Control Engineering Theory and Applications

The book provides general knowledge of automatic control engineering and its applications.

Providing an overview of control theory and systems, the chapters introduce transfer functions, modeling of control systems, automatic control systems, block diagrams, and signal flow graphs. While control system analysis and design are accompanied by root-locus methods and frequency response analyses, distributed control systems, nonlinearity in control systems including Z-transformation are also presented.

With straightforward demonstrations, examples, and multiple-choice questions, this book can be used as a reference textbook for electrical and electronics engineering, computer control engineering, automation engineering, mechatronics engineering, mechanics, robotics, AI control systems, hydraulics, process engineering, safety control engineering, aeronautical and aerospace engineering, auto-pilot system, decision-making system, and stock exchange, and will be suitable for majors, non-majors, and experts in the field of science and technology.

Jahangir Alam has been working as an associate professor at the School of Mechanical & Automotive Engineering in the South China University of Technology, P.R. China since 2014. He received his Ph.D. (Doctor of Engineering) degree in Mechatronic Engineering from Xiamen University, P.R. China. He has 6 patents and has published more than 30 papers, among which 16 papers are SCI/EI indexed. He has completed 19 scientific research projects as a principal investigator (PI) and participant. He has conducted the MOOC program for the course of Control Engineering Theory. His current interests in research and teaching are control engineering, robotic, machine visions, signals and image processing, biosignal and information processing, machine learning, biomaterials, biomedical engineering, etc.

Guoqing Hu is a scientist, an academician, a prominent pioneer professor, and a PhD supervisor at the School of Mechanical and Automotive Engineering in the South China University of Technology, Guangzhou, P.R. China. He was a visiting professor at The University of Queensland (Australia), The University of Nottingham (UK), The Chinese University of Hong Kong, and Case Western Reserve University (USA). He has published more than 200 journal papers (including SCI and EI indexed), authorized 30 patents, and completed more than 60 scientific research projects.

Hafiz Md. Hasan Babu is a prominent scientist, an academician, and a pioneer professor in the Department of Computer Science and Engineering, and the founding Chairman of the Department of Robotics and Mechatronics Engineering at the University of Dhaka, and a member of the Accreditation Council in the Ministry of Education in Bangladesh. He was also a Pro-Vice-Chancellor of National University and a member of the Prime Minister's ICT Task Force Committee in Bangladesh.

Huazhong Xu is an engineer and experimentalist at the School of Mechanical and Automotive in the South China University of Technology, Guangzhou, P.R. China. He has designed several industrial control systems, systems for testing the sealing structure of nuclear power gate, and systems for monitoring stress on the buried pipeline for natural gas transport.

Control Engineering Theory and Applications

Jahangir Alam, Guoqing Hu, Hafiz Md. Hasan Babu, and Huazhong Xu

CRC Press is an imprint of the
Taylor & Francis Group, an informa business

First edition published 2023
by CRC Press
6000 Broken Sound Parkway NW, Suite 300, Boca Raton, FL 33487-2742

and by CRC Press
4 Park Square, Milton Park, Abingdon, Oxon, OX14 4RN

CRC Press is an imprint of Taylor & Francis Group, LLC

ISBN: 978-1-032-27660-1 (hbk)
ISBN: 978-1-032-27734-9 (pbk)
ISBN: 978-1-003-29385-9 (ebk)

DOI: 10.1201/9781003293859

Typeset in Minion
by codeMantra

Contents

Preface

THIS BOOK HAS NINE CHAPTERS. Chapter 1 provides an overview of control theory and system. Chapter 2 introduces transfer function and modeling of control systems. Chapter 3 introduces automatic control systems, block diagrams, and signal flow graphs. Chapter 4 introduces transient response analyses in time domain. Chapter 5 introduces root-locus method: control system analysis and design. Chapter 6 introduces control system analysis and design by frequency response analyses. Chapter 7 introduces nonlinearity in control systems. Chapter 8 introduces distributed control system including Z-transformation to control system. Chapter 9 introduces applications with PID and motor control systems. This book will be useful to those in the fields of automation and control engineering, mechanical engineering, mechatronics engineering, robotics, control engineering, chemical engineering, stock exchange monitoring control system, automation control engineering, industrial engineering, processing engineering, and so on. Most of the examples in this book are illustrated using the MATLAB to provide clear understanding to students.

The chapters include examples and exercises as well as multiple-choice questions (MCQ). This book can be used as a reference textbook for mechanical, robotics, AI control systems, mechatronics, hydraulics, and other than control engineering majors, and it can also be useful to experts in the science and technology fields including business fields.

Acknowledgments

THE BOOK *CONTROL ENGINEERING THEORY AND APPLICATIONS* is an excellent textbook for undergraduate and graduate students in the fields of mechanics, robotics, AI control systems, mechatronics, hydraulics, and other than control engineering majors, and it can also be useful to experts in the science and technology fields including business fields.

We acknowledge the extraordinary help from Dr. Engr. Prof. Guoqing Hu for his excellence and inspiration in this book. We are very much pleased to receive support from South China University of Technology, China.

Finally, we acknowledge with gratitude the love and support from my family, students, readers, well-wishers, respected teachers, colleagues, friends, research mates, followers, and so on.

CHAPTER 1

Overview of Control Theory and System

AUTOMATIC CONTROL ENGINEERING IS learned mainly to solve the following two problems:

1. To analyze a given control system principle of work, component composition, and stability analysis.
2. To use the signal, machine, electricity, liquid, gas components, or equipment according to the actual needs of system design in order to achieve the required output.

1.1 BRIEF HISTORY AND DEVELOPMENTS OF CONTROL SYSTEMS

Control theory is one of the most important components of computer control engineering and mechatronics engineering (integrated with mechanical and electronic engineering). It was developed in the 1940s as a separate science and method due to rapid development of automatic control, electronic engineering, computer science and engineering, robotics, electrical engineering, decision science, economic processes, aerospace, and other multidisciplinary departments. In the late 1950s, the control theory has become more and more perfect due to the development of military and high technology in fields such as aviation, spaceflight, navigation, artillery, missile, electric power, and metallurgy. In 1954, Tsien hsue-Shen applied the thought and method of control theory, created the engineering control theory, applied the control theory to the field of engineering technology, and made it flourish. Control theory emerged in the late 1960s, namely, "classical control theory" and "modern control theory." The classical control theory is based on the transfer function, which mainly studies the analysis and design of the single input–single output (SISO) control system. The frequency-response and root-locus methods are the core of classical control theory. The system designed by these two methods are stable, and more or less meets a set of appropriate performance indicators. Generally speaking, these systems are satisfactory, but it is not the best system in a sense. Because modern devices with MIMO are becoming more and more complex, a large number

DOI: 10.1201/9781003293859-1

of equations are needed to describe modern control systems. Classical control theory involves SISO systems, and it is difficult to reveal the more profound characteristics of the system.

With the advent of the digital computer, the time-domain analysis of the complex system has become easier. The modern control theory develops at a historic moment. The state-space theory is one of the important properties of modern control theory, which was proposed by the famous American scholar R.E. Kalman to understand the control system. Modern control theory is based on state-space method in the control system. The researchers are adapting to the increasing complexity of modern equipment, mainly in the time domain; thus, the multiple input–multiple output (MIMO), variable parameters, variable structure, nonlinearity, high precision, high-efficiency design, and analysis of control system are studied to meet the strict requirements of the modern control theory, especially in the area of military, space technology, and industry application of precision, weight, and cost.

If 1950s–1960s years of the 20th century is the first revolution for the development of the control theory, the 1980s–1990s years of the 20th century is the second revolution of control theory due to invention and development of computer and software technology. In this period, control engineers have greatly improved the efficiency of the design of control systems by using MATLAB; however, the experts and scholars of control theory may be freed from the tedious calculation and digital simulation, and engaged in high technology and high-level research.

Without the participation of humans, an automatic control system automatically processes the production, or the physical quantity of the controlled object changes automatically according to the predetermined rules, such as hydraulic pressure system, flow rate, temperature, and pollution (P, Q, T, C) and voltage, current, power, resistance, capacitance, and inductance of electrical network (V, I, W, R, C, L).

According to the variation of a given program, a programmable controlled machine may be scheduled easily such as the processing of various predetermined shapes of the products, welding of various machine parts by an automatic robot or the process requirements, quenching process of insulation control, and so on. The common feature of these systems is in accordance with the provisions of the programing work and its modification.

In this unit, we will learn about the history and developments of control systems as well as its historical review. The application of automatic control system is believed to be in use even from the ancient civilizations. Several types of water clocks were designed and implemented to measure the time accurately from the 3rd century BC by Greeks and Arabs. But the first automatic system is considered the Watts Fly-ball Governor in 1788, which started the industrial revolution. The mathematical modeling of the Governor is analyzed by Maxwell in 1868. In the 19th century, Leonhard Euler, Pierre Simon Laplace, and Joseph Fourier developed different methods for mathematical modeling. The second system is considered Al Butz's Damper Flapper—a thermostat in 1885. He started the company, now named as Honeywell.

The beginning of the 20th century is known as the golden age of control engineering. During this time, classical control methods were developed at the Bell Laboratory by Hendrik Wade Bode and Harry Nyquist. Automatic controllers for steering ships were developed by Minorsky, Russian American Mathematician. He also introduced the concept of integral and derivative control in the 1920s. Meanwhile, the concept of stability was put forward by Nyquist and followed by Evans. The transforms were applied in control systems by Oliver Heaviside. Modern control methods were developed after the 1950s by Rudolf Kalman, to overcome the limitation of classical methods. Programmable logic controllers (PLCs) or programmable controllers were introduced in 1975.

1. *Liquid-level control system*
   Greeks began engineering feedback system around 300 BC. Some examples of liquid-level control system feedback system are shown in Figure 1.1.

2. *Steam pressure control system*
   Around 1681, application of steam pressure began with Denis Papin's safety valve.

3. *Temperature control system*
   In the 17th century, Cornelis Drebbel in Holland invented purely mechanical temperature control system for hatching eggs.

4. *Speed control system*
   In 1745, Edmund Lee applied control theory to windmills. In 1809, William Cubitt improved movable louvers. In the 18th century, James Watt invented fly-ball speed

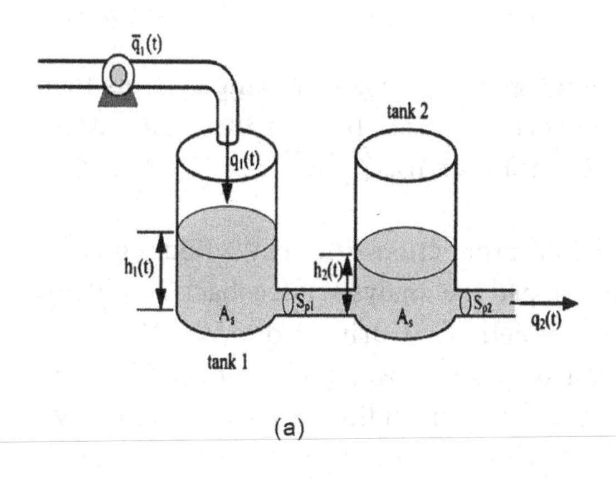

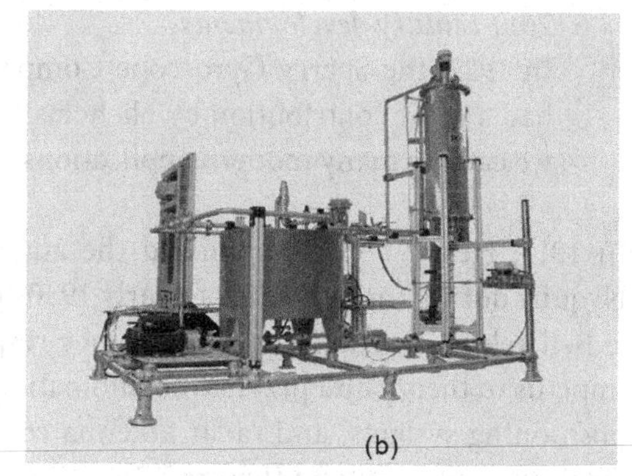

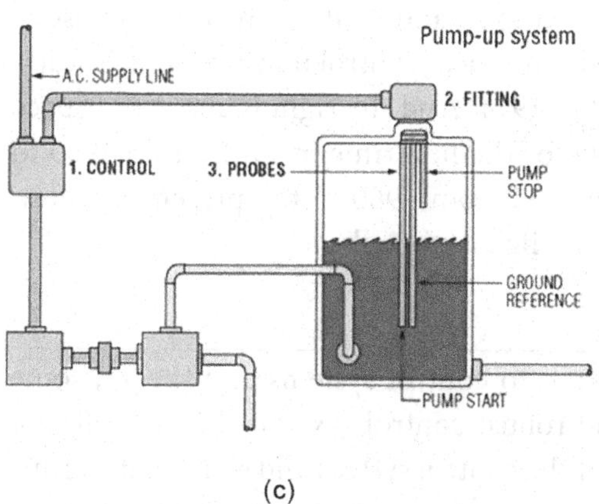

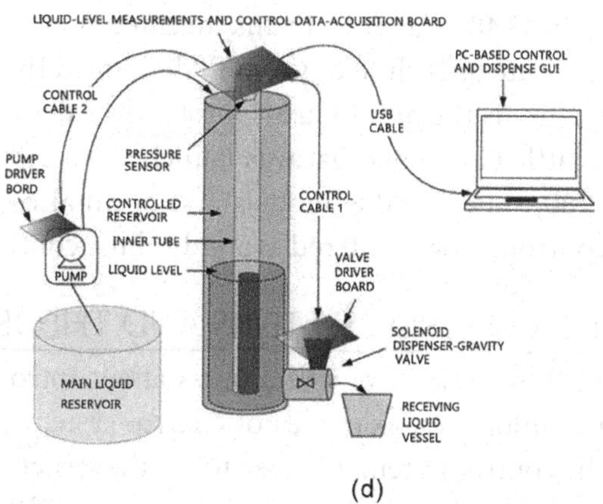

FIGURE 1.1 Liquid-level control system.

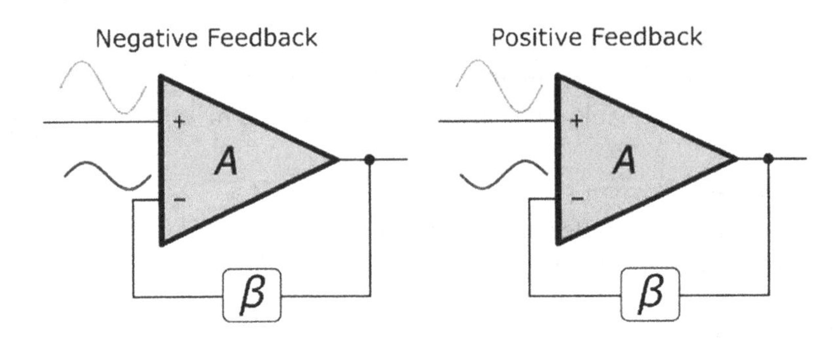

FIGURE 1.2 Feedback amplifiers.

governor to control the speed of steam engines. In 1877, Adams Prize developed "The Criterion of Dynamical Stability."

5. *Stability, stabilization, and steering development*
   James Watt's fly-ball governor for the speed control of a steam engine was invented in 1769. In 1868, James Clerk Maxwell published the stability criterion for a third-order-based coefficient of differential equation. In 1874, Edward John Routh extended the stability to fifth-order systems toward "The Criterion of Dynamical Stability."

6. *20th-century developments*
   In 1922, the Sperry Gyroscope Company installed automatic steering systems that had a great contribution by Nicholas Minorsky, a Russian born in 1885. Nowadays, we can see many modern applications of control systems.

In 1913, Henry Ford mechanized the automobile production line. H.W. Bode and H. Nyquist during the late 1920s to early 1930s developed the analysis of feedback amplifiers at Bell Telephone Lab by using frequency-response methods. During World War II, a large impetus to theory and practice of automatic control occurred which was auto-pilots, gun-positioning systems, and radar antenna control systems including negative and positive feedback control system (Figure 1.2).

In 1948, Walter R. Evans invented root-locus methods, and in 1952, the Massachusetts Institute of Technology (MIT) developed the first numerical control machine tool. The left picture is the first industrial robot was designed in 1954. And the right is about the Space Shuttle Columbia that was launched successfully for the first time in 1981. From 1960 to 1980, state-variable models and optimal control and from 1980 to the present, modern control theory centered around robust control, intelligent control, etc.

1.2 CONTROL SYSTEMS AND THEORY

In this section, we will discuss about introduction to control systems and theory, some terminologies, examples of control systems, and robust control systems. Before defining the control system, we need to see the structure of the control system and some of its terminologies. Figure 1.3 shows the structure of the control system. The structure of the control system consists of mechanical systems, electrical systems, mathematics, processing systems, electronic systems, materials properties, structures, and computation and decision.

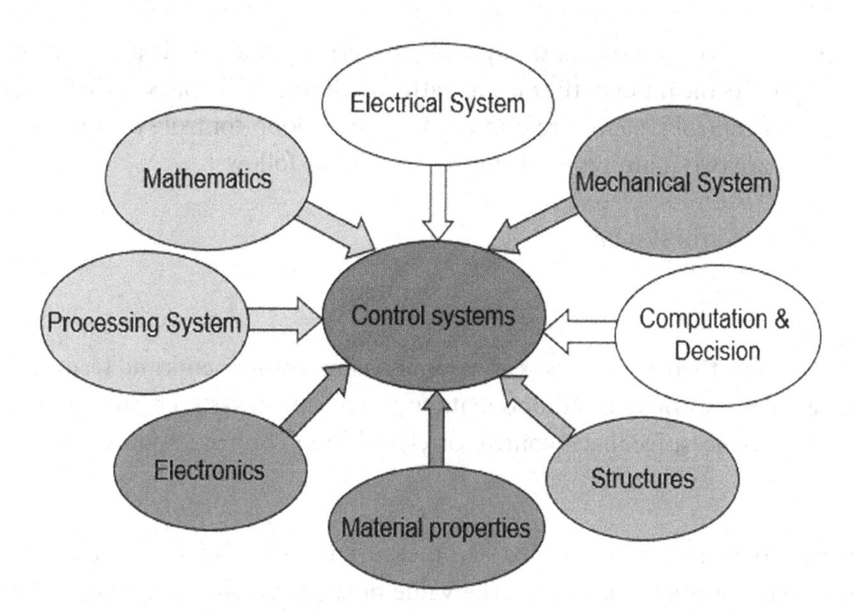

FIGURE 1.3 Structure of control system.

1.2.1 Control Theory and Engineering

There are various types of control systems, but all of them are created to control outputs. The system used for controlling the position, velocity, acceleration, temperature, pressure, voltage and current, etc., are examples of control systems.

Let us take an example of the simple temperature controller of the room, to clear the concept. Suppose there is a simple heating element, which is heated up as long as the electric power supply is switched on. As long as the power supply switch of the heater is on, the temperature of the room rises, and after achieving the desired temperature of the room, the power supply is switched off. Again, due to ambient temperature, the room temperature falls, and then manually the heater element is switched on to achieve the desired room temperature again. In this way, one can manually control the room temperature at the desired level. This is an example of manual control system.

This system can further be improved by using a timer switching arrangement of the power supply where the supply to the heating element is switched on and off at a predetermined interval to achieve desired temperature level of the room. There is another improved way of controlling the temperature of the room. Here, one sensor measures the difference between the actual temperature and the desired temperature. If there is any difference between them, the heating element functions to reduce the difference, and when the difference becomes lower than a predetermined level, the heating elements stop functioning.

Both forms of the system are automatic control systems. In the former one, the input of the system is entirely independent of the output. The temperature of the room (output) increases as long as the power supply switch is kept on. That means the heating element produces heat as long as the power supply is kept on, and the final room temperature does not have any control over the input power supply of the system. This system is referred to as open-loop control system. But in the latter case, the heating elements of the system function depend upon the difference between the actual temperature and the desired temperature.

This difference is called the error of the system. This error signal is fed back to the system to control the input. As the input to the output path and the error feedback path create a closed loop, this type of control system is referred to as a closed-loop control system.

Hence, there are two main types of control systems as follows:

1. Open-loop control system
2. Closed-loop control system

To study the control theory, we must know some system components and terminologies. A control system consists of some components such as control, systems, subsystems, plants, processes, disturbances, feedback control or closed-loop control systems, and open-loop control systems.

1. *Control*: Control is an action that can measure the value of a controlled variable. It means that the control measures the value of the controlled variable of the system, and it can be applied to the control signal of the system to correct or limit the deviation of the measured value from the desired value.
2. *Control variable*: The control variable is the quantity or condition that is measured and controlled in the system; however, the controlled variable is the response or output of the system.
3. *Control signal or manipulated variable*: A control signal or manipulated variable is the quantity or condition that is varied by the controller in the control system, and it affects the value of the controlled variable.
4. *System*: A system is a combination of components used in the system. In a system, a combination of components act together and perform a certain objective of the system, e.g., an automatic water temperature control; however, a system need not be a physical system such as economics and decision system.
5. *Subsystem*: A subsystem is a combination of subcomponents in the system.
6. *Plants*: Plants are a set of machine parts in the system. It may be a piece of equipment in the system. However, a set of machine parts functioning together to perform a particular operation or to control any physical object are called plants in a control system. Some examples of plants are mechanical devices, heating furnaces, chemical reactors, spacecraft, etc.
7. *Processes*: Processes can be defined as any operation to be controlled in the system, e.g., chemical, economic, and biological processes, whereas the system may be affected by disturbances; thus, the disturbance(s) are the affected signal to the output of a system or in the input or even in internal effects during the processing.
8. *Disturbances*: Disturbances are signals that tend to adversely affect the value of the output of a system. If a disturbance is generated within the system, it is called internal

disturbance, while an external disturbance is generated outside the system and may happen with an input.

9. *Feedback control or closed-loop control systems*: If the system is feedback or loop system with output to the input, that is called feedback control or closed-loop control systems; it can be summarized as follows:

    - The difference between output and some reference input of a system,
    - Unpredictable (random) disturbances are specified, and
    - Predictable (expected) or known disturbances can always be compensated within the system by using compensator(s).

10. *Open-loop control systems*: A control system in which the control action is totally independent of the output of the system is called open-loop control system. A manual control system is also an open-loop control system. The figure below shows a control system block diagram of an open-loop control system in which process output is totally independent of the controller action. If the system has no loop to compare the response with input, then the system is called open-loop control system. To define the control system, first we should look at Figure 1.4.

Therefore, we can say, the key point of a control system is the combination of the above systems, and it is the heart of any system. To understand the classification of control theory, control engineering has its own categorization depending on the different methodologies used. The main types of control engineering include

1. Classical control engineering,
2. Modern control engineering,
3. Robust control engineering,
4. Optimal control engineering,
5. Adaptive control engineering,
6. Nonlinear control engineering, and
7. Game theory.

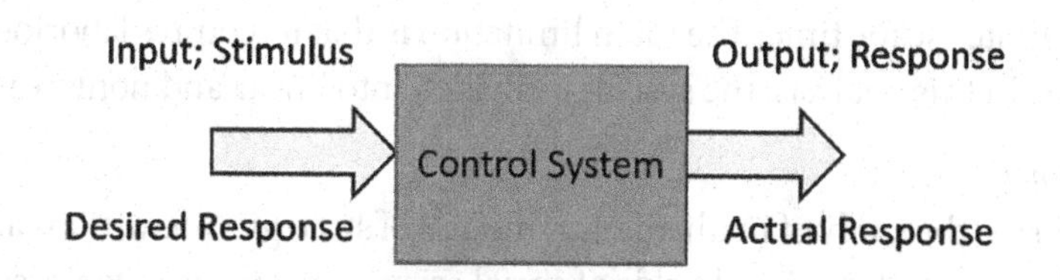

FIGURE 1.4 Block diagram of open-loop control system.

1. *Classical control engineering*
   The systems are usually represented by using ordinary differential equations. In classical control engineering, these equations are transformed and analyzed in a transformed domain. Laplace transform, Fourier transform, and z transform are examples. This method is commonly used in SISO systems.

2. *Modern control engineering*
   In modern control engineering, higher order differential equations are converted to first-order differential equations. These equations are solved very similar to the vector method. By doing so, many complications dealt while solving higher order differential equations are solved. These are applied in MIMO systems where analysis in the frequency domain is not possible. Nonlinearities with multiple variables are solved by modern methodology. State-space vectors, Eigenvalues, and Eigenvectors belong to this category. State variables describe the input, output, and system variables.

3. *Robust control engineering*
   In robust control methodology, the changes in the performance of the system with the change in parameters are measured for optimization. This aids in widening the stability and performance, and also in finding alternate solutions. Hence, in robust control, the environment, internal inaccuracies, noises, and disturbances are considered to reduce the fault in the system.

4. *Optimal control engineering*
   In optimal control engineering, the problem is formulated as a mathematical model of the process, physical constraints, and performance constraints, to minimize the cost function. Thus, optimal control engineering is the most feasible solution for designing a system with minimum cost.

5. *Adaptive control engineering*
   In adaptive control engineering, adaptive controllers are employed in which parameters are made adaptive by some mechanism. The block diagram given below shows an adaptive control system.

   In this kind of controllers, an additional loop for parameter adjustment is present in addition to the normal feedback of the process (Figure 1.5).

6. *Nonlinear control engineering*
   Nonlinear control engineering focuses on the nonlinearities that cannot be represented by using linear ordinary differential equations (i.e., they are not linear control systems). This system will exhibit multiple isolated equilibrium points, limit cycles, and bifurcations with finite escape time. The main limitation is that it requires laborious mathematical analysis. In this analysis, the system is divided into linear and nonlinear parts.

7. *Game theory*
   Game theory is the study of mathematical models of strategic interactions among rational agents. It has applications in all fields of social science, as well as in logic, systems science and computer science. For example, it addressed two-person zero-sum games (by John

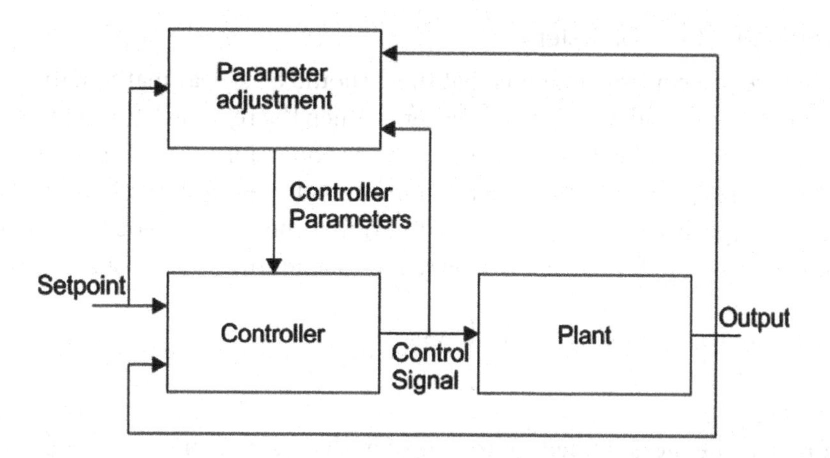

FIGURE 1.5 Parameter adjustment to the normal feedback process.

von Neumann), in which each participant's gains or losses are exactly balanced by those of other participants. In the 21st century, game theory applies to a wide range of behavioral relations, mathematical economics, business, as well as statistics to treat decision-making under uncertainty; it is now an umbrella term for the science of logical decision making in humans, animals, engineering and technology as well as computers, mechanical, electrical, aeronautics, stock market, business, and so on.

1.2.2 Definition of a Control System

Figure 1.4 is a block diagram of a control system. In the center is the control block. The system accepts the input signal or value by the center of the block which is called stimulus or desired response; after doing the process itself, the control system gives an output value or actual response signal, which is called the response of the system. Therefore, we can define control system as follows:

A control system consists of *subsystems* and *processes* (or *plants*) assembled for the purpose of obtaining a desired *output* with desired *performance*, given a specified *input*.

A control system is a system of devices or a set of devices that manage, command, direct, or regulate the behavior of other devices or systems to achieve desired results. In other words, the definition of a control system can be simplified as a system that controls other systems. As the human civilization is being modernized day by day, the demand for automation is increasing accordingly. Automation highly requires the control of devices. In recent years, control systems have played a central role in the development and advancement of modern technology and civilization. Practically every aspect of our day-to-day life is affected more or less by some control system. A bathroom toilet tank, a refrigerator, an air conditioner, a geezer, an automatic iron, and an automobile are all control systems. These systems are also used in industrial processes for more output. We find control systems in the quality control of products, weapons system, transportation systems, power system, space technology, robotics, and many more. The principles of control theory are applicable to both engineering and nonengineering fields.

1.2.3 Features of a Control System

The main feature of a control system is that there should be a clear mathematical relationship between the input and output of the system. When the relation between the input and output of the system can be represented by linear proportionality, the system is called a linear control system. Again, when the relationship between input and output cannot be represented by single linear proportionality, but rather the input and output are related by some nonlinear relation, the system is referred to as a nonlinear control system.

1.2.4 Requirements of a Good Control System

1. *Accuracy*
   Accuracy is the measurement of tolerance of the instrument, and it defines the limits of the errors made when the instrument is used in normal operating conditions. Accuracy can be improved by using feedback elements. To increase the accuracy of any control system, an error detector should be present in the control system.

2. *Sensitivity*
   The parameters of a control system are always changing with the change in surrounding conditions, internal disturbance, or any other parameters. This change can be expressed in terms of sensitivity. Any control system should be insensitive to such parameters but sensitive to input signals only.

3. *Noise*
   An undesired input signal is known as noise. A good control system should be able to reduce the noise effect for better performance.

4. *Stability*
   It is an important characteristic of the control system. For the bounded input signal, the output must be bounded, and if the input is zero, then the output must be zero. Such a control system is called a stable system.

5. *Bandwidth*
   An operating frequency range decides the bandwidth of the control system. Bandwidth should be as large as possible for the frequency response of a good control system.

6. *Speed*
   It is the time taken by the control system to achieve its stable output. A good control system possesses high speed. The transient period for such a system is very small.

7. *Oscillation*
   A small number of oscillations or constant oscillations of output tend to indicate the system to be stable.

1.2.5 Advantages of Control Systems or Why Do We Need Control Systems?

The advantages of control systems are mainly power amplification, remote control, convenience of input form, and compensation for disturbances. We can see some examples of control systems such as speed control system, temperature control system, automatic water temperature control system, engineering business organization system, and robust control system.

1. *Speed control system*
   A speed control system has a control valve to control the on/off switch. This control switch can control the speed of the fuel in the engine and produce the power level for the power cylinder (Figure 1.6).

2. *Temperature control system*
   Temperature control system is another example of a control system. Here, Figure 1.7 shows the block diagram of a temperature control system that has an electric furnace

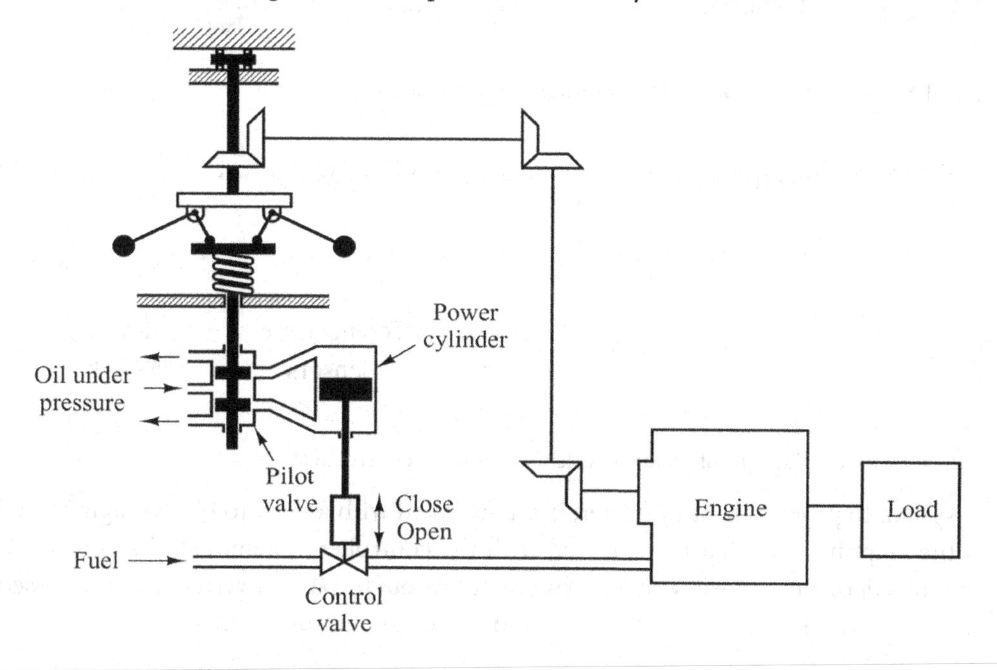

FIGURE 1.6 Speed control system.

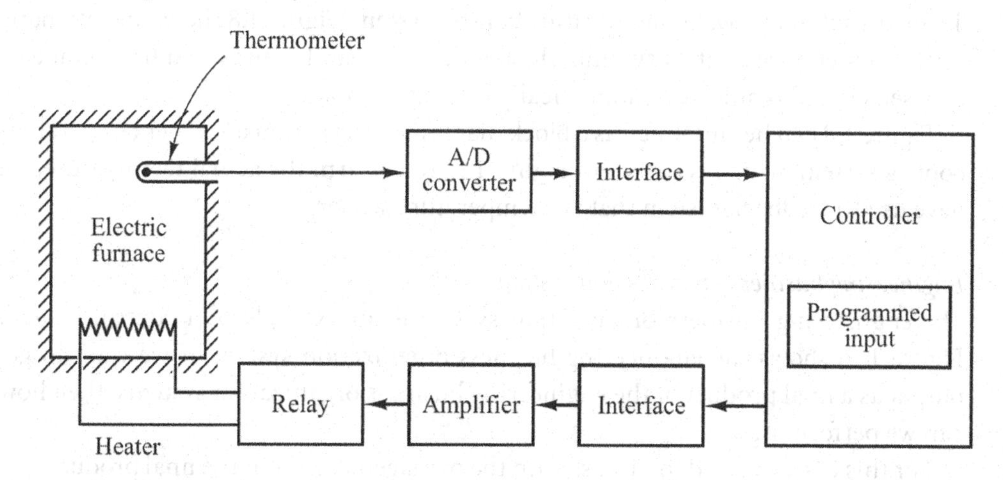

FIGURE 1.7 Temperature control system.

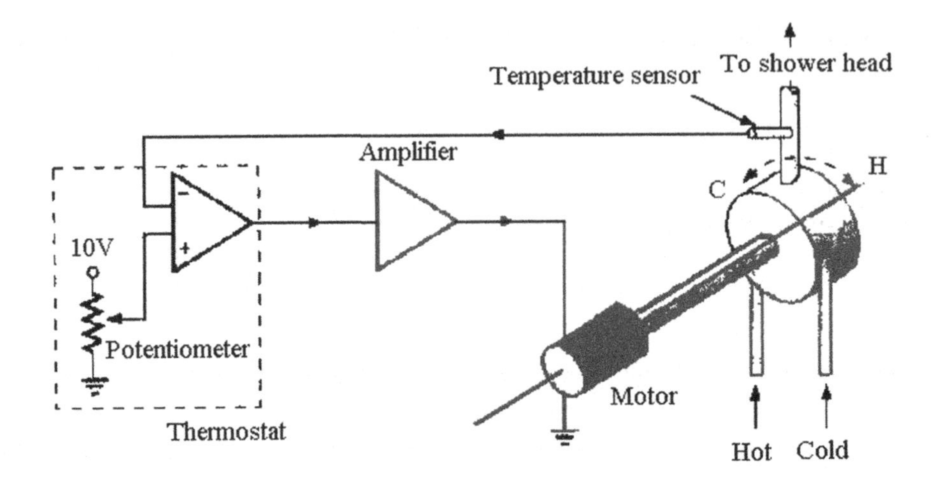

FIGURE 1.8 An automatic water temperature control system.

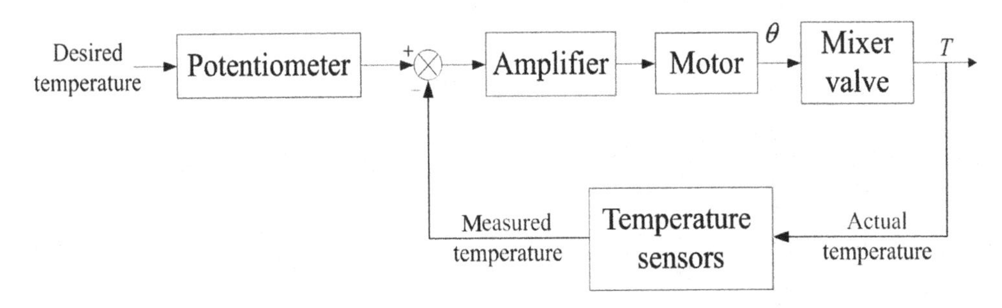

FIGURE 1.9 Block diagram of an automatic temperature control system.

system to produce heat by heater; it is connected with a relay to receive signals from the amplifier and that is interfaced with the controller system. Another side of the controller is interfaced with a thermometer through A/D converter. This is a closed-loop control system to maintain or control the temperature in this system.

3. *Automatic water temperature control system*

   This is an automatic water temperature control system. Figure 1.8 shows an automatic water temperature control system. The temperature can be measured by a temperature sensor and is adjusted automatically by using a motor.

   Figure 1.8 can be simplified as a block diagram of the automatic water temperature control system, which is shown in Figure 1.9. To confirm the actual temperature, we have used a feedback system that is a temperature sensor.

4. *Engineering business organization system*

   The engineering business organization system is an example of a control system. Figure 1.10 shows the engineering business organization system. If we want to get output as a final product of the engineering business organization systems, then how can we perform it?

   For this case, we need to discuss with the management about the final product that we want to produce. After approval by the management, research and development

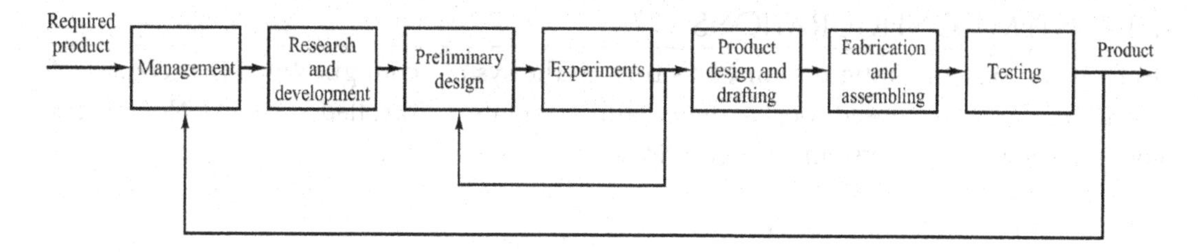

FIGURE 1.10 Engineering business organization systems.

are necessary. Then we can finish the preliminary design of the product. If the test results of the experiments are not satisfactory enough, then we should start the process from preliminary design and then experiment again; if the tested results are satisfactory, then we may finish the product design, draft for fabrication, and assemble the product. To ensure the final output of the product, it is needed to perform the final test to confirm the final product; if the tested results are not satisfactory enough with the requirements, we must check it from the beginning of the process.

5. *Robust control system*
   To understand a robust control system, the following must be considered:
   - Design a mathematical model of a plant or control object;
   - Include an error in the modeling process;
   - Plant differs from the used model;
   - Ensure the designed controller is satisfactory;
   - Uncertainty or error between actual plant and mathematical model;
   - Include uncertainty or error in the design process of the control system.

Therefore, more specifically, robust control system is measured by the robust stability and robust performance. The robust stability can be measured by this inequality:

Robust stability:

$$\left\| \frac{W_m(s)K(s)G(s)}{1+K(s)G(s)} \right\|_\infty < 1 \tag{1.1}$$

The robust performance is measured by this inequality in Eq. (1.1).

Robust performance:

$$\left\| \frac{W_s(s)}{1+K(s)G(s)} \right\|_\infty < 1 \tag{1.2}$$

where $W_m(s)$ is the chosen transfer function, $K(s)$ is the controller, $G(s)$ is the transfer function, and $W_s(s)$ is the selective transfer function. Remember carefully, from the robust stability, the controller $K(s)$ guarantees internal stability of all the systems in a group of systems that include the system with the actual plant. However, the specified performance can be satisfied in all systems that belong to the group.

1.3 SYSTEM CONFIGURATIONS

To focus on system configurations, let's turn to the system configurations that consist of open-loop systems, closed-loop systems, and computer-controlled systems. Now, I am going to show you the system configurations one by one briefly.

1.3.1 Open-Loop Control System

The open-loop control systems are those systems in which the output does not affect the control action. The system can accept input and can produce output, but the output is not connected with input; however, it is unable to compare the response(s) with input value.

1.3.1.1 Practical Examples of Open-Loop Control System

1. *Electric hand drier*—Hot air (output) comes out as long as you keep your hand under the machine, irrespective of how much your hand is dried.
2. *Automatic washing machine*—This machine runs according to the pre-set time irrespective of whether washing is completed or not.
3. *Bread toaster*—This machine runs as per adjusted time irrespective of whether toasting is completed or not.
4. *Automatic tea/coffee maker*—These machines also function for preadjusted time only.
5. *Timer-based clothes drier*—This machine dries wet clothes for preadjusted time, but it does not matter how much the clothes are dried.
6. *Light switch*—Lamps glow whenever the light switch is on irrespective of whether light is required or not.
7. *Volume on stereo system*—Volume is adjusted manually irrespective of output volume level.

Let's see another example of mixing hot and cold water by using a system that is shown in Figure 1.11a. It is a linear and open-loop control system that uses a mixer valve. Here, θ indicates the input and T indicates the output of the system, but the output is not connected to the input; thus, it is called an open-loop system. Look at the second Figure 1.11b that shows a light control and water tap control system with linear on/off control switch that can be understood as another example of an open-loop control system. Therefore, it is clear and concise to show a simple block diagram of an open-loop control system in Figure 1.12, and the closed-loop control system has been shown in Figure 1.13.

1.3.1.2 Advantages and Disadvantages of Open-Loop Control System

Let's see the open-loop system carefully and you can find that it has some excellent advantages, such as

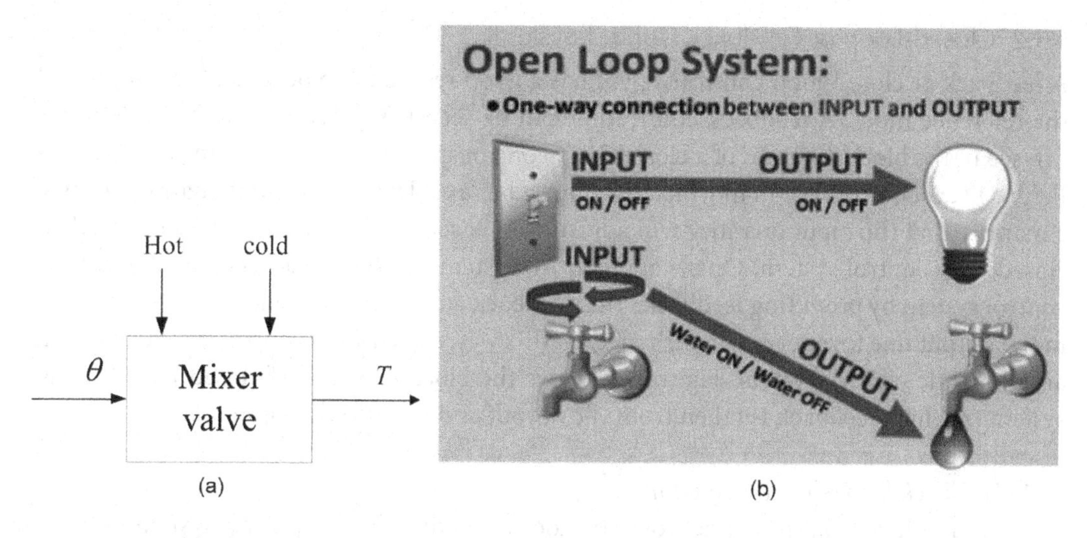

FIGURE 1.11 (a) Mixing hot and cold water and (b) light control and water tap control system.

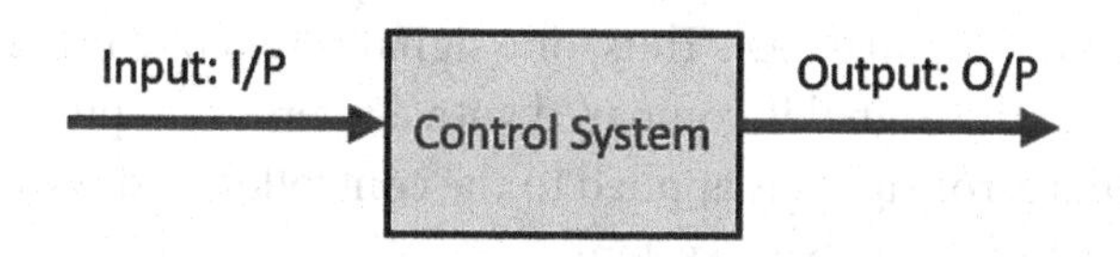

FIGURE 1.12 Block diagram of open-loop system.

- Simple construction,
- Less expensive,
- Easy to maintain,
- Generally stable and no stability problem, and
- Convenient to use as output is difficult to measure.

However, the open-loop control system has also some disadvantages, such as

- They are inaccurate,
- Unreliable,
- Any change in output cannot be corrected automatically,
- It has greater sensitivity to disturbances, and
- Inability to correct the disturbances.

1.3.2 Closed-Loop or Feedback Control System

A feedback or closed-loop control system has a good relationship between the output and the reference input. A feedback connection can be seen from a closed-loop system as we can see in the block diagram of a closed-loop control system or feedback control system in Figure 1.9. A control system in which the output has an effect on the input quantity in such a manner that the input quantity will adjust itself based on the output generated is called closed-loop control system. Open-loop control system can be converted into closed-loop control system by providing feedback. This feedback automatically makes suitable changes in the output due to external disturbance. In this way, closed-loop control system is called an automatic control system. Figure 1.9 shows the block diagram of closed-loop control system in which feedback is taken from the output and fed into the input.

1. *Feedback loop of control system*

   Feedback is a common and powerful tool when designing a control system. A feedback loop is a tool that takes the system output into consideration and enables the system to adjust its performance to meet a desired result of the system.

   In any control system, the output is affected by changes in environmental conditions or any kind of disturbance. Thus, one signal is taken from the output and is fed back to the input. This signal is compared with a reference input, and the error signal is generated. This error signal is applied to the controller, and the output is corrected. Such a system is called feedback system.

   When the feedback signal is positive, the system is called a positive feedback system. For a positive feedback system, the error signal is the addition of reference input signal and a feedback signal. When the feedback signal is negative, the system is called a negative feedback system. For a negative feedback system, the error signal is given by the difference between the reference input signal and the feedback signal. Effects of feedback are as follows:

   - Error between system input and system output is reduced;
   - System gain is reduced by a factor $1/(1 \pm GH)$;

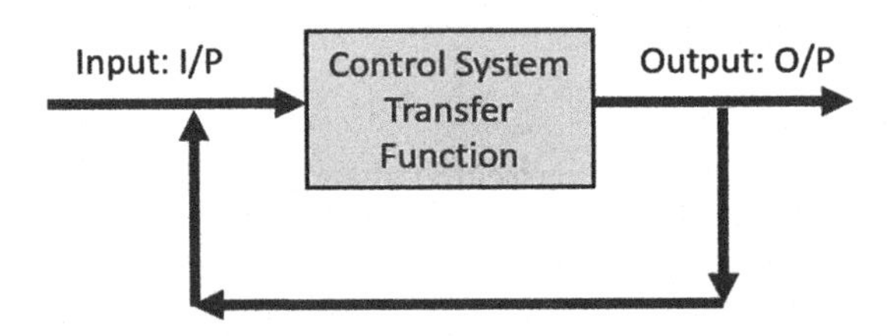

FIGURE 1.13 Block diagram of closed-loop system.

- Improvement in sensitivity;
- Stability may be affected; and
- Improve the speed of response.

2. *Practical examples of closed-loop control system*

    1. *Automatic electric iron*—Heating elements are controlled by the output temperature of the iron.
    2. *Servo voltage stabilizer*—Voltage controller operates depending upon the output voltage of the system.
    3. *Water level controller*—Input water is controlled by the water level of the reservoir.
    4. *Missile launched and autotracked by radar*—The direction of a missile is controlled by comparing the target and position of the missile.
    5. *An air conditioner*—An air conditioner functions depending upon the temperature of the room.
    6. *Cooling system in car*—It operates depending upon the temperature that it controls.
    7. Let's see a general example of a closed-loop control system or feedback control system in Figure 1.13. In this system, the signal generator acts as an input system that generates the input signal to the control system. The controller receives this specific signal from the signal generator and passes it to the actuator through an amplifier for further processing in the plant system. After good processing, it gives the output, which is called system response or system output. The output or response can be sensed by the sensor in the comparison unit of the input system to compare the results with the input value. Until getting a satisfactory result, the loop will work for better results. This is called a closed-loop control system.

Now, let's see another example of a closed-loop control system which is temperature control system shown in Figure 1.14. It is a typical automatic temperature control system which is not classified as the closed-loop control system. The desired temperature can be

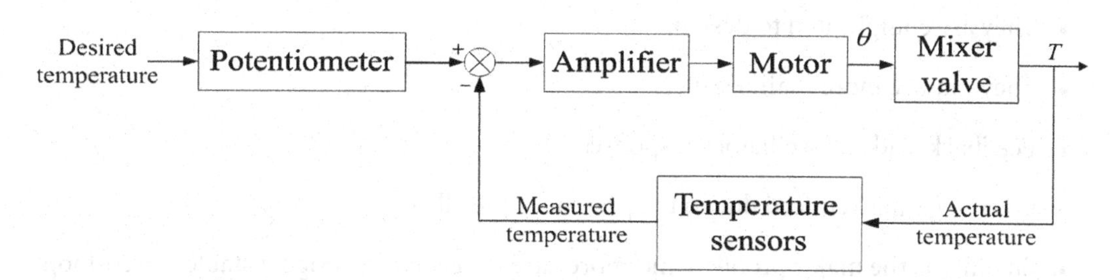

FIGURE 1.14 Temperature control system.

measured by a potentiometer; however, it can transfer to the system by amplifying through the amplifier. The motor can rotate to adjust the mixture valve to control the temperature in the system. A temperature sensor, which is a thermometer, can measure the temperature of the system. If the measured temperature is not fit for the actual temperature, then the process must be performed repeatedly until reaching the targeted temperature; thus, it is classified as a closed-loop control system process.

The *closed-loop control* system is often referred to as

1. Positive feedback and
2. Negative feedback.

1.3.2.1 Advantages and Disadvantages of Closed-Loop Control System

However, the output of the closed-loop control system has an effect(s) on control action. There are some advantages and disadvantages of the closed-loop control systems compared with open-loop control systems. The advantages are as follows:

- Less sensitive to external disturbances and internal variations,
- It has greater accuracy,
- Closed-loop control systems are more accurate even in the presence of nonlinearity,
- Highly accurate as any error arising is corrected due to the presence of feedback signal,
- Bandwidth range is large,
- It facilitates automation,
- The sensitivity of the system may be made small to make the system more stable, and
- This system is less affected by noise.

However, the disadvantages are obviously higher in cost and power and more complex in structure. Thus, we can say

- They are costlier.
- They are complicated to design.
- They require more maintenance.
- Feedback leads to oscillatory response.
- Overall gain is reduced due to the presence of feedback.
- Stability is the major problem and more care is needed to design a stable closed-loop system.

1.3.3 Comparison of Closed-Loop and Open-Loop Control Systems

SN	Open-Loop Control System	Closed-Loop Control System
1	The feedback element is absent.	The feedback element is always present.
2	An error detector is not present.	An error detector is always present.
3	It is a stable one.	It may become unstable.
4	Easy to construct.	Complicated construction.
5	It is economical.	It is costly.
6	Having small bandwidth.	Having large bandwidth.
7	It is inaccurate.	It is accurate.
8	Less maintenance.	More maintenance.
9	It is unreliable.	It is reliable.
10	Examples: hand drier, tea maker.	Examples: servo voltage stabilizer and perspiration.

1.3.4 Components of Closed-Loop System

We must learn about the components of the closed-loop control system. The typical closed-loop control system has the following:

- Signal generator,
- Measurement elements or sensors,
- Comparison elements,
- Amplifier elements,
- Actuator,
- Controller or compensator, and
- Controlled plant.

1.4 APPLICATION OF CONTROL THEORY IN ENGINEERING AND TECHNOLOGY

1.4.1 Control System Engineering

Control system engineering is the branch of engineering that deals with the principles of control theory, to design a system that yields the desired behavior in a controlled manner. Hence, although control engineering is often taught within electrical engineering at university, it is an interdisciplinary topic.

Control system engineers analyze, design, and optimize complex systems that consist of highly integrated coordination of mechanical, electrical, chemical, metallurgical, electronic, or pneumatic elements. Thus, control engineering deals with a diverse range of dynamic systems that include human and technological interfacing. These systems are broadly referred to as control systems.

Control system engineering focuses on the analysis and design of systems to improve the speed of response, accuracy, and stability of the system.

The two methods of control systems are classical and modern. The mathematical model of the system is set up as the first step followed by analysis, designing, and testing. Necessary conditions for the stability are checked and finally, optimization follows.

In the classical method, mathematical modeling is usually done in the time domain, frequency domain, or complex domain. The step response of a system is mathematically modeled in time-domain differential analysis to find its settling time, percentage of overshoot, etc. Laplace transforms are most commonly used in the frequency domain to find the open-loop gain, phase margin, bandwidth, etc. of the system. The concept of the transfer function, Nyquist stability criteria, sampling of data, Nyquist plot, poles and zeros, Bode plots, and system delays all come under the umbrella of classical control engineering stream.

Modern control engineering deals with MIMO systems, state-space approach, Eigenvalues, and Elgenvectors, etc. Instead of transforming complex ordinary differential equations, modern approach converts higher order equations to first-order differential equations, solved by the vector method.

Automatic control systems are most commonly used as it does not involve manual control. The controlled variable is measured and compared with a specified value to obtain the desired result. As a result of automated systems for control purposes, the cost of energy or power, as well as the cost of the process, will be reduced increasing its quality and productivity.

1.4.2 Differential Equation of a System

The dynamics of a system can be attributed to the systems of differential equations. In mechanical systems according to the stress analysis, we can list the differential equations of the system; in the electrical network system according to the characteristics of the electrical network, we can list the differential equations of the electrical network. Figure 1.15 shows the mechanical system; the differential equations of this mechanical system are obtained by force analysis as follow:

$$\begin{cases} m\ddot{x}(t)+f\dot{x}(t)+kx(t)=F(t) \\ x(0)=x_0 \qquad ;\dot{x}(0)=v_0 \end{cases} \tag{1.3}$$

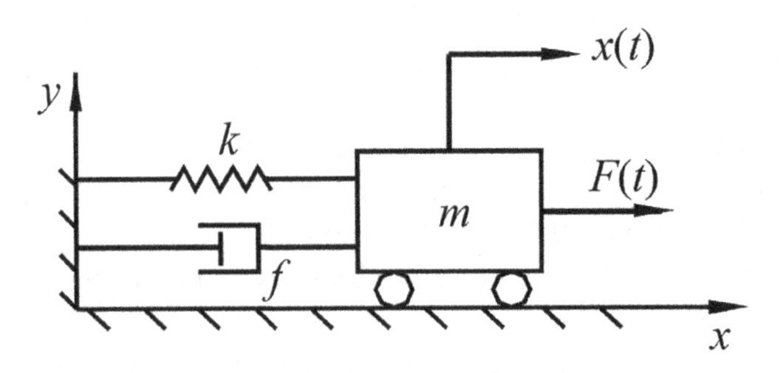

FIGURE 1.15 The Mass-spring-damper system.

The characteristic of Eq. (1.3) can be expressed as follows:

$$mp^2 + fp + k = 0 \tag{1.4}$$

where $p = \dfrac{d}{dt}$, which is related to the structure and parameters of the system itself, whereas it reflects the characteristics of the system itself independently.

Here, $F(t)$ is known as a system of external input and $x(t)$ is known as the output of the system. From the perspective of the generalized kinetic analysis, Figure 1.15 shows the physical model of the mechanical system, electrical system, and any other control system, which can describe the relationship of three functions such as the system, input, and output.

From the above analysis, it can be obtained from the control engineering point of view, which can be summarized as follows:

1. When the system input is known, the output (known as response) of the system through the output (response) is performed to study the various problems of the system itself to analyze the system.
2. To determine the system, the input forms are important to study the optimal control problem so that the output as far as possible meets the optimal requirements of the system.
3. If the output or input is known, the best possible requirement is that the input is given, which is the optimal design problem.
4. To determine the system, if the output is known, the recognition of the input information is important to perform filtering and prediction problems.
5. If both input and output are known, it can determine the structure and parameters of the system to establish a mathematical model that is known as the system identification problem.

1.4.3 Physical Model and System Block Diagram

The *Function recorder* is a kind of universal automatic recording instrument, which can automatically draw the function relationship between the two measurement functions on the paper. At the same time, the recorder also has a paper feeding mechanism, which is used to describe the function of time. The recorder adopts the principle of negative feedback. Its structure is comprised of attenuator, measuring circuit, amplifying device, servo motor, speed measuring set, gear system, and rope wheel as shown in Figure 1.16. The system of the input signal is the voltage to be recorded, object to the recording pen, and displacement. The function recorder control system is a displacement control recording pen that draws on the paper to record voltage signal curve.

In Figure 1.16, the measuring circuit is a bridge circuit that is comprised of two potentiometers (R_P and R_M); the recording pen is fixed on the brush of the potentiometer (R_M); and the output voltage of the measuring circuit (U_P) is proportional to the displacement of the recording pen. When the input signal is u_r, the bias voltage ($\Delta u = u_r - u_p$) is obtained

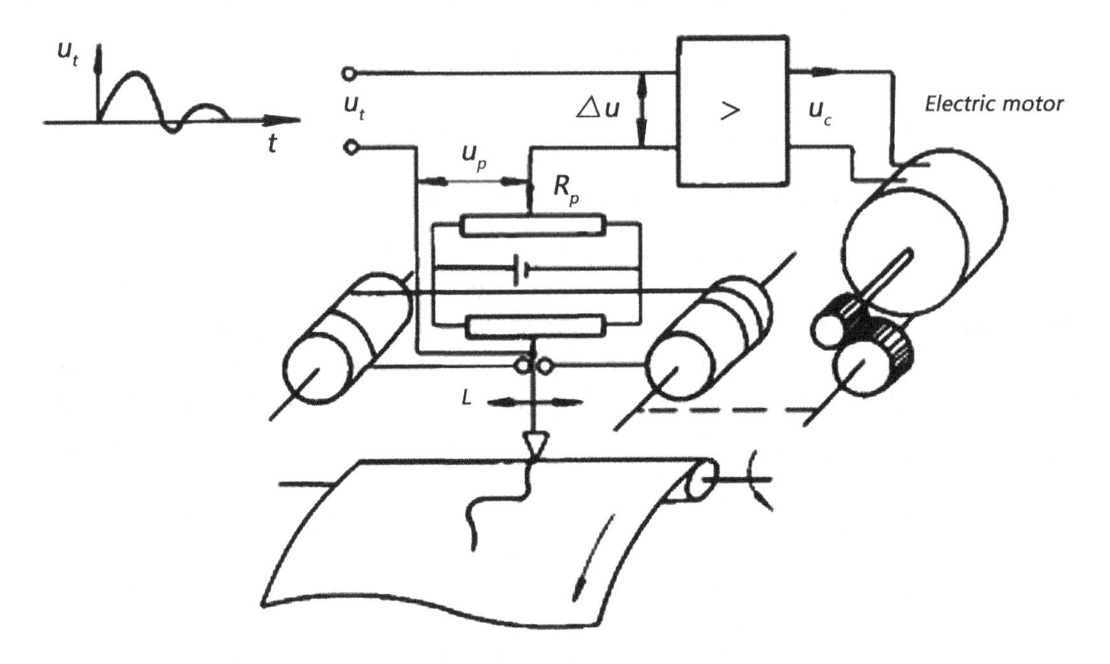

FIGURE 1.16 Schematic diagram of function recorder.

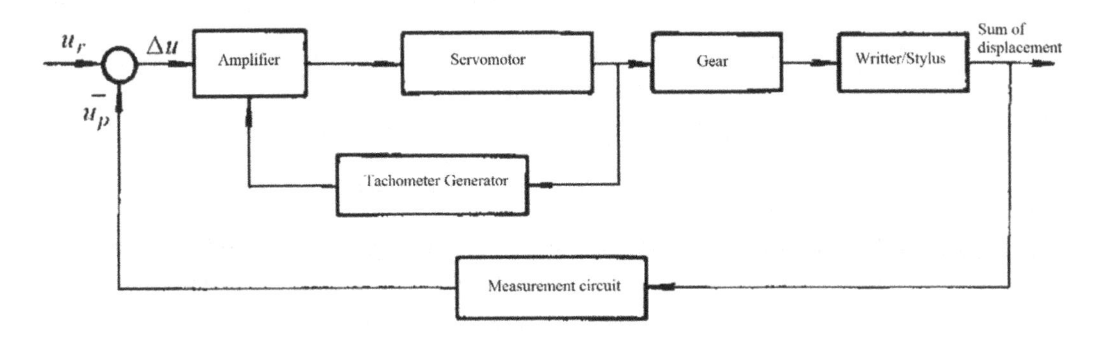

FIGURE 1.17 Block diagram of function recorder system.

at the input port of the amplifier, whereas when the servo motor is amplified by the signal, the recording pen is driven to move through the gear system and the rope wheel is to reduce the deviation of the voltage. When the deviation is $u = 0$ as well as when the motor stops rotating, the pen is not moving, but when the input signal changes with time, the pen draws continuous curve on the paper by changing with respect to time.

The block diagram of the function recorder control system is shown in Figure 1.17. A tacho-generator is a feedback proportional to the rotation speed of the motor voltage signal to increase the damping ratio of the system and improve the performance of the system.

The function of recorder controls the recording pen to record the output voltage signal. The input voltage signal can be any function of time. The components of the system are composed of mechanical, electrical, and other components. The voltage and displacement transmitted signals in the system are continuous signals with time variation. Thus, the function recorder system is a follow-up system with time variation.

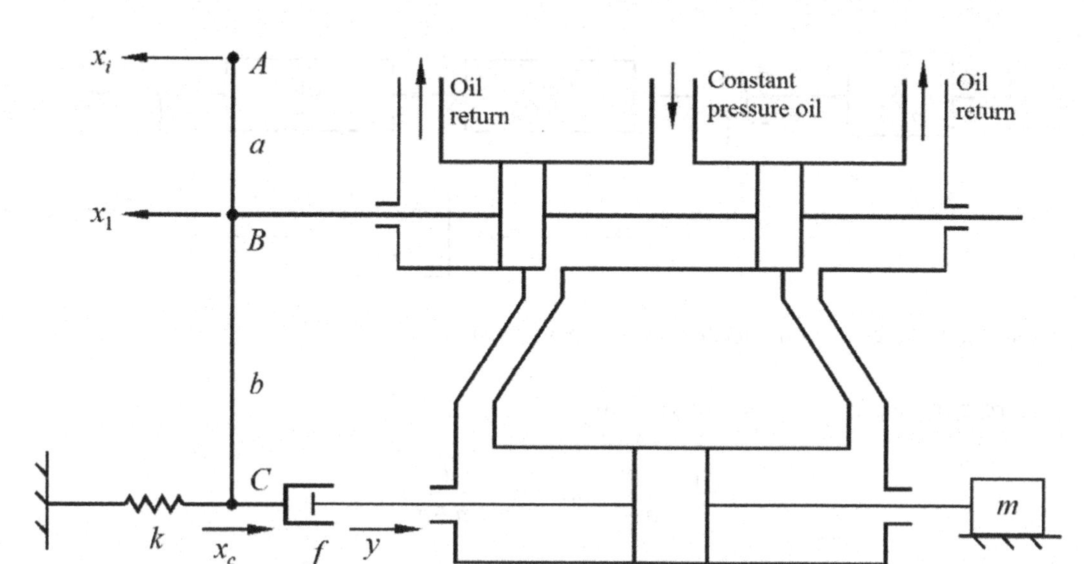

FIGURE 1.18 Hydraulic system.

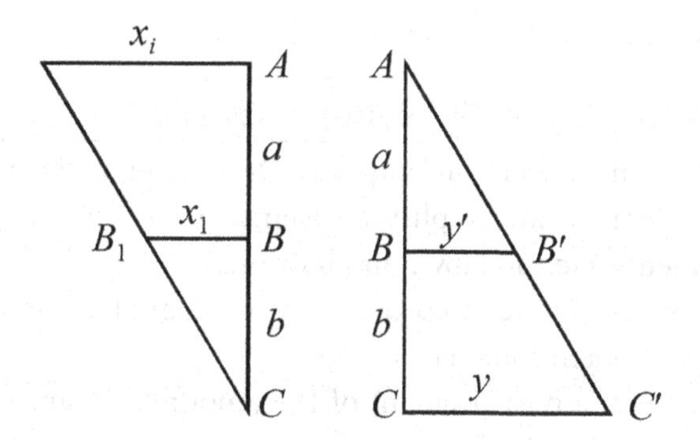

FIGURE 1.19 Relationship of spool displacement.

A simple hydraulic system is shown in Figure 1.18. Figure 1.18 shows a servo system—x_i is the input displacement and y is the output displacement. We want to determine the system block diagram in this system.

According to the working principle of the hydraulic system, when in a point of entry to the left displacement of x_i, namely by B to the spool input always left displacement, then open the control valve, after that high-pressure oil will pass into the hydraulic cylinder on the left side; whereas, it pushes the piston to moves right direction. Piston rod through the C driven at B point moves to the right; B point through the stem-driven spools is pushed back to the midpoint position; however, the plug valve and hydraulic cylinder will stop the motion. When A points enter to the right to a displacement B, x_i points move to the right, that is, when the spool input displaces to the right, the movement process will move to the opposite direction. Thus, you can get the B point for the system comparison point. The relationship system block diagram of spool displacement is shown in Figure 1.19, and the block diagram of hydraulic servo system is shown in Figure 1.20.

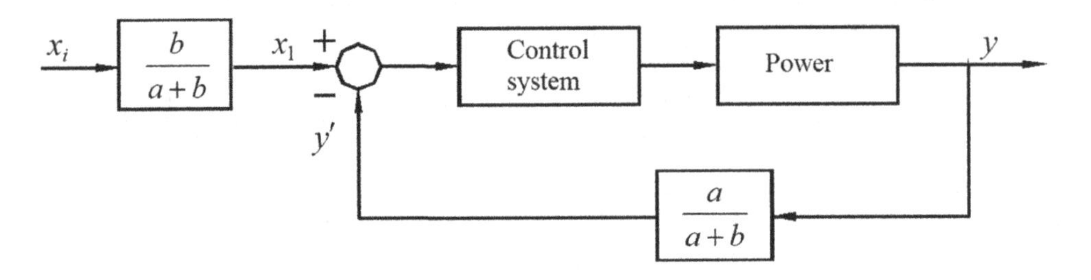

FIGURE 1.20 Block diagram of a hydraulic servo system.

Here, from Figure 1.19, we can say that

$$\frac{x_1}{x_i} = \frac{b}{a+b} \ldots\ldots\ldots x_1 = \frac{b}{a+b} x_i$$

$$\frac{y'}{y} = \frac{a}{a+b} \ldots\ldots\ldots y' = \frac{a}{a+b} y$$

Thus, the total displacement of the spool is $x_1 - y'$.

1.5 BASIC ELEMENTS OF THE CONTROL SYSTEM

A typical control system includes input/output (I/O) element, a given element, a feedback element, comparison element, an amplifying element, the actuator, control object elements, correction elements, etc., as shown in Figure 1.21.

A given element: It is mainly used to produce a given signal or input signals such as the speed control system of a potentiometer.

Feedback element: It is the measurement of the amount of transfer or output and the feedback signal. There is a definite functional relationship between the input and output signals (usually measured in a percentage %), such as the speed motor driving system. The feedback element is also known as the measuring element.

Comparison element: It is used to compare the deviation between the input signal and the feedback signal. Addition and subtraction are made in the input and output of a system and presented the deviation (error) signal; this role is often performed by the integrated circuit or the measuring element; it can also be a differential circuit but it is often not a dedicated physical component, but sometimes it shows the more links or system error.

Amplifier or amplifying element: The amplifying element carries on the signal amplification and the power expansion to the deviation signal. It causes the output quantity to have enough power or the requested physical quantity; for example, servo power amplifier and electro-hydraulic servo valve.

Actuator: The actuator directs the manipulation of the components of the control object. According to the deviated signal amplification, the object is performed to a task, and it is controlled to produce a consistent output; for example, hydraulic cylinder, hydraulic motor, and actuator motor.

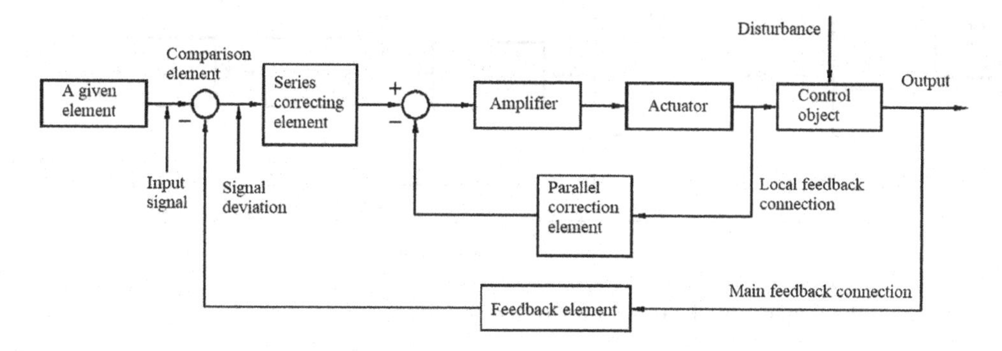

FIGURE 1.21 Block diagram of a typical feedback control system.

Control object: Control object is the part of a control system in a plant(s). Its output is the system variable (or control); for example, machine tools, working platforms, equipment, and production line.

Correcting element: It is also known as the correction device. For stability control system and to improve the system performance, it is convenient to adjust the parameter(s) and structure of the components. We can see two types of correction forms—main feedback correction and series correction.

1.6 BASIC TYPES OF AUTOMATIC CONTROL SYSTEMS

There are many basic types of automatic control systems, but only the main difference is the structure and the task of the different control systems. We can see different types of basic automatic control systems below.

1.6.1 According to the Motion Law of Quantity

a. The key point of the constant control system (such as voltage source, temperature control system) is to overcome the disturbance effect of the controlled variable.
b. Program control system is shown in Figure 1.22, such as computer numerical control systems. It is the known function of time for the input of the control system.
c. Dynamic system, which gives the time of the unknown function, that is, the quantitative transformation of the law cannot be accurately determined in advance, but the output can quickly and accurately reproduce the quantitative transformation. It can be used to be aimed at the enemy system such as cannon, hydraulic profiling tool and slide servo system.

1.6.2 According to the System Response Characteristics

a. Continuous control system can be divided into linear and nonlinear control systems. Linear control system can be described by linear differential equations, whereas nonlinear control system is not described by linear differential equation. Nonlinear

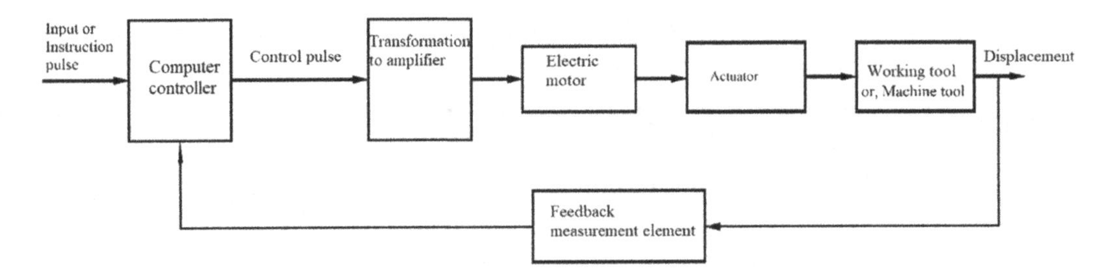

FIGURE 1.22 Block diagram of a program control system.

control system can be used as piecewise function in differential equations to describe the control system.

b. Digital control system, that is, the discrete control system, is represented by the digital quantity of all the amount of the system. The digital quantity is not continuous, but it is performed with computer control.

1.6.3 According to Implementation of the Components of the Physical Properties

a. Electric control system,
b. Hydraulic control system,
c. Mechanical control system,
d. Electromechanical integrated control system, and
e. Thermal energy control system.

1.6.4 Classification According to the Type of Control

a. Open-loop control system, that is, there is no feedback loop between the output and input of the system, whereas the output signal has no effect in the input signal of the control system. The characteristic of open-loop control system indicates that the structure is simple, the cost is low, and it is usually used for the stability of the system structure without interference or the interferences of the system.

b. Closed-loop control system, that is, there is a feedback loop between the output and input of the system, whereas the output signal has an effect on the input system of the control system; thus, the closed-loop control system improves the control accuracy of the system.

It is important to mention that it does not matter what kind of control system, but the control system must be *stable*, *speedy*, and *accurate*. **Stability** is an important keyword to study control systems or control engineering; stability is the most basic requirement of the control system. The system refers to the dynamic process of oscillation trend or enables the system to restore the balance ability and the output disturbance signals with respect to the time of the equilibrium state. **Speed** is the stable condition in a control system. The system needs to be affected by the disturbance signal and the output quantity; however, it is required that the deviation quantity between the output quantity and the given quantity which can be eliminated with time after the system is influenced by the disturbance signal. **Accuracy** is measured from the deviation that is between the output quantity and the given quantity at the end of the adjustment process; it is also known as the steady-state accuracy. Accuracy is one of the important indicators to measure the performance of the system. Stability, speed, and accuracy of the control system are mutually restricted in the same control system; for excellent speed, there may be a strong oscillation; if we improve the stability, the control process may slow down and the accuracy may also be reduced. These problems are important issues; thus, we must solve this problem by studying in the field of control engineering.

1.7 DESIGN AND ANALYSIS PROCESS

In this section, we focus on control system performance and design process as well as the test waveforms used in the control system. Now, I am going to talk about the design and compensation of control systems. Performance specifications, system compensation, and design procedures are the key features of the design and compensation of control systems.

1.7.1 Design and Compensation of Control Systems

Performance specifications of the control systems are performed by the following procedures:

a. System compensation,

b. Design procedures, and

c. Performance specifications.

In design and compensation of a control system, performance specifications play an important role; they can be performed by stability, quickness or rapidness, and accuracy. The system is stable or unstable that can be determined by analyzing the system stability; the speed of the system response is analyzed by the quickness or rapidness; however, the system errors or accuracy or the higher or lower accuracy also play important roles that can be performed by system accuracy analysis (Figure 1.23).

1.7.2 Control Systems Analysis Process

The control systems analysis process is important to analyze the system. Therefore, the objectives of the analyses and design of the control system are performed by

a. Transient response analyses,

b. Steady-state response analyses, and

c. Stability analyses.

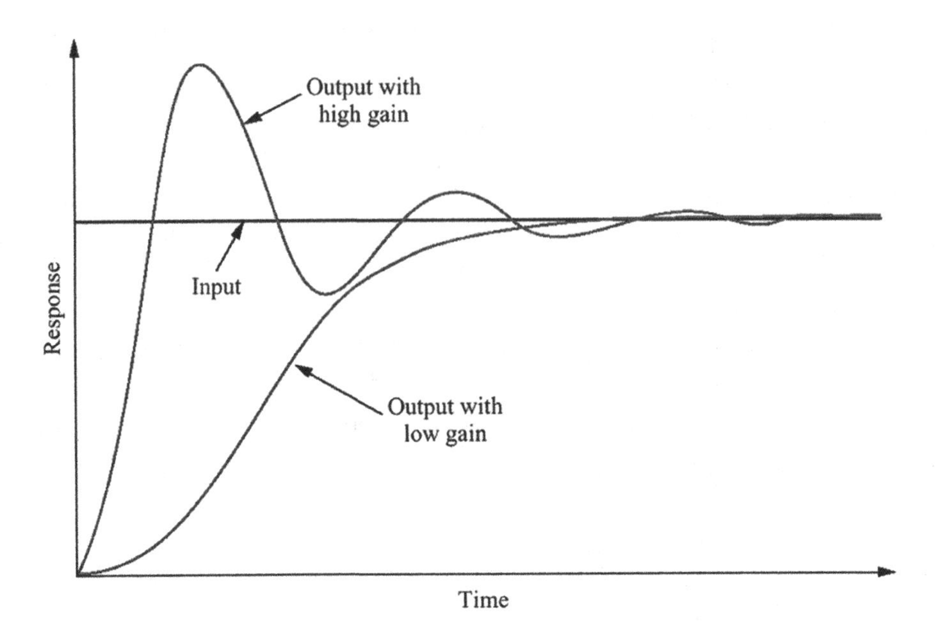

FIGURE 1.23 Performance analysis curve with high and low gain outputs of control system.

1. *Control systems analysis*

   We will see how to analyze the control system; Figure 1.23 is excellent for control systems analysis. Here, it can be seen that X-axis shows time and Y-axis shows response of a control system. The X–Y coordinate system shows a fixed straight input line that is parallel to the X-axis; however, there are two curves that can also be seen from this figure—one of the curves is existed output with high gain value and another curve is existed output with low gain value with respect to time. If the time value is increased, the amplitude of the output or response is decreased along with input value. As it also can be seen, the time is closed to infinity or equal to infinity, then the amplitude of the output or response are almost closed to the input value but the output with high gain curve is more oscillation than the output with low gain curve; however, the output with low gain curve is slower than the output with high gain curve and the output with low gain curve reaches closed to input gain before the output with higher gain curve.

   Let's see how to analyze the performance of a control system. Figure 1.24 shows an example of performance analysis of control system. It shows that the curve has X-axis which is indicated by time and the Y-axis shows the output of the system. Here, we can see the input command for input value and there is one response curve with low gain. As the starting time varies from initial point zero to initial condition zero, the response curve shows transient response until its final value of time; the response is almost in steady state from final value to infinity time; however, the steady-state response is almost parallel with the input command. The gap or difference between steady-state and input command is called steady-state error. To analyze the system performance of a control system, this transient response, steady-state response, and steady-state error are the main factors to study control system.

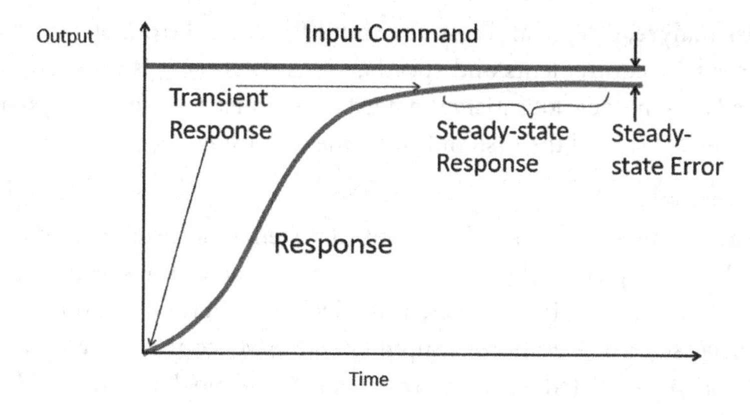

FIGURE 1.24 Performance analysis curve of control system.

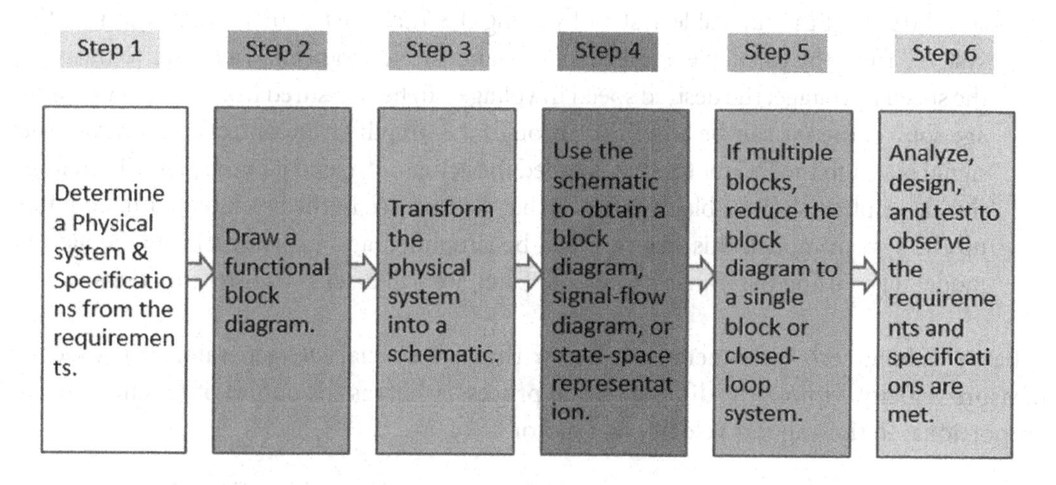

FIGURE 1.25 Control systems design process.

2. *Control systems design process*

To learn the control systems design process, we may follow the following six steps to design a control system, as shown in Figure 1.25.

Step 1: First, you should determine a physical system and specifications from the requirements.

Step 2: After getting all the specifications from Step 1, draw a functional block diagram of your specific system.

Step 3: We should transform the physical system into a schematic diagram for a more specific understanding.

Step 4: By using the schematic, you should obtain a block diagram, signal-flow diagram, or state-space representation of the control system.

Step 5: If there are multiple blocks and connections, you must reduce the block diagram to a simplified block diagram or a single block or closed-loop system so that you can easily build up a mathematical model or a simple model to analyze it easily.

Step 6: To analyze, design, and test, the simplified model from Step 6 can be used to observe the requirements and specifications whether it is satisfied or not. If the desired design does not satisfy the needs, you must redesign the system and must redo Steps 1–6 until the system fulfills the requirements.

3. *Design examples*
Let's see an example of a closed-loop control system of a speed turntable. Figure 1.26 shows a closed-loop control system of a speed turntable. The speed turntable is connected with DC motor, Tachometer, and a DC amplifier; DC amplifier is connected with a speed setting operational amplifier (OP-AM) connected with a battery. The OP-AM is also connected with Tachometer; and the block diagram of Figure 1.26 is shown in Figure 1.27.

How can you design the block diagram of this system? We want to measure the actual speed (rpm) of the turntable system by using the *Tachometer*; this is our output of the system. Thus, the turntable is called a plant and the Tachometer is a sensor to measure the speed in voltage; the desired speed in voltage can be measured from a battery or voltage source, then it can be amplified through the amplifier in controller. However, the signal is sent to the DC motor that is called the actuator to send it to the plant. Therefore, this description shows a block diagram that you can see in the block diagram. To build up the system model, this diagram can be simplified and expressed in mathematical modeling by using the transfer function. Later, we will discuss transfer functions.

What is tachometer? Tachometer is a sensor that is a special DC generator. It is used to measure velocity without the differentiation process, whereas the output of a tachometer is proportional to the angular velocity of a motor.

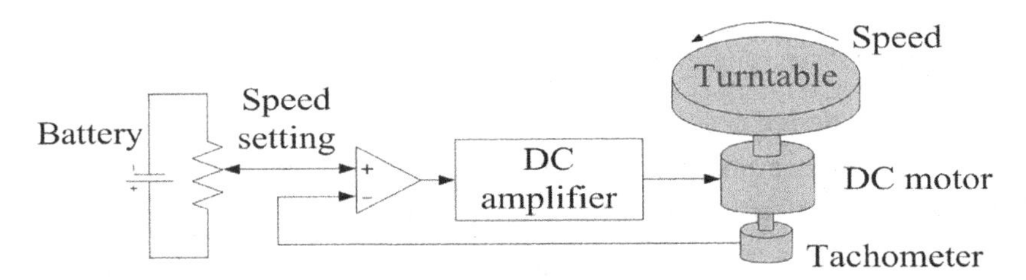

FIGURE 1.26 Closed-loop control system of a speed turntable.

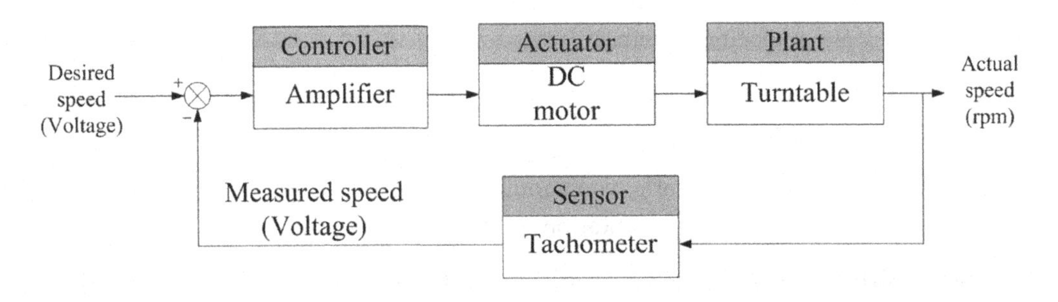

FIGURE 1.27 Block diagram of a speed turntable.

1.8 TEST WAVEFORMS USED IN CONTROL SYSTEMS

We will discuss the test waveforms used in the control system because it is very important to discuss for the control systems. Impulse signal, step signal, ramp signal, parabola signal, and sinusoid signal are usually the most useful test waveforms for the control systems, as shown in Table 1.1.

1. *Impulse signal*
   It is denoted by $\delta(t)$; $\delta(t)$ is equal to infinity for the ranges in the limit of negative zero to positive zero; and $\delta(t)$ is equal to zero for otherwise; or the integral value of $\delta(t)$ is equal to unity for ranges in the limit of negative zero to positive zero. The diagram of the impulse signal can be seen from the figure where the impulse can be seen only at the origin with amplitude. The impulse signal can be used for transient response modeling.

2. *Step signal*
   It is denoted by $u(t)$; $u(t)$ is equal to unity for the ranges of t is greater than zero; and $u(t)$ is equal to zero for the ranges of t is less than zero. The diagram of the step signal can be seen where the step can be seen as a DC line or a straight line that is parallel to the X-axis in the positive coordinate. The step signal can be used for transient response modeling and steady-state error analysis.

TABLE 1.1 Test Waveforms Used in Control Systems

Input	Function	Description	Sketch	Use
Impulse	$\delta(t)$	$\delta(t)=\begin{cases}\infty, \text{ for } 0^- < t < 0^+ \\ 0, \text{ otherwise}\end{cases}$ $\int_{0^-}^{0^+} \delta(t)dt = 1$	$f(t)$, $\delta(t)$, t	Transient response modeling
Step	$u(t)$	$u(t)=\begin{cases}1, \text{ for } t>0 \\ 0, \text{ for } t<0\end{cases}$	$f(t)$, t	Transient response, steady-state error.
Ramp	$tu(t)$	$tu(t)=\begin{cases}t, \text{ for } t\geq 0 \\ 0, \text{ otherwise}\end{cases}$	$f(t)$, t	Steady-state error.
Parabola	$\frac{1}{2}t^2u(t)$	$\frac{1}{2}t^2u(t)=\begin{cases}\frac{1}{2}t^2, \text{ for } t\geq 0 \\ 0, \text{ otherwise}\end{cases}$	$f(t)$, t	Steady-state error.
Sinusoid	$\sin\omega t$		$f(t)$, t	Transient response modeling, and steady-state error

3. *Ramp signal*
   It is denoted by $tu(t)$; $tu(t)$ is equal to unity for the ranges of t that is greater than or equal to zero; and $tu(t)$ is equal to zero for the other ranges of t. The diagram of the ramp signal can be seen from the figure in which the ramp can be seen as a DC line or a straight line that is passing through origin in the positive coordinate. The ramp signal can be used for steady-state error analysis.

4. *Parabola signal*
   It is denoted by $\frac{1}{2}t^2u(t)$; $\frac{1}{2}t^2u(t)$ is equal to $\frac{1}{2}t^2$ for the ranges of t is greater than or equal to zero; and $tu(t)$ is equal to zero for the other ranges of t. The diagram of the parabola signal can be seen from the figure where the parabola can be seen as a curve that is passing through its origin in the positive coordinate. The ramp signal can be used for steady-state error analysis.

5. *Sinusoid signal*
   It is denoted by $\sin\omega t$. The diagram of the sinusoid signal can be seen from the figure where the sinusoid can be seen as a sinusoid wave. The sinusoid signal can be used for transient response modeling and steady-state error analysis.

EXAMPLES

In this chapter, the students require to understand the basic concepts of the control system, the difference between the open-loop control and closed-loop control, and the basic principles and components of the closed-loop control system. The learning abstract should be displayed in a simple block diagram of a schematic diagram.

Example 1.1

In a refrigerator, the refrigeration system works as shown in Figure 1.28. Explain the working principle of the system and point out the controlled object, the amount and the quantity of the system, and draw the system block diagram.

Solution

The task is to keep the temperature in the refrigerator (T_c) equal to the set temperature (T_r). The refrigerator is the controlled object, the temperature is the controlled quantity in the box or cabinet, and the output voltage of the potentiometer is set by the controller knob (corresponding to the desired temperature T_r value), which is the given quantity.

The bimetallic temperature sensor (measuring element) in the temperature controller senses the temperature in the refrigerator and converts it into a voltage signal. It is compared with the output voltage of the potentiometer (given element) set by the controller knob (corresponding to the desired temperature T_r). The relay is controlled by the deviation voltage (Δu representing the deviation between actual temperature, T_c and desired temperature, T_r). When the deviation voltage (Δu) reaches a certain

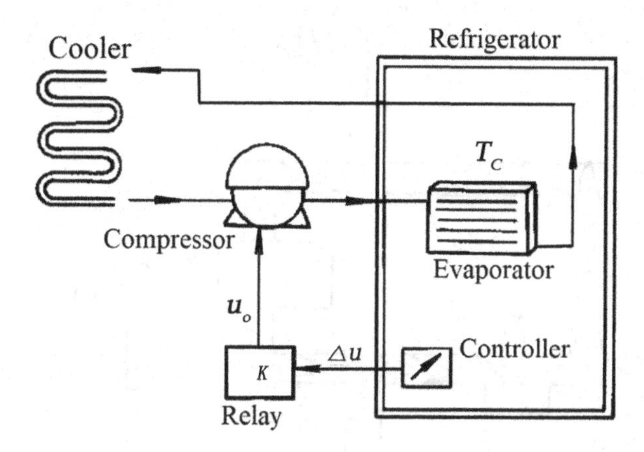

FIGURE 1.28 Schematic diagram of refrigerator cooling system.

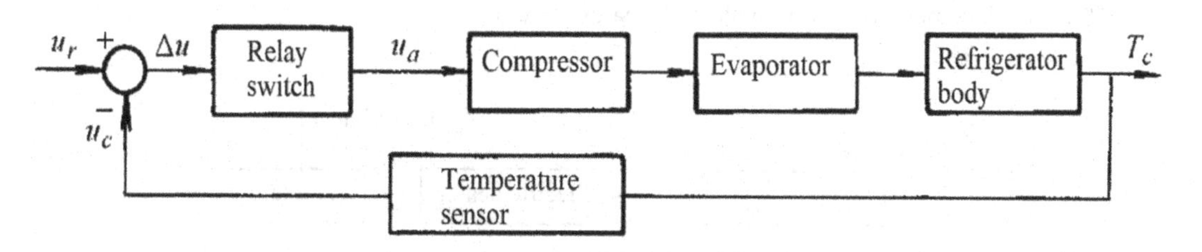

FIGURE 1.29 Block diagram of refrigerator cooling system.

value, the relay switching connects to start the compressor, whereas the high-temperature and low-pressure gaseous refrigerating liquid is transferred in the evaporator through the cooler to dissipate the heat.

After cooling, the low temperature and low-pressure refrigerating liquid are compressed into a low temperature and the high-pressure liquid to the evaporator. Decompression diffuses into gas rapidly, absorbs heat from the tank, the cabinet holds the lower the temperature. The high-temperature and low-pressure refrigerant are drawn into the cooler. By such kind of circulation flow, the refrigerator performs to achieve the effect of refrigeration. The relay, the compressor, the evaporator, and the cooler constitute the actuator of the system to complete the cooling function. The system block diagram of the refrigerator cooling system is shown in Figure 1.29.

Example 1.2

In this example, an electric hot water heater is provided as shown in Figure 1.30. In order to maintain the desired temperature, the power supply of the electric heater is switched on or off by the temperature control switch. When hot water is used, the water in the water tank is allowed to flow out of the water tank and the cold water is added. Explain the principle of the system and draw the schematic diagram of the system.

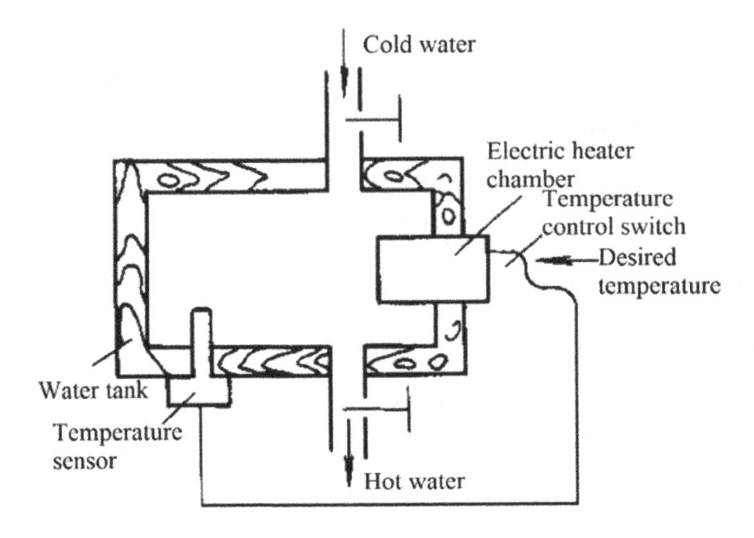

FIGURE 1.30 Schematic diagram of electric heater system.

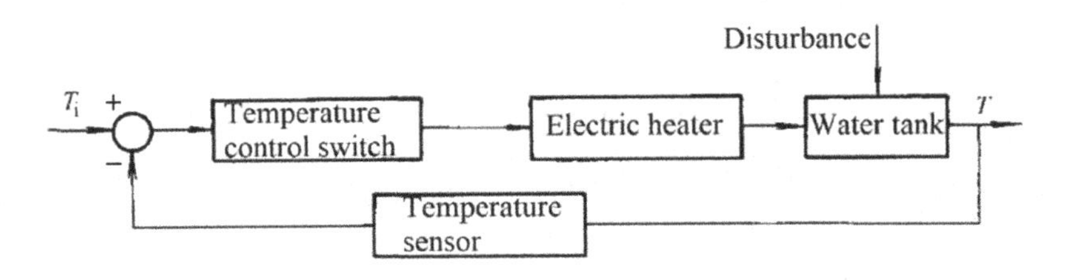

FIGURE 1.31 Block diagram of electric heater system.

Solution

In the electric water heater system, the input is the predetermined desired temperature (given value) which is set as T_i (°C); the output is the actual water temperature of the tank, set as T (°C); the control object is the water tank; disturbance signal is mainly caused by the cooling effect of releasing hot water and injecting cold water.

When $T_i = T$, the electric heater does not work, but the water temperature is kept as the desired temperature in the water tank. When hot water is used, the actual temperature drops due to disturbance. The temperature measuring element senses the change of $T < T_i$. This change is converted into an electrical signal to make the temperature switch work in temperature. The temperature control switch is powered on to heat up the water in the tank to reach at $T_i = T$. The block diagram of the system principle is shown in Figure 1.31.

Example 1.3

Figure 1.32 is the schematic block diagram of the automatic liquid-level control system. In any case, the liquid level (c) is expected to remain unchanged. Explain the working principle of the system and draw the block diagram of the system principle.

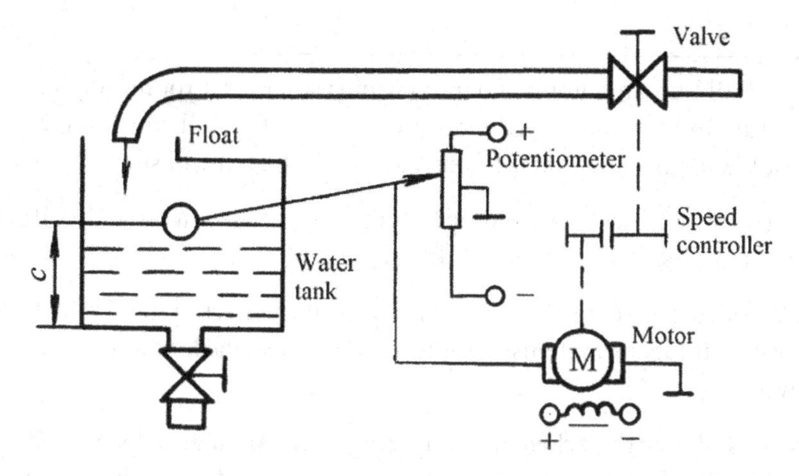

FIGURE 1.32 Schematic diagram of automatic control system of liquid-level control.

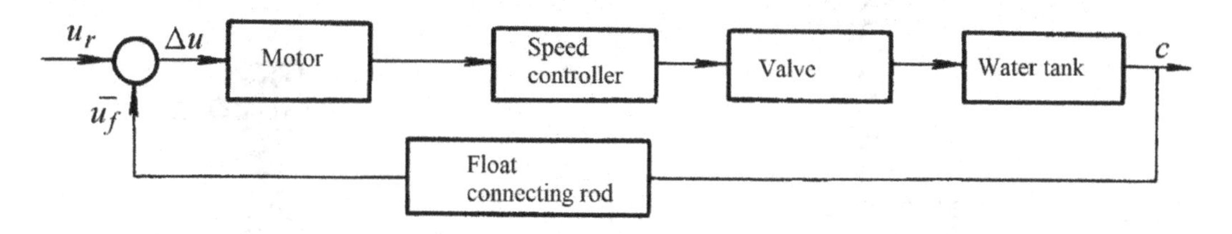

FIGURE 1.33 Block diagram of liquid-level automatic control system.

Solution

The control task of the system is to keep the liquid level at a constant. The water tank is the controlled object, the water level is the controlled quantity in the tank, and the voltage of the potentiometer is set to u_r (the desired value of the liquid level is c_r), which is the given quantity.

When the potentiometer brush is located in the midpoint (corresponding to u_r), the motor will not move, the control valve will be opened to a certain extent so that the inflow of the liquid is equal to the outflow of the liquid in the tank to maintain the liquid level at the desired height, c_r. Once the inflow or outflow of liquid changes, for example, when the liquid level is increased, the float position is also increased, and the brush of the potentiometer moves downward from the midpoint position through the lever; thus, it is provided a certain control voltage to the motor. Then the motor will drive to open the valve automatically by controlling through the speed controller or reducer. To reduce the flow or to control the speed of the flow to the liquid tank, the reducer or the speed controller is used to operate this automatic system. At this time, the liquid level of the tank drops and the float position drops accordingly until the potentiometer brush returns to the midpoint position; the system is in equilibrium again and the liquid level returns to the given height. On the contrary, if the liquid level of the tank drops, the system will automatically increase the speed of the speed controller to open the valve rapidly as well as increase the inflow of the liquid. The block diagram of the system principle is shown in Figure 1.33.

EXERCISES

1. Can you build up an automatic fan control system for a room? Suppose that if the temperature in the room is greater than 30°C, the fan will run automatically, but if the temperature in the room is lower than 30°C, the fan will stop.

2. There are many closed-loop and open-loop control systems in daily life for several specific examples. Explain their work principle.

3. What is feedback system? What are the effects of feedback system? What are negative and positive feedback systems? Why is the main feedback of a stable system always negative?

4. Figure 1.34 shows the schematic diagram of the angular velocity control system. The shaft of the governor rotates ω at an angular speed through the reduction gear.

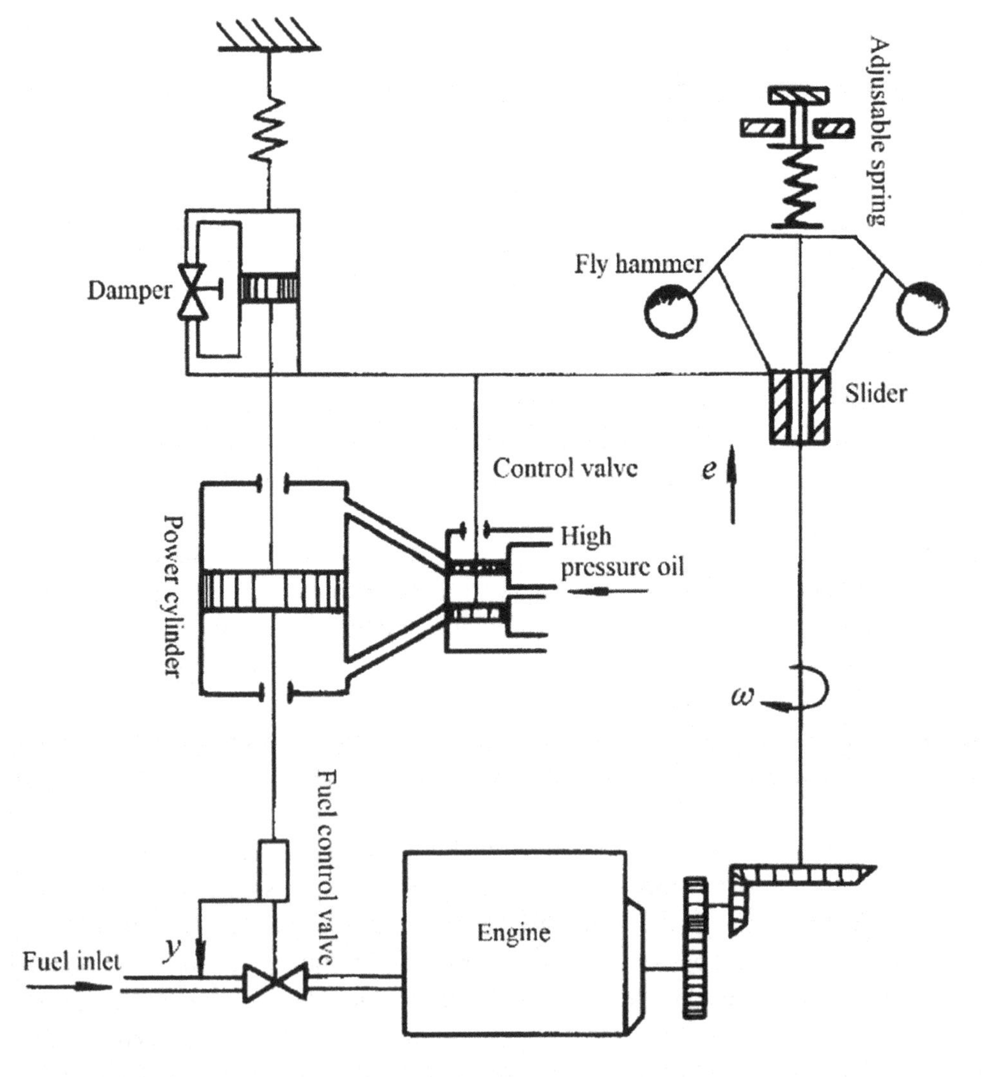

FIGURE 1.34 Schematic diagram of an angular velocity control system.

The centrifugal force is generated by the rotating hammer that is offset by the spring force. The required speed is adjusted by the spring prestresses. Draw the functional block diagram of the angular velocity control system.

5. Figure 1.35 is the block diagram of the automatic door control system of the warehouse. Explain the working principle of the automatic door control system and draw the system block diagram.

6. Figure 1.36 shows a schematic diagram of a hydraulic servo system with calibration. It consists of a servo valve, hydraulic cylinder, differential lever, and damper. The input is the displacement signal $r(t)$ at the end of the differential lever. The servo valve

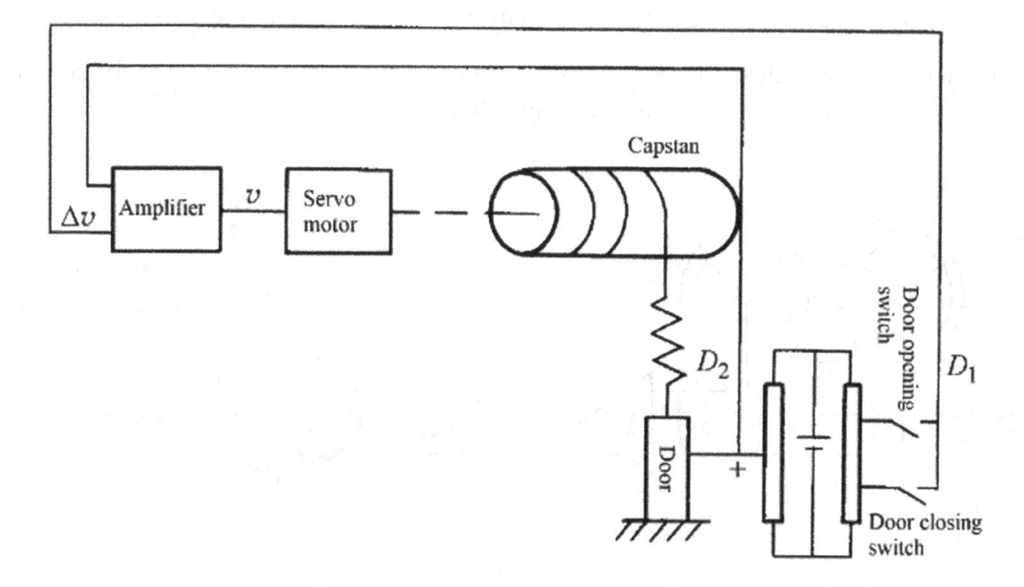

FIGURE 1.35 Schematic diagram of warehouse automatic door control system.

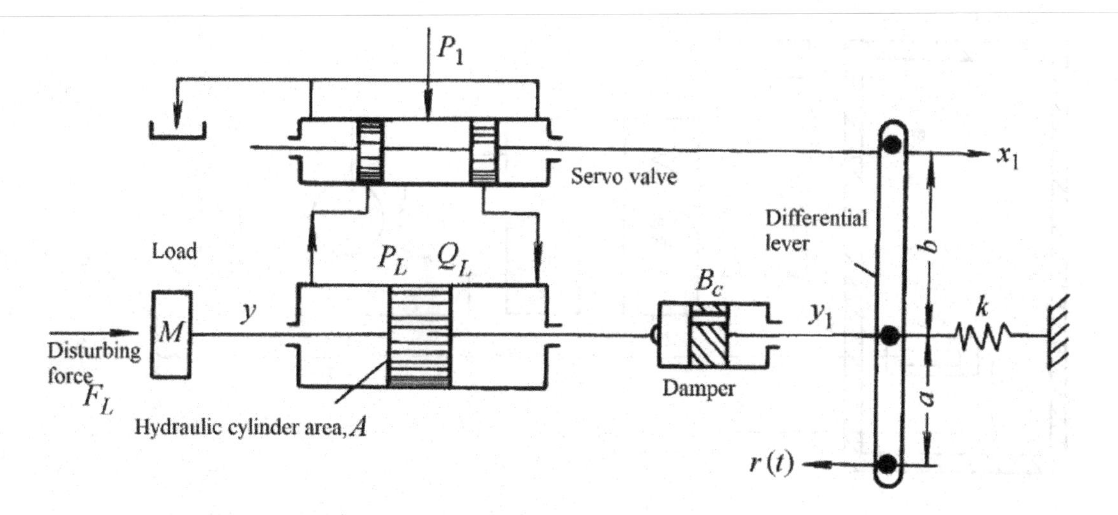

FIGURE 1.36 Schematic diagram of a hydraulic servo system.

is open, the hydraulic oil flows into the hydraulic cylinder, the piston moves, feedback is passed through the damper, and the valve is reduced to zero by the differential lever and the input displacement. Draw the block diagram of the system.

7. Figure 1.37 shows the position control system of a servo system. The angular displacement error detection device of the system is comprised of two potentiometers (instruction potentiometer and feedback potentiometer). Describe the working principle and draw the block diagram of the system.

8. Figure 1.38 is the schematic diagram of the electric furnace temperature control system. Analyze the working process of the system keeping the temperature of the electric furnace at constant, point out the controlled object, controlled quantity and the function of each part of the system, and draw the block diagram of the system.

9. There are many closed-loop and open-loop control systems in daily life for several specific examples. Explain their work principle.

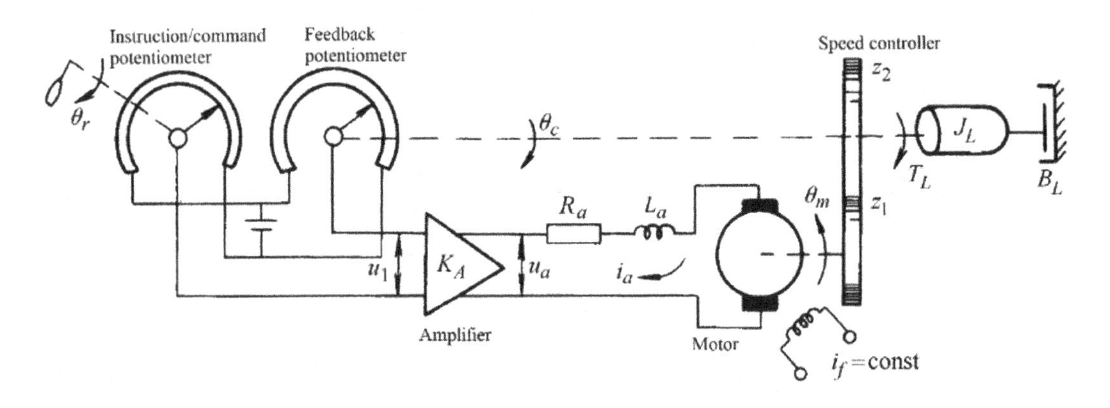

FIGURE 1.37 Schematic diagram of position control system of a servo.

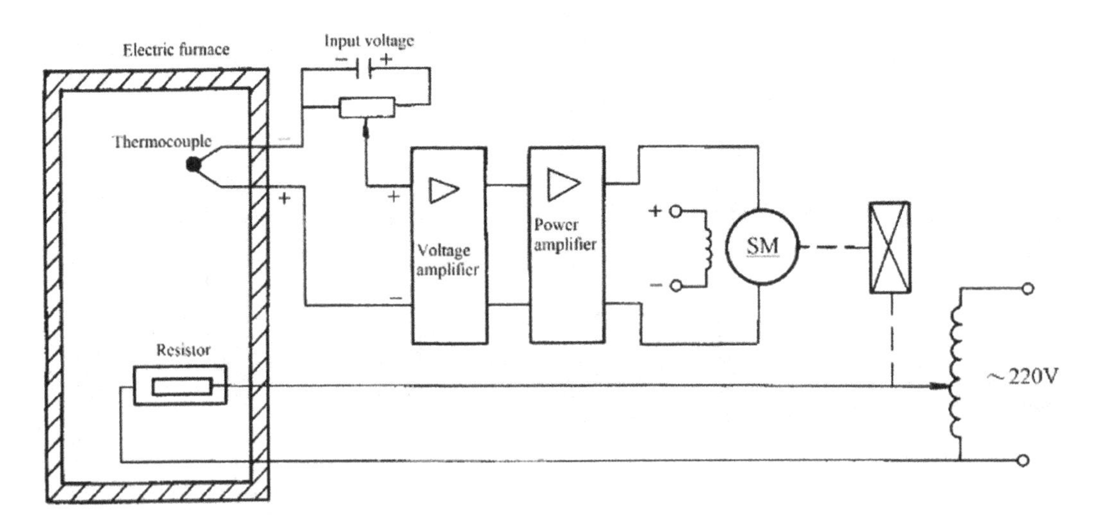

FIGURE 1.38 Schematic diagram of electric furnace temperature control system.

10. Figure 1.34 shows the schematic diagram of the angular velocity control system. The shaft of the governor rotates with ω at an angular speed through the reduction gear. The centrifugal force is generated by the rotating hammer, which is offset by the spring force. The required speed is adjusted by the spring prestresses. Draw the functional block diagram of the angular velocity control system.

11. Figure 1.35 is the block diagram of the automatic door control system of the warehouse. Explain the working principle of the automatic door control system and draw the system block diagram.

12. Figure 1.36 shows a schematic diagram of a hydraulic servo system with calibration. It consists of a servo valve, hydraulic cylinder, differential lever, and damper. The input is the displacement signal $r(t)$ at the end of the differential lever. The servo valve is open, the hydraulic oil flows into the hydraulic cylinder, the piston moves, feedback is passed through the damper, and the valve is reduced to zero by the differential lever and the input displacement. Draw the block diagram of the system.

13. Figure 1.37 shows the position control system of a servo system. The angular displacement error detection device of the system is composed of two potentiometers (instruction potentiometer and feedback potentiometer). Describe the working principle and draw the block diagram of the system.

14. Figure 1.38 is the schematic diagram of the electric furnace temperature control system. Analyze the working process of the system keeping the temperature of the electric furnace at constant, point out the controlled object, controlled quantity, and the function of each part of the system, and draw the block diagram of the system.

MCQ AND TRUE/FALSE QUESTIONS

1. The influence of the output signal of the system on the control action has
   a. open loop
   b. closed loop
   c. not both (a) and (b)
   d. both (a) and (b)

2. For system antiinterference ability
   a. open loop is stronger
   b. closed loop is stronger
   c. both are strong
   d. both are not strong

3. As a system

   a. open-loop oscillation

   b. closed-loop nonoscillation

   c. open loop must be oscillation

   d. closed loop must be oscillation

4. Control is an action that can measure the value of a controlled variable. It means that control measures the value of the controlled variable of the system.
   Answer: True/False

5. Control variable is the quantity or condition that is measured and controlled in the system, and controlled variable is the **response or output** of the system.
   Answer: True/False

6. A system is a combination of components and machines that are used in the system.
   Answer: True/False

7. How you can compensate for the predictable disturbances within the system?

   a. By using compensator(s).

   b. By using operational amplifiers.
   Answer: (a) True (b) False

8. The first significant work in automatic control was James Watt's centrifugal governor for the speed control of a steam engine that was invented in the

   a. 20th century

   b. 18th century

   c. 17th century

   d. 16th century

9. In 1922, Minorsky worked on automatic controllers for steering ships and showed how stability could be determined from the differential equations describing the system.
   Answer: True/False

10. In 1932, Nyquist developed a relatively simple procedure for determining the stability of closed-loop systems on the basis of open-loop response to steady-state sinusoidal inputs.
    Answer: True/False

11. In 1934, Hazen, who introduced the term *servomechanisms* for position control systems, discussed the design of relay servomechanisms capable of closely following a changing input.
    Answer: True/False

12. During the decade of the 1940s, frequency-response methods (especially the Bode diagram methods due to Bode) made it possible for engineers to design linear closed-loop control systems that satisfied performance requirements.
    Answer: True/False

13. Many industrial control systems in the 1900s and 1950s used PID controllers to control pressure, temperature, etc.
    Answer: True/False

14. In the early 1940s, Ziegler and Nichols suggested rules for tuning PID controllers, called Ziegler–Nichols tuning rules. From the late 1940s to the 1950s, the root-locus method was fully developed by Evans.
    Answer: True/False

15. From 1960 to 1980, optimal control of both deterministic and stochastic systems, as well as adaptive and learning control of complex systems, were fully investigated.
    Answer: True/False

16. When was the modern control theory centered around robust control and associated topics?

    a. From the 1980s to the 1990s

    b. From the 1970s to the 1979s

    c. From the 1960s to the 1970s

    d. From the 1940s to the 1950s

17. Modern control theory is based on the time-domain analysis of differential equation systems.
    Answer: True/False

18. Modern control theory made the design of control systems simpler because the theory is based on a model of an actual control system. However, the system's stability is sensitive to the error between the actual system and its model.
    Answer: True/False

19. The temperature in the electric furnace is measured by a thermometer, which is an

    a. analog device

    b. digital device

20. The analog temperature is converted to a digital temperature by a converter.

    a. A/D

    b. D/A

21. The digital temperature is fed to a controller through an interface. This digital temperature is compared with the programmed input temperature, and if there is any discrepancy (error), the controller sends out a signal to the heater, through an interface, amplifier, and relay, to bring the furnace temperature to the desired value.
    Answer: True/False

22. In the 17th century, Cornelis Drebbel in Holland invented a purely mechanical temperature control system for hatching eggs.
    Answer: True/False

23. "The criterion of Dynamical Stability" was invented in

    a. 1877 (by Adams Prize)

    b. 1868 (by James Clerk Maxwell)

24. The situation of stability, stabilization, and steering development was published as stability criterion for third-order-based coefficient of differential equation in and extended stability to fifth-order systems in

    a. 1868 (by James Clerk Maxwell); 1874 (by Edward John Routh)

    b. 1877 (by Adams Prize); 1868 (by Edward John Routh)

25. In which year, Henry Ford's mechanized automobile production line.

    a. 1913

    b. 1940

    c. 1868

    d. 1877

26. H.W. Bode and H. Nyquist developed the analysis of at by using frequency-response methods.

    a. feedback amplifiers; Bell Telephone Lab

    b. feedback system; AT&T Lab

27. In 1948, Walter R. Evans invented root-locus methods, and in 1952, the Massachusetts Institute of Technology (MIT) developed the first

a. control machine tool

b. numerical control machine tool

c. robot hand tool

d. above all the tools

28. Space Shuttle Columbia was launched successfully for the first time in 1981.
    Answer: True/False

29. The open-loop system has some excellent advantages such as

    a. Simple construction and less expensive

    b. No stability problem

    c. Convenient to use

    d. All of them

30. The open-loop control system has also some disadvantages such as it has lower sensitivity to disturbances and the inability to correct the disturbances.
    Answer: True/False

31. The *closed-loop control* system is often referred to as *Positive feedback and Negative feedback system;* however, the output of the closed-loop control system affects (s) on control action.
    Answer: True/False

32. The advantages and disadvantages of the closed-loop control systems compared with open-loop control systems. The advantages can be seen such as less sensitivity to external disturbances and internal variations, and it has greater accuracy. In addition, the disadvantages are obviously higher in cost and power and more complex in structure.
    Answer: True/False

33. It is necessary for us to learn about the components of closed-loop control system. The typical closed-loop control system consists of a signal generator, measurement elements or sensors, comparison elements, amplifier elements, actuator, controller or compensator, and controlled plant.
    Answer: True/False

34. To design the compensation of a control system, the performance specifications are an important role; it can be performed by stability, quickness or rapidness, and accuracy.
    Answer: True/False

35. The control systems' analysis process is important to analyze the system. Therefore, the objectives of the analyses and design of the control system are performed by

    a. Transient response analyses

    b. Steady-state response analyses

    c. Stability analyses

    d. All of them

36. A special DC generator is used to measure velocity without a differentiation process. The output of a tachometer is proportional to the linear velocity of the motor.
    Answer: True/False

37. As the starting time varies from initial point zero to initial condition zero, the response curve shows transient response until its final value of time; the response is almost steady state from the final value to infinity time; however, the steady-state response is no parallel with the input command.
    Answer: True/False

38. The gap or difference between steady-state and input command is called steady-state error.
    Answer: True/False

39. To determine a physical system and specifications from the requirements of the control system is step

    a. 6

    b. 3

    c. 1

    d. 2

40. To measure the actual speed of the turntable system by using the Tachometer, what unit should be considered?

    a. rpm

    b. prm

    c. no need unit

    d. rmp

41. The impulse signal can be used for transient response modeling and steady-state error analysis.
    Answer: True/False

42. The step signal can be used for transient response modeling and steady-state error analysis

    Answer: True/False

43. The ramp signal can be used for steady-state error analysis.

    Answer: True/False

44. The ramp signal can be used for steady-state error analysis

    Answer: True/False

45. The sinusoid signal can be used for transient response modeling, and steady-state error analysis.

    Answer: True/False

CHAPTER 2

Transfer Function and Modeling of Control Systems

2.1 MATHEMATICAL MODELING OF THE CONTROL SYSTEM

Before going to discuss the transfer function, we can see the formation of the mathematical model for the transfer function, as the mathematical model is a set of equations that represent the dynamics of the system (Figure 2.1).

When we look at the organization of a mathematical model, we can see that the model is divided into frequency domain, time domain, block diagram, and signal flow graph; however, time domain is divided into two sections—the differential equation and the state space.

What do the signal in the time domain and frequency domain look like? Let's look at Figure 2.2; in this figure, we can see the time axis along the X-axis of the signal and the amplitude of the signal along the Y-axis. If we see carefully, we can observe that the signal shows wave(s) of harmonics in the time domain, whereas we can see only the single bar and different types of frequencies show their peak value of amplitude of the signal in the frequency domain. The block diagram and signal flow graph will be discussed later.

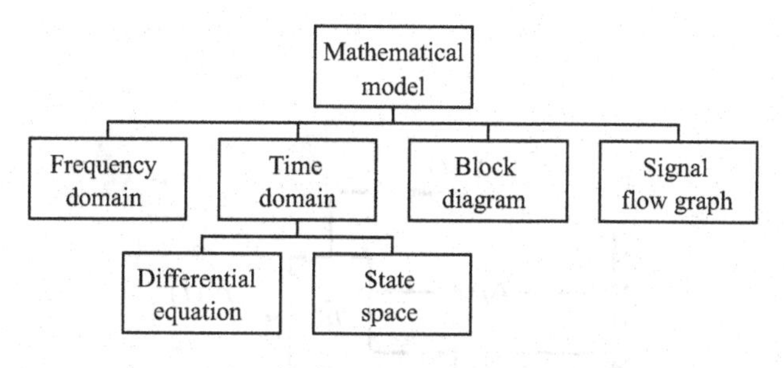

FIGURE 2.1 Organization of mathematical model of control system.

DOI: 10.1201/9781003293859-2

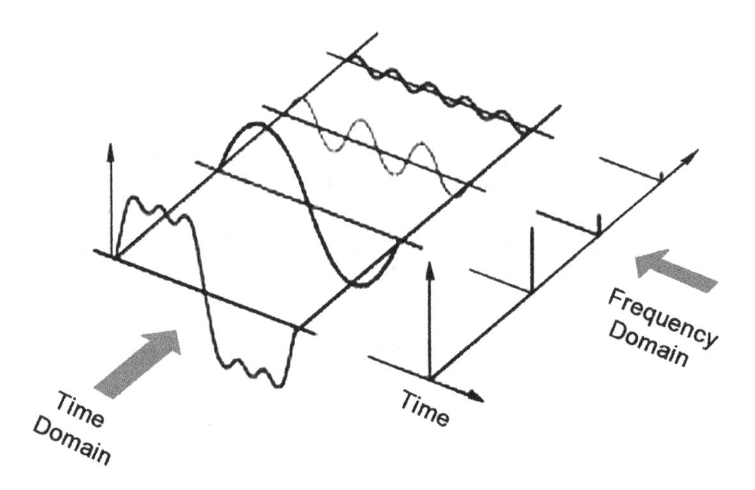

FIGURE 2.2 Signal in the time domain and frequency domain.

2.1.1 Differential Equation of the Control System

Example 2.1

Figure 2.3 shows the mass-spring-damping system, where f is the viscosity coefficient, m is the mass, k is the spring stiffness, $F(t)$ is the input force, and $y(t)$ is the output displacement.

Solution

According to Newton's second law, we can write

$$F(t) - f\dot{y}(t) - ky(t) = m\ddot{y}(t) \tag{2.1}$$

and

$$m\ddot{y}(t) + f\dot{y}(t) + ky(t) = F(t) \tag{2.2}$$

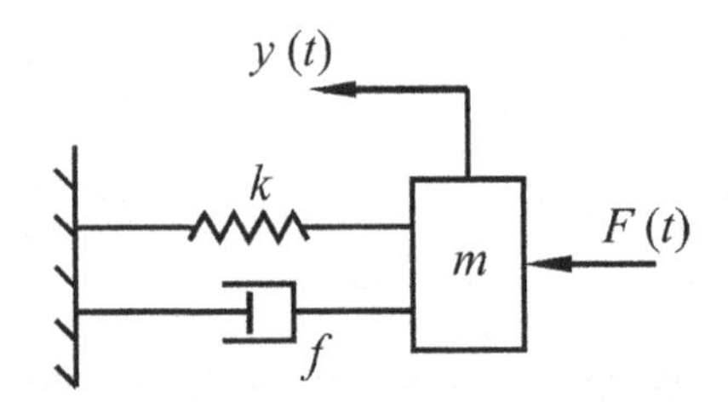

FIGURE 2.3 Mass-spring-damping system.

Example 2.2

Obtain the transfer functions $X_1(s)/U(s)$ and $X_2(s)/U(s)$ of the mechanical system shown in Figure 2.4a.

Solution

From the above mechanical system, it can be simplified that $U(s)$ is the input and $X_2(s)$ is the output, and $G(s)$ is the transfer function of the system. The system can be separated into two parts as the forces on m_1 due only to the motion of m_1 which is shown in Figure 2.4b and the forces on m_1 due only to the motion of m_2 which is shown in Figure 2.4c.

The equations of motion for the system can be found by using the Newton's law from Figure 2.4b the forces on m_1 due only to the motion of m_1 which is shown in

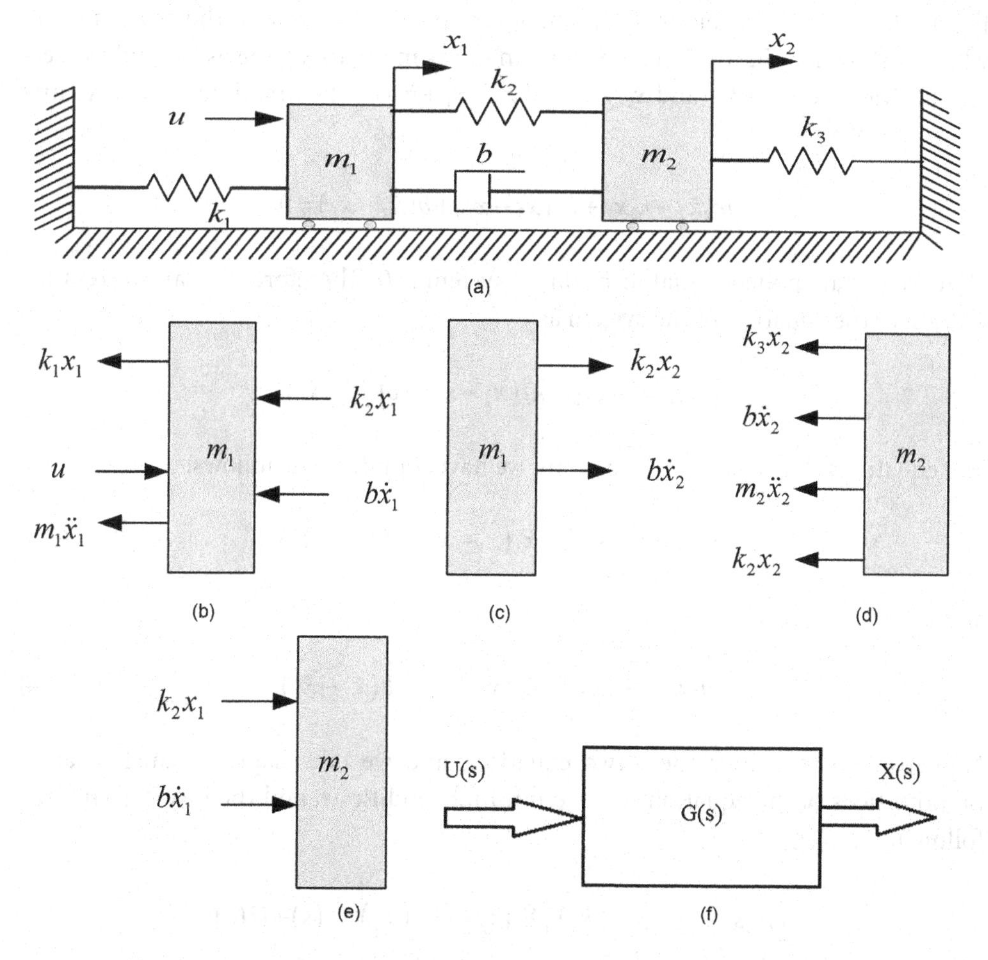

FIGURE 2.4 (a) Mass, spring, and damper system; (b) transformed free-body diagram due to motion of m_1; (c) transformed free-body diagram due to motion of m_2; (d) transformed free-body diagram due to motion of m_2; (e) transformed free-body diagram due to motion of m_1; and (f) transfer function of Figure 2.4a.

Figure 2.4b, and from Figure 2.4c, the forces on m_1 due only to the motion of m_2 which is shown in Figure 2.4c, then we can say the following:

$$m_1\ddot{x}_1 + k_1x_1 + k_2(x_1 - x_2) + b(\dot{x}_1 - \dot{x}_2) = u$$

or

$$m_1\ddot{x}_1 = u - k_1x_1 - k_2(x_1 - x_2) - b(\dot{x}_1 - \dot{x}_2)$$

whereas the system can be separated into other two parts as the forces on m_2 due only to the motion of m_2 which is shown in Figure 2.2d, and the forces on m_2 due only to the motion of m_1 which is shown in Figure 2.4e.

The equations of motion for the system can be found by using Newton's law from Figure 2.2d the forces on m_2 due only to the motion of m_2, which is shown in Figure 2.2d, and from Figure 2.2e, the forces on m_2 due only to the motion of m_1 which is shown in Figure 2.2e, then we can say as in the following as, m_2 and the second derivative of $x_2 + k_3$ and $x_2 + k_2$ and $x_3 - x_1 + b$ and the first derivative of x_3-first derivative of $x_1 = 0$.

$$m_2\ddot{x}_2 + k_3x_2 + k_2(x_2 - x_1) + b(\dot{x}_2 - \dot{x}_1) = 0$$

Here, the total applied initial force on m_2 system is 0. Therefore, we can express the motion of the equation of the system as

$$m_2\ddot{x}_2 = -k_3x_2 - k_2(x_2 - x_1) - b(\dot{x}_2 - \dot{x}_1)$$

The equations of motion for the system, we have found, are as follows:

$$m_1\ddot{x}_1 = u - k_1x_1 - k_2(x_1 - x_2) - b(\dot{x}_1 - \dot{x}_2) \tag{2.3}$$

and

$$m_2\ddot{x}_2 = -k_3x_2 - k_2(x_2 - x_1) - b(\dot{x}_2 - \dot{x}_1) \tag{2.4}$$

Now, we can rearrange these two equations and we take Laplace transformation on both sides of the equation with zero initial conditions, and then we can find the following:

$$\left[m_1s^2 + bs + (k_1 + k_2)\right]X_1(s) = (bs + k_2)X_2(s) + U(s)$$

and

$$\left[m_2s^2 + bs + (k_2 + k_3)\right]X_2(s) = (bs + k_2)X_1(s)$$

After rearranging these two equations, we can find the following expressions:

$$X_2(s)=\frac{(bs+k_2)X_1(s)}{\left[m_2s^2+bs+(k_2+k_3)\right]}$$

and

$$\left[m_1s^2+bs+(k_1+k_2)\right]X_1(s)=(bs+k_2)X_2(s)+U(s)$$

$$\left[m_1s^2+bs+(k_1+k_2)-\frac{(bs+k_2)^2}{m_2s^2+bs+(k_2+k_3)}\right]X_1(s)=U(s)$$

$$\left[\frac{(m_1s^2+bs+k_1+k_2)(m_2s^2+bs+k_2+k_3)-(bs+k_2)^2}{m_2s^2+bs+k_2+k_3}\right]X_1(s)=U(s)$$

and

$$\left[m_2s^2+bs+(k_2+k_3)\right]X_2(s)=(bs+k_2)X_1(s)$$

$$X_1(s)=\left[\frac{m_2s^2+bs+(k_2+k_3)}{(bs+k_2)}\right]X_2(s)$$

$$\left[\frac{(m_1s^2+bs+k_1+k_2)(m_2s^2+bs+k_2+k_3)-(bs+k_2)^2}{m_2s^2+bs+k_2+k_3}\right]X_1(s)=U(s)$$

$$\left[\frac{(m_1s^2+bs+k_1+k_2)(m_2s^2+bs+k_2+k_3)-(bs+k_2)^2}{bs+k_2}\right]X_2(s)=U(s)$$

Therefore, the transfer function can be found the following:

$$G_1(s)=\frac{X_1}{U}=\frac{m_2s^2+bs+k_2+k_3}{(m_1s^2+bs+k_1+k_2)(m_2s^2+bs+k_2+k_3)-(bs+k_2)^2}$$

and

$$G_2(s)=\frac{X_2}{U}=\frac{bs+k_2}{(m_1s^2+bs+k_1+k_2)(m_2s^2+bs+k_2+k_3)-(bs+k_2)^2}$$

Example 2.3

Figure 2.5 shows the mechanical model of a mass-damping system, where m is the mass, f_1, f_2 are the viscous damping coefficients, r is the input displacement, y is the output displacement.

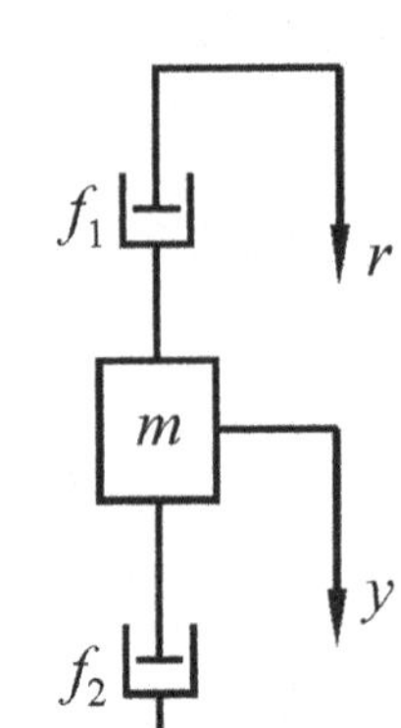

FIGURE 2.5 Mass-damping system.

Solution

According to Newton's second law, we can write

$$m\ddot{y}(t)+f_2\dot{y}(t)=f_1\left[\dot{r}(t)-\dot{y}(t)\right]$$

Simplifying Eq. (2.1), we get

$$m\ddot{y}(t)+\left(f_1+f_2\right)\dot{y}(t)=f_1\dot{r}(t)$$

Example 2.4

Figure 2.6a shows the passive *RC* circuit network; $u_i(t)$ is the input voltage of the network; $u_o(t)$ is the output voltage of the network; C is the capacitance; and R is the resistance.

Solution

According to Kirchhoff's law, we can write

$$u_i(t)=i(t)R+u_o(t) \tag{2.5}$$

$$u_o(t)=\frac{1}{C}\int i(t)dt \tag{2.6}$$

Simplifying Eqs. (2.5) and (2.6), we get

$$RC\frac{du_o(t)}{dt}+u_o(t)=u_i(t)$$

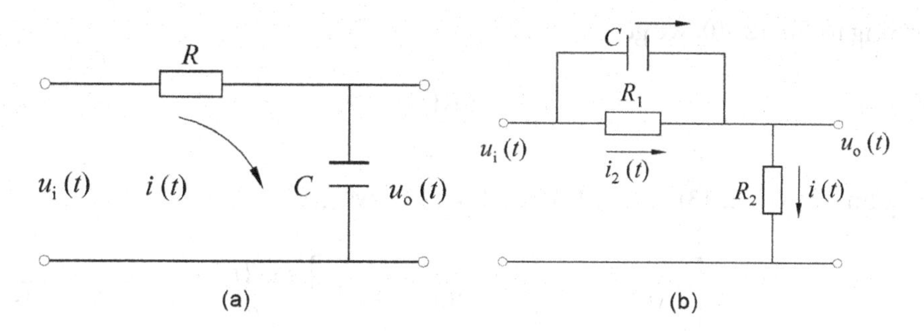

FIGURE 2.6 (a) RC networks and (b) passive complex circuitry networks.

Example 2.5

Figure 2.6b shows the passive *RC* circuit network; $u_i(t)$ is the input voltage of the network; $u_o(t)$ is the output voltage of the network; C is the capacitance; and R_1, R_2 is the resistance.

Solution

According to Kirchhoff's law, we can say

$$i_1(t)+i_2(t)=i(t) \tag{2.7}$$

$$u_i(t)=u_o(t)+R_1i_2(t) \tag{2.8}$$

$$\frac{1}{C}\int i_1(t)dt = R_1i_2(t) \tag{2.9}$$

and

$$u_o(t)=R_2i(t) \tag{2.10}$$

According to Eq. (2.8), we get

$$i_2(t)=\frac{u_i(t)-u_o(t)}{R_1} \tag{2.11}$$

According to Eq. (2.9), we get

$$i_1(t)=R_1C\frac{di_2(t)}{dt} \tag{2.12}$$

Putting Eq. (2.11) into Eq. (2.12), we get

$$i_1(t)=C\left[\dot{u}_i(t)-\dot{u}_o(t)\right] \tag{2.13}$$

According to Eq. (2.10), we get

$$i(t)=\frac{u_o(t)}{R_2} \tag{2.14}$$

Putting Eqs. (2.11), (2.13) and (2.14) into Eq. (2.7), we get

$$C\left[\dot{u}_i(t)-\dot{u}_o(t)\right]+\frac{1}{R_1}\left[u_i(t)-u_o(t)\right]=\frac{u_o(t)}{R_2} \tag{2.15}$$

$$R_1C\dot{u}_o(t)+\frac{R_1+R_2}{R_2}u_o(t)=R_1C\dot{u}_i(t)+u_i(t) \tag{2.16}$$

Example 2.6

Figure 2.7 shows the active network of the operational amplifier—$u_i(t)$ is the input voltage of the network, $u_o(t)$ is the output voltage of the network, C is the capacitance, R is the resistance, and K_o represents the multiplier amplification factor of the operational amplifier.

Solution

Let the feedback input at point A from the output $u_o(t)$ through a capacitance C of operational amplifier, K_o is generally very large multiplier amplification factor of the operational amplifier, and thus

$$u_o(t)=-K_ou_A(t) \tag{2.17}$$

It can be written as

$$u_A(t)=-\frac{u_o(t)}{K_o}\approx 0 \tag{2.18}$$

Since the input impedance is very high, we can say

$$i_1(t)\approx i_2(t) \tag{2.19}$$

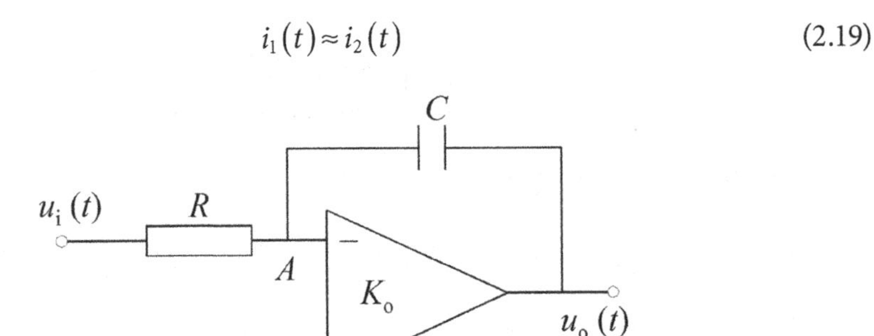

FIGURE 2.7 Active operational amplifier networks.

Thus, we can get

$$\frac{u_i(t)}{R} = -C\frac{du_o(t)}{dt} \tag{2.20}$$

Consequently,

$$RC\dot{u}_o(t) = -u_i(t) \tag{2.21}$$

Example 2.7

Figure 2.8 shows the armature-controlled dc type electric motor—$e_i(t)$ is the input voltage source in the armature-controlled motor system, $\omega_m(t)$ is the speed of the rotor system, R_a is the resistance of the armature winding, L_a is the inductance of the armature winding, $i_a(t)$ is the current which flows in the armature circuit, $e_m(t)$ is the voltage drops in the rotation system by rotor which is the induced electromotive force of the electric motor, $T(t)$ is the applied torque in the rotor system, $\theta_o(t)$ is the angular rotation or motion in the rotor system of this system network, f is the coefficient of the viscosity friction of the assuming electric motor and the load, J is the moment of inertia of the assuming electric motor and load, K_T is the coefficient of the electric motor torque, and K_e is the coefficient of the reverse-electric potential.

Solution

According to the Kirchhoff voltage law (KVL), the sum of all the voltages around a loop is equal to zero which means that the total applied voltage $e_i(t)$ is equal to the summation of the voltage drops (v_R, v_L, v_m) in resistor $R_a(t)$, inductor $L_a(t)$, and rotor system. Therefore, we get

$$e_i(t) = v_R + v_L + v_m$$

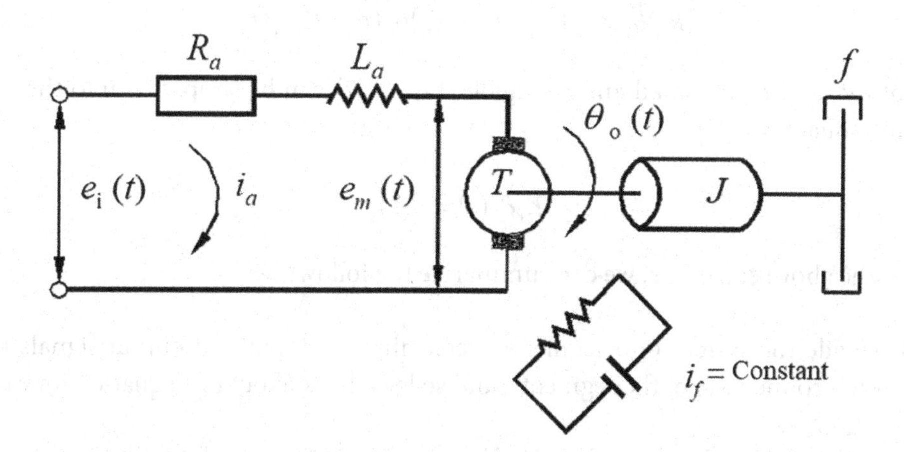

FIGURE 2.8 An armature-controlled direct-current type electric motor.

or

$$e_i(t) = i_a(t)R_a + L_a\frac{di_a(t)}{dt} + e_m(t) \tag{2.22}$$

According to the law of magnetic field effect to the load winding in the armature current circuit, the torque is developed by the motor which is proportional to the armature current $i_a(t)$; thus, we can say

$$T(t) = K_T i_a(t) \tag{2.23}$$

Here K_T is the motor torque constant.

According to the law of electromagnetic induction, we get

$$e_m(t) = K_e\dot{\theta}_o(t) \tag{2.24}$$

According to Newton's second law of motion, we get

$$T(t) - f\dot{\theta}_o(t) = J\ddot{\theta}_o(t) \tag{2.25}$$

Substituting Eq. (2.23) into Eq. (2.25), we get

$$i_a(t) = \frac{J}{K_{T_o}}\ddot{\theta}(t) + \frac{f}{K_T}\dot{\theta}_o(t) \tag{2.26}$$

Substituting Eqs. (2.24) and (2.26) into Eq. (2.22), we get

$$L_aJ\dddot{\theta}_o(t) + (L_af + R_aJ)\ddot{\theta}_o(t) + (R_af + K_TK_e)\dot{\theta}_o(t) = K_Te_i(t) \tag{2.27}$$

If L_a is small and negligible, the differential equation can be simplified into the following equation:

$$R_aJ\ddot{\theta}_o(t) + (R_af + K_TK_e)\dot{\theta}_o(t) = K_Te_i(t) \tag{2.28}$$

If both R_a and L_a are small and negligible, Eq. (2.27) can be simplified into the following equation:

$$K_e\dot{\theta}_o(t) = e_i(t) \tag{2.29}$$

From the above equations, we can summarize the following:

1. Divide the system into segments, determine the input and output signals for each connection of the segments, and so let's think about one equation for each segment.
2. According to the laws of physics or experimental methods, using the laws of physics, build primary equation of every segment and consider for simplification and linearization.

3. The equations of each segment are simultaneous; thus, we should cancel the intermediate variables, finally, the system differential equations of input and output variables and other parameters are obtained.
4. Linear differential equations with single input and single output can be arranged as follows:

$$a_0 \frac{d^n x_o(t)}{dt^n} + a_1 \frac{d^{n-1} x_o(t)}{dt^{n-1}} + a_2 \frac{d^{n-2} x_o(t)}{dt^{n-2}} + \cdots + a_{n-1} \frac{dx_o(t)}{dt} + a_n x_o(t) = b_0 \frac{d^m x_i(t)}{dt^m} +$$

$$b_1 \frac{d^{m-1} x_i(t)}{dt^{m-1}} + b_2 \frac{d^{m-2} x_i(t)}{dt^{m-2}} + \cdots + b_{m-1} \frac{dx_i(t)}{dt} + b_m x_i(t); \quad (n \geq m) \tag{2.30}$$

2.1.2 Linearization of the Differential Equation

In the actual control system, there are several kinds of nonlinear zone such as dead zone, saturated zone, gap, and other nonlinearities. Since the principle of linear superposition is not suitable for nonlinear systems, every system must be linearized. If the effect of the nonlinear factor is very small, the linearization is feasible. If the linearization is unfeasible, the nonlinear theory must be used. The methods of linearization are described below.

In general, the nonlinear characteristics of components or systems are shown in Figure 2.9, and the nonlinear functions are used as

$$y = f(x) \tag{2.31}$$

Description of the linearization method: The nonlinear function is developed into the Taylor series near the operating point x_0, and then the higher order term is omitted to obtain a linear function with increment variable $(x - x_0)$,

$$y = f(x) = f(x_0) + \left(\frac{df(x)}{dx}\right)_{x_0}(x - x_0) + \frac{1}{2!}\left(\frac{d^2 f(x)}{dx^2}\right)_{x_0}(x - x_0)^2 + \ldots \tag{2.32}$$

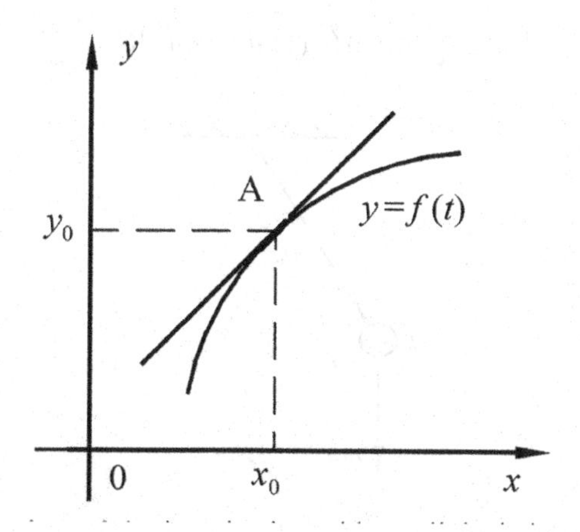

FIGURE 2.9 Linearization with small deviation.

If the increment variable $(x - x_0)$ is very small, we can omit its high-order term, and then we get

$$y - y_0 = f(x) - f(x_0) = \left(\frac{df(x)}{dx_0}\right)_{x_0} (x - x_0)$$

or

$$\Delta y = \left(\frac{df(x)}{dx_0}\right)_{x_0} \Delta x = k\Delta x \tag{2.33}$$

where $k = \left.\frac{df(x)}{dx}\right|_{x_0}$ is the proportional coefficient that represents the tangent gradient of the function $f(x)$ at point A. Similarly, we may say for the multivariable nonlinear function as in the following:

$$y = f(x_1, x_2, \ldots, x_n)$$

The closed working point of $(x_{10}, x_{20}, \ldots, x_{n0})$ has a new linear increment function as

$$\Delta y = \left.\frac{\partial f}{\partial x_1}\right|_{x_{10}, x_{20}, \ldots, x_{n0}} \Delta x_1 + \left.\frac{\partial f}{\partial x_2}\right|_{x_{10}, x_{20}, \ldots, x_{n0}} \Delta x_2 + \ldots + \left.\frac{\partial f}{\partial x_n}\right|_{x_{10}, x_{20}, \ldots, x_{n0}} \Delta x_n \tag{2.34}$$

By substituting the above linear increment equation into the system differential equation, the linear equation of the system can be obtained.

Example 2.8

Figure 2.10 shows a single pendulum system, where $T_i(t)$ is the output torque; $\theta_0(t)$ is the angle of the input pendulum, m is the mass of the pendulum, and l is the length of the pendulum.

Solution

According to Newton's second law, we get from the single pendulum system

$$T_i(t) - [mg \sin\theta_0(t)]l = (ml^2)\ddot{\theta}_0(t) \tag{2.35}$$

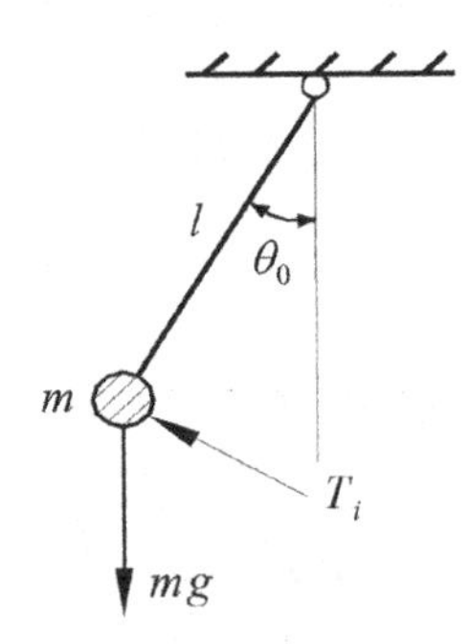

FIGURE 2.10 Single pendulum system.

Eq. (2.35) is a nonlinear differential equation, $\sin\theta_0 = 0$ at $\theta_0 = 0$, according to the Taylor expansion of $\sin\theta_0$, we get

$$\sin\theta_0 = \theta_0 - \frac{\theta_0^3}{3!} + \frac{\theta_0^5}{5!} - \cdots \tag{2.36}$$

If θ_0 is small enough, we can ignore its high-order infinitesimal, and thus

$$\sin\theta_0 \approx \theta_0 \tag{2.37}$$

Therefore, the linear equation of Eq. (2.35) can be expressed as

$$ml^2\ddot{\theta}_0(t) + mgl\theta_0(t) = T_i(t) \tag{2.38}$$

Example 2.9

Figure 2.11 shows a hydraulic pressure system. Here, p_s, p_1, p_2, p_L are the pressures; q_1, q_2, q_L are the flow rate; C_d is the flow coefficient; ρ is the fluid density; x is the displacement of the valve; w is the gradient of the area; and y is the displacement of the piston hydraulic cylinder.

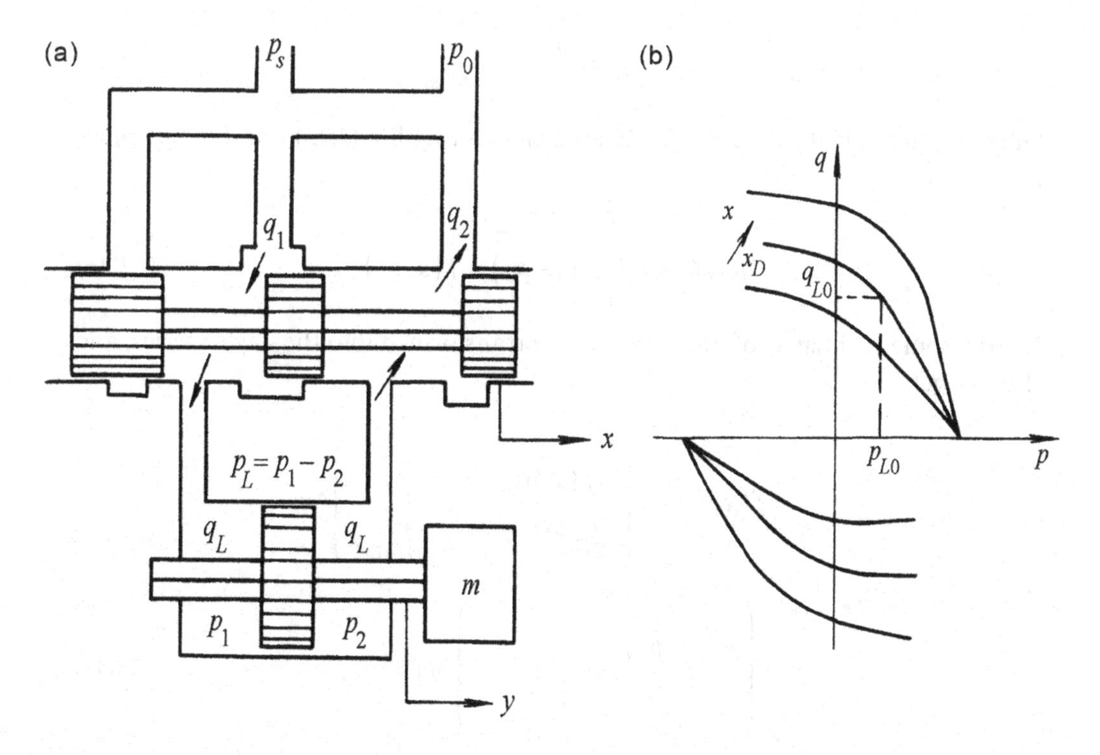

FIGURE 2.11 (a) Valve-controlled and (b) flow curve of the hydraulic pressure system.

Solution

According to the relationship of the hydrodynamics (Fluid dynamics), we get

$$q_1 = C_d wx\sqrt{\frac{2}{\rho}(p_s - p_1)} \tag{2.39}$$

$$q_2 = C_d wx\sqrt{\frac{2}{\rho}(p_2 - p_0)} \tag{2.40}$$

According to the continuity equation, we get

$$q_L = q_1 = q_2 \tag{2.41}$$

According to the force balance equation

$$p_s - p_1 = p_2 \tag{2.42}$$

$$p_L = p_1 - p_2 \tag{2.43}$$

Solving Eqs. (2.42) and (2.43), we get

$$p_1 = \frac{p_s + p_L}{2} \tag{2.44}$$

$$p_2 = \frac{p_s - p_L}{2} \tag{2.45}$$

Substituting Eq. (2.44) into Eq. (2.39) and substituting Eq. (2.45) into Eq. (2.40), we get

$$q_L = C_d wx\sqrt{\frac{1}{\rho}(p_s - p_L)} = f(x, p_L) \tag{2.46}$$

Nearby some point at x_0 of Eq. (2.46) can be transformed into the Taylor series, and then we get

$$q_L = f(x_0, p_{L0}) + \left(\frac{\partial f(x, p_L)}{\partial x} \bigg|_{\substack{x = x_0 \\ p_L = p_{L0}}} \right)\Delta x + \left(\frac{\partial f(x, p_L)}{\partial p_L} \bigg|_{\substack{x = x_0 \\ p_L = p_{L0}}} \right)\Delta p_L + \cdots \tag{2.47}$$

Here, we can define as

$$K_q = \frac{\partial f(x, p_L)}{\partial x}\Bigg|_{\substack{x=x_0 \\ p_L=p_{L0}}}$$, which is called the gain of the flow and

$$K_c = -\frac{\partial f(x, p_L)}{\partial p_L}\Bigg|_{\substack{x=x_0 \\ p_L=p_{L0}}}$$, which is called the coefficient of the flow–pressure

In Eq. (2.47), we get a linear equation by omitting the high-order infinitesimal as

$$\Delta q_L = K_q \Delta x - K_c \Delta p_L \tag{2.48}$$

Since the effect of hydraulic fluid flow is continuous, according to the continuity equation, we can get

$$\Delta q_L = A\frac{d(\Delta y)}{dt} \tag{2.49}$$

$$\Delta p_L A = m\frac{d^2 \Delta y}{dt^2} \tag{2.50}$$

where A represents the effective working area of the hydraulic cylinder. By solving Eqs. (2.48)–(2.50), we get

$$\frac{K_c}{A} m\frac{d^2 \Delta y}{dt^2} + A\frac{d\Delta y}{dt} = K_q \Delta x \tag{2.51}$$

From these examples, we should pay attention to the following matters when we will perform the linearization of the system equation:

1. Linearization is relative to a certain working point, but the working point may be different. With a different working point, the coefficients of the linear equation will be different.
2. The smaller the deviation, the higher the accuracy of its linearization.
3. Regarding the initial condition of the incremental or delta function, the initial condition must be set as 0. If the initial condition is non-zero, we should shift the generalized coordinate to the rated operating or working point.
4. Linearization is only used for the single-valued function without a discontinuity point or break point.

2.2 LAPLACE TRANSFORMATION AND INVERSE TRANSFORMATION

2.2.1 Applications of Laplace Transform

- Analysis of electrical and electronic circuits;
- Breaking down complex differential equations into simpler polynomial forms;
- Laplace transform gives information about steady as well as transient states;
- In machine learning, the Laplace transform is used for making predictions and making analyses in data mining; and
- The Laplace transform simplifies calculations in system modeling.

To determine the application of Laplace transforms in control systems, the control systems are usually designed to control the behavior of other devices. Examples of control systems can range from a simple home heating controller to an industrial control system that regulates the behavior of machinery. Generally, control engineers use differential equations to describe the behavior of various closed-loop functional blocks. Laplace transform is used here for solving these equations without the loss of crucial variable information.

The Laplace transform provides a useful method of solving certain types of differential equations when certain initial conditions are given, especially when the initial values are zero. The Laplace transform is also very useful in the area of mechanical and electrical circuit analysis. It is often easier to analyze the circuit in its Laplace form than to form it in differential equations. The techniques of Laplace transform are not only used in circuit analysis but are also used in

- Proportional-Integral-Derivative (PID) controllers;
- DC motor speed control systems;
- DC motor position control systems; and
- Second-order systems of differential equations (underdamped, overdamped, and critically damped).

Laplace transforms are frequently opted for signal processing. Along with the Fourier transform, the Laplace transform is used to study signals in the frequency domain. When there are small frequencies in the signal in the frequency domain then one can expect the signal to be smooth in the time domain. Filtering of a signal is usually done in the frequency domain for which Laplace acts as an important tool for converting a signal from the time domain to frequency domain.

2.2.2 Characterization of LTI Systems Using Laplace Transform

For a casual system region of convergence (ROC) associated with the system, the function is the right half-plane. A system is anticasual if its impulse response $h(t) = 0$ for $t > 0$.

If ROC of the system functions $H(s)$ includes the $j\omega$ axis, then the LTI of the system is called a stable system. If a casual system with rational system functions $H(s)$ has negative real parts for all of its poles, then the system is stable. Thus, the Laplace transform is a crucial tool in analyzing circuits. We can say as a stethoscope is to doctor Laplace transforms are to control engineer. What do you consider Laplace transforms as? In what way were they helpful to you?

2.2.3 Definition of Laplace Transformation

Define function $x(t)$ to be satisfied if and only if

1. $\begin{cases} x(t)=0; \forall t<0 \\ x(t) \textit{ is finite in the bounded area}; \forall t \geq 0 \end{cases}$

2. $\int_0^{\infty} x(t)\, e^{-st}\, dt < \infty$, $x(t)$ is a function of time t, and it is continuous or piecewise-continuous at its finite interval, so its Laplace transformation can be defined as

$$X(s)=\mathcal{L}[(x(t)]=\int_0^{\infty} x(t)e^{-st}\, dt$$

where s is the complex variable, and $s=\sigma+j\omega$; $x(t)$ is the primary function of $X(s)$; is the Laplace transformation of $x(t)$ (it is also called the transform function or image function); $\left[\frac{1}{t}\right]$ is the dimension of s; and $[x(t)][t]$ is the dimension of $X(s)$.

2.2.4 Laplace Transformation of Some General Function

In the control system, some typical functions are commonly used such as unit-step function, exponential function, impulse function, sine function, and cosine function.

1. *Unit-step function*, $1(t)$

$$1(t)=\begin{cases} 0 & ;\forall t<0 \\ 1 & ;\forall t \geq 0 \end{cases} \tag{2.52}$$

Thus,

$$X(s)=\mathcal{L}[(x(t)]=\mathcal{L}[1(t)]=\int_0^{\infty} 1(t)\, e^{-st} dt=-\frac{1}{s}\left[e^{-st}\right]_0^{\infty}=\frac{1}{s} \tag{2.53}$$

2. *Exponential function*, e^{at}

$$\mathcal{L}\left[e^{at}\right]=\int_0^{\infty} e^{at}e^{-st} dt=\int_0^{\infty} e^{-(s-a)t} dt=\frac{1}{s-a} \tag{2.54}$$

3. *Impulse function,* $\delta(t)$

$$\mathcal{L}[\delta(t)] = 1 \tag{2.55}$$

Proof: According to the definition of function $\delta(t)$,

$$\int_{-\infty}^{+\infty} x(t)\delta(t)dt = x(0)$$

Thus,

$$\mathcal{L}[\delta(t)] = \int_{0}^{+\infty} \delta(t)e^{-st}dt$$

$$= \int_{0^-}^{+\infty} \delta(t)e^{-st}dt$$

$$= \int_{-\infty}^{+\infty} \delta(t)e^{-st}dt$$

$$= e^{-st}\Big|_{t=0} = 1$$

Example 2.10

Obtain the Laplace transformation of $f(t) = e^{-\beta t}\delta(t) - \beta\, e^{-\beta t} \times 1(t)(\beta > 0)$.

Solution

According to the definition of Laplace transformation, we have

$$\mathcal{L}[f(t)] = \int_{0}^{+\infty} f(t)e^{-st}dt = \int_{0}^{+\infty} [e^{-\beta t}\delta(t) - \beta\, e^{-\beta t} \times 1(t)]e^{-st}dt$$

$$= \int_{0}^{+\infty} \delta(t)e^{-(s+\beta)t}dt - \beta\int_{0}^{+\infty} e^{-(s+\beta)t}dt$$

$$= e^{-(s+\beta)t}|t = 0 + \frac{\beta\, e^{-(s+\beta)t}}{s+\beta}\Bigg|_{0}^{+\infty} = 1 - \frac{\beta}{s+\beta} = \frac{s}{s+\beta}$$

4. $\sin\omega t, \cos\omega t$ *function*
   According to Euler's law

$$e^{j\theta} = \cos\theta + j\sin\theta \tag{2.56}$$

$$e^{-j\theta} = \cos\theta - j\sin\theta \tag{2.57}$$

From Eqs. (2.56) and (2.57), we have

$$\sin\theta = \frac{e^{j\theta} - e^{-j\theta}}{2j} \tag{2.58}$$

$$\cos\theta = \frac{e^{j\theta} + e^{-j\theta}}{2} \tag{2.59}$$

Taking the Laplace transformation to $\sin\omega t$ and $\cos\omega t$ function, we get

$$\mathcal{L}[\sin\omega t] = \int_0^\infty \sin\omega t \times e^{-st} dt$$

$$= \int_0^\infty \frac{e^{j\omega t} - e^{-j\omega t}}{2j} e^{-st} dt = \frac{1}{2j}\left(\frac{1}{s-j\omega} - \frac{1}{s+j\omega}\right) = \frac{\omega}{s^2+\omega^2} \tag{2.60}$$

$$\mathcal{L}[\cos\omega t] = \int_0^\infty \cos\omega t \times e^{-st} dt$$

$$= \int_0^\infty \frac{e^{j\omega t} + e^{-j\omega t}}{2} e^{-st} dt = \frac{1}{2}\left(\frac{1}{s-j\omega} + \frac{1}{s+j\omega}\right) = \frac{s}{s^2+\omega^2} \tag{2.61}$$

Example 2.11

Obtain the Laplace transformation of $\sin(\omega t + \theta)$.

Solution

$$\mathcal{L}[\sin(\omega t + \theta)] = \mathcal{L}[\sin\omega t\cos\theta + \cos\omega t\sin\theta] \ldots\ldots\ldots = \frac{\omega\cos\theta}{s^2+\omega^2} + \frac{s\sin\theta}{s^2+\omega^2}$$

$$= \frac{s\sin\theta + \omega\cos\theta}{s^2+\omega^2} \tag{2.62}$$

5. *The Laplace transformation of Power function t^n*

According to the power of Γ function,

$$\Gamma(a) = \int_0^\infty x^{a-1} e^{-x} dx \tag{2.63}$$

When a is a real and natural number of n, its recursion formula is expressed as

$$\Gamma(n+1) = n\Gamma(n) = n!$$

Defining

$$u = st \quad t = \frac{u}{s} \quad dt = \frac{1}{s} du$$

Thus,

$$\mathcal{L}\left[t^n\right] = \int_0^\infty t^n e^{-st} dt = \frac{1}{s^{n+1}} \int_0^\infty u^n e^{-u} du = \frac{n!}{s^{n+1}} \tag{2.64}$$

2.2.5 Property of the Laplace Transformation

1. *Addition property*

If

$$\mathcal{L}\left[f_1(t)\right] = F_1(s) \text{ and } \mathcal{L}\left[f_2(t)\right] = F_2(s)$$

then

$$\mathcal{L}\left[af_1(t) + bf_2(t)\right] = aF_1(s) + bF_2(s) \tag{2.65}$$

Proof:

$$\mathcal{L}\left[af_1(t) + bf_2(t)\right] = \int_0^\infty \left[af_1(t) + bf_2(t)\right] e^{-st} dt$$

$$\ldots\ldots = \int_0^\infty af_1(t) e^{-st} dt + \int_0^\infty bf_2(t) e^{-st} dt \ldots\ldots = aF_1(s) + bF_2(s)$$

$$= aF_1(s) + bF_2(s)$$

2. *Differential property*

$$\mathcal{L}\left[\frac{d}{dt}f(t)\right]=sF(s)-f(0) \tag{2.66}$$

Proof:

$$F(s)=\int_0^{\infty}f(t)e^{-st}dt$$

$$=\int_0^{\infty}f(t)\frac{1}{-s}de^{-st}$$

$$=f(t)\frac{e^{-st}}{-s}\Bigg|_0^{\infty}-\int_0^{\infty}\frac{e^{-st}}{-s}df(t)$$

$$=\frac{f(0)}{s}+\frac{1}{s}\int_0^{\infty}\left[\frac{d}{dt}f(t)\right]e^{-st}dt$$

Thus,

$$F(s)=\frac{f(0)}{s}+\frac{1}{s}L\left[\frac{d}{dt}f(t)\right]$$

Therefore,

$$\mathcal{L}\left[\frac{d}{dt}f(t)\right]=sF(s)-f(0)$$

Remember that

1. $\mathcal{L}\left[\frac{d^n}{dt^n}f(t)\right]=s^nF(s)-s^{n-1}f(0)-s^{n-2}\dot{f}(0)-\cdots-sf^{(n-2)}(0)-f^{(n-1)}(0)$
2. If $f(0)=\dot{f}(0)=\cdots=f^{(n-2)}(0)=f^{(n-1)}(0)=0$, then we may say as in the following:

$$\mathcal{L}\left[\frac{d^n}{dt^n}f(t)\right]=s^nF(s) \tag{2.67}$$

We can use the differential property to calculate the Laplace transformation of power function (t^n). To the power function t^n, its primary condition is

$$f(0)=\dot{f}(0)=\cdots=f^{(n-1)}(0)=0, \text{ and only for } f^{(n)}(t)=n!$$

Thus,

$$\mathcal{L}[n!]=\mathcal{L}\left[f^{(n)}(t)\right]=s^n\mathcal{L}\left[f(t)\right]-s^{n-1}f(0)-s^{n-2}f'(0)-\cdots-f^{(n-1)}(0)$$

Consequently,

$$\mathcal{L}[n!]=s^nL\left[t^n\right]$$

Hence,

$$\mathcal{L}[n!]=n!\mathcal{L}[1]=\frac{n!}{s}$$

Therefore,

$$\mathcal{L}\left[t^n\right]=\frac{n!}{s^{n+1}} \tag{2.68}$$

3. *Integral theorem*

$$\mathcal{L}\left[\int f(t)dt\right]=\frac{F(s)}{s}+\frac{f^{-1}(0)}{s};$$

where

$$f^{-1}(0)=\int f(t)dt\bigg|_{t=0} \tag{2.69}$$

Proof:

$$\begin{aligned}\mathcal{L}[\int f(t)dt]&=\int_0^\infty\left[\int f(t)dt\right]e^{-st}dt\\&=\int_0^\infty\left[\int f(t)dt\right]\frac{1}{-s}de^{-st}\\&=[\int f(t)dt]\frac{e^{-st}}{-s}\bigg|_0^\infty-\int_0^\infty\frac{e^{-st}}{-s}f(t)dt\\&=\frac{f^{-1}(0)}{s}+\frac{F(s)}{s}\end{aligned}$$

Interpretation:

1. $\mathcal{L}\left[\int\cdots\int f(dt)^n\right]=\frac{F(s)}{s^n}+\frac{f^{-1}(0)}{s^n}+\cdots+\frac{f^{-n}(0)}{s}$

   By using the formula, $f^{-n}(0)=\int\cdots\int f(t)(dt)^n|t=0$

2. If $f^{-1}(0)=f^{-2}(0)=\cdots=f^{-n+1}(0)=f^{-n}(0)=0,$

then

$$\mathcal{L}\left[\int\cdots\int f(dt)^n\right]=\frac{F(s)}{s^n}$$

Besides, we can also know the integration property of the transfer function from the Laplace transformation existing law. If $\mathcal{L}[f(t)]=F(s)$ and the integration $\int_s^{\infty} F(s)ds$ convergence, that we may say

$$\mathcal{L}\left[\frac{f(t)}{t}\right]=\int_s^{\infty} F(s)ds$$

Normally, we have

$$\mathcal{L}\left[\frac{f(t)}{t^n}\right]=\underbrace{\int_s^{\infty}ds\int_s^{\infty}ds\cdots\int_s^{\infty}}_{n}F(s)ds$$

And then it can be more simplified if the integration of $\int_0^{+\infty}\frac{f(t)}{t}dt$ exists, and if integration of $\int_0^{\infty}F(s)ds$ is convergent, then we have

$$\int_0^{+\infty}\frac{f(t)}{t}dt=\int_0^{+\infty}F(s)ds$$

Note that $F(s)=\mathcal{L}[f(t)]$ formula is often used to calculate some complex integrations in advanced mathematics.

Example 2.12

Obtain the integration of $\int_0^{\infty}\frac{\sin t}{t}dt$.

Solution

we have $\mathcal{L}[\sin t]=\frac{1}{s^2+1}$

Thus,

$$\int_0^{+\infty}\frac{\sin t}{t}dt=\int_0^{+\infty}\frac{1}{s^2+1}ds=\arctan s\Big|_0^{+\infty}=\frac{\pi}{2}$$

4. *Shifting theorem*

$$\mathcal{L}\left[e^{-at}f(t)\right]=F(s+a) \tag{2.70}$$

Proof:

$$\mathcal{L}\left[e^{-at}f(t)\right]=\int_0^{\infty}e^{-at}f(t)e^{-st}dt=\int_0^{\infty}f(t)e^{-(s+a)t}dt=F(s+a)$$

Example 2.13

Obtain the Laplace transformation of $f(t)=e^{-at}\cos\beta t$.

Solution

Since we know $\mathcal{L}[\cos\beta t]=\dfrac{s}{s^2+\beta^2}$

Thus, according to the shifting theorem, we can say

$$\mathcal{L}\left[e^{-at}\cos\beta t\right]=F(s+a)=\frac{s+a}{(s+a)^2+\beta^2}$$

5. *Delay theorem*

$$\mathcal{L}[f(t-a)\cdot 1(t-a)]=e^{-as}F(s)$$

Proof:

$$\mathcal{L}[f(t-a)\cdot 1(t-a)]=\int_a^{\infty} f(t-a)\cdot 1(t-a)e^{-st}dt$$

$$=\int_a^{\infty} f(t-a)e^{-st}dt$$

$$\underline{t-a=\tau}\int_0^{\infty} f(\tau)e^{-s(\tau+a)}d(\tau+a)$$

$$=e^{-as}\int_0^{\infty} f(\tau)e^{-s\tau}d\tau=e^{-as}F(s)$$

Example 2.14

Obtain the Laplace transformation of $f(t)=4\cos\left(2t-\dfrac{\pi}{3}\right)1\left(t-\dfrac{\pi}{6}\right)$

Solution

Since we know, $\mathcal{L}[f(t)]=\mathcal{L}\left[4\cos 2\left(t-\dfrac{\pi}{6}\right)\cdot 1\left(t-\dfrac{\pi}{6}\right)\right]$.

Thus, according to the delay theorem, we can get

$$\mathcal{L}[f(t)]=4\frac{se^{-\frac{\pi}{6}s}}{s^2+4}$$

6. *Initial value theorem*

$$\lim_{t \to 0} f(t) = \lim_{s \to \infty} sF(s) \tag{2.71}$$

Proof:

$$\mathcal{L}\left[\frac{df(t)}{dt}\right] = \int_0^\infty \frac{df(t)}{dt} e^{-st} dt$$

or

$$\mathcal{L}\left[\frac{df(t)}{dt}\right] = sF(s) - f(0)$$

Thus,

$$\int_0^\infty \frac{df(t)}{dt} e^{-st} dt = sF(s) - f(0)$$

Now, taking limits on both sides, we get

$$\lim_{s \to \infty} \int_0^\infty \frac{df(t)}{dt} e^{-st} dt = \lim_{s \to \infty} \left[sF(s) - f(0)\right]$$

Consequently,

$$0 = \lim_{s \to \infty} sF(s) - \lim_{s \to \infty} f(0)$$

Therefore,t

$$\lim_{t \to 0} f(t) = \lim_{s \to \infty} sF(s)$$

7. *Final value theorem*

$$\lim_{t \to \infty} f(t) = \lim_{s \to 0} sF(s) \tag{2.72}$$

Proof:

$$\mathcal{L}\left[\frac{df(t)}{dt}\right] = \int_0^\infty \frac{df(t)}{dt} e^{-st} dt$$

and

$$\mathcal{L}\left[\frac{df(t)}{dt}\right] = sF(s) - f(0)$$

Thus,

$$\int_0^\infty \frac{df(t)}{dt} e^{-st} dt = sF(s) - f(0)$$

Now, taking limits on both sides, we get

$$\lim_{s \to 0} \left[\int_0^\infty \frac{df(t)}{dt} e^{-st} dt \right] = \lim_{s \to 0} \left[sF(s) - f(0) \right]$$

Consequently,

$$\int_0^\infty \frac{df(t)}{dt} dt = \lim_{s \to 0} \left[sF(s) - f(0) \right]$$

Thus,

$$\lim_{t \to \infty} f(t) - f(0) = \lim_{s \to 0} sF(s) - f(0)$$

Therefore,

$$\lim_{t \to \infty} f(t) = \lim_{s \to 0} sF(s)$$

Special intention: The prerequisite of using the final value theorem is the existence of $\lim_{t \to \infty} f(t)$. The determinant of some functions such as sine function, $\sin \omega t$ and cosine function, $\cos \omega t$ must be between -1 and $+1$. Thus, the sine function, $\sin \omega t$ as well as the cosine function $\cos \omega t$ has not terminal value; therefore, the terminal value theorem is not suitable for sine function and cosine function.

8. *Scaling theorem*

$$\mathcal{L}\left[f\left(\frac{t}{a} \right) \right] = aF(as) \tag{2.73}$$

Proof:

Let, $\frac{t}{a} = t_1$, $as = s_1$; thus, we have

$$\mathcal{L}\left[f\left(\frac{t}{a} \right) \right] = \int_0^\infty f(t_1) e^{-s_1 t_1} d(at_1) = a \int_0^\infty f(t_1) e^{-s_1 t_1} dt_1 = aF(s_1)$$

Therefore,

$$\mathcal{L}\left[f\left(\frac{t}{a} \right) \right] = aF(as)$$

9. *Laplace transformation of time multiplication function*

$$\mathcal{L}[tf(t)] = -\frac{dF(s)}{ds} \tag{2.74}$$

Proof:
According to the Leibniz law,

$$\frac{dF(s)}{ds} = \frac{d}{ds}\int_0^\infty f(t)e^{-st}dt = \int_0^\infty \frac{\partial}{\partial s}\left[e^{-st}f(t)\right]dt = -\int_0^\infty tf(t)e^{-st}dt = -\mathcal{L}[tf(t)]$$

In the same way,

$$\mathcal{L}\left[t^n f(t)\right] = (-1)^n \frac{d^n F(s)}{ds^n}$$

Example 2.15

Obtain the Laplace transformation of function, $f(t) = t\sin\omega t$.

Solution

Since $\mathcal{L}[\sin\omega t] = \frac{\omega}{s^2+\omega^2}$, then according to the differential property of the transfer function, we know

$$\mathcal{L}[t\sin\omega t] = -\frac{d}{ds}\left[\frac{\omega}{s^2+\omega^2}\right] = \frac{2\omega s}{(s^2+\omega^2)^2} \tag{2.75}$$

10. *Laplace transformation of convolution integral*

$$\mathcal{L}[f(t) * g(t)] = F(s)G(s) \tag{2.76}$$

Here, $f(t) * g(t)$ is the mathematical expression of convolution integral, and its definition is

$$f(t) * g(t) = \int_0^t f(t-\tau)g(\tau)d\tau = g(t) * f(t) \tag{2.77}$$

Now we will prove the Laplace transformation of convolution integral.
Proof:

$$\mathcal{L}[f(t) * g(t)] = \mathcal{L}\left[\int_0^t f(t-\tau)g(\tau)d\tau\right] = \mathcal{L}\left[\int_0^\infty f(t-\tau)\cdot 1(t-\tau)g(\tau)d\tau\right]$$

$$= \int_0^{\infty}\left[\int_0^{\infty} f(t-\tau)\cdot 1(t-\tau) g(\tau) d\tau\right] e^{-st} dt$$

$$= \int_0^{\infty}\int_0^{\infty} f(t-\tau)\cdot 1(t-\tau) e^{-s(t-\tau)} dt \cdot g(\tau) e^{-s\tau} d\tau$$

Here, $\underline{t-\tau=\omega} \int_0^{\infty} f(\omega) e^{-s\omega} d\omega \int_0^{\infty} g(\tau) e^{-s\tau} d\tau = F(s)G(s)$

We can say that there are some problems about the Laplace transformation of some general engineering curves.

Example 2.16

Obtain the Laplace transformation of the square wave curve in Figure 2.12.

Solution

In Figure 2.12, the square wave function can be described as

$$f(t) = A\cdot 1(t) - 2A\cdot 1(t-T) + 2A\cdot 1(t-2T) - \cdots$$

Taking the Laplace transformation to the square wave function in the above equation, we get

$$\mathcal{L}[f(t)] = \frac{A}{s} - \frac{2A}{s} e^{-sT} + \frac{2A}{s} e^{-2sT} - \frac{2A}{s} e^{-3sT} + \cdots$$

When Re(s) > 0, we have $\left|e^{-sT}\right| < 1$; thus, we obtain

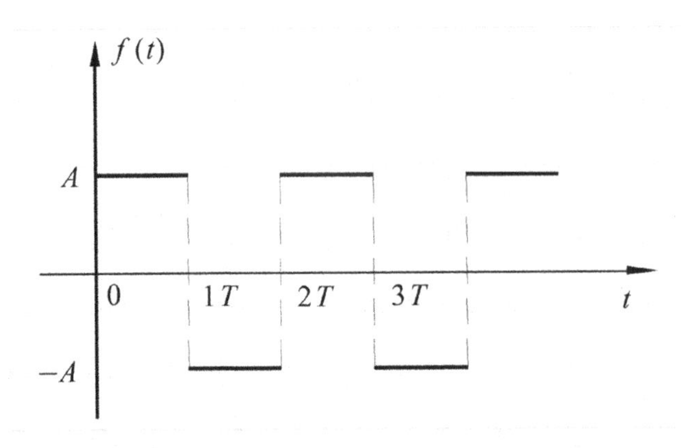

FIGURE 2.12 Square wave function.

$$\mathcal{L}[f(t)]=\frac{A}{s}-\frac{2A}{s}e^{-sT}\left(1-e^{-sT}+e^{-2sT}-e^{-3sT}+e^{-4sT}+\cdots\right)$$

$$=\frac{A}{s}-\frac{2A}{s}\frac{e^{-sT}}{1+e^{-sT}}=\frac{A}{s}\left(\frac{1-e^{-sT}}{1+e^{-sT}}\right)=\frac{A}{s}\tanh\left(\frac{sT}{2}\right)$$

Example 2.17

Obtain the Laplace transformation of the staircase curve in Figure 2.13.

Solution

According to Figure 2.13, we may write the expression of staircase function as

$$f(t)=A\left[1(t)+1(t-T)+1(t-2T)+1(t-3T)+\ldots\right]$$

Taking the Laplace transformation to the staircase function $f(t)$, we get

$$\mathcal{L}[f(t)]=A\left(\frac{1}{s}+\frac{1}{s}e^{-sT}+\frac{1}{s}e^{-2sT}+\frac{1}{s}e^{-3sT}+\cdots\right)=\frac{A}{s}\left(1+e^{-sT}+e^{-2sT}+e^{-3sT}+\cdots\right)$$

When $\mathrm{Re}(s)>0$, we have $\left|e^{-sT}\right|<1$; thus, we can rewrite as in the following:

$$\mathcal{L}[f(t)]=\frac{A}{s}\frac{1}{1-e^{-sT}}=\frac{A}{s}\frac{1}{1-e^{-\frac{sT}{2}}}\frac{1}{1+e^{-\frac{sT}{2}}}=\frac{A}{2s}\left(1+\coth\frac{sT}{2}\right)$$

Example 2.18

Obtain the Laplace transformation of the periodical triangle wave function in Figure 2.14, where

$$f(t)=\begin{cases} t & ;0\le t<A \\ 2A-t & ;A\le t<2A \end{cases}\quad \text{and } f(t+2A)=f(t) \text{ for } t>0.$$

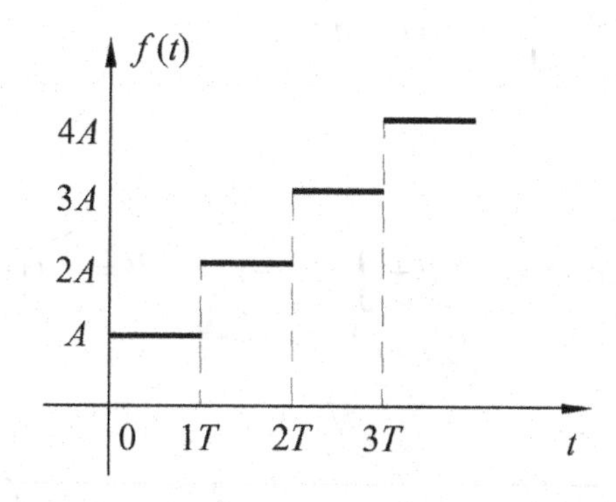

FIGURE 2.13 Staircase curve.

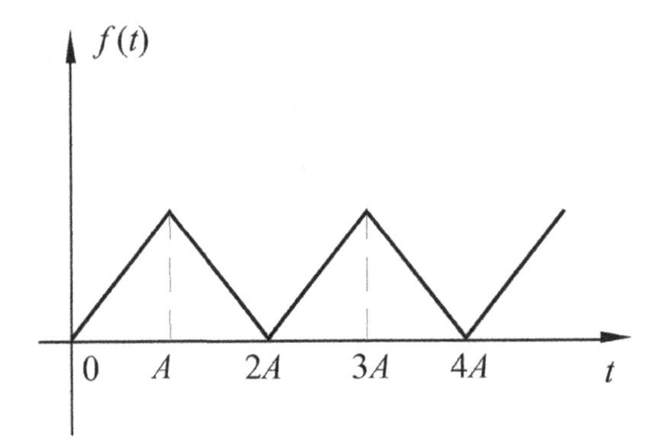

FIGURE 2.14 Periodical triangle wave function.

Solution

According to the definition of the Laplace transformation, we have

$$\mathcal{L}[f(t)]=\int_0^{+\infty} f(t)e^{-st}dt$$

$$=\int_0^{2A} f(t)e^{-st}dt+\int_{2A}^{4A} f(t)e^{-st}dt+\cdots+\int_{2kA}^{2(k+1)A} f(t)e^{-st}dt+\cdots$$

$$=\sum_{k=0}^{+\infty}\int_{2kA}^{2(k+1)A} f(t)e^{-st}dt$$

Let, $t=\tau+2kA$, and then

$$\int_{2kA}^{2(k+1)A} f(t)\,e^{-st}dt=\int_0^{2A} f(\tau+2kA)e^{-s(\tau+2kA)}d\tau$$

$$=e^{-2kAs}\int_0^{2A} f(\tau)\,e^{-s\tau}d\tau$$

Hence,

$$\int_0^{2A} f(t)e^{-st}dt=\int_0^{A} te^{-st}dt+\int_A^{2A}(2A-t)\,e^{-st}dt=\frac{1}{s^2}(1-e^{-As})^2$$

Consequently,

$$\mathcal{L}[f(t)]=\sum_{k=0}^{+\infty}e^{-2kAs}\int_0^{2A} f(t)\,e^{-st}dt=\int_0^{2A} f(t)\,e^{-st}dt\left(\sum_{k=0}^{+\infty}e^{-2kAs}\right)$$

When $\text{Re}(s) > 0$, it has $\left|e^{-2As}\right| < 1$

Thus,

$$\sum_{k=0}^{+\infty}\left(e^{-2kAs}\right) = \frac{1}{1-e^{-2As}}$$

Therefore,

$$\mathcal{L}[f(t)] = \frac{1}{1-e^{-2As}}\int_0^{2A} f(t)e^{-st}dt$$

$$= \frac{1}{1-e^{-2As}}(1-e^{-As})^2 \cdot \frac{1}{s^2}$$

$$= \frac{1}{s^2}\cdot\frac{(1-e^{-As})^2}{\left(1-e^{-As}\right)\left(1+e^{-As}\right)}$$

$$= \frac{1}{s^2}\cdot\frac{1-e^{-As}}{1+e^{-As}} = \frac{1}{s^2}\tanh\frac{As}{2}$$

In general case, those functions with a period of T, i.e., $f(t+2A) = f(t)$ for $t>0$, if $f(t)$ is

$$\mathcal{L}[f(t)] = \frac{1}{1-e^{-sT}}\int_0^{T} f(t)e^{-st}dt; \qquad (\text{Re}(s) > 0)$$

Note that this is the Laplace transformation formula of a periodic function.

2.2.6 Laplace Transformation Table

Let's look at the Laplace transformation table. The second column of the table shows the function of the linear system, but the third column shows the Laplace function in terms of s function that we can get by using the Laplace transformation of the linear function in the second column.

Laplace transforms are often applied to determine how to solve problems involving partial differential equations. Directly using the differential equation to calculate the inverse Laplace transformation is very complex; thus, we try to avoid it and use Laplace transformation (Table 2.1).

2.2.7 Graphs of Hyperbolic Functions

The relationship between general functions and hyperbolic functions are shown in Table 2.2. Figure 2.15 shows the graphs of hyperbolic functions which are performed by the following functions,

$y = \sinh x$ $\quad y = \tanh x$ $\quad y = \text{sech}\, x$

$y = \cosh x$ $\quad y = \coth x$ $\quad y = \text{cosech}\, x$

TABLE 2.1 Laplace Transformation Table

No.	$f(t)$	$F(s)$
1	$1(t)$	$\frac{1}{s}$
2	$\delta(t)$	1
3	e^{at}	$\frac{1}{s-a}$
4	$t^m,\ (m>-1)$	$\frac{\Gamma(m+1)}{s^{m+1}}$
5	$t^m e^{at},\ (m>-1)$	$\frac{\Gamma(m+1)}{(s-a)^{m+1}}$
6	$\sin\omega t$	$\frac{\omega}{s^2+\omega^2}$
7	$\cos\omega t$	$\frac{s}{s^2+\omega^2}$
8	$\sinh\omega t$	$\frac{\omega}{s^2-\omega^2}$
9	$\cosh\omega t$	$\frac{\omega}{s^2-\omega^2}$
10	$t\sin\omega t$	$\frac{2\omega s}{(s^2+\omega^2)^2}$
11	$t\cos\omega t$	$\frac{s^2-\omega^2}{(s^2+\omega^2)^2}$
12	$t\sinh\omega t$	$\frac{2\omega s}{(s^2-\omega^2)^2}$
13	$t\cosh\omega t$	$\frac{s^2+\omega^2}{(s^2-\omega^2)^2}$
14	$t^m \sin\omega t(m>-1)$	$\frac{\Gamma(m+1)}{2j(s^2+\omega^2)^{m+1}}[(s+j\omega)^{m+1}-(s-j\omega)^{m+1}]$
15	$t^m \cos\omega t(m>-1)$	$\frac{\Gamma(m+1)}{2(s^2+\omega^2)^{m+1}}[(s+j\omega)^{m+1}+(s-j\omega)^{m+1}]$
16	$e^{-at}\sin\omega t$	$\frac{\omega}{(s+a)^2+\omega^2}$
17	$e^{-at}\cos\omega t$	$\frac{s+a}{(s+a)^2+\omega^2}$
18	$e^{-at}\sin(\omega t+\varphi)$	$\frac{(s+a)\sin\varphi+\omega\cos\varphi}{(s+a)^2+\omega^2}$
19	$\sin^2\omega t$	$\frac{1}{2}\left(\frac{1}{s}-\frac{s}{s^2+4\omega^2}\right)$
20	$\cos^2\omega t$	$\frac{1}{2}\left(\frac{1}{s}+\frac{s}{s^2+4\omega^2}\right)$
21	$\sin\omega t\sin\beta t$	$\frac{2\omega\beta s}{\left[s^2+(\omega+\beta)^2\right]\left[s^2+(\omega-\beta)^2\right]}$

(Continued)

TABLE 2.1 (*Continued*) Laplace Transformation Table

No.	$f(t)$	$F(s)$
22	$e^{at}-e^{bt}$	$\dfrac{a-b}{(s-a)(s-b)}$
23	$ae^{at}-be^{bt}$	$\dfrac{(a-b)s}{(s-a)(s-b)}$
24	$\dfrac{1}{\omega}\sin\omega t-\dfrac{1}{\beta}\sin\beta t$	$\dfrac{\beta^2-\omega^2}{(s^2+\omega^2)(s^2+\beta^2)}$
25	$\cos\omega t-\cos\beta t$	$\dfrac{(\beta^2-\omega^2)s}{(s^2+\omega^2)(s^2+\beta^2)}$
26	$\dfrac{1}{\omega^2}(1-\cos\omega t)$	$\dfrac{1}{s(s^2+\omega^2)}$
27	$\dfrac{1}{\omega^3}(\omega t-\sin\omega t)$	$\dfrac{1}{s^2(s^2+\omega^2)}$
28	$\dfrac{1}{\omega^4}(\cos\omega t-1)+\dfrac{1}{2\omega^2}t^2$	$\dfrac{1}{s^3(s^2+\omega^2)}$
29	$\dfrac{1}{\omega^4}(\cosh\omega t-1)-\dfrac{1}{2\omega^2}t^2$	$\dfrac{1}{s^3(s^2-\omega^2)}$
30	$\dfrac{1}{2\omega}(\sin\omega t+\omega t\cos\omega t)$	$\dfrac{s^2}{(s^2+\omega^2)^2}$
31	$\dfrac{1}{2\omega^3}(\sin\omega t-\omega t\cos\omega t)$	$\dfrac{1}{(s^2+\omega^2)^2}$
32	$(1-at)e^{-at}$	$\dfrac{s}{(s+a)^2}$
33	$t\left(1-\dfrac{a}{2}t\right)e^{-at}$	$\dfrac{s}{(s+a)^3}$
34	$\dfrac{(1-e^{-at})}{a}$	$\dfrac{1}{s(s+a)}$
35	$\dfrac{1}{ab}+\dfrac{1}{b-a}\left(\dfrac{e^{-bt}}{b}-\dfrac{e^{-at}}{a}\right)$	$\dfrac{1}{s(s+a)(s+b)}$
36	$\dfrac{e^{-at}}{(b-a)(c-a)}+\dfrac{e^{-bt}}{(a-b)(c-b)}+\dfrac{e^{-ct}}{(a-c)(b-c)}$	$\dfrac{1}{(s+a)(s+b)(s+c)}$
37	$\dfrac{ae^{-at}}{(a-b)(c-a)}+\dfrac{be^{-bt}}{(a-b)(b-c)}+\dfrac{ce^{-ct}}{(c-a)(b-c)}$	$\dfrac{s}{(s+a)(s+b)(s+c)}$
38	$\dfrac{a^2e^{-at}}{(b-a)(c-a)}+\dfrac{b^2e^{-bt}}{(a-b)(c-b)}+\dfrac{c^2e^{-ct}}{(a-c)(b-c)}$	$\dfrac{s^2}{(s+a)(s+b)(s+c)}$
39	$\dfrac{e^{-at}-e^{-bt}[1-(a-b)t]}{(a-b)^2}$	$\dfrac{1}{(s+a)(s+b)^2}$

(*Continued*)

TABLE 2.1 (*Continued*) Laplace Transformation Table

No.	$f(t)$	$F(s)$
40	$\dfrac{e^{-at}-e^{-bt}\left[1-(a-b)t\right]}{(a-b)^2}$	$\dfrac{1}{(s+a)(s+b)^2}$
41	$e^{-at}-e^{\frac{at}{2}}\left(\cos\dfrac{\sqrt{3}at}{2}-\sqrt{3}\sin\dfrac{\sqrt{3}at}{2}\right)$	$\dfrac{3a}{s^3+a^3}$
42	$\sin at\cosh at-\cos at\sinh at$	$\dfrac{4a^3}{s^4+4a^4}$
43	$\dfrac{1}{2a^2}\sin at\sinh at$	$\dfrac{s}{s^4+4a^4}$
44	$\dfrac{1}{2a^3}(\sinh at-\sin at)$	$\dfrac{1}{s^4-a^4}$
45	$\dfrac{1}{2a^2}(\cosh at-\cos at)$	$\dfrac{s}{s^4-a^4}$
46	$\dfrac{1}{a}\left[a_0-(a_0-a)e^{-at}\right]$	$\dfrac{s+a_0}{s(s+a)}$
47	$\dfrac{a_0t}{a}+\left(\dfrac{a_0}{a^2}-t\right)\left(e^{-at}-1\right)$	$\dfrac{s+a_0}{s^2(s+a)}$
48	$\dfrac{1}{a^2}\left[a_0at+a_1a-a_0+\left(a_0-a_1a+a^2\right)e^{-at}\right]$	$\dfrac{s^2+a_1s+a_0}{s^2(s+a)}$
49	$\dfrac{\sqrt{a^2+\omega^2}}{\omega}\sin(\omega t+\varphi),\varphi=\tan^{-1}\dfrac{\omega}{a}$	$\dfrac{s+a}{s^2+\omega^2}$
50	$\dfrac{\sqrt{(b-a)^2+\omega^2}}{\omega}e^{-at}\sin(\omega t+\varphi),\varphi=\tan^{-1}\dfrac{\omega}{b-a}$	$\dfrac{s+b}{(s+a)^2+\omega^2}$
51	$\dfrac{1}{ab}\left[a_0-\dfrac{b(a_0-a)}{b-a}e^{-at}+\dfrac{a(a_0-b)}{b-a}e^{-bt}\right]$	$\dfrac{s+a_0}{s(s+a)(s+b)}$
52	$\dfrac{1}{b-a}\left[(a_0-a)e^{-at}-(a_0-b)e^{-bt}\right]$	$\dfrac{s+a_0}{(s+a)(s+b)}$
53	$\dfrac{a_0}{ab}+\dfrac{a^2-aa_1-1+a_0}{a(a-b)}e^{-at}-\dfrac{b^2-a_1b+a_0}{b(a-b)}e^{-bt}$	$\dfrac{s^2+a_1s+a_0}{s(s+a)(s+b)}$
54	$\dfrac{1}{a^2b^2}\left[abt-a-b+\dfrac{1}{a-b}\left(a^2e^{-bt}-b^2e^{-bt}\right)\right]$	$\dfrac{1}{s^2(s+a)(s+b)}$
55	$\dfrac{1}{ab}(1+a_0t)-\dfrac{a_0(a+b)}{a^2b^2}+\dfrac{1}{a-b}\left[\left(\dfrac{a_0-b}{b^2}\right)e^{-bt}-\left(\dfrac{a_0-a}{a^2}\right)e^{-at}\right]$	$\dfrac{s+a_0}{s^2(s+a)(s+b)}$
56	$\dfrac{a_0}{abc}-\dfrac{(a_0-a)e^{-at}}{a(b-a)(c-a)}-\dfrac{(a_0-b)e^{-bt}}{b(a-b)(c-b)}-\dfrac{(a_0-c)e^{-ct}}{c(a-c)(b-c)}$	$\dfrac{s+a_0}{s(s+a)(s+b)(s+c)}$
57	$\dfrac{e^{-at}}{a^2+\omega^2}+\dfrac{1}{\omega\sqrt{a^2+\omega^2}}\sin(\omega t-\varphi),\varphi=\tan^{-1}\dfrac{\omega}{a}$	$\dfrac{1}{(s+a)\left(s^2+\omega^2\right)}$

(*Continued*)

TABLE 2.1 (*Continued*) Laplace Transformation Table

No.	$f(t)$	$F(s)$
58	$\frac{1}{a^2+\omega^2}+\frac{1}{\omega\sqrt{a^2+\omega^2}}e^{-at}\sin(\omega t-\varphi), \varphi=\tan^{-1}\frac{\omega}{-a}$	$\frac{1}{s[(s+a)^2+\omega^2]}$
59	$\frac{a_0}{a^2+\omega^2}+\frac{1}{\omega}\sqrt{\frac{(a_0-a)^2+\omega^2}{a^2+\omega^2}}e^{-at}\sin(\omega t+\varphi)$	$\frac{s+a_0}{s[(s+a)^2+\omega^2]}$
60	$\frac{e^{-ct}}{(c-a)^2+\omega^2}+\frac{e^{-at}\sin(\omega t-\varphi)}{\omega\sqrt{(c-a)^2+\omega^2}}, \varphi=\tan^{-1}\frac{\omega}{c-a}$	$\frac{1}{(s+c)[(s+a)^2+\omega^2]}$
61	$\frac{1}{\omega_n\sqrt{1-\xi^2}}e^{-\xi\omega_n t}\sin\omega_n\sqrt{1-\xi^2}t$	$\frac{1}{s^2+2\xi\omega_n s+\omega_n^2}$
62	$\frac{-1}{\sqrt{1-\xi^2}}e^{-\xi\omega_n t}\sin(\omega_n\sqrt{1-\xi^2}t-\tan^{-1}\frac{\sqrt{1-\xi^2}}{\xi})$	$\frac{s}{s^2+2\xi\omega_n s+\omega_n^2}$
63	$\frac{\omega_n}{\sqrt{1-\xi^2}}e^{-\xi\omega_n t}\sin\omega_n\sqrt{1-\xi^2}t$	$\frac{\omega_n^2}{s^2+2\xi\omega_n s+\omega_n^2}$
64	$1-\frac{1}{\sqrt{1-\xi^2}}e^{-\xi\omega_n t}\sin(\omega_n\sqrt{1-\xi^2}t+\tan^{-1}\frac{\sqrt{1-\xi^2}}{\xi})$	$\frac{\omega_n^2}{s(s^2+2\xi\omega_n s+\omega_n^2)}$
65	$\frac{T\omega_n}{1+T^2\omega_n^2}e^{-\frac{t}{T}}+\frac{1}{\sqrt{1+T^2\omega_n^2}}\sin(\omega_n t-\tan^{-1}\omega_n T)$	$\frac{\omega_n^2}{(1+Ts)(s^2+\omega_n^2)}$
66	$\frac{T\omega_n^2 e^{-\frac{t}{T}}}{1-2\xi\omega_n T+T^2\omega_n^2}+\frac{\omega_n e^{-\xi\omega_n t}\sin(\omega_n\sqrt{1-\xi^2}t-\varphi)}{\sqrt{(1-\xi^2)(1-2\xi T\omega_n-T^2\omega_n^2)}}$	$\frac{\omega_n^2}{(1+Ts)(s^2+2\xi\omega_n s+\omega_n^2)}$
67	$\left[t-\frac{2a}{a^2+\omega_n^2}+\frac{1}{\omega_n}e^{-at}\sin\left(\omega_n t+2\tan^{-1}\frac{\omega_n}{a}\right)\right]\frac{1}{a^2+\omega_n^2}$	$\frac{1}{s^2[(s+a)^2+\omega_n^2]}$

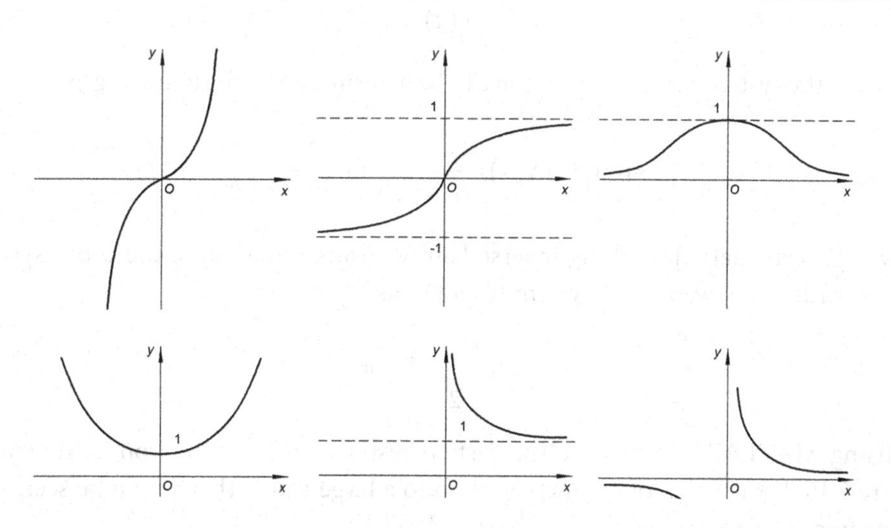

FIGURE 2.15 Graphs of hyperbolic functions.

TABLE 2.2 Definition of Hyperbolic Functions

	General Function	Hyperbolic Functions
1	$\sin x$	$\sin hx = \frac{e^x - e^{-x}}{2}$
2	$\cos x$	$\cos hx = \frac{e^x + e^{-x}}{2}$
3	$\tan x$	$\tan hx = \frac{e^x - e^{-x}}{e^x + e^{-x}}$
4	$\cot x$	$\cot hx = \frac{e^x + e^{-x}}{e^x - e^{-x}}$
5	$\sec x$	$\sec hx = \frac{2}{e^x + e^{-x}}$
6	$\mathrm{cosec}\, x$	$\mathrm{cosec}\, hx = \frac{2}{e^x - e^{-x}}$

Example 2.19

Find the Laplace transform of the given a differential equation,

$$\frac{dy(t)}{dt} + 2y(t) = x(t)$$

Solution

From the given differential equation, we can convert the differential by using the Laplace transformation

$$sY(s) + 2Y(s) = X(s)$$

Therefore, from above equation, transfer function can be written as

$$G(s) = \frac{Y(s)}{X(s)} = \frac{1}{s+2}$$

However, the system response function can be determined by the following expression

$$Y(s) = G(s)X(s) = \frac{1}{s(s+2)} = \frac{1/2}{s} - \frac{1/2}{s+2}$$

Now, it is clear that after taking inverse Laplace transformation of the above system response function, we get the system response as

$$y(t) = \frac{1}{2} - \frac{1}{2}e^{-2t}$$

By using MATLAB, we can see the system response of the function as shown in Figure 2.16. If the value of t is increased up to a large value, then it can be seen that the response curve reaches the steady-state level.

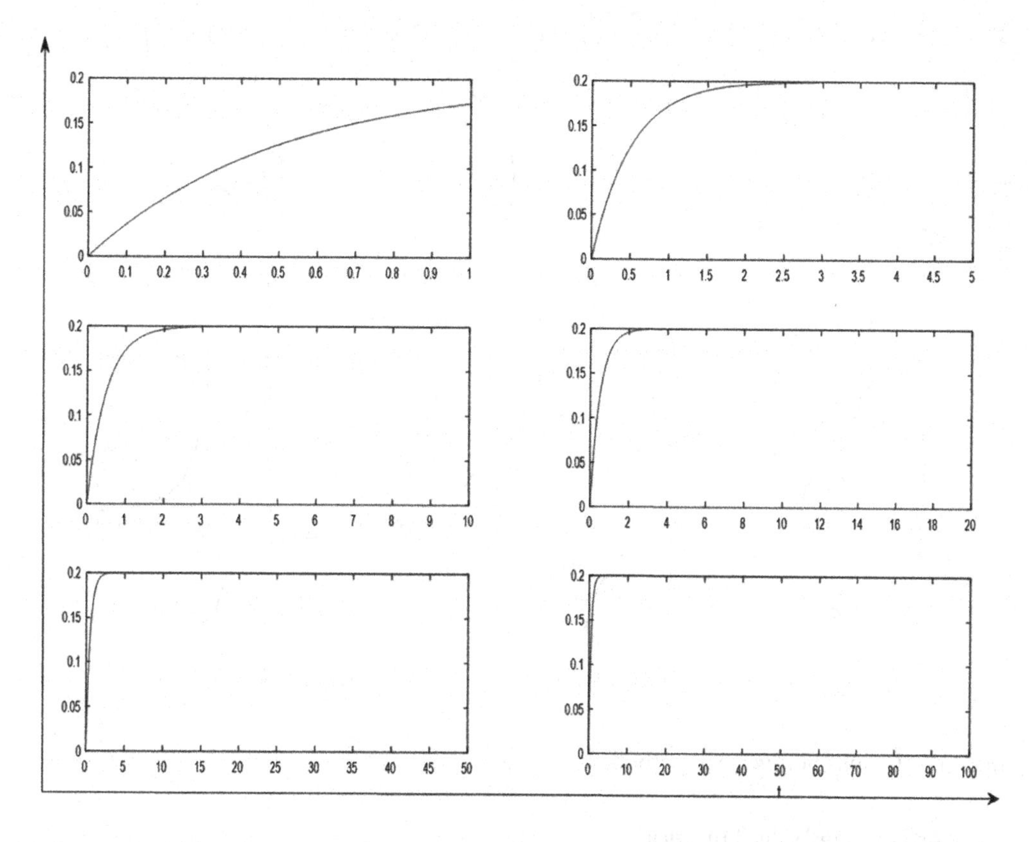

FIGURE 2.16 The system response of the function.

MATLAB code:

```
t=0:0.001:1;
y=(1/5)-(1/5)*exp(-2*t);
subplot(3,2,1)
plot(t, y)
```

2.2.8 Inverse Laplace Transformation

The formula of inverse Laplace transformation is

$$f(t)=\frac{1}{2\pi j}\int_{a-j\infty}^{a+j\infty}F(s)e^{st}ds \tag{2.78}$$

The simplified form can be expressed as

$$f(t)=\mathcal{L}^{-1}\left[F(s)\right]$$

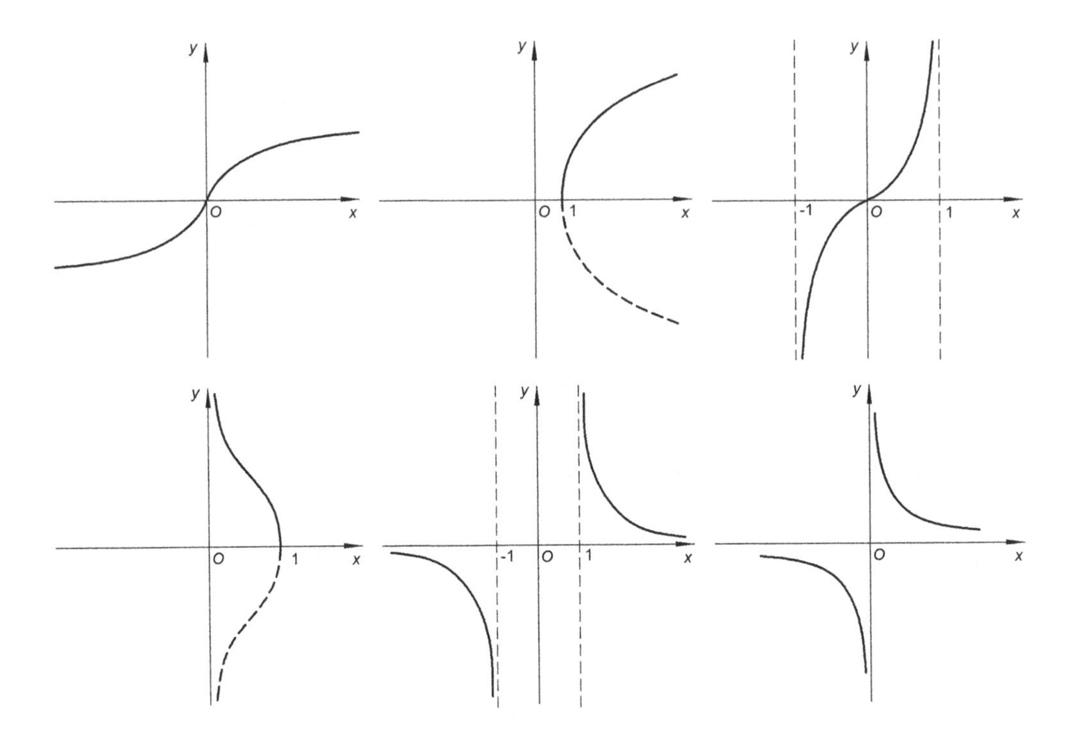

FIGURE 2.17 Graphs of inverse hyperbolic functions.

2.2.9 Inverse Hyperbolic Functions

If $x = \sinh y$, then $y = \sinh^{-1} x$ a is called the *inverse hyperbolic sine* of x. Similarly, we define the other inverse hyperbolic functions. The inverse hyperbolic functions are multiple-valued and as in the case of inverse trigonometric functions, we restrict ourselves to principal values for which they can be considered single-valued. The following list shows the principal values [unless otherwise indicated] of the inverse hyperbolic functions expressed in terms of logarithmic functions, which are taken as real-valued (Figure 2.17).

$$\sinh^{-1} x = \ln\left(x + \sqrt{x^2 + 1}\right); \quad -\infty < x < \infty$$

$$\cosh^{-1} x = \ln\left(x + \sqrt{x^2 - 1}\right); \quad x \geq l \text{ [}\cosh^{-1} x > 0 \text{ is principal value]}$$

$$\tanh^{-1} x = \frac{1}{2}\ln\left(\frac{1+x}{1-x}\right); -1 < x < 1$$

$$\coth^{-1} x = \frac{1}{2}\ln\left(\frac{x+1}{x-1}\right); x > 1 \text{ or } x < -1$$

$$\operatorname{sec h}^{-1} x = \ln\left(\frac{1}{x} + \sqrt{\frac{1}{x^2} - 1}\right); 0 < x \leq l \text{ [}\operatorname{sec h}^{-1} \; x > 0 \text{ is principal value]}$$

$$\operatorname{cosech}^{-1} x = \ln\left(\frac{1}{x} + \sqrt{\frac{1}{x^2} + 1}\right); x \neq 0$$

2.2.10 Graphs of Inverse Hyperbolic Functions

$y=\sinh^{-1}x$	$y=\tanh^{-1}x$	$y=\text{sech}^{-1}x$
$y=\cosh^{-1}x$	$y=\coth^{-1}x$	$y=\text{cosech}^{-1}x$

Example 2.20

Obtain the inverse Laplace transformation of $F(s)=\dfrac{s}{s^2+2s+5}$.

Solution

We may simplify the function as

$$F(s)=\frac{s}{s^2+2s+5}=\frac{(s+1)-1}{(s+1)^2+2^2}$$

Expanding partial fraction, we get

$$=\frac{s+1}{(s+1)^2+2^2}-\frac{1}{2}\frac{2}{(s+1)^2+2^2}$$

Taking inverse Laplace transformation on both sides, we get

$$f(t)=\mathcal{L}^{-1}\left[F(s)\right]=\mathcal{L}^{-1}\left[\frac{s+1}{(s+1)^2+2^2}-\frac{1}{2}\frac{2}{(s+1)^2+2^2}\right]$$

$$=e^{-t}\cos 2t-\frac{1}{2}e^{-t}\sin 2t.$$

In the control system, we use rational polynomial to express the transfer function as

$$F(s)=\frac{M(s)}{N(s)}=\frac{b_0s^m+b_1s^{m-1}+\cdots+b_{m-1}s+b_m}{s^n+a_1s^{n-1}+\cdots+a_{n-1}s+a_n};\quad \text{where, } n\geq m$$

According to the engineering mathematical concept, the s is a variable that makes the denominator to reach zero; that is called "pole"; a point for s which makes the numerator become zero is called "zero." According to the real coefficient of the polynomial, if the denominator has a polynomial of degree n, it must have n roots. Thus, $F(s)$ can be analyzed as

$$F(s)=\frac{b_0s^m+b_1s^{m-1}+\cdots+b_{m-1}s+b_m}{(s+p_1)^{r_1}(s+p_2)^{r_2}\cdots(s+p_r)^{r_l}(s^2+c_1s+d_1)^{k_1}\cdots(s^2+c_gs+d_g)^{k_g}}$$

Thereafter,

$$r_1+r_2+\cdots+r_l+2\left(k_1+k_2+\cdots+k_g\right)=n$$

To find $F(s)$, we normally use the partial fraction expansion method; then, we take an inverse Laplace transformation.

1. *For single-pole case:*

$$F(s)=\frac{b_0s^m+b_1s^{m-1}+\cdots+b_{m-1}s+b_m}{s^n+a_1s^{n-1}+\cdots+a_{n-1}s+a_n}=\frac{b_0s^m+b_1s^{m-1}+\cdots+b_{m-1}s+b_m}{(s+p_1)(s+p_2)\cdots(s+p_n)}$$

or

$$F(s)=\frac{a_1}{s+p_1}+\frac{a_2}{s+p_2}+\cdots+\frac{a_n}{s+p_n}$$

Here, a_k is a constant, the residue at the pole $s=-p_k$ is

$$a_k=\left[F(s)(s+p_k)\right]_{s=-p_k}$$

Taking a_k into the expression of *F*(*s*) and taking the inverse Laplace transformation, we get

$$f(t)=\mathcal{L}^{-1}\left[F(s)\right]=\left(a_1e^{-p_1t}+a_2e^{-p_2t}+\cdots a_ne^{-p_nt}\right)$$

Example 2.21

Obtain the inverse Laplace transformation of $F(s)=\dfrac{s+3}{s^2+3s+2}$.

Solution

We may simplify the function

$$F(s)=\frac{s+3}{s^2+3s+2}=\frac{s+3}{(s+1)(s+2)}=\frac{a_1}{s+1}+\frac{a_2}{s+2}$$

In the equation, we may determine a_1 and a_2 as in the following:

$$a_1=\left[\frac{s+3}{(s+1)(s+2)}\times(s+1)\right]_{s=-1}=2$$

$$a_2=\left[\frac{s+3}{(s+1)(s+2)}\times(s+2)\right]_{s=-2}=-1$$

Substituting the value of a_1 and a_2, we get

$$F(s)=\frac{2}{s+1}+\frac{-1}{s+2}$$

Taking the inverse Laplace transformation, we get

$$f(t)=\mathcal{L}^{-1}\left[F(s)\right]=\mathcal{L}^{-1}\left[\frac{2}{s+1}-\frac{1}{s+2}\right]=2e^{-t}-e^{-2t}.$$

2. *For conjugate complex pole:*

$$F(s)=\frac{M(s)}{N(s)}=\frac{a_1s+a_2}{(s+\sigma+j\beta)(s+\sigma-j\beta)}+\frac{a_3}{s+p_3}+\cdots+\frac{a_n}{s+p_n}$$

Let the steps of obtaining a_1, a_2 are the two sides of the equation simply multiply by $(s+\sigma+j\beta)(s+\sigma-j\beta)$, and define $s=-\sigma-j\beta$ (or $s=-\sigma+j\beta$), we get

$$(a_1s+a_2)\big|_{s=-\sigma-j\beta}=[F(s)(s+\sigma+j\beta)(s+\sigma-j\beta)]\big|_{s=-\sigma-j\beta}$$

Therefore, we get a_1, a_2, and $\frac{a_1s+a_2}{(s+\sigma)^2+\beta^2}$, and if it can be formulated as a transfer function of sine or cosine function, then we can take an inverse Laplace transformation of it.

Example 2.22

Obtain the inverse Laplace transformation of $F(s)=\frac{s+1}{s^3+s^2+s}$.

Solution

We may simply the function as

$$F(s)=\frac{s+1}{s^3+s^2+s}=\frac{a_1s+a_2}{s^2+s+1}+\frac{a_3}{s}$$

The term denominator (s^2+s+1) may be expressed as in the following:

$$s^2+s+1=\left(s+\frac{1}{2}+j\frac{\sqrt{3}}{2}\right)\left(s+\frac{1}{2}-j\frac{\sqrt{3}}{2}\right)$$

If both sides of $F(s)$ are multiplied by (s^2+s+1) and then evaluated at $s=-\frac{1}{2}-j\frac{\sqrt{3}}{2}$, we get

$$F(s)\times(s^2+s+1)=\frac{s+1}{s^3+s^2+s}\times(s^2+s+1)\bigg|_{s=-\frac{1}{2}-j\frac{\sqrt{3}}{2}}$$

Thus,

$$\frac{s+1}{s}\bigg|_{s=-\frac{1}{2}-j\frac{\sqrt{3}}{2}}=(a_1s+a_2)\big|_{s=-\frac{1}{2}-j\frac{\sqrt{3}}{2}}$$

or

$$\frac{1}{2}+j\frac{\sqrt{3}}{2}=-\frac{1}{2}a_1+a_2+j\left(-\frac{\sqrt{3}}{2}a_1\right)$$

Equating the above equation on both sides, we get

$$\begin{cases} \frac{1}{2} = -\frac{1}{2}a_1 + a_2 \\ \frac{\sqrt{3}}{2} = -\frac{\sqrt{3}}{2}a_1 \end{cases}$$

Solving this equation, we get

$$\begin{cases} a_1 = -1 \\ a_2 = 0 \end{cases}$$

Now if we can evaluate that $s = 0$, then we get

$$a_3 = \left[\frac{s+1}{s(s^2+s+1)} \times s \right]_{s=0} = 1$$

Consequently, if we substitute a_1, a_2, and a_3, we get

$$F(s) = \frac{-s}{s^2+s+1} + \frac{1}{s} = \frac{-\left(s+\frac{1}{2}\right) + \frac{\sqrt{3}}{3}\frac{\sqrt{3}}{2}}{(s+\frac{1}{2})^2 + (\frac{\sqrt{3}}{2})^2} + \frac{1}{s}$$

Taking the inverse Laplace transformation, we get

$$f(t) = \mathcal{L}^{-1}[F(s)] = -e^{-\frac{1}{2}t} \cos\frac{\sqrt{3}}{2}t + \frac{\sqrt{3}}{3}e^{-\frac{1}{2}t} \sin\frac{\sqrt{3}}{2}t + 1.$$

3. *The inverse Laplace transformation of multiple poles:*

$$F(s) = \frac{M(s)}{N(s)} = \frac{M(s)}{(s+p_1)^r(s+p_2)\cdots(s+p_k)}$$

$$= \frac{a_r}{(s+p_1)^r} + \frac{a_{r-1}}{(s+p_1)^{r-1}} + \cdots + \frac{a_1}{(s+p_1)} + \frac{b_2}{(s+p_2)} + \cdots + \frac{b_k}{(s+p_k)}$$

According to the residue theorem, we can calculate a_i,

$$a_r = [F(s)(s+p_1)^r]_{s=-p_1}$$

$$a_{r-1} = \left\{ \frac{d}{ds}[F(s)(s+p_1)^r] \right\}_{s=-p_1}$$

$$a_{r-2} = \frac{1}{2!}\left\{ \frac{d^2}{ds^2}[F(s)(s+p_1)^r] \right\}_{s=-p_1} \quad \vdots$$

$$a_{r-i} = \frac{1}{i!}\left\{\frac{d^i}{ds^i}\left[F(s)(s+p_1)^r\right]\right\}_{s=-p_1} \qquad \vdots$$

$$a_1 = \frac{1}{(r-1)!}\left\{\frac{d^{r-1}}{ds^{r-1}}\left[F(s)(s+p_1)^r\right]\right\}_{s=-p_1}$$

According to the inverse Laplace transformation (Table 2.1),

$$\mathcal{L}^{-1}\left[\frac{1}{(s+p_1)^i}\right] = \frac{t^{i-1}}{(i-1)!}e^{-p_1 t}$$

Therefore, we can get the inverse Laplace transformation of the multiple poles.

Example 2.23

Obtain the inverse Laplace transformation of $F(s) = \frac{s^2+2s+5}{(s+2)^3}$.

Solution

We may simply the function as

$$F(s) = \frac{s^2+2s+5}{(s+2)^3} = \frac{a_3}{(s+2)^3} + \frac{a_2}{(s+2)^2} + \frac{a_1}{(s+2)}$$

In the equation, we may determine a_1, a_2 and a_3 as follows:

$$a_3 = \left[\frac{s^2+2s+5}{(s+2)^3}(s+2)^3\right]_{s=-2} = 5$$

$$a_2 = \frac{d}{ds}\left[\frac{s^2+2s+5}{(s+2)^3}(s+2)^3\right]_{s=-2} = -2$$

$$a_1 = \frac{d^2}{2!ds^2}\left[\frac{s^2+2s+5}{(s+2)^3}(s+2)^3\right]_{s=-2} = 1$$

Substituting the value of a_1, a_2 and a_3, we get

$$F(s) = \frac{5}{(s+2)^3} - \frac{2}{(s+2)^2} + \frac{1}{(s+2)}$$

Taking the inverse Laplace transformation, we get

$$f(t) = \left(\frac{5}{2}t^2 - 2t + 1\right)e^{-2t}$$

4. *The inverse Laplace transformation of multiple poles can also be obtained by the Heaviside formula:*

$$f(t)=\frac{1}{(r-1)!}\left\{\frac{d^{r-1}}{ds^{r-1}}\left[F(s)(s+p_1)^r e^{st}\right]\right\}_{s=-p_1}+\sum_{i=2}^{k} b_i e^{-p_i t} \tag{2.79}$$

If we express *F*(*s*) as a fraction, then we get

$$F(s)=\frac{M(s)}{N(s)}$$

Let $p_i(i=2,3,\cdots,k)$ are single poles, then the Heaviside formula can be rewritten as

$$f(t)=\frac{1}{(r-1)!}\left\{\frac{d^{r-1}}{ds^{r-1}}\left[F(s)(s+p_1)^r e^{st}\right]\right\}_{s=-p_1}+\sum_{i=2}^{k}\left.\frac{M(s)}{\frac{dN(s)}{ds}}\right|_{s=-p_i} -e^{-p_i t}$$

Example 2.24

Obtain the inverse Laplace transformation of $F(s)=\frac{s^2+2s+5}{(s+2)^3}$.

Solution

We may simplify the function by using the Heaviside formula as follows:

$$\begin{aligned}
f(t)&=\frac{1}{2!}\frac{d^2}{ds^2}\left[(s+2)^3\frac{s^2+2s+5}{(s+2)^3}e^{st}\right]_{s=-2}\\
&=\frac{1}{2}\frac{d^2}{ds^2}\left[\left(s^2+2s+5\right)e^{st}\right]_{s=-2}\\
&=\frac{1}{2}\left[2e^{st}+(4s+4)te^{st}+\left(s^2+2s+5\right)t^2e^{st}\right]_{s=-2}\\
&=\frac{5}{2}t^2e^{-2t}-2te^{-2t}+e^{-2t}
\end{aligned}$$

2.3 CONSTANT COEFFICIENT LINEAR DIFFERENTIAL FUNCTION USING THE LAPLACE TRANSFORMATION

Example 2.25

Obtain the solution of function $\ddot{y}+2\dot{y}-3y=e^{-t}$ with the initial condition $y(0)=1,\ \dot{y}(0)=0$.

Solution

Taking the Laplace transformation on both sides and then setting the initial condition into the equation, we get

$$\left[s^2Y(s)-sy(0)-\dot{y}(0)\right]+2\left[sY(s)-y(0)\right]-3Y(s)=\frac{1}{s+1}$$

Substituting the initial condition into the above equation and obtaining the algebraic equation with an uncertain constant to analyze may be expressed as

$$Y(s)=\frac{s^2+3s+3}{(s+1)(s-1)(s+3)}$$

Obtaining the inverse Laplace transformation of $Y(s)$, then expanding $Y(s)$ as fraction form as in the following:

$$Y(s)=\frac{s^2+3s+3}{(s+1)(s-1)(s+3)}=\frac{\frac{7}{8}}{(s-1)}+\frac{-\frac{1}{4}}{(s+1)}+\frac{\frac{3}{8}}{(s+3)}$$

Taking inverse Laplace transformation of $Y(s)$, we get

$$y(t)=\frac{1}{8}\left(7e^{t}-2e^{-t}+3e^{-3t}\right)$$

This is the solution of the differential function, $\ddot{y}+2\dot{y}-3y=e^{-t}$.

Example 2.26

Obtain the solution of function under the initial condition $y(0)=1,\ \ \dot{y}(0)=0$.

$$\begin{cases} \ddot{y}-\ddot{x}+\dot{x}-y=e^{t}-2 \\ 2\ddot{y}-\ddot{x}-2\dot{y}+x=-t \end{cases}$$

Solution

Taking the Laplace transformation on both sides and then setting the initial condition into the equation, we get

$$\begin{cases} s^2Y(s)-s^2X(s)+sX(s)-Y(s)=\frac{1}{s-1}-\frac{2}{s} \\ 2s^2Y(s)-s^2X(s)-2sY(s)+X(s)=-\frac{1}{s^2} \end{cases}$$

By simplifying and solving the equations, we get

$$\begin{cases} X(s)=\dfrac{2s-1}{s^2(s-1)^2} \\ Y(s)=\dfrac{1}{s(s-1)^2} \end{cases}$$

According to the Heaviside formula, we can take the inverse Laplace transforms of $X(s)$ and $Y(s)$. Since $X(s)=\dfrac{2s-1}{s^2(s-1)^2}$ has four poles as $s=0$, 0, 1, 1, according to Heaviside formula, we have

$$x(t)=\lim_{s\to 0}\frac{d}{ds}\left[\frac{2s-1}{(s-1)^2}e^{st}\right]+\lim_{s\to 1}\frac{d}{ds}\left[\frac{2s-1}{s^2}e^{st}\right]$$

$$=\lim_{s\to 0}\left[te^{st}\frac{2s-1}{(s-1)^2}-\frac{2s}{(s-1)^3}e^{st}\right]+\lim_{s\to 1}\left[te^{st}\frac{2s-1}{s^2}+\frac{2(1-s)}{s^3}e^{st}\right]$$

$$=-t+te^t$$

However, $Y(s)=\dfrac{1}{s(s-1)^2}$ has 3 poles as $s=0$, 1, 1; according to Heaviside formula, we have

$$y(t)=\frac{1}{3s^2-4s+1}e^{st}\bigg|_{s=0}+\lim_{s\to 1}\frac{d}{ds}\left[(s-1)^2\frac{1}{s(s-1)^2}e^{st}\right]$$

$$=1+\lim_{s\to 1}\frac{d}{ds}\left[\frac{1}{s}e^{st}\right]$$

$$=1+\lim_{s\to 1}\left(\frac{t}{s}e^{st}-\frac{1}{s^2}e^{st}\right)$$

$$=1+te^t-e^t$$

Therefore, the solution of the equations is

$$\begin{cases} x(t)=-t+te^t \\ y(t)=1+te^t-e^t \end{cases}$$

From above, we can draw a conclusion about the steps of using the Laplace transform to find the result of the differential equations as

a. Obtain the Laplace transform function of the differential equations and consider the initial conditions.

b. Solve the algebraic equations about the Laplace transforms of the dependent variables.
c. Solve the equations of the inverse Laplace transforms of the dependent variables, which will help to obtain the result of the differential equations.

2.4 TRANSFER FUNCTION

The transfer function is established on the basis of the Laplace transform, and it describes the functional relationship between the input and output of a linear time-invariant (LTI) system in terms of the structural parameters of the system itself. It is determined by the internal parameters which are inherent in the system itself, and it is independent of input and output or driver functions. The dimension depends on the input and output. It does not represent the physical structure or the physical properties of the system. Many different physical systems have the same transfer function, and this is just as the same differential equation can describe the physical systems of many different structures. To study transfer function, we have to learn about LTI system that we will discuss in detail.

The transfer function is established on the basis of zero initial conditions, that is, the initial conditions of system input and output are all zero. The transfer function is defined as the ratio of the Laplace transform of the output to the Laplace transform of the input, i.e.,

$$G(s)=\frac{\mathcal{L}[x(t)]}{\mathcal{L}[y(t)]}=\frac{X(s)}{Y(s)}=\frac{\text{The Laplace transform of the output}}{\text{The Laplace transform of the input}}$$

It is assumed that the differential equation of the LTI system is

$$\begin{aligned}&a_o x^{(m)}(t)+a_1 x^{(m-1)}(t)+\ldots+a_{m-1}\dot{x}(t)+a_m x(t)\\&=b_o y^{(n)}(t)+b_1 y^{(n-1)}(t)+\ldots+b_{n-1}\dot{y}(t)+b_n y(t)\end{aligned} \tag{2.80}$$

In Eq. (2.80), $x(t)$ is the output of the system and $y(t)$ is the input, whereas $a_i\,(i=0,1,\ldots,m)$ and $b_j\,(j=0,1,\ldots,n)$ are the constant coefficients of the system's structure. It is assumed that the initial condition is zero, taking the Laplace transform of Eq. (2.80) and then we obtain the algebraic equation of s,

$$\begin{aligned}&(a_o s^m+a_1 s^{m-1}+\ldots+a_{m-1}s+a_m)X(s)\\&=(b_o s^n+b_1 s^{n-1}+\ldots+b_{n-1}s+b_n)Y(s)\end{aligned} \tag{2.81}$$

Therefore, the transfer function of the system is defined as in the following:

$$G(s)=\frac{X(s)}{Y(s)}=\frac{b_o s^n+b_1 s^{n-1}+\ldots+b_{n-1}s+b_n}{a_o s^m+a_1 s^{m-1}+\ldots+a_{m-1}s+a_m}=\frac{M(s)}{N(s)} \tag{2.82}$$

From Eq. (2.82), we can express

$$M(s) = b_o s^n + b_1 s^{n-1} + \ldots + b_{n-1} s + b_n$$

and

$$N(s) = a_o s^m + a_1 s^{m-1} + \ldots + a_{m-1} s + a_m$$

The implication of the zero initial condition in Eq. (2.82) is summarized as

1. The input does not act on the system until $t = 0$, so the input of the system and the values of all derivatives are zero when $t = 0$.
2. Before the input is applied to the system, the system is relatively stationary, so the derivatives of the output of the system are also zero when $t = 0$; therefore, the transfer function can fully describe the dynamic characteristics of the system.

2.4.1 The Advantages of the Transfer Function

1. Transfer function is simpler to describe the system than the differential equation. By the Laplace transform, the differential equation is simplified into an algebraic equation.
2. The transfer function is a rational fractional function of the complex variable(s). The order (m) of the numerator polynomial is lower than the order (n) of the denominator polynomial; and all the coefficients $a_i\ (i = 0, 1, \ldots, m)$ and $b_j\ (j = 0, 1, \ldots, n)$ are all real numbers; however, if the system has inertia, then the system energy is limited; thus, it must be satisfied by $n \geq m$.
3. The transfer function only depends on the structure of the system and its components; however, it is independent of external action.
4. The zero-pole configuration of the transfer function determines the transient response of the system.
5. When the system input(s) a typical signal, the output of the system is related to the transfer function, for example, when the input $x(t) = \delta(t)$, then $X(s) = 1$; thus, the output $Y(s) = G(s)$ is the transfer function of the system.
6. Let $s = j$ in the transfer function, and then we can analyze the frequency response of the system.

Therefore, the transfer function has I / O relationships with system components or systems; the transfer function can be described by linear, time-invariant, and differential equations with initial conditions zero such that $y(t) = y(0-) = 0$ as well as the following condition can be satisfied as

$$\left.\frac{d^{n-1}y}{dt^{n-1}}\right|_{t=0} = \left.\frac{d^{n-2}y}{dt^{n-2}}\right|_{t=0} = \ldots = \left.\frac{dy}{dt}\right|_{t=0} = y(0) = 0 \tag{2.83}$$

whereas we can say, the *transfer function* is the function of a linear (*L*), time-invariant (*TI*), differential equation system. It can be defined as the ratio of the Laplace transform of the output (response function) to the Laplace transform of the input (driving function) under the assumption that all initial conditions are zero.

$$\text{Transfer function}, G(s) = \left.\frac{\mathcal{L}[\text{output}]}{\mathcal{L}[\text{input}]}\right|_{\text{zero initial conditions}}$$

2.4.2 Mathematical Representation of the Transfer Function

Let's consider an LTI nth-order differential equation, so the system dynamics with mth order 'a' coefficient differential equation of input variable x of the equation on the left side is called input system; however, the system dynamics with nth-order 'b' coefficient differential equation of output variable y of the equation in the left side is called output system. Here, you must remember that n is greater or equal to m; i.e., $n \geq m$.

$$a_m \frac{d^m x(t)}{dt^m} + a_{m-1} \frac{d^{m-1} x(t)}{dt^{m-1}} + \ldots + a_1 \frac{dx(t)}{dt} + a_0\ x(t)$$

$$= b_n \frac{d^n y(t)}{dt^n} + b_{n-1} \frac{d^{n-1} y(t)}{dt^{n-1}} + \ldots + b_1 \frac{dy(t)}{dt} + b_0\ y(t) \tag{2.84}$$

From the LTI system dynamics, we can find the transfer function, $G(s)$ is equal to the Laplace transformation of output system divided by the Laplace transformation of input system. Therefore, the transfer function, $G(s)$ is equal to $Y(s)$ divided by $X(s)$ which gives the nth-order 'b' coefficient in s and divided by mth order 'a' coefficient in s. In addition, the system dynamics is represented by algebraic equations in s. Moreover, if $n = m$ or if the highest power of s in input system in the denominator of the transfer function is equal to n, the system is called nth-order system.

$$G(s) = \frac{\mathcal{L}\left\{b_n \dfrac{d^n y}{dt^n} + b_{n-1} \dfrac{d^{n-1} y}{dt^{n-1}} + \ldots + b_1 \dfrac{dy}{dt} + b_0 y\right\}}{\mathcal{L}\left\{a_m \dfrac{d^m x}{dt^m} + a_{m-1} \dfrac{d^{m-1} x}{dt^{m-1}} + \ldots + a_1 \dfrac{dx}{dt} + a_0 x\right\}}$$

and

$$G(s) = \frac{Y(s)}{X(s)} = \frac{b_n s^n + b_{n-1} s^{n-1} + \ldots + b_1 s + b_0}{a_m s^m + a_{m-1} s^{m-1} + \ldots + a_1 s + a_0} \tag{2.85}$$

2.4.3 Comments and Keynotes

The mathematical model of a transfer function is an operational method of expressing the differential equation that relates the output variable to the input variable; the property of a system itself, independent of the magnitude and nature of the input; including the units necessary to relate the input to the output. However, it does not provide any information concerning the physical structure of the system. If the transfer function of a system is **known**, the output or response can be studied for various forms of inputs to understand the nature of the system. If the transfer function of a system is **unknown**, it may be established experimentally by introducing known inputs and studying the output of the system.

Example 2.27

Find the transfer function represented by

$$\frac{d^2y}{dt^2}+4\frac{dy}{dt}+8y=2x+\frac{dx}{dt};y=y(t),\, x=x(t)$$

Solution

Taking the Laplace transform of both sides, assuming zero initial conditions, we have

$$s^2Y+4sY+8Y=2X+sX$$

Therefore, the transfer function is

$$G(s)=\frac{Y}{X}=\frac{s+2}{s^2+4s+8}$$

2.5 FUNCTION TRANSFORMATIONS

In this section, we will focus on the convolution integral of the LTI system, impulse response function (IRF), Laplace transform theorems and table, and function transformations. What is convolution integral of LTI system?

We have

$$G(s)=\frac{Y(s)}{X(s)}$$

From here, we can rearrange the above transfer function as

$$Y(s)=G(s)X(s)$$

Now, we say

$$y(t)=\int_0^t x(\tau)g(t-\tau)d\tau\equiv\int_0^t g(\tau)x(t-\tau)d\tau$$

where $g(t)$ and $x(t)$ are 0 for $t>0$.

2.5.1 Impulse Response Function (IRF)

Let, unit-impulse input, $x(t)=1$. Now taking the Laplace operator on both sides of this input, we get $X(s)=1$. We know the output $Y(s)=G(s)\ X(s)$, so the output for this case we can say $Y(s)=G(s)$ because $X(s)=1$.

$$x(t)=1$$

Thus,

$$\mathcal{L}\{x(t)\}=1$$

Hence,

$$X(s)=1$$

Now, consider the IRF which can be denoted by $g(t)$, and this $g(t)$ is the response of a linear system to unit-impulse input when the initial conditions are zero. However, the Laplace transform of IRF gives a transfer function such that Laplace of $g(t)$ will give us the transfer function, $G(s)$, whereas the inverse Laplace of transfer function $G(s)$ provides IRF, $g(t)$.

$$[g(t)]=G(s)\equiv\mathcal{L}^{-1}[G(s)]=g(t)$$

2.5.2 Laplace Transformation Theorems

Look at the table, we can see some definitions of the Laplace transform theorems. Here, we can see the definition, linearity, frequency shifting theorem, timeshift theorem, and scaling theorem (Table 2.3).

We also can see the differentiation theorem, integration theorem, final value theorem, and initial value theorem; the final value and initial value theorems are specially used in the control theory, so pay attention to remember these two theorems.

2.6 CONSTANT CONTROL ACTION OF TRANSFER FUNCTION

1. Proportional control of systems, $G(s)=K$ (K is a constant);
2. First-order system with inertia load, $G(s)=\dfrac{1}{Ts+1}$;
3. Derivative or differential control of systems;
4. Integral control of systems, $G(s)=\dfrac{K}{s}$ (K is a constant); and
5. Second-order oscillatory control, $G(s)=\dfrac{1}{T^2s^2+2\xi Ts+1}$; $(0<\xi<1)$.

TABLE 2.3 Laplace Transform Theorems

Item No.	Theorem	Name
1	$\mathcal{L}[f(t)] = F(s) = \int_{0^-}^{\infty} f(t)e^{-st}dt$	Definition
2	$\mathcal{L}[kf(t)] = kF(s)$	Linearity theorem
3	$\mathcal{L}[f_1(t) + f_2(t)] = F_1(s) + F_2(s)$	Linearity theorem
4	$\mathcal{L}[e^{-at}f(t)] = F(s+a)$	Frequency shift theorem
5	$\mathcal{L}[f(t-T)] = e^{-sT}F(s)$	Timeshift theorem
6	$\mathcal{L}[f(at)] = \frac{1}{a}F\left(\frac{s}{a}\right)$	Scaling theorem
7	$\mathcal{L}\left[\frac{df}{dt}\right] = sF(s) - f(0-)$	Differentiation theorem
8	$\mathcal{L}\left[\frac{d^2f}{dt^2}\right] = s^2F(s) - sf(0-) - f(0-)$	Differentiation theorem
9	$\mathcal{L}\left[\frac{d^nf}{dt^n}\right] = s^nF(s) - \sum_{k=1}^{n} s^{n-k}f^{k-1}(0-)$	Differentiation theorem
10	$\mathcal{L}\left[\int_{0-}^{t} f(\tau)d\tau\right] = \frac{F(s)}{s}$	Integration theorem
11	$f(\infty) = \lim_{s\to 0} sF(s)$	Final value theorem
12	$f(0+) = \lim_{s\to\infty} sF(s)$	Initial value theorem

1. *Proportional control of systems,* $G(s) = K$ *(K is a constant)*
   In the time domain, the input is proportional to the output and the system is expressed as a zero-order differential equation, i.e.,

$$y(t) = Kx(t) \tag{2.86}$$

Taking the Laplace transform of Eq. (2.86) under the zero initial condition, we obtain

$$Y(s) = KX(s)$$

Therefore, the transfer function can be expressed as $G(s)$ which can be denoted in terms of proportional constant action as K, i.e.,

$$G(s) = \frac{Y(s)}{X(s)} = K \tag{2.87}$$

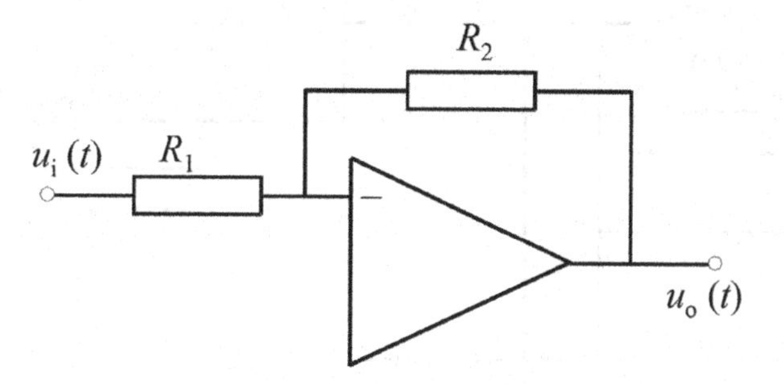

FIGURE 2.18 Operational amplifier with negative feedback.

Example 2.28

Figure 2.18 shows an operational amplifier, where $u_i(t)$ is the input voltage, $u_o(t)$ is the output voltage, R_1 and R_2 are two resistors.

Solution

According to KVL, we obtain

$$u_o(t) = -\frac{R_2}{R_1} u_i(t) \tag{2.88}$$

Taking the Laplace transformation of Eq. (2.88), we obtain

$$U_o(s) = -\frac{R_2}{R_1} U_i(s)$$

Therefore, the transfer function of the operational amplifier is

$$G(s) = \frac{U_o(s)}{U_i(s)} = -\frac{R_2}{R_1} = K \tag{2.89}$$

Example 2.29

Figure 2.19 shows a hydraulic displacement amplifier, where x_i and x_o are the displacement of the big piston and the small piston, respectively, and A_i and A_o are the proportional constant of the big piston and the small piston, respectively.

Solution

According to the equation of continuity, we obtain

$$A_i \dot{x}_i(t) = A_o \dot{x}_o(t) = q \tag{2.90}$$

Taking the integral in both sides of Eq. (2.90), we obtain

$$A_i x_i(t) = A_o x_o(t) = \int q\,dt \tag{2.91}$$

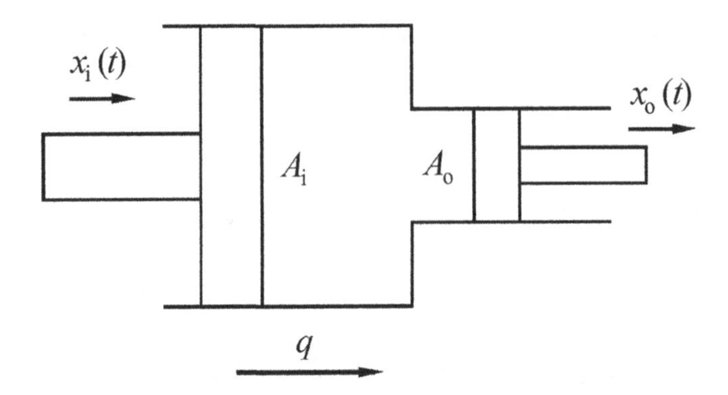

FIGURE 2.19 Hydraulic displacement amplifier.

Taking the Laplace transform of the first two terms of Eq. (2.91), rearrange it. Then we can obtain the transfer function of the system, i.e.,

$$G(s)=\frac{X_o(s)}{X_i(s)}=\frac{A_i}{A_o}=K$$

Example 2.30

Figure 2.20 shows a pair of gears, where $n_i(t), n_o(t)$ are the rotational speed of the input shaft and the output shaft, respectively; and Z_i and Z_o are numbers of teeth input gear and output gear, respectively.

Solution

According to theory of dynamics, we know that

$$n_i(t)\cdot Z_i = n_o(t)\cdot Z_o \tag{2.92}$$

Taking the Laplace transform of Eq. (2.92), we obtain the transfer function,

$$G(s)=\frac{N_o(s)}{N_i(s)}=\frac{Z_i}{Z_o}=K \tag{2.93}$$

2. *First-order system with inertia load,* $G(s)=\dfrac{1}{Ts+1}$

In the time domain, the input and output functions can be expressed as the following differential equation:

$$T\dot{x}_o(t)+x_o(t)=x_i(t) \tag{2.94}$$

Then we can take the Laplace transform of Eq. (2.94) under the zero initial condition, and we get

$$TsX_o(s)+X_o(s)=X_i(s)$$

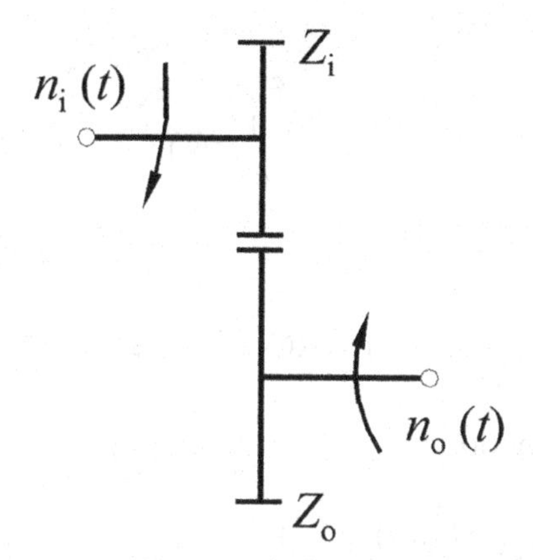

FIGURE 2.20 A pair of gears.

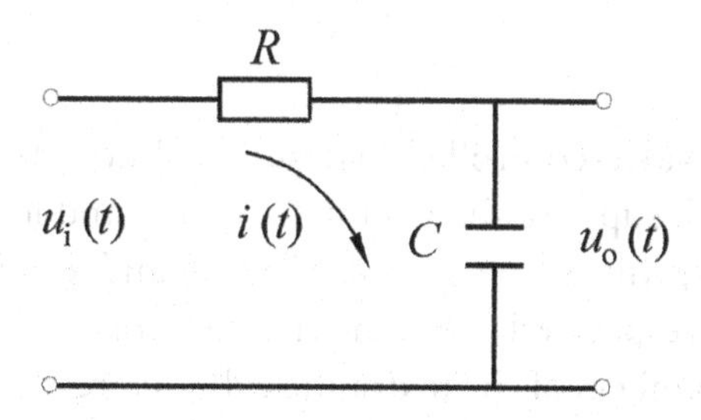

FIGURE 2.21 Nonelectric-source circuit.

Therefore, the transfer function is

$$G(s)=\frac{Y(s)}{X(s)}=\frac{1}{Ts+1}; \text{ where } T \text{ is a constant} \tag{2.95}$$

Example 2.31

Figure 2.21 shows a nonelectric-source circuit. $u_i(t)$ and $u_o(t)$ are the input and output voltage; R is a resistor; and C is a capacitor.

Solution

According to KVL, we obtain

$$u_i(t)=i(t)R+\frac{1}{C}\int i(t)dt \tag{2.96}$$

$$u_o(t)=\frac{1}{C}\int i(t)dt \tag{2.97}$$

Taking the Laplace transform of Eqs. (2.96) and (2.97), we obtain

$$U_i(s)=I(s)R+\frac{1}{Cs}I(s) \tag{2.98}$$

Thus,

$$U_o(s)=\frac{1}{Cs}I(s) \tag{2.99}$$

Eliminating $I(s)$, we get

$$U_i(s)=(RCs+1)U_o(s)$$

Therefore, the transfer function is

$$G(s)=\frac{U_o(s)}{U_i(s)}=\frac{1}{RCs+1}(\text{where } RC=T) \tag{2.100}$$

Example 2.32

Figure 2.22 shows the valve-controlled hydraulic cylinder type power mechanism system with spring load, suppose the volumetric effect and leakage are ignored and only the action of the spring is considered. Here, x_i and x_o are the displacement of servo valve and spring, respectively; A is the effective area of the piston; and p_i and p_o are pressures of two chambers of the hydraulic cylinder, respectively.

Solution

According to the flow equation, we obtain

$$q_L(t)=K_q x_i(t)-K_c p_L(t) \tag{2.101}$$

$$q_L(t)=A\dot{x}_o(t) \tag{2.102}$$

The power balance equation is

$$Ap_L(t)=k_s x_o(t) \tag{2.103}$$

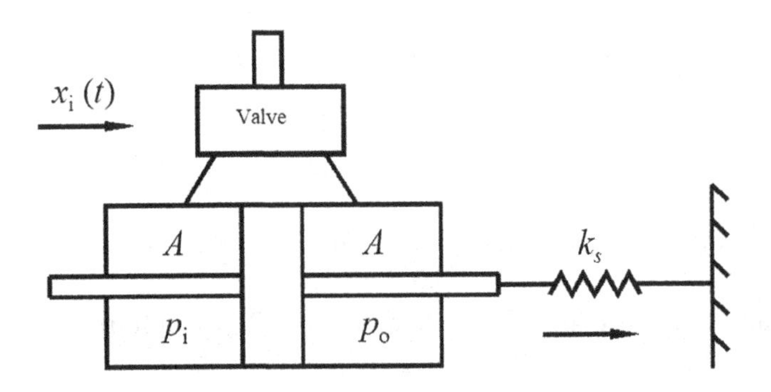

FIGURE 2.22 Hydraulic power mechanism with spring load.

Taking the Laplace transform of Eqs. (2.101)–(2.103) and eliminating q_L and p_L we get

$$G(s)=\frac{X_o(s)}{X_i(s)}=\frac{\frac{K_qA}{K_ck_s}}{\frac{A^2}{K_ck_s}s+1}=\frac{K}{Ts+1} \tag{2.104}$$

where $K=\frac{K_qA}{K_ck_s}$, $T=\frac{A^2}{K_ck_s}$

Example 2.33

Figure 2.23 shows a spring-damper system, where x_i and x_o are displacements of the input and output, k is the stiffness of the spring, and f is the damping coefficient.

Solution

As shown in Figure 2.23, it can be obtained according to the dynamics as

$$k\left[x_i(t)-x_o(t)\right]=f\dot{x}_o(t) \tag{2.105}$$

Taking the Laplace transform of Eq. (2.105), we obtain the transfer function, i.e.,

$$G(s)=\frac{X_o(s)}{X_i(s)}=\frac{1}{\frac{f}{k}s+1}=\frac{1}{Ts+1}\quad (f/k=T) \tag{2.106}$$

3. *Derivative or differential control of systems*

For ideal derivative control, $G(s)=Ks$ (K is a constant). The output variable is proportional to the first-order derivative of the input variable, that is,

$$x_o(t)=K\dot{x}_i(t) \tag{2.107}$$

Thus, taking the Laplace transform of Eq. (2.107), we obtain

$$X_o(s)=KsX_i(s)$$

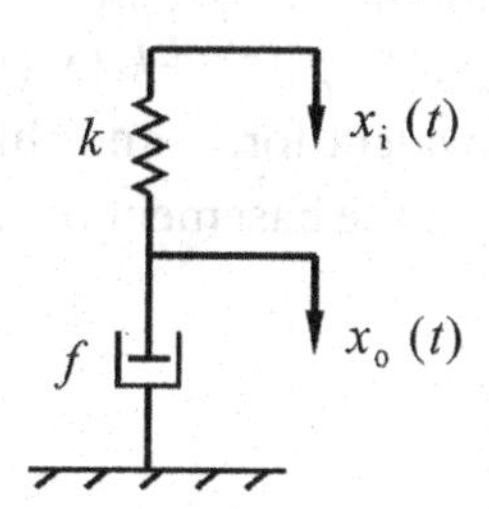

FIGURE 2.23 Spring-damper system.

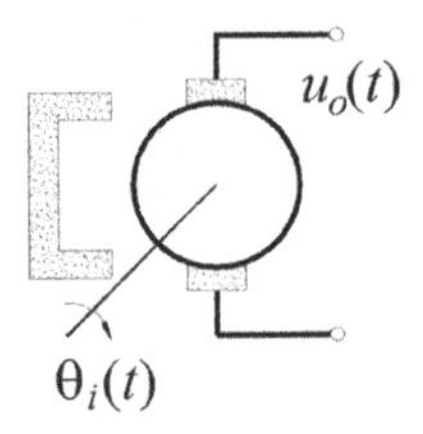

FIGURE 2.24 Permanent magnet DC motor.

Therefore, the transfer function is

$$G(s) = \frac{X_o(s)}{X_i(s)} = Ks \tag{2.108}$$

Example 2.34

Figure 2.24 shows a permanent magnetic type DC motor, where $\theta_i(t)$ and $u_o(t)$ are the input angle and the output voltage and K is the constant.

Solution

According to Faraday's law, we obtain

$$u_o(t) = K\frac{d\theta_i(t)}{dt} \tag{2.109}$$

Taking the Laplace transform of Eq. (2.89) and simplifying it, we obtain the transfer function of the system. i.e.,

$$G(s) = \frac{U_o(s)}{\Theta_i(s)} = Ks \tag{2.110}$$

Note that actually it is difficult to realize the ideal differential or derivative constant control; thus, we usually use $G(s) = \frac{KTs}{Ts+1}$ (T and K are constants) to access the approximated derivative or differential control constant action of the systems.

Example 2.35

Figure 2.25 shows a hydraulic differential system. The force on the piston under the action of an external force is $A\left(p_i(t) - p_o(t)\right) + k_1\left(x_i(t) - x_o(t)\right)$. The force on the cylinder body is opposite to its equivalent force, and this force is balanced by the connecting springs to the cylinder and the basement bed.

Solution

The force balance equation is

$$A\left(p_i(t) - p_o(t)\right) + k_1\left(x_i(t) - x_o(t)\right) = k_s x_o(t) \tag{2.111}$$

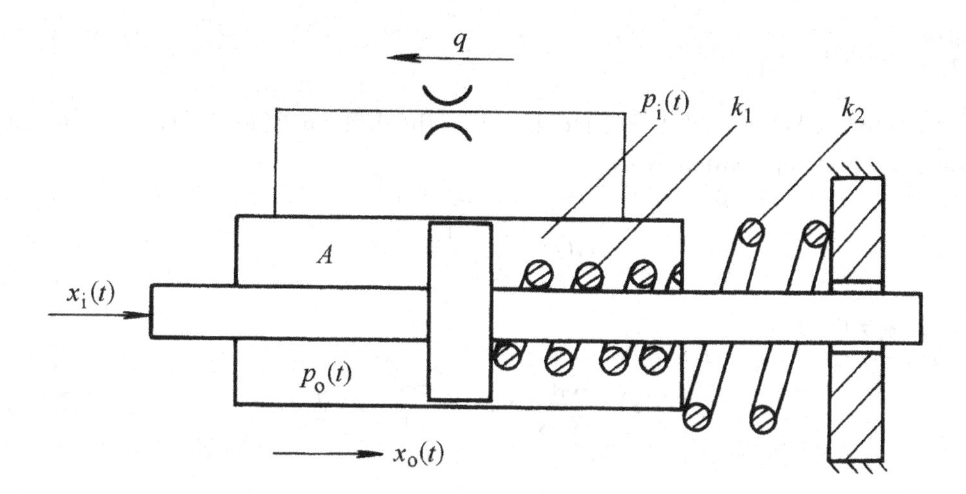

FIGURE 2.25 Hydraulic differential system.

The balance equation of the flow is

$$q = C_d\left(p_i(t) - p_o(t)\right) = A\left(\dot{x}_i(t) - \dot{x}_o(t)\right) \tag{2.112}$$

In Eq. (2.112), C_d is a stress-flow constant. By solving Eq. (2.111), we can obtain

$$p_i(t) - p_o(t) = \frac{k_1 + k_s}{A} x_o(t) - \frac{k_1}{A} x_i(t) \tag{2.113}$$

By substituting Eq. (2.113) into (2.111), we have

$$\frac{k_1 + k_s}{A} x_o(t) - \frac{k_1}{A} x_i(t) = \frac{A}{C_d}\left(\dot{x}_i(t) - \dot{x}_o(t)\right) \tag{2.114}$$

Taking the Laplace transform of Eq. (2.114) and simplifying it, we obtain the transfer function as

$$G(s) = \frac{X_o(s)}{X_i(s)} = \frac{\frac{A}{C_d}s + \frac{k_1}{A}}{\frac{A}{C_d}s + \frac{k_1 + k_s}{A}} \tag{2.115}$$

When $k_1 = 0$, which means there are no springs in the cylinder, Eq. (2.115) becomes

$$G(s) = \frac{X_o(s)}{X_i(s)} = \frac{\frac{A}{C_d}s}{\frac{A}{C_d}s + \frac{k_s}{A}} = K\frac{s}{Ts + 1} \tag{2.116}$$

where $K = \dfrac{A^2}{C_d k_s} = T$.

Note that, when $T \ll 1$, Eq. (2.116) becomes the differential or derivative constant control action approximately as

$$G(s) = \frac{X_o(s)}{X_i(s)} = Ks \tag{2.117}$$

When $k_1 \neq 0$

$$G(s) = \frac{X_o(s)}{X_i(s)} = k_A \frac{\tau s + 1}{Ts + 1}, \quad \tau > T \tag{2.118}$$

where

$$K_A = \frac{k_1}{k_1 + k_s}, \quad \mathrm{T} = \frac{A^2}{C_d(k_1 + k_s)}, \quad \tau = \frac{A^2}{C_d k_1}$$

Example 2.36

Figure 2.26 shows a passive differential circuit, where R is a resistor, C is a capacitor, $u_i(t)$ is the input voltage, and $u_o(t)$ is the output voltage.

Solution

According to KVL, we obtain

$$u_i(t) = \frac{1}{C}\int i(t)dt + i(t)R \tag{2.119}$$

$$u_o(t) = i(t)R \tag{2.120}$$

Taking the Laplace transforms of Eqs. (2.119) and (2.120) and eliminating $I(s)$, we obtain the transfer function as

$$G(s) = \frac{U_o(s)}{U_i(s)} = \frac{RCs}{RCs + 1} = \frac{Ts}{Ts + 1}\left(K = 1, T = RC\right) \tag{2.121}$$

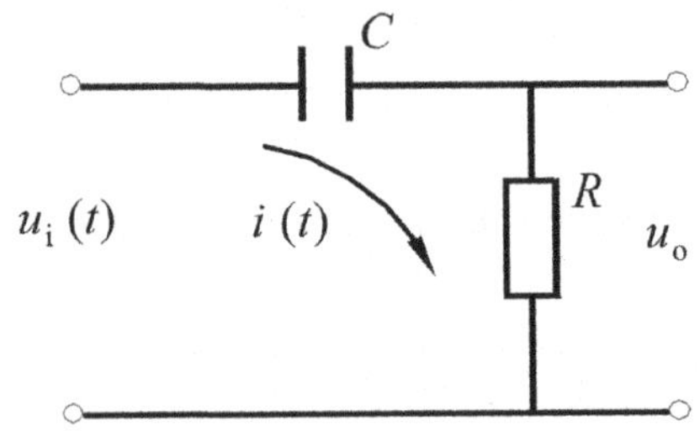

FIGURE 2.26 Passive differential circuit.

4. *Integral control of systems,* $G(s)=\frac{K}{s}$ *(K is a constant)*

The output variable is proportional to the integral of the input variable, that is,

$$x_o(t)=K\int x_i(t)dt \tag{2.122}$$

Taking the Laplace transform of Eq. (2.122), we obtain

$$X_o(s)=K\frac{X_i(s)}{s}$$

Therefore, the transfer function is

$$G(s)=\frac{X_o(s)}{X_i(s)}=\frac{K}{s} \tag{2.123}$$

Example 2.37

Figure 2.27 shows an active integral circuit. R is the resister. C is the capacitor. $u_i(t), u_o(t)$ are the input voltage and the output voltage.

Solution

According to KVL, we obtain

$$\frac{u_i(t)}{R}=-C\dot{u}_o(t) \tag{2.124}$$

Taking the Laplace transform of Eq. (2.124) and simplifying it, we obtain the transfer function as

$$G(s)=\frac{U_o(s)}{U_i(s)}=\frac{-\frac{1}{RC}}{s}=\frac{K}{s};\left(K=-\frac{1}{RC}\right) \tag{2.125}$$

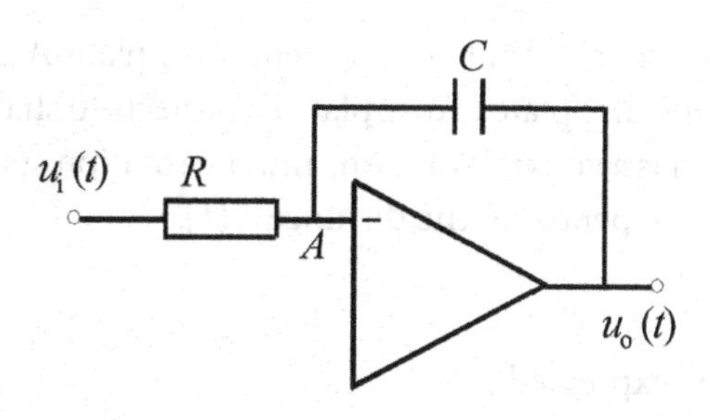

FIGURE 2.27 Active integral circuit.

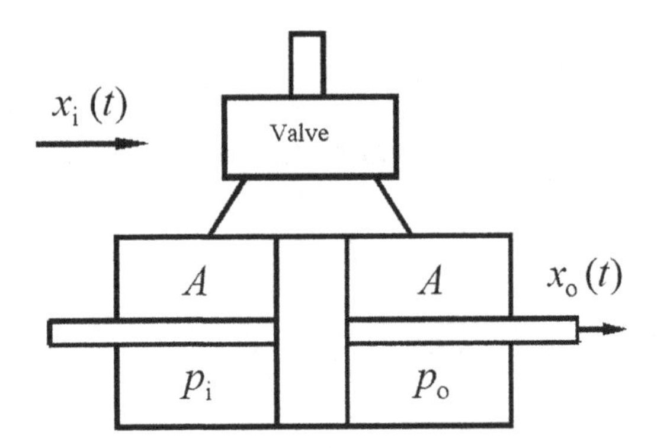

FIGURE 2.28 Valve control hydraulic cylinder actuating unit without load.

Example 2.38

Figure 2.28 shows the valve-controlled hydraulic cylinder type power mechanism without load; suppose the volume effect and leakage are ignored; here, x_i and x_o are the displacement of servo valve and hydraulic cylinder piston; A is the effective area of the piston; and p_i and p_o are pressures of two chambers of a hydraulic cylinder.

Solution

According to the continuity equation of flow, we have

$$q_L(t) = A\dot{x}_o(t) = K_q x_i(t) - K_c p_L(t) \tag{2.126}$$

Since there is no load, we get

$$p_L(t) = p_i(t) - p_o(t) = 0 \tag{2.127}$$

Taking the Laplace transforms of Eqs. (2.126) and (2.127) and simplifying it, we obtain

$$G(s) = \frac{X_o(s)}{X_i(s)} = \frac{K_q}{A}\frac{1}{s} = \frac{K}{s} \tag{2.128}$$

Example 2.39

Figure 2.29 shows a mechanical integral system; disc plate A rotates at a constant speed and drives to rotate disc plate B; disc plate B connects to shaft = 1 * ROMAN I with a sliding key which is a coaxial rotation; and the rotation speed of disc plate B (or axis I) and disc plate A depends on the distance $e_i(t)$.

Solution

The relationship can be expressed as

$$n(t) = Ke_i(t) \tag{2.129}$$

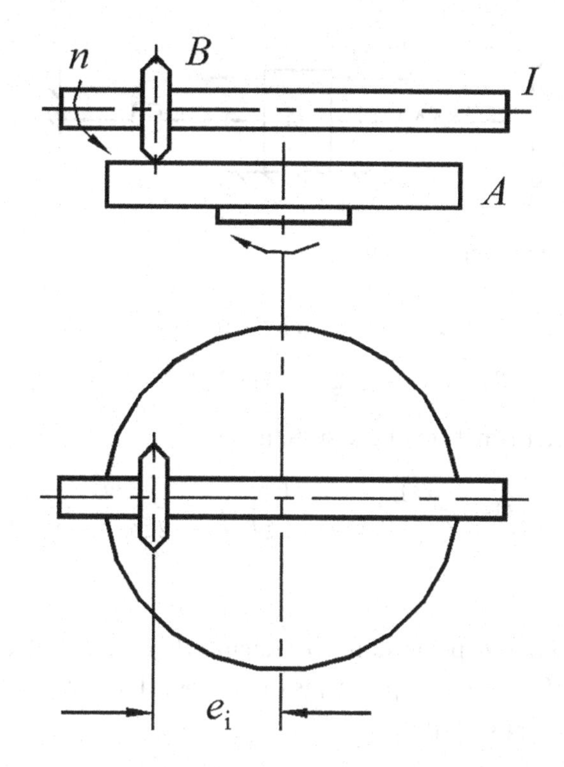

FIGURE 2.29 Mechanical integral device.

where $e_i(t)$ is the input of the distance, $\theta_o(t)$ is the angle and output of shaft I, and $n(t)$ is the rotation speed of shaft I.

According to the mechanism, we have

$$n(t)=\dot{\theta}_o(t) \tag{2.130}$$

Substituting Eq. (2.129) into Eq. (2.130), we have

$$\dot{\theta}_o(t)=Ke_i(t) \tag{2.131}$$

By taking the Laplace transform of Eq. (2.131), we obtain

$$G(s)=\frac{\Theta_o(s)}{E_i(s)}=\frac{K}{s} \tag{2.132}$$

Second-order oscillatory control, $G(s)=\dfrac{1}{T^2s^2+2\xi Ts+1};(0<\xi<1)$

If the input and output functions can be used as the following second-order differential equation,

$$T^2\ddot{x}_o(t)+2T\xi\dot{x}_o(t)+x_o(t)=x_i(t) \tag{2.133}$$

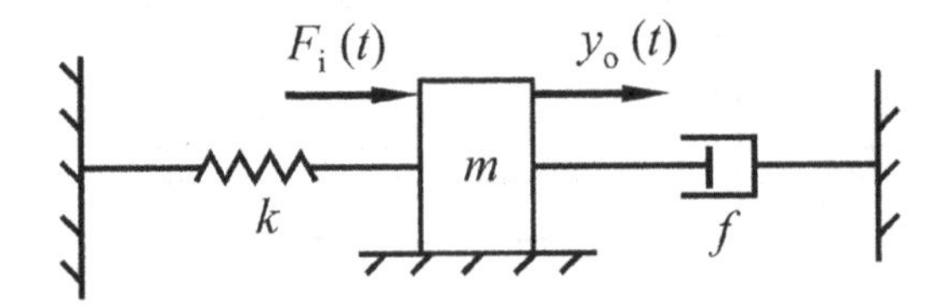

FIGURE 2.30 Mass-spring-damper system.

Taking the Laplace transform of Eq. (2.133), we have

$$T^2s^2X_o(s)+2T\xi sX_o(s)+X_o(s)=X_i(s) \tag{2.134}$$

Therefore, the transfer function of the system is

$$G(s)=\frac{X_o(s)}{X_i(s)}=\frac{1}{T^2s^2+2\xi Ts+1}=\frac{\omega_n^2}{2\xi\omega_n s+\omega_n^2} \tag{2.135}$$

Example 2.40

Figure 2.30 shows a mass-spring-damper system, where $F_i(t)$ is the input force, m is the mass, k is the stiffness of the spring, f is the viscous damping coefficient, and $y_o(t)$ is the displacement of the output.

Solution

According to Newton's laws, we have

$$F_i(t)-f\dot{y}_o(t)-ky_o(t)=m\ddot{y}_o(t) \tag{2.136}$$

Take the Laplace transform of Eq. (2.136). Simplifying it, we obtain

$$G(s)=\frac{Y_o(s)}{F_i(s)}=\frac{1}{ms^2+fs+k}=\frac{1/k}{(\sqrt{\frac{m}{k}})^2s^2+2\times\frac{f}{2\sqrt{mk}}\times\sqrt{\frac{m}{k}}s+1} \tag{2.137}$$

where $T=\sqrt{\frac{m}{k}}, \xi=\frac{f}{2\sqrt{mk}}$.

Example 2.41

Figure 2.31 shows an *R-L-C* passive circuit, where $u_i(t), u_o(t)$ are the voltage of the input and output, R is the resister, C is the capacitor, and L is the inductance.

Solution

According to KVL, we obtain

$$u_i(t)=L\frac{di(t)}{dt}+Ri(t)+\frac{1}{C}\int i(t)dt \tag{2.138}$$

$$u_o(t)=\frac{1}{C}\int i(t)dt \tag{2.139}$$

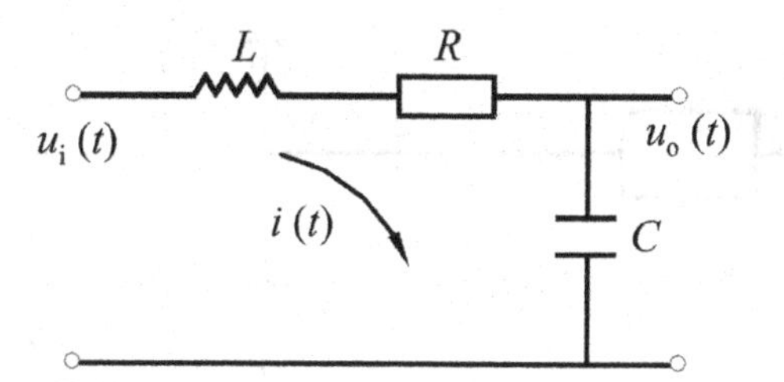

FIGURE 2.31 *RLC* passive circuit.

By taking the Laplace transforms of Eqs. (2.138) and (2.139) and rearranging them, we obtain the transfer function as

$$G(s)=\frac{1}{LCs^2+RCs+1}=\frac{1}{(\sqrt{LC})^2s^2+2\times\frac{RC}{2\sqrt{LC}}\times\sqrt{LC}s+1} \tag{2.140}$$

where $T=\sqrt{LC},\xi=\frac{RC}{2\sqrt{LC}}$.

Example 2.42

Figure 2.32 shows a gear drive. Assume that there are no gaps and deformation in this mechanism. Try to obtain the converted moment of inertia of the driving shaft, the equivalent viscous friction coefficient, and the transfer function.

Solution

The moment equation of shaft 1 is expressed as

$$J_1\ddot{\theta}_1+f_1\dot{\theta}_1+M_1=M \tag{2.141}$$

The moment equation of shaft 2 is expressed as

$$M_2=J_2\ddot{\theta}_2+f_2\dot{\theta}_2 \tag{2.142}$$

Taking the Laplace transforms of Eqs. (2.139) and (2.140), we get

$$J_1s^2\Theta_1(s)+f_1s\Theta_1(s)+M_1(s)=M(s) \tag{2.143}$$

$$M_2(s)=J_2s^2\Theta_2(s)+f_2s\Theta_2(s) \tag{2.144}$$

$$\frac{r_2}{r_1}=\frac{z_2}{z_1}=i \tag{2.145}$$

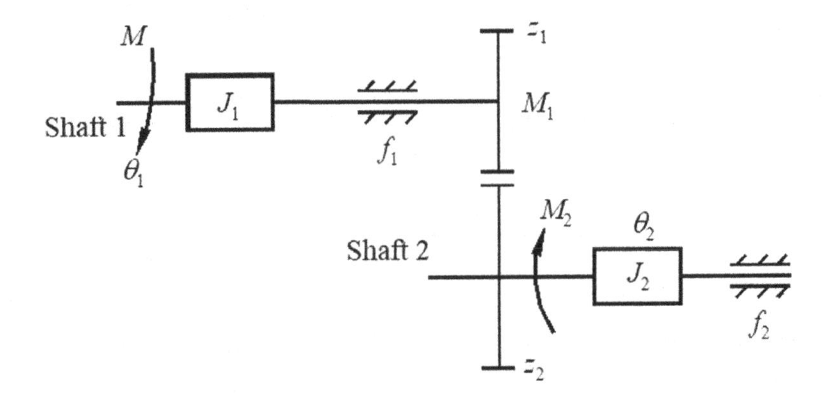

FIGURE 2.32 Gear driving system.

In Eq. (2.145), z_1 and z_2 are the number of teeth of gear 1 and gear 2. In Eqs. (2.143) and (2.142), M_1 means the resisting moment of gear 1 and M_2 means the driving moment of gear 2. Since the two gears have the same linear velocity at the engagement point P, $v_1 = v_2$, and the circumferential forces acting on the circumferential meshing points are equal; thus, $F_1 = F_2$, and therefore, the same magnitude of peripheral force at meshing point.

Thus,

$$F = \frac{M}{r}, \ v = \omega r.$$

Hence,

$$\frac{\omega_1}{\omega_2} = \frac{r_2}{r_1} = \frac{z_2}{z_1} \tag{2.146}$$

Consequently,

$$\frac{M_2}{M_1} = \frac{r_2}{r_1} = \frac{z_2}{z_1} \tag{2.147}$$

By substituting Eq. (2.147) into Eqs. (2.143) and (2.144) and eliminating M_1 and M_2, we have

$$M\frac{z_1}{z_2} = \left[J_1 s^2 + J_2\left(\frac{z_1}{z_2}\right)^2 s^2 + f_1 s + f_2\left(\frac{z_1}{z_2}\right)^2 s\right]\Theta_2(s) \tag{2.148}$$

Therefore, the transfer function of the system is

$$G(s) = \frac{\Theta_2(s)}{M(s)} = \frac{z_1 / z_2}{\left[J_1 + J_2\left(\frac{z_1}{z_2}\right)^2\right]s^2 + \left[f_1 + f_2\left(\frac{z_1}{z_2}\right)^2\right]s}$$

$$=\frac{1/i}{s\left[\left(J_1+\frac{J_2}{i^2}\right)s+\left(f_1+\frac{f_2}{i^2}\right)\right]}=\frac{1/i}{s(Js+f)} \tag{2.149}$$

The rotating inertia is converted to the motor shaft as $J = J_1 + \frac{J_2}{i^2}$ and the viscous friction coefficient as $f = f_1 + \frac{f_2}{i^2}$.

2.7 CONTROL SYSTEM MODELING

Example 2.43

Figure 2.33 shows the mass-spring-damping system, where $\boldsymbol{F_i(t)}$ is the input force, $\boldsymbol{m}$ is weight, $\boldsymbol{k}$ is the spring stiffness, $\boldsymbol{f}$ is the viscous damping coefficient, and $\boldsymbol{y_o(t)}$ is the displacement of the output. Draw the block diagram of the control system as shown in Figure 2.33.

Solution

List the differential equations of the control system from Figure 2.33.

$$F_1(t)=ky_o(t) \tag{2.150}$$

$$F_i(t)-F_1(t)-F_2(t)=m\frac{d^2y_o(t)}{dt^2} \tag{2.151}$$

$$F_2(t)=f\frac{dy_o(t)}{dt} \tag{2.152}$$

Taking the Laplace transforms of Eqs. (2.150)–Eq. (2.152), we get

$$F_1(s)=kY_o(s) \tag{2.153}$$

$$F_i(s)-F_1(s)-F_2(s)=ms^2Y_o(s) \tag{2.154}$$

$$F_2(s)=fsY_o(s) \tag{2.155}$$

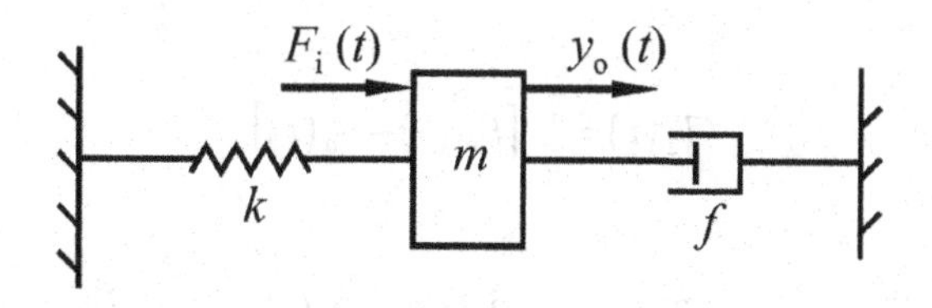

FIGURE 2.33 Mass-spring-damping system.

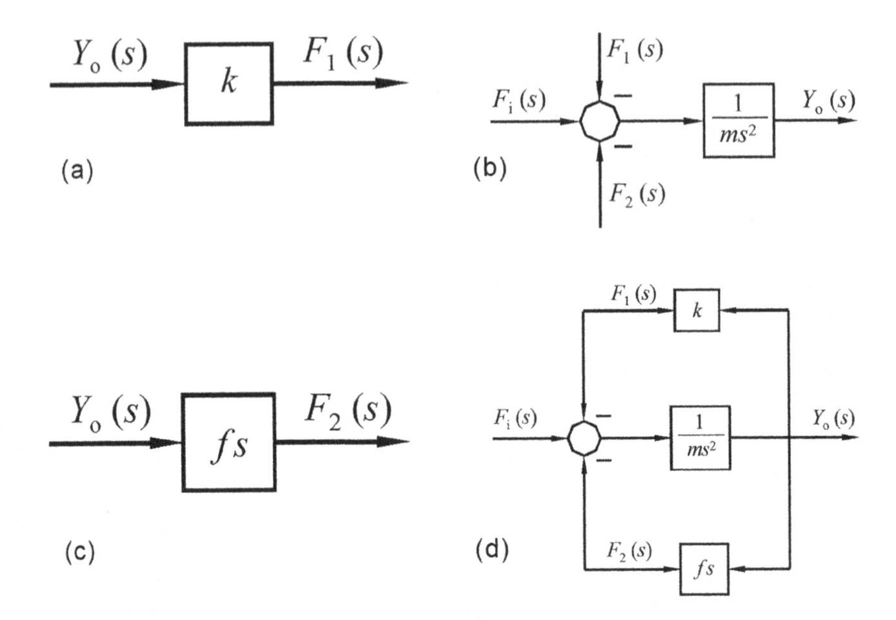

FIGURE 2.34 Block diagram of mass-spring-damping system.

Now, we can draw the block diagram of Eqs. (2.153)–(2.155) which is shown in Figure 2.34a–c, respectively, whereas Figure 2.34d shows the block diagram of the whole system which is combination of three blocks (Figure 2.34a–c).

Example 2.44

Figure 2.35 shows the moment of inertia-spring-damper system, where $\theta_i(t)$ $\theta_o(t)$ are rotational angles of the input and output; k_1, k_2 are coefficient stiffnesses of spring; J_1, J_2 are moments of inertia; T_1, T_2 are torques; and f is the viscous friction.

Solution

Set $\theta_m(t)$ in the middle of the moment of inertia J_1 and J_2, and the following equations are given according to the principle of dynamics:

$$T_1(t) = k_1\left[\left(\theta_i(t) - \theta_m(t)\right)\right] \tag{2.156}$$

$$T_1(t) - T_2(t) = J_1\ddot{\theta}_m(t) \tag{2.157}$$

$$T_2(t) = k_2\left[\theta_m(t) - \theta_o(t)\right] \tag{2.158}$$

$$T_2(t) = J_2\ddot{\theta}_o(t) + f\dot{\theta}_o(t) \tag{2.159}$$

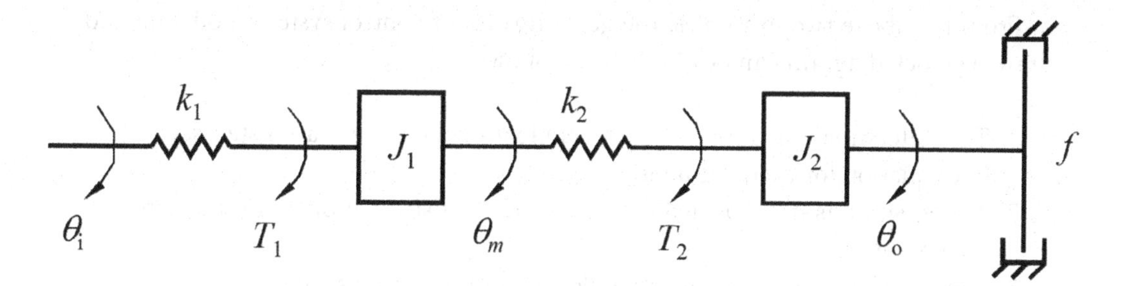

FIGURE 2.35 Moment of inertia-spring-damping system.

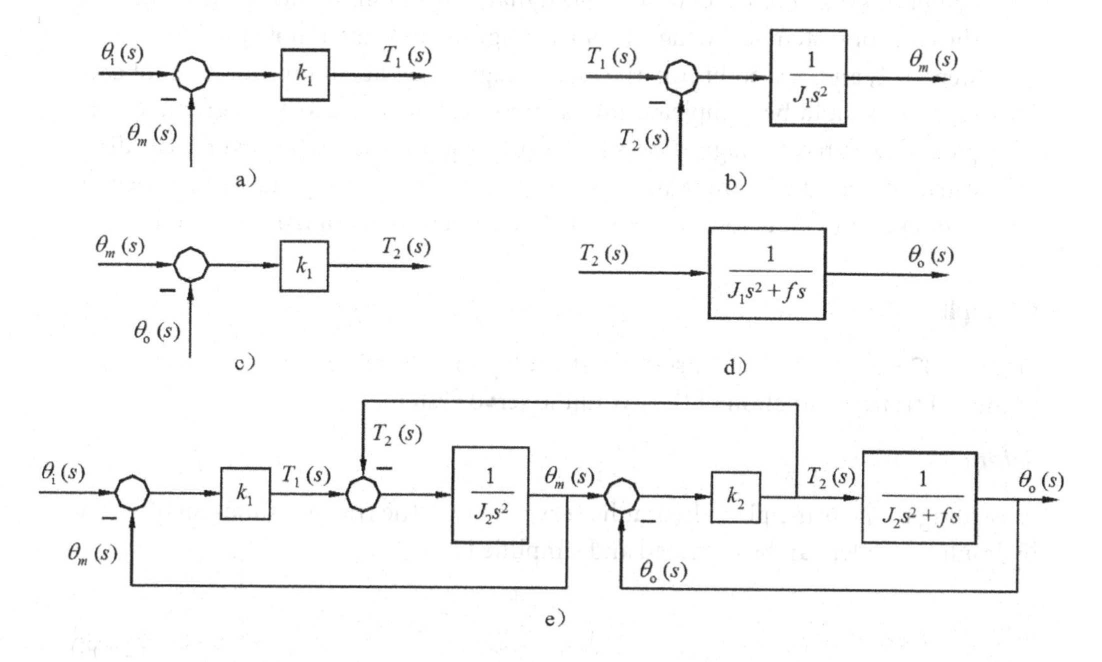

FIGURE 2.36 Block diagram of the moment of an inertia-spring-damper system of Figure 2.35.

Taking the Laplace transforms of Eqs. (2.156)–(2.159) and using zero initial conditions, we obtain

$$T_1(s) = k_1\left[\theta_i(s) - \theta_m(s)\right] \tag{2.160}$$

$$T_1(s) - T_2(s) = J_1 s^2 \theta_m(s) \tag{2.161}$$

$$T_2(s) = k_2\left[\theta_m(s) - \theta_o(s)\right] \tag{2.162}$$

$$T_2(s) = J_2 s^2 \theta_o(s) + fs\theta_o(s) \tag{2.163}$$

From Eqs. (2.160) to (2.163), we obtain the elements of the block diagram that are shown in Figure 2.36a–d. By connecting signals properly, we can construct a block diagram, as shown in Figure 2.36e.

From the above two examples, the general rules of control system modeling and drawing block diagram can be obtained as follows:

1. The differential equations of each segment of the system are listed separately, one equation for each segment.
2. The equation is transformed by the Laplace transformation assuming zero initial conditions.
3. Draw the block diagram or signal flow diagram of each segment.
4. Marge each link of the block diagram or signal flow graph into one, to form a complete system block diagram and signal flow graph. If you are familiar with the control system, drawing the block diagram and signal flow graph can omit steps 1, 2, and 3. Combining the block diagram or signal flow diagram of each segment should be completed into a simplified general block diagram; a complete system block diagram or signal flow diagram can be formed for the dedicated system. If you are familiar with the control system, you can omit steps 1, 2, or even 3 by drawing the complete block diagram or signal flow graph.

Example 2.45

Figure 2.37 shows a hydraulic servo system; try to determine the system block diagram and transfer function of the hydraulic servo system.

Solution

According to the principle of hydraulic servo control, the transfer function of servo-hydraulic cylinder can be obtained and simplified as

$$\frac{Y(s)}{X_v(s)} = \frac{\frac{K_q}{A}}{s} \tag{2.164}$$

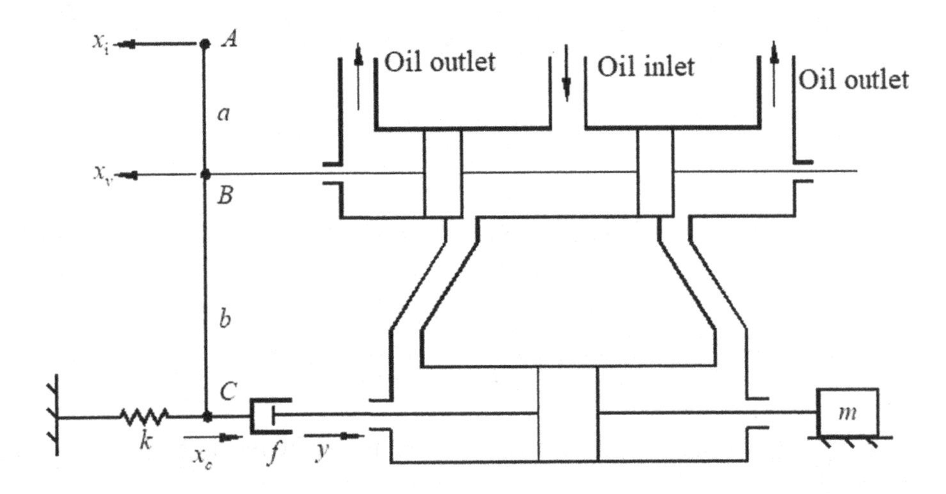

FIGURE 2.37 A hydraulic servo system.

In Eq. (2.164), A is a piston area (m^2) and K_q is the gain for flow (m^3/s). The transfer function relation of the feedback system is

$$X_v(s)=\frac{b}{a+b}X_i(s)-\frac{a}{a+b}X_c(s) \tag{2.165}$$

The transfer function of springs and dampers is defined as

$$F_f(s)=fs\left[X_c(s)-Y(s)\right] \tag{2.166}$$

In Eq. (2.166), F_f is the damping force of the damper; it's going in the opposite direction of the velocity; f is the damping coefficient of the damper, whereas spring restoring force is F_k and X_c is the transfer function of the deformation, then we get,

$$F_k(s)=kX_c(s) \tag{2.167}$$

Note that F_k has the same direction with F_f. The force balance equation can be expressed as

$$F_f(s)+F_k(s)=0 \tag{2.168}$$

According to Eqs. (2.164)–(2.168), the system transfer function is expressed as follows:

$$\frac{Y(s)}{X_i(s)}=\frac{b}{a+b}\times\frac{\frac{\frac{K_q}{A}}{s}}{1+\frac{a}{a+b}\times\frac{\frac{K_q k_f}{A}}{k_f s+1}} \tag{2.169}$$

In Eq. (2.169), $k_f=\frac{f}{k}$ and Figure 2.38 shows the block diagram of Figure 2.37.

However, for the circuit network system, using the complex impedance principle and avoiding differential equations, we can draw the block diagram of a system or signal flow graph to get the transfer function easily. The basic characteristics of the complex impedance are shown in Table 2.4.

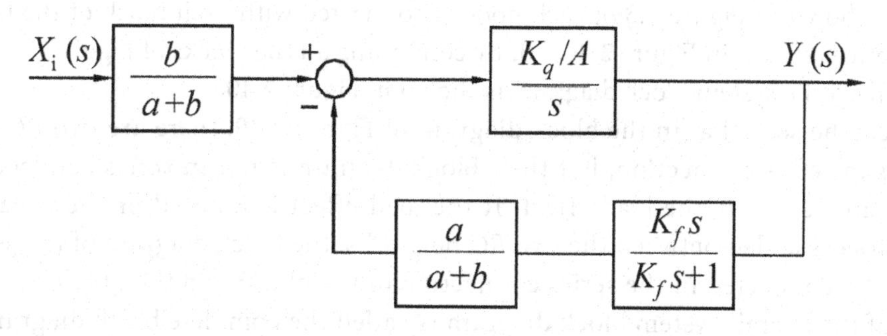

FIGURE 2.38 Block diagram of Figure 2.45.

TABLE 2.4 Basic Control Relationship of Typical Circuit Component

	A Typical Circuit	The Differential Equation	Laplace Transfer	The Complex Impedance (Transfer Function)
Resistive load	$u(t)$, $i(t)$, R	$u(t)=i(t)\times R$	$U(s)=I(s)\times R$	R
Capacitive load	$u(t)$, $i(t)$, C	$u(t)=\frac{1}{C}\int_0^t i(t)dt$	$U(s)=I(s)\times\frac{1}{Cs}$	$\frac{1}{Cs}$
Inductive load	$u(t)$, $i(t)$, L	$u(t)=L\frac{di(t)}{dt}$	$U(s)=I(s)\times Ls$	Ls

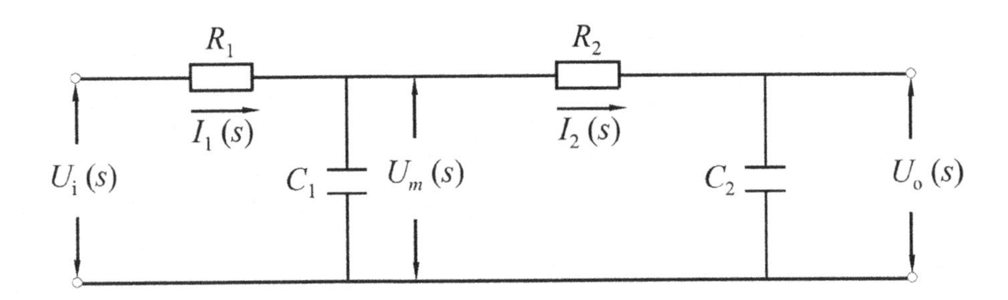

FIGURE 2.39 Passive filter network.

Example 2.46

Figure 2.39 shows a passive filter network in cascaded form with *RC* system. Draw the block diagram of the cascaded passive filter network.

Solution

Suppose that you set the intermediate variable in the middle of the cascaded network as U_m (shown in Figure 2.39); each node is connected with each block of the block diagram as shown in Figure 2.40a–d. By combining all the blocks of Figure 2.40a–d, we can get the system block diagram as shown in Figure 2.40e.

It can be seen that in the block diagram of Figure 2.40, there are two *RC* networks in series connection, but their block diagram is not in series connection; the main reason is the load effect. If the load effect is ignored or the isolation amplifier is added between the two *RC* networks, the block diagram of the series circuit is connected in the series connection of the block; and then the composition of the overall system block diagram is called the complete block diagram of the system.

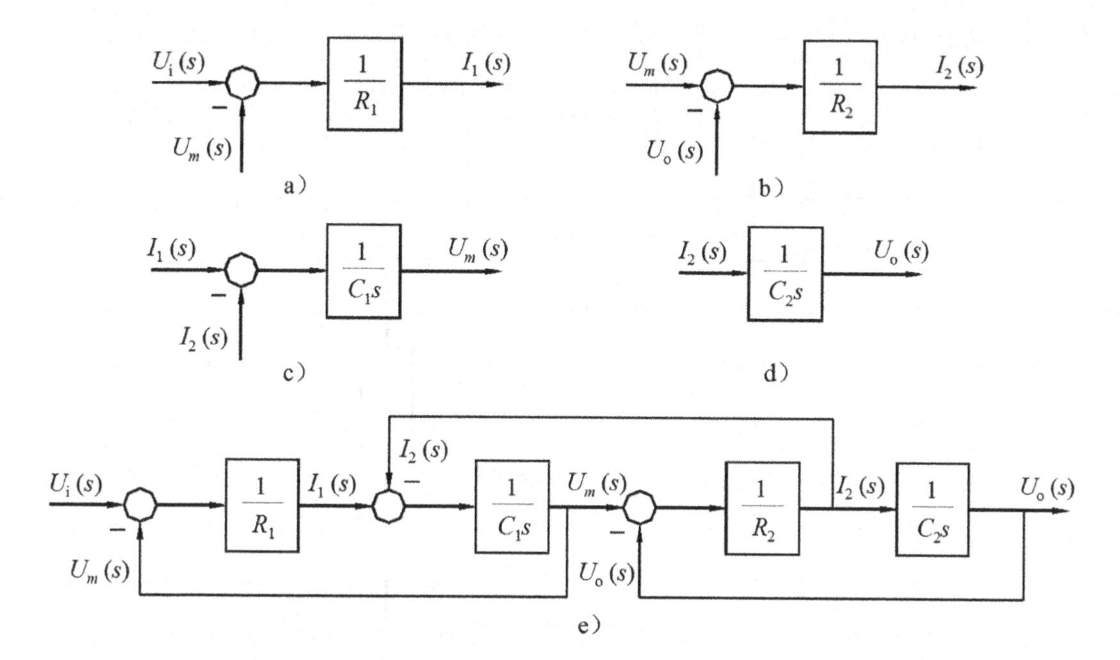

FIGURE 2.40 Block diagram of Figure 2.39.

Examples 2.47

Figure 2.41 shows a passive electric *RC* network, where $u_o(t)$ is the output voltage, $u_i(t)$ is the input voltage, R_1 and R_2 are the resistance, and C_1 and C_2 are the capacitance. (i) Draw block diagram; and (ii) find the transfer function.

Solution

1. According to Figure 2.40, we can determine the equation as follows
   The input circuits

$$u_i(t) = u_o(t) + R_1 i_2(t)$$

The output circuits

$$u_o(t) = \frac{1}{C_2}\int [i_1(t) + i_2(t)]dt + [i_1(t) + i_2(t)]R_2$$

The middle loop

$$\frac{1}{C_1}\int i_1(t)dt = R_1 i_2(t)$$

Taking the Laplace transforms of the above equations and drawing the block diagram of each connection of the network, as shown in Figure 2.42a–c and

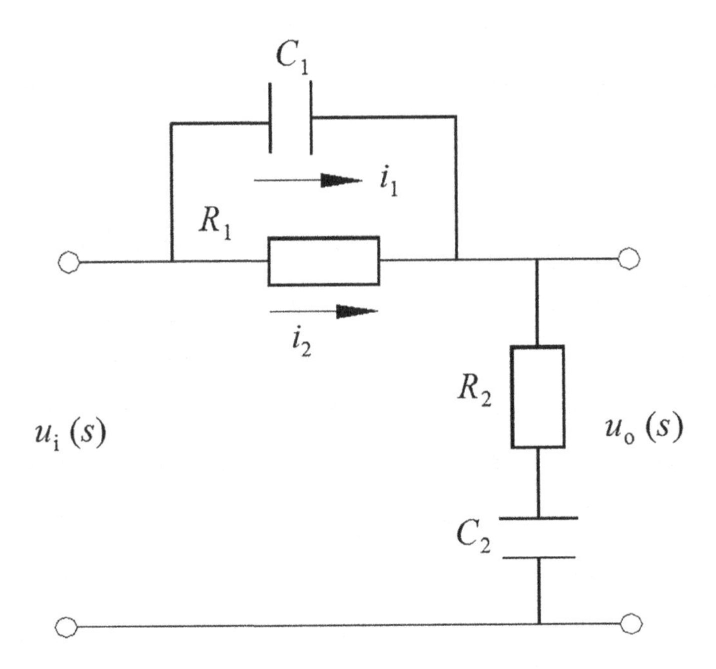

FIGURE 2.41 Passive electric network.

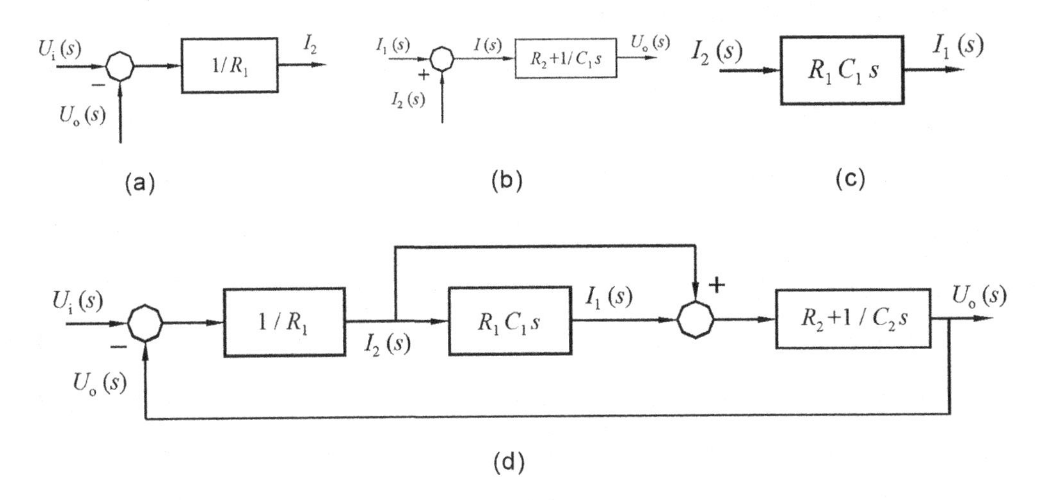

FIGURE 2.42 Complete system block diagram of Figure 2.40.

combining the block diagram of Figure 2.42a–c, we can get the complete system block diagram as shown in Figure 2.42d.

$$U_i(s) - U_o(s) = I_2(s)R_1$$

$$U_o(s) = (\frac{1}{C_2 s} + R_2)[(I_1(s) + I_2(s)]$$

$$\frac{1}{C_1 s} I_1(s) = R_1 I_2(s)$$

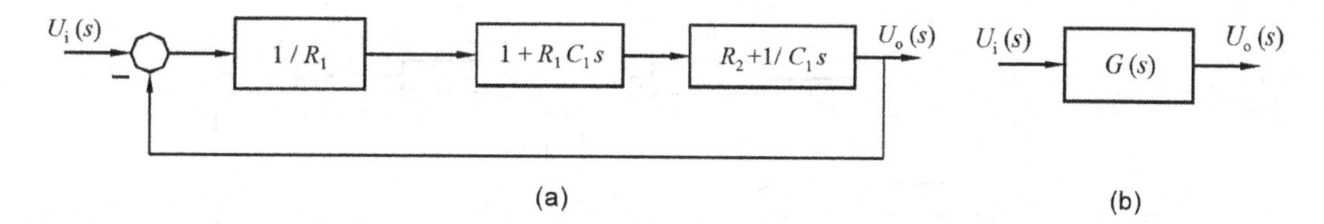

FIGURE 2.43 Simplified block diagram of Figure 2.42.

2. Simplify the block diagram of Figure 2.43 to get the system transfer function. Eliminate the parallel circuit as shown in Figure 2.43a the feedback loop as shown in Figure 2.43b.

   Therefore, the system transfer function of the system is

$$G(s)=\frac{U_o(s)}{U_i(s)}=\frac{R_1R_2C_1C_2s^2+(R_1C_1+R_2C_2)s+1}{R_1R_2C_1C_2s^2+(R_1C_1+R_2C_2+R_1C_2)s+1}$$

Example 2.48

Find the transfer function of the active circuit network as shown in Figure 2.44, where $u_o(t)$ is the output voltage; $u_i(t)$ is the input voltage; R_1, R_2, , R_4 and R_5 are resistances; C is the capacitance.

Solution

As shown in Figure 2.44, R_2, R_4 and R_5 are connected at the intermediate point $u_A(t)$. According to the theory of the complex impedance, C is the capacitance and $\frac{1}{Cs}$ is the complex impedance. According to the characteristics of an operational amplifier and the KVL, we can determine the following equations:

$$\begin{cases}\dfrac{U_i(s)}{R_1}=-\dfrac{U_A(s)}{R_2}\\[2ex]-\dfrac{U_A(s)}{R_2}=\dfrac{U_A(s)-U_o(s)}{R_5}+\dfrac{U_A(s)}{R_4+\dfrac{1}{Cs}}\end{cases}$$

Eliminating the intermediate variable $U_A(s)$, the transfer function that we can get is

$$\frac{U_o(s)}{U_i(s)}=-\frac{R_2+R_5}{R_1}\frac{\dfrac{R_2R_4+R_2R_5+R_4R_5}{R_2+R_5}Cs+1}{R_4Cs+1}$$

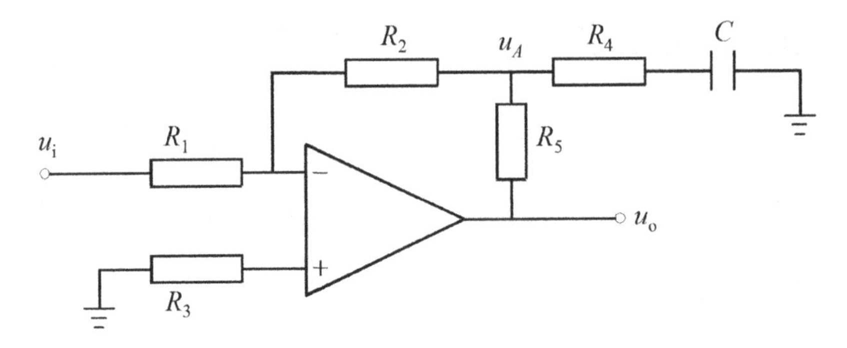

FIGURE 2.44 Active circuit network with operational amplifier (OP-AM).

2.8 MECHANICAL SYSTEM MODELING OF CONTROL SYSTEM

Here, I will focus on translational mechanical system transfer functions, rotational mechanical system transfer functions, and servo-system equation of motion by inspection.

2.8.1 Translational Mechanical System Transfer Functions

Look at the table of the translational mechanical system transfer functions. The first column of the table shows mechanical components and their representation of spring system, viscous damper, and mass system. The third column shows their force-velocity and force displacement. The last column shows their impedance (Table 2.5).

The spring system is represented by K, and if the force is $f(t)$ and the displacement is $x(t)$ along the X-axis, then the force-velocity we can see in the second column, and the force displacement that we can see in the third column is

$$f(t) = Kx(t) \tag{2.170}$$

Now, taking the Laplace transformation in both sides, we get

$$\mathcal{L}\{Kx(t)\} \leftrightarrow F(s) = KX(s)$$

However, the impedance which is called the transfer function of the spring can be found from the third column as

$$Z_M(s) = \frac{F(s)}{X(s)}$$

Here we have $F(s) = KX(s)$; therefore, we get

$$Z_M(s) = \frac{KX(s)}{X(s)} = K$$

TABLE 2.5 Transfer Functions of Translational Mechanical System

Component	Force-Velocity	Force Displacement	Impedance $Z_M(s)=F(s)/X(s)$
Spring $T(t)$ $\theta(t)$ K	$f(t)=K\int_0^t v(\tau)d\tau$	$f(t)=Kx(t)$	K
Viscous Damper $x(t)$ $f(t)$ f_v	$f(t)=f_v v(t)$	$f(t)=f_v\dfrac{dx(t)}{dt}$	$f_v s$
Mass $x(t)$ $f(t)$	$f(t)=M\dfrac{dv(t)}{dt}$	$f(t)=M\dfrac{d^2x(t)}{dt^2}$	MS^2

Note: Here the unit of $f(t)$ is N (Newtons), $v(t)$ is m/s (meter/second), $x(t)$ is m (meter), K is N/m (Newton/meter), f_v is N-s/m (Newton-second/meter), M is kg (kilogram = newton-second²/meter).

K is the transfer function of the spring system. However, from the second column, the force displacement of the viscous damper is $f(t)=f_v\dfrac{dx(t)}{dt}$ and the force displacement of mass is

$$f(t)=M\frac{d^2x(t)}{dt^2} \tag{2.171}$$

Therefore, similarly, the transfer function of the viscous damper is $f_v s$, and the transfer function of mass is Ms^2, respectively. Here, the mentioned symbols $f(t)$, $x(t)$, K, $f_v(t)$, and M are the force that is measured by N (Newtons), displacement which is measured by m (meters), $K=\text{N}/\text{m}$, $f(t)=N-s/m$, and $M=\text{kg}$, respectively.

2.8.2 Rotational Mechanical System Transfer Functions

Transfer functions of rotational mechanical system are similar to transfer functions of translational mechanical system, but transfer functions of rotational mechanical system create torque-angular velocity due to torque and angular velocity of spring, viscous damper, and inertia system, which is shown in the second column in the table. Similarly, from the third column of the table, the torque-angular displacement of spring, viscous damper, and inertia are $T(t)=K\theta(t)$, $T(t)=D\dfrac{d\theta(t)}{dt}$, and $T(t)=J\dfrac{d^2\theta(t)}{dt^2}$, respectively. However, from the fourth column of the table, the transfer function or impedance of spring is K, the transfer function or impedance of the viscous damper is Ds, and the transfer function or impedance of the inertia is Js^2 (Table 2.6).

To find the transfer function, $X(s)/F(s)$, consider the mechanical system as shown in Figure 2.45, where M is the mass system, K is the spring system, D is the damper system, $x(t)$ is the displacement, and $f(t)$ is the force system.

TABLE 2.6 Transfer Functions of Rotational Mechanical System

Component	Torque-angular velocity	Torque-angular displacement	Impedance $Z_M(s)=T(s)/\theta(s)$
Vicous $T(t)$ $\theta(t)$ Damper D	$T(t)=K\int_0^t \omega(\tau)d\tau$	$T(t)=K\theta(t)$	K
$T(t)$ $\theta(t)$ Inertia J	$T(t)=D\omega(t)$	$T(t)=D\frac{d\theta(t)}{dt}$	Ds
F(s) → $\frac{1}{Ms^2+Ds+K}$ → X(s)	$T(t)=J\frac{d\omega(t)}{dt}$	$T(t)=J\frac{d^2\theta(t)}{dt^2}$	JS^2

Note: Here the unit of $T(t)$ is N-m (Newtons-meters), $\theta(t)$ is rad (radians), $\omega(t)$ is rad/s (radian/second), K is N-m/rad (Newtons-meter/radian), D is N-m-s/rad (Newton-meters-seconds/radian), J is $\text{Kg}-\text{m}^2$ (kilograms-meters^2=newton-meters-seconds^2/radian).

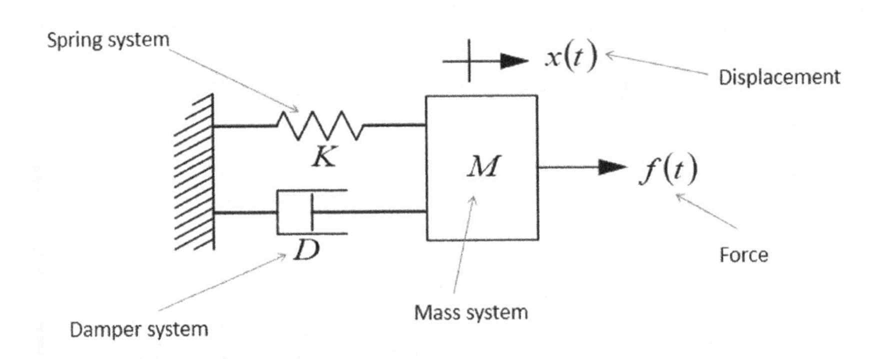

FIGURE 2.45 Mass, spring, and damper system.

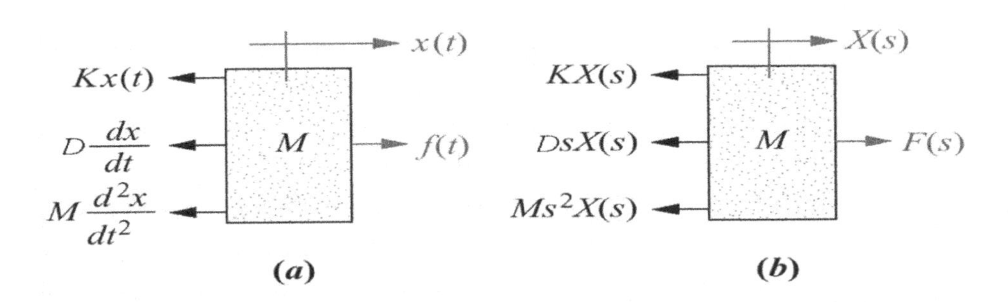

FIGURE 2.46 (a)-(b) Transformed free-body diagram, and (c) transfer function of Figure 2.45.

To solve this problem, we can separate the system into two parts as shown in Figure 2.46; however, the mass, spring, and damper system are representing in terms of force displacement function which is shown in Figure 2.46a; and then, it can be transformed in transfer function as a free-body diagram by using the Laplace transformation of mass, spring, and damper system which is shown in Figure 2.46b.

According to Newton's law, the total force is equal to the sum of all of the forces shown in the mass system, and then we have

$$M\frac{d^2x(t)}{dt^2}+D\frac{dx(t)}{dt}+Kx(t)=f \tag{2.172}$$

Now, taking the Laplace transform with zero initial conditions on both sides of the above equation, we get

$$\left(Ms^2+Ds+K\right)X(s)=F(s)$$

Solving for the transfer function yields

$$G(s)=\frac{X(s)}{F(s)}=\frac{1}{Ms^2+Ds+K} \tag{2.173}$$

Therefore, the $X(s)$ is the output and the $F(s)$ is the input, so the transfer function $G(s)$ can be shown in a block diagram as shown in the following.

2.8.3 Equation of Motion by Inspection

Now, turn to learn equation of motion by inspection for a servo system. There are three inertia systems as J_1, J_2, and J_3 which are connected with a spring system K as well as with damper system as D_1, D_2, and D_3. Let's see the Figure 2.47 that shows the mass, spring and damper system with inertia load. is directly connected with a damper system D_1 and a spring system K; the inertia J_1 contains the motion $\theta_1(t)$; The inertia J_2 is directly connected with a spring system K and a damper system D_2; the inertia J_2 contains the motion $\theta_2(t)$; and $\theta_3(t)$ with torque $T(t)$; however, the inertia J_3 is directly connected with two damper systemsD_2 and D_3; the inertia J_3 contains the motion $\theta_3(t)$

Here, it is mentioned that the damper system D_1 and the damper system D_3 are not directly connected to each other, so there are no effects on these both damper systems directly, and these two dampers are fixed with two separate fixed body systems. Now there is the question, how can you establish and compute the equation of motion in this system?

Let's see carefully!

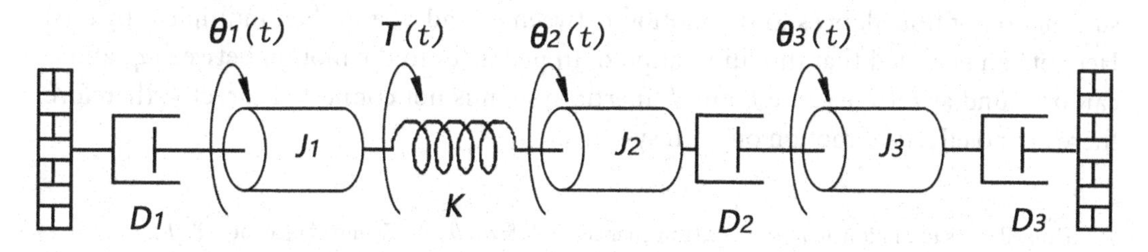

FIGURE 2.47 Mass, spring, and damper system with inertia system.

The summation of applied torques at $q_1(t)$ can be distributed or transmitted which is equal to the summation of impedances to the motion at q_1 and then multiplied by $q_1(s)$-the summation of impedances to the motion between q_1 and q_2 and then multiplied by $q_2(s)$—summation of impedances to the motion between q_1 and q_3 and then multiplied by $q_3(s)$.

It can be noted that the summation of impedances to the motion between q_1 and q_2 can be found only at the spring system K, because it has only one component which is spring K between J_1 and J_2 system, i.e.,

Σ(Impedances to motion at θ_1) $\theta_1(s)$ − Σ(Impedances at θ_1 & θ_2) $\theta_2(s)$ − Σ(Impedances at θ_1 & θ_3) $\theta_3(s)$ = Σ(applied torques at θ_1)

Therefore, if the total applied torques at $q_1(t)$ are denoted by $T(s)$, then the equation of motion can be written as

$$T(s) = \left(J_1 s^2 + D_1 s + K\right)\theta_1(s) - K\theta_2(s) - 0\theta_3(s) \tag{2.174}$$

The summation of applied torques at q_2 can be distributed or transmitted, which is equal to the summation of impedances to the motion at q_2 and then multiplied by $q_2(s)$ – the summation of impedances to the motion between q_1 and q_2 and then multiplied by $q_2(s)$ – the summation of impedances to the motion between q_2 and q_3 and then multiplied by $q_3(s)$. Here, it can be noted that the summation of impedances to the motion between q_2 and q_3 can be found only at the damper system D_2 because it has only one component which is damper D_2 between J_2 and J_3 inertia system, i.e.,

Σ(Impedances to motion at θ_2) $\theta_2(s)$ − Σ(Impedances at θ_1 & θ_2) $\theta_2(s)$ − Σ(Impedances at θ_2 & θ_3) $\theta_3(s)$ = Σ(applied torques at θ_2)

Therefore, the applied torques $T(s)$ is absent due to no-load condition at $q_1(t)$, and then the equation of motion can be written as

$$\left(J_2 s^2 + D_2 s + K\right)\theta_2(s) - K\theta_1(s) - D_2 s\theta_3(s) = 0 \tag{2.175}$$

The summation of applied torques at q_3 can be distributed or transmitted which is equal to the summation of impedances to the motion at q_3 and then multiplied by $q_3(s)$-the summation of impedances to the motion between q_1 and q_3 and then multiplied by $q_1(s)$-summation of impedances to the motion between q_2 and q_3 and then multiplied by $q_2(s)$. Here, it can be noted that the summation of impedances to the motion between q_1 and q_3 can be found at zero because J_1 and J_3 inertia system is not connected directly; therefore, there are no effects of motion on both systems.

Σ(Impedances to motion at θ_3) $\theta_3(s)$ − Σ(Impedances at θ_1 & θ_3) $\theta_1(s)$ − Σ(Impedances at θ_2 & θ_3) $\theta_2(s)$ = Σ(applied torques at θ_3)

Therefore, the applied torques $T(s)$ is absent due to no-load condition at $q_3(t)$, and then the equation of motion can be written as

$$\left(J_3 s^2 + D_3 s + D_2 s\right)\theta_3(s) - 0\theta_1(s) - D_2 s\theta_2(s) = 0 \tag{2.176}$$

2.9 MATHEMATICAL MODELING OF ELECTRICAL AND ELECTRONIC SYSTEMS

Basic laws governing electrical circuits are Kirchhoff's current law and voltage law. Kirchhoff's current law (node law) states that the algebraic sum of all currents entering and leaving a node is zero. (This law can also be stated as follows: The sum of currents entering a node is equal to the sum of currents leaving the same node.) Kirchhoff's voltage law (loop-law) states that at any given instant, the algebraic sum of the voltages around any loop in an electrical circuit is zero. (This law can also be stated as follows: The sum of the voltage drops is equal to the sum of the voltage rises around a loop.) A mathematical model of an electrical circuit can be obtained by applying one or both of Kirchhoff's laws to it. In this section, we will study electrical and electronic systems, operational amplifiers, and electromechanical system.

2.9.1 Electrical and Electronic Systems

Now let's look at Table 2.7, which shows the voltage-current and impedance relationships for capacitors, resistors, and inductors. In the table, the last two column shows the impedance and admittance for capacitor, resistor, and inductor. The impedance, as well as admittance, is useful for the transfer function of the system modeling. Therefore, the impedance of the capacitor, resistor, and inductor can be expressed as in the following that we can see in the last second column of the table by using

$$Z(s) = \frac{V(s)}{I(s)}$$

TABLE 2.7 Transfer Functions of Electrical Elements

Component	Voltage-Current, $v(t) \sim i(t)$	Current-Voltage, $i(t) \sim v(t)$	Voltage-Charge, $v(t) \sim q(t)$	Impedance $Z_M(s) = V(s)/I(s)$	Admittance $Y(s) = I(s)/V(s)$
Capacitor	$v(t) = \frac{1}{C}\int_0^t i(\tau)d\tau$	$i(t) = C\frac{dv(t)}{dt}$	$v(t) = \frac{1}{C}q(t)$	$\frac{1}{Cs}$	Cs
Registor	$v(t) = Ri(t)$	$i(t) = \frac{1}{R}v(t)$	$v(t) = R\frac{dq(t)}{dt}$	R	$\frac{1}{R} = G$
Inductor	$v(t) = L\frac{di(t)}{dt}$	$i(t) = \frac{1}{L}\int_0^t v(\tau)d\tau$	$v(t) = L\frac{d^2q(t)}{dt^2}$	Ls	$\frac{1}{Ls}$

Note: Here, the unit—$v(t)$ is V (volts), $i(t)$ is A (amps), $q(t)$ is Q (coulombs), C is F (Farads), R is Ω (ohms), G is ℧ (mhos), L is H (Henries).

Then, the impedance for capacitor is

$$Z(s)=\frac{1}{Cs}$$

The impedance for resistor is

$$Z(s)=R$$

The impedance for inductor is

$$Z(s)=Ls$$

whereas the admittance of capacitor, resistor, and inductor can be expressed as in the following that we can see in the last column of the table by using

$$Y(s)=\frac{I(s)}{V(s)}$$

Then, the admittance for the capacitor is

$$Y(s)=Cs$$

The admittance for the resistor is

$$Y(s)=\frac{1}{R}=G$$

And the admittance for the inductor is

$$Y(s)=\frac{1}{Ls}$$

1. *RLC circuits:*

   Now let's see an example. Find the transfer function relating the voltage $U_o(s)$ and the input voltage $U_i(s)$ from the given Figure 2.48.

   In this example, this is an *RLC* circuit network diagram, where *R* is a resistor, *L* is an inductor, *C* is a capacitor, *i*(*t*) is the current, $u_i(t)$ is the input voltage, and $u_o(t)$ is the output voltage in this network. Can you find the transfer function of this network? How can we develop a system model for this RLC network? According to Kirchhoff's voltage law (KVL), the sum of all the voltages around a loop is equal to zero which means that the total applied voltage $u_i(t)$ is equal to the summation of the voltage drops in a resistor ($u_R(t)$), inductor $u_L(t)$, and capacitor $u_C(t)$. Therefore, we can say

$$u_i(t)=u_R(t)+u_L(t)+u_C(t) \tag{2.177}$$

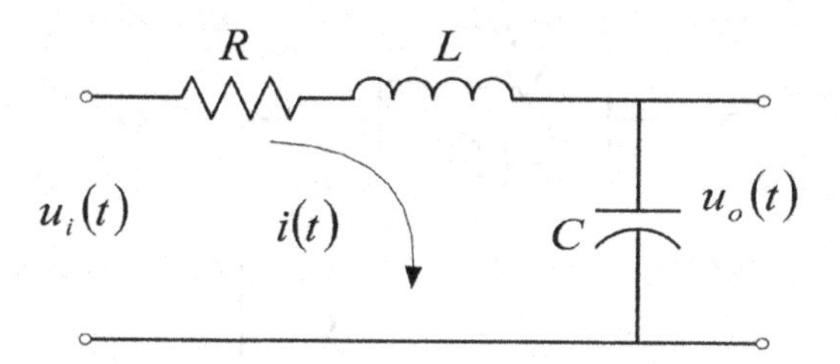

FIGURE 2.48 RLC network with series connection.

Then,

$$u_i(t) = Ri(t) + L\frac{di(t)}{dt} + \frac{1}{C}\int i(t)dt \tag{2.178}$$

and here

$$u_o(t) = \frac{1}{C}\int i(t)dt \tag{2.179}$$

Now, taking the Laplace transformation on both sides of the above equation with assuming zero initial condition, we can get

$$U_i(s) = LsI(s) + RI(s) + \frac{1}{Cs}I(s) \tag{2.180}$$

and

$$U_o(s) = \frac{1}{Cs}I(s) \tag{2.181}$$

Now, what should be the transfer function of the system? So, the transfer function of the RLC network can be expressed as

$$\frac{U_o(s)}{U_i(s)} = \frac{\frac{1}{Cs}I(s)}{LsI(s) + RI(s) + \frac{1}{Cs}I(s)} \tag{2.182}$$

$$\frac{U_o(s)}{U_i(s)} = \frac{\frac{1}{Cs}I(s)}{\left(LCs^2 + RCs + 1\right)\frac{1}{Cs}I(s)} \tag{2.183}$$

$$\frac{U_o(s)}{U_i(s)} = \frac{1}{LCs^2 + RCs + 1} \tag{2.184}$$

Now let's see another example of the system modeling for the transfer function. Obtain the transfer function of the following electrical system, as shown in Figure 2.49.

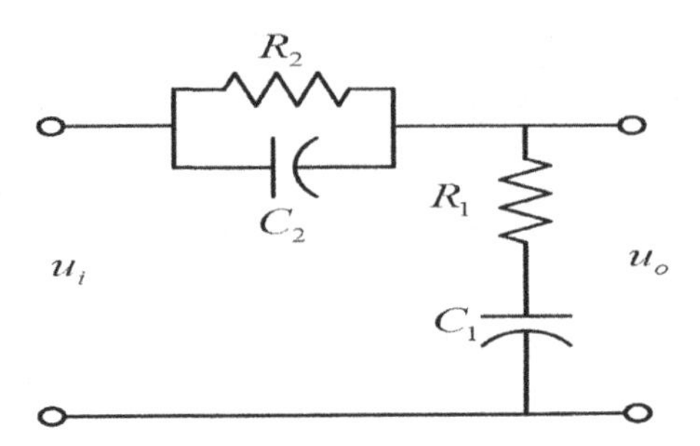

FIGURE 2.49 RLC network with parallel connection.

In this example, this is an *RC* circuit mesh network diagram, where R_1 and R_2 are resistors, C_1 and C_2 are capacitors, $i(t)$ is the current, $u_i(t)$ is the input voltage, and $u_o(t)$ is the output voltage in this mesh network. Can you find the transfer function of this analogous electrical network? How can we develop a system model for this analogous electrical network? There are two scenarios that we can see in this mesh network as R_2 and C_2 will be the same.are parallel; therefore, the voltage across/drops in mesh circuit R_2 and C_2 will be the same. After applying the Laplace transformation, the voltage-current relationship of the R_2 parallel with C_2mesh can be found from this mesh as

$$U_{R_1}(s)=\left(\frac{1}{\frac{1}{R_2}+C_2s}\right)I(s) \tag{2.185}$$

However, R_1 and C_1 resistors are series networks, so after applying the Laplace transformation of the output mesh network, the voltage-current relationship of the R_1 series with C_1mesh can be found as

$$U_o(s)=\left(R_1+\frac{1}{Cs_1}\right)I(s) \tag{2.186}$$

According to Kirchhoff's voltage law, the sum of all the voltages around a loop is equal to zero which means that the total applied voltage $u_i(t)$ is equal to the summation of the voltage drops in resistor R_1, resistor R_2, and capacitor $u_C(t)$. Therefore, we can say as

$$u_i(t)=u_{R_1}(t)+u_{R_2}(t)+u_C(t) \tag{2.187}$$

After taking the Laplace transformation, we get

$$U_i(s)=U_{R_1}(s)+U_{R_2}(s)+U_C(s) \tag{2.188}$$

After rearranging all, we get

$$U_i(s) = \left[\frac{1}{(1/R_2) + C_2 s} + R_1 + \frac{1}{Cs_1}\right] I(s)$$

Now we can find the transfer function of the analogous electrical mesh network as

$$\frac{U_o(s)}{U_i(s)} = \frac{R_1 + \dfrac{1}{Cs_1}}{\dfrac{1}{(1/R_2) + C_2 s} + R_1 + \dfrac{1}{Cs_1}}$$

Thus,

$$\frac{U_o(s)}{U_i(s)} = \frac{(R_1 C_1 s + 1)(R_2 C_2 s + 1)}{(R_1 C_1 s + 1)(R_2 C_2 s + 1) + R_2 C_1 s} \tag{2.189}$$

2. *Inverting operational amplifier:*
   Consider the inverting operational amplifier shown in Figure 2.50. How can you find the transfer function $U_o(s)/U_i(s)$ of the inverting operational amplifiers?

   In this system, $U_i(s)$ is the input function, $U_1(s)$ is the feedback input voltage from low output impedance, $U_o(s)$ is the output function; $Z_1(s)$ is high input impedance, and $Z_2(s)$ is the low output impedance; however, $I_1(s)$ is the input current, $I_2(s)$ is the output current, and $I_a(s)$ is the input feedback current passed from $Z_2(s)$ low output impedance.

   Under ideal conditions, the input feedback current $I_a(s)$ and the feedback input voltage $U_1(s)$ are received from $Z_2(s)$ low output impedance which can be found as

$$I_a(s) = 0$$

$$U_1(s) = 0$$

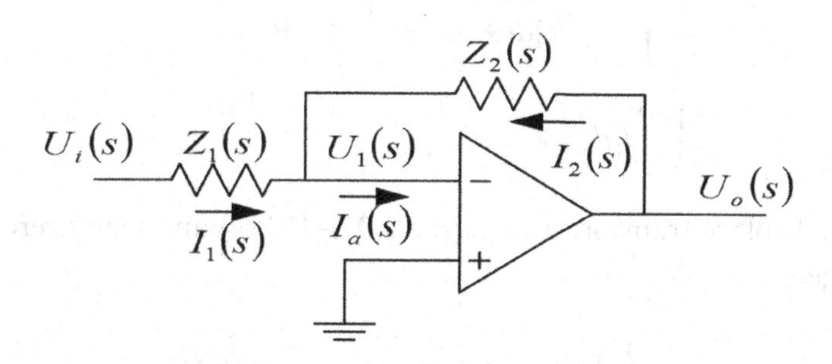

FIGURE 2.50 Inverting operational amplifier network.

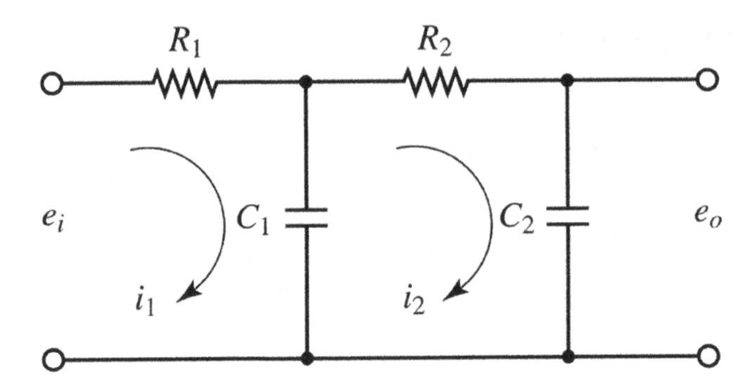

FIGURE 2.51 Cascaded electrical networks.

We may write the node equation at $U_1(s)$, and then we can express as

$$\frac{U_i(s)-U_1(s)}{Z_1(s)}-\frac{U_1(s)-U_o(s)}{Z_2(s)}=0 \tag{2.190}$$

After rearranging and simplifying the above equation, we can say

$$\frac{U_o(s)}{U_i(s)}=-\frac{Z_2(s)}{Z_1(s)} \tag{2.191}$$

3. *Transfer functions of cascaded elements:*
   Many feedback systems have components that load each other. Consider the system shown in Figure 2.51. Assume that e_i is the input and e_o is the output. The capacitances C_1 and C_2 are not charged initially.

   It will be shown that the second stage of the circuit (R_2C_2 portion) produces a loading effect on the first stage (R_1C_1 portion). The equations for this system are

$$\frac{1}{C_1}\int(i_1-i_2)dt+R_1i_1=e_i \tag{2.192}$$

and

$$\frac{1}{C_1}\int(i_2-i_1)dt+R_2i_2+\frac{1}{C_2}\int i_2dt=0 \tag{2.193}$$

$$\frac{1}{C_2}\int i_2dt=e_o \tag{2.194}$$

Taking the Laplace transforms of Eqs. (2.192)–(2.194) and using zero initial conditions, we get

$$\frac{1}{C_1s}\left[I_1(s)-I_2(s)\right]+R_1I_1(s)=E_i(s) \tag{2.195}$$

and

$$\frac{1}{C_1 s}\left[I_2(s)-I_1(s)\right]+R_2 I_2(s)+\frac{1}{C_1 s}I_2(s)=0 \tag{2.196}$$

$$\frac{1}{C_2 s}I_2(s)=E_o(s) \tag{2.197}$$

Eliminating $I_1(s)$ from Eqs. (2.195) to (2.197) and writing $E_i(s)$ in terms of $I_2(s)$, we find the transfer function between $E_o(s)$ and $E_i(s)$ to be

$$\frac{E_o(s)}{E_i(s)}=\frac{1}{(R_1C_1s+1)(R_2C_2s+1)+R_1C_2s}$$

$$=\frac{1}{R_1C_1R_2C_2s^2+(R_1C_1+R_2C_2+R_1C_2)s+1} \tag{2.198}$$

The term R_1C_2s in the denominator of the transfer function represents the interaction of two simple RC circuits. Since $(R_1C_1+R_2C_2+R_1C_2)^2>4R_1C_1R_2C_2$, the two roots of the denominator of Eq. (2.198) are real. The present analysis shows that if two RC circuits are connected in a cascade, the output from the first circuit is the input to the second, and the overall transfer function is not the product of $\frac{1}{(R_1C_1s+1)}$ and $\frac{1}{(R_2C_2s+1)}$. The reason for this is that, when we derive the transfer function for an isolated circuit, we implicitly assume that the output is unloaded. In other words, the load impedance is assumed to be infinite, which means that no power is being withdrawn at the output. When the second circuit is connected to the output of the first, a certain amount of power is withdrawn, and thus the assumption of no loading is violated. Therefore, if the transfer function of this system is obtained under the assumption of no loading, then it is not valid. The degree of the loading effect determines the amount of modification of the transfer function.

4. *Electronic controllers:*
   In what follows, we shall discuss electronic controllers using operational amplifiers. We begin by deriving the transfer functions of simple operational amplifier circuits. Then we derive the transfer functions of some of the operational amplifier controllers. Finally, we give operational amplifier controllers and their transfer functions in the form of a table as shown in Figure 2.52.

5. *Operational amplifiers:*
   Operational amplifiers, often called op-amps, are frequently used to amplify signals in sensor circuits. Op-amps are also frequently used in filters used for compensation purposes. Figure 2.53 shows an op-amp. It is a common practice to choose the ground as 0 V and measure the input voltages e_1 and e_2 relative to the ground. The input e_1 to the minus terminal of the amplifier is inverted and the input e_2 to the plus

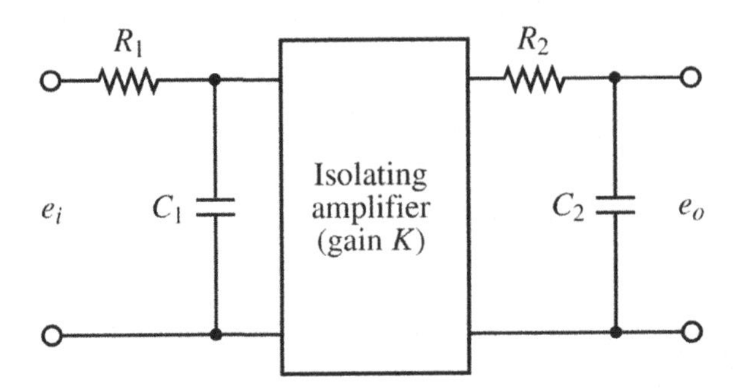

FIGURE 2.52 Electronic controllers.

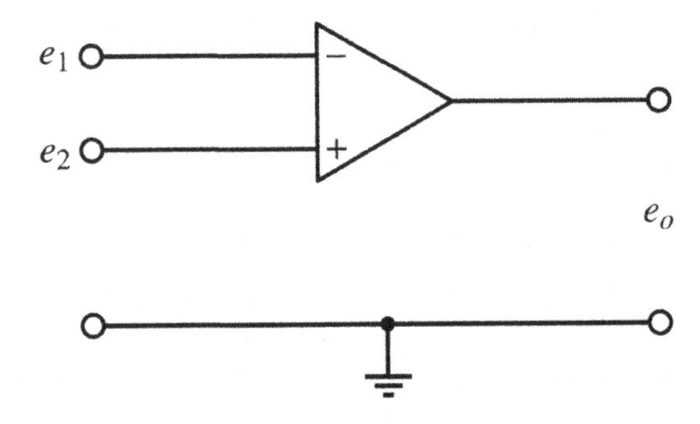

FIGURE 2.53 Operational amplifiers without amplifiers feedback load.

terminal is not inverted. The total input to the amplifier thus becomes $e_2 - e_1$. Hence, for the circuit shown in Figure 2.53, we have

$$e_o = K(e_2 - e_1) = -K(e_1 - e_2)$$

where the inputs e_1 and e_2 may be DC or AC signals and K is the differential gain (voltage gain).

The magnitude of K is ~10^5–10^6 for dc and ac signals with frequencies less than approximately 10 Hz. (The differential gain K decreases with the signal frequency and becomes about unity for frequencies of 1–50 MHz.) Note that the op-amp amplifies the difference in voltages e_1 and e_2. Such an amplifier is commonly called a differential amplifier. Since the gain of the op-amp is very high, it is necessary to have negative feedback from the output to the input to make the amplifier stable. (The feedback is made from the output to the inverted input so that the feedback is a negative.) In the ideal op-amp, no current flows into the input terminals, and the output voltage is not affected by the load connected to the output terminal. In other words, the input impedance is infinity and the output impedance are zero. In an actual op-amp, a very small amount of current (almost negligible) flows into an input terminal and the output cannot be loaded too much. In our analysis here, we make the assumption that the op-amps are ideal.

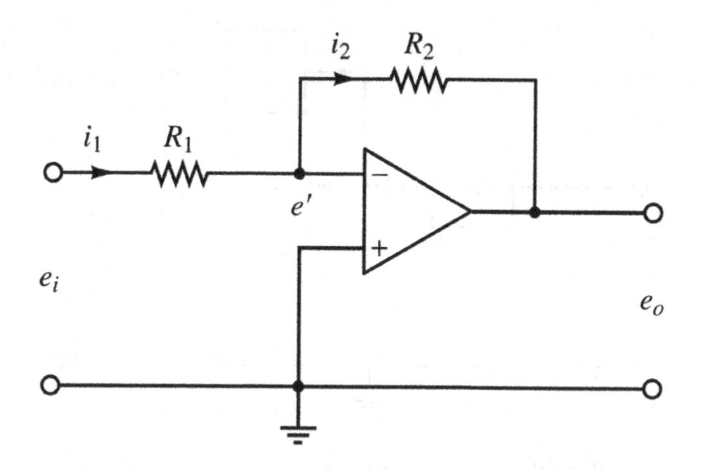

FIGURE 2.54 Inverting amplifier.

6. *Inverting amplifier:*
Consider the operational amplifier circuit shown in Figure 2.54. Let us obtain the output voltage e_o.
The equation for this circuit can be obtained as follows: Define

$$i_1 = \frac{e_i - e'}{R_1}, i_2 = \frac{e' - e_o}{R_2}$$

Since only a negligible amount of current flows into the amplifier, the current i_1 must be equal to current i_2. Thus,

$$\frac{e_i - e'}{R_1} = \frac{e' - e_o}{R_2}$$

Since $K\left(0 - e'\right) = e_o$ and $K \gg 1$, e' must be almost zero, or $e' \cong 0$. Hence, we have

$$\frac{e_i}{R_1} = \frac{e_o}{R_2}$$

or

$$e_o = \frac{R_2}{R_1} e_i$$

Thus, the circuit shown is an inverting amplifier. If $R_1 = R_2$, then the op-amp circuit shown acts as a sign inverter.

7. *Impedance approach to obtaining transfer functions:*
Consider the op-amp circuit shown in Figure 2.55. Similar to the case of electrical circuits we discussed earlier, the impedance approach can be applied to op-amp

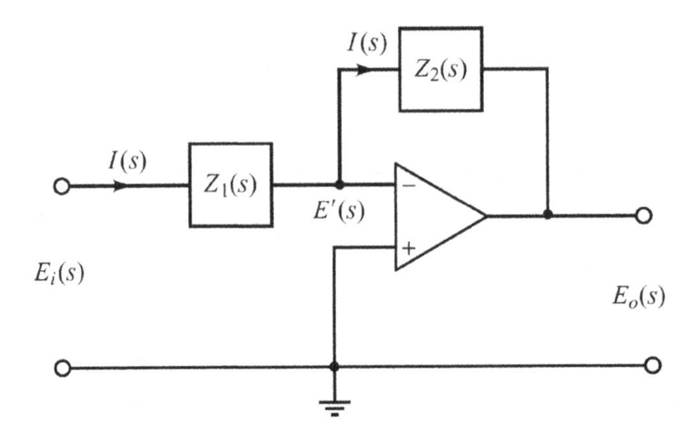

FIGURE 2.55 Operational amplifier circuit with load.

circuits to obtain their transfer functions. For the circuit shown in Figure 2.55, we have

$$\frac{E_i(s)-E'(s)}{Z_1}=\frac{E'(s)-E_o(s)}{Z_2}$$

Since $E'(s)\cong 0$, we have

$$\frac{E_o(s)}{E_i(s)}=-\frac{Z_2(s)}{Z_1(s)} \tag{2.199}$$

8. *Lead or lag networks using operational amplifiers:*
   Figure 2.56a shows an electronic circuit using an operational amplifier. The transfer function for this circuit can be obtained as follows: Define the input impedance and feedback impedance as Z_1 and Z_2, respectively. Then

$$Z_1=\frac{R_1}{R_1C_1s+1},Z_2=\frac{R_2}{R_2C_2s+1}$$

Hence, referring to Eq. (2.206), we have

$$\frac{E_o(s)}{E_i(s)}=-\frac{Z_2(s)}{Z_1(s)}=\frac{R_2}{R_1}\left(\frac{R_1C_1s+1}{R_2C_2s+1}\right)=-\frac{C_1}{C_2}\left(\frac{s+\frac{1}{R_1C_1}}{s+\frac{1}{R_2C_2}}\right) \tag{2.200}$$

Notice that the transfer function in Eq. (2.200) contains a minus sign. Thus, this circuit is sign inverting. If such a sign inversion is not convenient in the actual application, a sign inverter may be connected to either the input or the output of the circuit of Figure 2.56a.

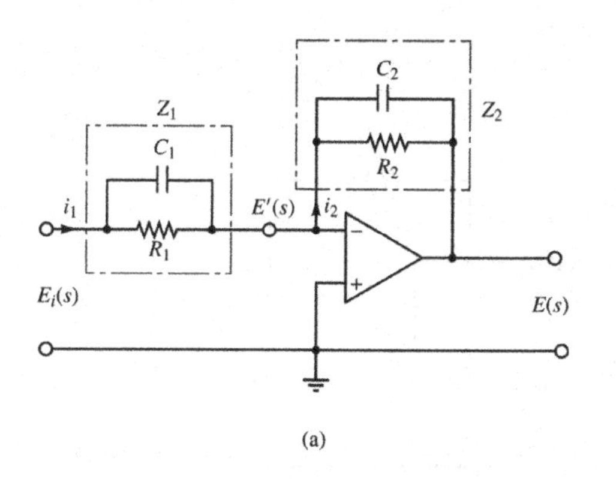

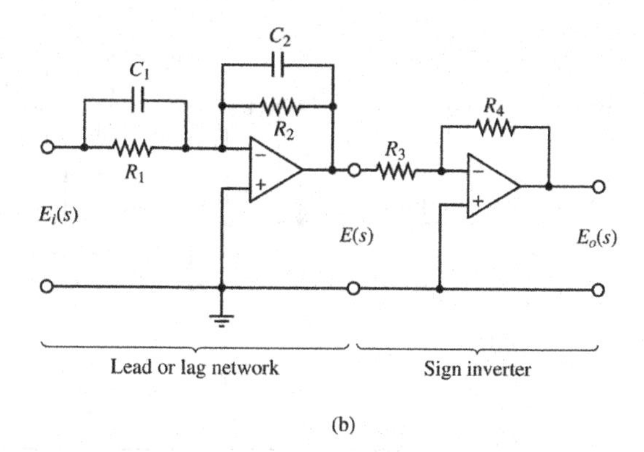

FIGURE 2.56 (a) Operational amplifier circuit and (b) operational amplifier circuit used as a lead or lag compensator.

An example is shown in Figure 2.56b. The sign inverter has the transfer function of

$$\frac{E_o(s)}{E_i(s)} = -\frac{R_4}{R_3}$$

The sign inverter has the gain of $-\frac{R_4}{R_3}$. Hence, the network shown in Figure 2.56b has the following transfer function:

$$\frac{E_o(s)}{E_i(s)} = \frac{R_2 R_4}{R_1 R_3}\left(\frac{R_1 C_1 s + 1}{R_2 C_2 s + 1}\right) = -\frac{R_4 C_1}{R_3 C_2}\left(\frac{s + \frac{1}{R_1 C_1}}{s + \frac{1}{R_2 C_2}}\right)$$

or

$$\frac{E_o(s)}{E_i(s)} = K_c \alpha \left(\frac{Ts + 1}{\alpha Ts + 1}\right) = K_c \left(\frac{s + \frac{1}{T}}{s + \frac{1}{\alpha T}}\right) \tag{2.201}$$

where

$$T = R_1 C_1, \alpha T = R_2 C_2, K_c = \frac{R_4 C_1}{R_3 C_2}$$

whereas

$$K_c \alpha = \frac{R_4 C_1}{R_3 C_2}\frac{R_2 C_2}{R_1 C_1} = \frac{R_2 R_4}{R_1 R_3}, \alpha = \frac{R_2 C_2}{R_1 C_1}$$

This network has a dc gain of

$$K_c \alpha = \frac{R_2 R_4}{R_1 R_3}$$

Note that this network, whose transfer function is given by Eq. (2.201), is a lead network if $R_1 C_1 > R_2 C_2$, or $\alpha < 1$. It is a lag network if $R_1 C_1 < R_2 C_2$.

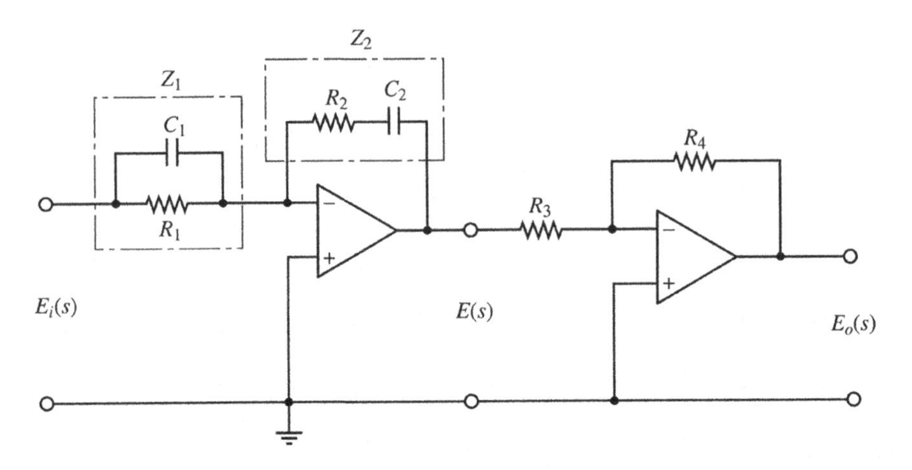

FIGURE 2.57 PID controller using operational amplifiers.

9. *PID controller using operational amplifiers:*
Figure 2.57 shows an electronic proportional-plus-integral-plus-derivative controller (a PID controller) using operational amplifiers. The transfer function $\frac{E_o(s)}{E_i(s)}$ is given by

$$\frac{E_o(s)}{E_i(s)} = -\frac{Z_2(s)}{Z_1(s)}$$

where

$$Z_1 = \frac{R_1}{R_1C_1s+1}, Z_2 = \frac{R_2C_2s+1}{C_2s}$$

Thus,

$$\frac{E_o(s)}{E_i(s)} = -\left(\frac{R_2C_2s+1}{C_2s}\right)\left(\frac{R_1C_1s+1}{R_1}\right)$$

whereas

$$\frac{E_o(s)}{E_i(s)} = -\frac{R_4}{R_3}$$

Consequently,

$$\frac{E_o(s)}{E_i(s)} = \frac{E_o(s)}{E(s)}\frac{E(s)}{E_i(s)} = \frac{R_2R_4}{R_1R_3}\left(\frac{R_2C_2s+1}{C_2s}\right)\left(\frac{R_1C_1s+1}{R_2}\right)$$

or

$$\frac{E_o(s)}{E_i(s)} = \frac{R_2R_4}{R_1R_3}\left(\frac{R_1C_1+R_2C_2}{R_2C_2} + \frac{1}{R_2C_2s} + R_1C_1s\right)$$

Therefore,

$$\frac{E_o(s)}{E_i(s)} = \frac{R_4(R_1C_1 + R_2C_2)}{R_1R_3C_2}\left[1 + \frac{1}{(R_1C_1 + R_2C_2)s} + \frac{R_1C_1R_2C_2}{R_1C_1 + R_2C_2s}s\right] \tag{2.202}$$

Notice that the second operational amplifier circuit acts as a sign inverter as well as a gain adjuster.

When a PID controller is expressed as

$$\frac{E_o(s)}{E_i(s)} = K_p\left(1 + \frac{T_i}{s} + T_d s\right)$$

K_p is the proportional gain, T_i is the integral time, and T_d is the derivative time. From Eq. (2.202), we obtain the proportional gain K_p, integral time T_i, and derivative time T_d to be

$$K_p = \frac{R_4(R_1C_1 + R_2C_2)}{R_1R_3C_2}$$

$$T_i = \frac{1}{R_1C_1 + R_2C_2}$$

and

$$T_d = \frac{R_1C_1R_2C_2}{R_1C_1 + R_2C_2s}$$

When a PID controller is expressed as

$$\frac{E_o(s)}{E_i(s)} = K_p + \frac{K_i}{s} + K_d s$$

K_p is the proportional gain, K_i is the integral gain, and K_d is the derivative gain. For this controller,

$$K_p = \frac{R_4(R_1C_1 + R_2C_2)}{R_1R_3C_2}$$

$$K_i = \frac{R_4}{R_1R_3C_2}$$

and

$$T_d = \frac{R_4R_2C_1}{R_3}$$

Table 2.8 shows a list of operational amplifier circuits that may be used as controllers or compensators.

TABLE 2.8 Operational amplifier circuits that may be used as compensators

	Control Action	$\frac{E_o(s)}{E_i(s)}$	Operational Amplifier Circuits
1	P	$\frac{R_4}{R_3}\frac{R_2}{R_1}$	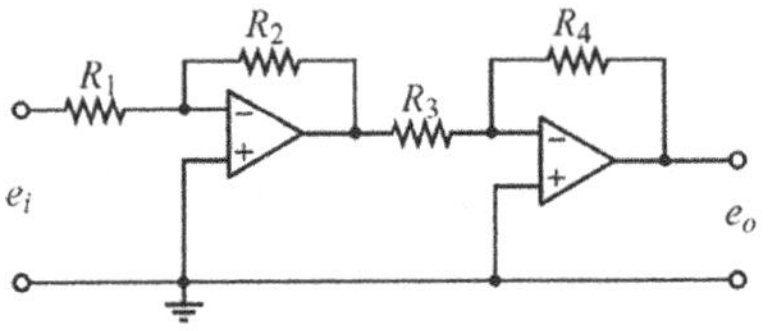
2	I	$\frac{R_4}{R_3}\frac{1}{R_1C_2s}$	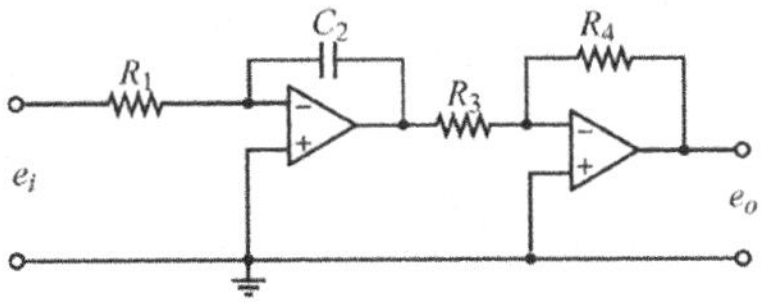
3	PD	$\frac{R_4}{R_3}\frac{R_2}{R_1}(R_1C_1s+1)$	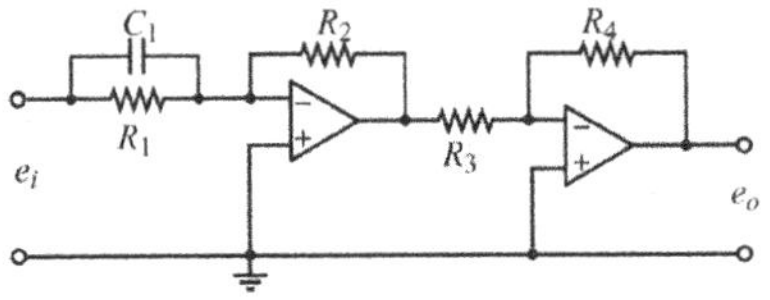
4	PI	$\frac{R_4}{R_3}\frac{R_2}{R_1}\left(\frac{R_2C_2s+1}{R_2C_2s}\right)$	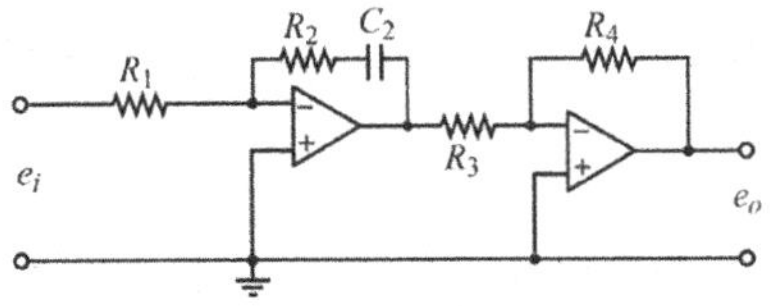
5	PID	$\frac{R_4}{R_3}\frac{R_2}{R_1}\frac{(R_1C_1s+1)(R_2C_2s+1)}{R_2C_2s}$	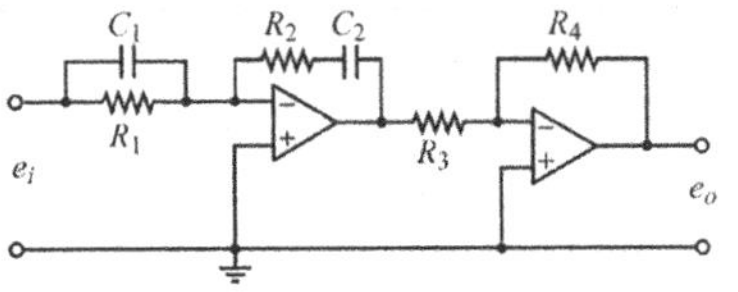
6	Lead or lag	$\frac{R_4}{R_3}\frac{R_2}{R_1}\frac{R_1C_1s+1}{R_2C_2s+1}$	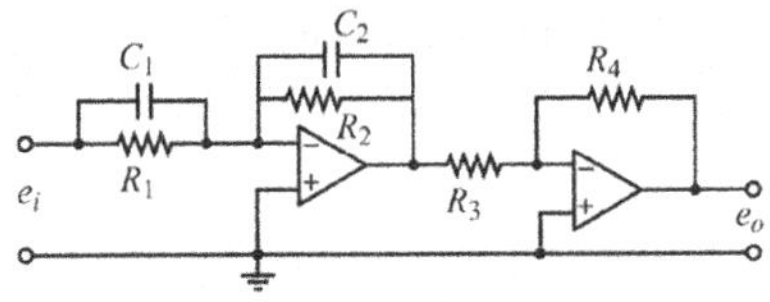
7	Lag-lead	$\frac{R_4}{R_3}\frac{R_2}{R_1}\frac{\left[(R_1+R_3)C_1s+1\right](R_2C_2s+1)}{(R_1C_1s+1)\left[(R_2+R_4)C_2s+1\right]}$	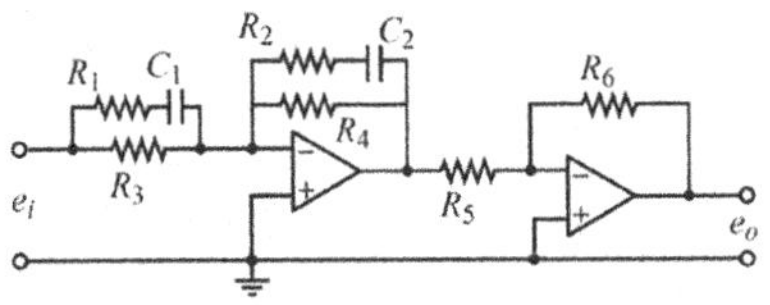

2.9.2 Electromechanical System

To discuss electromechanical system. Let's see an example of an armature-controlled DC electric motor system. In this example of an armature-controlled DC electric motor system as shown in Figure 2.58, how can you find the transfer function between e_a and w_m?

Let interpret the terms from Figure 2.58, $e_a(t)$ is the input voltage source, $\omega_m(t)$ is the speed of the rotor system, R_a is the resistor, L_a is the inductor, $i_a(t)$ is the armature circuit current, $v_b(t)$ is the voltage drops in the rotation system by rotor, $T_m(t)$ is the applied torque in the rotor system, and $\theta_m(t)$ is the angular rotation or motion in the rotor system of this system network. The armature-controlled DC electric motor system is fixed with a fixed field.

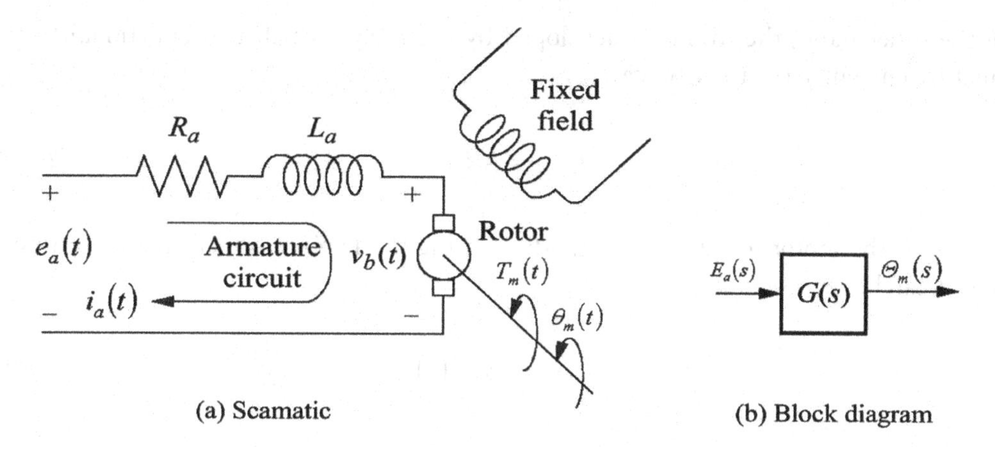

FIGURE 2.58 Armature-controlled motor.

However, this can be simplified into block diagram where $E_a(s)$ is the input source, $\Theta_m(s)$ is the speed in the output, and $G(s)$ is the system transfer function after taking the Laplace transformation of this system function.

Therefore, we have to find the transfer function as

$$\frac{\Theta_m(s)}{E_a(s)}$$

To solve this problem, we can say that the current is carried in the armature circuit; as a result, the current is rotating in a magnetic field as well as it is passing through the rotor system. Therefore, we can say that the voltage ($v_b(t)$) is proportional to speed $\omega_m(t)$. Thus, it can be expressed as

$$v_b(t) = K_b\omega_m(t) \tag{2.203}$$

$$v_b(t) = K_b\frac{d\theta_m(t)}{dt} \tag{2.204}$$

Here, the K_b is called back electromotive force EMF constant. After taking the Laplace transformation on both sides, it can be represented as

$$V_b(s) = K_b\Theta_m(s) \tag{2.205}$$

According to the KVL, the sum of all the voltages around a loop is equal to zero which means that the total applied voltage $e_a(t)$ is equal to the summation of the voltage drops in a resistor ($R_a(t)$), inductor $L_a(t)$, and rotor $v_b(t)$. Therefore, we can say

$$e_a(t) = R_a(t)e_a(t) + L_a(t)i_a(t) + v_b(t)i_a(t) \tag{2.206}$$

After taking the Laplace transformation on both sides, it can be represented as

$$R_aI_a(s) + L_asI_a(s) + V_b(s) = E_a(s) \tag{2.207}$$

On the other hand, the torque is developed by the motor which is proportional to the armature current, $i_a(t)$; thus, we can say

$$T_m(t) = K_t i_a(t) \tag{2.208}$$

Here, K_t is the motor torque constant. After taking the Laplace transformation on both sides, it can be represented

$$T_m(s) = K_t I_a(s) \tag{2.209}$$

Therefore,

$$I_a(s) = \frac{T_m(s)}{K_t} \tag{2.210}$$

However, the typical equivalent mechanical loading on a motor can be established where J_m is the inertia of the motor, $T_m(t)$ is the applied torque in the rotor system, $\theta_m(t)$ is the angular rotation or motion in the rotor system, $\omega_m(t)$ is the speed of the motor, and D_m is the damper system as shown in the Figure 2.59.

Therefore, the applied torque can be expressed as

$$\omega_m(t) = \frac{d\theta_m(t)}{dt} \tag{2.211}$$

Therefore,

$$T_m(t) = J_m \frac{d\omega_m(t)}{dt} + D_m \omega_m(t)$$

After taking the Laplace transformation on both sides, it can be represented as

$$T_m(s) = (J_m s + D_m)\Theta_m(s)$$

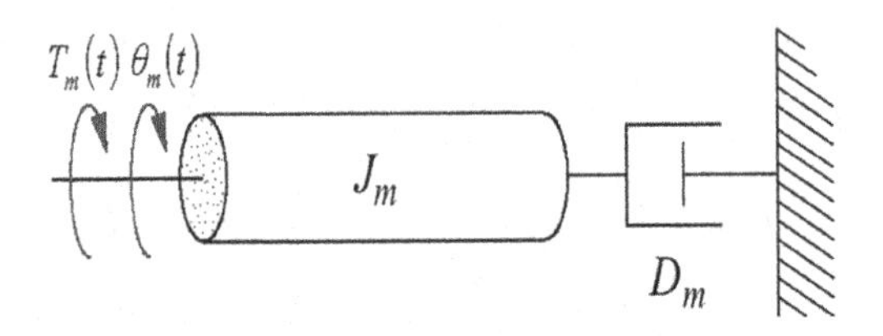

FIGURE 2.59 Typical equivalent mechanical loading on a motor.

Since we have the relation as

$$I_a(s)=\frac{T_m(s)}{K_t}$$

Therefore, we get

$$I_a(s)=\frac{(J_m s+D_m)\Theta_m(s)}{K_t}$$

We can substitute the value of $I_a(s)$ and $V_b(s)$ in the following equation, which we can simplify as

$$R_a I_a(s)+L_a s I_a(s)+V_b(s)=E_a(s) \tag{2.212}$$

Then

$$(R_a+L_a s)\frac{(J_m s+D_m)\Theta_m(s)}{K_t}+K_b\Theta_m(s)=E_a(s)$$

Therefore, we can establish the transfer function as

$$\frac{\Theta_m(s)}{E_a(s)}=\frac{K_t}{(R_a+L_a s)(J_m s+D_m)+K_t K_b}=\frac{K_t}{J_m L_a s^2+J_m R_a s+D_m R_a s+R_a D_m+K_t K_b}$$

or

$$\frac{\Theta_m(s)}{E_a(s)}=\frac{K_t/J_m L_a}{s^2+\left(\frac{R_a}{L_a}+\frac{D_m}{J_m}\right)s+\frac{R_a D_m+K_t K_b}{J_m L_a}} \tag{2.213}$$

2.10 LIQUID-LEVEL SYSTEMS AND THERMAL SYSTEM

2.10.1 Liquid-Level Systems

Consider the system shown in Figure 2.60a. The variables are defined as follows:

$\bar{Q}$ = steady-state flow rate (before any change), m^3/second

q_i = small deviation of inflow rate from its steady-state value, m^3/second

q_o = small deviation of outflow rate from its steady-state value, m^3/second

$\bar{H}$ = steady-state head (before any change), m

h = small deviation of head from its steady-state value, m

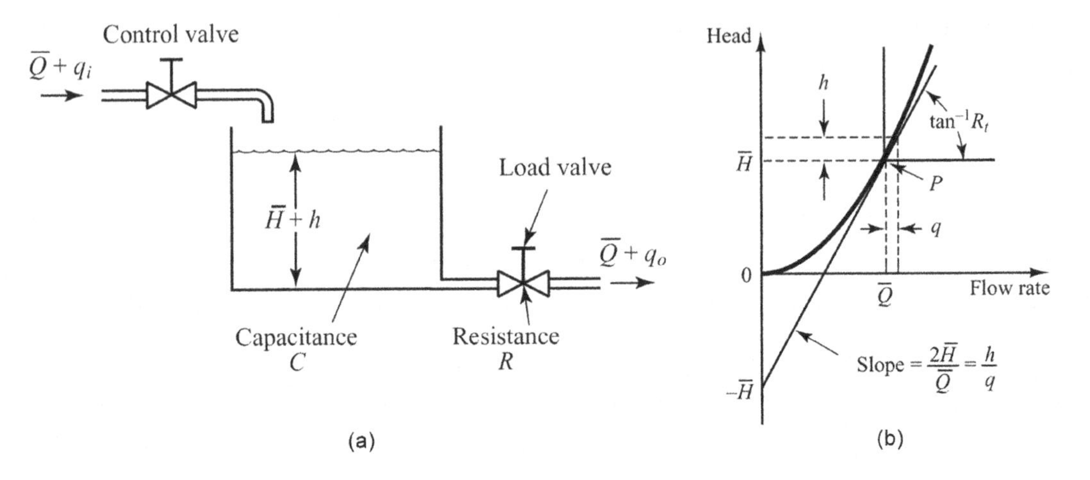

FIGURE 2.60 (a) Liquid-level system and (b) head-versus-flow-rate curve.

As stated previously, a system can be considered linear if the flow is laminar. Even if the flow is turbulent, the system can be linearized if changes in the variables are kept small. Based on the assumption that the system is either linear or linearized, the differential equation of this system can be obtained as follows: since the inflow minus outflow during the small-time interval dt is equal to the additional amount stored in the tank, we see that

$$C\,dh = \left(q_i - q_o\right)dt$$

From the definition of resistance, the relationship between q_o and h is given by

$$q_o = \frac{h}{R}$$

The differential equation for this system for a constant value of R becomes

$$RC\frac{dh}{dt} + h = Rq_i \tag{2.214}$$

Note that RC is the time constant of the system. Taking the Laplace transforms of both sides of Eq. (2.214), assuming the zero initial condition, we obtain

$$(RCs+1)H(s) = RQ_i(s)$$

Where,

$$H(s) = \mathcal{L}[h] \text{ and } Q_i(s) = \mathcal{L}[q_i]$$

If q_i is considered the input and h the output, the transfer function of the system is

$$\frac{H(s)}{Q_i(s)} = \frac{R}{RCs+1}$$

If, however, q_o is taken as the output, the input being the same, then the transfer function is

$$\frac{Q_o(s)}{Q_i(s)} = \frac{1}{RCs+1}$$

where we have used the relationship

$$Q_o(s) = \frac{1}{R}H(s)$$

2.10.2 Liquid-Level Systems with Interaction

Consider the system shown in Figure 2.61. In this system, the two tanks interact. Thus, the transfer function of the system is not the product of two first-order transfer functions.

$\bar{Q}$ = steady-state flow rate (before any change), m^3/second

q = small deviation of inflow rate from its steady-state value, m^3/second

q_2 = small deviation of outflow rate from its steady-state value, m^3/second

$\bar{H}_1$ = steady-state liquid level of tank 1 (before any change), m

$\bar{H}_2$ = steady-state liquid level of tank 2 (before any change), m

In the following, we shall assume only small variations of the variables from the steady-state values. Using the symbols as defined in Figure 2.61, we can obtain the following equations for this system:

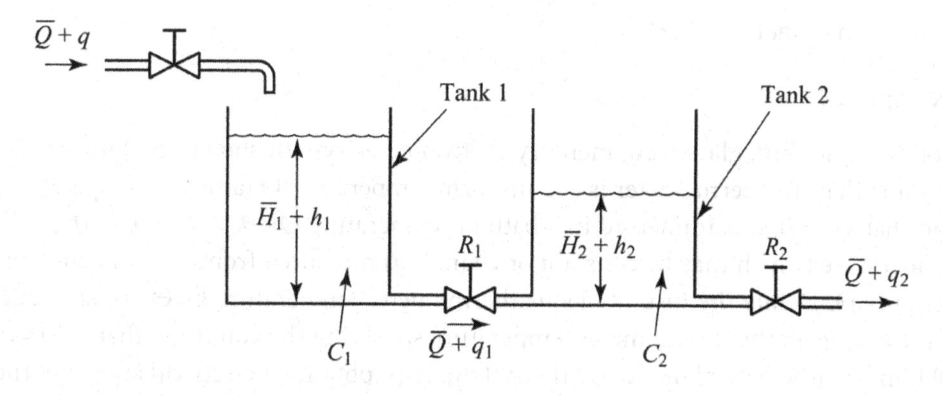

FIGURE 2.61 Liquid-level systems with interaction.

$$\frac{h_1 - h_2}{R_1} = q_1 \tag{2.215}$$

$$C_1 \frac{dh_1}{dt} = q - q_1 \tag{2.216}$$

$$\frac{h_2}{R_2} = q_2 \tag{2.217}$$

$$C_2 \frac{dh_2}{dt} = q_1 - q_2 \tag{2.218}$$

If q is considered the input and q_2 the output, the transfer function of the system is

$$\frac{Q_2(s)}{Q(s)} = \frac{1}{R_1C_1R_2C_2s^2 + (R_1C_1 + R_2C_2 + R_2C_1)s + 1} \tag{2.219}$$

It is instructive to obtain Eq. (2.219), the transfer function of the interacted system, by block diagram reduction. From Eqs. (2.215) to (2.218), we obtain the elements of the block diagram, as shown in Figure 2.62a. By connecting signals properly, we can construct a block diagram, as shown in Figure 2.62b. This block diagram can be simplified, as shown in Figure 2.62c. Further simplifications result in Figure 2.62d and e. Figure 2.62e is equivalent to Eq. (2.219).

Notice the similarity and difference between the transfer function given in the transfer function of cascaded elements section (see Eq. 2.205) and that given by

$$\frac{E_o(s)}{E_i(s)} = \frac{1}{R_1C_1R_2C_2s^2 + (R_1C_1 + R_2C_2 + R_1C_2)s + 1}$$

The term R_2C_1s that appears in the denominator of Eq. (2.215) exemplifies the interaction between the two tanks. Similarly, the term R_1C_2s in the denominator represents the interaction between the two *RC* circuits shown in the transfer function of cascaded elements section (see Figure 2.51).

2.10.3 Thermoelectrical System

Example 2.49

Consider the thin, glass-wall, mercury thermometer system shown in Figure 2.63a. Assume that the thermometer is at a uniform temperature $\bar{\Theta}$ (ambient temperature) and that at $t = 0$, it is immersed in a bath of temperature $\bar{\Theta} + \theta_b$ where θ_b is the bath temperature (which may be constant or changing) measured from the ambient temperature $\bar{\Theta}$. Define the instantaneous thermometer temperature by $\bar{\Theta} + \theta$. so that θ is the change in the thermometer temperature satisfying the condition that $\theta(0) = 0$. Obtain a mathematical model for the system. Also obtain an electrical analog of the thermometer system.

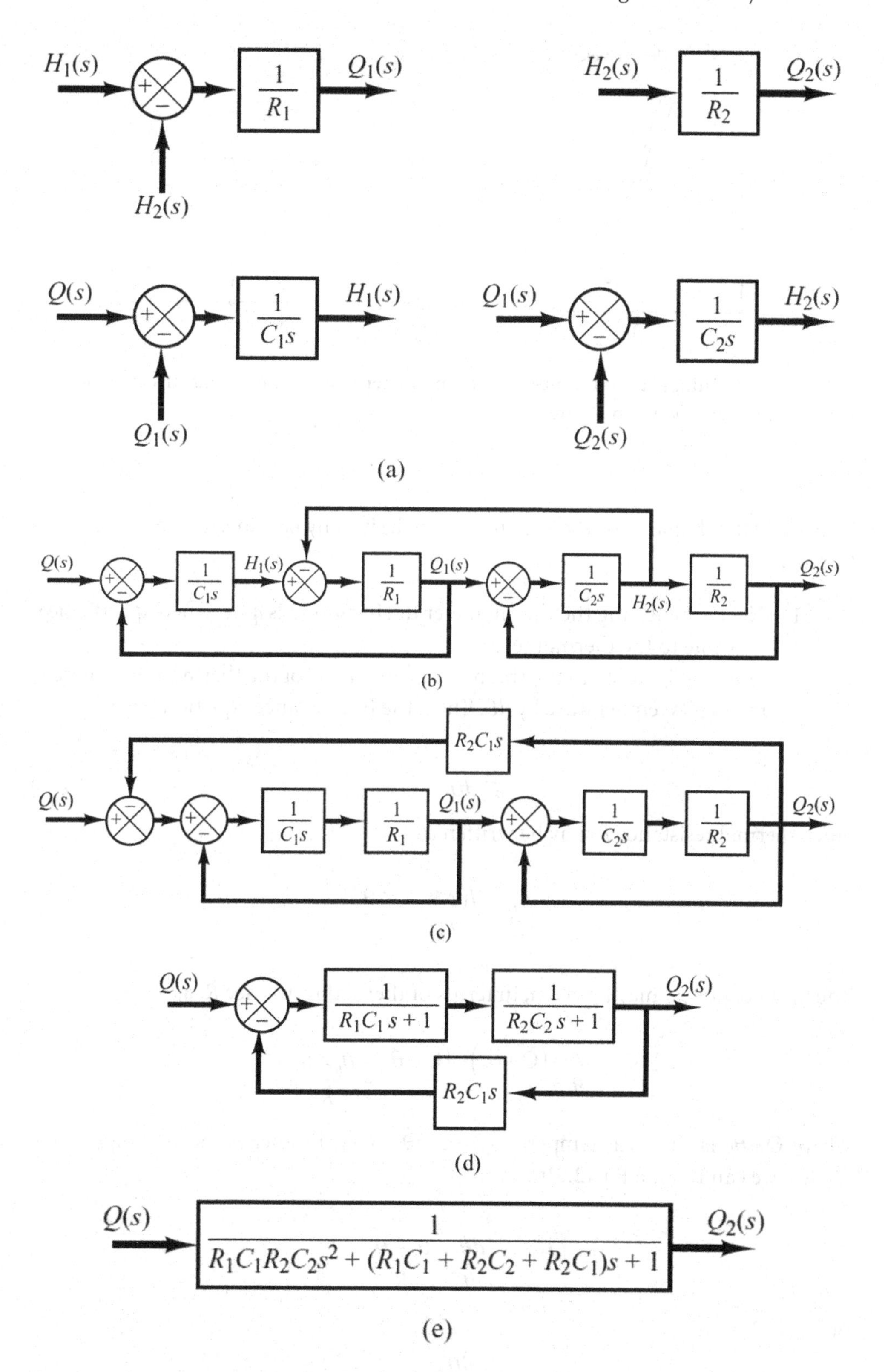

FIGURE 2.62 (a) Elements of the block diagram of the system, (b) block diagram of the system; and (c)–(e) successive reductions of the block diagram.

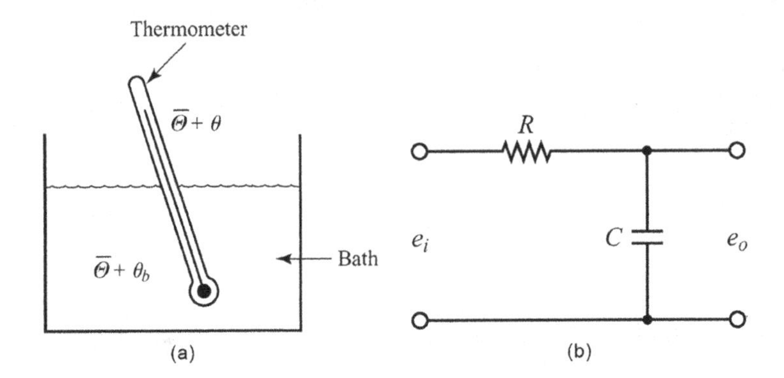

FIGURE 2.63 (a) Thin, glass-wall, mercury thermometer system and (b) Electrical analog of the thermometer system shown in Figure 2.60.

Solution

A mathematical model for the system can be derived by considering heat balance as follows:

1. The heat entering the thermometer during dt sec is $q\ dt$, where q is the heat flow rate to the thermometer.
2. This heat is stored in the thermal capacitance C of the thermometer, thereby raising its temperature by $d\theta$. Thus, the heat balance equation is

$$C\ d\theta = q\ dt \tag{2.220}$$

Since thermal resistance R may be written as

$$R = \frac{d(\Delta\theta)}{dq} = \frac{\Delta\theta}{q}$$

The heat flow rate q may be given, in terms of thermal resistance R, as

$$q = \frac{(\bar{\Theta}+\theta_b)-(\bar{\Theta}+\theta)}{R} = \frac{\theta_b-\theta}{R}$$

where $\bar{\Theta}+\theta_b$ is the bath temperature and $\bar{\Theta}+\theta$ is the thermometer temperature. Hence, we can rewrite Eq. (2.216) as

$$C\frac{d\theta}{dt} = \frac{\theta_b-\theta}{R}$$

or

$$RC\frac{d\theta}{dt} + \theta = \theta_b \tag{2.221}$$

Eq. (2.221) is a mathematical model of the thermometer system. Referring to Eq. (2.221), an electrical analog for the thermometer system can be written as

$$RC\frac{d\theta}{dt}+e_o=e_i$$

An electrical circuit represented by this last equation is shown in Figure 2.63b.

EXAMPLES

1. Figure 2.64b shows a schematic diagram of an automobile suspension system. As the car moves along the road, the vertical displacements at the tires act as the motion excitation to the automobile suspension system. The motion of this system consists of a translational motion of the center of mass and a rotational motion about the center of mass. Mathematical modeling of the complete system is quite complicated.

   A very simplified version of the suspension system is shown in Figure 2.64c. Assuming that the motion x_i at point P is the input to the system and the vertical motion x_o of the body is the output, obtain the transfer function $X_o(s)/X_i(s)$. (Consider the motion of the body only in the vertical direction.) Displacement x_o is measured from the equilibrium position in the absence of input x_i.

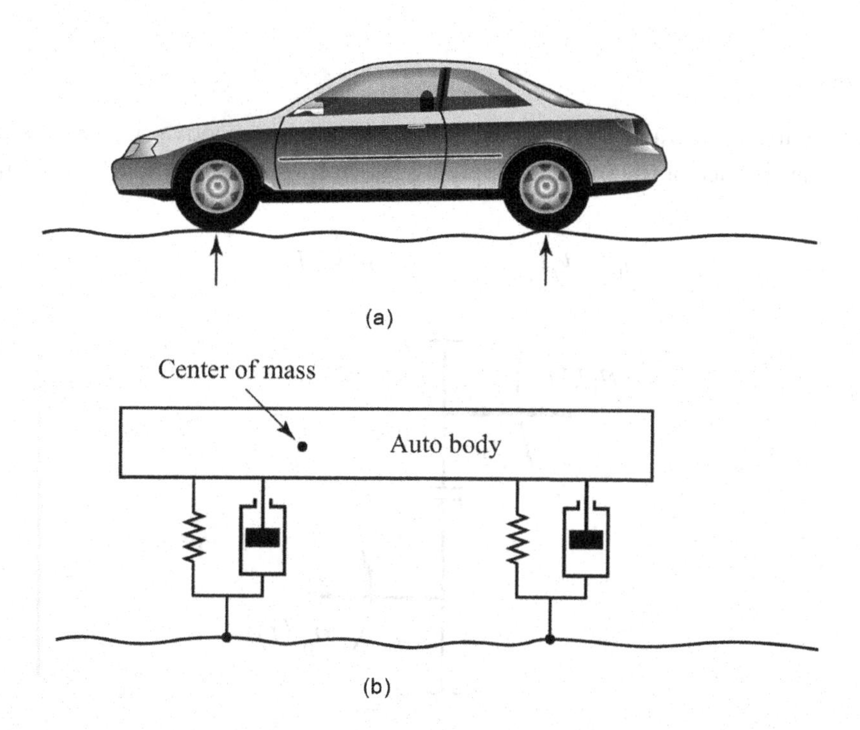

FIGURE 2.64 (a) Automobile system, (b) automobile suspension system, and (c) simplified suspension system.

Solution

The equation of motion for the system shown in Figure 2.64c is

$$m\ddot{x}_o + b(x_o - \dot{x}_i) + k(x_o - x_i) = 0$$

or

$$m\ddot{x}_o + b\dot{x}_o + kx_o = b\dot{x}_i + kx_i$$

Taking the Laplace transform of this last equation, assuming zero initial conditions, we obtain

$$(ms^2 + bs + k)X_o(s) = (bs + k)X_i(s)$$

Hence the transfer function $X_o(s)/X_i(s)$ is given by

$$\frac{X_o(s)}{X_i(s)} = \frac{bs + k}{ms^2 + bs + k}$$

2. Obtain the transfer function $Y(s)/U(s)$ of the system shown in Figure 2.65. The input u is a displacement input.

Solution

Assume that displacements x and y are measured from respective steady-state positions in the absence of the input u. Applying Newton's second law to this system, we obtain

$$m_1\ddot{x} = k_2(y - x) + b(\dot{y} - \dot{x}) + k_1(u - x)$$

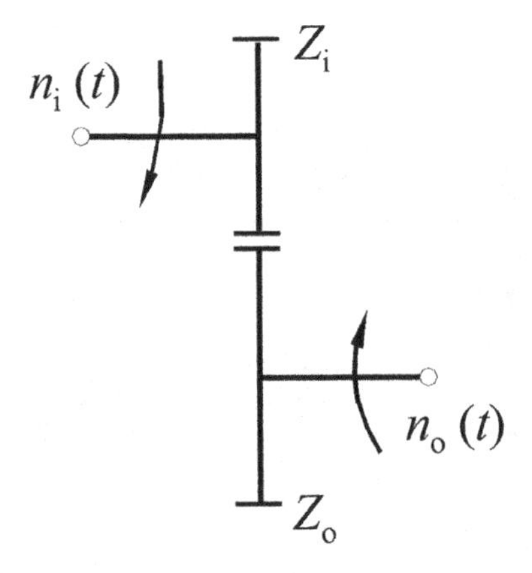

FIGURE 2.65 Motorcycle suspension system.

and

$$m_2 \ddot{y} = -k_2(y-x) - b(\dot{y}-\dot{x})$$

Hence, we have

$$m_1 \ddot{x} + b\dot{x} + (k_1 + k_2)x = b\dot{y} + k_2 y + k_1 u$$

$$m_2 \ddot{y} + b\dot{y} + k_2 y = b\dot{x} + k_2 x$$

Taking Laplace transforms of these two equations and assuming zero initial conditions, we obtain

$$\left[m_1 s^2 + bs + (k_1 + k_2)\right] X(s) = (bs + k_2)Y(s) + k_1 U(s)$$

$$[m_2 s^2 + bs + k_2]Y(s) = (bs + k_2)X(s)$$

Eliminating $X(s)$ from the last two equations, we have

$$\left[m_1 s^2 + bs + (k_1 + k_2)\right] \frac{m_2 s^2 + bs + k_2}{bs + k_2} Y(s) = (bs + k_2)Y(s) + k_1 U(s)$$

Thus,

$$\frac{Y(s)}{U(s)} = \frac{bs + k_2}{m_1 m_2 s^4 + (m_1 + m_2)bs^3 + [k_1 m_2 (m_1 + m_2)k_2]s^2 + k_1 bs + k_1 k_2}$$

EXERCISES

1. Two-mechanical systems are shown in Figure 2.66a and b. Find out their respective differential equations, where x_i is the displacement of input and x_o is the displacement of output, and assuming that the end of the output has no load effect.
2. Take Laplace transform of the following function.
    1. $f(t) = (4t+5)(t) + (t+2)\cdot 1(t)$
    2. $f(t) = [4\cos(2t - \pi/3)]\cdot 1(t - \pi/6) + e^{-5}t \cdot 1(t)$
    3. $f(t) = (15t^2 + 4t + 6)(t) + 1(t-2)$
    4. $f(t) = 6\sin(3t - 135°)\cdot 1(t - \pi/4)$
3. Take the Laplace transform of the following function.
    1. $F(s) = \dfrac{s+1}{(s+2)(s+3)}$

2. $F(s)=\dfrac{e^{-s}}{s-1}$
3. $F(s)=\dfrac{s}{(s+1)^2(s+2)}$
4. $F(s)=\dfrac{4}{s^2+s+4}$

4. Use the Laplace transform method to solve the following differential equations:

1. $\dfrac{d^2x(t)}{dt^2}+6\dfrac{dx(t)}{dt}+8x(t)=1(t);x(0)=1,\left.\dfrac{dx(t)}{dt}\right|_{t=0}=0$

2. $\dfrac{dx(t)}{dt}+10x(t)=2;x(0)=0$

3. $\dfrac{dx(t)}{dt}+100x(t)=300;\left.\dfrac{dx(t)}{dt}\right|_{t=0}=50$

4. $\dfrac{d^2x(t)}{dt^2}+\dfrac{dx(t)}{dt}+x(t)=\delta(t);x(0)=0,\left.\dfrac{dx(t)}{dt}\right|_{t=0}=0$

5. $\dfrac{d^2x(t)}{dt^2}+2\dfrac{dx(t)}{dt}+x(t)=1(t);x(0)=0,\left.\dfrac{dx(t)}{dt}\right|_{t=0}=0$

5. Find the transfer function of the passive network respectively in Figure 2.67.
6. Obtain the transfer function of the active network in Figure 2.68.
7. System of differential equations is as follows:

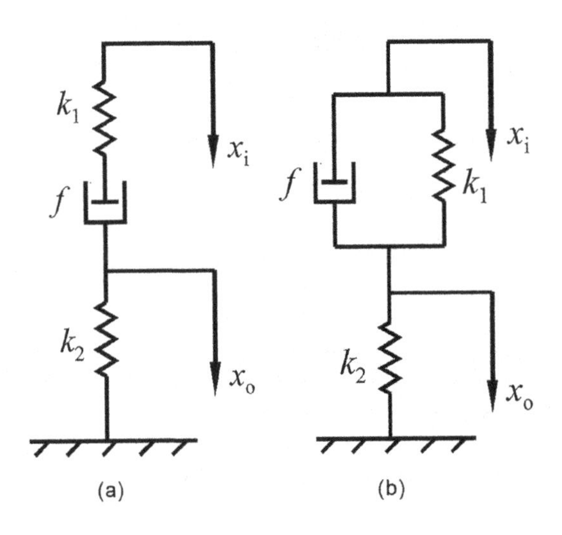

FIGURE 2.66 Mechanical system with damper and spring system.

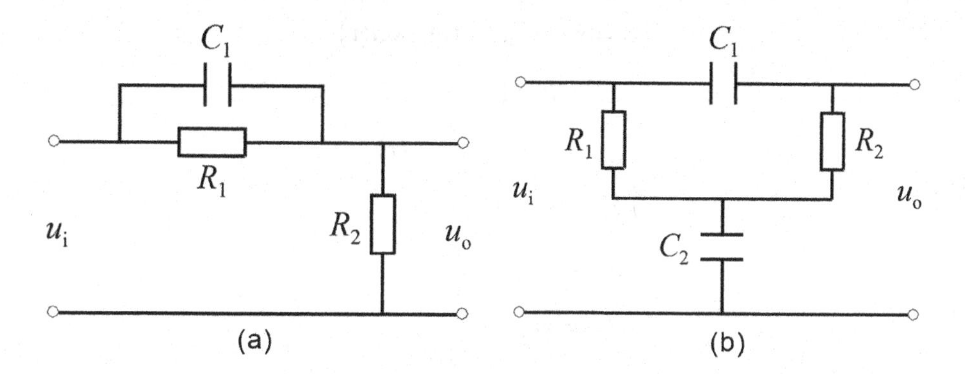

FIGURE 2.67 Passive network with load R and C.

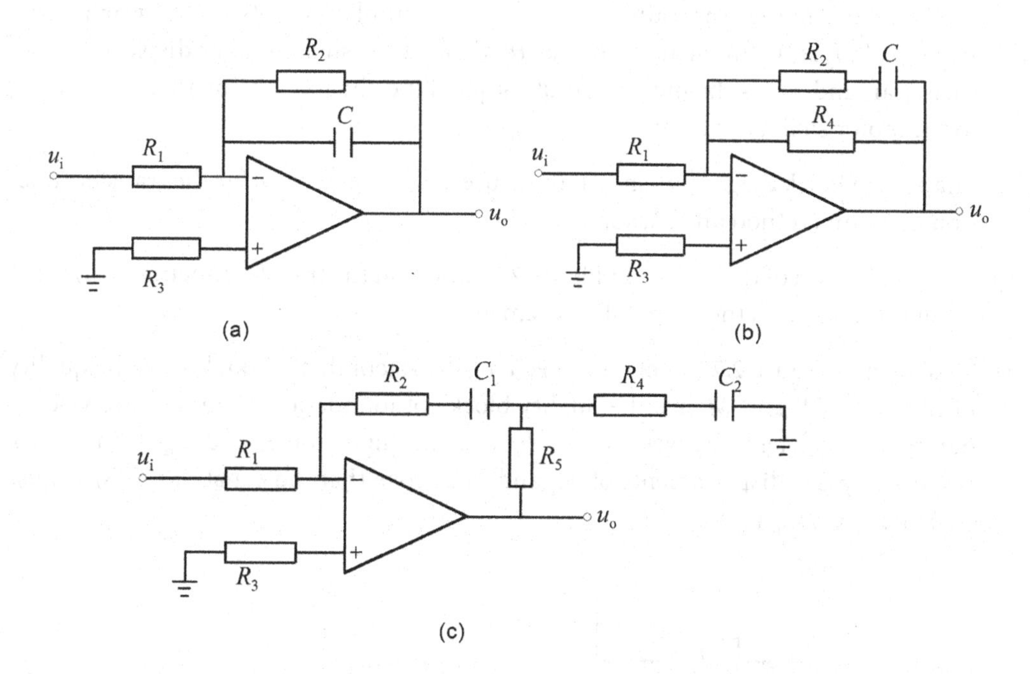

FIGURE 2.68 Active network with load R and C.

$$\dot{x}_1 = k_1\left[x_i(t) - x_o(t) - \beta x_3\right]$$

$$x_2 = \tau\ \dot{x}_1(t)$$

$$T\dot{x}_3 + x_3 = x_1 + x_2$$

$$\dot{x}_o(t) = k_2 x_3$$

where $x_i(t)$ is the input; $x_o(t)$ is the output; x_1, x_2, and x_3 are the intermediate variable; and τ, β, k_1, and k_2 are the constants. Draw the dynamic structure of the system and find the transfer function $X_o(s)/X_i(s)$ and $X_o(s)/n(s)$.

$$x_1 = x_i(t) - \tau\, \dot{x}_o(t) + k_1 n(t)$$

$$x_2(t) = k_o x_1(t)$$

$$x_3(t) = x_2 - n(t) - x_5(t)$$

$$T\dot{x}_4(t) = x_3(t)$$

$$x_5(t) = x_4(t) - x_o(t)$$

$$\dot{x}_o(t) = x_5(t) - x_o(t)$$

where $x_i(t)$ is the input; $x_o(t)$ is the total output; $n(t)$ are inputs; and k_o, k_1, T, and τ are constants.

8. In Figure 2.69 in the gear train, z_1, z_2, z_3, z_4 is the number of teeth on the gear, respectively, J_1, J_2, J_3 is the moment of inertia; θ_1, θ_2, θ_3 is the shaft angular displacement in each gear; and M_m is the motor output torque. Find the motion equation converting to the motor shaft gear train.

9. The system block diagram is shown in Figure 2.70, respectively, for the transfer function diagram method and Mason rules.

10. Draw a diagram of the system in Figure 2.71 and find the transfer function. $F_i(t)$ is the input power; $x_o(t)$ is the output displacement.

11. As shown in Figure 2.72, there is a mechanical system that is suitable for the quality of the mass. M_1 and M_2 are the quality blocks of foundation, μ_1 and μ_2 are viscous damping coefficients between mass; $F_i(t)$ is the input force; and $x_{o1}(t)$ and $x_{o2}(t)$, respectively, are displacements of M_1 and M_2. Prove that the equations $G_1(s) = x_{o1}(s)/F_i(s)$ and $G_2(s) = x_{o2}(s)/F_i(s)$.

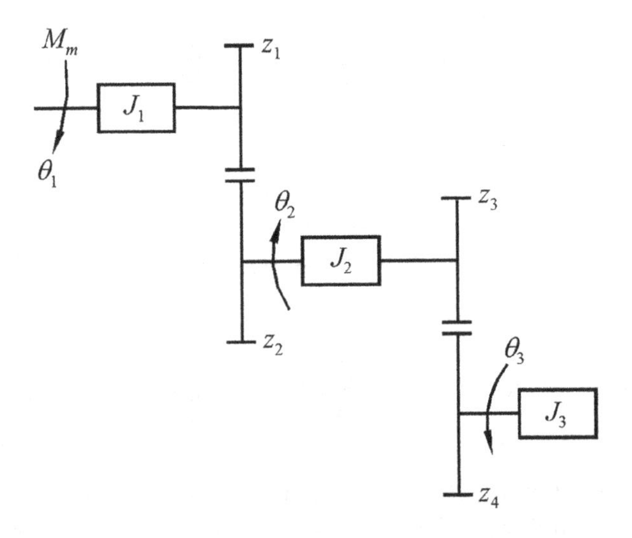

FIGURE 2.69 Gear train.

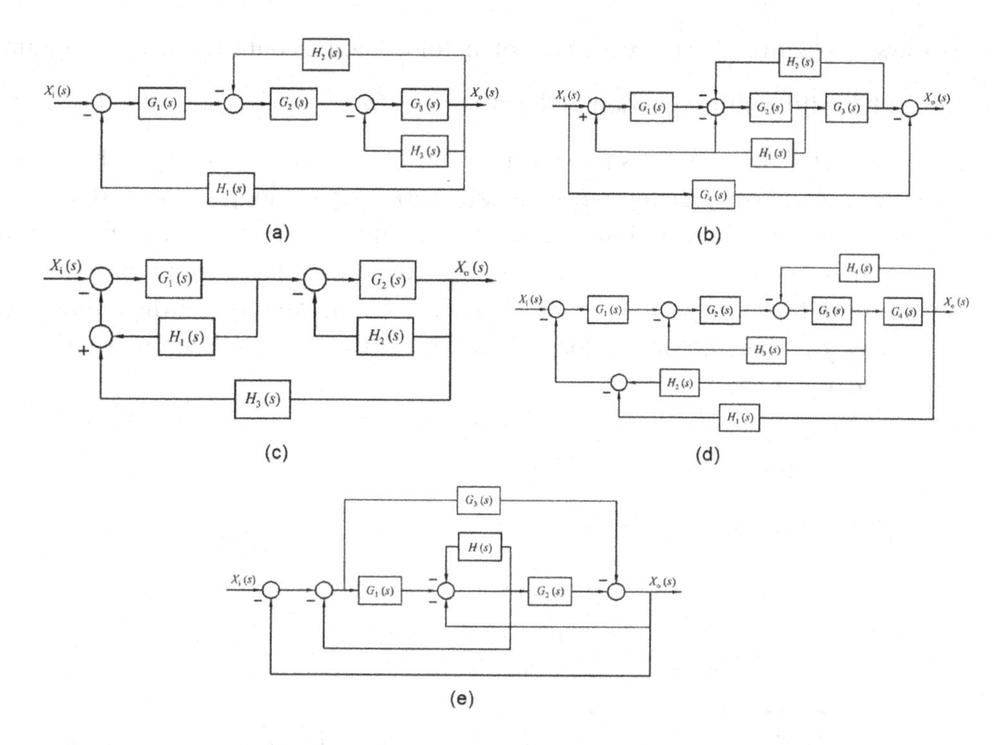

FIGURE 2.70 System block diagram and its simplification.

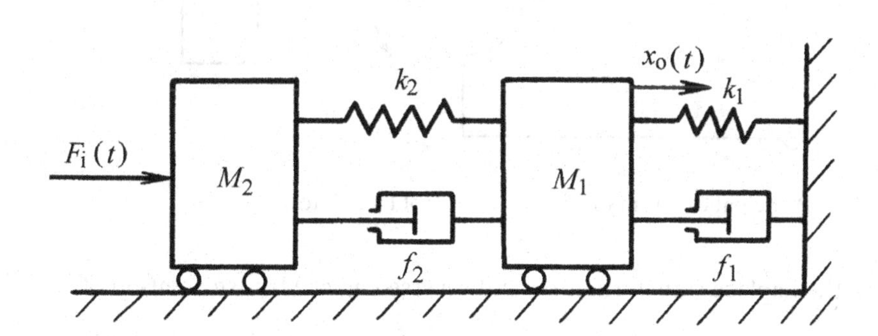

FIGURE 2.71 Mechanical system with friction system, and two mass system M_1 and M_2.

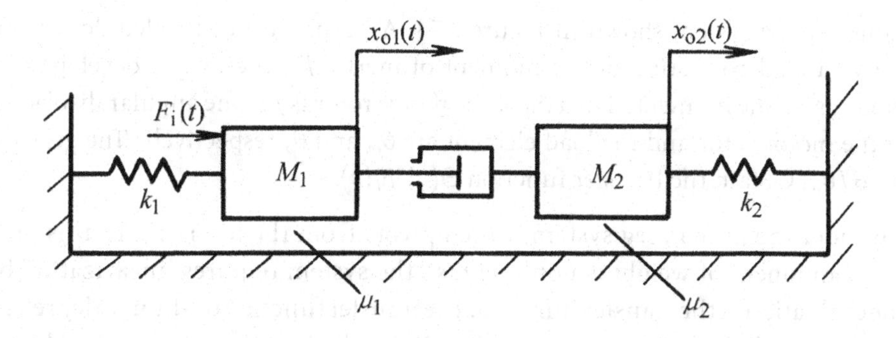

FIGURE 2.72 Mechanical system with two mass system M_1 and M_2 without friction.

12. As shown in Figure 2.73, there is a type of motor armature control principle diagram.
    1. Find the differential equation of the motor and the transfer function of the motor.
    2. Obtain $\Omega_M(s)/U_a(s)$. R_a is the resistance in the armature loop; L_a is the inductance in the armature loop; i_a is the armature current loop; $u_a(t)$ is applied to the armature voltage, as input of the system; E_b is armature counter electromotive force as the excitation current; $\omega_M(t)$ is the rotational speed of the motor, as output of the system; J_M is for motor load to the motor shaft of the moment of inertia. f_M is the motor and load equivalent to the viscous damping coefficient on the motor shaft.

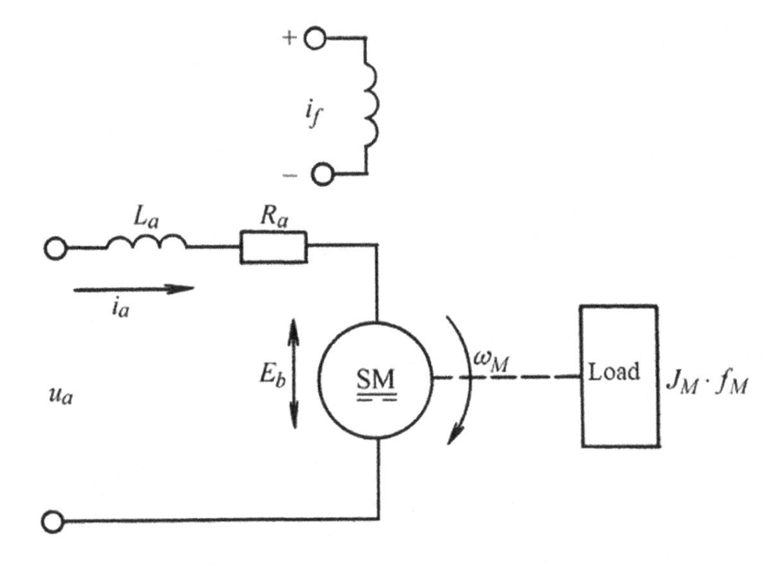

FIGURE 2.73 Principal diagram of armature control DC motor.

13. Find the Laplace transformation and inverse Laplace transformation of

$$\frac{d^2y}{dt^2} + 12\frac{dy}{dt} + 32y = 32u(t)$$

14. Consider the system shown in Figure 2.74. An armature-controlled dc servomotor drives a load consisting of the moment of inertia J_L. The torque developed by the motor is T. The moment of inertia of the motor rotor is J_m. The angular displacements of the motor rotor and the load element are θ_m and θ, respectively. The gear ratio is $n = \theta / \theta_m$. Obtain the transfer function $\Theta(s)/E_i(s)$.

15. Consider the human leg system, which pivots from the hip joint. In this problem the component of weight is nonlinear, so the system requires linearization before the evaluation of the transfer function. The transfer function of a human leg relates the output angular rotation about the hip joint to the input torque supplied by the leg

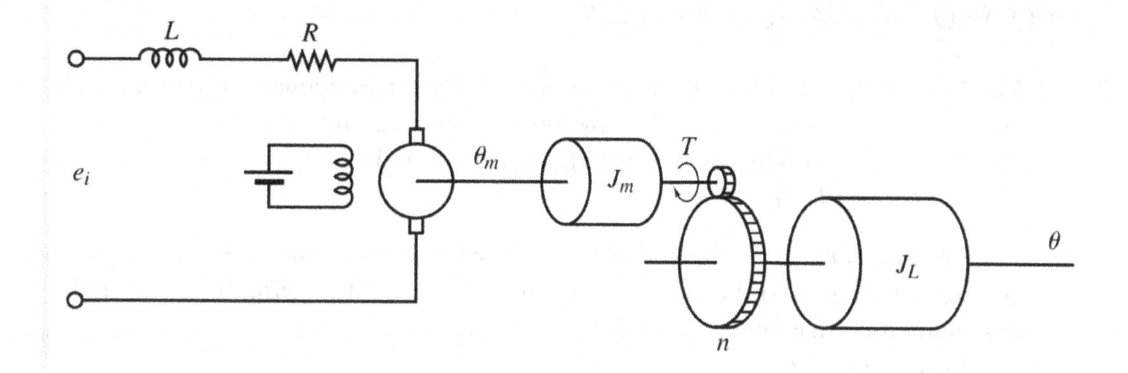

FIGURE 2.74 Armature-controlled dc servomotor system.

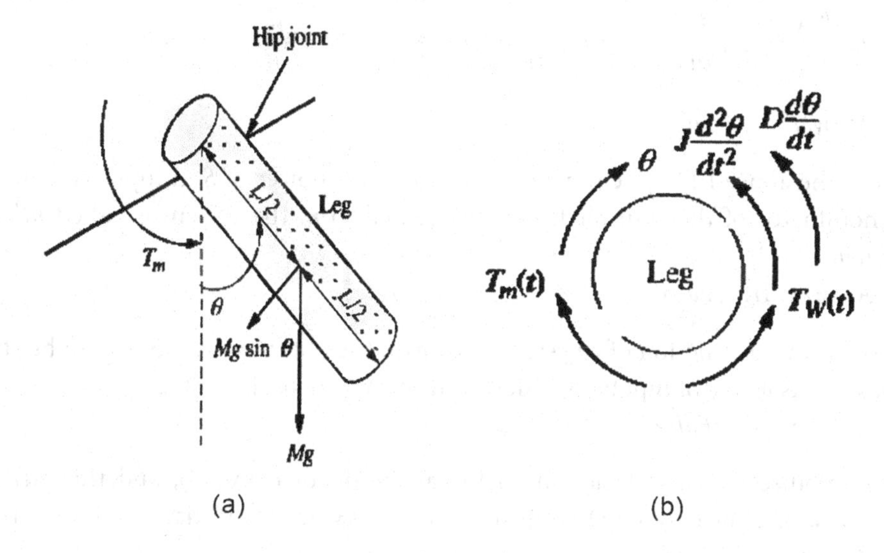

FIGURE 2.75 Human leg system.

muscle. A simplified model for the leg is shown in Figure 2.75. The model assumes an applied muscular torque/muscle torque, $T_m(t)$, viscous damping, D, at the hip p joint, and inertia, J, around the hip joint. Also, a component of the weight of the leg, M_g, where M is the mass of the leg and g is the acceleration due to gravity, creates a nonlinear torque. If we assume that the leg is of uniform density, the weight can be applied at $L/2$, where L is the length of the leg (Milsum, 1966). Do the following:

a. Evaluate the nonlinear torque,

b. Find the transfer function $q(s)/T_m(s)$, for small angles of rotation, where $q(s)$ is the angular rotation of the leg about the hip joint.

MCQ AND TRUE/FALSE QUESTIONS

1. The transfer function has I/O relationships of system components or systems; the transfer function can be described by linear, time-invariant, and differential equations with initial conditions zero such that $y(t) = y(0-) = 1$.

   Answer: True/False

2. *Transfer Function* can be defined as the ratio of the Laplace transform of the output (response function) to the Laplace transform of the input (driving function) under the assumption that all initial conditions are zero.

   Answer: True/False

3. From the following LTI differential equation, you must remember that n is less or equal to m.

   $$a_m \frac{d^m x}{dt^m} + a_{m-1} \frac{d^{m-1} x}{dt^{m-1}} + \ldots + a_1 \frac{dx}{dt} + a_0 x = b_n \frac{d^n y}{dt^n} + b_{n-1} \frac{d^{n-1} y}{dt^{n-1}} + \ldots + b_1 \frac{dy}{dt} + b_0 y$$

   Answer: True/False

4. From the above LTI system, if $n = m$ or if highest power of S in input system in the denominator of the transfer function is equal to n, the system is called *nth-order system*.

   Answer: True/False

5. If the transfer function of a system is known, the output or response can be studied for various forms of inputs to understand the nature of the system.

   Answer: True/False

6. The impulse response function (IRF) can be denoted by $g(t)$, and this $g(t)$ is the response of a linear system to unit-impulse input when the initial conditions are zero.

   Answer: True/False

7. The Laplace transform of impulse response function gives transfer function such that, The Laplace of g(t) will give us the transfer function, G(s), whereas the inverse Laplace of transfer function G(s) provides impulse response function (IRF), g(t).

   Answer: True/False

8. The system response of $\frac{dy(t)}{dt} + 2y(t) = x(t)$ is $y(t) = \frac{1}{2} - \frac{1}{2} e^{-t}$.

   Answer: True/False

9. The transfer function of the viscous damper is $f_v s$, and the transfer function of mass is $Ms^2 + 1$

   Answer: True/False

10. Find the motion of the equation at the moment of inertia at J_2 shown in Figure 2.47.

1. $\left(J_1s^2+D_1s+K\right)\theta_1(s)-K\theta_2(s)-0\theta_3(s)=T(s)$
2. $\left(J_2s^2+D_2s+K\right)\theta_2(s)-K\theta_1(s)-D_2s\theta_3(s)=0$
3. $\left(J_3s^2+D_3s+D_2s\right)\theta_3(s)-0\theta_1(s)-D_2\theta_2(s)=0$
4. All of them

11. Choose the correct impedance and admittance of capacitor, resistor, and inductor that are denoted by

1. Impedance: $Z(s)=\frac{1}{Cs}$, $Z(s)=R$, $Z(s)=Ls$; Admittance: $Y(s)=Cs$, $Y(s)=\frac{1}{R}=G$, $Y(s)=\frac{1}{Ls}$
2. Admittance: $Z(s)=\frac{1}{Cs}$, $Z(s)=R$, $Z(s)=Ls$; Impedance: $Y(s)=Cs$, $Y(s)=\frac{1}{R}=G$, $Y(s)=\frac{1}{Ls}$

12. Find the transfer function of the following electrical RLC network shown in Figure 2.48.

1. $\frac{U_o(s)}{U_i(s)}=\frac{1}{LCs^2+RCs+1}$
2. $\frac{U_o(s)}{U_i(s)}=\frac{1}{LCs^2+RCs+LRC+1}$

13. Electric, a hydraulic, or a pneumatic motor or valve are examples of the actuator.
    Answer: True/False

14. An actuator is a type of AC motor that is responsible for moving or controlling a mechanism or system.
    Answer: True/False

15. Actuator produces input to the plant so that the output signal will approach the reference input signal.
    Answer: True/False

16. Sensor can be used to compare the output to the reference input signal.
    Answer: True/False

17. Automatic control systems can be represented by block/blocks or a set of block diagrams which is contained the input, output, and transfer function. However, a block diagram of a system is a pictorial representation of the functions performed by each component and of the flow of signals.
    Answer: True/False

18. Summing point is a junction or symbol in a block diagram of a system; a signal can be added or subtracted or both added and subtracted in the summing point or junction;

quantities of the signal being added or subtracted have the same dimension and the same units.

Answer: True/False

19. Any linear control system may be represented by a block diagram consisting of blocks, summing points, and branch points. Remember that the feedback connection has no block, so the feed transfer function value is unity here.

    Answer: True/False

20. After taking the Laplace transformation of $\frac{d^2x(t)}{dt^2}+\frac{dx(t)}{dt}+x(t)=y(t)$, we can get

    1. $s^2X(s)+sX(s)+X(s)=Y(s)$
    2. $s^2x(s)+sx(s)+x(s)=y(s)$

    Choose the right answer.

CHAPTER 3

Automatic Control Systems, Block Diagrams, and Signal Flow Graphs

3.1 AUTOMATIC CONTROL SYSTEMS

An automatic controller compares the actual value of the plant output with the reference input (which is called the set point or desired value) and determines the deviation or error detection by using the comparator. The feedback and reference input signals can be compared to determine the deviation that provides error detection by actuating the error signal. However, the amplifier is the main element of the automatic controller that is connected with the actuator to perform the process and the output of the plant, which can be connected with a sensor to make sense of the automatic system for the expected output (Figure 3.1).

It is needed to mention that the sensor may be a displacement measurement sensor, pressure measurement sensor, voltage measurement sensor, etc. The automatic controller produces a control signal called control action that will reduce the deviation to zero or a smaller value. Now, there is one question. What is an actuator in the system?

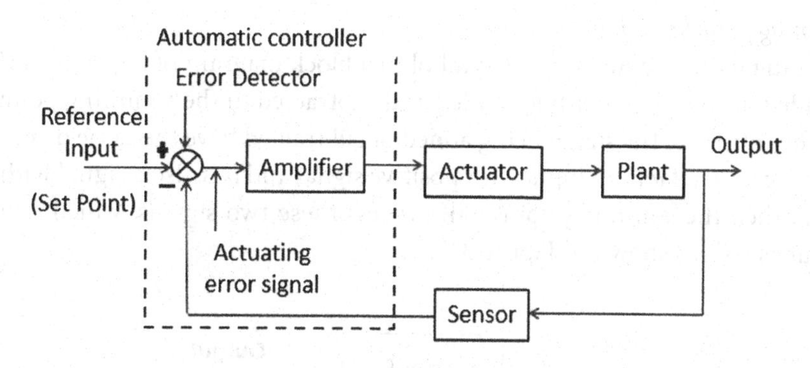

FIGURE 3.1 Automatic control system.

DOI: 10.1201/9781003293859-3

An **actuator** is a type of motor that is responsible for moving or controlling a mechanism or system in the plant. Once the actuator gets the control action signal, then it works. Examples of actuators are electric, a hydraulic, or a pneumatic motor or valve.

3.1.1 Functions of Automatic Controllers

Some of the functions of automatic controllers are detecting the actuating error signal at a very low power level and amplifying it to a sufficiently high level. The output of the automatic controller is fed to an actuator. Actually, an actuator is a power device that produces the input to the plant according to the control action signal so that the output signal will approach the reference input signal. The sensor converts the output variable into another suitable variable such as displacement measurement sensor, pressure measurement sensor, and voltage measurement sensor. The sensor can be used to compare the output to the reference input signal. The set point is converted to a reference input as the feedback signal from the sensor or measuring element.

3.1.2 Automatic Control Systems Representation

How can we represent automatic control systems? Automatic control systems can be represented combined with

1. Block diagrams;
2. Summing point/junction;
3. Branch point; and
4. Pickoff point/junction.

1. *Block diagrams*:
   Automatic control systems can be represented by block/blocks or a set of block diagrams which contain the input, output, transfer function, etc. However, a block diagram of a system is a pictorial representation of the functions performed by each component and of the flow of signals, as shown in Figure 3.2.

2. *Summing point/junction*:
   Summing point is a junction or symbol in a block diagram of a system; a signal can be added or subtracted or both added and subtracted in the summing point or junction; quantities of the signal being added or subtracted have the same dimension and the same units; if a is an input with positive signal and b is also a signal with negative signal, then the summing point will process these two signals, which will give the output as $a–b$, as shown in Figure 3.3a.

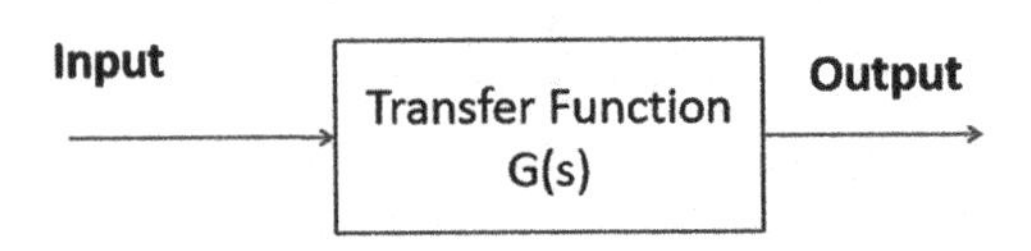

FIGURE 3.2 Block diagram of transfer function.

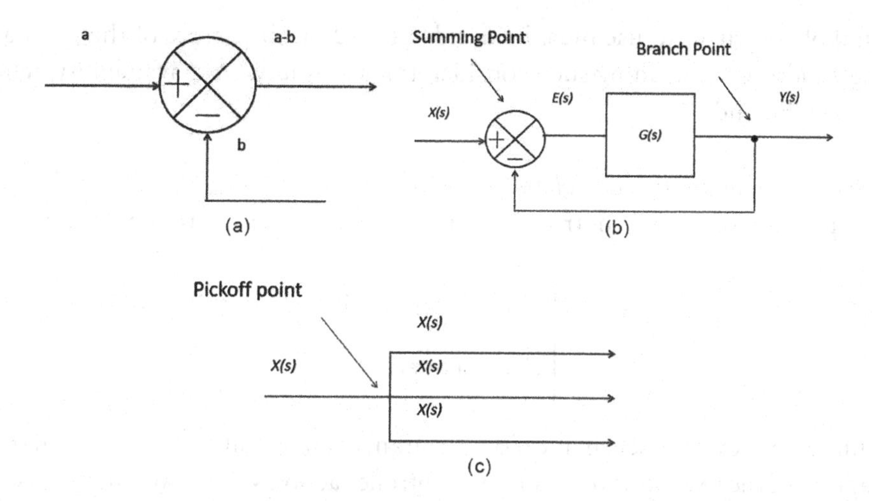

FIGURE 3.3 (a) Summing point/junction, (b) branch point, and (c) pickoff point/junction.

3. *Branch point*:
   A branch point is a point from which the signal goes concurrently to the other blocks or summing points from a block. The branch point must hold a connection point to pass the signal. The summing point and the branch point are not the same point and same processing function, as shown in Figure 3.3b.
4. *Pickoff point/junction*:
   A pickoff point distributes the input signal to the output points. The pickoff point must not be undiminished, as shown in Figure 3.3c.

3.1.3 Classifications of Automatic or Industrial Controllers

There are several kinds of industrial controllers which can be classified as

1. Two-position or on–off controllers;
2. Proportional controllers;
3. Integral controllers;
4. Proportional-plus-integral controllers, that is called PI controllers;
5. Proportional-plus-derivative controllers, that is called PD controllers; and
6. Proportional-plus-integral-plus-derivative controllers, that is called PID controllers;

Most industrial controllers use electricity or pressurized fluid such as oil or air as power sources. Therefore, controllers may also be classified according to the kind of power employed in the operation, such as

1. Pneumatic controllers;
2. Hydraulic controllers; or
3. Electronic controllers.

What kind of controller to use must be decided based on the nature of the plant and the operating conditions, including such considerations as safety, cost, availability, reliability, accuracy, weight, and size.

1. *Two-position or on–off controllers*:
   Two-position or on–off control action is denoted by $u(t)$. Therefore, we can say

$$u(t)=\begin{cases} U_1, & \text{For } e(t)>0 \\ U_2, & \text{For } e(t)<0 \end{cases} \tag{3.1}$$

   Figure 3.4a shows the schematic block diagram of an on–off controller. From the block diagram of the two-position or on–off controller action switch, we can see that e is the actuating error signal that can be said as $e(t)$. For $e(t)>0$, the switch goes to the on or upper position that is indicated by U_1 state in the controller part, and the switch goes to the off or down position that is indicated by U_2 state in the controller part.

   In a two-position control system, the actuating element has only two fixed positions, which are, in many cases, simply on and off. Two-position or on–off control is relatively simple and inexpensive and, for this reason, is very widely used in both industrial and domestic control systems. Let the output signal from the controller be $u(t)$ and the actuating error signal be $e(t)$. In two-position control, the signal $u(t)$ remains at either a maximum or minimum value, depending on whether the actuating error signal is positive or negative.

   Please note that U_1 and U_2 are constants. The minimum value U_2 is usually either zero or $-U_1$. Two-position controllers are generally electrical devices, and an electric solenoid-operated valve is widely used in such controllers. Pneumatic proportional controllers with very high gains act as two-position controllers and are sometimes called pneumatic two-position controllers.

   From Figure 3.4b, the range through which the actuating error signal must move before the switching occurs is called the *differential gap*. A differential gap is shown in Figure 3.4b. Such a differential gap causes the controller output $u(t)$ to keep its present value until the actuating error signal has moved slightly beyond the zero value. In some cases, the differential gap is a result of unintentional friction and lost motion;

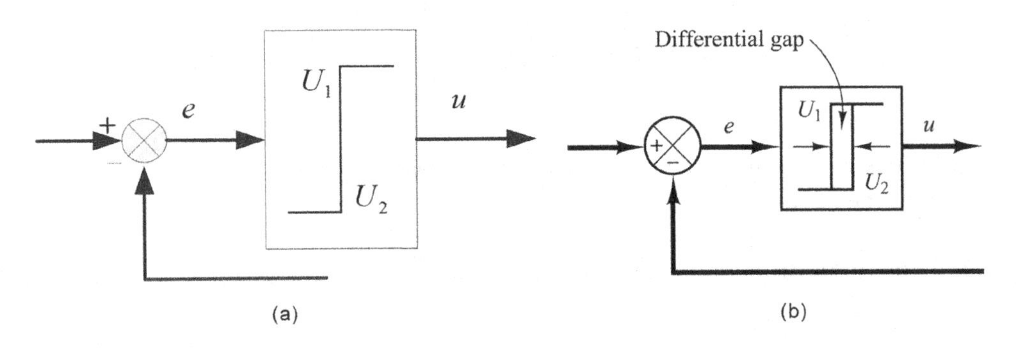

FIGURE 3.4 Schematic diagram of an on–off the controller.

however, quite often it is intentionally provided in order to prevent too-frequent operation of the on–off mechanism.

Consider the liquid-level control system shown in Figure 3.5a, where the electromagnetic valve shown in Figure 3.5b is used for controlling the inflow rate. This valve is either open or closed. With this two-position control, the water inflow rate is either a positive constant or zero. As shown in Figure 3.5c, the output signal continuously moves between the two limits required to cause the actuating element to move from one fixed position to the other. Note that the output curve follows one of two exponential curves, one corresponding to the filling curve and the other to the emptying curve. Such output oscillation between two limits is a typical response characteristic of a system under two-position control.

2. *Proportional controllers*:
   For a controller with proportional control action, the relationship between the output of the controller $u(t)$ and the actuating error signal $e(t)$ is

$$u(t)=K_p e(t)$$

or in Laplace-transformed quantities,

$$\frac{U(s)}{E(s)}=K_p$$

where K_p is the proportional gain. Whatever the actual mechanism may be and whatever the form of the operating power, the proportional controller is essentially an amplifier with an adjustable gain.

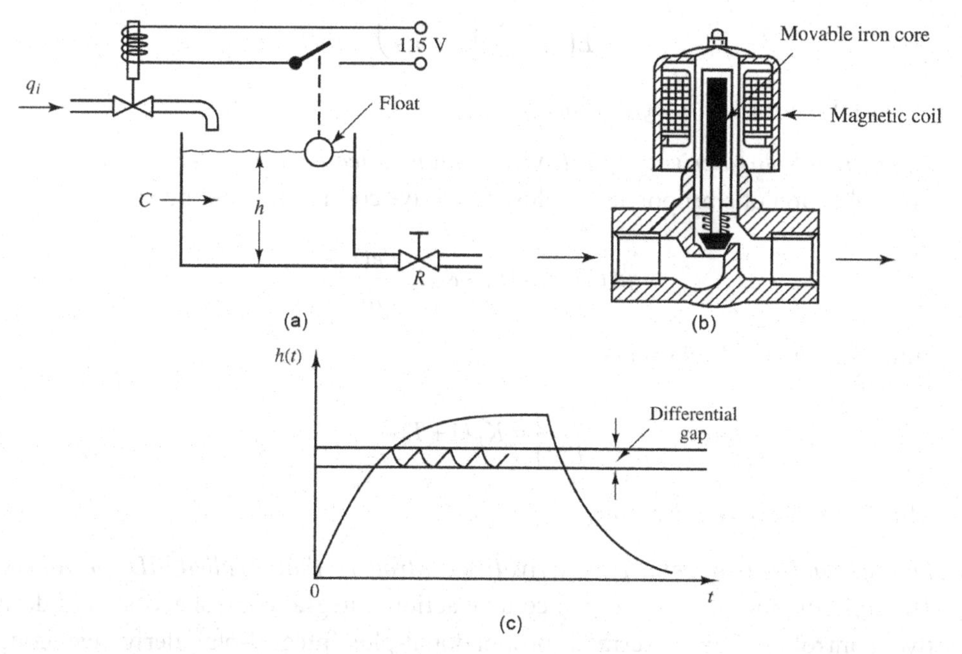

FIGURE 3.5 (a) Liquid-level control system, (b) electromagnetic valve, and (c) level $h(t)$-versus-t curve for the system.

3. *Integral controllers*:
   In a controller with integral control action, the value of the controller output *u(t)* is changed at a rate proportional to the actuating error signal *e(t)*. That is,

$$\frac{du(t)}{dt} = K_i e(t)$$

   Or, we can say

$$u(t) = K_i \int_0^t e(t)dt$$

   where K_i is an adjustable constant. The transfer function of the integral controller is expressed as

$$\frac{U(s)}{E(s)} = \frac{K_i}{s}$$

4. *Proportional-plus-integral controllers, that is called PI controllers*:
   The control action of a proportional-plus-integral controller is defined by

$$u(t) = K_p e(t) + \frac{K_p}{T_i} \int_0^t e(t)dt$$

   or the transfer function of the controller is expressed as

$$\frac{U(s)}{E(s)} = K_p \left(1 + \frac{1}{T_i s}\right)$$

   where T_i is called the *integral time.*

5. *Proportional-plus-derivative controllers, that is called PD controllers*:
   Control action of a proportional-plus-derivative controller is defined by

$$u(t) = K_p e(t) + K_p T_d \frac{de(t)}{dt}$$

   Thus, the transfer function is

$$\frac{U(s)}{E(s)} = K_p (1 + T_d s)$$

   where T_d is the *derivative time.*

6. *Proportional-plus-integral-plus-derivative controllers, that is called PID controllers*:
   The combination of proportional control action, integral control action, and derivative control action is termed proportional-plus-integral-plus-derivative control action. It has the advantages of each of the three individual control actions. The equation of a controller with this combined action is given by

TABLE 3.1 Transfer Function of Industrial Controllers

Components	Differential Equation	Transfer Function $\frac{U(s)}{E(s)}$
Proportional	$u(t)=K_pe(t)$	K_p
Proportional+integral	$u(t)=K_pe(t)+\frac{K_p}{T_i}\int e(t)dt$	$K_p\left(1+\frac{1}{T_is}\right)$
Proportional+derivative	$u(t)=K_pe(t)+K_pT_d\frac{de(t)}{dt}$	$K_p(1+T_ds)$
Proportional+integral+derivative	$u(t)=K_pe(t)+\frac{K_p}{T_i}\int_0^t e(t)dt+K_pT_d\frac{de(t)}{dt}$	$K_p\left(1+\frac{1}{T_is}+T_ds\right)$

$$u(t)=K_pe(t)+\frac{K_p}{T_i}\int_0^t e(t)dt+K_pT_d\frac{de(t)}{dt}$$

Or, the transfer function is expressed as

$$\frac{U(s)}{E(s)}=K_p\left(1+\frac{1}{T_i}+T_ds\right)$$

where K_p is the proportional gain, T_i is the integral time, and T_d is the derivative time. The block diagram of a proportional-plus-integral-plus-derivative controller is shown in Figure 2.10.

3.1.4 Transfer Function of Automatic or Industrial Controllers

In Table 3.1, we can see the component of the industrial controllers in the first column, differential equation of the controller is shown in the second column, and the transfer function of the industrial controllers are shown in the third column.

3.2 BLOCK DIAGRAM OF CONTROL SYSTEM AND ITS SIMPLIFICATION

In the control system, a block diagram is a pictorial diagram of mathematical representation that describes the relationship of signals between the components of a system. A block diagram is a structural diagram that represents the causal relationship between the input and output variables of the system. However, it represents the operation of the variables in the system. It is a very simple method to describe a complex system in a control system.

3.2.1 Components of the Block Diagram

The block diagram of the control system consists of multiple sets of signals, but it consists of four basic units as follows:

1. *Directed signal line*: This is a straight line with an arrow indicating the direction in which the signal is being transmitted. The signal is a time function or an image function on the line marks, as shown in Figure 3.6a.

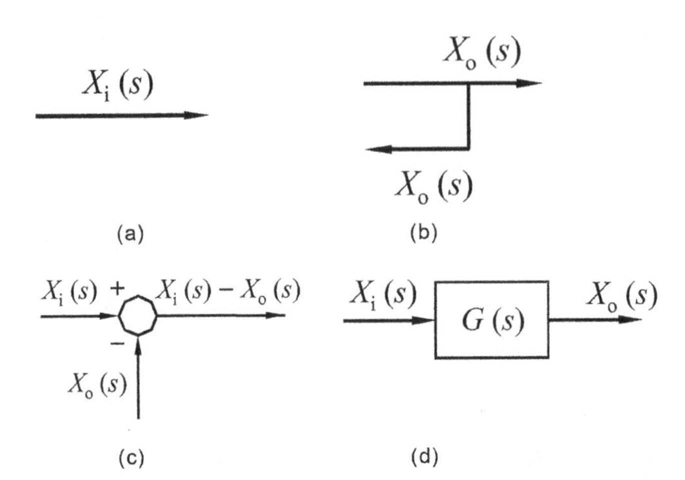

FIGURE 3.6 Four basic units of block diagram.

2. *Outlet point (measuring point)*: Measuring point represents the position where the signal is extracted or measured. The signal characteristics derived from the same position are exactly the same, as shown in Figure 3.6b.

3. *Comparing point (summing point)*: Summing point measures the addition or subtraction of two or more signals; it is also a point where we can compare the signal in this point or junction. "+" means the signals addition together. Sometimes "+" can be negligible. "−" means signals subtraction together, as shown in Figure 3.6c.

4. *Block (link)*: Block represents the mathematical transformation of signal. The transfer function of components or system is written in the square box, as shown in Figure 3.6d. Obviously, the output variable of the square box is equal to the product of the input variable and the transfer function, i.e.

$$X_o(s) = G(s)X_i(s) \tag{3.2}$$

3.2.2 Equivalent Transformation of the Block Diagram

1. *Equivalent transformation of cascaded (series) block diagram*:
   The *cascaded* or series block diagram is shown in Figure 3.7b, which can be equivalently converted to the product of the transfer function of each cascaded (series) block (link), i.e.

$$G(s) = G_1(s)G_2(s)\ldots G_{n-1}(s)G_n(s) \tag{3.3}$$

2. *Equivalent transformation of the parallel system*:
   Figure 3.8a shows a parallel block diagram, and its equivalent block diagram is shown in Figure 3.8b. The total transfer function of the parallel block diagram is equal to the sum of the transfer functions of each block diagram.

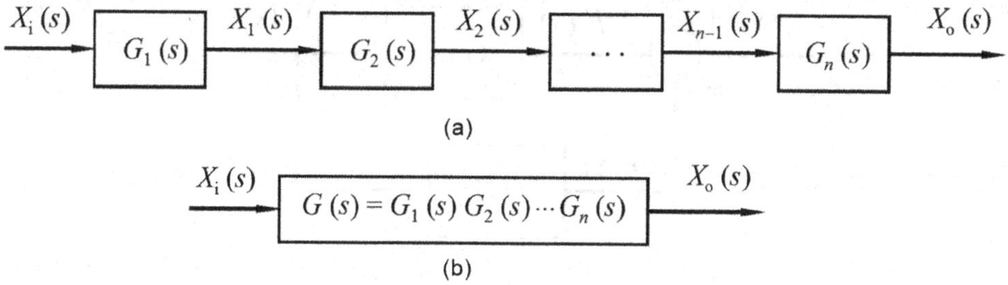

FIGURE 3.7 Block diagram of cascaded system.

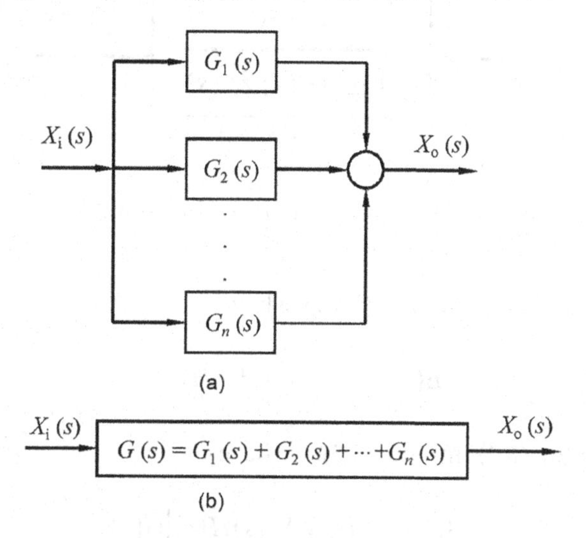

FIGURE 3.8 Block diagram of parallel system.

The transfer function of the parallel block diagram is defined as

$$G(s)=G_1(s)+G_2(s)+\ldots+G_{n-1}(s)+G_n(s) \tag{3.4}$$

3. *Equivalent block diagram of feedback system*:
Figure 3.9a shows a feedback block diagram system, and its equivalent block diagram is shown in Figure 3.9b. The total transfer function of the feedback block diagram system is equal to the sum of the transfer functions of each block diagram.
The relation between input and output is expressed as

$$E(s)=X_i(s)\mp B(s) \tag{3.5}$$

and

$$X_o(s)=E(s)G(s) \tag{3.6}$$

On the other hand,

$$B(s)=X_o(s)H(s) \tag{3.7}$$

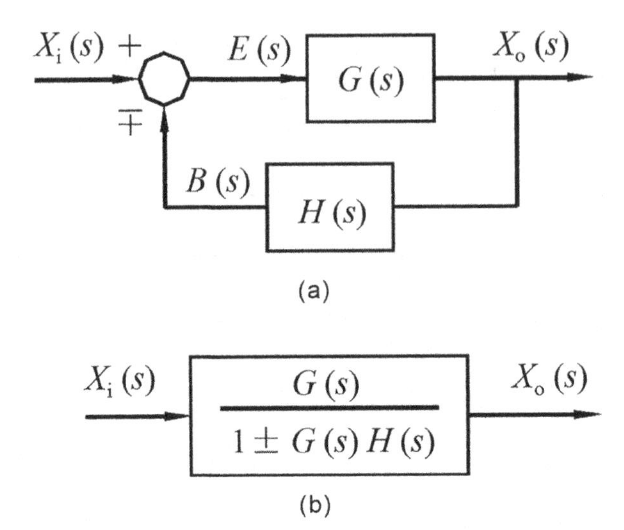

FIGURE 3.9 Block diagram of feedback connection/closed-loop system.

By substituting Eq. (3.6) into Eq. (3.4), we have

$$E(s) = X_i(s) \mp X_o(s)H(s) \tag{3.8}$$

By substituting Eq. (3.7) into Eq. (3.5), we have

$$X_o(s) = \left[X_i(s) \mp X_o(s)H(s)\right]G(s) \tag{3.9}$$

Simplifying Eq. (3.8), we have

$$\frac{X_o(s)}{X_i(s)} = \frac{G(s)}{1 \pm G(s)H(s)} \tag{3.10}$$

In Eq. (3.9), "+" means negative feedback system and "–" means positive feedback system. They are contrasted to the "+" and "–" in Eq. (3.4). In Eq. (3.4), "–" means negative feedback system, and "+" means positive feedback system. However, $G(s)$ means feed forward transfer function, and $G(s)H(s)$ means the open-loop transfer function of the closed-loop system.

3.2.3 Rules of Equivalence for Block Diagrams

1. The transfer function is the product of each forward channel or link remains unchanged and
2. The product of transfer functions in each loop remains unchanged.

The transfer rules of common blocks are shown in Table 3.2.

TABLE 3.2 Transfer Rules of Block Diagrams

No.	Original Block Diagrams	Equivalent Block Diagrams	Comments
1	A, G_1, AG_1, G_2, AG_1G_2	A, G_1G_2, AG_1G_2	Cascaded equivalence
2	A, G_1, AG_1, G_2, AG_1G_2	A, G_2, AG_2, G_1, AG_1G_2	Multiplication exchange
3	A, G_1, G_2, $\pm$, $AG_1 \pm AG_2$	A, $G_1 \pm G_2$, $AG_1 \pm AG_2$	Parallel equivalence
4	A, G_1, G_2, $\pm$, $AG_1 \pm AG_2$	A, G_2, $\frac{1}{G_2}$, G_1, $\pm$, $AG_1 \pm AG_2$	Equivalently converted to unit paralleling
5	A, $\pm$, G_1, G_2, B	A, $\frac{G_1}{1 \mp G_1G_2}$, B	Feedback equivalence
6	A, $-$, G_1, G_2, B	A, $\frac{1}{G_2}$, $-$, G_2, G_1, B	Equivalently converted to unit feedback
7	A, $\pm$, P, $A \pm P$, $A \pm P$	P, $\pm$, $A \pm P$, A, $\pm$, P, $A \pm P$	Exchange the comparing point and the exit point
8	A, $\pm$, P, $A \pm P$, $\pm$, Q, $A \pm P \pm Q$	A, $\pm$, Q, $A \pm Q$, $\pm$, P, $A \pm P \pm Q$	Exchange comparing points
9	A, $\pm$, Q, $A \pm Q$, P, $\pm$, $A \pm P \pm Q$	P, $\pm$, A, $\pm$, Q, $A \pm P \pm Q$	Combine the comparing points
10	P, $\pm$, A, $\pm$, Q, $A \pm P \pm Q$	A, $\pm$, Q, $A \pm Q$, P, $\pm$, $A \pm P \pm Q$	Desperate comparing points
11	A, G, $\pm$, P, $AG \pm P$	A, $\pm$, G, $AG \pm P$, $\frac{1}{G}$, P	Move the comparing points forward
12	A, $\pm$, P, G, $AG \pm PG$	A, G, $\pm$, $AG \pm PG$, P, G	Move the comparing points backward
13	A, G, AG, A	A, G, AG, $\frac{1}{G}$, A	Move the exit points backward
14	A, G, AG, AG	A, G, AG, G, AG	Move the exit points forward

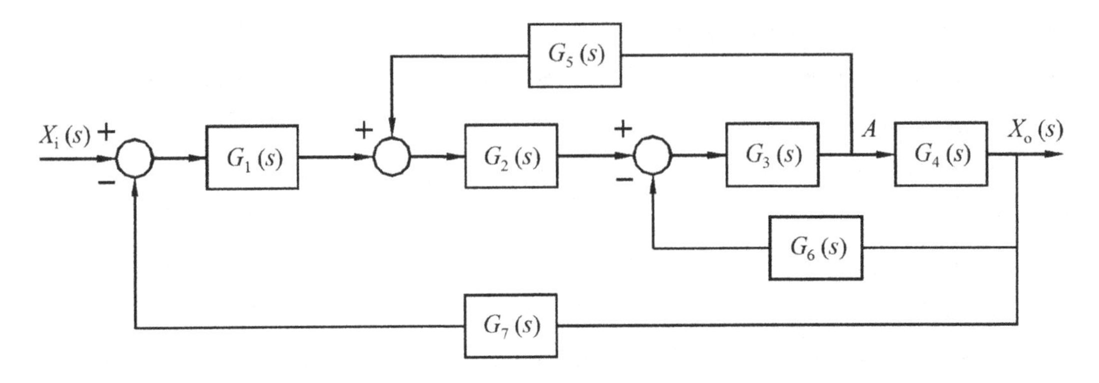

FIGURE 3.10 Block diagram of a negative feedback system.

Example 3.1

Simplify the block diagram shown in Figure 3.10 and obtain the transfer function.

Solution

1. First, we move point A to the front of the block $G_4(s)$, and then we can get Figure 3.11a.
2. Eliminating $G_3(s), G_4(s)$, and $G_6(s)$ loops, the result is shown in Figure 3.11b.
3. Eliminating $G_2(s)$, $\frac{G_3(s)G_4(s)}{1+G_3(s)G_4(s)G_6(s)}$, $\frac{1}{G_4(s)}$, and $G_5(s)$ loops, the result is shown in Figure 3.11c.
4. Eliminating all loops, we get the result as shown in Figure 3.11d.

   Therefore, the transfer function is

$$G(s)=\frac{X_o(s)}{X_i(s)}=\frac{G_1(s)G_2(s)G_3(s)G_4(s)}{1+G_2(s)G_3(s)G_5(s)+G_3(s)G_4(s)G_6(s)+G_1(s)G_2(s)G_3(s)G_4(s)G_7(s)}$$

3.3 DIFFERENT KINDS OF BLOCK DIAGRAM

In this section, we will focus on different kinds of block diagrams, such as

1. Closed-loop system;
2. Open-loop transfer function and feedforward transfer function;
3. Closed-loop transfer function;
4. Transfer function of nonloading cascaded elements; and
5. Parallel transfer function.

1. *Closed-loop system*:
   Block diagram of closed-loop system is shown in Figure 3.12. The output $Y(s)$ is fed back to the summing point, where it is compared with the reference input $X(s)$.

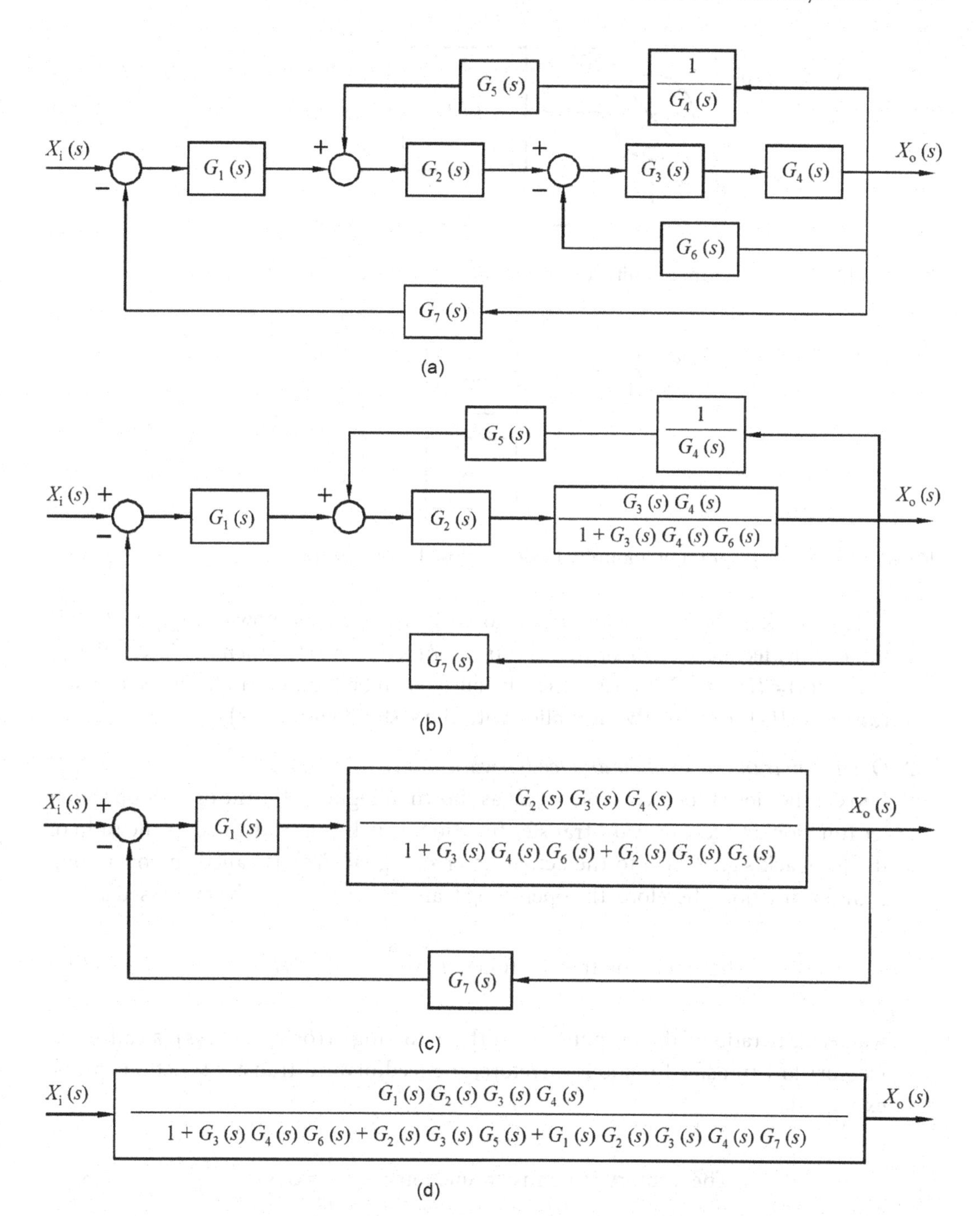

FIGURE 3.11 (a) Block diagram of a negative feedback system, (b)–(d) step by step simplified diagram.

The output of the block, $Y(s)$ in this case, is obtained by multiplying the transfer function $G(s)$ by the input to the block, $E(s)$. Any linear control system may be represented by a block diagram consisting of blocks, summing points, and branch points. Remember here, the feedback connection has no block; thus, the feed-transfer function value is unity in this closed-loop system.

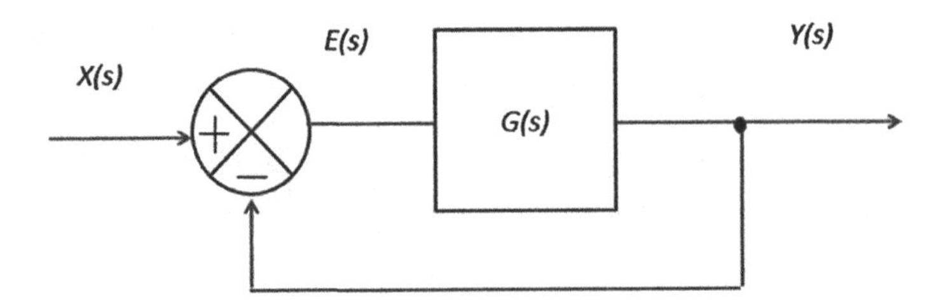

FIGURE 3.12 Block diagram of unity feedback closed-loop system.

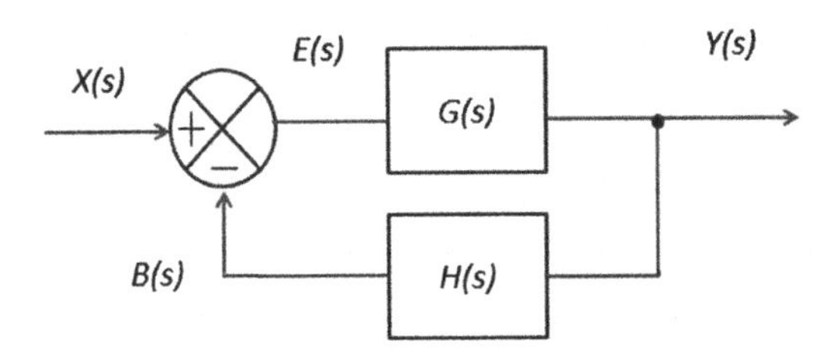

FIGURE 3.13 Block diagram of nonunity feedback closed-loop system.

Now, look at the block diagram of closed-loop system as shown in Figure 3.13. If there is any feedback element whose transfer function is $H(s)$, then the feedback signal is $B(s)=H(s)Y(s)$; it means that the output can be found from feedback transfer function $H(s)$ and directly multiplied with the system output $Y(s)$.

2. *Open-loop transfer function and feedforward transfer function*:
   Look at the closed-loop control system as shown in Figure 3.13. The open-loop transfer function and feedforward transfer function, which can be defined as the ratio of the feedback signal $B(s)$ to the actuating error signal $E(s)$, is called the open-loop transfer function; therefore, the open-loop transfer function can be expressed as

$$\text{The open-loop transfer function} = \frac{B(s)}{E(s)} = G(s)H(s) \tag{3.11}$$

   whereas the ratio of the output $Y(s)$ to the actuating error signal $E(s)$ is called the feedforward transfer function; therefore, the feedforward transfer function can be expressed as

$$\text{The feedforward transfer function} = \frac{Y(s)}{E(s)} = G(s) \tag{3.12}$$

3. *Closed-loop transfer function*:
   Here, if the feedback transfer function $H(s)$ is unity or $H(s)=1$ from Eqs. (3.11) and (3.12), then the ratio can be expressed as

$$\frac{B(s)}{E(s)} = \frac{Y(s)}{E(s)} \tag{3.13}$$

We can find the closed-loop transfer function in Figure 3.13. The system shown in Figure 3.13 shows the relations between output $Y(s)$ and input $X(s)$:

$$E(s) = X(s) - B(s) \tag{3.14}$$

Since we know that

$$B(s) = H(s)Y(s) \tag{3.15}$$

Therefore, by substituting Eq. (3.15) into Eq. (3.14), it can be rewritten as

$$E(s) = X(s) - H(s)Y(s) \tag{3.16}$$

The transfer function relating output $Y(s)$ to input $X(s)$, which is called the closed-loop transfer function, can be displayed as

$$Y(s) = G(s)E(s) \tag{3.17}$$

Now, we can substitute Eq. (3.16) into Eq. (3.17), and then we can rewrite the output $Y(s)$ as

$$Y(s) = G(s)E(s) = G(s)\left[X(s) - H(s)Y(s)\right] \tag{3.18}$$

After rearranging Eq. (3.18), we get the closed-loop transfer function as

$$\frac{Y(s)}{X(s)} = \frac{G(s)}{1 + G(s)H(s)} \tag{3.19}$$

However, we can transform input $X(s)$ on the right-hand side of Eq. (3.19), and thus we can say

$$Y(s) = \left[\frac{G(s)}{1 + G(s)H(s)}\right]X(s) \tag{3.20}$$

4. *Transfer function of nonloading cascaded elements*:
   The output $Y(s)$ of the closed-loop system clearly depends on both the feedforward transfer function $G(s)$ and the nature of the input $X(s)$. After simplifying the closed-loop block diagram, we can get the result as shown in Figure 3.14.
   Now, let's see the transfer function of nonloading cascaded elements. The transfer function of element 1 can be defined as

$$G_1(s) = \frac{X_2(s)}{X_1(s)} \tag{3.21}$$

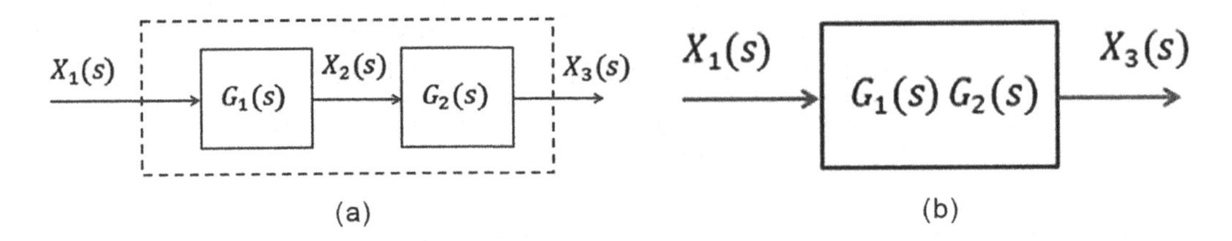

FIGURE 3.14 Nonloading cascaded elements.

Therefore,

$$X_2(s)=G_1(s)X_1(s) \tag{3.22}$$

whereas the transfer function of element 2 can be defined as

$$G_2(s)=\frac{X_3(s)}{X_2(s)} \tag{3.23}$$

Thus,

$$X_3(s)=G_2(s)X_2(s) \tag{3.24}$$

If the input impedance of the second element is infinite, the output of the first element is not affected by the connection to the second element. The whole cascaded form can be written as

$$G(s)=\frac{X_3(s)}{X_1(s)} \tag{3.25}$$

Here, if we multiply by $X_2(s)$ in the numerator and denominator, then we get

$$G(s)=\frac{X_2(s)X_3(s)}{X_1(s)X_2(s)} \tag{3.26}$$

Rearranging the above expression, we can say

$$G(s)=\left(\frac{X_2(s)}{X_1(s)}\right)\left(\frac{X_3(s)}{X_2(s)}\right) \tag{3.27}$$

Therefore,

$$G(s)=G_1(s)G_2(s) \tag{3.28}$$

5. *Transfer function of parallel connection*:
   From this figure, the transfer function of the parallel elements can be expressed as (Figure 3.15)

$$Y(s) = \pm G_1(s)X(s) \pm G_2(s)X(s) \quad (3.29)$$

Therefore, the simplified transfer function can be found here:

$$G(s) = \frac{Y(s)}{X(s)} = \pm G_1(s) \pm G_2(s) \quad (3.30)$$

Example 3.2

A closed-loop system subjected to a disturbance. Determine the output $C(s)$ to the two inputs $R(s)$ and $D(s)$ for the system shown in Figure 3.16.

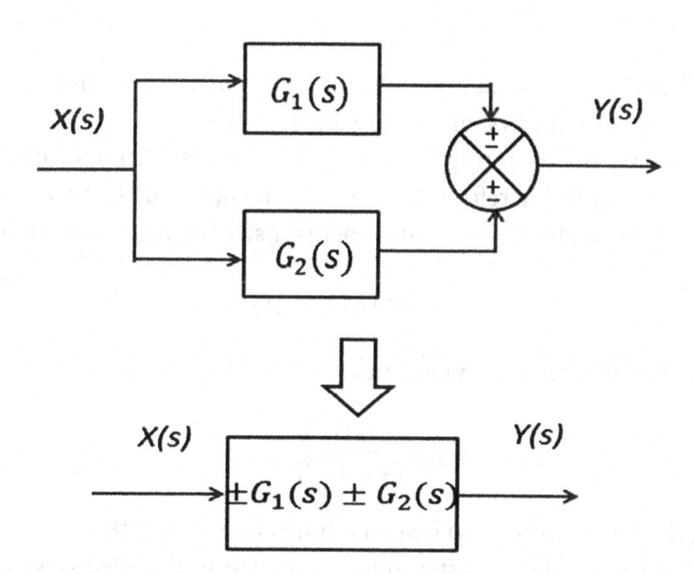

FIGURE 3.15 Parallel connection elements.

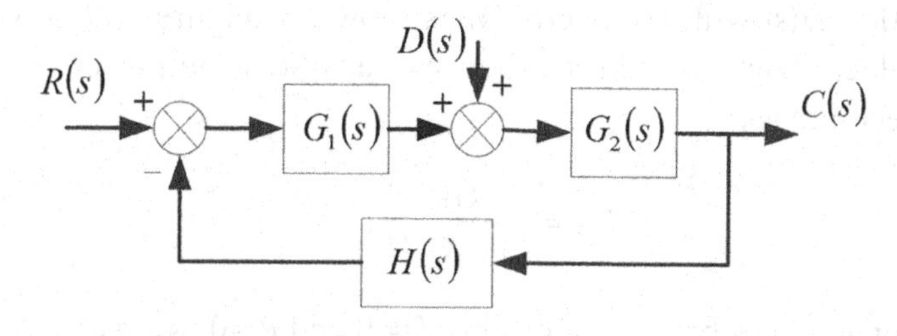

FIGURE 3.16 A closed-loop system with disturbance (D).

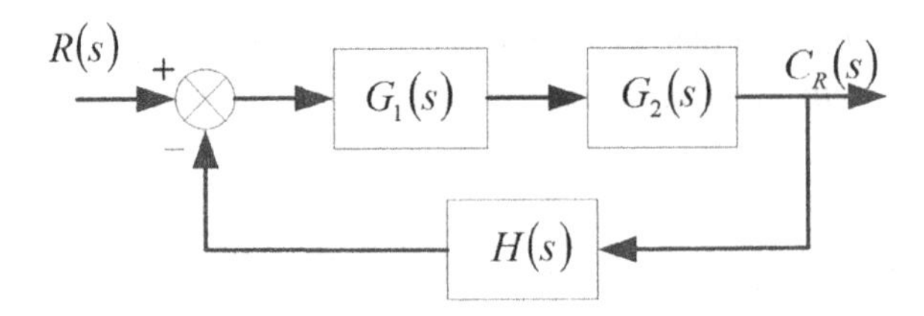

FIGURE 3.17 A closed-loop system without disturbance (D=0).

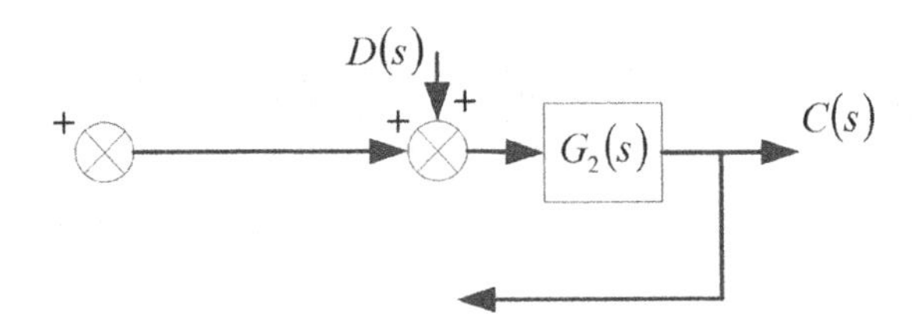

FIGURE 3.18 System diagram with R=0.

Solution

How can we find the solution of the problem? First, setting $D=0$ and rearranging the block diagram produce the system (Figure 3.17).

Since we set $D=0$, the summing point should be removed from the diagram, which will make it very simple. Therefore, we can see that $G_1(s)$ and $G_2(s)$ are the series connected so these can be multiplied in the inner transfer function and can be written as

$$G(s)=G_1(s)G_2(s)$$

The output of closed-loop system becomes

$$C_R=\frac{G_1G_2}{1+G_1G_2H}R(s)$$

Now, setting $R=0$, we can see two cases as shown in Figure 3.18.

Since $R=0$, the first inner transfer function $G_1(s)$ as well as feedback transfer function $H(s)$ are inactive to perform, but the inner transfer function $G_2(s)$ will work properly because $D(s)$ is working properly. Once we get output $C(s)$, the feedback transfer function $H(s)$ will work actively, as shown in Figure 3.19. Therefore, rearranging the block diagram then it produces the system output of the closed-loop system that becomes as,

$$C_D=\frac{G_2}{1+G_1G_2H}D(s)$$

Now summing up from the two conditions, $D=0$ and $R=0$, we get

$$C=C_R+C_D$$

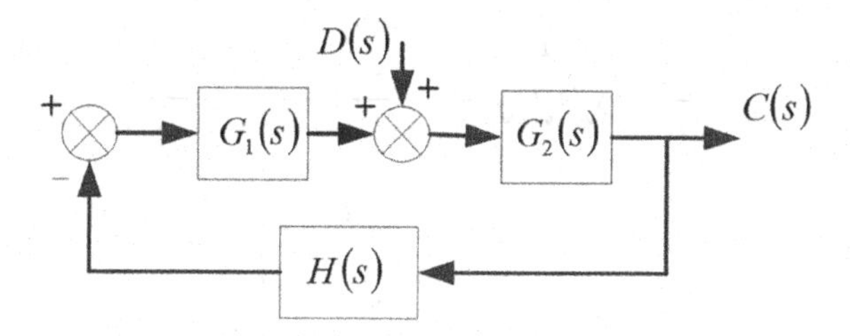

FIGURE 3.19 A closed-loop system with R=0.

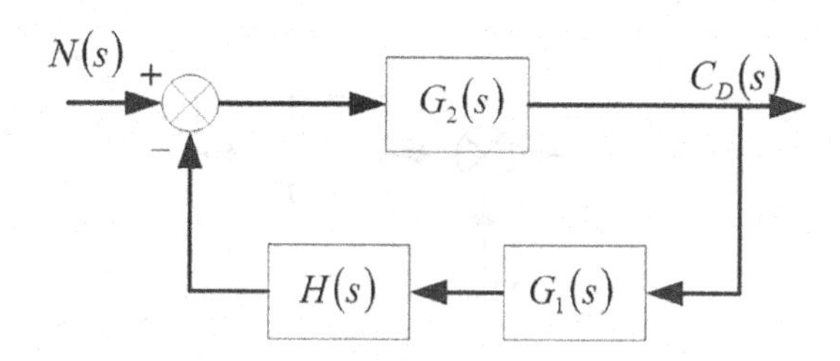

FIGURE 3.20 Simplified closed-loop system of Figure 3.16.

We have already found C_R and C_D, so if we can substitute C_R and C_D, then we get

$$C = C_R + C_D = \frac{G_1 G_2}{1 + G_1 G_2 H} R + \frac{G_2}{1 + G_1 G_2 H} D$$

However, if we want to draw the simplified system diagram from the above expression, we can redesign our system as shown in Figure 3.20, where *N(s)* is the input.

3.4 REDUCTION OF BLOCK DIAGRAMS

In this section, we will study the procedures for drawing block diagrams and their reduction. The following are procedures for drawing block diagrams:

- Write the equations that describe the dynamic behavior of each component;
- Assuming zero initial conditions, take the Laplace transforms of these equations;
- Represent each Laplace-transformed equation individually in a block form; and
- Assemble the elements into a completed block diagram.

Let's see an example of an RC circuit as shown in Figure 3.21.

Therefore, the equations for the RC circuit can be written as follows:

$$i = \frac{e_i - e_o}{R} \tag{3.31}$$

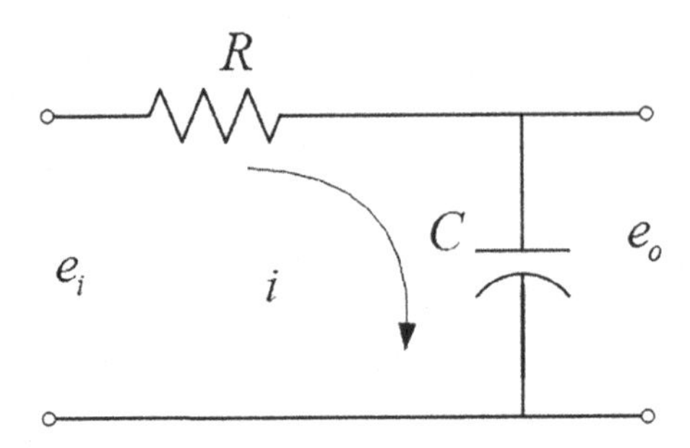

FIGURE 3.21 RC networks.

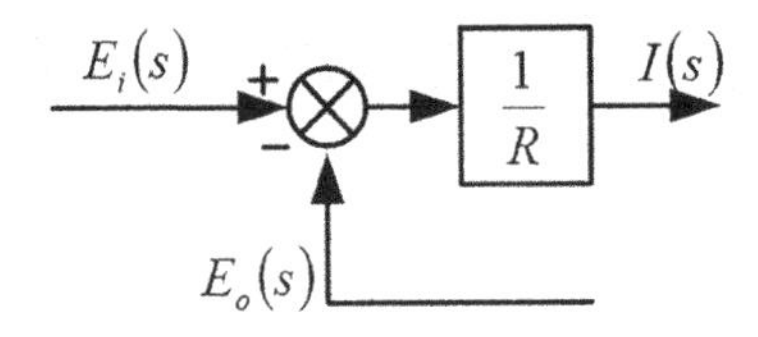

FIGURE 3.22 Block diagram of Eq. (3.33).

and

$$e_o = \frac{\int i dt}{C} \tag{3.32}$$

Now, applying the Laplace transformation on both sides of the equations with zero initial conditions, we can find the following equations:

$$I(s) = \frac{E_i(s) - E_o(s)}{R} \tag{3.33}$$

and

$$E_o(s) = \frac{1}{Cs} I(s) \tag{3.34}$$

From Eq. (3.33), we can see that $I(s)$ is the output current, $E_i(s)$ is the positive input/supplying source voltage, $E_o(s)$ is the negative feedback voltage which is actually comes from output branch, and $1/R$ is the system transfer/response signal of the system which is called $G(s)$. Thus, we can see the block diagram of Eq. (3.33) as shown in Figure 3.22.

Similarly, from Eq. (3.34), we can see $E_o(s)$ is the output voltage, $I(s)$ is the input current, and $1/C_s$ is the transfer/system response from the capacitor; thus, the block diagram of Eq. (3.34) system equation can be seen in Figure 3.23.

Now, we can combine these two block diagrams into one block diagram, and then we can see the block diagram of the above RC circuit, which is shown in Figure 3.24.

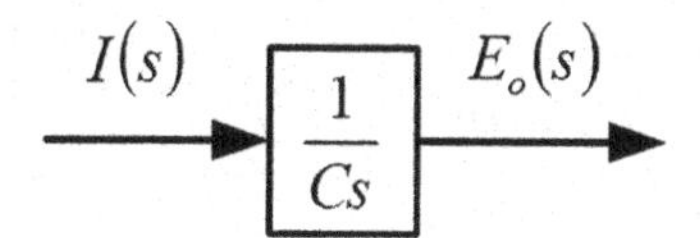

FIGURE 3.23 Block diagram of Eq. (3.34).

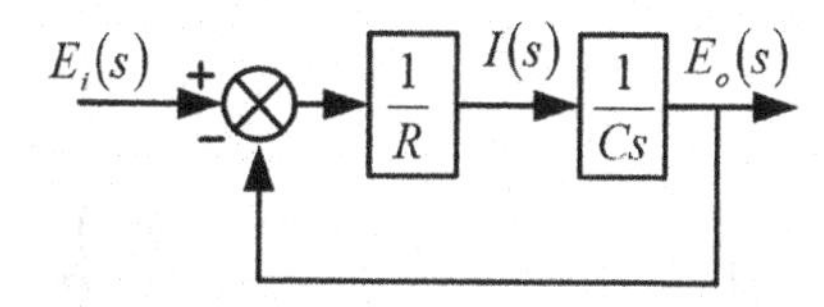

FIGURE 3.24 Block diagram of RC network.

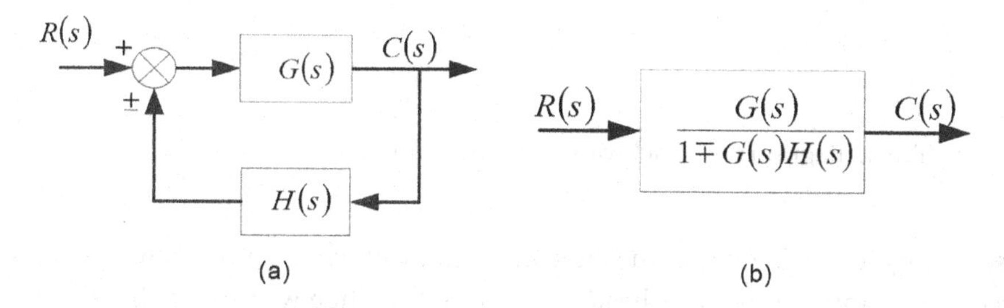

FIGURE 3.25 A closed-loop system.

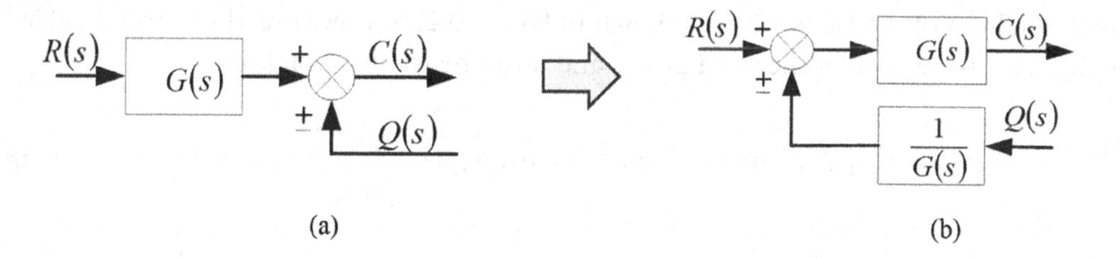

FIGURE 3.26 Simplified block diagram.

Look at Figure 3.24 carefully and you can see that there are two transfer functions in the block diagram as $G_R(s) = 1/R$ and $G_C(s) = 1/C_s$. These two transfer functions are connected with a series. Therefore, can you simplify these block diagram? And, how can we simplify the block diagram?

To simplify the block diagram or for reduction of block diagram, let's consider a closed-loop/feedback system (Figure 3.25). Here, $R(s)$ is the input, $C(s)$ is the output, $H(s)$ is the feed-transfer function, and $G(s)$ is the transfer function. Thus, this closed-loop system can be simplified as shown in Figure 3.26b, and the transfer function of the system can be written as

$$G_c(s) = \frac{G(s)}{1 \mp G(s)H(s)} \tag{3.35}$$

FIGURE 3.27 Ahead Bblock diagram.

FIGURE 3.28 Block diagram with backward summing junction.

With a summing junction on the left side and from the block diagram shown in Figure 3.27, we moved the summing junction ahead of a block $G(s)$. Then we can see that the input $R(s)$ and the output $C(s)$ as well as $Q(s)$ are not changed, but a reciprocal function of $G(s)$ can be seen as $1/G(s)$, which is a feedback transfer function due to movement of the summing junction ahead of the block $G(s)$, as shown in Figure 3.27b. Therefore, the output $C(s)$ will be the same from both of the two figures, and it can be written as follows:

$$C(s)=R(s)G(s)\pm Q(s)=R(s)G(s)\pm\frac{Q(s)}{G(s)}G(s) \tag{3.36}$$

However, let's see this block diagram with a branch point on the left side as shown in Figure 3.28, and from this block diagram, we moved the branch point ahead of a block $G(s)$. Then we can see that the input $R(s)$ and the output $C(s)$ as well as $Q(s)$ are not changed, but a transfer function as $G(s)$ can be seen due to moving the branch point ahead of the block $G(s)$, as we can see in the right figure. Therefore, the output $C(s)$ and $Q(s)$ will be the same from both of the two figures and it can be written as follows:

$$Q(s)=C(s)=R(s)G(s) \tag{3.37}$$

Now, let's see a block diagram with a summing junction on the left side of Figure 3.29. If moved the summing junction to the past of a block $G(s)$, then we can see that the input $R(s)$ and the output $C(s)$ as well as $Q(s)$ are not changed, but a transfer function of $G(s)$ can be seen as $G(s)$, which is the same as the transfer function $G(s)$ due to moving the summing junction to the past of the block $G(s)$ that we can see in Figure 3.29. Therefore, the output $C(s)$ will not change, and it can be written as follows:

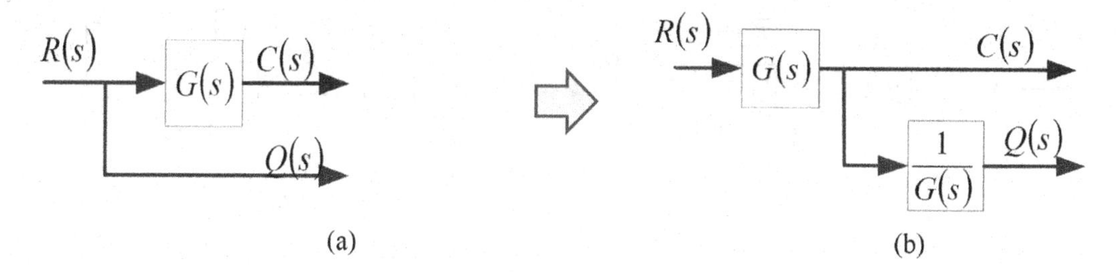

FIGURE 3.29 Block diagram with past connection of transfer function but no summing junction.

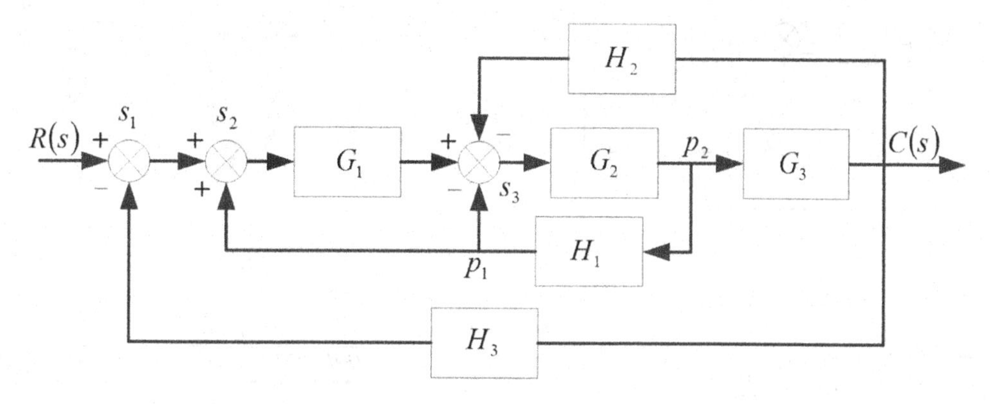

FIGURE 3.30 Multiple loop system block diagram.

$$C(s)=[R(s)\pm Q(s)]G(s)=R(s)G(s)\pm Q(s)G(s) \tag{3.38}$$

Now, let's see a block diagram with a branch point on the left side. From this block diagram, if we move the branch point to the past of a block $G(s)$, we can see that the input $R(s)$ and the output $C(s)$ as well as $Q(s)$ are not changed, but a reciprocal function of $G(s)$ can be seen as $1/G(s)$ due to moving the branch point to the past of the block $G(s)$ that we can see in the right figure. Therefore, the output $Q(s)$ and the input $R(s)$ will be the same, but $C(s)$ will be unchanged. It can be written as follows:

$$Q(s)=R(s)=R(s)G(s)\frac{1}{G(s)} \tag{3.39}$$

Example 3.3

Let's see this example, simplify the multiple-loop system block diagram, and determine the closed-loop transfer function as shown in Figure 3.30.

Solution

Let's move pickoff point P_2 to the right of block G_3; then we obtain the result as shown in Figure 3.31a.

Now, we can merge two summing junctions into one summing junction because these two summing junctions are cascaded, and if we simplify these as one summing

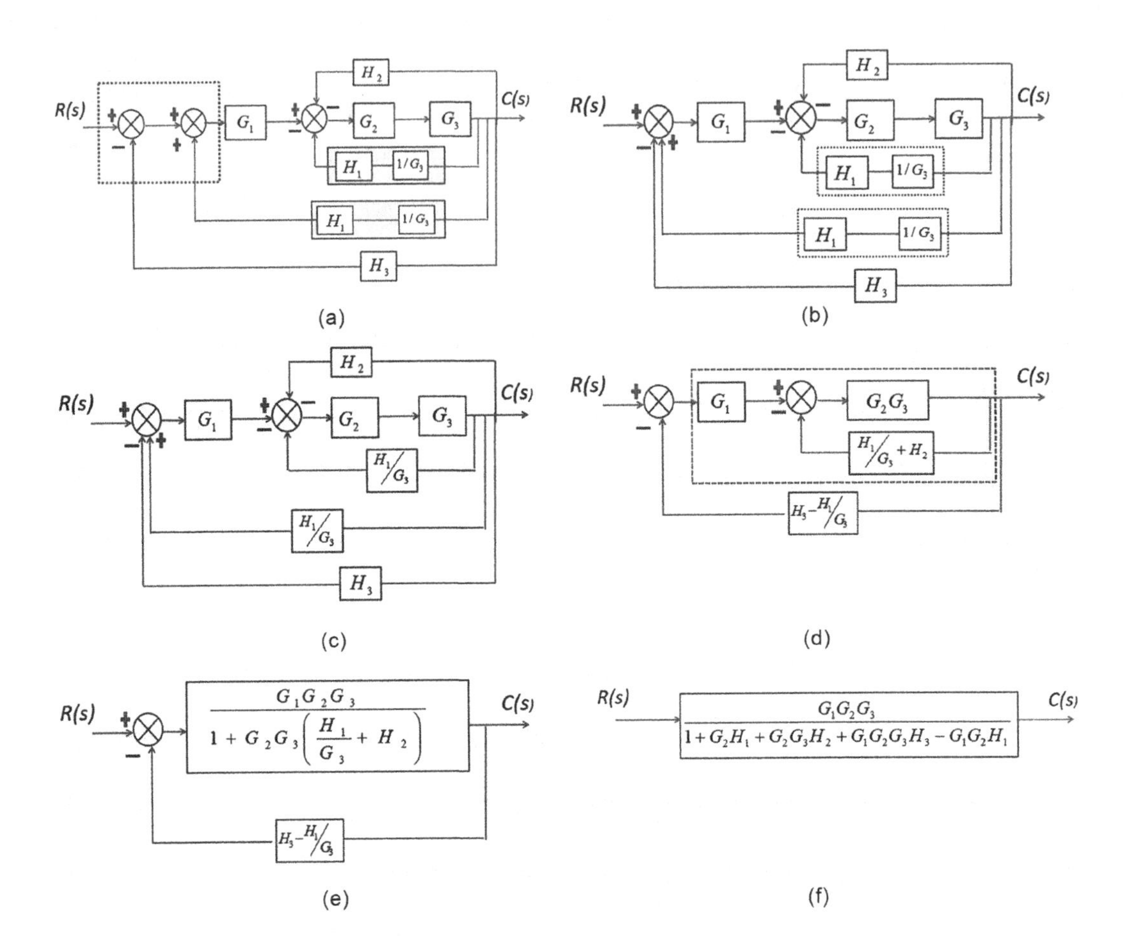

FIGURE 3.31 Simplified system of Figure 3.29.

junction, then the system will be simple, but the value will not be changed, as shown in Figure 3.31b. Here, we can see that there are two feedback systems, and the feedback transfer function of both H_1 and $1/G_3$ are series connected, which can be simplified as we can see in Figure 3.31c. However, we can see that there are two feedback systems that transfer function H_1/G_3 and H_3 are parallel connected which can be simplified as we can see in Figure 3.31d. Now, these block diagrams can be simplified, and we can get the result as shown in Figure 3.31e. Thus, finally, the simplified block diagram and transfer function of the given system can be seen as shown in Figure 3.31f.

3.5 SIGNAL FLOW GRAPH AND MASON'S RULES

In this section, we will study signal flow graphs, terms of signal flow graph, signal flow graph algebra, signal flow graph of linear systems, and signal flow graphs of control systems.

3.5.1 Signal Flow Graphs

The control system in engineering is often composed of multiple circuits that intersected with each other. In this case, it is very troublesome and error susceptible to simplify the block diagram of each other. If the block diagram is changed into the signal flow diagram, Mason's formula is used to calculate the transfer function, which is very convenient.

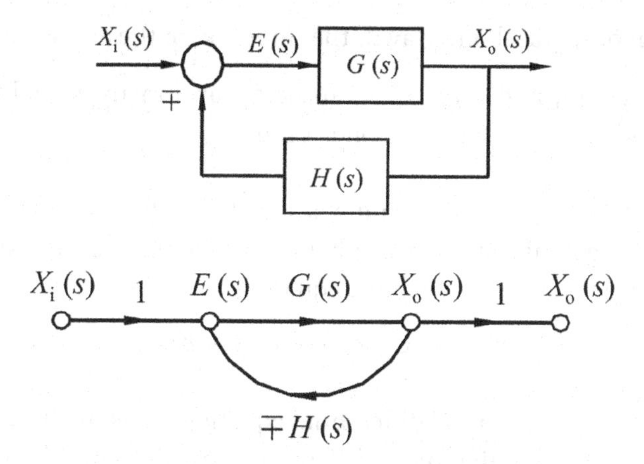

FIGURE 3.32 Simplified signal flow graph.

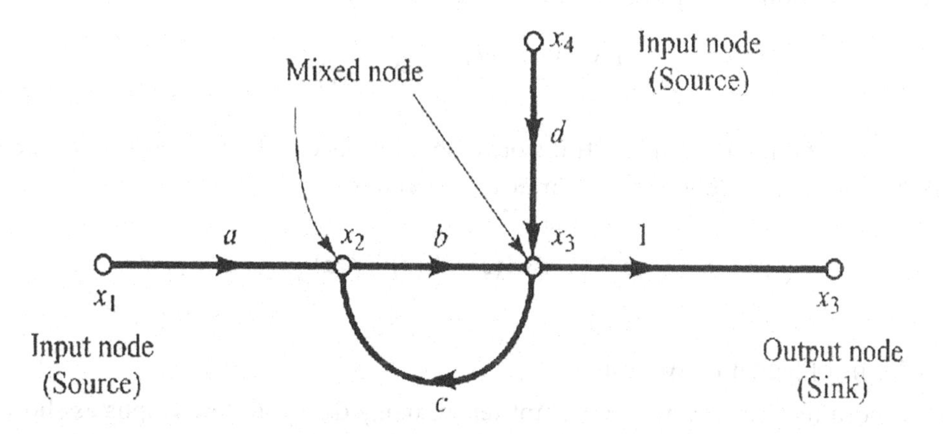

FIGURE 3.33 General signal flow graph.

First, look at the relationship between the block diagram as shown in Figure 3.9a and the corresponding signal flow diagram as shown in Figure 3.32.

The block diagram as shown in is Figure 3.9a transformed into the signal flow diagram in Figure 3.32; it can be seen that the signal flow diagram in the network is composed of some directional line segments connecting nodes with an arrow. The basic properties of a signal flow diagram are as follows:

1. Node(s) represent(s) variables or signals of the system. Usually, nodes are set from left to right. The signal of each node is the algebraic sum of all signals passing through the node; the signals that flow to each branch of the same node are represented by the signals of the node. Figure 3.33 shows a signal flow graph with nodes; here, $X_i(s)$, $E(s)$, $X_o(s)$ is indicated as node in the signal flow graph with circle "o";
2. The branch is equivalent to the multiplier. When the signal flows through the branch, the signal flowing into the branch is multiplied by the gain of the branch input flow which is equal to the signal flowing out of the branch. In Figure 3.32, we may say $X_o(s) = G(s)E(s)$;

3. The signal can only travel one-way in the direction of the arrow on the branch; and
4. In the same system, node variables can be set arbitrarily, signal flow graph is not unique, but the final transfer function is unique.

Therefore, we may define a signal flow graph as a pictorial diagram that represents a set of simultaneous linear algebraic equations. What is a linear algebraic equation?

Let's see an example of a linear algebraic equation

$$x_1 = a_{11}x_1 + a_{12}x_2 + a_{13}x_3 + b_1u_1 \tag{3.40}$$

In control systems, firstly, the linear differential equations must be transformed into algebraic equation in s. What are the linear differential equations and how to transform the linear differential equations into algebraic equation in s?

Let's see an example of linear differential equations as

$$\frac{d^2x(t)}{dt^2} + \frac{dx(t)}{dt} + x(t) = y(t) \tag{3.41}$$

However, after taking the Laplace transformation, in this equation on both sides then we can represent as in the following algebraic equation in s.

$$s^2X(s) + sX(s) + X(s) = Y(s) \tag{3.42}$$

3.5.2 Terms of Signal Flow Graph

Now we need to discuss some important terminology of signal flow graphs as shown in Figure 3.33, such as

1. Node;
2. Transmittance;
3. Branch;
4. Input node or source;
5. Output node or sink
6. Mixed node;
7. Loop;
8. Loop gain;
9. Path;
10. Nontouching loops;
11. Forward path; and
12. Forward path gain.

Description the Terms of Signal Flow Graph:

1. *Node*: A node is a point representing a variable or signal.

2. *Transmittance*: Transmittance may be a real gain or complex gain, which can be expressed in the transfer function between two nodes.

3. *Branch*: Branch is the directed line segment joining two nodes, and the gain of a branch is a transmittance.

4. *Input node or source*: It corresponds to an independent variable that will be connected to the outgoing branches or node(s).

5. *Output node or sink*: It corresponds to a dependent variable that will be connected to the incoming branches.

   Figure 3.33 shows a general signal flow graph, where x_1 is the input node or source which is an independent variable, and x_2 is the output node or sink which is a dependent variable. The signal is incoming from x_1 to x_2 node; thus, x_1 to x_2 has a directed graph, so it is a branch. However, "a" is the transmittance.

6. *Mixed node*: It is connected with both incoming and outgoing branches. Look at Figure 3.33 carefully, and we can see that x_2 and x_3 are mixed node.

7. *Loop*: Loop is a closed path. Look at Figure 3.33 carefully, and we can see that node x_2 to node x_3 and then node x_3 to node x_2 are closed path, so this is a loop in this signal flow graph.

8. *Loop gain*: Loop gain is the product of branch transmittances of a loop. Look at Figure 3.33 carefully, and we can see that the loop gain of the signal flow graph is bc.

9. *Path*: A path is a traversal way of connected branches in the direction of the branch arrows.

   - *Open path*: If no node is crossed more than once, it is called an open path;
   - *Closed path*: If the path ends at the same node from which it began and does not cross any other node more than once, it is called a closed path; and
   - *Open nor closed path*: If the path ends at the different node from which it began and crosses some node more than once, it is called an open nor closed path.

10. *Nontouching loops*: If the loops do not possess any common nodes, then it is called nontouching loops.

11. *Forward path*: The path from the input node to an output node that does not cross more than once, it is called a forward path. Please look at the signal flow graph and we can see that the path from the input node $R(s)$ to the output node $C(s)$ did not cross more than once; thus, this is called a forward path.

12. *Forward path gain*: The forward path gain is the product of the branch transmittances of a forward path in the signal flow graph(s). Please look at the signal flow graph in Figure 3.33, and it can be seen that the forward path gain from the input node $R(s)$ to the output node $C(s)$ is $P = ab$.

3.5.3 Signal Flow Graph Algebra

Signal flow graph algebra is important to discuss about Mason's rules, which will be discussed next, whereas the I/O relationship is determined by Mason's rules as

- The value of a node with one incoming branch is shown in the figure; thus, the linear algebra can be written from the signal flow graph as $x_2 = ax_1$.
- If the branches are in the cascaded form, then the total transmittance of cascaded branches given as from the first branch it will be $x_2 = ax_1$ and from the second branch it will be $x_3 = ax_2$; therefore, the simplified linear equation will be $x_3 = abx_1$, which has been shown in Figure 3.40.
- Similarly, the parallel branches may be combined by adding the transmittances as $x_2 = (a+b)x_1$; and
- A mixed node may be eliminated as shown in the second figure of Figure 3.34.

However, a loop may be eliminated as shown in Figure 3.35.

That's why it can be seen the simplified linear algebra of a loop as

$$x_3 = \frac{ab}{1-bc}x_1 \tag{3.43}$$

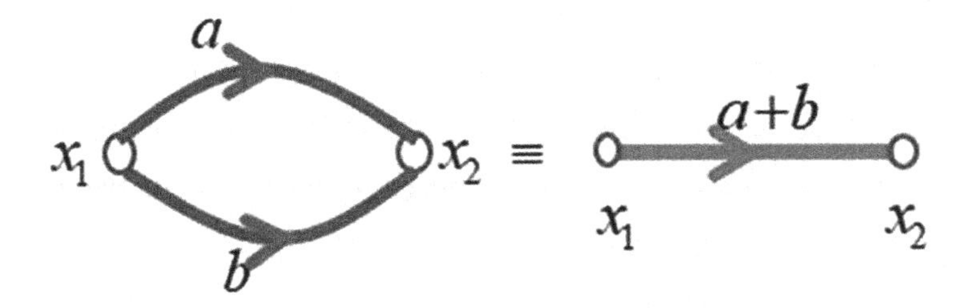

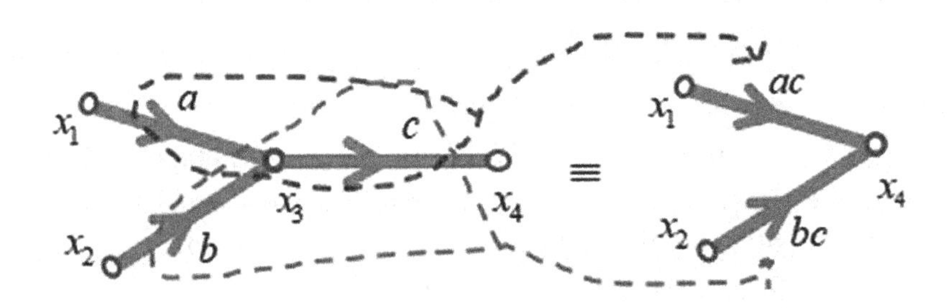

FIGURE 3.34 Signal flow graph algebra.

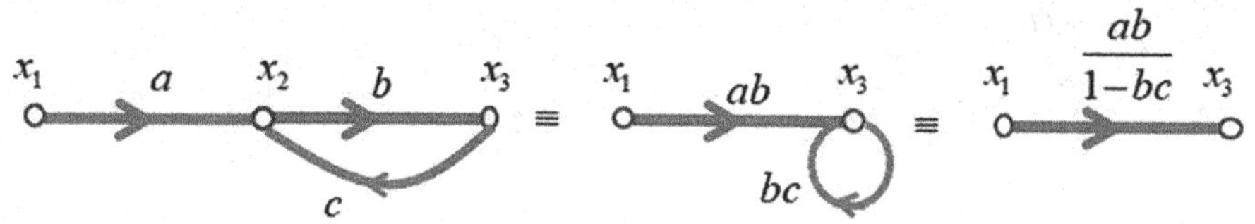

FIGURE 3.35 Loop in a signal flow graph.

3.5.4 Signal Flow Graph of Linear System

Now, we will study about signal flow graph of linear systems. How can we represent linear system(s) into signal flow graph? If u_1 and u_2 are the input variables; x_1, x_2 and x_3 are the output variables; and a_{ij} is the transmittance of linear systems, then let's see the following examples of linear systems. How can we construct the signal flow graph of these linear systems? And how these linear systems look like if combined?

$$x_1 = a_{11}x_1 + a_{12}x_2 + a_{13}x_3 + b_1u_1 \tag{3.44}$$

$$x_2 = a_{21}x_1 + a_{22}x_2 + a_{23}x_3 + b_2u_2 \tag{3.45}$$

$$x_3 = a_{31}x_1 + a_{32}x_2 + a_{33}x_3 \tag{3.46}$$

To sketch the signal flow graph of linear system, first, locate the output variables from each of the linear systems and after that

Step 1. Add the input variables;

Step 2. Put the transmittance; and

Step 3. Draw the signal flow graph.

The individual signal flow graph of the linear systems from Eqs. (3.44) to (3.46) are shown in Figure 3.36a–c. After that, we can add all of the three signal flow graphs into one signal flow graph as shown in Figure 3.36d.

3.5.5 Signal Flow Graph of a Control System

To represent the signal flow graph of a control system, it is very important to study the signal flow graph and Mason's rules for the control systems. The signal flow graphs of control systems are shown in Figure 3.37.

Example 3.4

I am showing an example of a control system and how can we convert the block diagram of a control system to a signal flow graph of Figure 3.38.

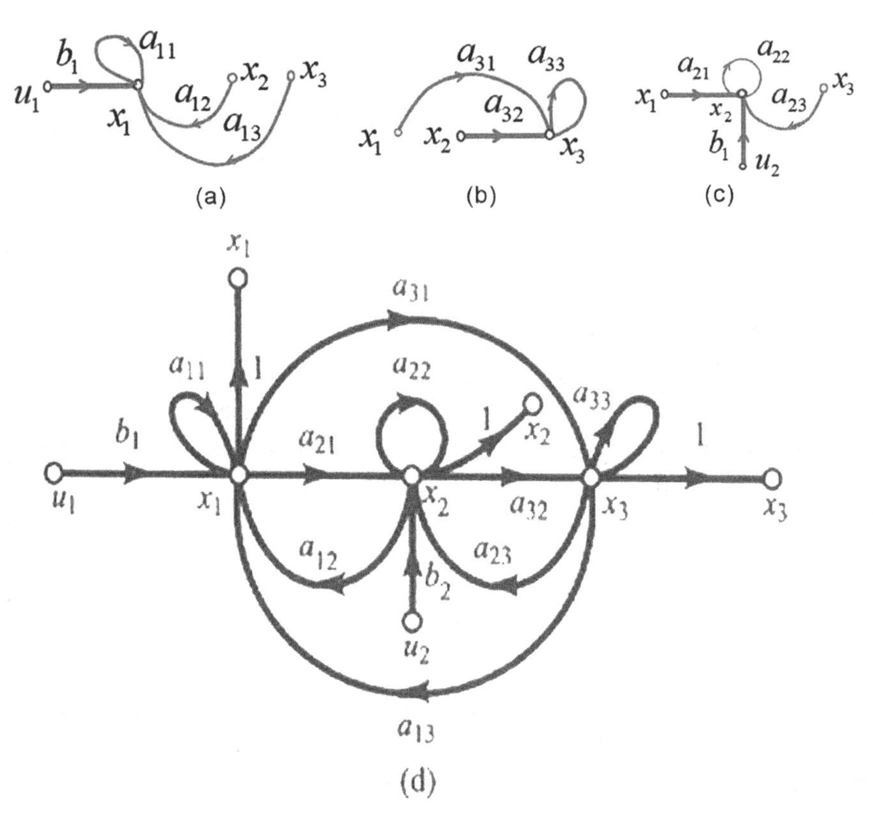

FIGURE 3.36 Signal flow graphs of a linear system: (a) $x_1 = a_{11}x_1 + a_{12}x_2 + a_{13}x_3 + b_1u_1$, (b) $x_2 = a_{21}x_1 + a_{22}x_2 + a_{23}x_3 + b_2u_2$, (c) $x_3 = a_{31}x_1 + a_{32}x_2 + a_{33}x_3$, and (d) linear system of all together of (a), (b), and (c) in a signal flow graph.

Solution

Look carefully at Figure 3.38. First, we can determine the node instead of the input source, output, and branch point; the node is denoted by a circle in the graph shown in Figure 3.39a. And then we can connect the node and put transmittances on branches. Thus, we can get the signal flow graph of a control system shown in Figure 3.39b, whereas we can simplify the signal flow graph for further use, as shown in Figure 3.39c and d.

3.6 MASON'S RULES AND MASON'S GAIN FORMULA

In this section, we will study Mason's gain formula for signal flow graphs and some important examples for the applications of Mason's gain formula. What is Mason's gain formula for the signal flow graph?

- Mason's gain formula is a relationship between an input variable and an output variable of signal flow graphs.

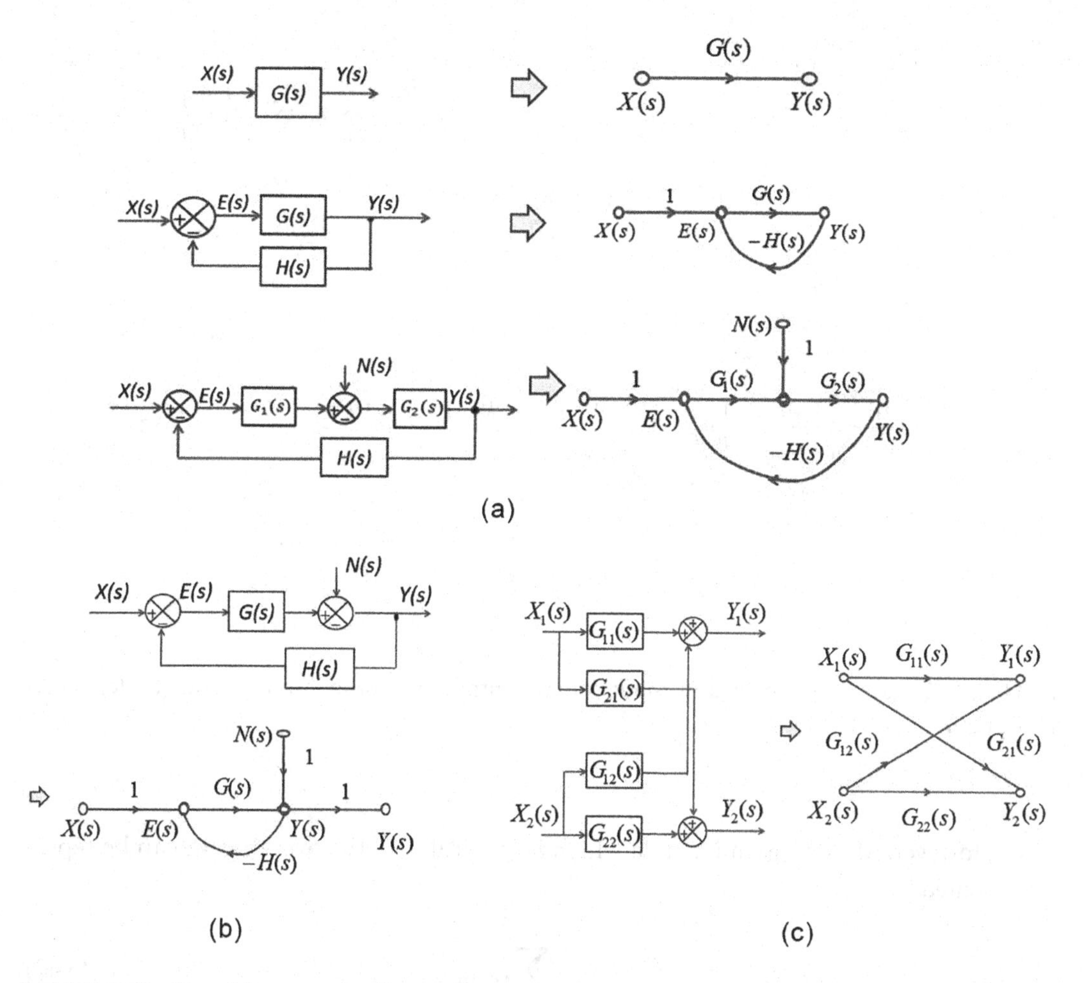

FIGURE 3.37 Signal flow graph of control systems.

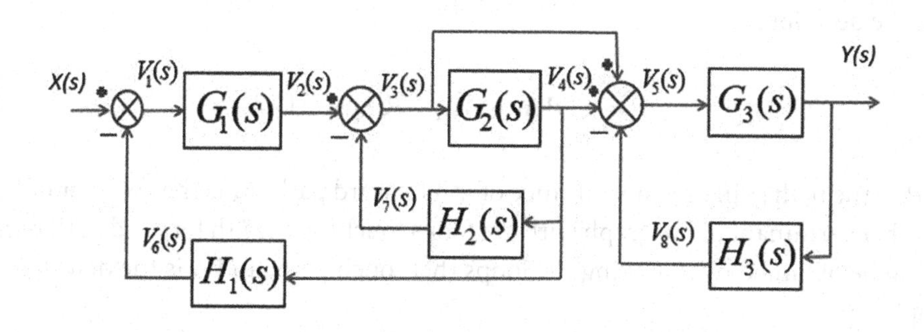

FIGURE 3.38 Block diagram of a control system.

- Overall gain or overall transmittance is the main factor to evaluate Mason's gain formula. The transmittance can be measured from between an input node and an output node.

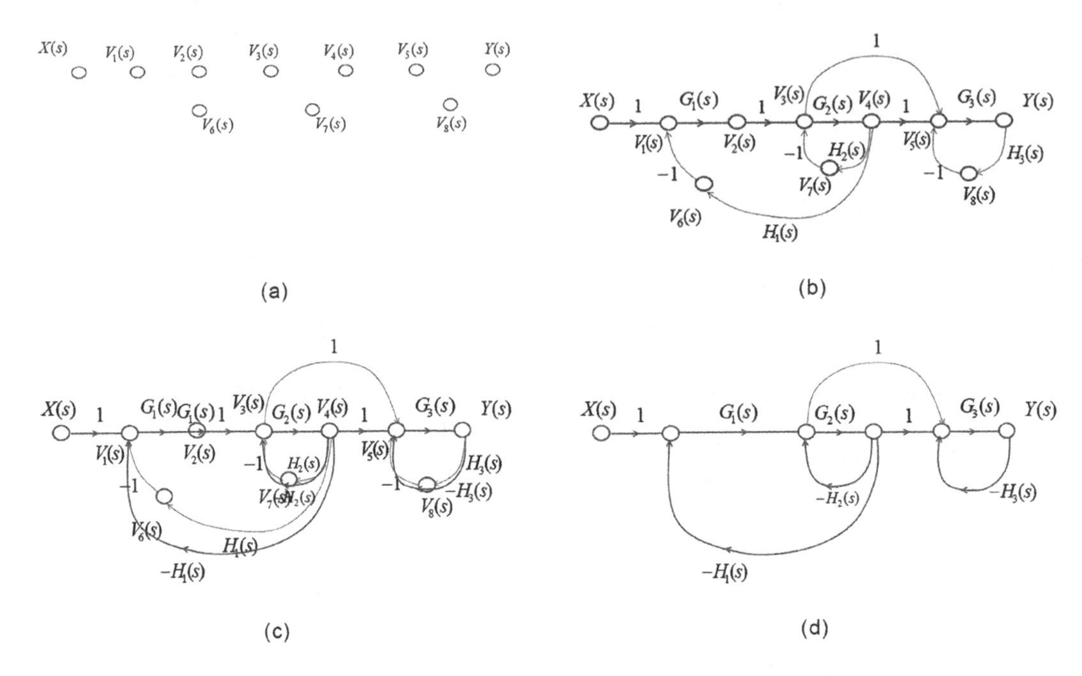

FIGURE 3.39 (a) Nodes instead of input source, output, and branch point, and (b)–(c) signal flow graph.

- However, Mason's gain formula which is applicable to the overall gain can be represented by

$$P = \frac{1}{\Delta}\sum_{k} P_k \Delta_k \tag{3.47}$$

or it can be described as

$$P = \frac{1}{\Delta}(P_1\Delta_1 + P_2\Delta_2 + P_3\Delta_3 + \ldots) \tag{3.48}$$

where P_k is the path gain or transmittance of kth forward path; Δ_k is the cofactor of kth forward path determinant of the graph with the loops touching the kth forward path removed, and it can be obtained by removing the loops that touch path; and Δ is the determinant of the graph.

However,

$$\Delta = 1 - \sum_{a} L_a + \sum_{b,c} L_b L_c - \sum_{d,e,f} L_d L_e L_f + \ldots + (-1)^m \sum \ldots + \ldots \tag{3.49}$$

where $\sum_{a} L_a$ is evaluated from individual single variable loop gains of the signal flow graph; $\sum_{b,c} L_b L_c$ is evaluated from nontouching loops gains taken two variables at a time; $\sum_{d,e,f} L_d L_e L_f$ is evaluated from nontouching loops gains taken three variables at a time; and so on.

Therefore, this equation can be written as follows:

$$\Delta = 1 - \sum_{a} L_a + \sum_{b,c} L_b L_c - \sum_{d,e,f} L_d L_e L_f + \ldots + (-1)^m \sum \ldots + \ldots \tag{3.50}$$

Thus,

$$\Delta = 1 - (L_1 + L_2 + L_3 + \ldots) + (L_1 L_2 + L_2 L_3 + \ldots) - (L_1 L_2 L_3 + L_2 L_3 L_4 + \ldots) + \ldots \tag{3.51}$$

Example 3.5

Consider the following system as a signal flow graph shown in Figure 3.40. Let us obtain the closed-loop transfer function $C(s)/R(s)$ by use of Mason's gain formula.

Solution

To solve this problem, we need to find the forward path gain. Thus, we can see only one forward path between I/O. Thus, the forward path gain is

$$P_1 = G_1 G_2 G_3$$

Now, we need to convert the control system into a signal flow graph shown in Figure 3.41.

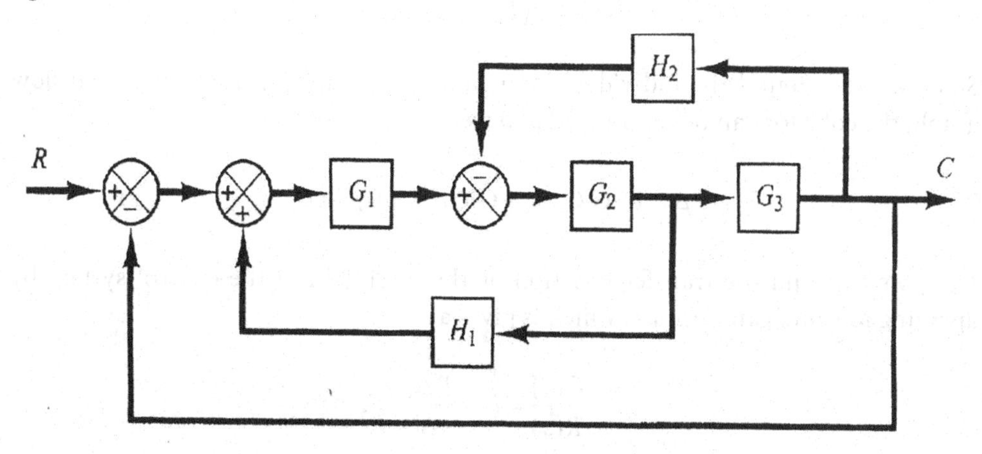

FIGURE 3.40 Closed-loop control system.

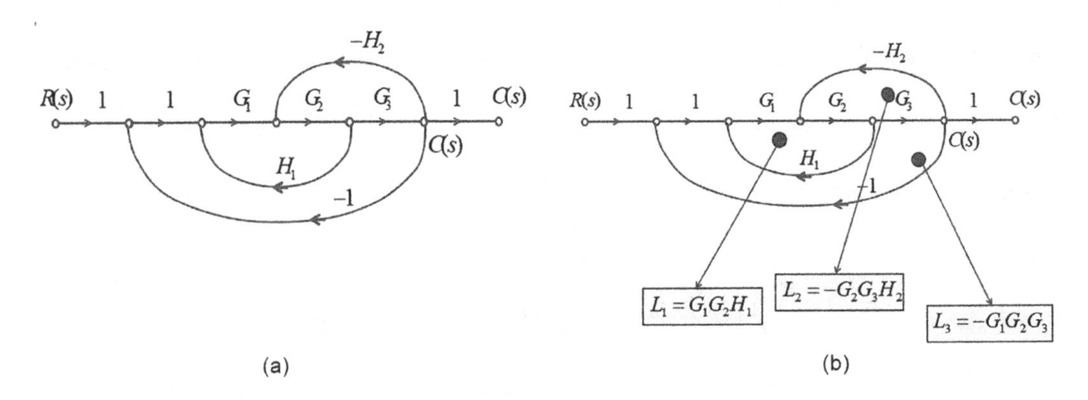

FIGURE 3.41 (a) Signal flow graph of Figure 3.45 and (b) individual loop gain of the signal flow graph.

Here, we can see that there are three loops in the signal flow graphs; thus, the gains of the three individual paths or loops can be written as

$$L_1 = G_1G_2H_1$$

$$L_2 = -G_2G_3H_2$$

and

$$L_3 = -G_1G_2G_3,$$

whereas we can see the nontouching common branch loop in the signal flow graph; therefore, the nontouching common branch loop determinant can be obtained as

$$\Delta = 1 - (L_1 + L_2 + L_3)$$

Since we have found the individual loop gains L_1, L_2, and L_3 from the signal flow graph, the cofactor can be rearranged and we get

$$\Delta = 1 - G_1G_2H_1 + G_2G_3H_2 + G_1G_2G_3$$

Now, we can find the transfer function or the total gain of the control system by applying Mason's gain formula which is gives as

$$\frac{C(s)}{R(s)} = P = \frac{P_1\Delta_1}{\Delta}$$

Thus,

$$P = \frac{P_1\Delta_1}{\Delta} = \frac{G_1G_2G_3\Delta_1}{1 - G_1G_2H_1 + G_2G_3H_2 + G_1G_2G_3}$$

Since the path touches all the three loops, the cofactor delta 1 of the determinant can be obtained from delta by removing the all loops, and then the value of delta 1 is unity; thus, we can get the total gain of the signal flow graph as follows:

$$P = \frac{G_1G_2G_3}{1 - G_1G_2H_1 + G_2G_3H_2 + G_1G_2G_3}$$

Example 3.6

Now I am going to show you another example of a signal flow graph to obtain the closed-loop transfer function by applying Mason's gain formula (Figure 3.42).

Solution

Consider the following system as a signal flow graph. Let us obtain the closed-loop transfer function $C(s)/R(s)$ by the use of Mason's gain formula. To solve this problem, we can see there are three forward paths between I/O as shown in Figure 3.43, which are

$$P_1 = G_1G_2G_3G_4G_5$$

$$P_2 = G_1G_6G_4G_5$$

and,

$$P_3 = G_1G_2G_7$$

However, the gains of the four individuals' loops are shown in Figure 3.44 as

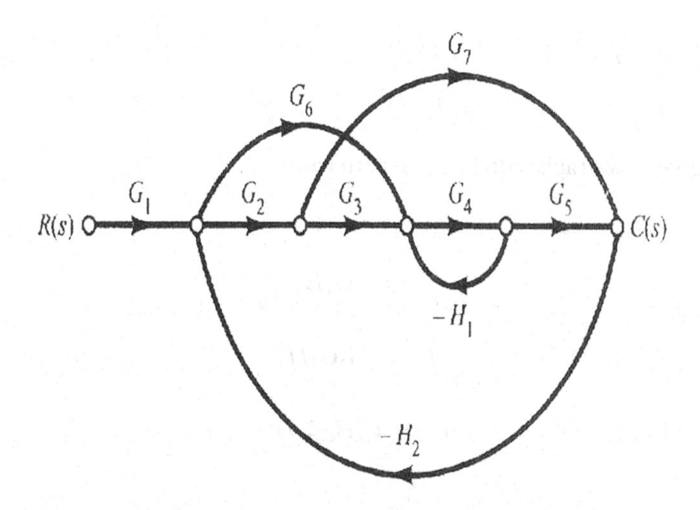

FIGURE 3.42 Signal flow graph of a closed-loop control system.

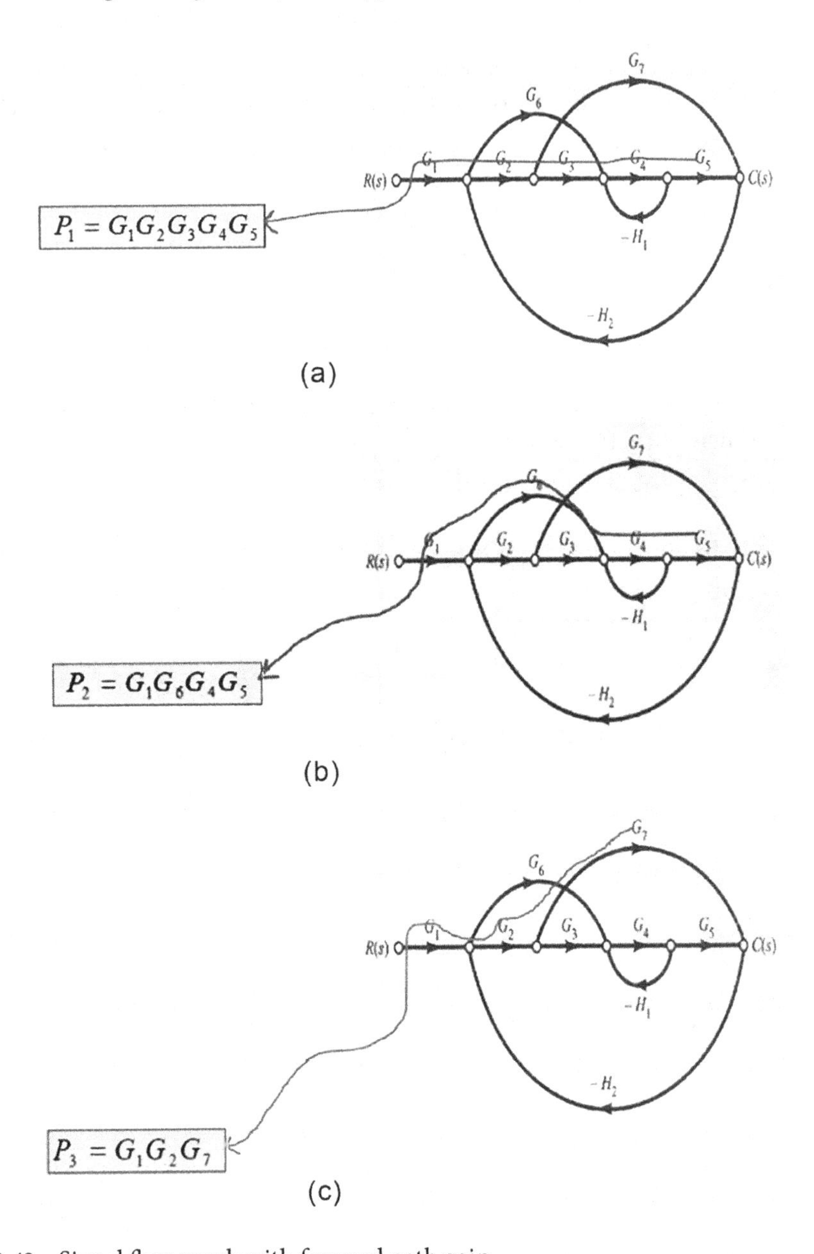

FIGURE 3.43 Signal flow graph with forward path gain.

$$L_1 = -G_4H_1$$

$$L_2 = -G_2G_7H_2$$

$$L_3 = -G_6G_4G_5H_2$$

and

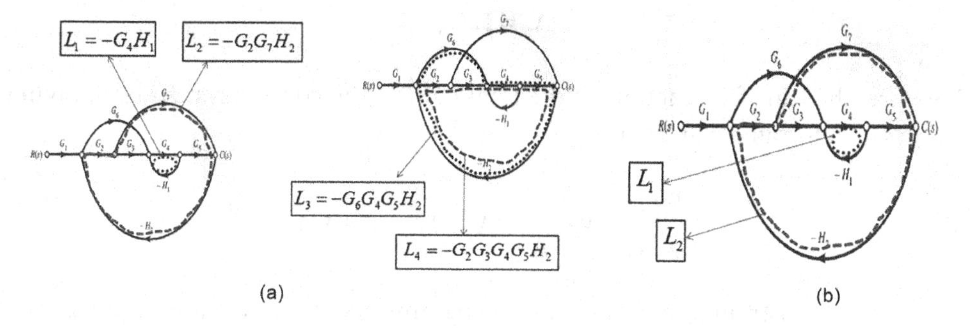

FIGURE 3.44 Signal flow graph with individual loop gain.

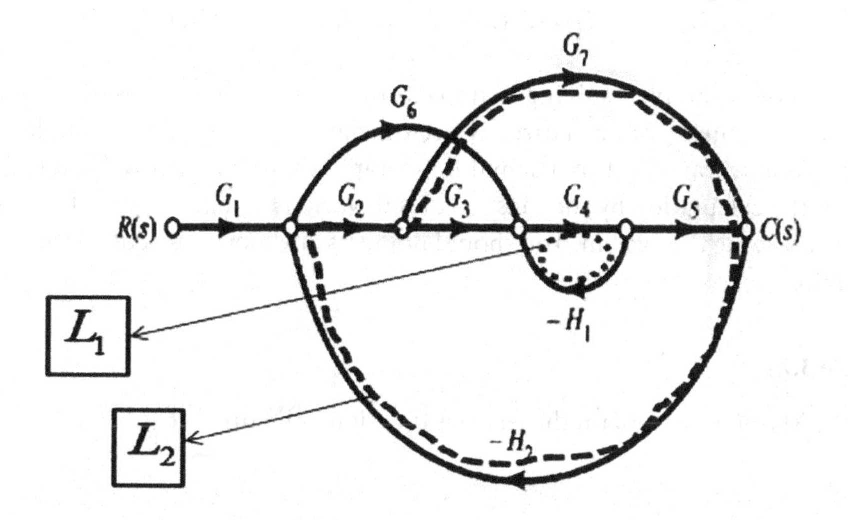

FIGURE 3.45 Signal flow graph with touch path and non-touch path.

$$L_4 = -G_2G_3G_4G_5H_2$$

The loop L_1 does not touch with loop L_2. Hence, the determinant Δ can be written as (Figure 3.45)

$$\Delta = 1 - (L_1 + L_2 + L_3 + L_4) + L_1L_2$$

To check the signal flow graph again, we can see that the forward path P_1 and P_2 are touch path with loop L_1, L_2, L_3, and L_4; therefore, the cofactor Δ_1 and Δ_2 can be found as

$$\Delta_1 = 1, \Delta_2 = 1$$

However, the forward path P_3 is a non-touch path with a loop L_1 only; therefore, the cofactor Δ_3 can be found as

$$\Delta_3 = 1 - L_1$$

Therefore, the transfer function or the total gain of the control system by applying Mason's gain formula can be given as

$$\frac{Y(s)}{X(s)} = P = \frac{1}{\Delta}(P_1\Delta_1 + P_2\Delta_2 + P_3\Delta_3)$$

Now, we can substitute the value of all the parameters in the formula, and then we can find the total gain or closed-loop transfer function as

$$P = \frac{G_1G_2G_3G_4G_5 + G_1G_6G_4G_5 + G_1G_2G_7(1 + G_4H_1)}{1 + G_4H_1 + G_2G_7H_2 + G_6G_4G_5H_2 + G_2G_3G_4G_5H_2 + G_4H_1G_2G_7H_2}$$

From the above examples, solving a simple transfer function in the system is easy, clear, and convenient; you can directly use the block diagram. For a complex system, you should transfer it to a signal flow graph and then use the Mason rules. Note that the calculation by the Mason rules to various signals should check carefully by observing the system. You should not miss any loop to calculate the transfer function.

Example 3.7

Using the Mason rules, obtain the transfer function of Figure 3.46.

Solution

To get the transfer function of the signal flow graph it can be determined by using Mason's rules. Thus, we get the forward path gain as

$$P_1 = G_1(s)G_2(s)G_3(s)G_4(s)$$

whereas

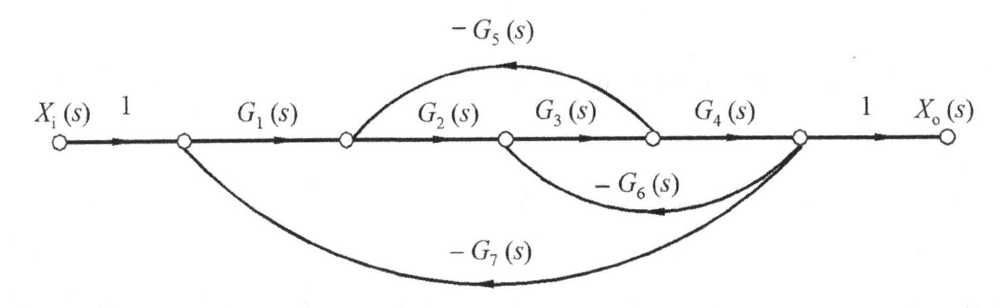

FIGURE 3.46 Signal flow graph.

$$\sum_a L_a = -G_2(s)G_3(s)G_5(s) - G_3(s)G_4(s)G_6(s) - G_1(s)G_2(s)G_3(s)G_4(s)G_7(s)$$

Hence,

$$\Delta = 1 - \sum_a L_a$$

or

$$\Delta = 1 + G_2(s)G_3(s)G_5(s) + G_3(s)G_4(s)G_6(s) + G_1(s)G_2(s)G_3(s)G_4(s)G_7(s)$$

In this case,

$$\Delta_1 = 1$$

Since we have, the transfer function or the total gain of the control system we may write as,

$$P = \frac{1}{\Delta} P_1 \Delta_1$$

Therefore,

$$P = \frac{G_1(s)G_2(s)G_3(s)G_4(s)}{1 + G_2(s)G_3(s)G_5(s) + G_3(s)G_4(s)G_6(s) + G_1(s)G_2(s)G_3(s)G_4(s)G_7(s)}$$

Example 3.8

Use the signal flow graph and Mason's rules to obtain the transfer function from Figure 3.47.

Solution

First, transfer the block diagram (a) to signal flow graph (b), and according to Mason's rules, we can get

$P_1 = G_1G_2G_3$ and $P_2 = G_1G_4$

whereas

$$\sum_a L_a = -G_1(s)G_2(s)G_3(s) - G_1(s)G_4(s) - G_1(s)G_2(s)H_1(s) - G_2(s)G_3(s)H_2(s) - G_4(s)H_2(s)$$

Hence,

$$\Delta = 1 - \sum_a L_a$$

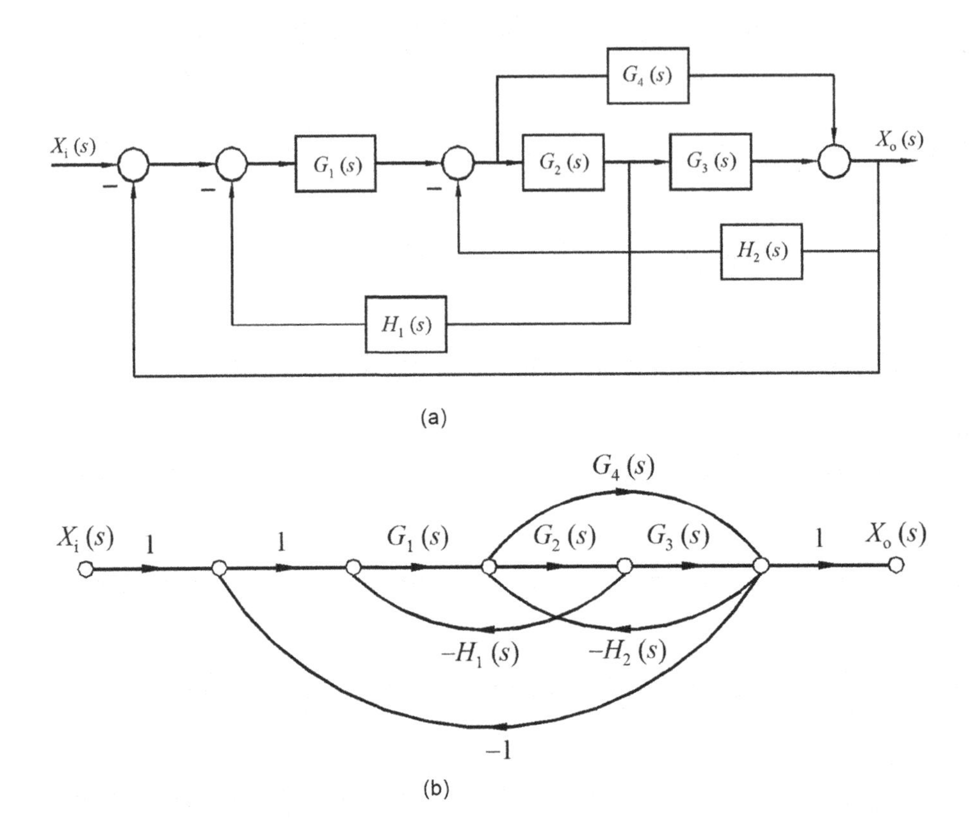

FIGURE 3.47 Block diagram and signal flow graph.

or

$$\Delta = 1 + G_1(s)G_2(s)G_3(s) + G_1(s)G_4(s) + G_1(s)G_2(s)H_1(s) + G_2(s)G_3(s)H_2(s) + G_4(s)H_2(s)$$

In this case,

$$\Delta_1 = 1, \text{ and } \Delta_2 = 1$$

The transfer function or the total gain of the control system as we get

$$P = \frac{1}{\Delta} P_1 \Delta_1$$

Therefore,

$$P = \frac{G_1G_2G_3 + G_1G_4}{1 + G_1G_2G_3 + G_1G_4 + G_1G_2H_1 + G_2G_3H_2 + G_4H_2}$$

From the above examples, solving a simple transfer function in the system is easy, and you can directly use the block diagram, both clear and convenient. For a complex system, transfer it to a signal flow graph and then use the Mason rules.

EXERCISES

1. A system of signal flow graph is shown in Figure 3.48. Find the transfer function.
2. A signal control system flow graph is shown in Figure 3.49. Comprehensive test expressions derived output speed O and the relationship between load torque and input speed I and T_L. A system of signal flow diagram is shown in Figure 3.50. Find the transfer function.

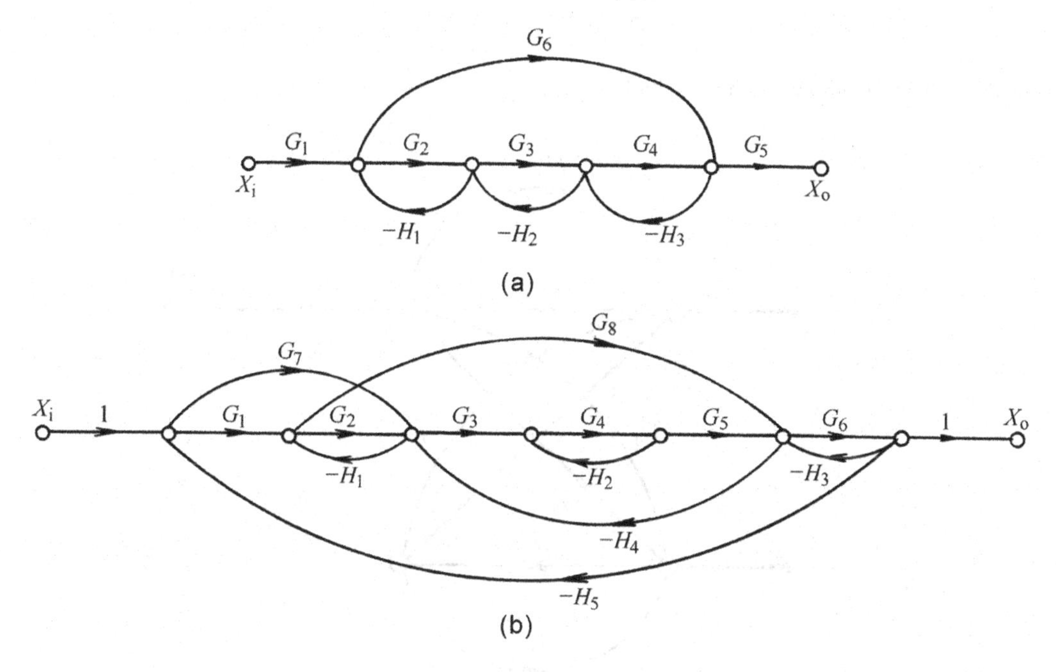

FIGURE 3.48 Signal flow diagram of a system.

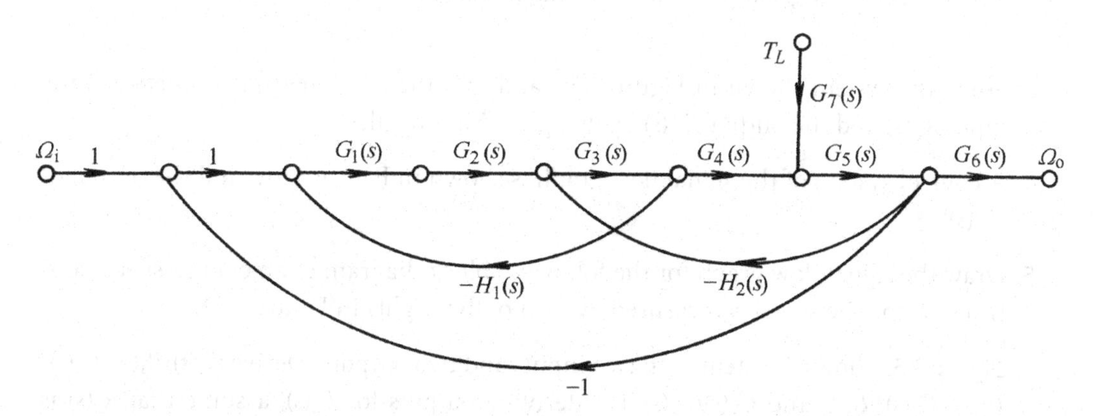

FIGURE 3.49 Signal flow diagram of a system with external input signal.

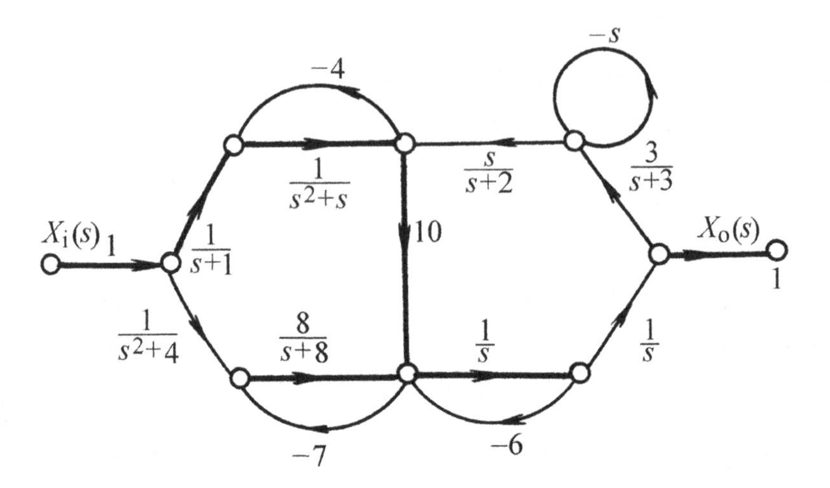

FIGURE 3.50 Signal flow graph of multiple flow.

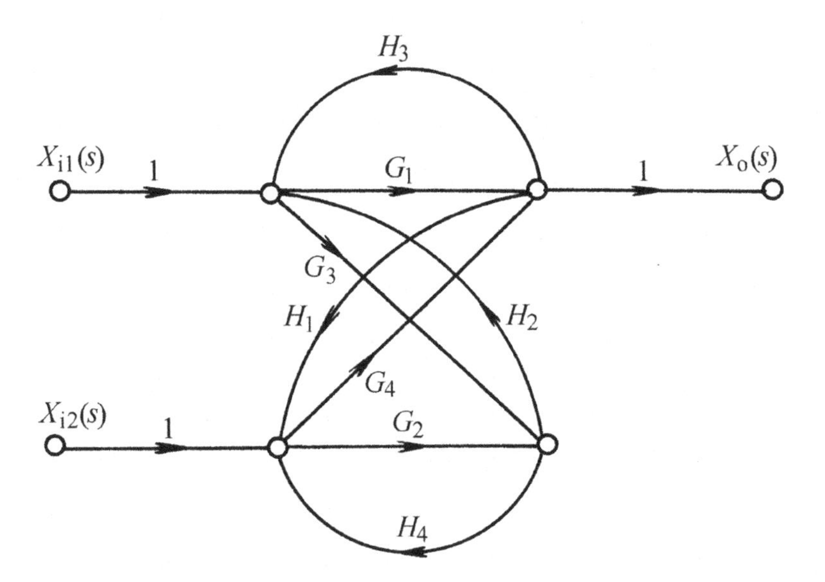

FIGURE 3.51 Signal flow graph of multiple flow with two inputs.

3. Study the system shown in Figure 3.51, and find the transfer function between the input $X_i(s)$ and the output $X_o(s)$ by using the Mason rules.

4. A control system of the signal flow graph is shown in Figure 3.51. Find the output of $X_o(s)$.

5. Draw the signal flow graph for the following block diagram of an engine-speed control system. The speed is measured by a set of flyweights in Figure 3.52.

6. Figure 3.53 shows a system with two inputs and two outputs. Derive $C_1(s)/R_1(s)$, $C_1(s)/R_2(s)$, $C_2(s)/R_1(s)$, and $C_2(s)/R_2(s)$. (In deriving outputs for $R_1(s)$, assume that $R_2(s)$ is zero, and vice versa.)

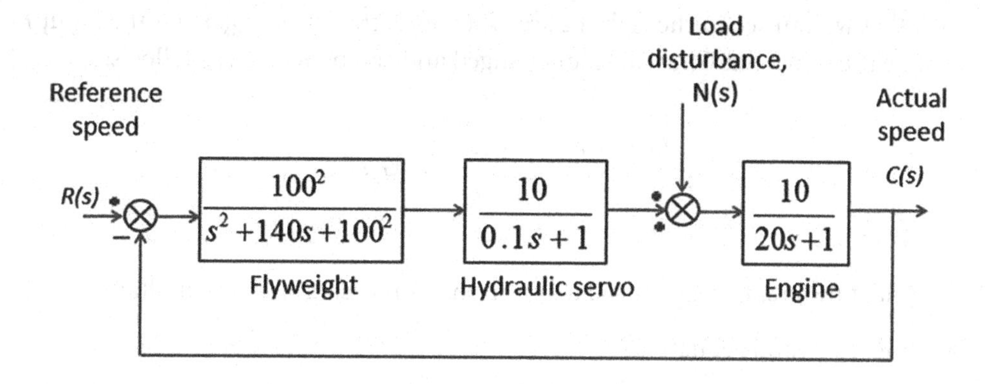

FIGURE 3.52 Signal flow graph of engine speed control system.

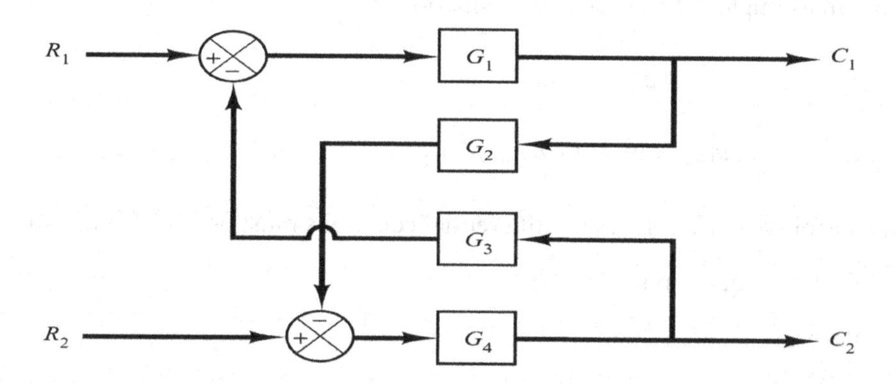

FIGURE 3.53 Signal flow graph of two input and two output systems.

MCQ AND TRUE/FALSE QUESTIONS

1. Please comment on the following figures.

   A block diagram with a summing junction on the left side, and from this block diagram, we moved the summing junction to the past of a block $G(s)$, and then we can see that the input $R(s)$ and the output $C(s)$ as well as $Q(s)$ are not changed, but a transfer function of $G(s)$ can be seen as $G(s)$ which is the same as transfer function $G(s)$ due to moving the summing junction to past of the block $G(s)$ as seen in the right figure. Then,

$$C(s)=\left[R(s)\pm Q(s)\right]G(s)=R(s)G(s)\pm Q(s)G(s)$$

   Answer: True/False

2. Please comment on the following figures.

   A block diagram with a branch point on the left side, and from this block diagram, we moved the branch point to the past of a block $G(s)$, and then we can see that the input $R(s)$ and the output $C(s)$ as well as $Q(s)$ are not changed, but a reciprocal function of $G(s)$ can be seen as $1/G(s)$ due to moving the branch point to past of the block

$G(s)$ that we can see in the right figure. Therefore, the output $Q(s)$ and the input $R(s)$ will be the same, but $C(s)$ will be unchanged and can be written as follows:

$$Q(s) = R(s) = R(s)G(s)\frac{1}{G(s)}$$

Answer: True/False

3. A signal flow graph is a pictorial diagram that represents a set of simultaneous.
   a. linear algebraic equations
   b. linear equations
4. Give an example of linear algebraic equation.
   a. $x_1 = a_{11}x_1 + a_{12}x_2 + a_{13}x_3 + b_1u_1$
   b. $x_1 = a_{11}x_1^2 + a_{12}x_2 + a_{13}x_3 + b_1u_1$
5. In control systems, first, linear differential equations must be transformed into
   a. algebraic equation in s
   b. algebraic equation in s^2
6. Loop gain is the product of branch transmittances of an open loop.
   Answer: True/False
7. The forward path gain is the product of the branch transmittances of a forward path in the signal flow graph(s).
   Answer: True/False
8. Mason's gain formula has a relationship between an input variable and an output variable of signal flow graphs.
   Answer: True/False
9. Overall gain or overall transmittance is not the main factor to evaluate Mason's gain formula. The transmittance can be measured from between an input node and an output node.
   Answer: True/False

CHAPTER 4

Transient Response Analyses in Time Domain

IN THIS CHAPTER, WE will focus on typical signal analysis in the time domain, control systems analysis, and analyzing process, transient response of first-order systems, transient response of second-order systems, second-order systems and transient response specifications, transient response of high-order systems, stability analyses and criterion, and MATLAB with MATLAB SIMULINK.

4.1 TYPICAL SIGNALS ANALYSIS IN TIME DOMAIN

The commonly used typical input signals are step functions, ramp functions (e.g., speed or velocity), parabolic functions (e.g., acceleration), impulse functions, sinusoidal functions, and white noise. Any of the typical input signals should be selected as a function of time, whereas a typical input signal depends on the system and working conditions of the system. For example,

- If the control quantity is a function that changes with time, such as radar antenna, artillery, and temperature control device, it is more appropriate to choose ramp function.
- If the control system is shocked with input, such as missile launch and satellite launch, then the choice of an impulse function is more appropriate.
- If the input of the control system works forward and backward directions as round-trip motion with respect to time, such as machine vibration, it is better to select the positive sine function (cosine).
- If the input quantity of the control system is a suddenly changed such as suddenly power on or power off, then the choice of step signal is appropriated. It is worth noting that the performance index of the response in the time domain is defined by selecting the step signal as the input.

DOI: 10.1201/9781003293859-4

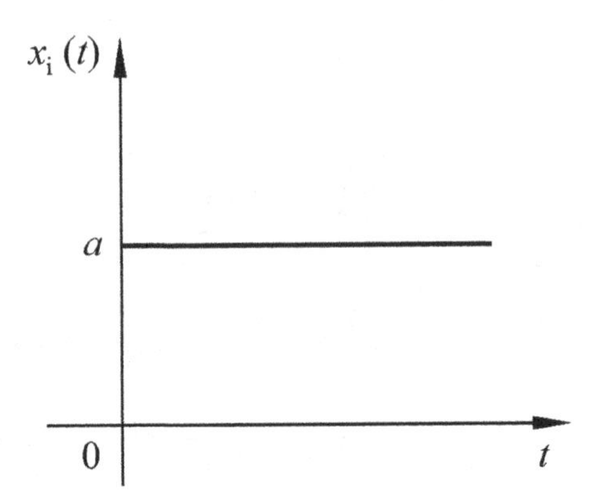

FIGURE 4.1 Step function.

4.1.1 Step Function and Unit-Step Function

Step function refers to system input when the input quantity of the control system has suddenly changed. Figure 4.1 shows the step function. If the quantity of input is connected suddenly or disconnected suddenly or stopped suddenly, etc., then the mathematical expression of step function is defined as

$$x_i(t) = \begin{cases} a & \forall t \quad 0 \\ 0 & \forall t < 0 \end{cases} \tag{4.1}$$

where $i = 0, 1, \ldots\ldots, n$; and a is a constant; if $a = 1$, then it is called unit-step function. It is denoted by $u(t)$, and it can be defined as

$$u(t) = \begin{cases} 1 & ; \forall t \geq 0 \\ 0 & ; \forall t < 0 \end{cases} \tag{4.2}$$

The diagram of the step signal can be seen in Figure 4.1, where the step can be seen as a constant straight line or a straight line that is parallel to the X-axis in the positive coordinate. The step signal can be used for transient response modeling and steady-state error (SSE) analysis.

4.1.2 Ramp Function and Unit-Ramp Function

Figure 4.2 shows the ramp function. The diagram of the ramp signal can be seen in Figure 4.2, where the ramp can be seen as a line or a straight line that is passing through the origin in the positive coordinate system. The ramp signal can be used for SSE analysis. If the input is changed constantly, then we can see the ramp function characteristics in the system.

Thus, the mathematical expression of the ramp function can be defined as

$$x_i(t) = \begin{cases} at & \forall t \quad 0 \\ 0 & \forall t < 0 \end{cases} \tag{4.3}$$

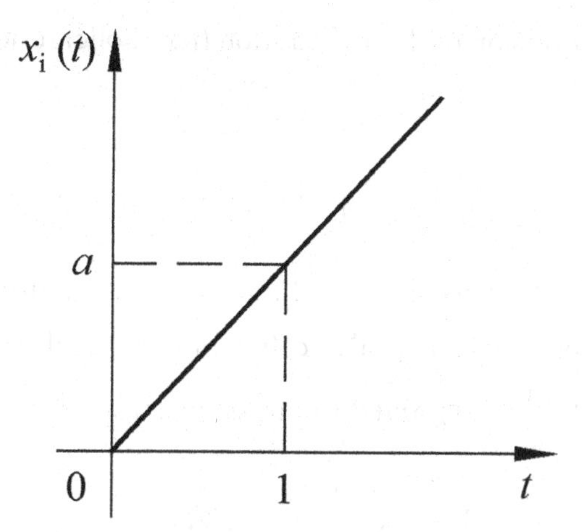

FIGURE 4.2 Ramp function.

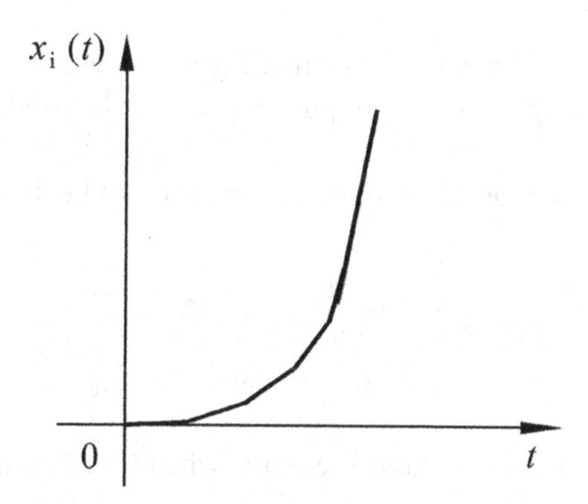

FIGURE 4.3 Parabolic (acceleration) function.

where $i = 0,1,\ldots..,n$, and a is a constant; if $a = 1$, then it is called unit-ramp function or it is also called only ramp function. It is denoted by $tu(t)$, and it can be defined as

$$tu(t) = \begin{cases} t & ; \forall t \geq 0 \\ 0 & ; \forall t < 0 \end{cases} \tag{4.4}$$

4.1.3 Parabolic (Acceleration) Function

Parabolic (acceleration) function refers to the input quantity when the function changes to the acceleration (parabolic) curve, as shown in Figure 4.3; the diagram of the parabola signal can be seen in Figure 4.3, where the parabola can be seen as a curve that is passing through origin in the positive coordinate. The ramp signal can be used for SSE analysis.

The mathematical expression of the acceleration (parabolic) function can be defined as

$$x_i(t)=\begin{cases} at^2 & \forall t \;\; 0 \\ 0 & \forall t<0 \end{cases} \tag{4.5}$$

where $i=0,1,\ldots..,n$, and a is a constant; if $a=1$ then it is called unit-parabolic (acceleration) function; when $a=\frac{1}{2}$ sometimes it is also called only parabolic (acceleration) function. Normally, it is denoted by $\frac{1}{2}t^2u(t)$, and it can be defined as

$$\frac{1}{2}t^2u(t)=\begin{cases} t^2 & ;\forall t\geq 0 \\ 0 & ;\forall t<0 \end{cases} \tag{4.6}$$

4.1.4 Impulse Function

The diagram of the impulse signal can be seen in Figure 4.4, where the impulse can be seen only at the origin with amplitude. The impulse signal can be used for transient response modeling.

The mathematical expression of the impulse function can be defined as

$$x_i(t)=\begin{cases} \lim\limits_{t_0\to 0}\frac{a}{t_0} & ;\forall 0\leq t\leq t_0 \\ 0 & ;\forall 0\langle t \text{ or } t\rangle t_0 \end{cases} \tag{4.7}$$

where $i=0,1,\ldots..,n$, and a is a constant; therefore, when $0\leq t\leq t_0$, then the function is ∞, and impulse height $\frac{a}{t_0}$ is ∞ (infinite), whereas the interval t_0 is very small (infinitesimal), the impulse area is a; usually, the pulse intensity is measured in the area a; thus, when the

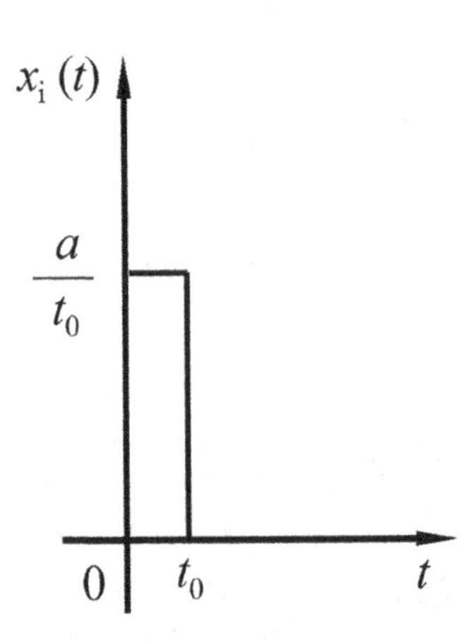

FIGURE 4.4 Impulse function.

area $a=1$, it is called unit-impulse function which is also called δ function, as shown in Figure 4.4. Thus, a unit-impulse function is denoted by $\delta(t)$, and it is defined as

$$\delta(t)=\begin{cases} \lim\limits_{t_0\to 0}\dfrac{1}{t_0} & ;\forall 0\le t\le t_0 \\ 0 & ;\forall 0\rangle t \text{ or } t\rangle t_0 \end{cases} \tag{4.8}$$

or we can say as

$$\delta(t)=\begin{cases} \int_{0^-}^{0^+}\delta(t)dt=1 & ;\forall 0\le t\le t_0 \\ 0 & ;\forall 0\rangle t \text{ or } t\rangle t_0 \end{cases} \tag{4.9}$$

The important property of $\delta(t)$ is that the Laplace transform of $\delta(t)$ is equal to 1, i.e.,

$$\mathcal{L}\{\delta(t)\}=1 \tag{4.10}$$

Proof

According to the properties of step function, we obtain

$$\delta(t)=\lim_{t_0\to 0}\left[\frac{1(t)}{t_0}-\frac{1(t-t_0)}{t_0}\right]=\lim_{t_0\to 0}\frac{1}{t_0}\left[1(t)-1(t-t_0)\right]$$

$$\mathcal{L}\{\delta(t)\}=\lim_{t_0\to 0}\frac{1}{t_0}\left[\frac{1}{s}-\frac{1}{s}e^{-t_0 s}\right]=\lim_{t_0\to 0}\frac{1}{t_0 s}\left[1-\left(1-t_0 s+\frac{1}{2!}t_0^2 s^2-\cdots\right)\right]$$

$$=\lim_{t_0\to 0}\frac{1}{t_0 s}\left[t_0 s-\frac{1}{2!}t_0^2 s^2+\cdots\right]$$

$$=1$$

4.1.5 Sinusoid Function

The diagram of the sinusoid signal can be seen in Figure 4.5, where the sinusoid can be seen as a sinusoid wave. The sinusoid signal can be used for transient response modeling, and SSE analysis. Generally, sinusoid function is denoted by $\sin\omega t$.

The mathematical definition of the sinusoid function is written as

$$x_i(t)=\begin{cases} a\sin\omega t & ;\forall t\ \ge 0 \\ 0 & ;\forall t<0 \end{cases} \tag{4.11}$$

where $i=0,1,\ldots.,n$ and a is a constant.

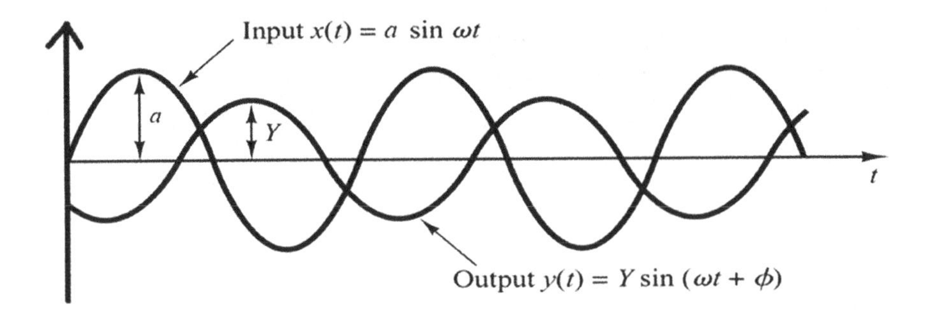

FIGURE 4.5 Sinusoid signal.

4.2 CONTROL SYSTEMS ANALYSIS AND ANALYZING PROCESS

We will see how to analyze the control system. Figure 4.6 is excellent for control systems analysis. Here, it can be seen that X-axis shows time and Y-axis shows the response of a control system. In the X–Y coordinate system, it shows a fixed straight input line that is parallel to the X-axis; however, two curves also can be seen in this Figure 4.6, one of the curves is existed output with a high gain value, and the other curve is existed output with low gain value with respect to time. If the time value is increased, the amplitude of the output or response is decreased along with the input value. As it also can be seen that the time is closed to infinity or equal to infinity, and then the amplitude of the output or response is almost closed to the input value, but the output with a high gain curve is more oscillation than the output with a low gain curve. However, the output with a low gain curve is slower than the output with high gain curve, and the output with low gain curve reaches close to input gain before the output with higher gain curve.

The control systems analyze process is important to analyze any system. Therefore, the objectives of the analyses and design of the control system are performed by

1. Transient response analyses,
2. Steady-state response analyses, and
3. Stability analyses.

4.2.1 Transient Response and Steady-State Response

Due to varying time starting from initial point zero with initial condition zero, the response cure shows transient response until its final value of time; the response is almost steady-state from final value to infinity time. However, the steady-state response is almost parallel with the input command. The gap or difference between steady-state and input command is called steady-state error, as shown in Figure 4.7. To analyze the system performance of a control system, these transient responses, steady-state responses, and SSE s are the main factors to study the control system. Thus, the time response of a control system consists of

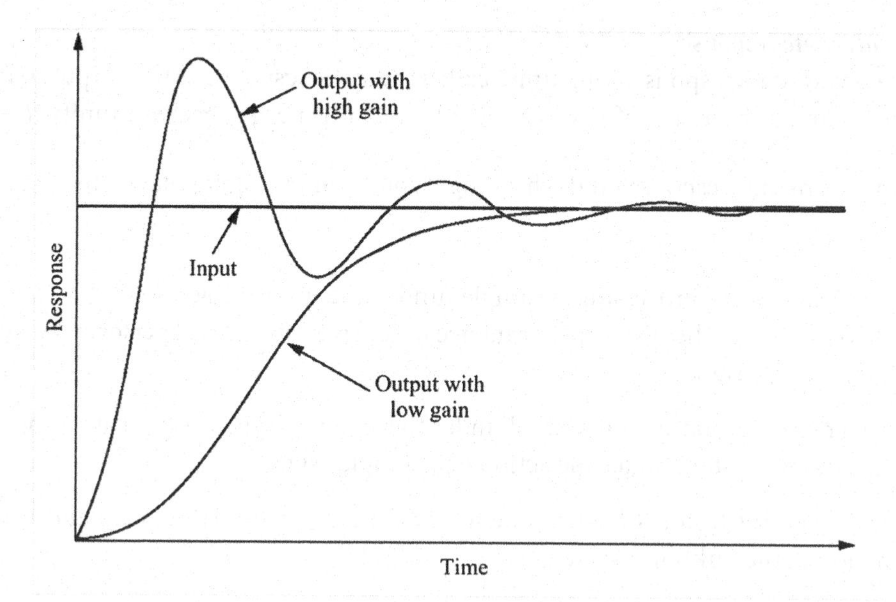

FIGURE 4.6 Control systems analysis curve.

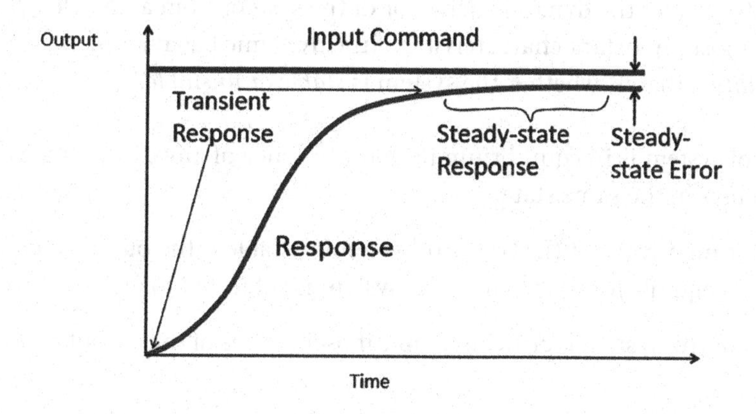

FIGURE 4.7 Transient response and steady-state response.

two parts: (a) the transient response and (b) the steady-state response. The system response is denoted by $c(t)$ which may be written as

$$c(t) = c_{tr}(t) + c_{ss}(t) \tag{4.12}$$

where $c_{tr}(t)$ is the transient response and $c_{ss}(t)$ is the steady-state response.

1. *Transient response:*
   This is the change of system output from initial state to stable state under the action of a typical signal input. Transient response is also called dynamic response or transient process or transient response.

2. *Steady-state response:*
   The steady-state response, sometimes called the static response, is the output state of a system under the action of a typical signal input as time approaches infinity ($t \to \infty$).

Analyzing transient response and choosing typical input signal have the following advantages:

1. The mathematical processing is simple; under the given typical signal action, it is easy to determine the system performance index (parameter) and perform the system analysis and design.
2. The transient response is often used under the action of typical signal as the basis for analyzing the system under the action of a complex signal.
3. It is convenient to identify the system and determine the parameters and transfer function of the unknown system.

4.2.2 Stability, Absolute Stability, Relative Stability and Steady-State Error

Now, I am going to discuss about **absolute stability.** In designing a control system, we must be able to predict the dynamic behavior of the system from a knowledge of the components. The most important characteristic of the dynamic behavior of a control system is *absolute stability* – that is, whether the system is *stable or unstable.*

- A control system is in equilibrium if, in the absence of any disturbance or input, the output stays in the same state.
- A linear time-invariant (LTI) control system is stable if the output eventually comes back to its equilibrium state when the system is subjected to an initial condition.
- An LTI control system is critically stable if oscillations of the output continue forever. And
- An LTI is unstable if the output diverges without being bound from its equilibrium state when the system is subjected to an initial condition.

Actually, the output of a physical system may increase to a certain extent but may be limited by mechanical "stops," or the system may break down or become nonlinear after the output exceeds a certain magnitude so that the linear differential equations no longer apply.

However, I am going to talk about *relative stability and SSE* that have important system behaviors (other than absolute stability) to which we must give careful consideration including relative stability and SSE. Since a physical control system involves energy storage when subjected to an input, the output of the system cannot follow the input immediately but exhibits a transient response before a steady state can be reached, as shown in Figure 4.7.

The transient response of a practical control system often exhibits damped oscillations before reaching a steady state. If the output of a system at steady state does not exactly agree with the input, the system is said to have SSE. This error is indicative of the accuracy of the system. In analyzing a control system, we must examine transient response behavior and steady-state behavior.

4.3 TRANSIENT RESPONSE OF FIRST-ORDER SYSTEMS

In this section, I will focus on first-order systems, characteristics of the exponential response curve, and important properties of LTI systems. Consider the first-order system shown in Figure 4.8a.

Physically, this system may represent an RC circuit, thermal system, or the like. A simplified block diagram is shown in Figure 4.8b. The system which can be described by a first-order differential equation is called first-order system; the transfer function or the input–output relationship of the given first-order system is given by

$$G(s)=\frac{C(s)}{R(s)}=\frac{1}{Ts+1} \tag{4.13}$$

We shall analyze the system responses to such inputs as the unit-step, unit-ramp, and unit-impulse functions, whereas the initial conditions are assumed to be zero.

4.3.1 Unit-Step Response of First-Order Systems

To analyze unit-step response of first-order systems, Eq. (4.13) can be rewritten as

$$C(s)=\left(\frac{1}{Ts+1}\right)R(s) \tag{4.14}$$

We have the unit-step function as

$$r(t)=1(t)$$

Taking the Laplace transform of the unit-step function, we obtain

$$R(s)=\frac{1}{s}$$

Substituting $R(s)=1/s$ in Eq. (4.14), we get

$$C(s)=\left(\frac{1}{Ts+1}\right)\left(\frac{1}{s}\right)$$

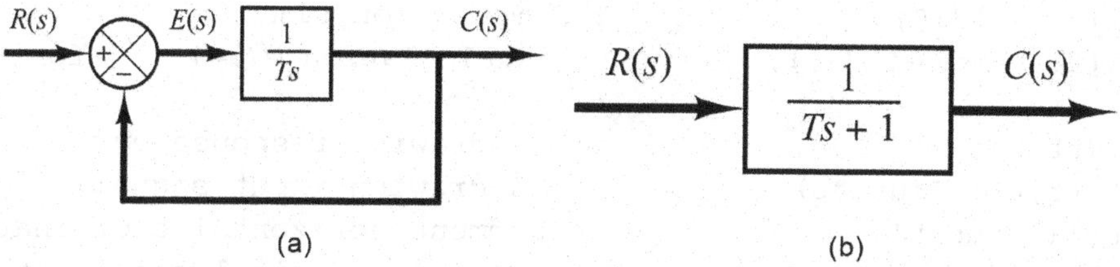

FIGURE 4.8 Block diagram of first-order systems.

TABLE 4.1 Unit-step response of first-order systems

t	0	T	$2T$	$3T$	$4T$	$5T$	...	∞
$c(t)$	0	0.632	0.865	0.95	0.982	0.993	...	1
$c(t)$	0	63.2%	86.5%	95%	98.2%	99.3%		100%

Expanding $C(s)$ by partial fractions, we get

$$C(s)=\frac{1}{s}-\frac{T}{Ts+1}=\frac{1}{s}-\frac{1}{s+\left(\frac{1}{T}\right)} \tag{4.15}$$

Taking the inverse Laplace transform of Eq. (4.14), we obtain

$$c(t)=1-e^{-\frac{t}{T}};\text{for}\quad t\geq 0 \tag{4.16}$$

Note that Eq. (4.16) is called unit-step response function of the first-order systems.

1. *Characteristics analysis of exponential response curve:*
   Now let's see, what are the characteristics of an exponential response curve? How can we generate this curve? According to Eq. (4.16), we obtain Table 4.1.
   To generate the unit-step response curve of the first-order system, we may recall Eq. (4.16), and in MATLAB software, by generating different values of t we obtain $c(t)$; however, we can plot the unit-step response curve of the first-order system in MATLAB, as shown in Figure 4.9.

```
MATLAB Code
T=10; ut=1;
for i1=1:2
i2(i1)=i1-1;
xot2(i1)=i1-1;
end
plot(i2,xot2)                                % drawing slope;
hold on
for i=1:51                          % iteration number with time
variable;
tt(i)=(i-1)/10;                     % Computation time 0~10T
xot3(i)=1-exp(-tt(i));                % calculation c(t)=1-e^(-t/T) for t≥0;
end
plot(tt,xot3)                           % drawing response curve;
plot(tt,ut+0*(ut+tt))                   % drawing input signal;
xlabel('Time t')                     % comment horizontal coordinates;
ylabel('Output response c(t)') % comment vertical coordinates;
grid;                                     % generate grid in curve;
```

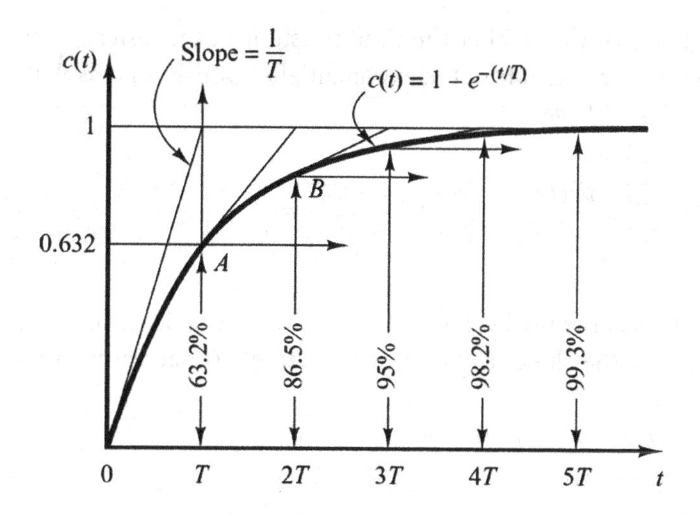

FIGURE 4.9 Unit-step response curve of the first-order system.

Now let's see the analysis of unit-step response from Figure 4.9. Look carefully and observe it. We already have seen the exponential response curve $c(t)$. The response function of the first-order systems states that initially the output $c(t)$ is zero and finally it becomes unity. One important characteristic of such an exponential response curve $c(t)$ is that at $t = T$, the value of $c(t)$ is 0.632, or the response $c(t)$ has reached 63.2% of its total change.

- If $t = 0$ to T, the exponential response curve has gone from 0% to 63.2% of the final value;
- If $t = 2T$, the response reaches 86.5% of the final value;
- At $t = 3T, 4T$, and $5T$, then the response reaches 95%, 98.2%, and 99.3%, respectively, of the final value; and
- Thus, if $t \geq 4T$, the response remains within 2% of the final value.

As seen from *unit-step response function of the first-order systems*, the steady state is reached mathematically only after an infinite time. In practice, however, a reasonable estimate of the response time is the length of time the unit-step response curve needs to reach and stay within the 2% line of the final value or four-time constants.

The first-order inertial system is always stable without vibration. At $t = 3T \sim 4T$, the response curve reaches the steady-state value of 95%–98%, which is called the settling time; thus, the settling time usually takes

$$t_s = (3 \sim 4)T$$

Note that the smaller the time constant T, the faster the system responds. Another important characteristic of the exponential response curve is that the slope of the tangent line at $t=0$ is $1/T$. Note that the smaller the time constant T, the faster the system responds. Another important characteristic of the exponential response curve is that the slope of the tangent line at $t=0$ is $1/T$, since

$$\left.\frac{dc(t)}{dt}\right|_{t=0}=\left.\frac{d}{dt}\left(1-e^{-\frac{t}{T}}\right)\right|_{t=0}=\left.\frac{1}{T}e^{-\frac{t}{T}}\right|_{t=0}=\frac{1}{T}$$

The output would reach the final value at $t=T$ if it maintained its initial speed of response. However, we see that the slope of the response curve $c(t)$ decreases monotonically from $1/T$ at $t=0$ to 0 at $t=\infty$.

Now, we recall Eq. (4.16), and then get

$$e^{-\frac{t}{T}}=1-c(t),\text{ for } t\geq 0$$

Taking logarithm on both sides, we obtain

$$\left[-\frac{1}{T}\log e\right]t=\log\left[1-c(t)\right] \tag{4.17}$$

Here, $\log[1-c(t)]$ is the ordinary longitudinal in the coordinates system along with Y-axis and the abscissa time t is in the X-axis; then we get a straight line through the origin, as shown in Figure 4.10. To measure the unit-step response $c(t)$ of the system, $\log[1-c(t)]$ is the ordinary longitudinal in the coordinates system along with Y-axis, but the time t is along with the X-axis in the semi-logarithmic paper; if we get a straight line passing through the origin (similar to Figure 4.10), then we can identify the first-order system.

4.3.2 Unit-Ramp Response of First-Order System

To analyze unit-ramp response of first-order systems, we have the unit-ramp function as

$$r(t)=t$$

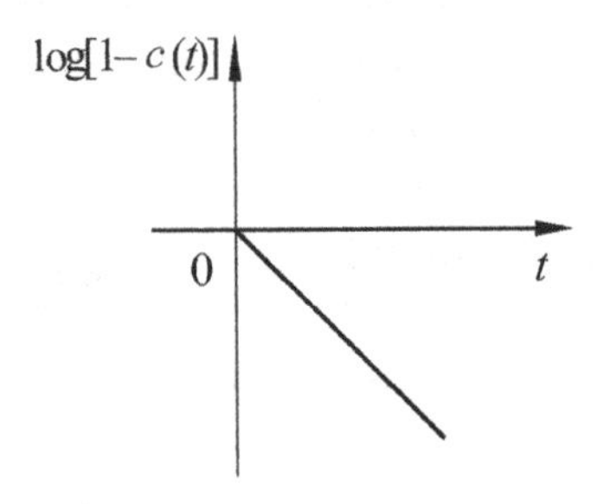

FIGURE 4.10 Logarithm identification curve of the first-order system.

Taking the Laplace transform of the unit-ramp function, we obtain

$$R(s)=\frac{1}{s^2}$$

Substituting $R(s)=\frac{1}{s^2}$ in Eq. (4.14), we get

$$C(s)=\left(\frac{1}{Ts+1}\right)\left(\frac{1}{s^2}\right)$$

Expanding $C(s)$ by partial fractions, we get

$$C(s)=\frac{1}{Ts+1}\frac{1}{s^2}=\frac{1}{s^2}-\frac{T}{s}+\frac{T}{s+\frac{1}{T}} \tag{4.18}$$

Taking the inverse Laplace transform of Eq. (4.18), we obtain

$$c(t)=t-T+Te^{-\frac{1}{T}t}\text{; for } t\geq 0 \tag{4.19}$$

Note that Eq. (4.19) is called unit-ramp response function of the first-order systems, whereas the error is

$$e(t)=r(t)-c(t)=t-\left[t-T+Te^{-\frac{1}{T}t}\right]=T\left(1-e^{-\frac{1}{T}t}\right) \tag{4.20}$$

In Eq. (4.20), when t approaches to infinity, i.e., $t\infty$, $e{-}t/T$ approaches zero, and the error signal $e(t)$ approaches to T or $e(\infty)=T$; then $e_{ss}=e(\infty)=T$, as shown in Figure 4.11.

If we choose different values of t, then we can draw the unit-ramp response signal $c(t)$ as well the error signal $e(t)$ with respect to unit-ramp input; the unit-ramp input and the system output $c(t)$ are shown in Figure 4.11. The error signal $e(t)$ in following the unit-ramp input is equal to T for sufficiently large t. Therefore, when input is a unit-ramp function, the SSE of the first-order system is T. Obviously, if the time constant T is smaller, then the SSE is also smaller. The smaller the time constant T, the smaller the SSE in the ramp input. If T is increased then the SSE is larger.

To generate the unit-step response curve of the first-order system, we may recall Eq. (4.19), and in MATLAB software, by generating different values of t we obtain $c(t)$; however, we can plot the unit-step response curve of the first-order system in MATLAB, as shown in Figure 4.11.

```
MATLAB Code
T=10;
rt=t;
t=0:50;                     % iteration number with time  variable;
```

```
hold on
plot(t/5,t*T/50)
a=t/T;                          % computing a=t/T, where T=10;
ct=t-T+T*exp(-a);               % calculation c(t)=(1/T)e^(-(1/T)t) for t≥0;
hold on
plot(t/5,ct*10.5/50)            % drawing response curve;
xlabel('Time t')                % comment horizontal coordinates;
ylabel('Output response c(t)')  % comment vertical
                                  coordinates;
grid;                           % generate grid in curve;
```

4.3.3 Unit-Impulse Response of First-Order System

To analyze unit-impulse response of first-order systems, we have the unit-impulse function as

$$r(t)=\delta(t)$$

Taking the Laplace transform of the unit-impulse function, we obtain

$$R(s)=1$$

Substituting $R(s)=1$ in Eq. (4.14), we get

$$C(s)=\left(\frac{1}{Ts+1}\right)(1)$$

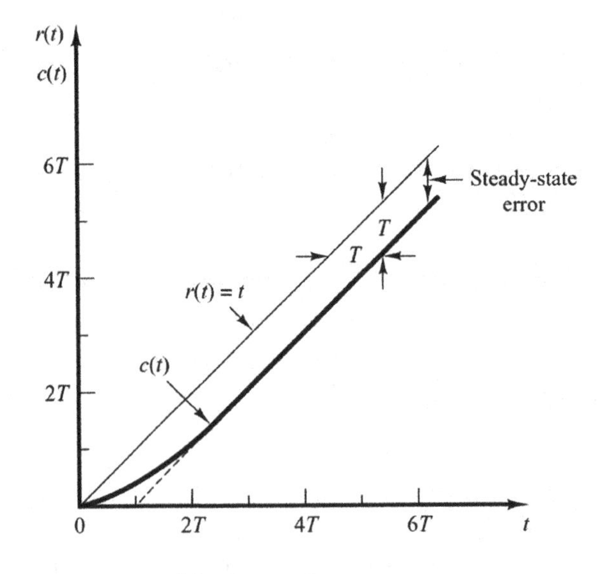

FIGURE 4.11 Unit-ramp response of the first-order system.

Rearranging right-hand side of $C(s)$, we get

$$C(s)=\frac{\frac{1}{T}}{s+\frac{1}{T}}=\frac{1}{T}\left(\frac{1}{s+\frac{1}{T}}\right) \tag{4.21}$$

Taking the inverse Laplace transform of Eq. (4.21), we obtain

$$c(t)=\frac{1}{T}e^{-\frac{1}{T}t};\ \text{for } t\geq 0 \tag{4.22}$$

Note that Eq. (4.22) is called unit-impulse response function of the first-order systems. If we consider the value of t and T, then by using the MATLAB software, we can sketch the unit-impulse response curves of the first-order system, as shown in Figure 4.12. The X-axis shows the time, and the Y-axis shows the $c(t)$ of unit-impulse response curve.

To generate the unit-impulse response curve of the first-order system, we may recall Eq. (4.22), and in MATLAB software, by generating different values of t, we obtain $c(t)$; however, we can plot the unit-impulse response curve of the first-order system in MATLAB, as shown in Figure 4.12.

```
MATLAB Code
T=10;
t=0:50;                              % iteration number with time
variable;
a=t/T;                            % computing a=t/T, where T=10;
ct=exp(-a);                          % calculation c(t)=(1/T)e^(-(1/T)t) for t≥0;
plot(t/5,ct)                      % drawing response curve;
xlabel('Time t')                     % comment horizontal coordinates;
ylabel('Output response c(t)')                % comment vertical
coordinates;
grid;                                   % generate grid in curve;
```

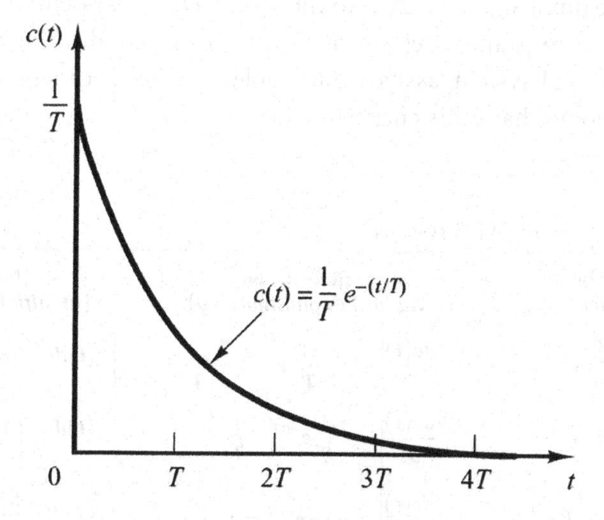

FIGURE 4.12 Unit-impulse response of the first-order system.

4.3.4 Important Property of LTI Systems

From these above analyses, we can obtain the relationship between unit-step function, unit-ramp function, and unit-impulse function, i.e.,

$$\delta(t)=\frac{d}{dt}[1(t)] \tag{4.23}$$

$$1(t)=\frac{d}{dt}[t] \tag{4.24}$$

We recall Eqs. (4.16), (4.19), and (4.22) of unit-step function, unit-ramp function, and unit-impulse function, respectively, and then rewrite as

$$c_{\text{step}}(t)=1-e^{-\frac{t}{T}};\text{ for } t\geq 0 \tag{4.25}$$

$$c_{\text{ramp}}(t)=t-T+Te^{-\frac{1}{T}t};\text{ for } t\geq 0 \tag{4.26}$$

$$c_{\text{impulse}}(t)=\frac{1}{T}e^{-\frac{1}{T}t};\text{ for } t\geq 0 \tag{4.27}$$

According to Eqs. (4.25)–(4.27), we obtain

$$c_{\text{impulse}}(t)=\frac{dc_{\text{step}}(t)}{dt} \tag{4.28}$$

$$c_{\text{step}}(t)=\frac{dc_{\text{ramp}}(t)}{dt} \tag{4.29}$$

Thus, the response of the system to the derivative of the input signal can be obtained by taking the derivative of the system to the input signal. Similarly, the response of the system to the integral of the input signal is equal to the integral of the system to the original signal response. The integral constant is determined by the initial conditions, which is an important condition of the LTI system, as shown in Table 4.2. Linear time-varying systems and nonlinear systems do not have this characteristic.

TABLE 4.2 Important property of LTI systems

Input	Output ($for\ t\geq 0$)	Differentiation ($at\ initial\ condition\ t=0$)	Integration ($at\ initial\ condition\ t=0$)
Unit-Impulse	$c(t)=\frac{1}{T}e^{-t/T}$	$\left.\frac{dc(t)}{dt}\right\vert_{t=0}=-\frac{1}{T^2}e^{-t/T}=-\frac{1}{T^2}$	$\int_{t=0} c(t)dt=-e^{-t/T}=-1$
Unit-Step	$c(t)=1-e^{-t/T}$	$\left.\frac{dc(t)}{dt}\right\vert_{t=0}=\frac{1}{T}e^{-t/T}=\frac{1}{T}$	$\int_{t=0} c(t)dt=t+Te^{-t/T}=T$
Unit-Ramp	$c(t)=t-T+Te^{-t/T}$	$\left.\frac{dc(t)}{dt}\right\vert_{t=0}=1-e^{-t/T}=0$	$\int_{t=0} c(t)dt=\frac{t^2}{2}-Tt-T^2e^{-t/T}=-T^2$

To discuss about an important property of LTI systems, the unit-impulse, unit-step, and unit-ramp inputs of the first-order system are important to analyze the responses $c(t)$; we have already discussed responses $c(t)$ which are shown in the first and second columns of Table 4.2, whereas in the third column shows the differential value with initial condition zero of the response $c(t)$, and the fourth column shows the integral value with initial condition zero of the response $c(t)$. Comparing the system responses of unit-step, unit-ramp, and unit-impulse inputs of the first-order system clearly indicates that

1. The response to the derivative of an input signal can be obtained by differentiating the response of the system to the original signal;
2. It can also be seen that the response to the integral of the original signal can be obtained by integrating the response of the system to the original signal; and
3. By determining the integration constant from the zero-output initial condition.

This is a property of LTI systems, whereas remember that linear time-varying systems and nonlinear systems do not possess this property.

4.3.5 Time Response of First-Order System

Let's show you the time response of the first-order systems can be represented as shown in Table 4.3. The impulse signal, unit-step signal, and ramp signal are shown in the first column; time-domain representations of the signals are shown in the second column; frequency-domain representations of the signals are shown in the third column, and time-response representations of the signals are shown in the fourth column.

Example 4.1

Let's discuss about this example. A small DC motor is used in a speed control system for a computer tape unit. It is required that the time constant of the system be less than 0.1 s. By investigating the possibility of using open- or closed-loop configurations, design a suitable control system given that the transfer function of the motor from input voltage to output speed is given by

$$\frac{\Omega}{V_i}=\frac{2.5}{s+5}$$

TABLE 4.3 Time response of first-order system

	Input Signal		
	Time Domain	Frequency Domain	**Time Response**
Impulse signal	δ(t)	1	$\frac{1}{\tau}e^{-\frac{t}{T}},(t\geq 0)$
Unit-step signal	1(t)	1/s	$1-e^{-\frac{t}{T}},(t\geq 0)$
Ramp signal	t	$1/s^2$	$t-T+Te^{-\frac{t}{T}},(t\geq 0)$

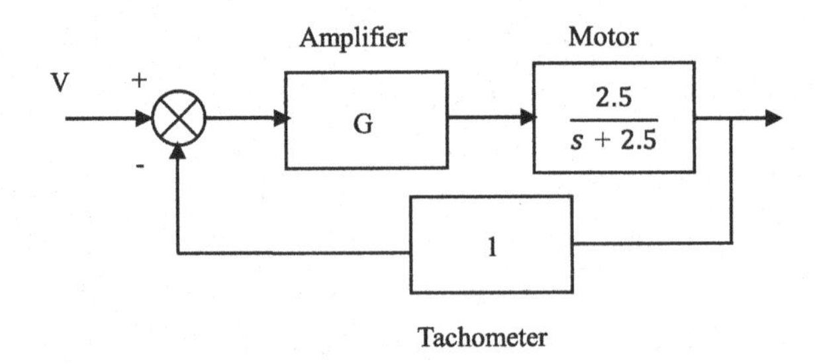

FIGURE 4.13 The closed-loop control system of the small DC motor.

Solution

To solve this problem, the expression can be rearranged, and then we are using an open-loop system. The transfer function is as follows:

$$\frac{\Omega}{V_i}=\frac{2.5}{s+5}=\frac{0.5}{0.2s+1}$$

Then the time constant is

$$\tau=0.2\,\text{s}$$

The transfer function of the closed-loop system is as follows:

$$\frac{\Omega}{V_i}=\frac{2.5G}{s+5+2.5G}=\frac{K'}{1+\tau s}$$

where $K'=\dfrac{2.5G}{5+2.5G}$ $\tau=\dfrac{1}{5+2.5G}$

In order for the closed-loop system to meet the design requirement of $\tau<0.1$ s that we can see in the closed-loop control system in the block diagram shown in Figure 4.13, we get

$$\tau=\frac{1}{5+2.5G}<0.1\leftrightarrow G>2$$

4.4 TRANSIENT RESPONSE OF SECOND-ORDER SYSTEMS

In this section, I will focus on second-order systems and servo system as well as step response of second-order system. Now, the question is what is second-order system? In this section, we shall obtain the response of a typical second-order control system to a step input, ramp input, and impulse input. Here we consider a servo system as an example of a second-order system.

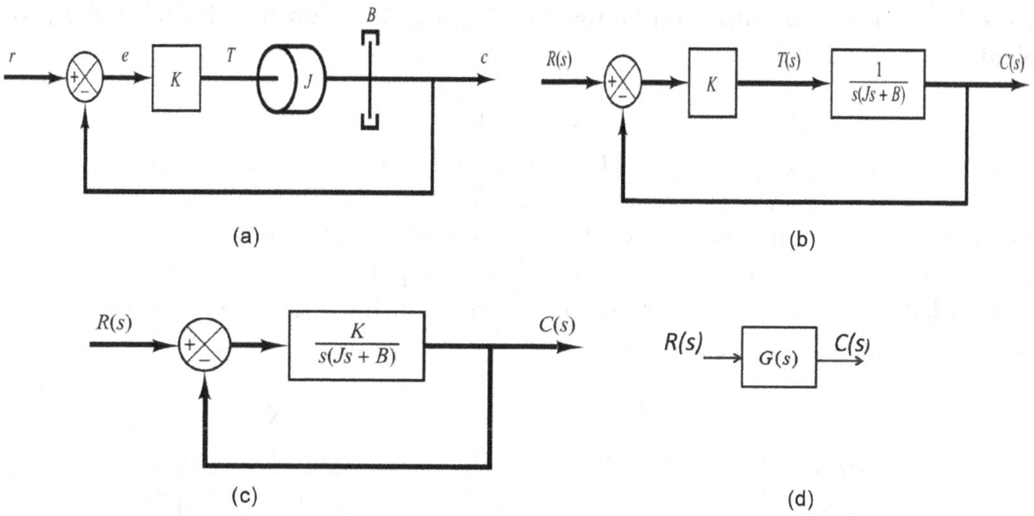

FIGURE 4.14 Block diagram of servo control system.

The systems can be described by second-order differential equations that are called second-order systems. Physically speaking, the second-order system always contains two energy storage components; the energy conversion causes the system between the two components to tend to reciprocating oscillation. When the damping ratio is not sufficiently large, the system presents oscillation characteristics; thus, the second-order system is also called second-order oscillation system.

Now let's see as an example, the **servo system,** shown in Figure 4.14a. The servo system consists of a proportional controller gain (K), actuating error signal (e), torque (T), inertia (J), viscous-friction elements (B), and load elements (inertia (J) and viscous-friction elements (B)). Suppose that we wish to control the output position c in accordance with the input position r.

The equation for the load elements is defined as

$$J\ddot{c}+B\dot{c}=T \tag{4.30}$$

or

$$J\frac{d^2c}{dt^2}+B\frac{dc}{dt}=T \tag{4.31}$$

where T is the torque produced by the proportional controller whose gain is K. By taking Laplace transforms on both sides of Eq. (4.31) with assuming the zero initial conditions, we obtain

$$Js^2C(s)+BsC(s)=T(s) \tag{4.32}$$

Thus, from the transfer function between $C(s)$ and $T(s)$ as shown in Figure 4.14(b), we obtain

$$\frac{C(s)}{T(s)} = \frac{1}{s(Js+B)} \tag{4.33}$$

By using this transfer function of Eq. (4.33), we can redraw the second-order system as we can see in Figure 4.14c. It can be modified and the simplified open-loop control system of this block diagram can be seen as shown in Figure 4.14d. The closed-loop transfer function of the servo system is then obtained as

$$G(s) = \frac{C(s)}{R(s)} = \frac{\dfrac{K}{s(Js+B)}}{1+\dfrac{K}{s(Js+B)}} = \frac{K}{Js^2+Bs+K} = \frac{\dfrac{K}{J}}{s^2+\left(\dfrac{B}{J}\right)s+\left(\dfrac{K}{J}\right)} \tag{4.34}$$

Such a system where the closed-loop transfer function possesses two poles that are called a second-order system, whereas second-order systems may involve one or two zeros that we will discuss this later. We can rewrite Eq. (4.34), then we get as

$$\frac{C(s)}{R(s)} = \frac{K/J}{\left[s+\dfrac{B}{2J}+\sqrt{\left(\dfrac{B}{2J}\right)^2-\dfrac{K}{J}}\right]\left[s+\dfrac{B}{2J}-\sqrt{\left(\dfrac{B}{2J}\right)^2-\dfrac{K}{J}}\right]} \tag{4.35}$$

Therefore, we can see that the closed-loop poles are complex conjugates if the discriminant, $B^2-4JK<0$, and they are real if the discriminant, $B^2-4JK\geq 0$. In the transient response analysis, it is convenient to rewrite as

$$\frac{K}{J} = \omega_n^2 \tag{4.36}$$

$$\frac{B}{J} = 2\xi\omega_n = 2\sigma \tag{4.37}$$

where ω_n is called *undamped natural frequency*, σ is called the *attenuation*, and ζ is the *damping ratio* of the system. However, the damping ratio ζ is the ratio of the actual damping B to the critical damping B_c, where the critical damping,

$$B_c^2 - 4JK = 0$$

or

$$B_c = 2\sqrt{JK}$$

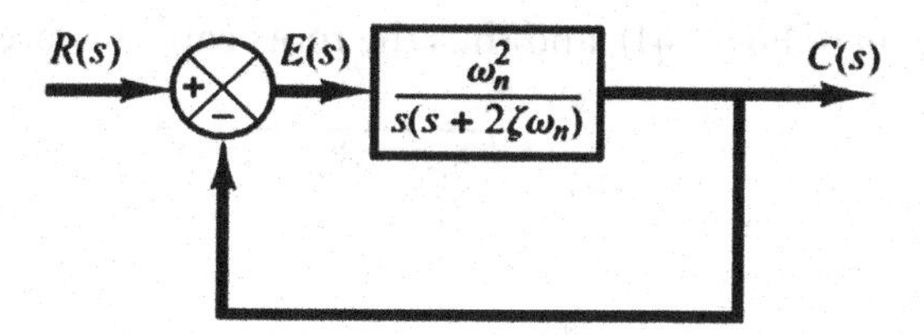

FIGURE 4.15 Closed-loop control system of second-order system.

Thus,

$$\xi = \frac{B}{B_c} = \frac{B}{2\sqrt{JK}} \tag{4.38}$$

Therefore, Eq. (4.34) can be revised as in the following which we may look at the modified closed-loop control system block diagram as shown in Figure 4.15 where we can see the standard form of the second-order system; thus, the closed-loop transfer function of the standard form of the second-order system can be written as

$$\frac{C(s)}{R(s)} = \frac{\omega_n^2}{s^2 + 2\xi\omega_n s + \omega_n^2} \tag{4.39}$$

Eq. (4.39) is the standard form of the second-order system that is very important in further discussion of the control system analysis and discussions.

The typical transfer function of second-systems also can be written as

$$G(s) = \frac{C(s)}{R(s)} = \frac{1}{T^2 s^2 + 2\xi Ts + 1} \tag{4.40}$$

where $T = \frac{1}{\omega_n}$.

Now, we may discuss the dynamic behavior of the second-order system that can then be described in terms of two parameters ζ and ω_n.

4.4.1 Unit-Step Response of Second-Order Systems

1. *Underdamped case* ($0 < \xi < 1$)
   When $0 < \xi < 1$, it is called *underdamped case.* In underdamped state, the poles of the second-order system are a pair of conjugate complex roots, i.e., the second-order system has two poles. We recall Eq. (4.39), and we can find the characteristics equation from the denominator of Eq. (4.39) then, Eq. (4.34) can be modified and we get

$$s^2 + 2\xi\omega_n s + \omega_n^2 = 0 \tag{4.41}$$

There are two roots of Eq. (4.41), and thus the roots can be expressed as

$$s_{1,2} = \frac{-2\xi\omega_n \pm \sqrt{(2\xi\omega_n)^2 - 4.1.\omega_n^2}}{2.1}$$

$$= -\sigma \pm \sqrt{(\xi\omega_n)^2 - \omega_n^2}$$

$$= -\sigma \pm \sqrt{(\xi\omega_n)^2 - \omega_n^2}$$

$$= -\sigma \pm \omega_n\sqrt{\xi^2 - 1}$$

$$= -\sigma \pm j\omega_n\sqrt{1-\xi^2}$$

$$= -\sigma \pm j\omega_d$$

Thus, we can say the two poles or two roots as

$$s_{1,2} = -\sigma \pm j\omega_d \tag{4.42}$$

i.e.

$$s_1 = -\sigma \pm j\omega_d = -\xi\omega_n + j\omega_d$$

and

$$s_2 = -\sigma \pm j\omega_d = -\xi\omega_n - j\omega_d$$

where

$$\omega_d = \omega_n\sqrt{1-\xi^2} \tag{4.43}$$

or

$$\omega_n = \frac{\omega_d}{\sqrt{1-\xi^2}} \tag{4.44}$$

Here, ω_d is the damped natural frequency or damped oscillation frequency or damped frequency.

Therefore, the closed-loop poles are complex conjugates and lie in the left-half s-plane, as shown in Figure 4.16. The system is then called underdamped. Thus, Eq. (4.39) can be expressed as

$$\frac{C(s)}{R(s)} = \frac{\omega_n^2}{(s+\xi\omega_n + j\omega_d)(s+\xi\omega_n - j\omega_d)} \tag{4.45}$$

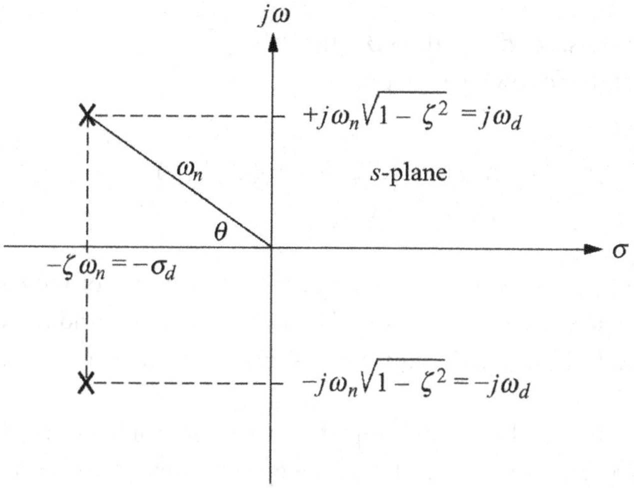

FIGURE 4.16 Poles or roots of the second-order system for underdamped case.

For the unit-step response, then we substitute $R(s)=\frac{1}{s}$ in Eq. (4.45), we get

$$C(s)=\frac{\omega_n^2}{\left(s+\xi\omega_n+j\omega_d\right)\left(s+\xi\omega_n-j\omega_d\right)}\cdot\frac{1}{s}$$

$$=\frac{1}{s}-\frac{s+\xi\omega_n}{\left(s+\xi\omega_n\right)^2+\omega_d^2}-\frac{\xi\omega_n}{\left(s+\xi\omega_n\right)^2+\omega_d^2} \tag{4.46}$$

$$=\frac{1}{s}-\frac{s+\xi\omega_n}{\left(s+\xi\omega_n\right)^2+\omega_d^2}-\frac{\xi}{\sqrt{1-\xi^2}}\cdot\frac{\omega_d}{\left(s+\xi\omega_n\right)^2+\omega_d^2} \tag{4.47}$$

Taking the inverse Laplace transform of Eq. (4.47), we obtain

$$c(t)=1-e^{-\xi\omega_n t}\left(\cos\omega_d t+\frac{\xi}{\sqrt{1-\xi^2}}\sin\omega_d t\right) \tag{4.48}$$

Here,

$$\mathcal{L}^{-1}\left\{\frac{s+\xi\omega_n}{\left(s+\xi\omega_n\right)^2+\omega_d^2}\right\}=e^{-\xi\omega_n t}\cos\omega_d t;\mathcal{L}^{-1}\left\{\frac{\omega_d}{\left(s+\xi\omega_n\right)^2+\omega_d^2}\right\}=e^{-\xi\omega_n t}\sin\omega_d t$$

Thus,

$$c(t)=1-\frac{e^{-\xi\omega_n t}}{\sqrt{1-\xi^2}}\left(\sqrt{1-\xi^2}\cos\omega_d t+\xi\sin\omega_d t\right);\ \text{for } t\geq 0 \tag{4.49}$$

$$c(t)=1-\frac{e^{-\xi\omega_n t}}{\sqrt{1-\xi^2}}\sin\left(\omega_d t+\tan^{-1}\frac{\sqrt{1-\xi^2}}{\xi}\right);\ \text{for } t\geq 0 \tag{4.50}$$

Let $\sin\beta=\sqrt{1-\xi^2}$; thus, $\cos\beta=\xi$; then $\beta=\tan^{-1}\frac{\sqrt{1-\xi^2}}{\xi}$.

Therefore, we get the following equation

$$c(t)=1-\frac{e^{-\xi\omega_n t}}{\sqrt{1-\xi^2}}\sin(\omega_d t+\beta) \tag{4.51}$$

According to Eq. (4.50), when $0<\xi<1$, the unit-step response of second-order system is oscillating with the attenuation of the angular frequency in ω_d, and the response curve of the system is shown in Figure 4.17. Figure 4.17 shows that the amplitude is increasing with ξ decreasing.

According to Eq. (4.51), the underdamped unit-step response is divided into two parts: (a) the part of steady-state system under unit-step function, there is no SSE and (b) transient part that is damped sinusoidal oscillation; the oscillation frequency is ω_d, so it is called damped the oscillation frequency and its value and damped ratio are related. The second-order oscillation system is less damped state from ξ=0.2, 0.4, 0.6, and 0.8; the curve is shown in Figure 4.17.

%The MATLAB program for Figure 4.17a are as follows:

```
T=10;
ut=1;
Wn=0.10;
t=0:100;
hold on
plot(t, ut+0*(ut+t))
for Z=[0,0.1,0.2,0.3,0.4,0.5,0.6,0.7,0.8,1,2]
y=exp(-Z.*Wn.*t);
Wd=Wn.*(sqrt(1-Z.*Z));
B=atan((sqrt(1-Z.*Z))./Z);
ct=1-(y./(sqrt(1-Z.*Z))).*sin(Wd.*t+B);
hold on
plot(t, ct)
end
xlabel('t')
ylabel('c(t)')
title('Underdamped step responses of second-order system
oscillation')
grid
```

%The MATLAB program for Figure 4.17b are as follows:

```
for j=1:4
Z(j)=j*2/10;
beta=atan((sqrt(1-Z(j)*Z(j)))/Z(j));
for i=1:200
```

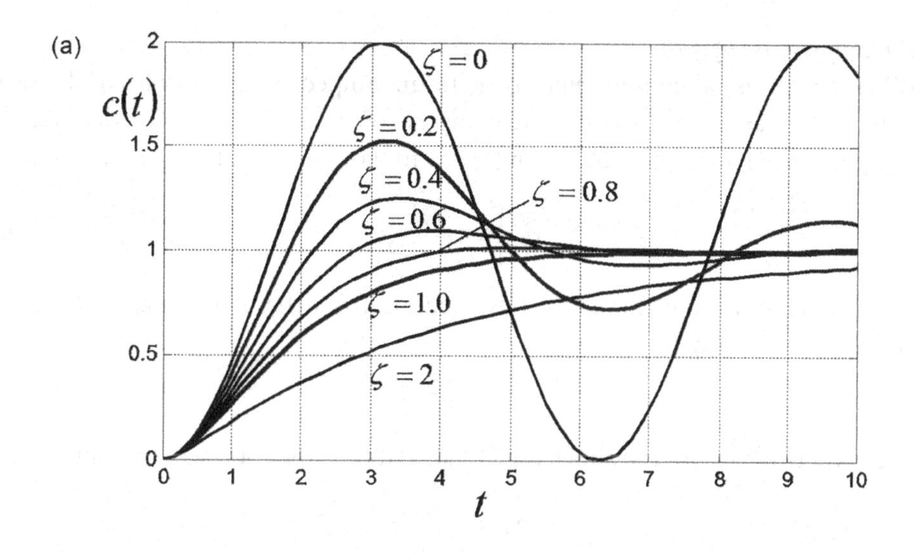

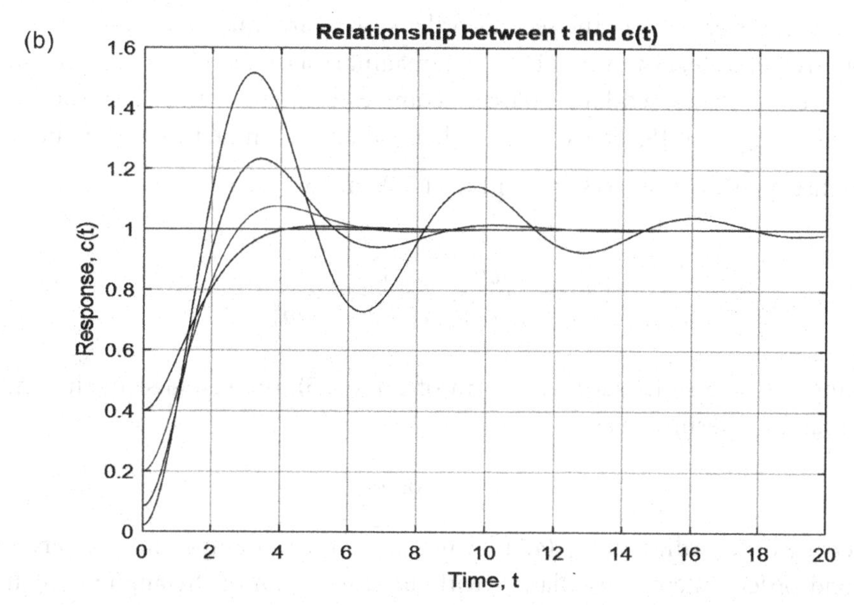

FIGURE 4.17 Unit-step response of the second-order system for underdamped case.

```
wntss(i, j)=(i-1)/10;
xos(i, j)=1-exp(-wntss(i, j)*Z(j))*sin(wntss(i,
j)*sqrt(1-Z(j)*Z(j))+beta);
xo1s(i, j)=1;
end
end
plot(wntss, xos, wntss, xo1s)
grid
xlabel('Time, t')
ylabel('Response, c(t)')
title('Oscillation relationship between t and c(t)')
```

2. *Undamped case ($\xi = 0$)*
   When $\xi = 0$, it is called *undamped case.* In undamped state, the poles of the second-order system are a pair of conjugate complex roots, i.e., the second-order system has two poles. We recall Eq. (4.39), then for the undamped case it can be expressed as

$$\frac{C(s)}{R(s)} = \frac{\omega_n^2}{s^2 + \omega_n^2} \tag{4.52}$$

Thus, we can find the characteristics equation from the denominator of Eq. (4.52), and then we can say

$$s^2 + \omega_n^2 = 0 \tag{4.53}$$

There are two roots of the above Eq. (4.53), and thus the roots can be expressed as

$$s_{1,2} = \pm j\omega_n \tag{4.54}$$

The two roots or poles of the second-order system are imaginary roots and same but roots are opposite; however, it is on the imaginary axis, as shown in Figure 4.18.

Therefore, the closed-loop poles are complex conjugates and lie on the s-plane, as shown in Figure 4.18. The system is then called undamped case. For the unit-step response, we substitute $R(s) = \frac{1}{s}$ in Eq. (4.52), and we get

$$C(s) = \frac{\omega_n^2}{s^2 + \omega_n^2} \cdot \frac{1}{s} = \frac{1}{s} - \frac{1}{s^2 + \omega_n^2} \tag{4.55}$$

Taking the inverse Laplace transform of Eq. (4.55), the response of the undamped case can be determined as

$$c(t) = 1 - \cos(\omega_n t) \tag{4.56}$$

where $t \geq 0$; according to Eq. (4.56), when $\xi = 0$, the undamped unit-step response of second-order system is oscillating without attenuation of the angular frequency in ω_d, and the response curve of the system is shown in Figure 4.19. Figure 4.19 shows that the amplitude is not increasing with ξ. The response curve of the system is a

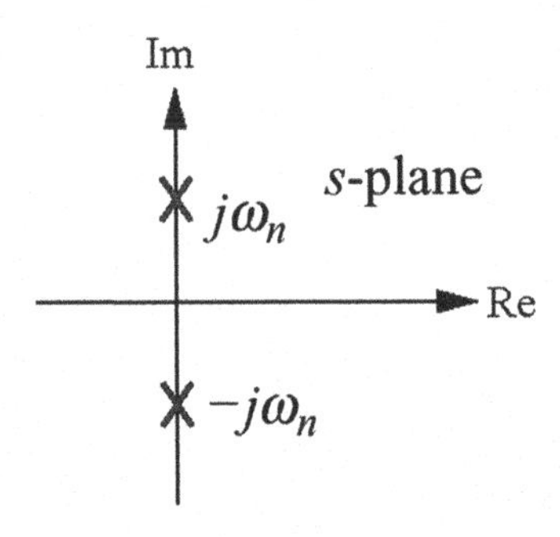

FIGURE 4.18 Poles or roots of the second-order system for undamped case.

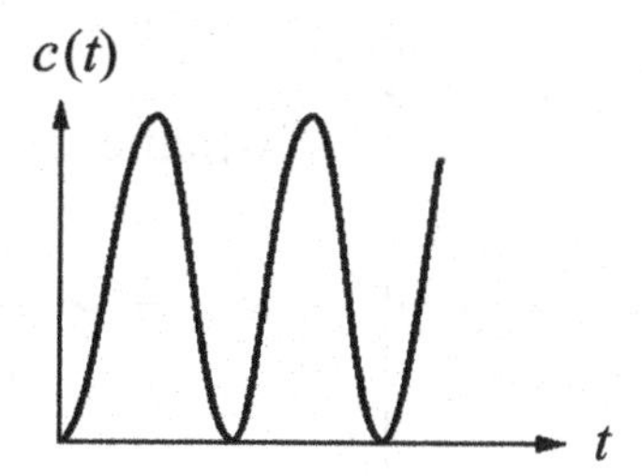

FIGURE 4.19 Unit-step response of the second-order system for undamped case.

nondamped oscillation. ω_n is the nondamped oscillation frequency, the value of ω_n is determined by the structural parameters of the system itself; thus, ω_n is called the natural frequency.

```
%The MATLAB program are as follows:
T=10;
ut=1;
Wn=0.10;
t=0:100;
hold on
plot(t, ut+0*(ut+t))
Z=0;
ct=1-cos(Wn*t);
hold on
plot(t, ct)
xlabel('t')
ylabel(' Unit response, c(t)')
title('Undamped step responses of second-order system
oscillation')
grid
```

3. *Critically damped case ($\xi = 1$)*

When $\xi = 1$, it is called *critically damped case.* In a critically damped case, the poles of the second-order system are a pair of real roots, i.e., the second-order system has two real poles. We recall Eq. (4.39), and then for the critically damped case, it can be expressed as

$$\frac{C(s)}{R(s)} = \frac{\omega_n^2}{s^2 + 2\omega_n s + \omega_n^2} \tag{4.57}$$

Thus, we can find the characteristics equation from the denominator of Eq. (4.57), and then we can say

$$s^2 + 2\omega_n s + \omega_n^2 = 0 \tag{4.58}$$

There are two roots of the above Eq. (4.58), and thus the roots can be expressed as

$$s_1 = s_2 = -\omega_n \tag{4.59}$$

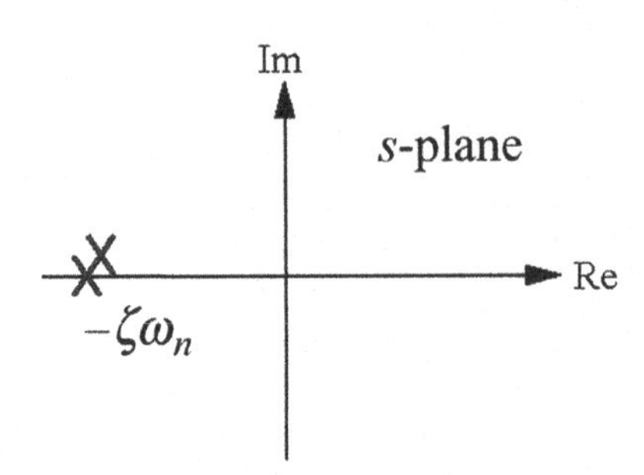

FIGURE 4.20 Poles or roots of the second-order system for critically damped case.

The two roots or poles of the second-order system are real roots or real poles and the same roots or poles; however, it is on the real axis at the same point, as shown in Figure 4.18.

Therefore, the closed-loop poles are real and lie on the s-plane, as shown in Figure 4.20. The system is then called a critically damped case. For the unit-step response, we substitute $R(s)=\frac{1}{s}$ in Eq. (4.57), we get

$$C(s)=\frac{\omega_n^2}{(s+\omega_n)^2}.\frac{1}{s}=\frac{1}{s}-\frac{1}{s+\omega_n}-\frac{\omega_n}{(s+\omega_n)^2} \quad (4.60)$$

Taking the inverse Laplace transform of Eq. (4.60), the response of the undamped case can be determined as

$$c(t)=1-e^{-\omega_n t}-\omega_n te^{-\omega_n t}=1-(1+\omega_n t)e^{-\omega_n t} \quad (4.61)$$

where $t\geq 0$; according to Eq. (4.61), when $\xi=1$, the critically damped for unit-step response of second-order system is not well oscillating without attenuation of the angular frequency in ω_d; there is no overshoot in the system; and the response curve of the system is shown in Figure 4.21. Figure 4.21 shows that the amplitude is not increasing with ξ and the response is steady at a certain value of t.

The critically damped is a nonperiodic rising process with a steady-state value of 1. The rate of change of critically damped response can be found from Eq. (4.61), and then we get

$$\frac{d}{dt}[c(t)]=\omega_n^2 te^{-\omega_n} \quad (4.62)$$

When $t=0$, the rate of change of critically damped response is zero; when $t>0$, the rate of change of critically damped response is positive, whereas $c(t)$ is monotonically increasing, when the rate of change of critically damped response ($t\infty$) approaches zero; thus, the response characteristics of the critically damped case are not oscillating at the steady state, as shown in Figure 4.21.

```
%The MATLAB program are as follows:
T=10;
```

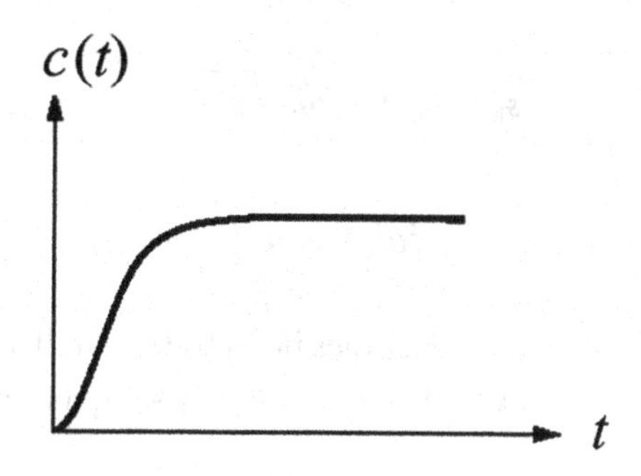

FIGURE 4.21 Unit-step response of the second-order system for critically damped case.

```
ut=1;
Wn=0.10;
t=0:100;
hold on
plot(t, ut+0*(ut+t))
Z=1;
ct=1-(1+Wn*t).*exp(-Wn*t);
hold on
plot(t, ct)
xlabel('t')
ylabel('c(t)')
title('Critically damped step responses of second-order system')
grid
```

4. Overdamped case ($\xi > 1$)

When $\xi > 1$, it is called *overdreamed case.* In overdamped case, the poles of the second-order system are a pair of negative real roots, i.e., the second-order system has two negative real poles. We recall Eq. (4.39), and we can find the characteristics equation from the denominator of Eq. (4.39) and then the roots can be expressed as

$$s_{1,2} = \frac{-2\xi\omega_n \pm \sqrt{(2\xi\omega_n)^2 - 4.1.\omega_n^2}}{2.1}$$

$$= -\sigma \pm \sqrt{(\xi\omega_n)^2 - \omega_n^2}$$

$$= -\sigma \pm \sqrt{(\xi\omega_n)^2 - \omega_n^2}$$

$$= -\sigma \pm \omega_n\sqrt{\xi^2 - 1}$$

Thus, we can say the two poles or two roots as

$$s_{1,2} = -\sigma \pm \omega_n\sqrt{\xi^2 - 1} \tag{4.63}$$

i.e.

$$s_1 = -\xi\omega_n + \omega_n\sqrt{\xi^2 - 1}$$

and

$$s_2 = -\xi\omega_n - \omega_n\sqrt{\xi^2 - 1}$$

Therefore, the closed-loop negative real poles lie in the left-half s-plane, as shown in Figure 4.22. The system is then called overdamped case. Thus, Eq. (4.39) can be expressed as

$$\frac{C(s)}{R(s)} = \frac{\omega_n^2}{\left(s+\xi\omega_n - \omega_n\sqrt{\xi^2-1}\right)\left(s+\xi\omega_n + \omega_n\sqrt{\xi^2-1}\right)} \tag{4.64}$$

For the unit-step response, then we substitute $R(s) = \frac{1}{s}$ in Eq. (4.45), and we get

$$C(s) = \frac{\omega_n^2}{\left(s+\xi\omega_n - \omega_n\sqrt{\xi^2-1}\right)\left(s+\xi\omega_n + \omega_n\sqrt{\xi^2-1}\right)} \cdot \frac{1}{s}$$

$$= \frac{1}{s} + \frac{\xi - \sqrt{\xi^2-1}}{2\sqrt{\xi^2-1}} \cdot \frac{1}{s+\xi\omega_n + \omega_n\sqrt{\xi^2-1}} + \frac{-\xi - \sqrt{\xi^2-1}}{2\sqrt{\xi^2-1}} \cdot \frac{1}{s+\xi\omega_n - \omega_n\sqrt{\xi^2-1}}$$

$$= \frac{1}{s} + \frac{1}{2\sqrt{\xi^2-1}\left(\xi+\sqrt{\xi^2-1}\right)} \cdot \frac{1}{s+\xi\omega_n + \omega_n\sqrt{\xi^2-1}} - \frac{1}{2\sqrt{\xi^2-1}\left(\xi-\sqrt{\xi^2-1}\right)} \cdot \frac{1}{s+\xi\omega_n - \omega_n\sqrt{\xi^2-1}}$$

$$= \frac{1}{s} + \frac{\omega_n}{2\sqrt{\xi^2-1}.\omega_n\left(\xi+\sqrt{\xi^2-1}\right)} \cdot \frac{1}{s+\omega_n\left(\xi+\sqrt{\xi^2-1}\right)} - \frac{\omega_n}{2\sqrt{\xi^2-1}.\omega_n\left(\xi-\sqrt{\xi^2-1}\right)} \cdot \frac{1}{s+\omega_n\left(\xi-\sqrt{\xi^2-1}\right)}$$

$$= \frac{1}{s} + \frac{\omega_n}{2\sqrt{\xi^2-1}.T1} \cdot \frac{1}{s+1} - \frac{\omega_n}{2\sqrt{\xi^2-1}.T2} \cdot \frac{1}{s+T2}$$

$$= \frac{1}{s} + \frac{\omega_n}{2\sqrt{\xi^2-1}}\left(\frac{1}{T_1} \cdot \frac{1}{s+T1} - \frac{1}{T_2} \cdot \frac{1}{s+T_2}\right) \tag{4.65}$$

where $T_1 = \omega_n\left(\xi+\sqrt{\xi^2-1}\right)$, and $T_2 = \omega_n\left(\xi-\sqrt{\xi^2-1}\right)$. Both terms have a time dimension that is called the second-order overdamped time constant, and obviously $T_1 \neq T_2$.

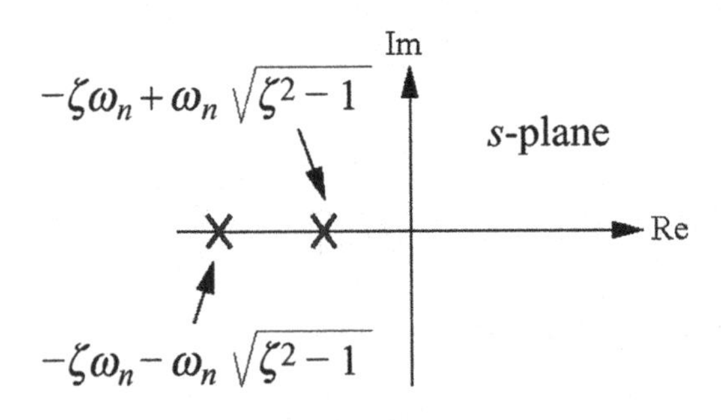

FIGURE 4.22 Poles or roots of the second-order system for overdamped case.

Taking the inverse Laplace transform of Eq. (4.65), we obtain

$$c(t)=1+\frac{\omega_n}{2\sqrt{\xi^2-1}}\left(\frac{1}{T_1}.e^{-T_1t}-\frac{1}{T_2}.e^{-T_2t}\right) \tag{4.66}$$

where $t\geq 0$; according to Eq. (4.66), when $\xi>1$, the overdamped unit-step response of second-order system is not well oscillating without attenuation of the angular frequency in ω_d; there is no overshoot in the system; and the response curve of the system is shown in Figure 4.23. Figure 4.23 shows that the amplitude is increasing with ξ and response is not steady after at a certain value of t.

```
%The MATLAB program are as follows:
ut=1; Wn=0.10;
t=0:100;
hold on
plot(t, ut+0*(ut+t))
Z=2;
T1=Wn.*(Z+sqrt(Z.*Z-1)); T2=Wn.*(Z-sqrt(Z.*Z-1));
A=Wn/(2*sqrt(Z.*Z-1));
ct=1+A*((1/T1)*exp(-T1*t)-(1/T2)*exp(-T2*t));
hold on
plot(t, ct)
xlabel('t')
ylabel('c(t)')
title('Critically damped step responses of second-order system')
grid
```

However, from Eq. (4.66), it is noted that e^{-T_1t} and e^{-T_2t} are called decaying exponential functions. If the damping ratio $\xi>1$, one of the two decaying exponentials decreases much faster than the other, so the faster-decaying exponential term (which corresponds to a smaller time constant) may be neglected. That is, if $-T_2$ is located very much closer to the $j\omega$ -axis than $-T_1$ (which means $|-T_2|<<|-T_1|$); then for an approximate solution, we may neglect $-T_1$. This is permissible because the effect of $-T_1$ on the response is much smaller than that of $-T_2$, since the term involving T_1 in Eq. (4.66); thus, T_1 term part decays much faster than the term involving T_2.

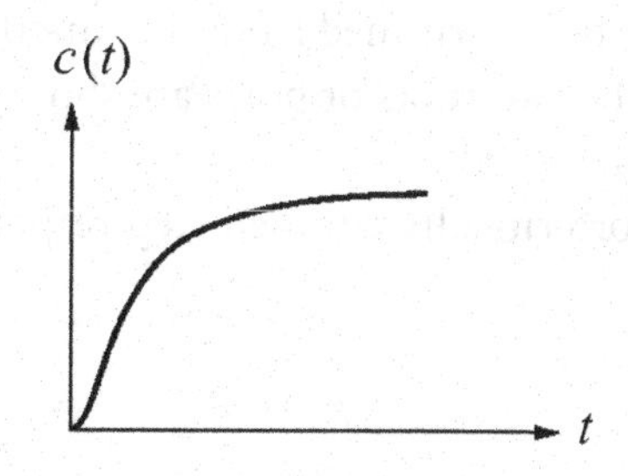

FIGURE 4.23 Unit-step response of the second-order system for overdamped case.

Once the faster-decaying exponential term has disappeared, the response is similar to that of a first-order system, and $C(s)/R(s)$ may be approximated by the transfer function as

$$\frac{C(s)}{R(s)}=\frac{\xi\omega_n-\omega_n\sqrt{\xi^2-1}}{s+\xi\omega_n-\omega_n\sqrt{\xi^2-1}}=\frac{T_2}{s+T_2} \tag{4.67}$$

Unit-step response curves of system can be represented as

$$C(s)=\frac{T_2}{s+T_2}.\frac{1}{s}=\frac{1}{s}-\frac{1}{s+T_2} \tag{4.68}$$

Taking the inverse Laplace transform of Eq. (4.68), we obtain

$$c(t)=1-e^{-T_2t};\text{ for } t\geq 0 \tag{4.69}$$

Therefore, the time response of the second-order system for an overdamped case can be seen from

$$c(t)=1-e^{-\left(\zeta-\sqrt{\zeta^2-1}\right)\omega_nt};\text{ for } t\geq 0 \tag{4.70}$$

Comments: In practical systems, it is usually with a certain damped ratio, so it is impossible to determine the natural frequency ω_n through the experimental method. Only measured damped frequency ω_d, because $\omega_d=\omega_n\sqrt{1-\xi^2}$, when $0<\xi<1$, the damped frequency is always lower than the natural frequency. When ξ increases and the damped oscillation frequency ω_d decreases, if $\xi\geq 1$, the damped frequency will not exist, and the system response does not have possesses oscillation. It can be drawn from the above analysis that for the second-order system, the response of $\xi=0$ is the oscillation amplitude, if damped is positive, the second-order system is stable. For the vast majority of mechanical and electrical control systems, the general second-order systems have positive damped, and the system is stable in that time frame.

As we can see in the sets of family curves as shown in Figure 4.17, if the damping ratio $0<\xi<1$, then the underdamped response is quickly oscillated with higher amplitude value that indicates the underdamped case; if the damping ratio $\xi=1$, then the critically response is slowly oscillating with lower amplitude value and which is under the input commands and very closed to the input line that indicates the critically damped case; and however, if the damping ratio $\xi>1$, then the overdamped response is very slowly responding with very low amplitude value which is always sluggish and that indicates the overdamped case of the second-order system.

As a summary, the step response $c(t)$ of the second-order system is shown in the third column of Table 4.4, and it can be determined from the function of ξ which can be seen in the first column of Table 4.4; the two roots or poles are shown on s-plane as we can see in the second column of Table 4.4.

%The MATLAB program for critically damped step responses of second-order system are as follows:

```
ut=1;
Wn=0.10;
t=0:100;
hold on
```

TABLE 4.4 Second-order step response as a function of ζ.

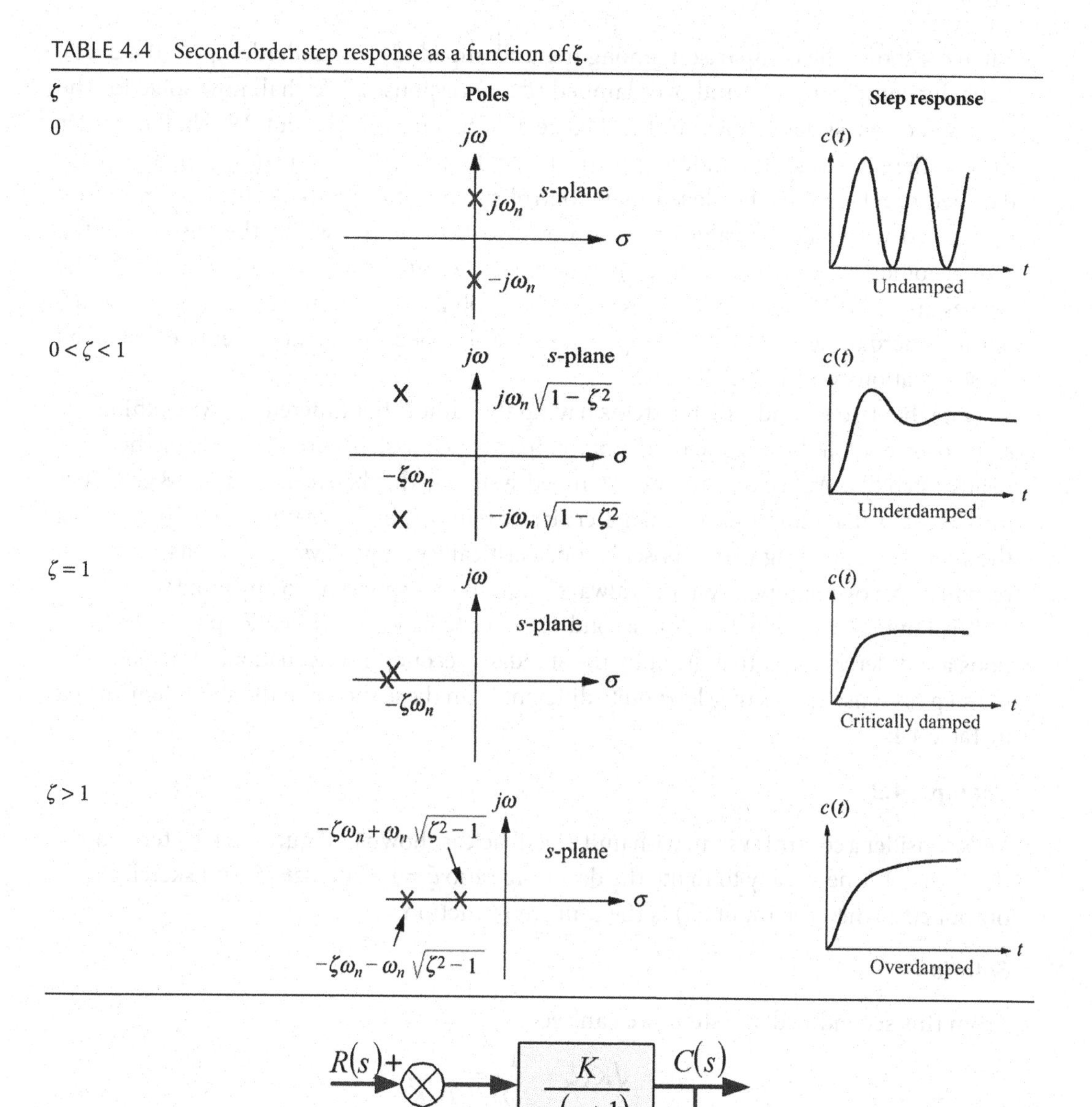

FIGURE 4.24 Control system with unity feedback.

```
plot(t, ut+0*(ut+t))
Z=2;
T2=Wn.*(Z-sqrt(Z.*Z-1));
ct=1+exp(-T2*t);
hold on
plot(t, ct)
xlabel('t')
ylabel('c(t)')
title('Critically damped step responses of second-order
system');grid
```

Table 4.4 shows the comparison among the undamped ($\xi = 0$), underdamped ($0 < \xi < 1$), critically damped ($\xi = 1$), and overdamped ($\xi > 1$) responses. We shall now solve for the response of the system shown in this Figure 4.24 to a unit-step input. We shall consider three different cases: the underdamped ($0 < \xi < 1$), critically damped ($\xi = 1$), and overdamped ($\xi > 1$) cases. The closed-loop control system of a family of unit-step response curves $c(t)$ with various values of ζ is shown in this Figure, where the abscissa is the dimensionless variable $\omega_n t$. *Unit-step response of second-order system for different* ξ. The curves are functions only of ξ. These curves are obtained from the system response $c(t)$ of the underdamped, critically damped, and overdamped cases. The system described by these equations was initially at rest.

Note that two second-order systems having the same ξ but different ω_n will exhibit the same overshoot and the same oscillatory pattern. Such systems are said to have the same relative stability. We see that an underdamped system with ξ between 0.5 and 0.8 gets close to the final value more rapidly than a critically damped or overdamped system. Among the systems responding without oscillation, a critically damped system exhibits the fastest response. An overdamped system is always sluggish in responding to any inputs.

It is important to note that, for second-order systems whose closed-loop transfer functions are different from that given by the standard second-order equation, remember that the step response curves may look quite different from those shown in the above figures and in Table 4.4.

Example 4.2

Let's consider a control system with unity feedback as shown in Figure 4.24. Determine the value of K necessary to make the damping ratio $\xi = 1$. Calculate K and sketch the output $c(t)$) when the input $r(t)$ is the unit step function.

Solution

From this second-order system, we can see

$$\omega_n = \sqrt{K}, \xi = \frac{1}{2\omega_n} = \frac{1}{2\sqrt{K}}$$

Since the damping ratio $\xi = 1$, $K = 1/4$.

If the input is the unit step function, the output may be written as this function.

$$G(s) = \frac{1/4}{s(s+1/2)^2} = \frac{1}{s} - \frac{1}{s+1/2} - \frac{(1/2)^2}{(s+1/2)^2}$$

Thus, the time response is therefore written as like this.

$$c(t) = 1 - e^{-\frac{1}{2}t} - \frac{1}{2}te^{-\frac{1}{2}t} = 1 - \left(1 + \frac{1}{2}t\right)e^{-\frac{1}{2}t}$$

Therefore, the time response $c(t)$ can be drawn from this equation by using MATLAB and it can be seen in Figure 4.25.

%The MATLAB program is as follows:

```
ut=1;
Wn=0.10;
```

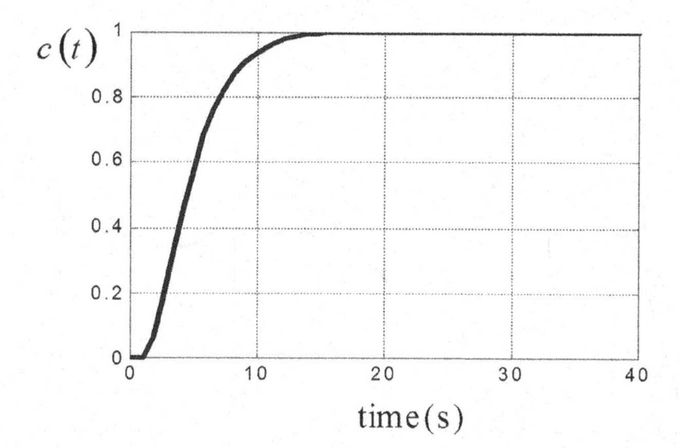

FIGURE 4.25 Time response.

```
t=0:40;
hold on
plot(t, ut+0*(ut+t))
ct=1-(1+0.5.*t).*exp(-0.5.*t);
hold on
plot(t, ct)
xlabel('t')
ylabel('c(t)')
title(step responses of second-order system')
grid
```

4.5 SECOND-ORDER SYSTEMS AND TRANSIENT RESPONSE SPECIFICATIONS

Frequently, the performance characteristics of a control system are specified in terms of the *transient response to a unit-step input*, since it is easy to generate and is sufficiently drastic. If the response to a step input is known, it is mathematically possible to compute the response to any input. The transient response of a system to a *unit-step input depends on the initial conditions.* For convenience in comparing transient responses of various systems, it is a common practice to use the standard initial condition that the system is at rest initially with the output and all-time derivatives thereof zero. Then the response characteristics of many systems can be easily compared. The transient response of a practical control system *often exhibits damped oscillations before reaching steady state.*

4.5.1 Definition of Transient Response Specifications

Stability is the most basic requirement of the control system. It can be meaningful to study the transient characteristics of the control system and it converges only after the transient response of the control system. The transient response specifications reflect the degree and speed of convergence of the control system. The transient characteristics of a system are usually measured or calculated in the time domain by the response under the action of a step function. In general, if the transient characteristics of the system under the action of step function can meet the requirements of the performance index or specifications, then

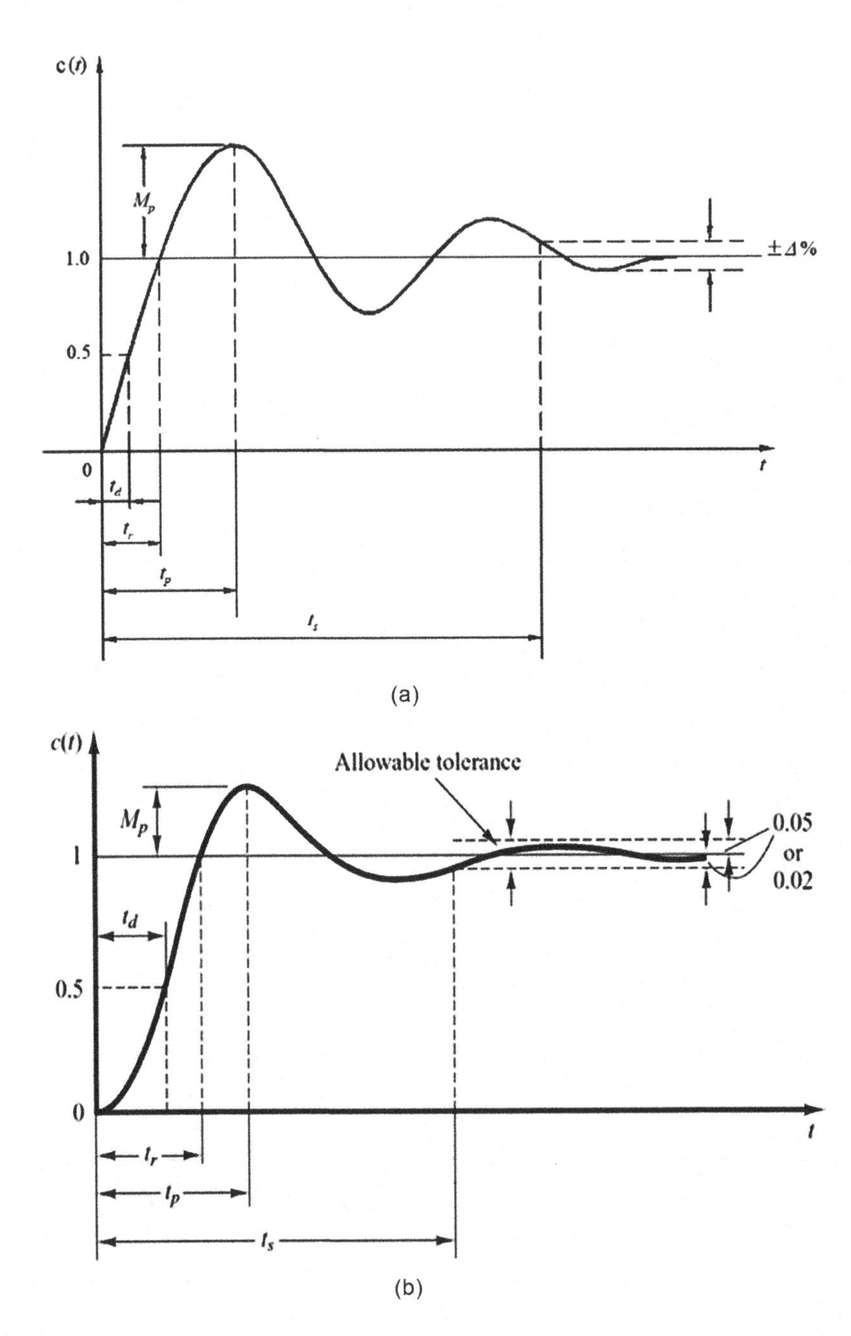

FIGURE 4.26 Specifications unit-step response curve.

the transient performance of the system under the action of other forms of function can also be satisfactory. As shown in Figure 4.26, the output response of the system is $c(t)$ under the action of the unit step function. It is assumed that the control system is in a relatively steady state before the action of the unit step function, and all the derivatives of the output amount are equal to zero; this assumption is realistic for most control systems.

In specifying the transient response characteristics of a control system to a unit-step input, it is common to specify the following specifications. The transient performance specifications can be described in the control system which is defined as

- Delay time, t_d;
- Rise time, t_r;
- Peak time, t_p;
- Maximum overshoot, M_p;
- Settling time, t_s;
- Number of the oscillation frequency, N.

1. *Delay time, t_d*
   The delay time is the time required for the response to reach half the final value the very first time—the time for the response to rise from 0% to 50% of its final value.
2. *Rise time, t_r*
   The rise time is the time required for the response to rise from 10% to 90%, 5% to 95%, or 0% to 100% of its final value. For underdamped second-order systems, the 0%–100% rise time is normally used. For overdamped systems, the 10%–90% rise time is commonly used.
3. *Peak time, t_p*
   The peak time is the time required for the response to reach the first peak of the overshoot.
4. *Maximum (percent) overshoot, M_p*
   The maximum overshoot is the maximum peak value of the response curve measured from unity. The difference between the maximum peak value of the response curve and the steady-state value is 1.

$$M_p = c(t_p) - c(\infty) \tag{4.71}$$

   If the final steady-state value of the response differs from unity, then it is common to use the maximum percent overshoot. It is defined by

$$M_p = \frac{c(t_p) - c(\infty)}{c(\infty)} \times 100\% \tag{4.72}$$

   The amount of the maximum (percent) overshoot directly indicates the relative stability of the system. Note that if the maximum output value $c(t_p)$ is less than steady-state value $c(\infty)$, then the response of system does not have maximum overshoot.
5. *Settling time, t_s*

The minimum time for the response curve to reach and stay within a range about the final value ±Δ%; as seen from Figure 4.26b, ±Δ% accepts ±2% or ±5%, and it may be determined from the objectives of the system design. Unless otherwise specified, the settling time takes ±5%. Settling time also called transition time and is related to the largest time constant of the control system. The settling time is the time required for the response curve to reach and stay within a range about the final value of size specified by absolute percentage of the final value (usually 2% or 5%). The settling time is related to the largest time constant of the control system. The percentage error criterion to use may be determined by the objectives of the system design.

4.5.2 Comments on Transient Response Specifications

Note that the oscillation times N, the settling time t_s and peak time t_p reflect the fast of the system; however, the maximum overshoot M_p (σ%) and oscillation times N reflect the steady state of the system.

The time-domain specifications just given are quite important, since most control systems are time-domain systems; that is, they must exhibit acceptable time responses. (This means that the control system must be modified until the transient response is satisfactory.)

Note that not all these transient response specifications necessarily apply to any given case. For example, for an overdamped system, the terms peak time and maximum overshoot do not apply. (For systems that yield SSEs for step inputs, this error must be kept within a specified percentage level. Detailed discussions of SSEs will be discussed later.)

Except for certain applications where oscillations cannot be tolerated, it is desirable that the transient response be sufficiently fast and be sufficiently damped. Thus, for a desirable transient response of a second-order system, the damping ratio must be between 0.4 and 0.8. Small values of ξ (that is, $\xi < 0.4$) yield excessive overshoot in the transient response, and a system with a large value of ξ (that is, $\xi > 0.8$) responds sluggishly.

We shall see later that the maximum overshoot (M_p) and the rise time (t_r) conflict with each other. In other words, both the maximum overshoot (M_p) and the rise time (t_r) cannot be made smaller simultaneously. If one of them is made smaller, the other necessarily becomes larger.

4.5.3 Transient Response Specifications of Second-Order Systems for Underdamped Case

Let's discuss how to derive transient response specifications of second-order systems. We shall obtain the rise time (t_r), peak time (t_p), maximum overshoot (M_p), and settling time (t_s) of the second-order system given by the standard second-order equation. These values will be obtained in terms of ξ and ω_n. Note that the system is assumed to be underdamped case ($0 < \xi < 1$).

1. *Rise time, t_r*

   We recall the underdamped case referring from Eq. (4.48) and We are letting the response of rise time, $c(t_r) = 1$, then we can simplify the following equation and we can rewrite it as follows:

$$c(t_r)=1=1-e^{-\xi\omega_n t}\left(\cos\omega_d t+\frac{\xi}{\sqrt{1-\xi^2}}\sin\omega_d t\right) \tag{4.73}$$

Since, $e^{-\zeta\omega_n t_r}\neq 0$, then we can find

$$\cos\omega_d t_r+\frac{\xi}{\sqrt{1-\xi^2}}\sin\omega_d t_r=0$$

or

$$\frac{\xi}{\sqrt{1-\xi^2}}\sin\omega_d t_r=-\cos\omega_d t_r$$

Thus, it can be simplified

$$\tan\omega_d t_r=\frac{\sqrt{1-\xi^2}}{-\xi}$$

or $\tan\omega_d t_r=\dfrac{\omega_n\sqrt{1-\xi^2}}{-\zeta\omega_n}$

Here, $\omega_n\sqrt{1-\xi^2}=\omega_d$ and $\xi\omega_n=\sigma$, as shown in Figure 4.27, then we have

$$\tan\omega_d t_r=-\frac{\omega_d}{\sigma}$$

Thus, the rise time, t_r, is

$$t_r=\frac{1}{\omega_d}\tan^{-1}\left(-\frac{\omega_d}{\sigma}\right)=\frac{1}{\omega_d}\tan^{-1}(-\beta)=\frac{1}{\omega_d}\tan^{-1}\left(\tan(\pi-\beta)\right)=\frac{\pi-\beta}{\omega_d}$$

$$t_r=\frac{\pi-\beta}{\omega_d} \tag{4.74}$$

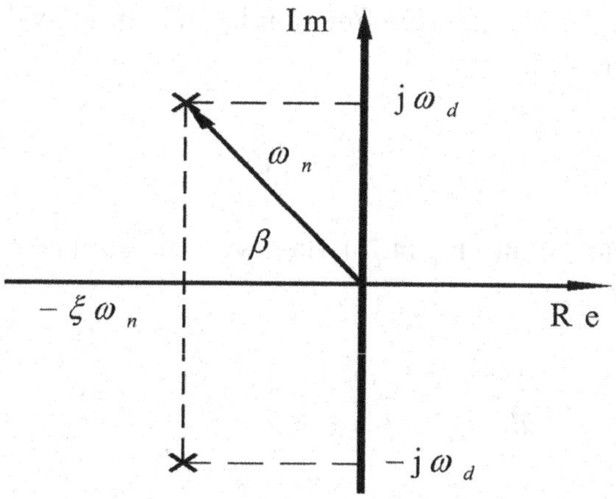

FIGURE 4.27 Definition of the angle.

where angle β is defined in Figure 4.27. Clearly, for a small value of rise time t_r, ω_d must be large.

2. *Peak time, t_p*
We recall the underdamped case referring to the system response signal of Eq. (4.48), we may obtain the peak time by differentiating $c(t)$ with respect to time and letting this derivative equal zero, then we can simplify the following equation and we can rewrite as follows

$$\frac{d}{dt}[c(t)]=\frac{d}{dt}\left[1-e^{-\xi\omega_n t}\left(\cos\omega_d t+\frac{\xi}{\sqrt{1-\xi^2}}\sin\omega_d t\right)\right]$$

or

$$\frac{dc(t)}{dt}=-\left[\left(\cos\omega_d t+\frac{\xi}{\sqrt{1-\xi^2}}\sin\omega_d t\right)\frac{d}{dt}e^{-\xi\omega_n t}+e^{-\xi\omega_n t}\frac{d}{dt}\left(\cos\omega_d t+\frac{\xi}{\sqrt{1-\xi^2}}\sin\omega_d t\right)\right]$$

Thus,

$$\frac{dc(t)}{dt}=\xi\omega_n e^{-\xi\omega_n t}\left(\cos\omega_d t+\frac{\xi}{\sqrt{1-\xi^2}}\sin\omega_d t\right)+e^{-\xi\omega_n t}\left(\omega_d\sin\omega_d t-\frac{\xi\omega_d}{\sqrt{1-\xi^2}}\cos\omega_d t\right)$$

i.e.,

$$\frac{dc(t)}{dt}=e^{-\xi\omega_n t}\left[\xi\omega_n\cos\omega_d t+\frac{\xi^2\omega_n}{\sqrt{1-\xi^2}}\sin\omega_d t+\omega_d\sin\omega_d t-\xi\omega_n\cos\omega_d t\right]$$

Therefore,

$$\frac{dc(t)}{dt}=\frac{\omega_n e^{-\zeta\omega_n t}}{\sqrt{1-\xi^2}}\sin\omega_d t \tag{4.75}$$

According to the definition, if the function has maximum value, and its unit-order differentiate is zero, i.e.,

$$\frac{dc(t)}{dt}=0$$

Substituting Eq. (4.76) into Eq. (4.75), then we evaluate at $t=t_p$ and we can simplify Eq. (4.75), we get

$$\left.\frac{dc(t)}{dt}\right|_{t=t_p}=\frac{\omega_n}{\sqrt{1-\xi^2}}e^{-\xi\omega_n t_p}\sin\omega_d t_p=0$$

Since $\frac{\omega_n}{\sqrt{1-\xi^2}}e^{-\xi\omega_n t_p}\neq 0$, then we get

$$\sin\omega_d t_p = 0 = \sin n\pi$$

Thus,

$$t_p = \frac{n\pi}{\omega_d} \tag{4.76}$$

Here, $n = 0,1,2,\ldots$; when $n = 1,3,5,\ldots$; which is an odd number, then there is a positive peak that is forward overshoot; while $n = 2,4,6,\ldots$; which is an even number, then the system appears negative to overshoot. The overshoot time is found in Figure 4.26a as

$$t_p = \frac{\pi}{\omega_n\sqrt{1-\xi^2}}, \frac{2\pi}{\omega_n\sqrt{1-\xi^2}}, \frac{3\pi}{\omega_n\sqrt{1-\xi^2}}, \ldots$$

However, if $n = 1$, then the peak time corresponds to the first peak overshoot, $\omega_d t_p = \pi$. Hence, we can define the peak time t_p as

$$t_p = \frac{\pi}{\omega_d} = \frac{\pi}{\omega_n\sqrt{1-\xi^2}} \tag{4.77}$$

Therefore, according to Eq. (4.77), the peak time tp corresponds to one-half cycle of the frequency of damped oscillation; in another word, the peak time is inversely proportional to the absolute value of the imaginary part of the closed-loop poles. When the damped ratio is certain, and distance between poles on the real axis distance farther, the peak time of the system is shorter.

3. *Maximum overshoot, M_p*

   To discuss maximum overshoot (M_p), the maximum overshoot occurs at the peak time or at $t = t_p = \pi/\omega_d$. Assuming that the final value of the output $c(\infty)$ is unity that is $c(\infty) = 1$, whereas maximum overshoot (M_p) is obtained from system response signal $c(t)$ of underdamped case. Thus, according to Eq. (4.49), we obtain the response of underdamped case as

$$c(t_p) = 1 - \frac{e^{-\xi\omega_n t_p}}{\sqrt{1-\xi^2}}\left(\sqrt{1-\xi^2}\cos\omega_d t_p + \xi\sin\omega_d t_p\right) \tag{4.78}$$

Substituting peak time, $t_p = \dfrac{\pi}{\omega_d}$ and then the maximum overshoot $M_p = c(t_p) - 1$ in Eq. (4.78). i.e.

$$M_p = c(t_p) - c(\infty) = c(t_p) - 1$$

i.e.,

$$M_p = \left[1 - \frac{e^{-\xi\omega_n\frac{\pi}{\omega_d}}}{\sqrt{1-\xi^2}}\left(\sqrt{1-\xi^2}\cos\omega_d\frac{\pi}{\omega_d} + \xi\sin\omega_d\frac{\pi}{\omega_d}\right)\right] - 1$$

or

$$M_p = -e^{-\xi\omega_n(\pi/\omega_d)}\left(\cos\pi + \frac{\xi}{\sqrt{1-\xi^2}}\sin\pi\right)$$

Thus,

$$M_p = e^{-\xi\omega_n(\pi/\omega_d)} = e^{-\left(\xi/\sqrt{1-\xi^2}\right)\pi}$$

i.e.,

$$M_p = e^{-\xi\omega_n(\pi/\omega_d)} \tag{4.79}$$

or

$$M_p = e^{-\left(\xi/\sqrt{1-\xi^2}\right)\pi} \tag{4.80}$$

Therefore, the maximum percent overshoot can be obtained as

$$M_p = e^{-\xi\omega_n(\pi/\omega_d)} \times 100\% \tag{4.81}$$

or

$$M_p = e^{-\left(\xi/\sqrt{1-\xi^2}\right)\pi} \times 100\% \tag{4.82}$$

If the final value $c(\infty)$ of the output is not unity, then we need to use the following equation as we have defined in Eq. (4.72), we can rewrite

$$M_p = \frac{c(t_p) - c(\infty)}{c(\infty)} \times 100\%$$

The changes for maximum overshoot M_p and damped ratio ξ are performed by MATLAB program which is shown in Table 4.5 as well as shown in Figure 4.28.

```
%The MATLAB program are as follows:
for i=1:99
Zt(i)=i/100;
wn=1;
mpp(i)=(exp(-Zt(i)/(sqrt(1-Zt(i).*Zt(i))).*pi))*100;
end
```

TABLE 4.5 Maximum overshoot of different damped ratio ξ

ξ	0	0.1	0.2	0.3	0.4	0.5	0.6	0.7	0.8	0.9	1.0
M_p	100	72.9	52	37.2	25.4	16.3	9.4	4.3	1.5	0.15	0

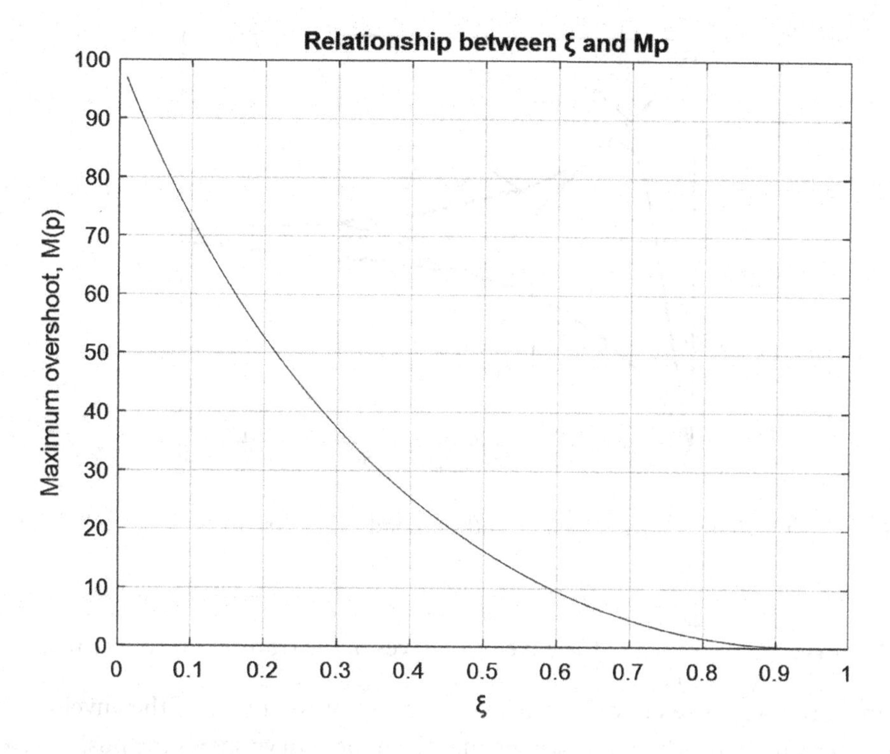

FIGURE 4.28 Relationship between M_p and ξ of second-order system.

```
plot(Zt, mpp)
grid
xlabel(' ξ ')
ylabel('Maximum overshoot, M(p)')
title('Relationship between ξ and Mp')
```

According to Eq. (4.80), Figure 4.28, and Table 4.5, we obtain the maximum overshoot which is only related to the damped ratio ξ; however, ξ is larger but M_p % is smaller.

4. *Settling time, t_s*

The transient response is obtained from the system response signal $c(t)$ of the underdamped case of second-order system.

Using Eq. (4.50) or (4.51) for the output curve of unit-step response is defined from

$$c(t)=1-\frac{e^{-\xi\omega_n t}}{\sqrt{1-\xi^2}}\sin\left(\omega_d t+\tan^{-1}\frac{\sqrt{1-\xi^2}}{\xi}\right);\text{ for } t\geq 0$$

or

$$c(t)=1-\frac{e^{-\xi\omega_n t}}{\sqrt{1-\xi^2}}\sin(\omega_d t+\beta)$$

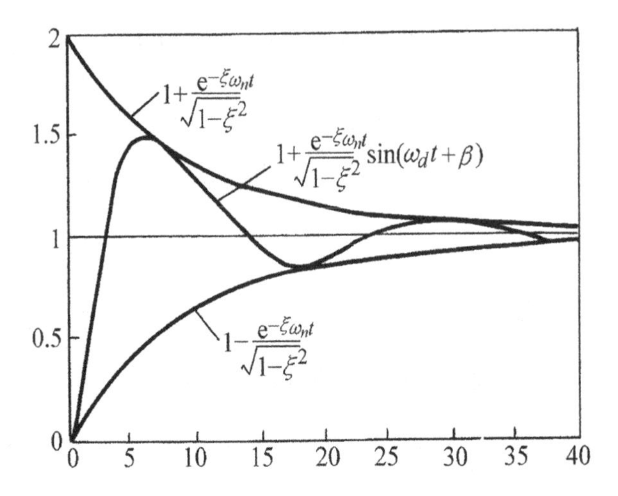

FIGURE 4.29 Unit-step response envelope of the second-order system.

The curves $1\pm\frac{e^{-\xi\omega_n t}}{\sqrt{1-\xi^2}}$ are the envelope curves of the transient response to a unit-step input. The response curve $c(t)$ always remains within a pair of the envelope curves, as shown in Figure 4.29. We can see that the upper curve shows the positive envelope of transient response which is lightly damped compared with negative envelope of transient response which is shown in a lower curve that is sluggish. The time constant of these envelope curves is $T=1/\xi\omega_n$. The speed of decay of the transient response depends on the value of the time constant $T=1/\xi\omega_n$. For a given ω_n, the settling time t_s is a function of the damping ratio ξ.

In Eq. (4.50) or (4.51), the unit-step response $c(t)$ of second-order system for different ξ, we see that for the same ω_n and for a range of ξ between 0 and 1, the settling time t_s for a very lightly damped system is larger than that for a properly damped system. For an overdamped system, the settling time t_s becomes large because of the sluggish response.

Response time $\omega_n t$

The envelope is symmetric with steady-state component on the transient response, obviously, the error range is defined as

$$\left|\frac{e^{-\xi\omega_n t}}{\sqrt{1-\xi^2}}\sin\left(\omega_d t+\tan^{-1}\frac{\sqrt{1-\xi^2}}{\xi}\right)\right|\le\frac{e^{-\xi\omega_n t_s}}{\sqrt{1-\xi^2}}=\Delta\% \tag{4.83}$$

To simplify Eq. (4.83), we obtain

$$t_s=-\frac{\ln\Delta+\ln\sqrt{1-\xi^2}}{\xi\omega_n} \tag{4.84}$$

If $\Delta = 5\%$, then we get the settling time as

$$t_s = -\frac{\ln 5\% + \ln\sqrt{1-\xi^2}}{\xi\omega_n} \tag{4.85}$$

When $\sqrt{1-\xi^2}$ is small, Eq. (4.85) can be simplified as

$$t_s = -\frac{\ln 0.05}{\xi\omega_n} = \frac{3}{\xi\omega_n} \tag{4.86}$$

Thus, when $t_s \geq \frac{3}{\xi\omega_n}$, the settling time corresponding to the error range ±5% tolerance band may be measured from the response of underdamped case for the second-order system. We obtain settling time is

$$t_s = \frac{3}{\xi\omega_n};(\text{for 5\% criterion}) \tag{4.87}$$

Similarly, when the settling time corresponding to the error range ±2% tolerance band, it may be measured from the response of underdamped case for the second-order system, we obtain settling time is

$$t_s = \frac{4}{\xi\omega_n};(\text{for 2\% criterion}) \tag{4.88}$$

The settling time corresponding to a ±2% or ±5% tolerance band may be measured in terms of the time constant $T = 1/\xi\omega_n$ from the curves of Figure 4.30 for different values of ξ. For $0<\xi<0.9$, if the 2% criterion is used, the settling time t_s is approximately

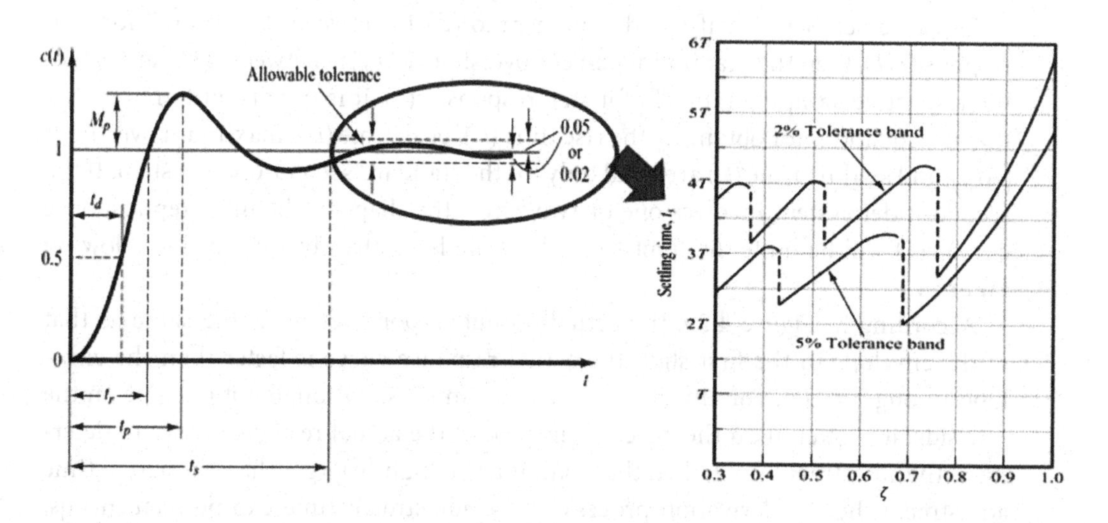

FIGURE 4.30 Relationship between settling time and ξ.

four times the time constant ($T = 1/\xi\omega_n$) of the system. If the 5% criterion is used, then the settling time t_s is approximately three times the time constant ($T = 1/\xi\omega_n$) of the system. Note that the settling time t_s reaches a minimum value around ξ = 0.76 (for the 2% criterion) or ξ = 0.68 (for the 5% criterion) and then increases almost linearly for large values of ξ.

Look at Figure 4.30b and we can see that the discontinuities in the curves have arisen because an infinitesimal change in the value of ξ can cause a finite change in the settling time (t_s). For convenience in comparing the responses of systems, we commonly refine Eqs. (4.86) and (4.88), and then the settling time (t_s) can be refined as

$$t_s = 4T = \frac{4}{\sigma} = \frac{4}{\xi\omega_n}; \ \left(\text{for 2\% criterion}\right) \tag{4.89}$$

and

$$t_s = 3T = \frac{3}{\sigma} = \frac{3}{\xi\omega_n}; \ \left(\text{for 5\% criterion}\right) \tag{4.90}$$

Note that the settling time (t_s) is inversely proportional to the product of the damping ratio (ξ) and the undamped natural frequency (ω_n) of the system.

Since the value of damping ratio (ξ) is usually determined from the requirement of permissible maximum overshoot (M_p), the settling time (t_s) is determined primarily by the undamped natural frequency (ω_n). This means that the duration of the transient period may be varied, without changing the maximum overshoot (M_p), by adjusting the undamped natural frequency (ω_n). From this analysis, it is evident that for rapid response, the undamped natural frequency (ω_n) must be large. To limit the maximum overshoot (M_p) and to make the settling time (t_s) small, the damping ratio (ξ) should not be too small. The relationship between the maximum percent overshoot (M_p) and the damping ratio (ξ) is presented in Figure 4.31.

Please remember that if the damping ratio (ξ) is between 0.4 and 0.7 for step response $c(t)$, then the maximum percent overshoot (M_p) is between 25%, and 4% of the respective damping ratio (ξ) for step response $c(t)$. It is important to remember that the equations for obtaining the rise time (t_r), peak time (t_p), maximum overshoot (M_p), and settling time (t_s) are valid only for the standard second-order system. If the second-order system involves one or two zeros, the shape of the unit-step response curve $c(t)$ will be quite different from those we have already seen in the following Figure 4.17.

According to Figure 4.29, the actual output response curve is the same as that of the envelope in the first shot; the actual response curve is faster than the envelope among ξ = 0.06. For any other ξ are in a similar situation, but it must be on the safe side; it is often used the envelope instead of the actual response curve to determine the adjustment time; then the result may be slightly larger than the actual time adjustment. By the derivation process of the adjustment time can be seen in Eqs. (4.89) and (4.90); when ξ is larger as an independent variable, then the dependent

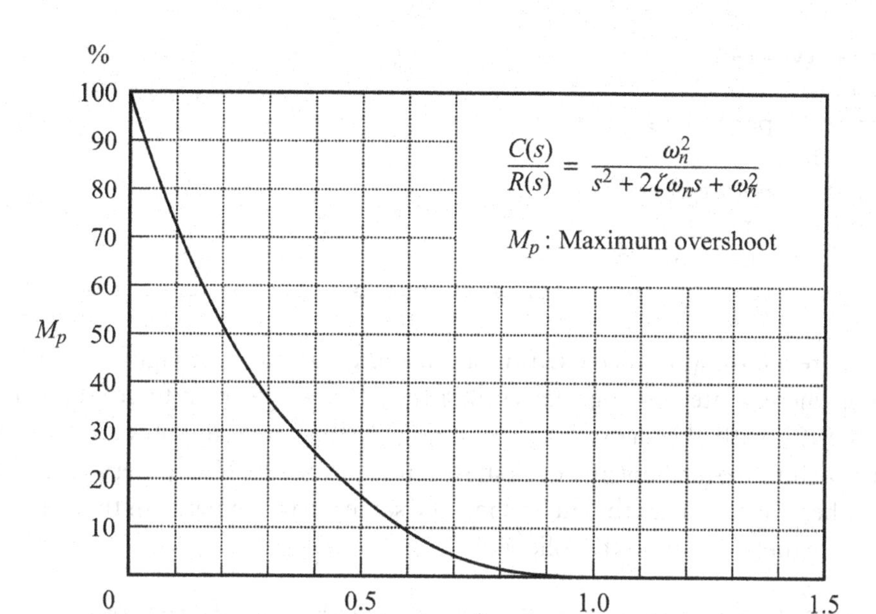

FIGURE 4.31 Relationship between M_p and ξ.

variable settling time (t_s) is smaller; when $\xi = 0.707$, the shortest time for the adjustment of the unit-step response of system namely the fastest response; when $\xi < 0.707$ and the smaller the ξ, then adjustment of the dependent variable settling time (t_s) is the longer; however, when $\xi > 0.707$ as the greater the ξ, then the regulating time is longer but adjustment of the dependent variable settling time (t_s) is the shorter. Adjusting time by the undamped natural frequency (ω_n) and damping ratio ξ relations is shown in Figure 4.31.

To find the envelop and the response curve as shown in Figure 4.29, we refer Eq. (4.50) or (4.51), and then we can write the program as follows:

```
%The MATLAB program are as follows:
T=10; ut=1;
Wn=0.2;
t=0:50;
hold on
plot(t, ut+0*(ut+t))
Z=1./(T*Wn);
A=exp(-Z.*Wn.*t)/sqrt(1-Z.*Z);
Wd=Wn.*(sqrt(1-Z.*Z));
B=atan((sqrt(1-Z.*Z))./Z);
ct=1-A.*sin(Wd.*t+B);
hold on
plot(t, ct)
Epositive=1+A;
```

```
Enegative=1-A;
hold on
plot(t, Epositive)
hold on
plot(t, Enegative)
xlabel('Time, t')
ylabel('Response, c(t)')
title('Unit-step responses envelops of second-order system');grid
```

To find the relationship between settling time and ξ as shown in Figure 4.32a, we refer Eq. (4.85), whereas after getting this settling time (t_s), we perform to get the maximum overshoot (M_p) with the same damping ratio (ξ), then we can show the relationship between maximum overshoot and ξ, as shown in Figure 4.32b; If we observe Figures 4.31b and 4.32, then we can say both figures show the same curve. To perform this relationship, we may write the program in the following.

```
%The MATLAB program for Figure 4.32(a) are as follows:
for i=1:99
Z(i)=i/100;
wn=1;
wnts(i)= -(log(0.05)+log(sqrt(1-Z(i)*Z(i))))/(Z(i)*wn);
end
figure
plot(Z, wnts)
hold on
grid
xlabel(' ξ ')
ylabel('Setling time, ts')
title('Relationship between ξ and ts')
```

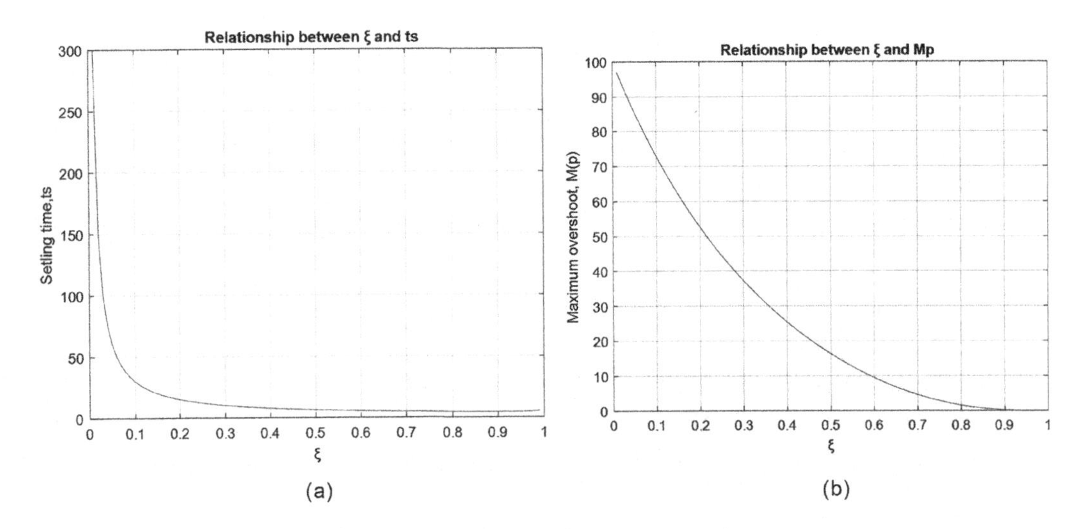

FIGURE 4.32 Relationship between (a) settling time and ξ and (b) M_p and ξ.

```
%The MATLAB program for Figure 4.32(b) are as follows:
for i=1:99
Zt(i)=i/100;
wn=1;
mpp(i)=(exp(-Zt(i)/(sqrt(1-Zt(i).*Zt(i))).*pi))*100;
end
plot(Zt, mpp)
grid
xlabel(' ξ ')
ylabel('Maximum overshoot, M(p)')
title('Relationship between ξ and Mp')
grid
```

Example 4.3

Consider the second-order system shown in Figure 4.15 which is again shown below. Let us obtain the rise time, peak time, maximum overshoot, and settling time when the system is subjected to a unit-step input. Where $\xi = 0.6$, $\omega_n = 5\,\text{rad/s}$.

Solution

From the given values of ξ and ω_n, we obtain $\omega_d = \omega_n\sqrt{1-\xi^2} = 4$ and $\sigma = \xi\omega_n = 3$. The rise time is t_r, then

$$t_r = \frac{\pi - \beta}{\omega_d} = \frac{3.14 - \tan^{-1}\dfrac{\omega_d}{\sigma}}{\omega_d} = \frac{3.14 - \tan^{-1}\dfrac{4}{3}}{4} = 0.55\,\text{s}$$

and the peak time is t_p, so

$$t_p = \frac{\pi}{\omega_d} = \frac{3.14}{4} = 0.78\,\text{s}$$

However, the maximum overshoot M_p is,

$$M_p = e^{-(\sigma/\omega_d)\pi} = e^{-(3/4)\times 3.14} = 0.095 = 9.5\%$$

and the settling time is t_s then,

$$t_s = \frac{4}{\sigma} = \frac{4}{3} = 1.33\,\text{s}; (\text{for } 2\% \text{ criterion})$$

$$t_s = \frac{3}{\sigma} = \frac{3}{3} = 1\,\text{s}; (\text{for } 5\% \text{ criterion})$$

5. *Delay time, t_d*

According to the definition of delay time, it is the time required for the unit step response to reach 50% of the steady-state level, i.e.,

$$c(t_d)=0.5 \tag{4.91}$$

According to Eq. (4.51), we get

$$c(t_d)=1-\frac{e^{-\xi\omega_n t_d}}{\sqrt{1-\xi^2}}\sin(\omega_d t_d+\beta) \tag{4.92}$$

Substituting Eq. (4.91) into Eq. (4.92), we get

$$1-\frac{e^{-\xi\omega_n t_d}}{\sqrt{1-\xi^2}}\sin(\omega_d t_d+\beta)=0.5 \tag{4.93}$$

Thus,

$$\omega_n t_d=\frac{1}{\xi}\ln\frac{\sin(\omega_d t_d+\beta)}{\sqrt{1-\xi^2}} \tag{4.94}$$

According to Eq. (4.94), where $\omega_n t_d$ accepts different values for the corresponding value ξ; then we can draw the curve with ξ vs t_d, as shown in Figure 4.33. According to Figure 4.33, the wide range of delay time (t_d) and the damping ratio (ξ) are related by the following equation as

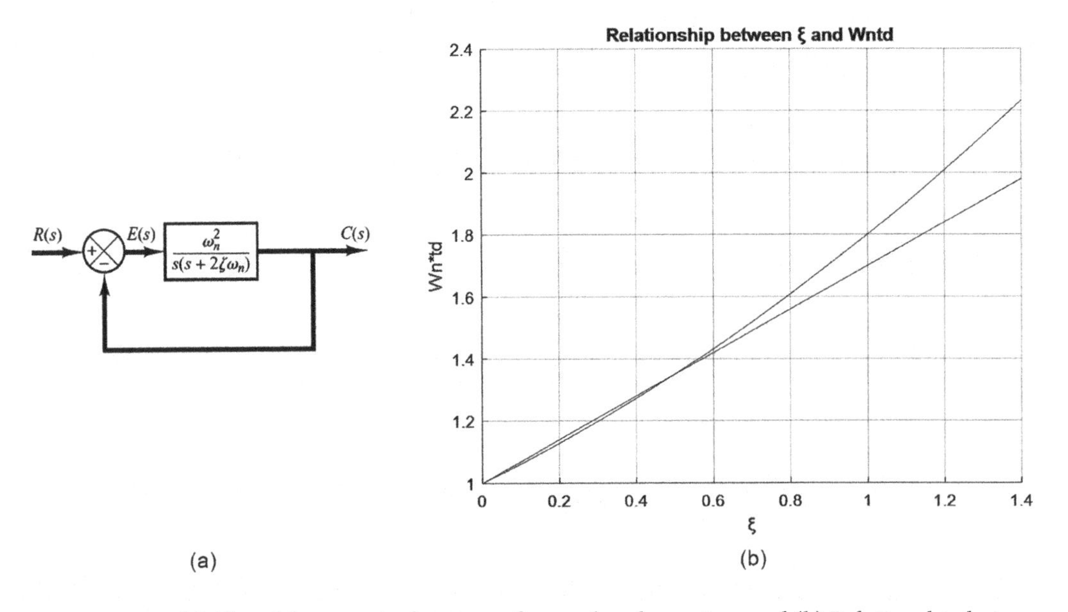

FIGURE 4.33 (a) Closed-loop control system of second-order system and (b) Relationship between ξ and t_d.

$$t_d = \frac{1+0.6\xi+0.2\xi^2}{\omega_n} \tag{4.95}$$

Satisfying the condition of underdamped case $(0<\xi<1)$, then Eq. (4.95) can be simplified to

$$t_d = \frac{1+0.7\xi}{\omega_n} \quad (0<\xi<1) \tag{4.96}$$

Eq. (4.94) shows that the increasing or decreasing the damped natural frequency (ω_d) can reduce the delay time (t_d); In other words, the closed-loop pole is far from the origin of the coordinate of s-plane; the system delay time (t_d) is shorter. However, according to Eqs. (4.95) and (4.96), when the natural frequency (ω_n) is constant, then the closed-loop poles are closed to the imaginary axis in the s-plane; therefore, the system delay time (t_d) is shorter.

6. *Number of the oscillation frequency, N*
By the definition, N is the number of oscillation frequencies in the time interval $0 \le t \le t_s$ and the system is considered under unit-step function, then the transient response frequency of oscillation is expressed as

$$N = \frac{t_s}{T_d} \tag{4.97}$$

where $T_d = \frac{2\pi}{\omega_d}$ is the damping period of the system.

According to the tolerance band $\Delta = \pm 2\%$, the number of oscillation frequency (N) is found by substituting $t_s = \frac{4}{\xi\omega_n}$ and $\omega_d = \omega_n\sqrt{1-\xi^2}$ in Eq. (4.97), and then we get

$$N = \frac{\frac{4}{\xi\omega_n}}{\frac{2\pi}{\omega_n\sqrt{1-\xi^2}}}$$

Thus,

$$N = \frac{2\sqrt{1-\xi^2}}{\pi\xi} \tag{4.98}$$

According to the tolerance band $\Delta = \pm 5\%$, the number of oscillation frequency (N) is found by substituting $t_s = \frac{3}{\xi\omega_n}$ and $\omega_d = \omega_n\sqrt{1-\xi^2}$ in Eq. (4.97), and then we get

$$N = \frac{\frac{3}{\xi\omega_n}}{\frac{2\pi}{\omega_n\sqrt{1-\xi^2}}}$$

Thus,

$$N = \frac{1.5\sqrt{1-\xi^2}}{\pi\xi} \tag{4.99}$$

We know the maximum overshoot, $M_p = e^{-\left(\frac{\xi}{\sqrt{1-\xi^2}}\right)\pi}$

or

$$\ln M_p = -\frac{\pi\xi}{\sqrt{1-\xi^2}} \tag{4.100}$$

Both Eqs. (4.98) and (4.99) will generate Eq. (4.100) and perform the relationship between the number of oscillation frequency (N) and the overshoot value (M_p), and thus we get

$$N = \begin{cases} -\dfrac{2}{\ln M_p} & ;\Delta = \pm 2\% \\ -\dfrac{1.5}{\ln M_p} & ;\Delta = \pm 5\% \end{cases} \tag{4.101}$$

The number of oscillation frequency (N) can also be obtained by the following conditions:

$$c(t) - c(\infty) = 0\text{ ; where, } t \le t_s \tag{4.102}$$

Therefore,

$$c(t) - c(\infty) = -\frac{e^{-\xi\omega_n t}}{\sqrt{1-\xi^2}}\sin\left[\omega_d t + \tan^{-1}\frac{\sqrt{1-\xi^2}}{\xi}\right] = 0\text{ ;}(\ t \le t_s) \tag{4.103}$$

Since $\dfrac{e^{-\xi\omega_n t}}{\sqrt{1-\xi^2}} \ne 0$, we get

$$\sin\left[\omega_d t + \tan^{-1}\frac{\sqrt{1-\xi^2}}{\xi}\right] = 0 \tag{4.104}$$

Consequently,

$$\omega_d t + \tan^{-1}\frac{\sqrt{1-\xi^2}}{\xi} = k\pi \quad ;(\text{for } k = 0,1,2,\cdots) \tag{4.105}$$

We choose $t = t_s$, and then we get

$$\omega_d t_s + \tan^{-1}\frac{\sqrt{1-\xi^2}}{\xi} = (m+d)\pi \tag{4.106}$$

where m is the integer number, d is the decimal numbers; at $t = t_s$, $c(t)$ is not exactly equal to $c(\infty)$; because the number of oscillation frequency is across twice at a time from $c(t)$ to $c(\infty)$, we get

$$N = \frac{m}{2}$$

Now, substituting $N = \frac{m}{2}$ into Eq. (4.106), we get

$$N = \frac{\omega_n \sqrt{1-\xi^2}\, t_s + \tan^{-1} \dfrac{\sqrt{1-\xi^2}}{\xi}}{2\pi} - \frac{d}{2} \tag{4.107}$$

We recall Eq. (4.84), and then we get

$$\omega_n t_s = -\frac{\ln \Delta + \ln \sqrt{1-\xi^2}}{\xi} \tag{4.108}$$

Substituting Eq. (4.108) into (4.107) and setting $d = 0$, we get

$$N = \left[-\frac{\sqrt{1-\xi^2}\left(\ln \Delta + \ln \sqrt{1-\xi^2}\right)}{2\pi\xi} + \frac{\tan^{-1} \dfrac{\sqrt{1-\xi^2}}{\xi}}{2\pi} \right] \tag{4.109}$$

Here, for given ξ and Δ, we can find out the corresponding oscillation number N. Figure 4.34 shows a relationship between N and ξ of $\Delta = 5\%$ tolerance band. By the above analysis, to make the second-order system satisfactory transient performance indicators, we must select the appropriate ξ and undamped natural frequency ω_n; by increasing the damping ratio ξ, we can improve the speed of the second-order system response and reduce the adjustment of the transition time; however, increasing the damping ratio ξ, we can reduce the oscillation performance of the system; thus, reducing the amount of overshoot M_p and the oscillation number N. When designing the system, increase the value, generally improve the open-loop amplification coefficient K is used to realize the system analysis; and improve the commonly used reduce open-loop amplification coefficient K. Obviously, the quickness and the oscillation behavior of systems between the insurmountable contradictions. The system is required not only to reduce oscillation, but also to have a certain degree of rapidity. The number of oscillation (N) with damping ratio (ξ) is shown in Figure 4.34.

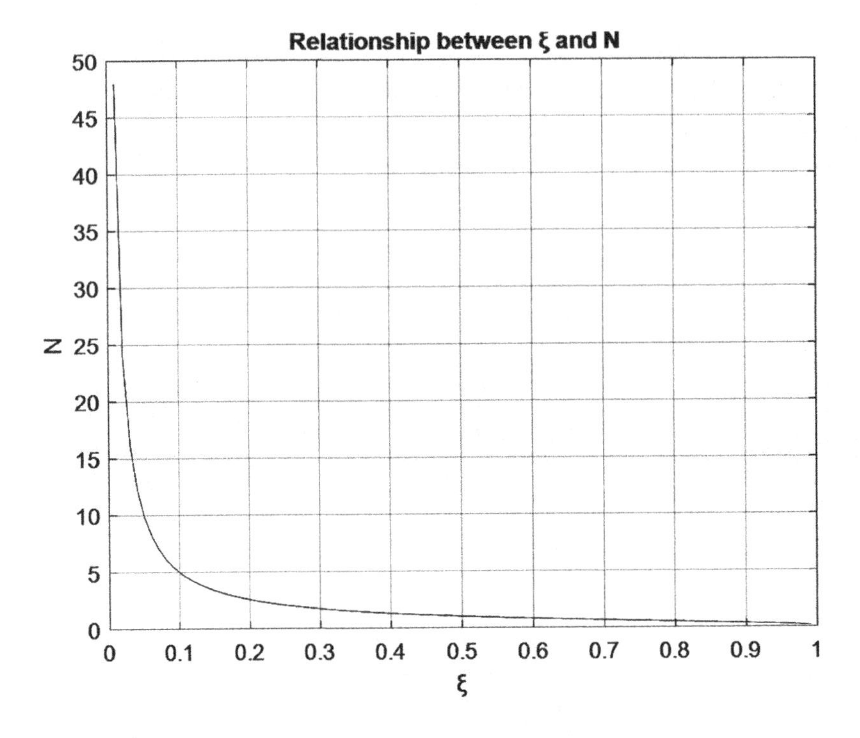

FIGURE 4.34 Relationship between ξ and N.

To find the relationship between N and ξ as shown in Figure 4.34, we refer Eq. (4.109). To perform this relationship, we may write the MATLAB program in the following.

```
%The MATLAB program are as follows:
for i=1:99
Z(i)=i/100;
A=sqrt(1-Z(i)*Z(i));
N(i)=-(A*(log(0.05)+log(A)))/(2*pi*Z(i))+(atan(A/Z(i)))/(2.*pi);
hold
end
plot(Z, N)
xlabel(' ξ ')
ylabel('N')
title('Relationship between ξ and N')
```

Example 4.4

We recall Figure 4.15 which is shown again below. This figure shows a second-order system, where $\xi = 0.5$, and $\omega_n = 4$ rad / s. When the input signal is a unit step function, obtain the transient performance index of the system.

Solution

According to the block diagram, we can find the transfer function of the system as

$$G(s)=\frac{\dfrac{\omega_n^2}{s(s+2\xi\omega_n)}}{1+\dfrac{\omega_n^2}{s(s+2\xi\omega_n)}}=\frac{\omega_n^2}{s^2+2\xi\omega_n s+\omega_n^2}$$

We obtain

$$\omega_d=\omega_n\sqrt{1-\xi^2}=4\sqrt{1-0.5^2}=3.46\text{ rad}$$

and $\sigma=\xi\omega_n=0.5\times 4\,\text{rad/s}=2.0\,\text{rad/s}$.

The rise time t_r:

$$t_r=\frac{\pi-\beta}{\omega_d}=\frac{\pi-\tan^{-1}\dfrac{\omega_d}{\sigma}}{\omega_d}=\frac{3.14-\tan^{-1}\dfrac{3.46}{2}}{3.46}=0.60\,\text{s}$$

The peak time t_p:

$$t_p=\frac{\pi}{\omega_d}=\frac{\pi}{3.46}=0.91\,\text{s}$$

The delay time t_d:

$$t_d=\frac{1+0.6\xi+0.2\xi^2}{\omega_n}=\frac{1+0.6\times 0.5+0.2\times 0.5^2}{4}=0.34\,\text{s}$$

The maximum overshoot M_p: $M_p\%=e^{-\frac{\pi\xi}{\sqrt{1-\xi^2}}}\times 100\%=16.3\%$

The settling or adjusting time t_s:

$$t_s=\begin{cases}\dfrac{4}{\xi\omega_n}=\dfrac{4}{0.5\times 4}=2\,\text{s};\Delta=\pm 2\% \\[2ex] \dfrac{3}{\xi\omega_n}=\dfrac{3}{0.5\times 4}=1.5\,\text{s};\Delta=\pm 5\%\end{cases}$$

The number of oscillation frequency N:

$$N=\begin{cases}\dfrac{2\sqrt{1-\xi^2}}{\pi\xi}=\dfrac{2\sqrt{1-0.5^2}}{\pi\times 0.5}=1.1\quad;\Delta=2\% \\[2ex] \dfrac{1.5\sqrt{1-\xi^2}}{\pi\xi}=\dfrac{1.5\sqrt{1-0.5^2}}{\pi\times 0.5}=0.83\quad;\Delta=5\%\end{cases}$$

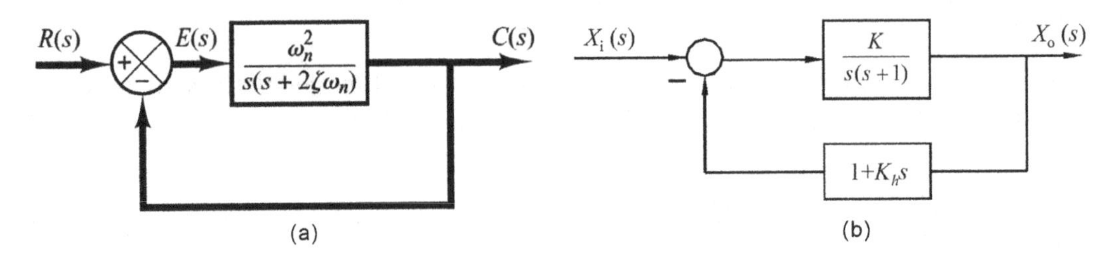

FIGURE 4.35 (a) Block diagram of a second-order system, and (b) Transfer function of a second-order system.

Example 4.5

Figure 4.35 shows a second-order system. The maximum overshoot of the system is 0.2, and the peak time is 1 s. Determine the gain K, the coefficient K_h, and other performance indexes of the system.

Solution

According to the question, we have the maximum overshoot as

$$M_p = e^{-\frac{\xi\pi}{\sqrt{1-\xi^2}}} = 0.2$$

Thus, the damping ratio, $\xi=0.456$.

The peak time t_p:

$$t_p = \frac{\pi}{\omega_d} = 1\,\text{s}$$

Thus, the damped natural frequency, $\omega_d = \pi\,\text{rad/s}$.

And the undamped natural frequency,

$$\omega_n = \frac{\omega_d}{\sqrt{1-\xi^2}} = \frac{\pi}{\sqrt{1-0.456^2}} = 3.53\,\text{rad/s}$$

The transfer function for the system,

$$G(s) = \frac{\frac{K}{s(s+1)}}{1+\frac{K}{s(s+1)}(1+K_h s)} = \frac{K}{s^2+(KK_h+1)s+K} = \frac{\omega_n^2}{s^2+2\xi\omega_n s+\omega_n^2}$$

Thus,

$$K = \omega_n^2 = 3.53^2 = 12.46\,\text{rad}^2/\text{s}^2$$

and

$$2\xi\omega_n = KK_h + 1$$

Thus,

$$K_h = \frac{2\xi\omega_n - 1}{K} = \frac{2\times0.456\times3.53-1}{12.46} = 0.178\,\text{s}$$

The delay time, $t_d = \dfrac{1+0.6\xi+0.2\xi^2}{\omega_n} = \dfrac{1+0.6\times0.456+0.2\times0.456^2}{3.53} = 0.373\,\text{s}$

The rise time, $t_r = \dfrac{\pi-\beta}{\omega_d} = \dfrac{\pi-\arccos 0.456}{\pi} = 0.65\,\text{s}$

The settling or adjusting time,

$$t_s = \begin{cases} \dfrac{4}{\xi\omega_n} = \dfrac{4}{0.456\times3.53} = 2.48\,\text{s} & ;\Delta = \pm2\% \\ \dfrac{3}{\xi\omega_n} = \dfrac{3}{0.456\times3.53} = 1.86\,\text{s} & ;\Delta = \pm5\% \end{cases}$$

The number of oscillation frequencies

$$N = \begin{cases} \dfrac{2\sqrt{1-\xi^2}}{\pi\xi} = \dfrac{2\sqrt{1-0.456^2}}{\pi\times0.456} = 1.24 & ;\Delta = 2\% \\ \dfrac{1.5\sqrt{1-\xi^2}}{\pi\xi} = \dfrac{1.5\sqrt{1-0.456^2}}{\pi\times0.456} = 0.93 & ;\Delta = 5\% \end{cases}$$

Example 4.6

Figure 4.36a shows a mass-spring-damper system, and Figure 4.36b shows the response curve of the system. The applied step force is 8.9 N in the system, and the recorded output time-response curve is shown in Figure 4.36b. Determine the system mass, spring stiffness, and damping coefficient f of the system.

Solution

According to Newton's second law,

$$F_i(t) - kx_o(t) - f\frac{dx_o(t)}{dt} = m\frac{dx_o^2(t)}{dt^2}$$

The transfer function of the system,

$$1G(s) = \frac{X_o(s)}{F_i(s)} = \frac{1}{ms^2+fs+k} = \frac{\frac{1}{k}\frac{k}{m}}{s^2+\frac{f}{m}s+\frac{k}{m}} = \frac{\frac{1}{k}\omega_n^2}{s^2+2\xi\omega_n s+\omega_n^2}$$

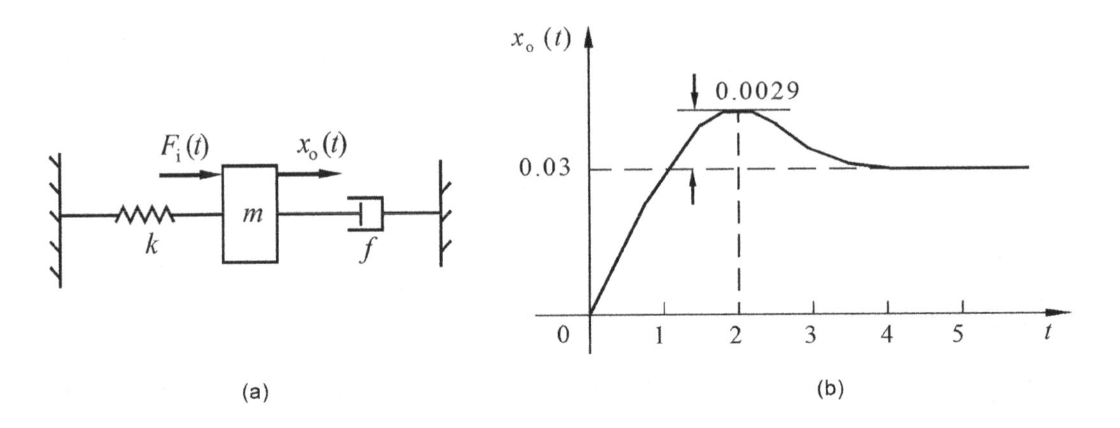

FIGURE 4.36 (a) A mass-spring-damper system, and (b) response curve.

Thus,

$$X_o(s) = \frac{1}{ms^2 + fs + k} F_i(s) = \frac{1}{ms^2 + fs + k} \cdot \frac{8.9}{s}$$

However,

$$x_o(\infty) = \lim_{s \to 0} s \cdot \frac{1}{ms^2 + fs + k} \frac{8.9}{s} = \frac{8.9}{k} = 0.03m$$

Thus, the spring stiffness,

$$k = \frac{8.9}{0.03} = 297\,\text{N/m}$$

According to the maximum overshoot from Figure 4.36b,

$$M_p = e^{-\frac{\xi\pi}{\sqrt{1-\xi^2}}} = \frac{0.0029}{0.03}$$

Thus, $\xi = 0.6$

The peak time,

$$t_p = \frac{\pi}{\omega_n \sqrt{1-\xi^2}}$$

Thus, the undamped natural frequency,

$$\omega_n = \frac{\pi}{t_p \sqrt{1-\xi^2}} = \frac{\pi}{2\sqrt{1-0.6^2}} = 1.96\,\text{rad/s}$$

The mass,

$$m=\frac{k}{\omega_n^2}=\frac{297}{1.96^2}=77.3\,\text{kg}$$

Therefore, the damping coefficient,

$$f=2\xi\omega_n m=2\times0.6\times1.96\times77.3=181.8\,\text{N}\cdot\text{m/s}$$

4.5.4 Transient Response Specifications of Second-Order Systems for Overdamped and Critically Damped Condition

Due to the slow response of overdamped systems, we generally do not intend for such kind of systems to be designed under the condition in an overdamped (>1) and critically damped ($=1$) case. However, for some special cases such as low gain and high inertia temperature control systems, it is needed to use the overdamping system. In addition, some systems do not allow overshooting of the time response but need a faster response such as the indication and recording meter systems, it is needed to use the critical damping system. In particular, the time response of some higher-order systems can often be approximated by the time response of overdamped second-order systems. Therefore, it is of great practical significance to study the method of estimating the overdamped transient performance of second-order systems.

When the damping ratio $\xi=1$, the step response of the system in with the zero initial conditions, we recall Eq. (4.61)

$$c(t)=1-e^{-\omega_n t}-\omega_n te^{-\omega_n t}=1-(1+\omega_n t)e^{-\omega_n t}$$

Where $t\geq0$ and $\xi=1$. When $\xi>1$, the step response of the system with the zero initial conditions, we recall Eq. (4.66), then we get

$$c(t)=1+\frac{\omega_n}{2\sqrt{\xi^2-1}}\left(\frac{1}{T_1}.e^{-T_1 t}-\frac{1}{T_2}.e^{-T_2 t}\right)$$

where $T_1=\omega_n\left(\xi+\sqrt{\xi^2-1}\right)$ and $T_2=\omega_n\left(\xi-\sqrt{\xi^2-1}\right)$. Both terms have a time dimension that is called the second-order overdamped time constant, and obviously $T_1\neq T_2$, $T1>T2$, $t\geq0$, and >1.

According to Eq. (3.71), among the performance indicators of transient response, only delay time (t_d), rise time (t_r), and the settling or adjusting time (t_s) are of practical significance in engineering. However, Eq. (4.66) is a transcendental equation, and it is impossible to get their exact solutions according to the definition of transient response performance index. At present, numerical solutions of dimensionless time under different conditions are mainly adopted by computers in engineering, and their approximate formulas are obtained by curve fitting.

7. *The delay time (t_d)*
   The fitting curve of the delay time (t_d) can be expressed by recalling Eq. (4.95) and the fitted curve was found in Figure 4.33. Thus, we say

$$t_d = \frac{1 + 0.6\xi + 0.2\xi^2}{\omega_n} \tag{4.110}$$

8. *The rise time (t_r)*
   As seen from Figures 4.33 to 4.37, there is no oscillation of the transient response characteristics; therefore, the required time is adopted for the response to rise from 10% to 90% of the steady-state value. We may use the numerical method to determine the relationship between the dimensionless rise time ($\omega_n t_r$) and the damping ratio ξ, as shown in Figure 4.33 which is described in the following approximated quadratic equation as

$$t_r = \frac{1 + 1.5\xi + \xi^2}{\omega_n} \tag{4.111}$$

9. *The settling time or adjusting time t_s*
   Adjust the time, according to Eq. (4.66), the ratio $\frac{T_1}{T_2}$ takes different values, and then we can solve the dimensionless time adjustment or settling time adjustment of $\frac{t_s}{T_1}$. As shown in Figure 4.37, the median error characteristic is chosen to adjust the settling

FIGURE 4.37 Relationship between ξ and $\omega_n t_r$.

time at 5%. As shown in Figure 4.37, we have chosen $\frac{T_1}{T_2}=1$, and then the equivalent relationship of critical damping and settling time or adjustment time can be found as

$$t_s = 4.75T_1 \qquad ;(\xi = 1) \tag{4.112}$$

When $\xi > 1$, and $T_1 \geq 4T_2$, i.e., when the value of the first closed-loop pole (T_1) is equal or four times larger than the value of the second closed-loop pole (T_2) of the over-damped second-order system, the system can be equivalent to the first-order system with a closed-loop pole $-\frac{1}{T_1}$. Thus, the settling time or the adjustment time can be written as

$$t_s = 3T_1 \tag{4.113}$$

The relative error of the system is not more than 10%. In Figure 4.37, the damping ratio ξ is a dependent parameter, because we may express as

$$s^2 + 2\xi\omega_n s + \omega_n^2 = \left(s+\frac{1}{T_1}\right)\left(s+\frac{1}{T_2}\right) = s^2 + \left(\frac{1}{T_1}+\frac{1}{T_2}\right)s + \frac{1}{T_1T_2} \tag{4.114}$$

Thus, we may write the relation between the damping ratio ξ and the independent variables $\frac{T_1}{T_2}$ as

$$\xi = \frac{1+T_1/T_2}{2\sqrt{T_1/T_2}} \tag{4.115}$$

Example 4.7

Figure 4.38 shows a servo motor control system with time constant T, and $T = 0.1\text{s}$; and K is an open-loop gain, the system requirement is under the unit step response with no overshoot and the adjusting time $t_s \leq 1\text{s}$. Determine the system open-loop gain K, delay time, and rise time.

Solution

In accordance with the requirements of the system under the unit step response with no overshoot, the damping ratio should be $\xi \geq 1$; and when $\xi = 1$, the system response

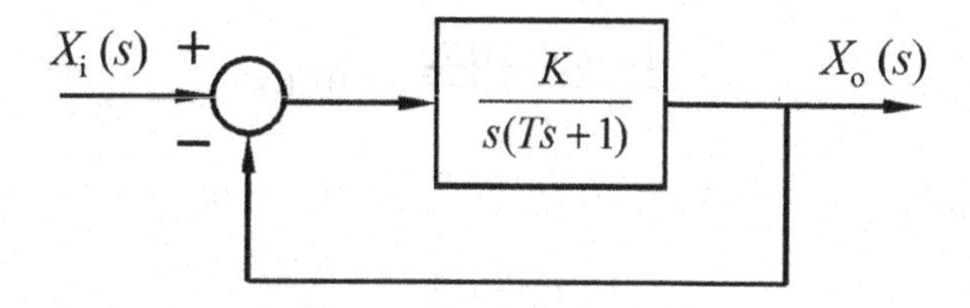

FIGURE 4.38 Servo motor control system.

is the fastest; thus, choosing $\xi = 1$, the settling time or the adjusting time can be found from Eq. (4.112), and then we get

$$t_s = 4.75T_1 \quad ;(\xi = 1)$$

According to Figure 4.38, the available characteristic equation of the closed-loop system can be written by using Eq. (4.114) as

$$s^2 + \frac{1}{T}s + \frac{K}{T} = 0 \tag{4.116}$$

The standard form of the closed-loop system for the characteristic equation is written as

$$s^2 + 2\xi\omega_n s + \omega_n^2 = 0 \tag{4.117}$$

Comparing Eqs. (4.116) with (4.117), we get

$$2\xi\omega_n = \frac{1}{T}$$

Thus,

$$\omega_n = \frac{1}{2\xi T} = \frac{1}{2\times1\times0.1} = 5\,\text{rad/s}$$

and

$$\omega_n = \sqrt{\frac{K}{T}} = \sqrt{\frac{K}{0.1}} = 5 \Rightarrow K = 2.5$$

However, we compare Eqs. (4.114) and (4.117), then we get,

$$\omega_n^2 = \frac{1}{T_1T_2}$$

When $\xi = 1$ and $T_1 = T_2$, then $T_1 = T_2 = 0.2\,\text{s}$; thus, $t_s = 4.75T_1 = 0.95\,\text{s}$.

Therefore, to meet the requirements of the settling time or the adjustment time and the performance of the settling time or the adjustment time, it satisfies $t_s \le 1\,\text{s}$.

Thus, the delay time

$$t_d = \frac{1 + 0.6\xi + 0.2\xi^2}{\omega_n} = 0.36\,\text{s}$$

And the rise time

$$t_r = \frac{1 + 1.5\xi + \xi^2}{\omega_n} = 0.7\,\text{s}$$

4.5.5 Servo System with Velocity-Feedback System

The derivative of the output signal can be used to improve system performance. Such kind of derivative can be described by this relationship as

$$u(t) = K_p T_d \frac{de(t)}{dt} \leftrightarrow U(s) = K_p T_d s \tag{4.118}$$

In obtaining the derivative of the output position signal, it is desirable to use a *tachometer* instead of physically differentiating the output signal. Please note that the differentiation amplifies noise effects. In fact, if discontinuous noises are present, differentiation amplifies the discontinuous noises more than the useful signal. For example, the output of a potentiometer is a discontinuous voltage signal because, as the potentiometer brush is moving on the windings, voltages are induced in the switchover turns and thus generate transients. The output of the potentiometer therefore should not be followed by a differentiating element. Now, the question, *what is tachometer?* Look at Figure 4.39a, it shows a tachometer which is a special direct current (dc) generator, and it is frequently used to measure velocity *without differentiation process.* The output of a tachometer is *proportional to the angular velocity of the motor, as shown in Figure 4.39b.* In any servo system, such a velocity signal can be easily generated by a tachometer.

Potentiometers are commonly used to control electrical devices. Let's see an example of a potentiometer that is shown in Figure 4.39c. Now, the question, *what is potentiometer?*

Potentiometers are commonly used to control electrical devices such as volume controls on audio equipment.

The output of a *potentiometer* is a discontinuous voltage signal because, as the potentiometer brush is moving on the windings, voltages are induced in the switchover turns and thus generate transients. The output of the potentiometer therefore should not be followed by a differentiating element.

Consider the servo system shown in Figure 4.40a. In this device, the velocity signal, together with the positional signal, is fed back to the input to produce the actuating error signal. In any servo system, such a velocity signal can be easily generated by a *tachometer.* The simplified block diagram is shown in Figure 4.40b.

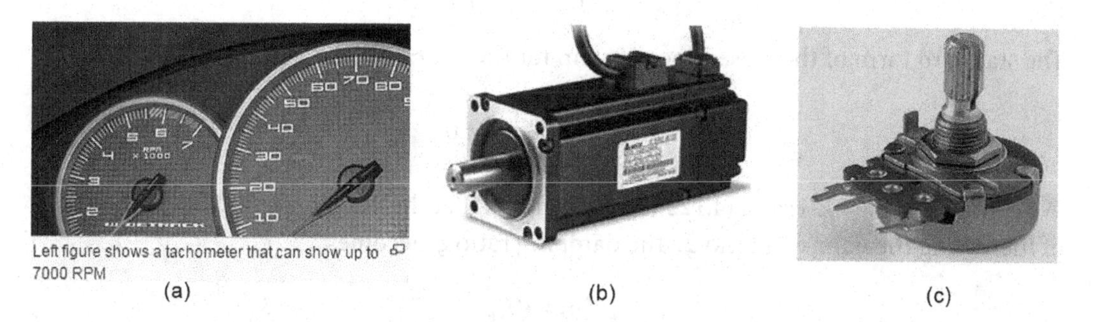

FIGURE 4.39 (a) Tachometer, (b) servo motor, and (c) potentiometer.

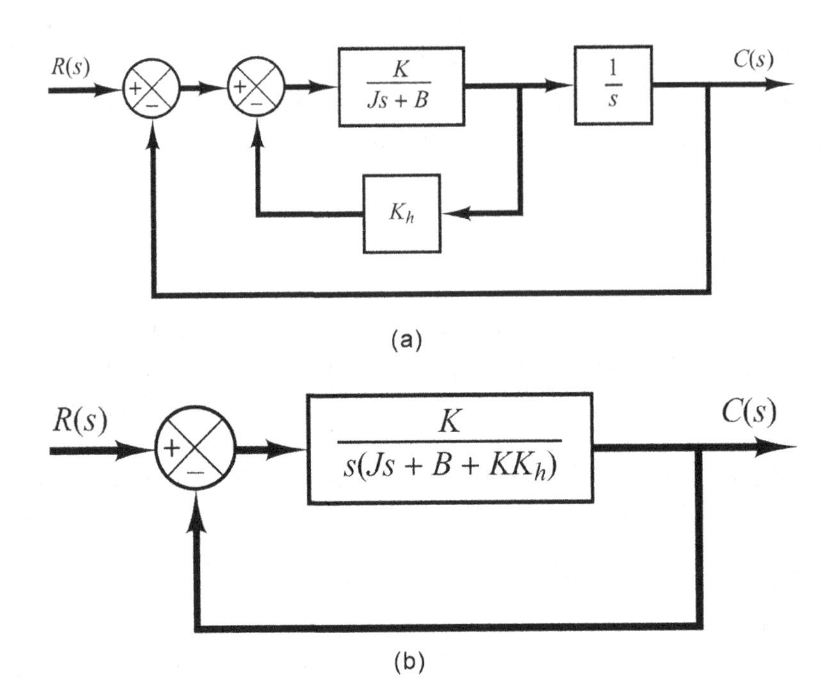

FIGURE 4.40 Closed-loop control system of a servo system.

The closed-loop transfer function between output $C(s)$ and input $R(s)$ in Figure 4.40b can be expressed as

$$\frac{C(s)}{R(s)} = \frac{K}{Js^2 + (B + KK_h)s + K} \tag{4.119}$$

The characteristic equation of Eq. (4.119) can be written as

$$Js^2 + (B + KK_h)s + K = 0 \tag{4.120}$$

or

$$s^2 + \frac{(B + KK_h)}{J}s + \frac{K}{J} = 0 \tag{4.121}$$

The standard form of the closed-loop system for the characteristic equation is written as

$$s^2 + 2\xi\omega_n s + \omega_n^2 = 0 \tag{4.122}$$

We compare Eqs. (4.121) and (4.122), and it is noted that the velocity feedback has the effect of increasing the damping ratio ξ. The damping ratio ξ becomes

$$\xi = \frac{B + KK_h}{2\sqrt{KJ}}$$

The undamped natural frequency $\omega_n = \sqrt{K/J}$ is not affected by velocity feedback K_h, but ξ is affected by velocity feedback. Noting that the maximum overshoot for a unit-step input can be controlled by controlling the value of the damping ratio ξ, we can reduce the maximum overshoot by adjusting the velocity-feedback constant K_h so that $0.4 \le \xi \le 0.7$; we can reduce the maximum overshoot by adjusting the velocity-feedback constant K_h so that $0.4 \le \xi \le 0.7$. It is important to remember that velocity feedback has the effect of increasing the damping ratio without affecting the undamped natural frequency of the system.

Example 4.8

Let's look at this example and consider the velocity-feedback system shown in Figure 4.40a. Determine the values of gain K and velocity-feedback constant K_h so that the maximum overshoot in the unit-step response is 0.2 and the peak time is 1 s. With these values of K and K_h, obtain the rise time and settling time. Assume that $J=1$ kg-m^2 and $B=1$ N-m/rad/s.

Solution

To determine of the values of K and K_h thus, the maximum overshoot M_p is given by the equation as

$$M_p = e^{-\left(\frac{\xi}{\sqrt{1-\xi^2}}\right)\pi}$$

Then substituting the value of maximum overshoot $M_p = 0.2$, we get

$$M_p = e^{-\left(\xi/\sqrt{1-\xi^2}\right)\pi} = 0.2$$

or

$$\frac{\xi\pi}{\sqrt{1-\xi^2}} = 1.61$$

Then,

$$\xi = 0.456$$

The peak time $t_p = 1\,\text{s}$, thus

$$t_p = \frac{\pi}{\omega_d} = 1$$

or

$$\omega_d = 3.14$$

Since $\xi = 0.456$, $w_n = \sqrt{K/J}$,

$$K = Jw_n^{\,2} = w_n^{\,2} = 12.5\,\text{N-m}$$

$$\sigma = \xi\omega_n = 0.456 \times 3.53 = 1.61$$

Then K_h is

$$K_h = \frac{2\sqrt{KJ}\xi - B}{K} = \frac{2\sqrt{K}\xi - 1}{K} = 0.178\,\text{s}$$

Rise time t_r: From equation

$$t_r = \frac{\pi - \beta}{\omega_d}$$

Thus,

$$t_r = \frac{\pi - \beta}{\omega_d} = \frac{3.14 - \tan^{-1}\dfrac{\omega_d}{\sigma}}{\omega_d} = \frac{3.14 - \tan^{-1}\dfrac{3.14}{1.61}}{4} = 0.65\,\text{s}$$

Settling time t_s: for the 2% criterion,

$$t_s = \frac{4}{\sigma} = \frac{4}{1.61} = 2.48\,\text{s}$$

For the 5% criterion,

$$t_s = \frac{3}{\sigma} = \frac{3}{1.61} = 1.86\,\text{s}$$

4.5.6 Unit-Impulse Response of the Second-Order System

We already have known the second-order standard equation, and thus we recall Eq. (4.39) as

$$\frac{C(s)}{R(s)} = \frac{\omega_n^2}{s^2 + 2\xi\omega_n s + \omega_n^2}$$

For a unit-impulse input $r(t)$, the corresponding Laplace transform is unity, i.e., $R(s) = 1$. The unit-impulse output $C(s)$ of the second-order system is written as

$$\frac{C(s)}{R(s)} = \frac{\omega_n^2}{\left(s^2 + \xi\omega_n + \omega_d\right)\left(s^2 + \xi\omega_n - \omega_d\right)} \tag{4.123}$$

or

$$C(s) = \frac{\omega_n^2}{\left(s^2 + \xi\omega_n + \omega_d\right)\left(s^2 + \xi\omega_n - \omega_d\right)}.R(s)$$

Thus,

$$C(s)=\frac{\omega_n^2}{\left(s^2+\xi\omega_n+\omega_d\right)\left(s^2+\xi\omega_n-\omega_d\right)}.1 \tag{4.124}$$

or

$$C(s)=\frac{\omega_n^2}{\left(s+\xi\omega_n+\omega_n\sqrt{\xi^2-1}\right)\left(s+\xi\omega_n-\omega_n\sqrt{\xi^2-1}\right)} \tag{4.125}$$

1. *Unit-impulse response of underdamped case* $(0<\xi<1)$
   To satisfy the condition of underdamped case, we rearrange Eq. (4.125) as follows:

$$C(s)=\frac{\frac{\omega_n}{\sqrt{1-\xi^2}}\left(\omega_n\sqrt{1-\xi^2}\right)}{\left(s+\xi\omega_n\right)^2+\left(\omega_n\sqrt{1-\xi^2}\right)^2} \tag{4.126}$$

   Satisfying with $0<\xi<1$ taking the inverse Laplace transform of Eq. (4.126), the time response c(t) is written as

$$c(t)=\frac{\omega_n}{\sqrt{1-\xi^2}}e^{-\xi\omega_n t}\sin\omega_n\sqrt{1-\xi^2}\;t;\text{ for }(t\geq 0) \tag{4.127}$$

2. *Unit-impulse response of critically damped case* $(\xi=1)$
   To satisfy the condition of critically damped case $(\xi=1)$, we rearrange Eq. (4.125) as follows:

$$C(s)=\frac{\omega_n^2}{\left(s+\omega_n\right)^2} \tag{4.128}$$

   Satisfying with $\xi=1$ taking the inverse Laplace transform of Eq. (4.128), the time response *c*(*t*) is written as

$$c(t)=\omega_n^2 te^{-\omega_n t};\text{ for }(t\geq 0) \tag{4.129}$$

3. *Unit-impulse response of overdamped case* $(\xi>1)$
   To satisfy the condition of critically damped case $(\xi>1)$, we rearrange Eq. (4.125) as follows:

$$C(s)=\frac{\omega_n}{2\sqrt{\xi^2-1}}\left[\frac{1}{s+\left(\xi-\sqrt{\xi^2-1}\right)\omega_n}-\frac{1}{s+\left(\xi+\sqrt{\xi^2-1}\right)\omega_n}\right] \tag{4.130}$$

Satisfying with $\xi > 1$ taking the inverse Laplace transform of Eq. (4.130), the time response $c(t)$ is written as

$$c(t) = \frac{\omega_n}{2\sqrt{\xi^2 - 1}}\left[e^{-\left(\xi - \sqrt{\xi^2 - 1}\right)\omega_n t} - e^{-\left(\xi + \sqrt{\xi^2 - 1}\right)\omega_n t}\right] \tag{4.131}$$

Similarly, in the case of overdamped $(\xi > 1)$, the unit-impulse curve of the second-order system can be based on the unit-step response of the overdamped the second-order system. Thus, we recall Eq. (4.66) of the unit-step response as we know

$$c(t) = 1 + \frac{\omega_n}{2\sqrt{\xi^2 - 1}}\left(\frac{1}{T_1}.e^{-T_1 t} - \frac{1}{T_2}.e^{-T_2 t}\right)$$

To get the impulse response, taking the first derivative of $c(t)$ w.r.t t, we get

$$\frac{dc(t)}{dt} = \frac{\omega_n}{2\sqrt{\xi^2 - 1}}\left(e^{-T_2 t} - e^{-T_1 t}\right)$$

whereas $T_1 = \omega_n\left(\xi + \sqrt{\xi^2 - 1}\right)$, and $T_2 = \omega_n\left(\xi - \sqrt{\xi^2 - 1}\right)$, and thus we get

$$c_\delta(t) = \frac{dc(t)}{dt} = \frac{\omega_n}{2\sqrt{\xi^2 - 1}}\left[e^{-\left(\xi - \sqrt{\xi^2 - 1}\right)\omega_n t} - e^{-\left(\xi + \sqrt{\xi^2 - 1}\right)\omega_n t}\right] \tag{4.132}$$

We can see that Eqs. (4.131) and (4.132) are the same unit-impulse responses. Note that without taking the inverse Laplace transform of $C(s)$, we can also obtain the time response $c(t)$ by differentiating the corresponding unit-step response, since the unit-impulse function is the time derivative of the unit-step function.

A family of unit-impulse response curves given by Eqs. (4.127), (4.129), and (4.131) with various values of ζ, as shown in Figure 4.41. Figure 4.41 shows the unit-impulse response of second-order system for (a) underdamped case, (b) critically damped case, (c) overdamped case, and (d) all together of underdamped, critically damped, and overdamped case. Using Eqs. (4.127), (4.129), and (4.131), we write the program in MATLAB software then we get the unit-impulse responses of underdamped case (shown in Figure 4.41a), critically damped case (shown in Figure 4.41b), and over-damped case (shown in Figure 4.41c), respectively. The curves $c(t)/\omega_n$ are plotted against the dimensionless variable $\omega_n t$, and thus they are functions only of ξ. For the critically damped and overdamped cases, the unit-impulse response is always positive or zero, that is, $c(t) \geq 0$. Thus, we recall Eq. (4.127) and we take the differentiation w.r.t t, and then we get

$$\frac{dc(t)}{dt} = \frac{\omega_n}{\sqrt{1 - \xi^2}}\frac{d}{dt}\left(e^{-\xi\omega_n t}\sin\omega_n\sqrt{1 - \xi^2}\,t\right) \tag{4.133}$$

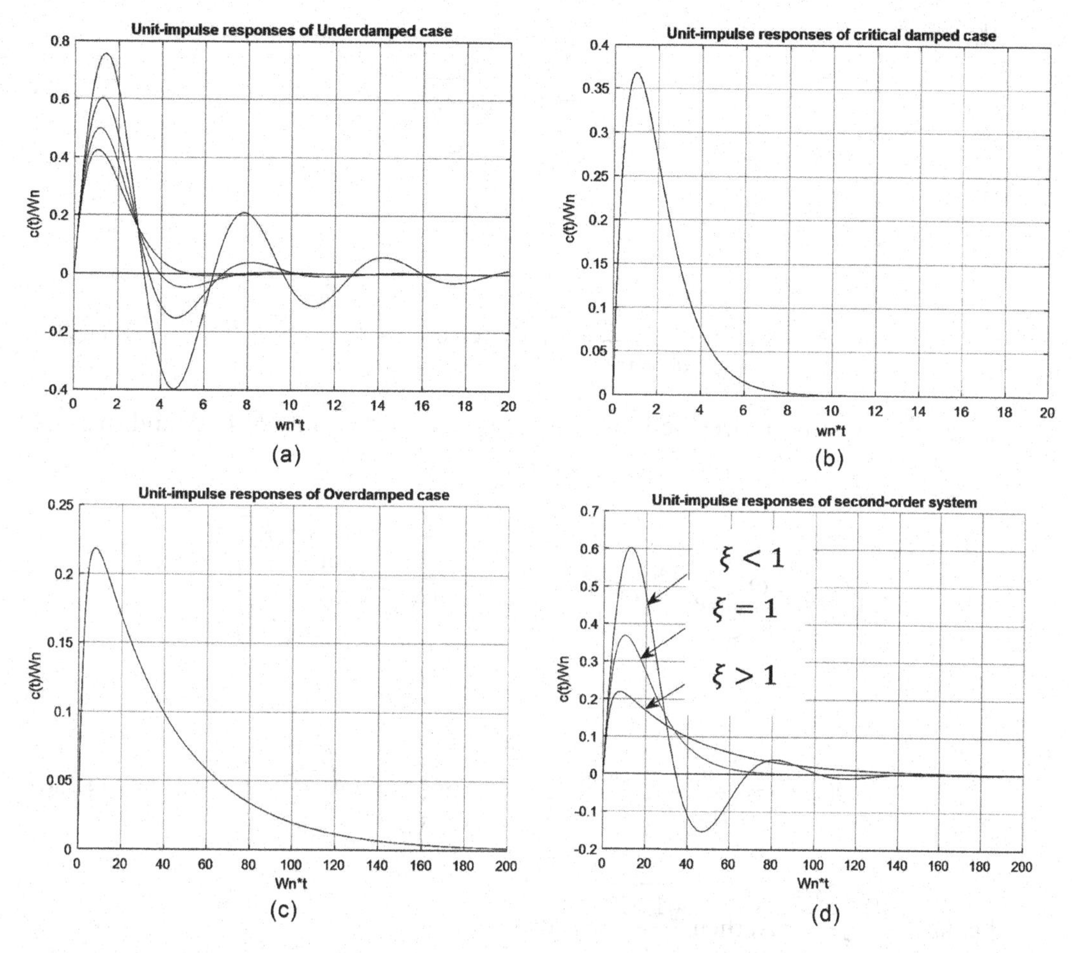

FIGURE 4.41 Unit-impulse response of the second-order system for (a) underdamped case, (b) critically damped case, (c) overdamped case, and (d) all together of underdamped, critically damped, and overdamped case.

Setting $\frac{dc(t)}{dt}=0$ in Eq. (4.133), we get

$$\frac{\omega_n e^{-\xi\omega_n t}}{\sqrt{1-\xi^2}}\left(\omega_n\sqrt{1-\xi^2}\cos\omega_n\sqrt{1-\xi^2}t-\xi\omega_n\sin\omega_n\sqrt{1-\xi^2}t\right)=0 \tag{4.134}$$

Since $\frac{\omega_n e^{-\xi\omega_n t}}{\sqrt{1-\xi^2}}\neq 0$, we get

$$\omega_n\sqrt{1-\xi^2}\cos\omega_n\sqrt{1-\xi^2}t-\xi\omega_n\sin\omega_n\sqrt{1-\xi^2}t$$

or

$$\omega_n\sqrt{1-\xi^2}=\xi\omega_n\tan\omega_n\sqrt{1-\xi^2}t$$

Thus,

$$\tan\omega_n\sqrt{1-\xi^2}\,t=\frac{\sqrt{1-\xi^2}}{\xi}$$

Therefore,

$$t=\frac{\tan^{-1}\dfrac{\sqrt{1-\xi^2}}{\xi}}{\omega_n\sqrt{1-\xi^2}};\text{ where, } 0<\xi<1 \tag{4.135}$$

To get the maximum overshoot, we substitute Eqs. (4.135) into (4.127), and then we get

$$c(t)_{\max}=\frac{\omega_n}{\sqrt{1-\xi^2}}e^{-\xi\omega_n\times\frac{\tan^{-1}\frac{\sqrt{1-\xi^2}}{\xi}}{\omega_n\sqrt{1-\xi^2}}}\cdot\sin\omega_n\sqrt{1-\xi^2}\times\frac{\tan^{-1}\dfrac{\sqrt{1-\xi^2}}{\xi}}{\omega_n\sqrt{1-\xi^2}}$$

or

$$c(t)_{\max}=\frac{\omega_n}{\sqrt{1-\xi^2}}e^{-\frac{\xi}{\sqrt{1-\xi^2}}.\tan^{-1}\frac{\sqrt{1-\xi^2}}{\xi}}\cdot\sin\left(\tan^{-1}\frac{\sqrt{1-\xi^2}}{\xi}\right) \tag{4.136}$$

Let $\tan^{-1}\dfrac{\sqrt{1-\xi^2}}{\xi}=A$, then $\dfrac{\sqrt{1-\xi^2}}{\xi}=\tan A$;

Thus, $\sin A=\sqrt{1-\xi^2}$.

Eq. (4.136) can be rewritten as

$$c(t)_{\max}=\frac{\omega_n}{\sqrt{1-\xi^2}}e^{-\frac{\xi}{\sqrt{1-\xi^2}}\cdot\tan^{-1}\frac{\sqrt{1-\xi^2}}{\xi}}\cdot\sqrt{1-\xi^2}$$

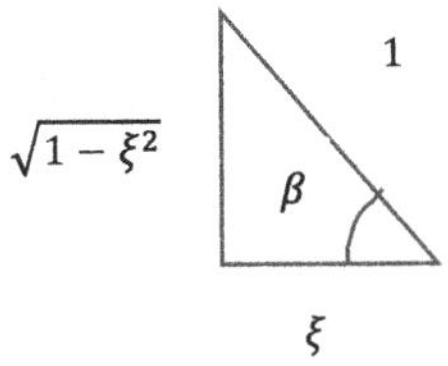

1

$\sqrt{1-\xi^2}$

β

ξ

Therefore,

$$c(t)_{\max}=\omega_n\,e^{-\frac{\xi}{\sqrt{1-\xi^2}}.\tan^{-1}\frac{\sqrt{1-\xi^2}}{\xi}};\text{(where, } 0<\xi<1) \tag{4.137}$$

Referring to Eq. (4.127) for the underdamped case, the unit-impulse response $c(t)$ oscillates about zero and takes both positive and negative values. From the foregoing analysis, we may conclude that if the impulse response $c(t)$ does not change the sign, the system is either critically damped or overdamped, in which case the corresponding step response does not overshoot but increases or decreases monotonically and approaches a constant value.

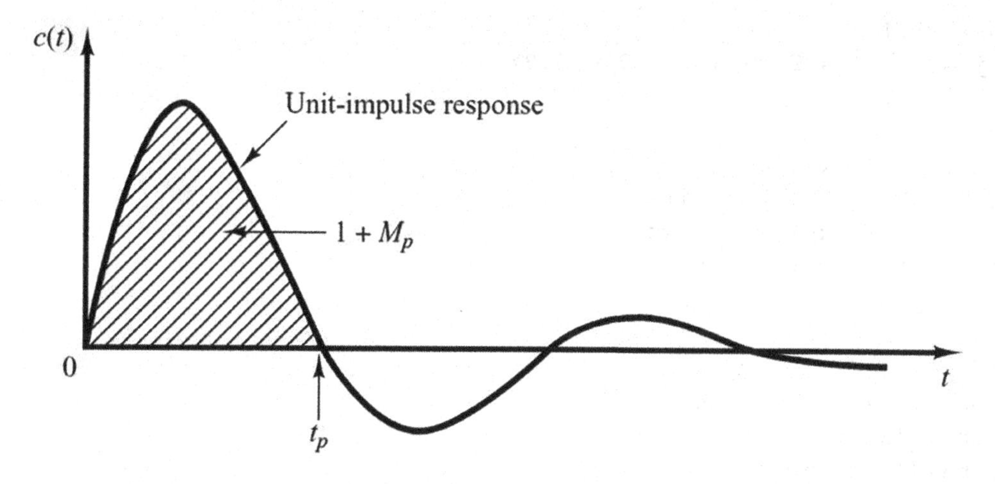

FIGURE 4.42 Unit-impulse response curve and area.

However, referring from Eq. (4.137) for the underdamped case, the maximum overshoot (M_p) for the unit-impulse response of the underdamped system occurs at the t which can be derived by differentiating and when $dc(t)/dt = 0$ from Eq. (4.133). Therefore, to find the maximum overshoot (M_p) and the response $c(t)_{max}$, the value of t can be substituted in Eq. (4.127). Therefore, the maximum overshoot (M_p) and the response $c(t)_{max}$ can be determined by using Eq. (4.137). Since the unit-impulse response function is the time derivative of the unit-step response function, the maximum overshoot (M_p) for the unit-step response can be found from the corresponding unit-impulse response. That is, the area under the unit-impulse response curve from $t = 0$ to the time of the first zero, as shown in this Figure 4.42.

In Figure 4.42, the area is shown as shaded by $1 + M_p$, where M_p is the maximum overshoot (for the unit-step response $c(t)$ from Eqs. 4.79 and 4.80) is given by

$$M_p = e^{-\xi\omega_n(\pi/\omega_d)} = e^{-\left(\xi/\sqrt{1-\xi^2}\right)\pi} \tag{4.138}$$

The peak time (t_p) (for the unit-step response $c(t)$) is given by Eq. (4.77) as

$$t_p = \frac{\pi}{\omega_d} \tag{4.139}$$

However, Eq. (4.77) corresponds to the time that the unit-impulse response first crosses the time axis, as shown in Figure 4.42.

To perform Figures 4.41 and 4.42, we may write the MATLAB program and we can analyze the curves. The MATLAB program is as below.

%The MATLAB program for the unit-impulse response of underdamped case $(0 < \xi < 1)$:

```
for j=1:4
Z(j)=j*0.2; % Z =0.2,0.4,0.6,0.8%
for i=1:201
wn(i, j)=1;
wt(i, j)=(i-1)/10;
sqrtt=sqrt(1-Z(j).*Z(j));
xown(i, j)=exp(-Z(j).*wt(i, j)).*(wn(i, j)/
sqrtt).*sin(sqrtt.*wt(i, j));
xo(i, j)=0;
end
end
plot(wt, xown, wt, xo)
xlabel('wn*t')
ylabel('c(t)/Wn')
title('Unit-impulse responses of Underdamped case'); grid
```

%The MATLAB program for unit-impulse response of critically damped case ($\xi = 1$):

```
for j=1:4
Z(j)=1+j*0; % Z =1%
for i=1:201
wn(i, j)=1;
wt(i, j)=(i-1)/10;
sqrtt=sqrt(1-Z(j).*Z(j));
xown(i, j)=wn(i, j)*wt(i, j)*exp(-wt(i, j));
xo(i, j)=0;
end
end
plot(wt, xown, wt, xo)
xlabel('wn*t')
ylabel('c(t)/Wn')
title('Unit-impulse responses of critical damped case'); grid
```

%The MATLAB program for unit-impulse response of overdamped case ($\xi > 1$):

```
Wn=0.10;
t=0:200;
hold on
plot(t, 0*(ut+t))
Z=2;
A=Wn./(2*(sqrt(Z.*Z-1)));
D1=(exp(-(Z-(sqrt(Z*Z-1))).*(Wn.*t)));
D2=(exp(-(Z+(sqrt(Z*Z-1))).*(Wn.*t)));
ct=A.*(D1-D2);
plot(t*Wn*10,ct/Wn)
xlabel('Wn*t')
ylabel('c(t)/Wn')
title('Unit-impulse responses of Overdamped case'); grid
```

%The MATLAB program for unit-impulse response of underdamped case ($0<\xi<1$), Critically damped case ($\xi=1$), and overdamped case ():

```
Wn=0.10;
t=0:200;
hold on
plot(t, 0*(ut+t))
for Z=[0.4,1,2]
if Z<1
A=Wn./(sqrt(1-Z.*Z));
y=exp(-Z.*Wn.*t);
B=Wn.*sqrt(1-Z*Z);
ct_underdamped=A.*y.*sin(B*t);
plot(t*Wn*10,ct_underdamped/Wn)
end
if Z==1
y=exp(-Z.*Wn.*t);
ct_critical=(Wn.*Wn).*t.*y;
plot(t*Wn*10,ct_critical/Wn)
end
if Z>1
V=Wn./(2.*(sqrt(Z.*Z-1)));
D1=(exp(-(Z-(sqrt(Z*Z-1))).*(Wn.*t)));
D2=(exp(-(Z+(sqrt(Z*Z-1))).*(Wn.*t)));
ct_overdamped=V.*(D1-D2);
plot(t*Wn*10,ct_overdamped/Wn)
end
end
xlabel('Wn*t')
ylabel('c(t)/Wn')
title('Unit-impulse responses of second-order system')
grid
```

%The MATLAB program for area of unit-impulse response from unit-step response:

```
Wn=0.10;
t=0:200;
hold on
plot(t, 0*(ut+t))
Z=0.4;
A=Wn./(sqrt(1-Z.*Z));
y=exp(-Z.*Wn.*t);
B=Wn.*sqrt(1-Z*Z);
ct=A.*y.*sin(B*t);
hold on
plot(t, ct)
D1=Z./(sqrt(1-Z.*Z));
D2=atan((sqrt(1-Z.*Z))./Z);
```

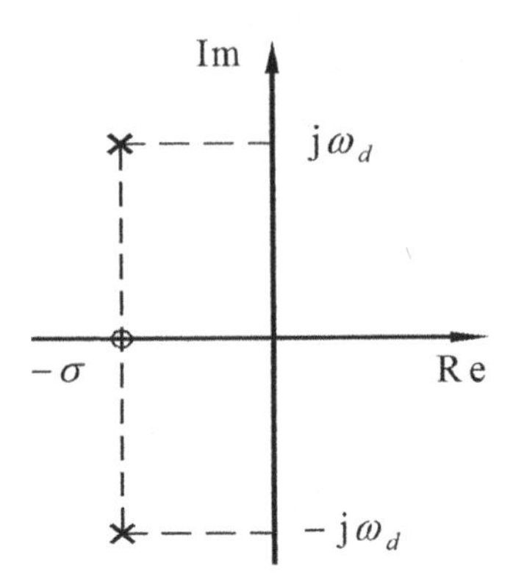

FIGURE 4.43 Pole-zero configuration of closed-loop control system.

```
ctmax=Wn.*exp(-D1.*D2);
hold on
plot(t, ctmax)
xlabel('t')
ylabel('c(t)')
title('Unit-impulse responses area')
grid
```

Example 4.9

A control system whose closed-loop poles and closed-loop zeros are located in a straight line parallel to the imaginary axis on the s-plane, as shown in Figure 4.43. Prove that the impulse response of the system is decaying with cosine function.

Solution

Proof: According to the pole-zero configuration of the closed-loop system as shown in Figure 4.43, the transfer function of the system can be obtained as

$$G(s)=\frac{C(s)}{R(s)}=\frac{K(s+\sigma)}{(s+\sigma+j\omega_d)(s+\sigma-j\omega_d)}=\frac{K(s+\sigma)}{(s+\sigma)^2+\omega_d^2}$$

The unit-impulse response, the input $C(s)=1$, then we get

$$C(s)=\frac{K(s+\sigma)}{(s+\sigma)^2+\omega_d^2}$$

Taking inverse Laplace transformation of $C(s)$, we get

$$c(t)=Ke^{-\sigma t}\cos\omega_d t$$

Therefore, the above equation shows that the impulse response of the system is decaying with cosine function

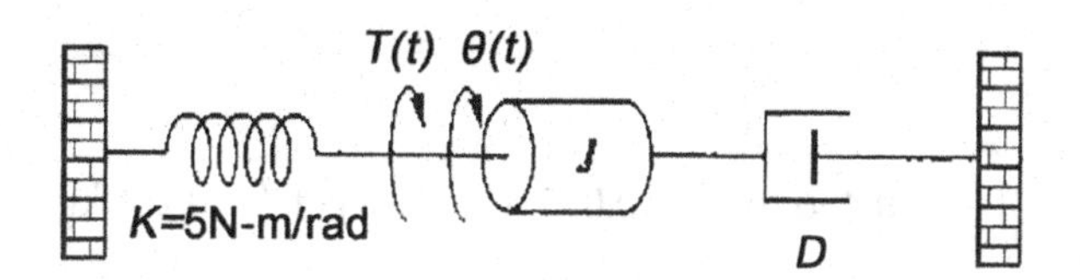

FIGURE 4.44 Inertia system with spring and damper system.

Example 4.10

Solve the following system to find the J and D to yield to 20% overshoot and settling time of 2 s of a step input of toque $T(t)$ (Figure 4.44).

Solution

The torque equation we may write as

$$T(t)=J\frac{d^2\theta(t)}{dt^2}+D\frac{d\theta(t)}{dt}+K\theta(t)$$

Taking the Laplace transformation, we get

$$T(s)=\left(Js^2+Ds+K\right)\theta(s)$$

Thus, the transfer function,

$$G(s)=\frac{\theta(s)}{T(s)}=\frac{1/J}{s^2+\frac{D}{J}s+\frac{K}{J}}$$

By comparing with the standard second-order equation, we can say

$$2\xi\omega_n=\frac{D}{J}\text{ and }\omega_n=\sqrt{\frac{K}{J}}$$

The settling time (t_s),

$$t_s=\frac{4}{\sigma}=\frac{4}{\xi\omega_n}=2$$

Thus, $\xi\omega_n=2$, and then $\frac{2}{\xi}=\sqrt{\frac{K}{J}}$.

Therefore,

$$\xi=2\sqrt{\frac{J}{K}}$$

The maximum overshoot,

$$M_p=e^{-\left(\xi\pi\sqrt{1-\xi^2}\right)}\text{ or }\xi=\frac{-\ln(M_p)}{\sqrt{\pi^2+\ln^2(M_p)}}\quad\xi=\frac{-\ln(20\%)}{\sqrt{\pi^2+\ln^2(20\%)}};\text{ then }\xi=0.456$$

Thus,

$$\xi = 2\sqrt{\frac{J}{K}} = 0.456, \text{ or } \frac{J}{5} = 0.052, \text{ then we get } J = 0.260\,\text{kg-m}^2$$

Since, $\xi\omega_n = 2$, then $2\xi\omega_n = 4$.

Therefore, $2\xi\omega_n = \frac{D}{J} = 4$. Thus, $\frac{D}{0.260} = 4$ or $D = 1.040\,\text{N-m-s/rad}$.

4.5.7 Unit-Ramp Response of the Second-Order System

When the input signal is a unit-ramp function, we recall Eq. (4.39), and the Laplace transform of the output of the second-order system is written as

$$C(s) = \frac{\omega_n^2}{s^2 + 2\xi\omega_n s + \omega_n^2} \cdot \frac{1}{s^2} = \frac{1}{s^2} - \frac{\frac{2\xi}{\omega_n}}{s} + \frac{\frac{2\xi}{\omega_n}(s + \xi\omega_n) + (2\xi^2 - 1)}{s^2 + 2\xi\omega_n s + \omega_n^2} \tag{4.140}$$

1. *Unit-ramp response of underdamped case* $(0 < \xi < 1)$
   Taking the inverse Laplace transform of Eq. (4.140), we can get the response $c(t)$ of the unit-ramp function for underdamped case as

$$c(t) = t - \frac{2\xi}{\omega_n} + \frac{1}{\omega_n\sqrt{1-\xi^2}} e^{-\xi\omega_n t} \sin(\omega_d t + 2\beta) \tag{4.141}$$

where $t \geq 0$, $\omega_d = \omega_n\sqrt{1-\xi^2}$, and $\beta = \tan^{-1}\frac{\sqrt{1-\xi^2}}{\xi}$.

Thus,

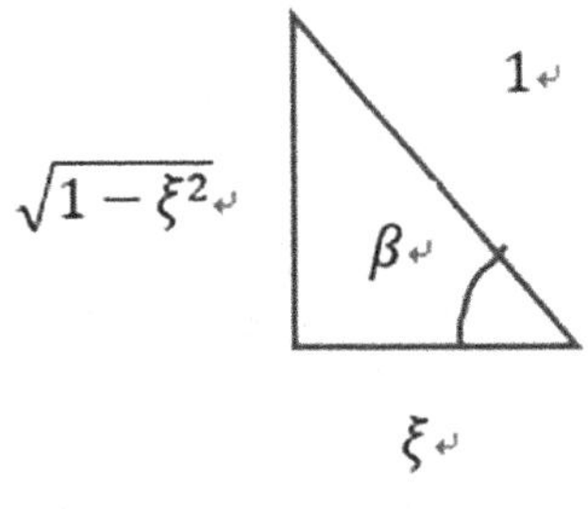

$$\tan 2\beta = \frac{2\tan\beta}{1 - \tan^2\beta} = \frac{2\xi\sqrt{1-\xi^2}}{2\xi^2 - 1}$$

1

$\sqrt{1-\xi^2}$

β

ξ

Hence,

$$c(t) = t - \frac{2\xi}{\omega_n} + \frac{1}{\omega_n\sqrt{1-\xi^2}} e^{-\xi\omega_n t} \sin\left(\omega_d t + \tan^{-1}\frac{2\xi\sqrt{1-\xi^2}}{2\xi^2 - 1}\right) \tag{4.142}$$

The unit-ramp response $c(t)$ of Eq. (4.142) is divided into the steady-state response and transient response. Thus, the steady-state response is defined as

$$c_{ss}(t) = t - \frac{2\xi}{\omega_n} \tag{4.143}$$

and the transient response is defined as

$$c_{tr}(t) = \frac{e^{-\xi\omega_n t}}{\omega_d} \sin(\omega_d t + 2\beta) \tag{4.144}$$

For the unit feedback of the second-order system, the error response of the second-order system is defined as

$$e(t)=r(t)-c(t) \tag{4.145}$$

When $t\to\infty$, the steady-state (ss) value of the error response is called the SSE of the system, and thus we get

$$e_{\text{SSE}}=e(\infty)=\lim_{t\to\infty}\left[r(t)-c(t)\right]=\frac{2\xi}{\omega_n} \tag{4.146}$$

The response curve of the second-order system for the underdamped system is shown in Figure 4.45.

Therefore, we get the error response $e(t)$ from Eqs. (4.142) to (4.145),

$$e(t)=\frac{2\xi}{\omega_n}-\frac{1}{\omega_n\sqrt{1-\xi^2}}e^{-\xi\omega_n t}\sin(\omega_d t+\tan^{-1}\frac{2\xi\sqrt{1-\xi^2}}{2\xi^2-1}) \tag{4.147}$$

Taking the derivative w.r.t t of Eq. (4.147) and set $\frac{de(t)}{dt}=0$, we get the peak time of unit-ramp error response as

$$t_p=\frac{\pi-\beta}{\omega_d}=\frac{\pi-\tan^{-1}\frac{\sqrt{1-\xi^2}}{\xi}}{\omega_n\sqrt{1-\xi^2}} \tag{4.148}$$

By using Eq. (4.147), the peak error of the system is

$$e(t_p)=\frac{2\xi}{\omega_n}\left(1+\frac{1}{2\xi}e^{-\xi\omega_n t_p}\right) \tag{4.149}$$

The maximum deviation of system is

$$e_{\max}=e(t_p)-e_{\text{SSE}}=\frac{1}{\omega_n}e^{-\xi\omega_n t_p} \tag{4.150}$$

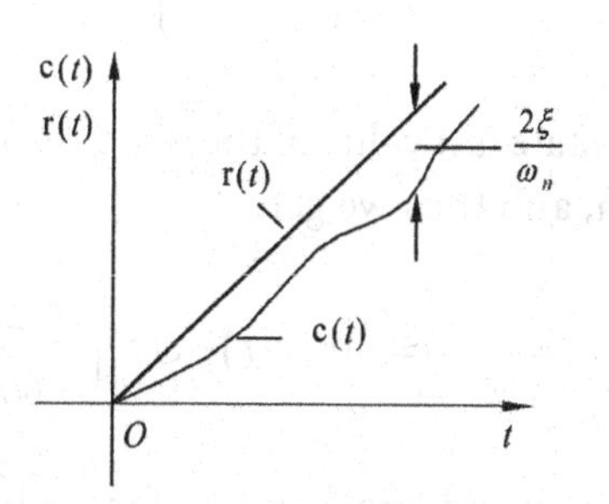

FIGURE 4.45 The unit-ramp response curve of the underdamped second-order system.

Similarly, according to the definition, the settling time or the adjustment time of the system can be obtained as

$$t_s = \frac{3}{\xi\omega_n}\ ; \text{for } \Delta = \pm 5\% \tag{4.151}$$

2. *Unit-ramp response of critical damping case* $(\xi = 1)$
To satisfy the condition of critically damped case ($\xi = 1$) for unit-ramp function, we rearrange Eq. (4.39) as follows:

$$C(s) = \frac{\omega_n^2}{s^2 + 2\omega_n s + \omega_n^2} \cdot \frac{1}{s^2} = \frac{1}{s^2} - \frac{\frac{2}{\omega_n}}{s} + \frac{1}{(s+\omega_n)^2} + \frac{\frac{2}{\omega_n}}{(s+\omega_n)} \tag{4.152}$$

Satisfying with $\xi = 1$ taking the inverse Laplace transform of Eq. (4.152), the time response $c(t)$ is written as

$$c(t) = t - \frac{2}{\omega_n} + te^{-\omega_n t} + \frac{2}{\omega_n}e^{-\omega_n t};\ (\text{for } t \geq 0) \tag{4.153}$$

The unit-ramp response $c(t)$ of Eq. (4.153) is divided into the steady-state response and transient response. Thus, the steady-state response is defined as

$$c_{ss}(t) = t - \frac{2}{\omega_n} \tag{4.154}$$

and the transient response is defined as

$$c_{tr}(t) = te^{-\omega_n t} + \frac{2}{\omega_n}e^{-\omega_n t} \tag{4.155}$$

For the unit feedback of the second-order system, the error response of the second-order system is defined as

$$e(t) = r(t) - c(t)$$

When $t \to \infty$, the steady-state (ss) value of the error response is called the steady-state error (SSE) of the system, and thus we get

$$e_{SSE} = e(\infty) = \lim_{t\to\infty}[r(t) - c(t)] = \frac{2}{\omega_n} \tag{4.156}$$

The response curve of the second-order system for the underdamped system is shown in Figure 4.46.

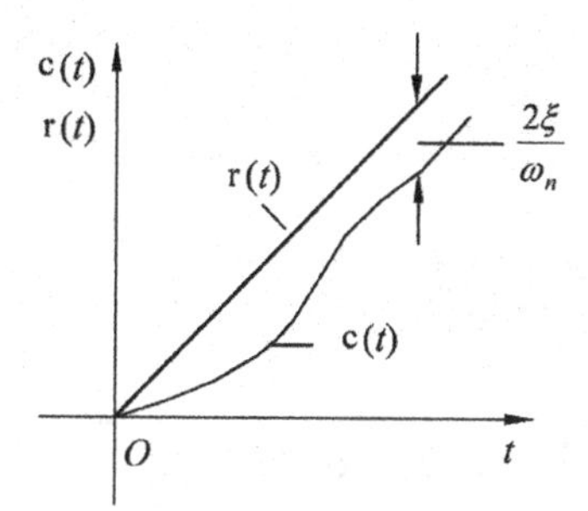

FIGURE 4.46 Critically damped second-order system unit-ramp response curve.

Therefore, we get the error response $e(t)$

$$e(t)=\frac{2}{\omega_n}\left[1-\left(1+\frac{1}{2}\omega_n t\right)e^{-\omega_n t}\right] \tag{4.157}$$

Taking the derivative, w.r.t t and set $\frac{de(t)}{dt}=0$, we get the peak time of unit-ramp error response as

$$\frac{2}{\omega_n}\left[-\left(\frac{1}{2}\omega_n\right)e^{-\omega_n t}-\left(\frac{1}{2}\omega_n t\right)(-\omega_n)e^{-\omega_n t}\right]=0 \text{ and thus, } t_p=\frac{1}{\omega_n} \tag{4.158}$$

The peak error of the system is

$$e(t_p)=1-\frac{1}{\omega_n}+\frac{2}{\omega_n}-\frac{1}{\omega_n}e^{-1}-\frac{2}{\omega_n}e^{-1}=\frac{\omega_n+1-3e^{-1}}{\omega_n} \tag{4.159}$$

The maximum deviation of the system is

$$e_{\max}=e(t_p)-e_{\text{SSE}}=\frac{\omega_n-1-3e^{-1}}{\omega_n} \tag{4.160}$$

Similarly, by the computer numerical solution according to the definition, the settling time or the adjustment time of the system can be obtained as

$$t_s=\frac{4.1}{\omega_n}\text{; for } \Delta=\pm 5\% \tag{4.161}$$

3. *Unit-ramp response of overdamped case* $(\xi>1)$
   To satisfy the condition of the overdamped case $(\xi>1)$ for unit-ramp function $\left(R(s)=\frac{1}{s^2}\right)$, we rearrange Eq. (4.39) as follows:

$$C(s)=\frac{\omega_n^2}{s^2+2\xi\omega_n s+\omega_n^2}\cdot\frac{1}{s^2}=\frac{\omega_n^2}{\left(s+\xi\omega_n+\omega_n\sqrt{\xi^2-1}\right)\left(s+\xi\omega_n-\omega_n\sqrt{\xi^2-1}\right)}\cdot\frac{1}{s^2}$$

or

$$C(s)=\frac{1}{s^2}-\frac{\frac{2\xi}{\omega_n}}{s}+\frac{\xi}{\omega_n\sqrt{\xi^2-1}}\left((T_1-1).\frac{1}{s+T_2}-(T_2-1).\frac{1}{s+T_1}\right) \tag{4.162}$$

where $T_1=\omega_n\left(\xi+\sqrt{\xi^2-1}\right)$, and $T_2=\omega_n\left(\xi-\sqrt{\xi^2-1}\right)$; then satisfying with $\xi>1$ taking the inverse Laplace transform of Eq. (4.162), the unit-ramp time response c(t) is written as

$$c(t)=t-\frac{2\xi}{\omega_n}+\frac{\xi}{\omega_n\sqrt{\xi^2-1}}\left((T_1-1)\cdot e^{-T_2t}-(T_2-1)\cdot e^{-T_1t}\right) \tag{4.163}$$

When $t\geq 0$, and $t\to\infty$, the SSE of unit-ramp system response e_{SSE} is

$$e_{\text{SSE}}=e(\infty)=\lim_{t\to\infty}\left[r(t)-c(t)\right]=\frac{2\xi}{\omega_n} \tag{4.164}$$

The system error response is

$$e(t)=\frac{2\xi}{\omega_n}\left[1-\frac{1}{2\sqrt{\xi^2-1}}\left((T_1-1)\cdot e^{-T_2t}-(T_2-1).e^{-T_1t}\right)\right] \tag{4.165}$$

Under the condition of the unit-ramp signal, the dynamic performance index of the overdamped second-order system can only be approximated by a computer. The output response of the second-order system of the overdamped case is shown in Figure 4.47.

To perform the unit-ramp responses, Figures 4.45–4.47 were sketched for underdamped, critically damped, and overdamped case, respectively, by using the response *c(t)* from Eqs. (4.142), (4.153), and (4.163), respectively. We may write the MATLAB program, and we can analyze the curves. The MATLAB program is as below.

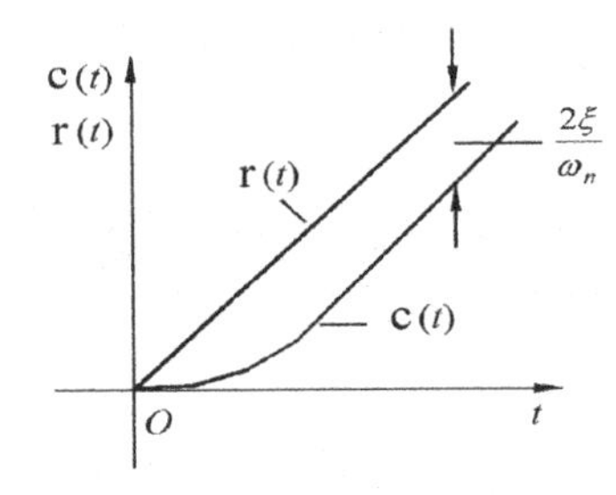

FIGURE 4.47 Unit-ramp response curve of overdamped for the second-order system.

%The MATLAB program for unit-ramp response of underdamped case ($0<\xi<1$):

```
rt=t;
Wn=0.10;
t=0:100;
hold on
plot(t, rt)
Z=0.2;
A=1./(Wn.*(sqrt(1-Z.*Z)));
y=exp(-Z.*Wn.*t);
Wd=Wn.*(sqrt(1-Z.*Z));
B=atan((sqrt(2.*Z.*sqrt(1-Z.*Z)))./(2.*Z.*Z-1));
ct_ramp=t-((2.*Z)./Wn)+A.*y.*sin(Wd.*t+B*pi/180);
hold on
plot(t, t.*ct_ramp./130)
xlabel('t')
ylabel('c(t)')
title('Unit-ramp responses of Underdamped case'); grid
```

%The MATLAB program for unit-ramp response of critically damped case ($\xi=1$):

```
rt=t;
Wn=0.10;
t=0:100;
hold on
plot(t, rt)
Z=1;
y=exp(-Wn.*t);
ct_ramp=t-(2/Wn)+t.*y+(2/Wn).*y;
hold on
plot(t, t.*ct_ramp/120)
xlabel('t')
ylabel('c(t)')
title('Unit-ramp responses of critical damped case')
grid
```

%The MATLAB program for unit-ramp response of overdamped case ($\xi>1$):

```
Wn=0.10;
t=0:100;
rt=t;
hold on
plot(t, rt)
for Z=[2]
A=Z./(Wn.*(sqrt(Z.*Z-1)));
T1=Wn.*(Z+sqrt(Z.*Z-1));
```

```
T2=Wn.*(Z-sqrt(Z.*Z-1));
E=(T1-1).*exp(-T2.*t)-(T2-1).*exp(-T1.*t);
ct_ramp=t-((2.*Z)./Wn)+A.*E;
hold on
plot(t, t.*ct_ramp/100)
end
xlabel('t')
ylabel('c(t)')
title('Unit-ramp responses of Overdamped case')
grid
```

Example 4.11

A control system is shown in Figure 4.48. The input signal is a unit-ramp function that determines the system unit-ramp response and error response expressions when K is 13.5, 200 and 1,500, respectively. If possible, please estimate the error performance of the system.

Solution

According to the open-loop transfer function of the system as shown in Figure 4.48, we may write

$$G(s)=\frac{5K}{s(s+34.5)}=\frac{\omega_n^2}{s(s+2\xi\omega_n)}$$

1. When $K=13.5$, the solution is $\xi=2.1$, and $\omega_n=8.2$. Substituting the parameters into Eqs. (4.163) and (4.165), we get

$$c(t)=t-0.51+0.51e^{-2.08t}-e^{-32.4t}$$

and

$$e(t)=0.51\left[1-e^{-2.08t}+0.004e^{-32.4t}\approx 0.51e^{-2.08t}\right]$$

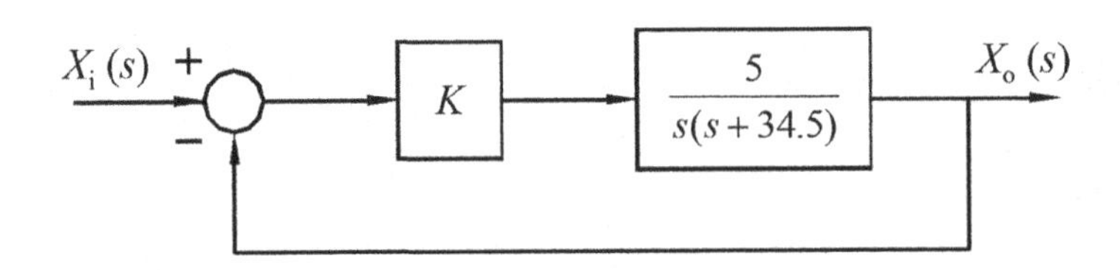

FIGURE 4.48 A closed-loop control system.

The system is equivalent to a first-order system, the equivalent time constant $T = \frac{1}{2.08} = 0.48\,\text{s}$, and thus we have $t_d = 0.33$ s, $t_r = 1.06$ s, $t_s = 1.44$ s, and $c_{SSE} = 0.51\,\text{rad}$

2. When $K = 13.5$, the solution is $\xi = 0.55$, and $\omega_n = 31.6$. Substituting the parameters into Eqs. (4.142) and (4.147), we get

$$c(t) = t - 0.035 + 0.038e^{-17.4t}\sin(26.4t + 113°)$$

and

$$e(t) = 0.035 - 0.038e^{-17.4t}\sin(26.4t + 113°)$$

However, we get from Eqs. (4.148), (4.150), and (4.151) as

$$t_p = \frac{\pi - \beta}{\omega_d} = 0.08\,\text{s},\ t_s = \frac{3}{\xi\omega_n} = 0.17\,\text{s},$$

$$c_{\max} = \frac{1}{\omega_n}e^{-\xi\omega_n t_p} = 0.008\,\text{rad, and } c_{SSE} = 0.035\,\text{rad}$$

3. When $K = 1{,}500$, the solution is $\xi = 0.2$, and $\omega_n = 86.6$. Substituting the parameters into Eqs. (4.142) and (4.147), we get

$$c(t) = t - 0.0046 + 0.12e^{-17.3t}\sin(84.9t + 157°)$$

and

$$e(t) = 0.0046 - 0.12e^{-17.3t}\sin(84.9t + 157°)$$

However, we get the system performance from Eqs. (4.148), (4.150), and (4.151) as

$$t_p = \frac{\pi - \beta}{\omega_d} = 0.02\,\text{s},\ t_s = \frac{3}{\xi\omega_n} = 0.17\,\text{s},$$

$$c_{\max} = \frac{1}{\omega_n}e^{-\xi\omega_n t_p} = 0.008\,\text{rad, and } c_{SSE} = 0.0046\,\text{rad}$$

The unit-ramp response of the system under different values of K is shown in Figure 4.49, and the error response curve is shown in Figure 4.50. By using the MATLAB program, we may perform this analysis.

The calculation of this example, Figure 4.49a shows the unit-ramp response curve and Figure 4.49b shows the error response curves of the unit-ramp response signal. In these analyses, if we increase the gain of the amplifier, the damping ratio (ξ) is decreased but it is reduced the SSE of the system, whereas it deteriorated the dynamic performance of the unit-ramp response.

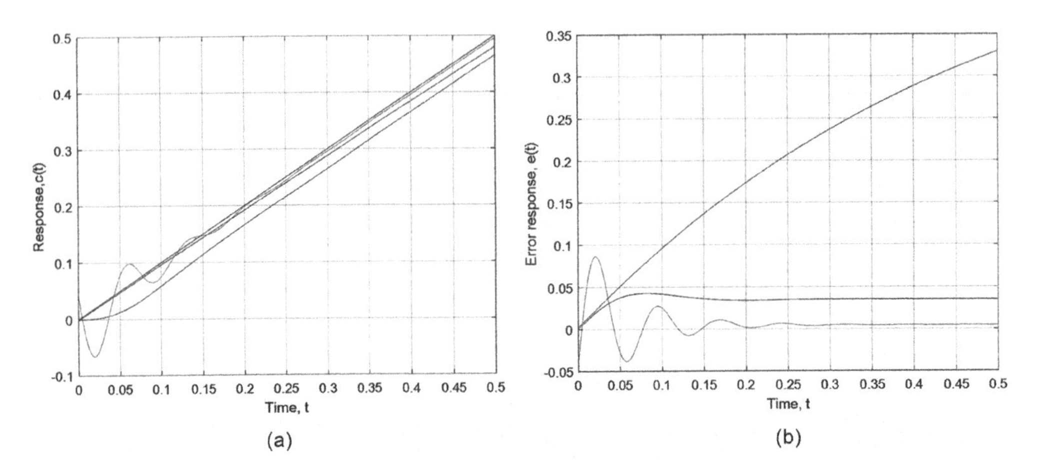

FIGURE 4.49 (a) Response curves of unit-ramp signal, and (b) errors of unit-ramp response curves.

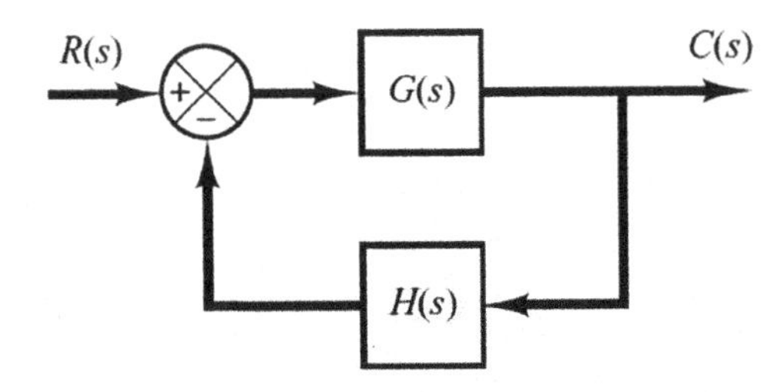

FIGURE 4.50 Closed-loop system with feedback transfer function.

Therefore, the damping ratio (ξ) should not be too small, and the value of ω_n must be large enough. But only the gain can be adjusted in the proportion of the control system, to satisfy both these two-characteristic difficulties at the same time; therefore, it is possible to adopt only by compromised solution (optimization).

%The MATLAB program for unit-ramp response and errors curve:

```
for j=1:500
t(j)=(j-1)/1000;
ct1(j)=t(j)-0.51+0.51*exp(-.08*t(j))-0.002*exp(-32.4*t(j));
e1(j)=0.51*(1-exp(-2.08*t(j))+0.004*exp(-32.4*t(j)));
ct2(j)=t(j)-0.035+0.038*exp(-17.4*t(j))*sin(26.4*t(j)+113/
180*pi);
e2(j)=0.035-0.038*exp(-17.4*t(j))*sin(26.4*t(j)+113/180*pi);
ct3(j)=t(j)-0.0046+0.12*exp(-17.3*t(j))*sin(84.9*t(j)+157/
180*pi);
e3(j)=0.0046-0.12*exp(-17.3*t(j))*sin(84.9*t(j)+157/180*pi);
ct(j)=t(j);
end
figure
plot(t, ct1,t, ct2,t, ct3,t, xt)
xlabel('Time, t')
ylabel('Response, c(t)')
grid
```

```
figure
plot(t, e1,t, e2,t, e3)
xlabel('Time, t')
ylabel('Error response, e(t)')
grid
```

4.6 TRANSIENT RESPONSE OF HIGH-ORDER SYSTEMS

The control system is usually a high-order system in practical engineering, for example, valve-controlled hydraulic cylinder, hydraulic booster, steering engine, air-controlled motor, electronic control robot, and so on are high-order systems. For an actual control system, it is difficult to find a low-order system; thus, it is very important to study the high-order system for the control system.

The closed-loop transfer function of the system as shown in Figure 4.50 is

$$\frac{C(s)}{R(s)}=\frac{G(s)}{1+G(s)H(s)}$$

In general, $G(s)$ and $H(s)$ are given as ratios of polynomials in s, or

$$G(s)=\frac{p(s)}{q(s)},\text{ and } H(s)=\frac{n(s)}{d(s)}$$

where $p(s)$, $q(s)$, $n(s)$ and $d(s)$ are polynomials in s. Thus, the closed-loop transfer function is written as

$$\frac{C(s)}{R(s)}=\frac{\dfrac{p(s)}{q(s)}}{1+\dfrac{p(s)}{q(s)}\cdot\dfrac{n(s)}{d(s)}}=\frac{p(s)d(s)}{q(s)d(s)+p(s)n(s)}$$

or

$$\frac{C(s)}{R(s)}=\frac{b_0s^m+b_1s^{m-1}+\cdots+b_{m-1}s+b_m}{a_0s^n+a_1s^{n-1}+\cdots+a_{n-1}s+a_n};\ m\le n \tag{4.166}$$

The higher-order system refers to the control system of the third order or above. Here, the LTI system with single input and single output will be mainly studied. The transfer function of the system can be described as

$$\frac{C(s)}{R(s)}=\frac{K\left(s^m+b_1s^{m-1}+\cdots+b_{m-1}s+b_m\right)}{s^n+a_1s^{n-1}+\cdots+a_{n-1}s+a_n} \tag{4.167}$$

The transient response of this system to any given input can be obtained by a computer simulation. If an analytical expression for the transient response is desired, then it is necessary to factor the denominator polynomial. MATLAB may be used for finding the roots of the denominator polynomial. Once the numerator and the denominator have been factored, $C(s)/R(s)$ can be written in the form as

$$\frac{C(s)}{R(s)}=\frac{K(s+z_1)(s+z_2)...(s+z_m)}{(s+p_1)(s+p_2)...(s+p_n)} \quad (4.168)$$

When $m \le n$ and $q+2r=n$, under the condition of the unit-step function, the output response of Eq. (4.168) is

$$C(s)=\frac{K(s+z_1)(s+z_2)...(s+z_m)}{(s+p_1)(s+p_2)...(s+p_n)}\cdot\frac{1}{s} \quad (4.169)$$

or

$$C(s)=\frac{K\left(s^m+b_1s^{m-1}+\cdots+b_{m-1}s+b_m\right)}{\sum_{j=1}^{q}(s+p_j)\sum_{k=1}^{r}\left(s^2+2\xi_k\omega_ks+\omega_k^2\right)}\cdot\frac{1}{s} \quad (4.170)$$

If the poles are different, Eq. (4.170) can be expanded into a partial fraction as

$$C(s)=\frac{a}{s}+\sum_{j=1}^{q}\frac{a_j}{s+p_j}+\sum_{k=1}^{r}\frac{\beta_k(s+\xi_k\omega_k)+\gamma_k\left(\omega_k\sqrt{1-\xi_k^2}\right)}{(s+\xi_k\omega_k)^2+(\omega_k\sqrt{1-\xi_k^2})^2} \quad (4.171)$$

Taking the inverse Laplace transform of Eq. (4.171), we get

$$c(t)=a+\sum_{j=1}^{q}a_je^{-p_jt}+\sum_{k=1}^{r}\beta_ke^{-\xi_k\omega_kt}\cos(\omega_k\sqrt{1-\xi_k^2})t+\sum_{k=1}^{r}\gamma_ke^{-\xi_k\omega_kt}\sin(\omega_k\sqrt{1-\xi_k^2})t \quad (4.172)$$

By Eq. (4.172), the transient response of the higher-order system is superposed by the response functions of the inertial and second-order oscillating elements; if every root has negative real part, the roots of the system are all in the left-half s-plane; except the constant term in Eq. (4.172), the other exponential-sinusoidal terms are decayed to zero with time $t \to \infty$, and then the system is stable. The unit-step response of the higher-order system is shown in Figure 4.51.

4.6.1 Real Poles and Pairs of Complex-Conjugate Poles of Higher-Order Systems

The real poles and pairs of complex-conjugate poles as well as dominant closed-loop poles. It will be seen that the response of a higher-order system is the sum of the responses of first-order and second-order systems. Consider the case where the poles of $C(s)$ consist of real poles and pairs of complex-conjugate poles. A pair of complex-conjugate poles yields a second-order term in s. We recall Eq. (4.171), it can be factored form of the higher-order characteristic equation consisting of first- and second-order terms. If the closed-loop poles involve multiple poles, $C(s)$ must have multiple-pole terms. We see that the response of

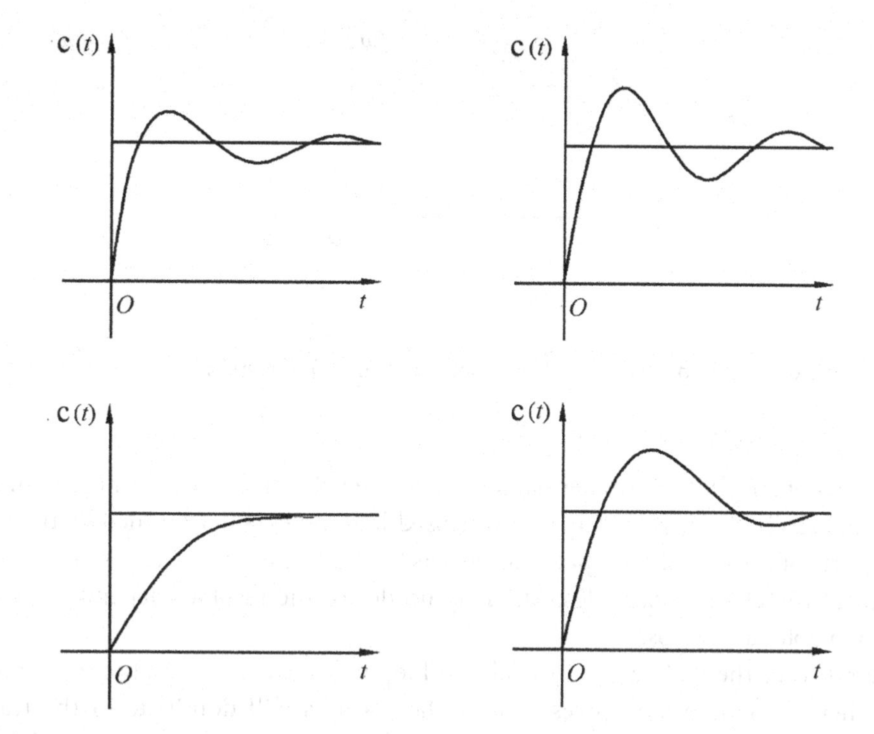

FIGURE 4.51 Unit-step response curve of the higher-order system.

a higher-order system is composed of a number of terms involving the simple functions found in the responses of first- and second-order systems.

Whereas, we recall Eq. (4.172), we can say that the response curve of a stable higher-order system is the sum of a number of exponential curves and damped sinusoidal curves. If all closed-loop poles lie in the left-half s-plane, then the exponential terms and the damped exponential terms will approach zero as time t increases to infinity. The steady-state output is then $c(\infty)=a$.

Let us assume that the system considered is a stable one. Then the closed-loop poles that are located far from the $j\omega$ axis have large negative real parts. The exponential terms that correspond to these poles decay very rapidly to zero, as shown in Figure 4.52. Note that the horizontal distance from a closed-loop pole to the $j\omega$ axis determines the settling time (t_s) of transients due to that pole. The smaller the distance, the longer the settling time. Therefore, we can say

$$t_s = \frac{1}{d} \tag{4.173}$$

Remember that the type of transient response is determined by the closed-loop poles, while the shape of the transient response is primarily determined by the closed-loop zeros. As we have seen earlier, the poles of the input $R(s)$ yield the steady-state response terms in the solution, while the poles of $C(s)/R(s)$ enter into the exponential transient response terms and/or damped sinusoidal transient response terms. The zeros of $C(s)/R(s)$ do not

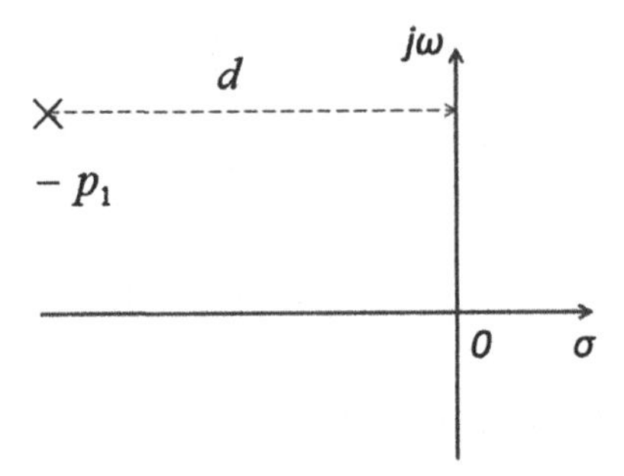

FIGURE 4.52 Relationship between pole and settling time (t_s) of response.

affect the exponents in the exponential terms, but they do affect the magnitudes and signs of the residues. The relative dominance of closed-loop poles is determined by the ratio of the real parts of the closed-loop poles, as well as by the relative magnitudes of the residues evaluated at the closed-loop poles. The magnitudes of the residues depend on both the closed-loop poles and zeros.

If the ratios of the real parts of the closed-loop poles exceed 5 and there are no zeros nearby, then the closed-loop poles nearest the $j\omega$ axis will dominate in the transient response behavior because these poles correspond to transient response terms that decay slowly. Those closed-loop poles that have dominant effects on the transient response behavior are called dominant closed-loop poles. Quite often the dominant closed-loop poles occur in the form of a complex-conjugate pair. The dominant closed-loop poles are the most important among all closed-loop poles. Some higher-order systems can be approximated by lower-order systems with appropriate simplification. There are two commonly simplification methods that are used to discuss in below.

1. *Dominant closed-loop pole method:* If the negative real part of the system poles is away from the imaginary axis, the transient response corresponding to the pole decays rapidly; the closest closed-loop pole has the slowest decaying transient response to the imaginary axis; the pole dominates the transient response; thus, it is called the dominant pole. For example, in a control engineering, the distance from one pole A to the imaginary axis is greater than five times the distance from another pole B to the imaginary axis and the influence of pole A on the system can be ignored.

2. *Dipole cancellation:* In the closed-loop transfer function, if the negative real part of the zero and the pole of the system are similar in numerical value, the zero and the pole can be eliminated together, to omit the influence of the pair of zero poles on the system, which is called dipole cancellation.

   Note that the gain of a higher-order system is often adjusted so that there will exist a pair of dominant complex-conjugate closed-loop poles. The presence of such poles in a stable system reduces the effects of such nonlinearities as dead zone, backlash, and coulomb-friction (Figure 4.53).

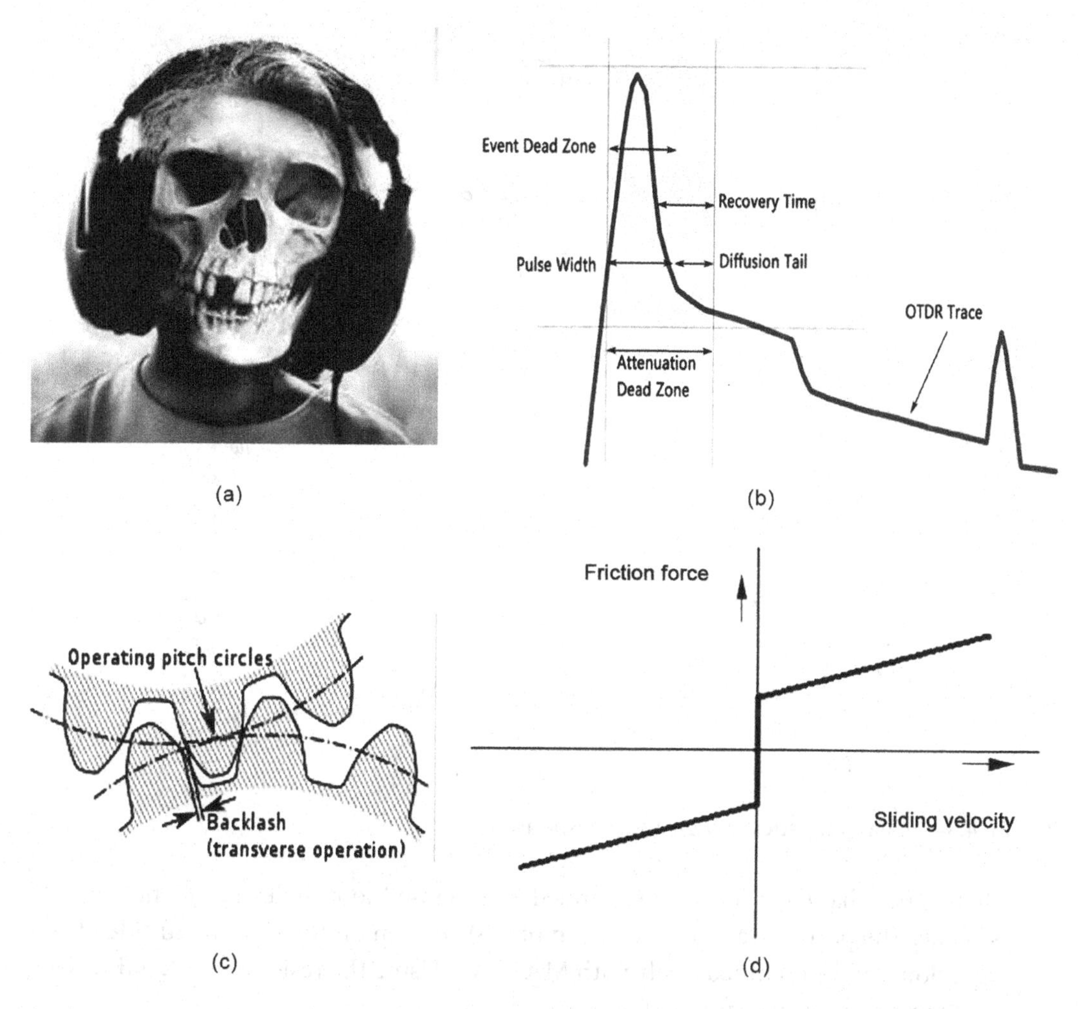

FIGURE 4.53 Effects of nonlinearities such as (a)–(b) dead zone, (c) backlash, and (d) coulomb-friction.

Let us examine the response behavior of the high-order response behavior of the system to a unit-step input. Consider first the case where the closed-loop poles are all real and distinct. For a unit-step input, Eq. (4.171) can be written as

$$C(s)=\frac{a}{s}+\sum_{j=1}^{q}\frac{a_j}{s+p_j} \tag{4.174}$$

where a_i is the residue of the pole at $s=-p_j$. If the system involves multiple poles, then $C(s)$ will have multiple-pole terms such as if the system has p_1, p_2, and p_3 poles and then we can see $C(s)$ from this system function.

$$C(s)=\frac{a}{s}+\sum_{j=1}^{q}\frac{a_j}{(s+p_1)(s+p_2)(s+p_3)} \tag{4.175}$$

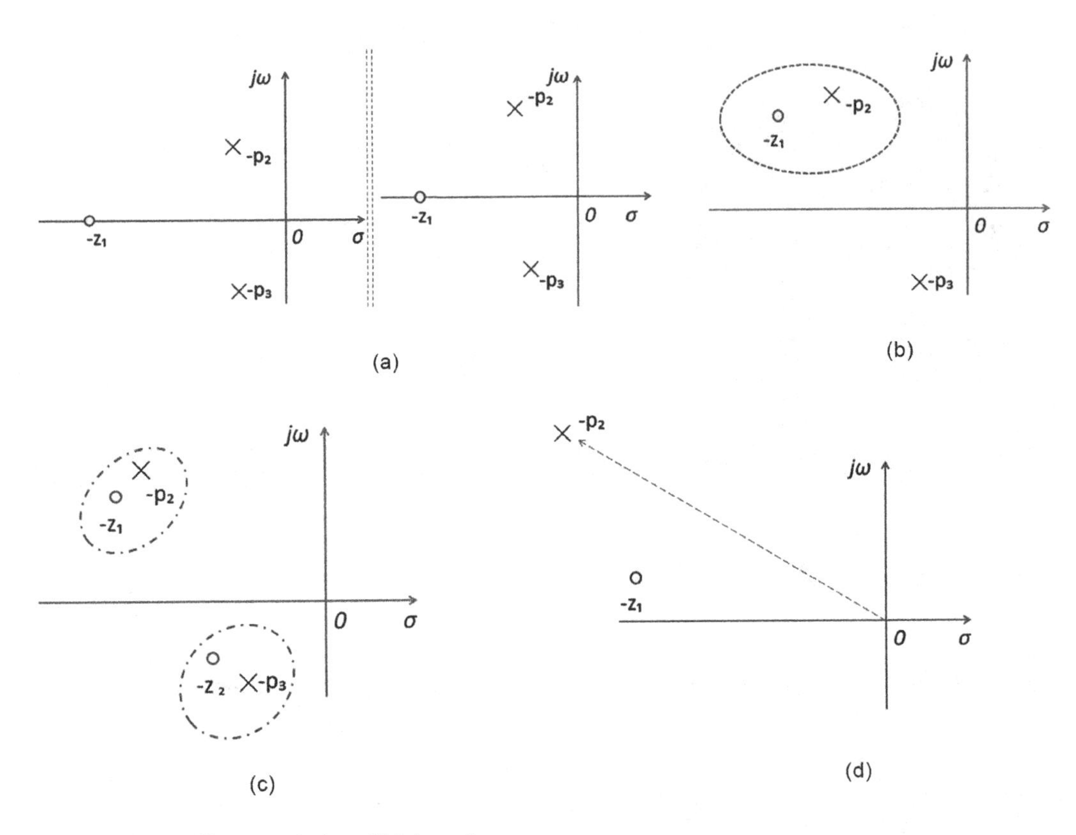

FIGURE 4.54 Characteristics of high-order response.

Remember that $C(s)$ must be expanded if all closed-loop poles lie in the left-half s-plane. The partial-fraction expansion of $C(s)$, as seen on the right-hand side of the equation, can be obtained easily with MATLAB. Using the residue command we can find the residue of the system. For example,

$$G(s)H(s)=\frac{K(s+z_1)}{(s+p_1)(s+p_2)(s+p_3)(s+p_4)} \tag{4.176}$$

In Eq. (4.176), we can see there are four poles p_1, p_2, p_3, and p_4 as well as one zero z_1 which has been shown in Figure 4.54a. The poles are indicated by the cross sign, and the zero is indicated by a circle. Here, p_2, and p_3 are complex-conjugate poles, as shown in Figure 4.54a.

If all closed-loop poles lie in the left-half s-plane, the relative magnitudes of the residues determine the relative importance of the components in the expanded form of $C(s)$. If there is a closed-loop zero close to a closed-loop pole, then the residue at this pole is small and the coefficient of the transient response term corresponding to this pole becomes small, as shown in Figure 4.54b. A pair of closely located poles and zeros will effectively cancel each other, as shown in Figure 4.54c. If a pole is located very far from the origin, the residue at this pole may be small, as shown in Figure 4.54d.

Therefore, we can see the characteristics of high-order response as the transients corresponding to such a remote pole are small and last a short time. Terms in the expanded form of $C(s)$ having very small residues contribute little to the transient response, and these terms may be neglected. If this is done, the higher-order system may be approximated by a lower-order one. Such an approximation often enables us to estimate the response characteristics of a higher-order system from those of a simplified one. For example, we may say as

$$C(s)=\frac{a}{s}+\frac{a_1}{s+p_1} \tag{4.177}$$

4.7 STABILITY ANALYSES AND CRITERION

In this section, we will learn *stability analysis and stability analysis in complex plane. Now the problem is why we need to investigate stability?* The answer is quite obvious, the issue of ensuring the *stability* of a closed-loop system is the most important to control system design. An *unstable* feedback system is of no practical value. Therefore, we may indicate that

- An LTI control system is in *equilibrium,* and it will departure from its equilibrium when it is subjected to transient disturbance;
- An LTI control system is *stable* if the output eventually comes back to its equilibrium state when the system is subjected to an initial condition;
- An LTI control system is *critically stable* if oscillations of the output continue forever; and
- It is *unstable* if the output diverges without being bound from its equilibrium state with the initial condition of the system.

As we can see in Figure 4.55a and b, it explains the balance of a pendulum, and Figure 4.55c shows the balance of a small ball. There is a necessary and sufficient condition for a feedback system to be stable is that all the poles of the system transfer function have negative

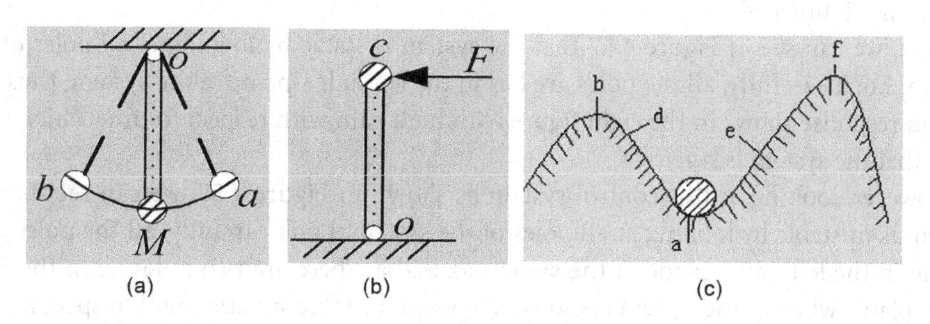

FIGURE 4.55 (a)–(b) The balance of a pendulum, and (c) The balance of a small ball.

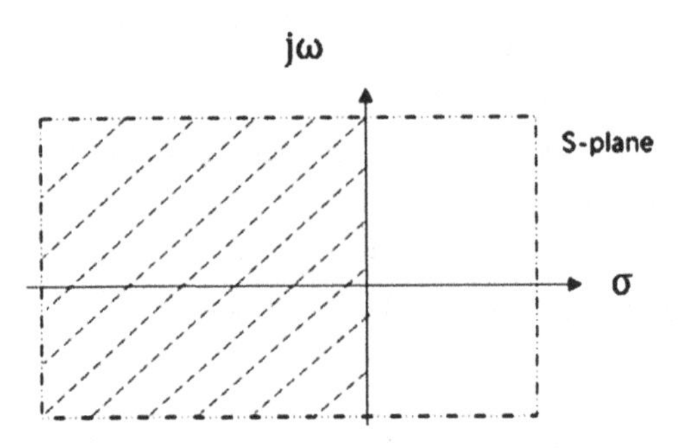

FIGURE 4.56 S-plane with a real and imaginary axis for pole-zero configuration.

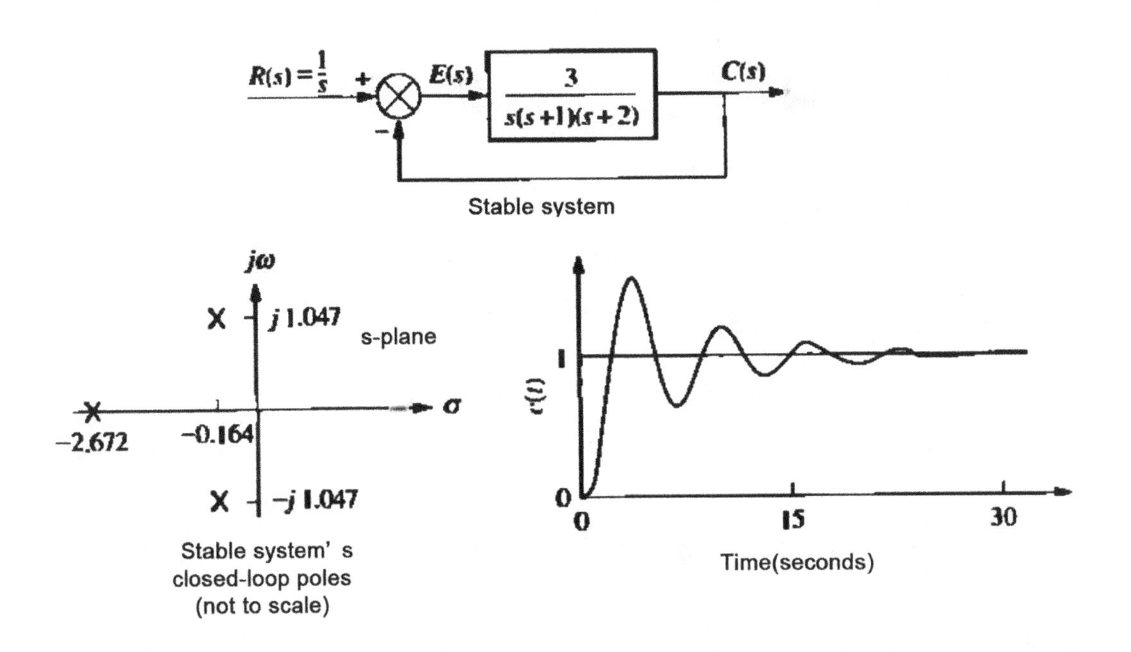

FIGURE 4.57 Stable system of closed-loop poles.

real parts. This means that all the poles of the system must lie in the left half s-plane as we can see in Figure 4.56.

Here, we can see in Figure 4.57 that the system is stable by looking at all poles of the system. Look carefully, all the poles are lies in the left half s-plane of the system; thus, the system response shows in the right figure with high gain with respect to time which indicates that the system is *stable*.

However, looking at this control system as shown in Figure 4.58, we can see that the system is unstable by looking at all poles of the system. Look carefully, all the poles does not lie in the left half s-plane of the system as well as there are two poles lie in the right half s-plane, whereas the system response shows an unstable situation with respect to time which indicates that the system is *unstable*.

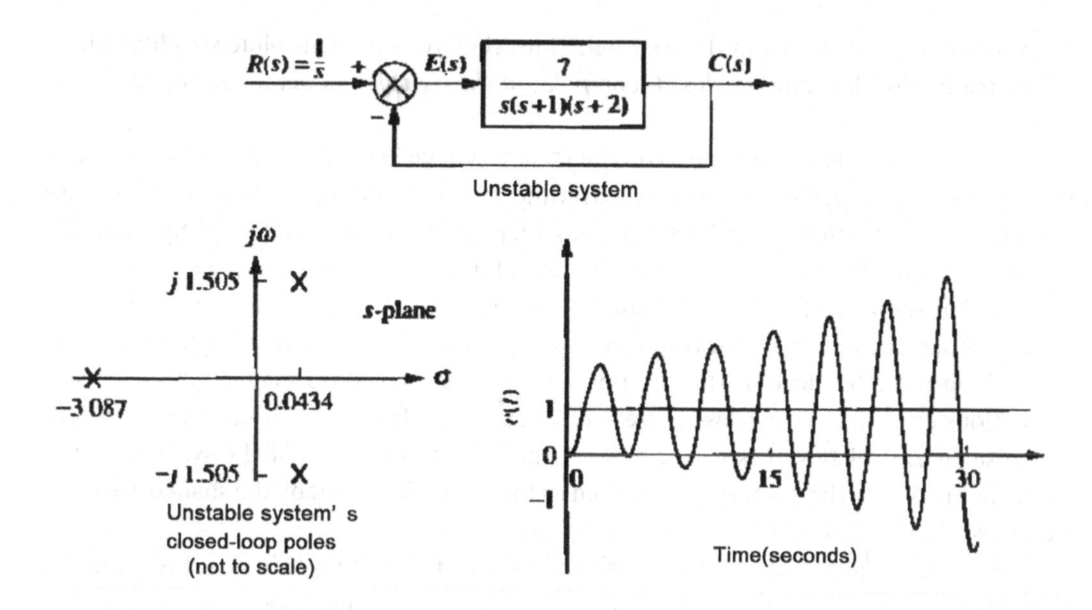

FIGURE 4.58 Unstable system of closed-loop poles.

4.7.1 Stability Analysis in Complex Plane

The stability of a linear closed-loop system can be determined from the location of the closed-loop poles in the *s-plane.* If any of these poles lie in the right-half s-plane, then with increasing time, they give rise to the dominant mode, and the transient response increases monotonically or oscillates with an increasing amplitude that we can see in the time response of Figure 4.58, which represents an *unstable system.* For such a system, as soon as the power is turned on, the output may increase with time. If no saturation takes place in the system and no mechanical stop is provided, then the system may eventually be subjected to damage and fail, since the response of a real physical system cannot increase indefinitely. Therefore, closed-loop poles in the right-half s-plane are not permissible in the usual linear control system.

However, if all closed-loop poles lie to the left of the *jω* axis, any transient response eventually reaches equilibrium. This represents a stable system. Stability is the most important problem in linear control systems. So, in this part, we will figure out the following problems.

- *What conditions will a system become unstable?*
- *If it is unstable, how stabilize the system?*
- *Why the system is stable if all closed-loop poles lie in the left-half s-plane?*

Whether a linear system is stable or unstable is a property of the system itself and does not depend on the input or driving function of the system. The poles of the input, or driving function, do not affect the property of stability of the system, but they contribute only to steady-state response terms in the solution. Thus, the problem of absolute stability can be solved readily by choosing *no closed-loop poles in the right-half s-plane, including the $j\omega$ axis.*

Mathematically, closed-loop poles on the $j\omega$ axis will yield oscillations, the amplitude of which is neither decaying nor growing with time. However, in practical cases where noise is present, the amplitude of oscillations may increase at a rate determined by the noise power level. Therefore, a control system should not have closed-loop poles on the $j\omega$ axis.

Note that the mere fact that all closed-loop poles lie in the left-half s-plane does not guarantee satisfactory transient response characteristics. If dominant complex-conjugate closed-loop poles lie close to the $j\omega$ axis, the transient response may exhibit excessive oscillations or may be very slow. Therefore, to guarantee fast, yet well-damped, transient response characteristics, it is necessary that the closed-loop poles of the system lie in a particular region in the complex plane, such as the region bounded by the shaded area that we can see in Figure 4.59.

In this figure, the region in the complex plane satisfying the conditions, the damping ratio, $\zeta > 0.4$ and settling time, $t_s < \frac{4}{\sigma}$. Since the relative stability and transient response performance of a closed-loop control system are directly related to the closed-loop pole-zero configuration in the s-plane, it is frequently necessary to adjust one or more system parameters in order to obtain suitable configurations. The effects of varying system parameters on the closed-loop poles will be discussed in detail later.

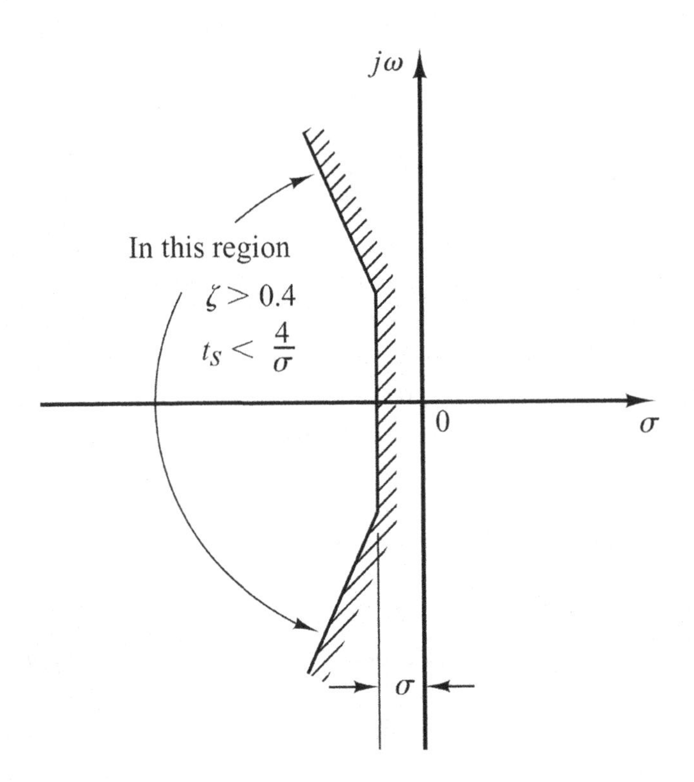

FIGURE 4.59 Condition of a stable and unstable system.

4.7.2 Routh's Stability Criterion of Stability Analyses

In this section, I will discuss Routh's stability criterion, relative stability analysis, and application of Routh's stability criterion.

1. *Routh's stability criterion*

   Since we have learned that the most important problem in linear control systems concerns stability. That is, under what conditions will a system become unstable? If it is unstable, how should we stabilize the system? It was stated that a control system is stable if and only if all closed-loop poles lie in the left-half s-plane. Most linear closed-loop systems have closed-loop transfer functions of the form as we can see the close-loop transfer functions as

$$\frac{C(s)}{R(s)}=\frac{b_0s^m+b_1s^{m-1}+...+b_{m-1}s+b_m}{a_0s^n+a_1s^{n-1}+...+a_{n-1}s+a_n}=\frac{B(s)}{A(s)},(m\leq n)$$

   where a and b are constants and $m\leq n$. A simple criterion, known as *Routh's stability criterion,* enables us to determine the number of closed-loop poles that lie in the right-half s-plane without having to factor the denominator polynomial.

   Routh's stability criterion informs us whether or not there are *unstable roots* in a polynomial equation without actually solving for them. This stability criterion applies to polynomials with only a finite number of terms. When the criterion is applied to a control system, information about *absolute stability* can be obtained directly from the coefficients of the characteristic equation. The procedure in Routh's stability criterion is as follows,

   Step 1: Write the polynomial in s in the following form like this.

$$a_0s^n+a_1s^{n-1}+...+a_{n-1}s+a_n=0$$

   where the coefficients a are real quantities. We assume that an $a_n\neq 0$; that is, any zero root has been removed.

   Step 2: If any of the coefficients are zero or negative in the presence of at least one positive coefficient, a root or roots exist that are imaginary or that have positive real parts. Therefore, in such a case, the system is *not stable.* If we are interested in only *absolute stability,* there is no need to follow the procedure further. *Note that all the coefficients must be positive.* Then we must stop here.

   Step 3: This is a necessary condition, as may be seen from the following argument:

   - A polynomial in s having real coefficients can always be factored into linear and quadratic factors, such as $(s+a)$ and (s^2+bs+c), where a, b, and c are real.

Therefore, from this quadratic as $s^2+bs+c=0$ then we can say to find the roots of this equation as like this.

$$s=\frac{-b\pm\sqrt{b^2-4ac}}{2a}$$

- The linear factors yield real roots and the quadratic factors yield complex-conjugate roots of the polynomial.
- The factor (s^2+bs+c) yields roots having negative real parts only if b and c are both positive.

For all roots to have negative real parts, the constants a, b, c, and so on in all factors must be positive. The product of any number of linear and quadratic factors containing only positive coefficients always yields a polynomial with positive coefficients. It is important to note that the condition that *all the coefficients be positive is not sufficient to assure stability.* Look at this equation.

$$a_0s^n+a_1s^{n-1}+...+a_{n-1}s+a_n=0$$

The necessary but not sufficient condition for stability is that the coefficients of this equation all be present and all have a positive sign. If all a's are negative, they can be made positive by multiplying both sides of the equation by −1. If all coefficients are positive, arrange the coefficients of the polynomial in rows and columns according to the following pattern:

$$\begin{array}{lllll}
s^n & a_0 & a_2 & a_4 & a_6\ldots \\
s^{n-1} & a_1 & a_3 & a_5 & a_7\ldots \\
s^{n-2} & b_1 & b_2 & b_3 & b_4\ldots \\
s^{n-3} & c_1 & c_2 & c_3 & c_4\ldots \\
s^{n-4} & d_1 & d_2 & d_3 & d_4\ldots \\
s^2 & e_1 & e_2 & & \\
s^1 & f_1 & & & \\
s^0 & g & & &
\end{array}$$

where b, c, d,… are continued until the remaining ones are all zero. The process of forming rows continues until we run out of elements (The total number of rows is $n+1$). The coefficients b_1, b_2, b_3, and so on, are evaluated as follows:

$$b_1=\frac{a_1a_2-a_0a_3}{a_1}$$

and we can find b_2 as follows:

$$b_2 = \frac{a_1 a_4 - a_0 a_5}{a_1}$$

and we can find b_3 as follows:

$$b_3 = \frac{a_1 a_6 - a_0 a_7}{a_1}$$

Similarly, the evaluation of the b's is continued until the remaining ones are all zero. The same pattern of cross-multiplying the coefficients of the two previous rows is followed in evaluating the c's, d's, e's, and so on. That is,

$$c_1 = \frac{b_1 a_3 - a_1 b_2}{b_1}$$

$$c_2 = \frac{b_1 a_5 - a_1 b_3}{b_1}$$

$$c_3 = \frac{b_1 a_7 - a_1 b_4}{b_1}$$

and

$$d_1 = \frac{c_1 b_2 - b_1 c_2}{c_1}$$

$$d_2 = \frac{c_1 b_3 - b_1 c_3}{c_1}, \ldots\ldots$$

This process is continued until the nth row has been completed. The complete array of coefficients is triangular. Note that in developing the array an entire row may be divided or multiplied by a positive number in order to simplify the subsequent numerical calculation without altering the stability conclusion.

Therefore, we can conclude that Routh's stability criterion states that the number of roots of this equation ($a_0 s^n + a_1 s^{n-1} + \ldots + a_{n-1} s + a_n = 0$) with positive real parts is equal to the *number of changes in the sign of the coefficients of the first column of the array.* It should be noted that the exact values of the terms in the *first column need not be known*; instead, only the *signs are needed.* The necessary and sufficient condition that all roots of this equation ($a_0 s^n + a_1 s^{n-1} + \ldots + a_{n-1} s + a_n = 0$) lie in the left-half s-plane is that all the coefficients of this equation ($a_0 s^n + a_1 s^{n-1} + \ldots + a_{n-1} s + a_n = 0$) be positive and all terms in the first column of the array have positive signs.

Let's see the example, consider the following polynomial:

$$s^4 + 2s^3 + 3s^2 + 4s + 5 = 0$$

Let us follow the procedure just presented and construct the array of coefficients. (The first two rows can be obtained directly from the given polynomial. The remaining terms are obtained from these. If any coefficients are missing, they may be replaced by zeros in the array.) Thus, in this polynomial, we can list as follows:

S^4	1	3	5
S^3	2	4	0
S^2	1	5	
S^1	−6		
S^0	5		

In this example, the number of changes in sign of the coefficients in the first column is 2. This means that there are two roots with positive real parts. However, note that the result is unchanged when the coefficients of any row are multiplied or divided by a positive number in order to simplify the computation as we can see below:

S^4	1	3	5	S^4	1	3	5
S^3	2	4	0	S^3	2	4	0
S^2	1	5		S^2	1	5	
S^1	−6			S^1	−3		
S^0	5			S^0	5		

2. *Relative stability analysis*
   Routh's stability criterion provides the answer to the question of absolute stability. This, in many practical cases, is not sufficient. We usually require information about the relative stability of the system. A useful approach for examining relative stability is to shift the s-plane axis and apply Routh's stability criterion. That is, we substitute

$$S = \hat{S} - \sigma$$

   where σ = constant into the characteristic equation of the system. Now, we may write the polynomial in terms of S and apply Routh's stability criterion to the new polynomial in the number of changes of sign in the first column of the array developed for the polynomial is equal to the number of roots that are located to the right of the vertical line $s = -\sigma$. Thus, this test reveals the number of roots that lie to the right of the vertical line $s = -\sigma$, as shown in Figure 4.60.

3. *Application of Routh's stability criterion to control system analysis*
   Now, I want to show you the application of Routh's stability criterion to control system analysis. Routh's stability criterion is of limited usefulness in linear control system analysis, mainly because it does not suggest how to improve relative stability or

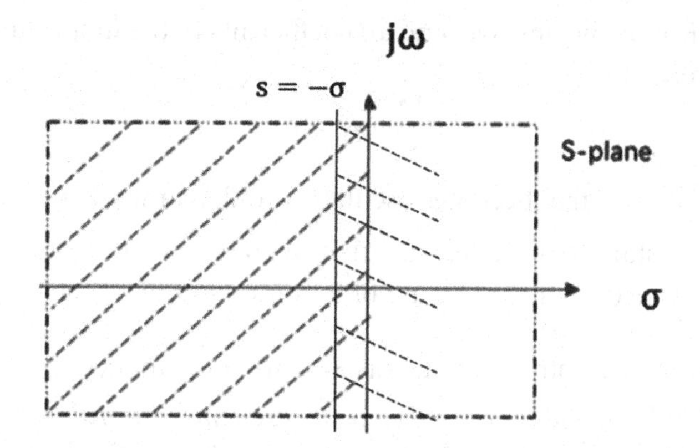

FIGURE 4.60 Relative stability.

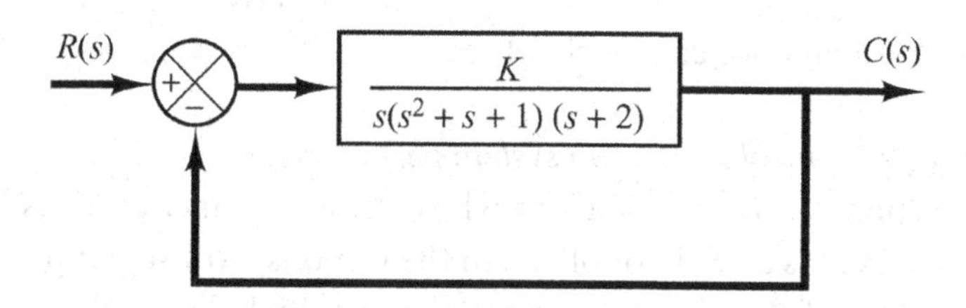

FIGURE 4.61 A closed-loop control system.

how to stabilize an unstable system. It is possible, however, to determine the effects of changing one or two parameters of a system by examining the values that cause instability. In the following, we shall consider the problem of determining the stability range of a parameter value.

Consider the system shown in Figure 4.61. Let us determine the range of K for stability. The closed-loop transfer function is

$$\frac{C(s)}{R(s)} = \frac{K}{s\left(s^2+s+1\right)\left(s+2\right)+K}$$

The characteristic equation is

$$s^4+3s^3+3s^2+2s+K=0$$

The array of coefficients becomes as follows:

$$\begin{array}{llll} s^4 & 1 & 3 & k \\ s^3 & 3 & 2 & 0 \\ s^2 & 7/3 & K & \\ s^1 & 2-\frac{9}{7}K & & \end{array}$$

For stability, K must be positive, and all coefficients in the first column must be positive. Therefore,

$$\frac{14}{9} > K > 0$$

When $K = \frac{14}{9}$, the system becomes oscillatory and, mathematically, the oscillation is sustained at constant amplitude. Note that the ranges of design parameters that lead to stability may be determined by use of *Routh's stability criterion.*

4.7.3 Effects of Integral Control Actions on System Performance

In this section, we will learn *proportional control of systems; integral control of systems; and response to torque disturbances.* You should remember that we shall investigate the effects of integral and proportional control on the system performance. Here we shall consider only simple systems so that the effects of integral, derivative, and proportional control actions on system performance can be clearly seen.

1. *Effects of integral control actions on system performance*

   In the proportional control of a plant whose transfer function does not possess an integrator 1/s, there is an SSE, or offset, in the response to a step input. Such an offset can be eliminated if the integral control action is included in the controller. In the *integral control of a plant*, the control signal—the output signal from the controller—at any instant is the area under the actuating-error-signal curve up to that instant, as shown in Figure 4.62a.

   We know that the integral control action is $u(t) = \frac{K_p}{T_i}\int e(t)dt$. From here, the plots of $e(t)$ and $u(t)$ curves show *nonzero control signal* when the *actuating error signal is zero* (that is called *integral control*). The control signal $u(t)$ can have a *nonzero value* when the *actuating error signal* $e(t)$ is zero. This is impossible in the case of the proportional controller since a nonzero control signal requires a nonzero actuating error

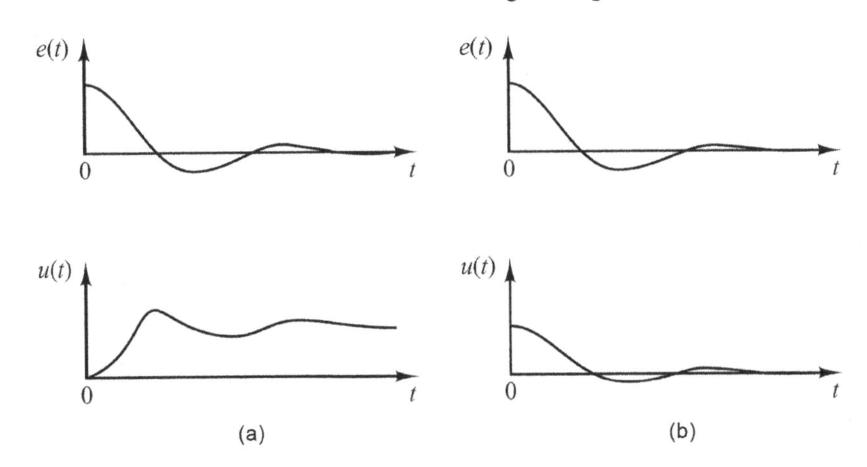

FIGURE 4.62 (a) Plots of $e(t)$ and $u(t)$ curves showing nonzero control signal when the actuating error signal is zero (integral control) and (b) plots of $e(t)$ and $u(t)$ curves showing zero control signal when the actuating error signal is zero (proportional control).

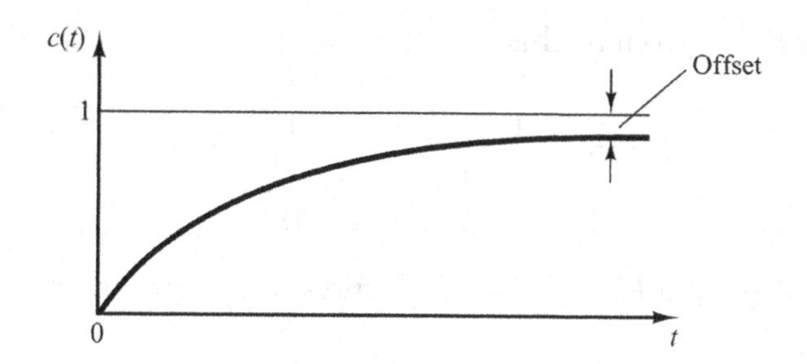

FIGURE 4.63 Unit-step response curve with offset.

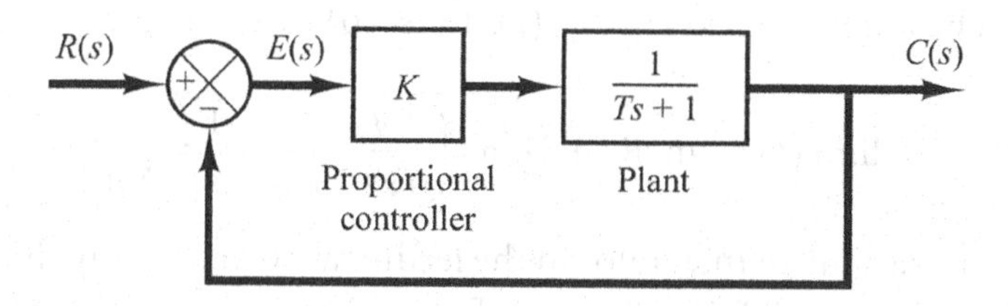

FIGURE 4.64 Proportional control of a system without an integrator.

signal. A nonzero actuating error signal at steady state means that there is an *offset*. This offset, is shown in Figure 4.63.

However, look at the proportional control actions in Figure 4.62b. We know that the *proportional control action* is $u(t) = K_p e(t)$. From here, the plots of *e(t)* and *u(t)* curves show *zero control signal* when the actuating *error signal is zero* (that means *proportional control*). The *curve e(t) vs. t* and the corresponding curve *u(t) vs. t* when the controller is of the proportional type.

Note that integral control action, while removing offset or SSE, may lead to oscillatory response of slowly decreasing amplitude or even increasing amplitude, both of which are usually undesirable.

2. *Proportional control of systems*

We shall show that the *proportional control of a system without an integrator* will result in an SSE with a step input. We shall then show that such an error can be eliminated if integral control action is included in the controller. Look at the proportional control system shown in Figure 4.64.

Considering this system, let us obtain the SSE in the *unit-step response* of the system. It can be defined as

$$G(s) = \frac{K}{Ts+1}$$

From the system, we can express it as

$$\frac{E(s)}{R(s)} = \frac{R(s)-C(s)}{R(s)} = 1 - \frac{C(s)}{R(s)} = 1 - \frac{\frac{K}{Ts+1}}{1+\frac{K}{Ts+1}.1} = \frac{1}{1+\frac{K}{Ts+1}} = \frac{1}{1+G(s)}$$

and the error $E(s)$ is given by this expression as

$$E(s)=\frac{1}{1+G(s)}R(s)=\frac{1}{1+\dfrac{K}{Ts+1}}.R(s)$$

For the unit-step input $R(s)=1/s$, we can express the error $E(s)$ as

$$E(s)=\left(\frac{Ts+1}{Ts+1+K}\right).\frac{1}{s}$$

The SSE can be found from $u(t)=K_p e(t)$. Thus, $e(t)$ can be expressed as

$$e_{ss}=\lim_{t\to\infty}e(t)=\lim_{s\to 0}sE(s)=\lim_{s\to 0}s.\left(\frac{Ts+1}{Ts+1+K}\right).\frac{1}{s}=\frac{1}{K+1}$$

Such a system without an integrator in the feedforward path always has an SSE in the step response. Such an SSE is called an *offset*, as shown in Figure 4.63.

3. *Integral control of systems*

Considering this system, the controller is integral, as shown in Figure 4.65.

The closed-loop transfer function of the system is as follows:

$$\frac{C(s)}{R(s)}=\frac{\dfrac{K}{s(Ts+1)}}{1+\dfrac{K}{s(Ts+1)}.1}=\frac{K}{s(Ts+1)+K}$$

or it can be simplified and we can rewrite as follows:

$$\frac{E(s)}{R(s)}=\frac{R(s)-C(s)}{R(s)}=\frac{s(Ts+1)}{s(Ts+1)+K}$$

From $u(t)=\frac{K_p}{T_i}\int e(t)dt$, the system is stable, the SSE for the unit-step response can be obtained by applying the final value theorem, then we can see the expression as

$$e_{ss}=\lim_{t\to\infty}e(t)=\lim_{s\to 0}sE(s)=\lim_{s\to 0}s.\left[\frac{s(Ts+1)}{Ts^2+s+K}\right]\frac{1}{s}=0$$

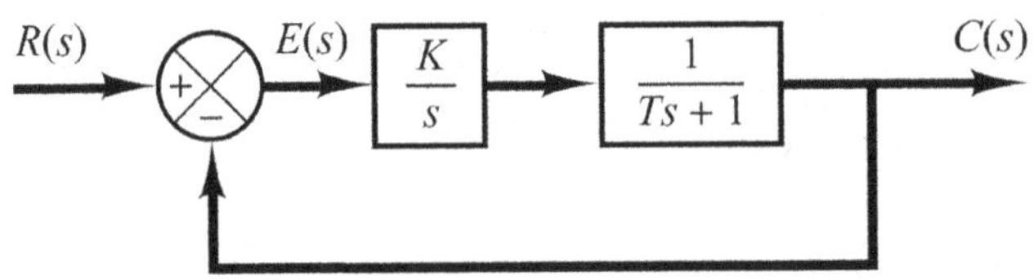

FIGURE 4.65 Integral control of systems.

The value of e_{ss} is equal to zero which indicates that an Integral control of the system thus *eliminates the* SSE in the response to the step input. This is an important improvement over the proportional control alone, which gives an *offset* that we can see in this response curve of Figure 4.63.

4. *Response to torque disturbances (proportional control)*
   Let us investigate the effect of a torque disturbance occurring at the load element from this figure. The proportional controller delivers torque T to position the load element, which consists of the moment of inertia and viscous friction. Torque disturbance is denoted by D.

   Assuming that the reference input is zero or $R(s)=0$, the transfer function between $C(s)$ and $D(s)$ is given by this equation as follows:

$$\frac{C(s)}{D(s)}=\frac{1}{Js^2+bs+K_p}$$

Hence,

$$\frac{E(s)}{D(s)}=-\frac{C(s)}{D(s)}=-\frac{1}{Js^2+bs+K_p}$$

or it can be expressed as

$$\frac{E(s)}{D(s)}=-\frac{C(s)}{D(s)}=-\frac{1}{Js^2+bs+K_p}$$

However, the expression can be seen in the form of $E(s)$ as

$$E(s)=-\frac{1}{Js^2+bs+K_p}.D(s)=-\frac{1}{Js^2+bs+K_p}.\frac{T_d}{s}$$

The SSE due to a step disturbance torque of magnitude T_d is given by this expression as

$$e_{ss}=\lim_{t\to\infty}e(t)=\lim_{s\to 0}sE(s)=\lim_{s\to 0}\left[\frac{-s}{Js^2+bs+K_p}.\frac{T_d}{s}\right]=\frac{T_d}{K_p}$$

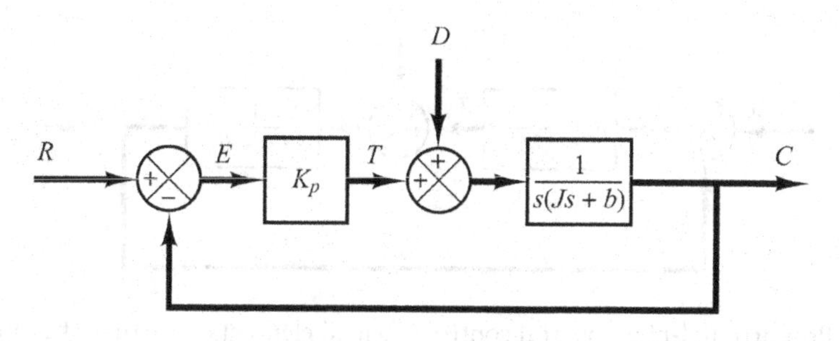

FIGURE 4.66 Control system with a torque disturbance.

At steady state, the proportional controller provides the torque $-T_d$, which is equal in magnitude but opposite in sign to the disturbance torque T_d. The steady-state output due to the step disturbance torque is written as

$$e_{ss} = -e_{ss} = \frac{T_d}{K_p}$$

The SSE can be reduced by increasing the value of the gain K_p. *Increasing* this value, however, will cause the system response to be *more oscillatory.*

5. *Response to torque disturbances (proportional + integral control)*
Now, I am going to discuss Response to Torque Disturbances for *Proportional-Plus-Integral Control* which is described as shown in following the mathematical expression:

$$u(t) = K_p e(t) + \frac{Kp}{Ti} \int e(t)dt$$

It can be shown in this control system. Please look at Figure 4.67 carefully.

To eliminate offset due to torque disturbance, the proportional controller may be replaced by a *proportional-plus-integral controller.* If *integral control action is added to the controller,* then, as long as there is *an error signal,* a torque is developed by the controller to reduce this error, provided the control system is a stable one.

The proportional-plus-integral control of the load element consists of a moment of inertia and viscous friction. The closed-loop transfer function between $C(s)$ and $D(s)$ can be defined as

$$\frac{C(s)}{D(s)} = \frac{s}{Js^3 + bs^2 + K_p s + \frac{K_p}{T_i}}$$

In the absence of the reference input, or $r(t) = 0$, the error signal is obtained from the following error signal $E(s)$.

$$E(s) = -\frac{s}{Js^3 + bs^2 + K_p s + \frac{K_p}{T_i}} D(s)$$

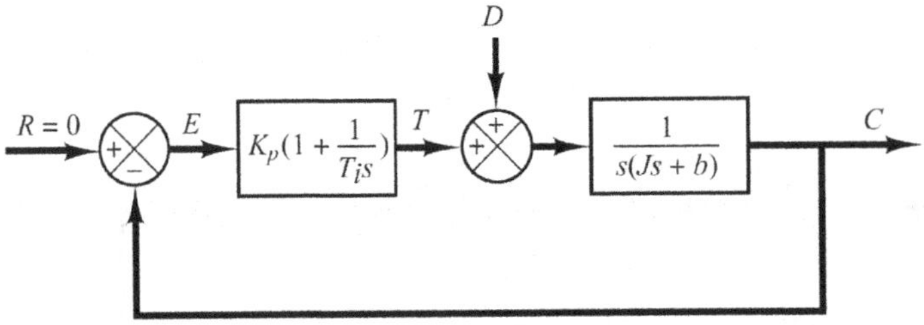

FIGURE 4.67 Proportional-plus-integral control of a load element consisting of a moment of inertia and viscous friction.

If this control system is stable—that is, if the roots of the characteristic equation can be seen from the following equation:

$$Js^3 + bs^2 + K_p s + \frac{K_p}{T_i} = 0$$

This equation has negative real parts—then the SSE in the response to a unit-step disturbance torque can be obtained by applying the final value theorem and then we can find

$$e_{ss} = \lim_{t\to\infty} e(t) = \lim_{s\to 0} sE(s) = \lim_{s\to 0}\left[\frac{-s^2}{Js^3 + bs^2 + K_p s + \frac{K_p}{T_i}}\right] \cdot \frac{1}{s} = 0$$

Thus SSE to the step disturbance torque can be eliminated if the controller is of the proportional-plus-integral type. Note that the integral control action added to the proportional controller has converted the originally second-order system to a third-order one. Hence, the control system may become *unstable for a large value of Kp*, since the roots of the characteristic equation may have *positive real parts*. Remember that *the second-order system is always stable if the coefficients in the system differential equation are all positive.*

It is important to point out that if the controller was an *integral controller*, as shown in Figure 4.68, then *the system always becomes unstable*, because the characteristic equation is always as

$$Js^3 + bs^2 + K = 0$$

and it will have roots with positive real parts. Such an unstable system cannot be used in practice. Note that in the system of Figure 4.68 the proportional control action tends to stabilize the system, while the integral control action tends to eliminate or reduce SSE in response to various inputs. Therefore, adding to a proportional controller provides high sensitivity.

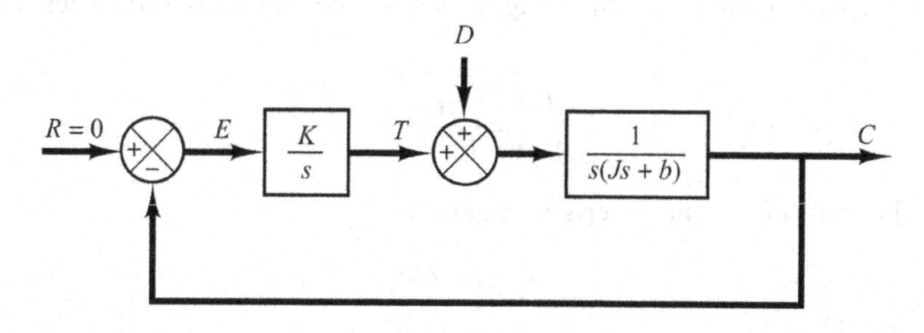

FIGURE 4.68 Integral control of a load element consisting of a moment of inertia and viscous friction.

4.7.4 Effects of Derivative Control Actions on System Performance

In this section, I will discuss *Derivative Control Action, Proportional Control of Systems with Inertia Load, and Proportional-plus-Deviation Control of Systems with Inertia Load.* Please remember that we shall investigate the effects of derivative control and proportional actions on the system performance. Here we shall consider only simple systems, so that the effects of derivative control and proportional control actions on system performance can be clearly seen.

1. *Derivative control action*
   Now I am going to discuss about derivative control action. Derivative control action, when added to a proportional controller, provides a means of obtaining a controller with high sensitivity.
   An advantage of using derivative control action is that it responds to the rate of change of the actuating error and can produce a significant correction before the magnitude of the actuating error becomes too large. Derivative control thus anticipates the actuating error, initiates an early corrective action, and tends to increase the stability of the system.
   Although derivative control does not affect the SSE directly, it adds damping to the system and thus permits the use of a larger value of the gain K, which will result in an improvement in the steady-state accuracy. Because derivative control operates on the rate of change of the actuating error and not the actuating error itself, this mode is never used alone. It is always used in combination with proportional or proportional-plus-integral control action.
2. *Proportional control of systems with inertia load*
   Although derivative control does not affect the *SSE* directly, it *adds damping to the system* and thus permits the use of a larger value of the gain K, which will result in an improvement in the steady-state accuracy. Because derivative control operates on the rate of change of the actuating error and not the actuating error itself, this mode is never used alone. It is always used in combination with proportional or proportional-plus-integral control action.
   Now, let's see the *proportional control of systems with Inertia Load.* Before we discuss further the effect of derivative control action on system performance, we shall consider the proportional control of an inertia load.
   Consider the system shown in Figure 4.69a. The closed-loop transfer function is obtained as

$$\frac{C(s)}{R(s)}=\frac{K_p}{Js^2+K_p}$$

   Since the roots of the characteristic equation

$$Js^2+K_p=0$$

   are imaginary, the response to a unit-step input continues to oscillate indefinitely which we can see in Figure 4.69b. Control systems exhibiting such response

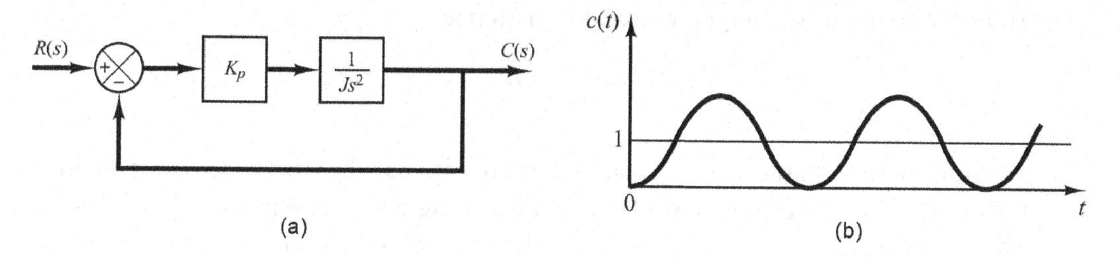

FIGURE 4.69 (a) Proportional control of a system with inertia load, and (b) response to a unit-step input.

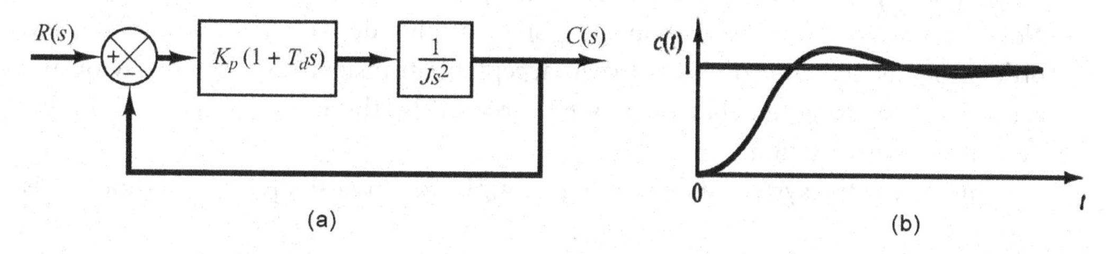

FIGURE 4.70 (a) Proportional-plus-derivative control of a system with inertia load, and (b) response to a unit-step input.

characteristics are not desirable. We shall see that the addition of derivative control will stabilize the system.

3. *Proportional-plus-derivative control of a system with inertia load*
Now, let's see the *Proportional-Plus-Derivative Control of a System with Inertia Load.* Let us modify the proportional controller to a proportional-plus-derivative controller whose transfer function is $K_p(1+T_d s)$ that we can see in Figure 4.70a.
The torque developed by the controller is proportional to $K_p(1+T_d \dot{e})$. Derivative control

- Is essentially anticipatory,
- Measures the instantaneous error velocity,
- Predicts large overshoot ahead of time; and
- Appropriately counteracts before too large an overshoot occurs.

Derivative control is essentially anticipatory, measures the instantaneous error velocity, predicts the large overshoot ahead of time, and produces an appropriate counteraction before too large an overshoot occurs, as shown in Figure 4.70b.

Considering the system shown in Figure 4.70a, the closed-loop transfer function is given by the following mathematical expression:

$$\frac{C(s)}{R(s)} = \frac{K_p(1+T_d s)}{Js^2 + K_p T_d s + K_p}$$

Thus, the characteristic equation can be written as

$$Js^2 + K_p T_d s + K_p = 0$$

The roots of this characteristic equation have negative real parts for positive values of J, K_p; thus, derivative control introduces a damping effect. A typical response curve $c(t)$ to a unit-step input is shown in this Figure 4.70b. It is clear that the response curve shows a *marked improvement over the original response curve* that we can see in Figure 4.70b.

4. *Proportional-plus-derivative control of the second-order system*
   Now, I am going to discuss about proportional-plus-derivative control of second-order systems. A compromise between acceptable transient response behavior and acceptable steady-state behavior may be achieved by the use of proportional-plus-derivative control action.
   Considering the system shown in Figure 4.71, the closed-loop transfer function is

$$\frac{C(s)}{R(s)} = \frac{K_p + K_d s}{Js^2 + (B + K_d)s + K_p}$$

The SSE for a unit-ramp input is

$$e_{ss} = \frac{B}{K_p}$$

The characteristic equation can be seen as

$$Js^2 + (B + K_d)s + K_p = 0$$

Thus, the effective damping coefficient of this system is thus $B + K_d$ rather than B. The damping ratio ξ of this system is

$$\xi = \frac{B + K_d}{2\sqrt{K_p J}}$$

How we can make SSE for ramp input and M_p for step input?

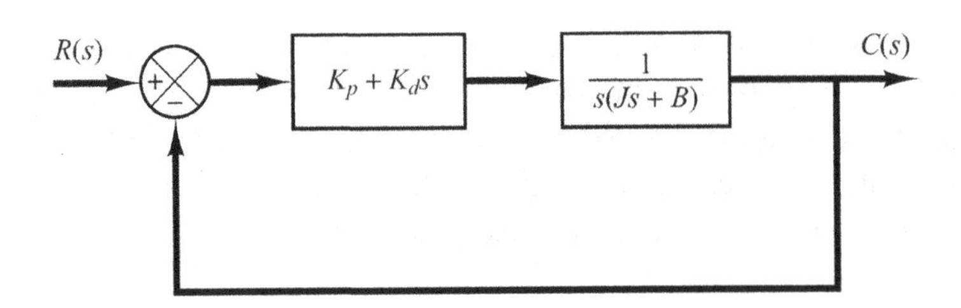

FIGURE 4.71 PD Control system.

It is possible to make both the SSE (e_{ss}) for a ramp input and the *maximum overshoot* (M_p) for a step input small by making B small, K_p large, and K_d large enough so that ξ is between 0.4 and 0.7.

4.7.5 Steady-State Errors in Unity-Feedback Control Systems

In this section, we will learn the classification of control systems, SSE in the unity-feedback control system, and SSEs.

1. *Preview of SSEs*
   Errors in a control system can be attributed to many factors. Changes in the reference input will cause unavoidable errors during transient periods and may also cause SSEs. Imperfections in the system components, such as static friction, backlash, and amplifier drift, as well as aging or deterioration, will cause errors at steady state. In this section, however, we shall not discuss errors due to imperfections in the system components. Rather, we shall investigate a type of SSE that is caused by the incapability of a system to follow particular types of inputs.
   Any physical control system inherently suffers SSE in response to certain types of inputs. A system may have no SSE to a step input, but the same system may exhibit nonzero SSE to a ramp input. Please remember that the only way we may be able to eliminate this error is to modify the system structure. Whether a given system will exhibit SSE for a given type of input depends on the type of open-loop transfer function of the system, to be discussed in what follows.

2. *Classification of control systems*
   Let's see the classification of control systems. Control systems may be classified according to their ability to follow

   - Step inputs,
   - Ramp inputs,
   - Parabolic inputs, and so on.

   This is a reasonable classification scheme, because actual inputs may frequently be considered combinations of such inputs. The magnitudes of the SSEs due to these individual inputs are indicative of the goodness of the system.

3. *Definition of system type*
   Let's see about *SSEs* in unity-feedback control system. Consider the unity-feedback control system with the following open-loop transfer function $G(s)$ as

$$G(s)=\frac{K(T_a s+1)(T_b s+1)...(T_m s+1)}{s^N (T_1 s+1)(T_2 s+1)...(T_p s+1)}$$

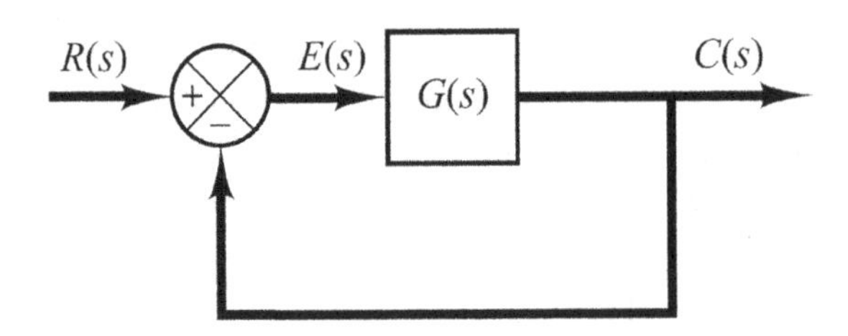

FIGURE 4.72 Unity control system.

It involves the term S^N in the denominator, representing *a pole of multiplicity N at the origin*. The present classification scheme is based on the number of integrations indicated by the *open-loop transfer function*. A system is called type 0 system if $N=0$; type 1 system if $N=1$; type 2 system if $N=2,\ldots,$ and so on. Note that this classification is different from that of the order of a system. As the type number is increased, the accuracy is improved; however, increasing the type number aggravates the stability problem. A compromise between *steady-state accuracy* and *relative stability* is always necessary.

We shall see later that if $G(s)$ is written so that each term in the numerator and denominator, except the term S^N, approaches unity as s approaches zero, then the open-loop gain K is directly related to the SSE.

4. *SSEs*
   To discuss SSEs, look at the closed-loop control system shown in Figure 4.72.

   Considering the system shown in Figure 4.72, the closed-loop transfer function is

$$\frac{C(s)}{R(s)}=\frac{G(s)}{1+G(s)}$$

The transfer function between the error signal $e(t)$ and the input signal $r(t)$ is

$$\frac{E(s)}{R(s)}=1-\frac{C(s)}{R(s)}=\frac{1}{1+G(s)}$$

where the error $e(t)$ is the difference between the input signal and the output signal. The final value theorem provides a convenient way to find the steady-state performance of a stable system. Since $E(s)$ is

$$E(s)=\frac{1}{1+G(s)}R(s)$$

the SSE is defined as

$$e_{ss}=\lim_{t\to\infty}e(t)=\lim_{s\to 0}sE(s)=\lim_{s\to 0}\frac{1}{1+G(s)}R(s)$$

where $e(t)$ is the difference between the input signal and the output signal.

5. *Static error constants definition*
The static error constants are defined in the next figures of merit of control systems. The higher the constants, the smaller the SSE. In a given system, the output may be the position, velocity, pressure, temperature, or the like. The physical form of the output, however, is immaterial to the present analysis. Therefore, in what follows, we shall call the output "position," the rate of change of the output "velocity," and so on. This means that in a temperature control system "position" represents the output temperature, "velocity" represents the rate of change of the output temperature, and so on.

6. *Static position error constant, K_p*
Now, let see about Static Position Error Constant K_p. The SSE of the system for a unit-step input $R(s) = 1/s$ is defined as

$$e_{ss} = \lim_{t\to\infty} e(t) = \lim_{s\to 0} sE(s) = \lim_{s\to 0} \frac{s}{1+G(s)} \cdot \frac{1}{s} = \frac{1}{1+G(0)}$$

The static position error constant K_p is defined by

$$K_p = \lim_{s\to 0} G(s) = G(0)$$

Thus, the SSE in terms of the static position error constant K_p is given by

$$e_{ss} = \frac{1}{1+K_p}$$

For a type 0 system if $N=0$ then we can see the static position error constant $K_p = K$, i.e.,

$$K_p = \lim_{s\to 0} \frac{K(T_a s+1)(T_b s+1)\ldots}{(T_1 s+1)(T_2 s+1)\ldots} = K$$

For a type 1 or higher system if $N=1$ or $N>1$, then we can see the static position error constant K_p=infinity, i.e.,

$$K_p = \lim_{s\to 0} \frac{K(T_a s+1)(T_b s+1)\ldots}{s^N (T_1 s+1)(T_2 s+1)\ldots} = \infty, \text{ for } N \geq 1$$

Hence, remember that for a type 0 system, the static position error constant K_p is finite, while for a type 1 or higher system, K_p is infinite. For a unit-step input, the SSE e_{ss} may be summarized as follows:

From

$$e_{ss} = \frac{1}{1+K_p}$$

We can say

$$e_{ss} = \begin{cases} \dfrac{1}{1+K}; & \text{for } N=0, \text{ type 0 system} \\ 0; \text{ for } N \geq 1, \text{ type 1 or higher system} \end{cases}$$

From the foregoing analysis, it is seen that the response of a feedback control system to a step input involves an SSE if there is no integration in the feedforward path. If small errors for step inputs can be tolerated, then:

- A type 0 system may be permissible, provided that the gain K is sufficiently large.
- If the gain K is too large, it is difficult to obtain reasonable relative stability.
- If zero SSE for a step input is desired, the type of the system must be one or higher.

7. *Static velocity error constant, K_v*
   Now, let's discuss about Static Velocity Error Constant K_v. The SSE of the system with a unit-ramp input $R(s) = 1/s^2$ is given by this expression as

$$e_{ss} = \lim_{t \to \infty} e(t) = \lim_{s \to 0} sE(s) = \lim_{s \to 0} \frac{s}{1+G(s)} \cdot \frac{1}{s^2} = \lim_{s \to 0} \frac{1}{sG(s)}$$

The static velocity error constant K_v is defined by

$$K_v = \lim_{s \to 0} sG(s)$$

Thus, the SSE in terms of the static velocity error constant K_v is given by

$$e_{ss} = \frac{1}{K_v}$$

- The term velocity error is used here to express the SSE for a ramp input.
- The dimension of the velocity error is the same as the system error.
- That is, velocity error is not an error in velocity, but it is an error in position due to a ramp input.

For a type 0 system, if $N=0$, then we can see the static velocity error constant $K_v=0$, i.e.,

$$K_v = \lim_{s \to 0} \frac{sK(T_a s+1)(T_b s+1)\ldots}{(T_1 s+1)(T_2 s+1)\ldots.} = 0$$

For a type 1, if $N=1$, then we can see the static velocity error constant $K_v=K$, i.e.,

$$K_v = \lim_{s\to 0} \frac{sK(T_a s+1)(T_b s+1)\ldots}{s(T_1 s+1)(T_2 s+1)\ldots.} = K$$

For a type 2 or higher system, if $N=2$ or $N>2$, then we can see the static velocity error constant K_v=infinity, i.e.,

$$K_v = \lim_{s\to 0} \frac{sK(T_a s+1)(T_b s+1)\ldots}{s^N(T_1 s+1)(T_2 s+1)\ldots.} = \infty, \text{ for } N \geq 2$$

The SSE e_{ss} for the unit-ramp input can be summarized as follows:

From,

$$e_{ss} = \frac{1}{K_v}$$

We can say

$$e_{ss} = \begin{cases} \dfrac{1}{K_v} = \infty; \text{ for } N=0, \text{ type 0 system} \\ \dfrac{1}{K_v} = \dfrac{1}{K}\text{ ; for } N=1, \text{ type 1 system} \\ \dfrac{1}{K_v} = 0\text{ ; for } N \geq 2, \text{ type 2 or higher system} \end{cases}$$

The foregoing analysis indicates that a type 0 system is incapable of following a ramp input in the steady state. The type 1 system with unity feedback can follow the ramp input with a finite error. In steady-state operation, the output velocity is exactly the same as the input velocity, but there is a positional error. This error is proportional to the velocity of the input and is inversely proportional to the gain K, as shown in Figure 4.73.

Figure 4.73 shows an example of the response of a type 1 system with unity feedback to a ramp input. The type 2 or higher system can follow a ramp input with zero error at steady state.

8. *Static acceleration error constant, K_a*

Now let's see about Static Acceleration Error Constant K_a. The SSE of the system with a unit-parabolic input (called acceleration input), which is defined by the expression as

$$r(t) = \begin{cases} \dfrac{t^2}{2}, \text{ for } t \geq 0 \\ 0, \text{ for } t < 0 \end{cases}$$

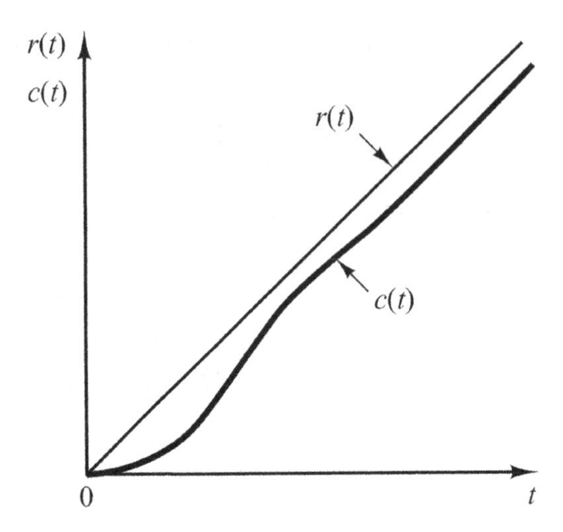

FIGURE 4.73 Response of type 1 unity-feedback system to a ramp input.

The SSE of the system for a unit-parabola input, $R(s) = 1/s^3$ as

$$e_{ss} = \lim_{t \to \infty} e(t) = \lim_{s \to 0} sE(s) = \lim_{s \to 0} \frac{s}{1+G(s)} \cdot \frac{1}{s^3} = \lim_{s \to 0} \frac{1}{s^2 G(s)}$$

The static acceleration error constant K_a is defined by the equation as

$$K_a = \lim_{s \to 0} s^2 G(s)$$

The SSE in terms of the static **acceleration** error constant K_v is given by

$$e_{ss} = \frac{1}{K_a}$$

Note that the acceleration error, the SSE due to a parabolic input, is an error in position.

1. *The values of K_a are obtained as follows:*
   For a type 0 system, if $N=0$, then we can see the static **acceleration** error constant $K_a=0$, i.e.,

$$K_a = \lim_{s \to 0} \frac{s^2 K(T_a s+1)(T_b s+1)\ldots}{(T_1 s+1)(T_2 s+1)\ldots.} = 0$$

   For a type 1, if $N=1$, then we can see the static **acceleration** error constant $K_a=0$, i.e.,

$$K_a = \lim_{s \to 0} \frac{s^2 K(T_a s+1)(T_b s+1)\ldots}{s(T_1 s+1)(T_2 s+1)\ldots.} = 0$$

For a type 2 system, if $N=2$, then we can see the static **acceleration** error constant K_a=infinity, i.e.,

$$K_a = \lim_{s \to 0} \frac{s^2 K(T_a s+1)(T_b s+1)...}{s^2 (T_1 s+1)(T_2 s+1)....} = K$$

For a type 3 or higher system, if $N=3$ or $N>3$, then we can see the static **acceleration** error constant K_a=infinity, i.e.,

$$K_a = \lim_{s \to 0} \frac{s^2 K(T_a s+1)(T_b s+1)...}{s^N (T_1 s+1)(T_2 s+1)....} = \infty, \; for\ N \geq 3$$

The SSE e_{ss} for the unit-parabolic input can be summarized as follows:

From

$$e_{ss} = \frac{1}{K_a}$$

We can say

$$e_{ss} = \begin{cases} \dfrac{1}{K_a} = \infty; \text{for type 0 and type 1 system} \\ \dfrac{1}{K_a} = \dfrac{1}{K}; \text{for type 2 system} \\ \dfrac{1}{K_a} = 0; \text{for type 3 or higher system} \end{cases}$$

Note that both type 0 and type 1 systems are incapable of following a parabolic input in the steady state. The type 2 system with unity feedback can follow a parabolic input with a finite error signal, as shown Figure 4.74.

Figure 4.74 shows an example of the response of a type 2 system with unity feedback to a parabolic input. The type 3 or higher system with unity feedback follows a parabolic input with zero error at steady state. Tables 4.6 and 4.7 summarize the SSEs for type 0, type 1, and type 2 systems when they are subjected to various inputs. The finite values for SSEs appear on the diagonal line. Above the diagonal, the SSEs are infinity; below the diagonal, they are zero.

Remember that the terms position error, velocity error, and acceleration error mean steady-state deviations in the output position. A finite velocity error implies that after transients have died out, the input and output move at the same velocity but have a finite position difference.

The error constants K_p, K_v, and K_a describe the ability of a unity-feedback system to reduce or eliminate SSE. Therefore, they are indicative of the steady-state performance. It is generally desirable to increase the error constants, while maintaining the transient response within an acceptable range. It is noted that to improve the steady-state performance we can increase the type of the system by adding an integrator or

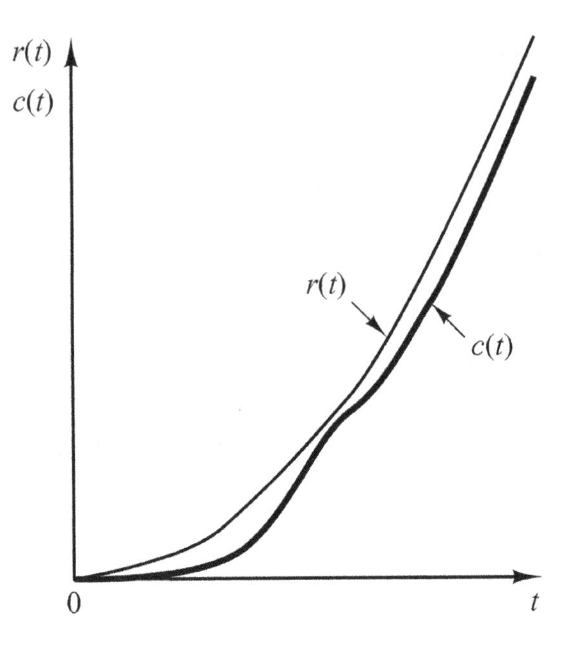

FIGURE 4.74 Response of type 2 unity-feedback system to a parabolic input.

TABLE 4.6 Summary of Steady-State Error in Terms of Gain K

	Step Input $r(t)=1$	**Ramp Input** $r(t)=t$	**Acceleration Input** $r(t)=\frac{1}{2}t^2$
Type 0 System	$\frac{1}{1+K}$	∞	∞
Type 1 System	0	$\frac{1}{K}$	∞
Type 2 System	0	0	$\frac{1}{K}$

TABLE 4.7 SSE Function of System Type and Input

		Type 0		**Type 1**		**Type 2**	
Input	**Steady-state error formula**	**Static error constant**	**Error**	**Static error constant**	**Error**	**Static error constant**	**Error**
Step	$\frac{1}{1+K_p}$	$K_p=K$	$\frac{1}{1+K}$	$K_p=\infty$	0	$K_p=\infty$	0
Ramp	$\frac{1}{K_v}$	$K_v=0$	∞	$K_v=\infty$	$\frac{1}{K}$	$K_v=\infty$	0
Parabola, Acceleration	$\frac{1}{K_a}$	$K_a=0$	∞	$K_a=0$	∞	$K_a=K$	$\frac{1}{K}$

integrators to the feedforward path. This, however, introduces an additional stability problem. The design of a satisfactory system with more than two integrators in series in the feedforward path is generally not easy.

4.7.6 Gear Train System and SSE for Disturbances

In this section, we will discuss about the gear train system and SSEs for disturbances.

1. *Gear train calculation*

   Gear trains are often used in servo systems

   - to reduce speed,
   - to magnify torque, or
   - to obtain the most efficient power transfer by matching the driving member to the given load.

   Consider the gear train system shown in Figure 4.75; a load (T_L) is driven by a motor through the gear train.

   - Assuming that the stiffness of the shafts of the gear train is infinite (please remember that there is neither backlash nor elastic deformation); and
   - That the number of teeth (N_1, N_2, N_3, and N_4) on each gear is proportional to the radius of the gear.

   Obtaining the equivalent moment of inertia and equivalent viscous-friction coefficient which is referred to the motor shaft and referred to the load shaft.

   - Thus, we can see that the numbers of teeth on gears 1, 2, 3, and 4 are N_1, N_2, N_3, and N_4, respectively.
   - The angular displacements of shafts 1, 2, and 3 are θ_1, θ_2, and θ_3, respectively.

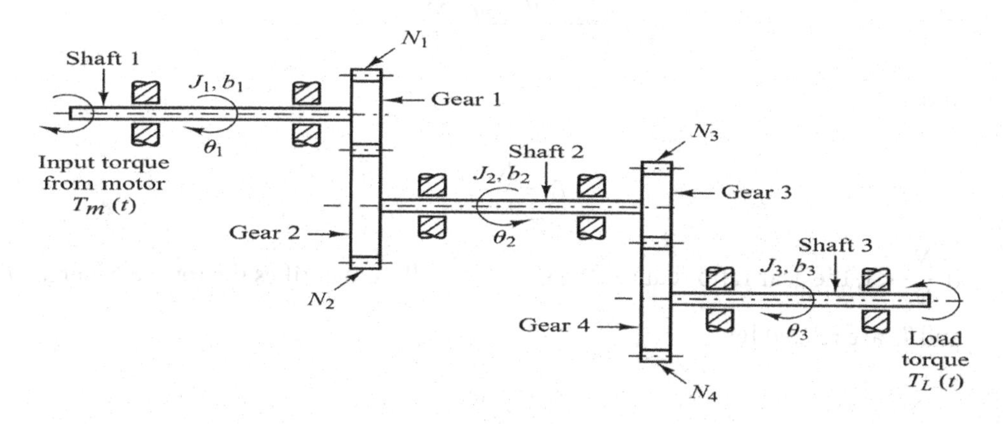

FIGURE 4.75 Gear train system.

- Thus, $\frac{\theta_2}{\theta_1} = \frac{N_1}{N_2}$ and $\frac{\theta_3}{\theta_2} = \frac{N_3}{N_4}$.
- The moment of inertia and viscous-friction coefficient of each gear train component are denoted by J_1, b_1; J_2, b_2; and J_3, b_3; respectively.
- Here, J_3 and b_3 include the moment of inertia and friction of the load.

For this gear train system, we can obtain the following equations:

For shaft 1, we can say as

$$J_1\ddot{\theta}_1 + b_1\dot{\theta}_1 + T_1 = T_m$$

where T_m is the torque developed by the motor and T_1 is the load torque on gear 1 due to the rest of the gear train.

For shaft 2, we can say as

$$J_2\ddot{\theta}_2 + b_2\dot{\theta}_2 + T_3 = T_2$$

where T_2 is the torque transmitted to gear 2 and T_3 is the load torque on gear 3 due to the rest of the gear train.

For shaft 3, we can say as

$$J_3\ddot{\theta}_3 + b_3\dot{\theta}_3 + T_L = T_4$$

where T_L is the load torque and T_4 is the torque transmitted to gear 4. Since the work done by gear 1 is equal to that of gear 2, we can say

$$T_1\theta_1 = T_2\theta_2$$

$$T_2 = \frac{T_1\theta_1}{\theta_2} = T_1\frac{N_2}{N_1},$$

and

$$T_1 = \frac{N_1}{N_2}T_2$$

If $\frac{N_1}{N_2} < 1$, the gear ratio reduces the speed as well as magnifies the torque, whereas T_3 and T_4 are related by

$$T_4 = T_3\frac{N_4}{N_3}$$

and θ_3 and θ_1 are related by this relation as

$$\theta_3 = \theta_2 \frac{N_3}{N_4} = \theta_1 \frac{N_1}{N_2} \cdot \frac{N_3}{N_4}$$

If $\frac{N_2}{N_1} < 1$, the gear ratio reduces the speed as well as magnifies the torque; then we can substitute the value of T_2 and we can get

$$J_1\ddot{\theta}_1 + b_1\dot{\theta}_1 + \frac{N_1}{N_2} T_2 = T_m$$

or

$$J_1\ddot{\theta}_1 + b_1\dot{\theta}_1 + \frac{N_1}{N_2}\left(J_2\ddot{\theta}_2 + b_2\dot{\theta}_2 + T_3\right) = T_m$$

Thus,

$$J_1\ddot{\theta}_1 + b_1\dot{\theta}_1 + \frac{N_1}{N_2}\left(J_2\ddot{\theta}_2 + b_2\dot{\theta}_2\right) + \frac{N_1}{N_2} T_3 = T_m$$

However, we can substitute the values of T_3 and T_4, and then we can get

$$J_1\ddot{\theta}_1 + b_1\dot{\theta}_1 + \frac{N_1}{N_2}\left(J_2\ddot{\theta}_2 + b_2\dot{\theta}_2\right) + \frac{N_1}{N_2} \cdot \frac{N_3}{N_4} T_4 = T_m$$

After rearranging and simplifying, we can say

$$J_1\ddot{\theta}_1 + b_1\dot{\theta}_1 + \frac{N_1}{N_2}\left(J_2\ddot{\theta}_2 + b_2\dot{\theta}_2\right) + \frac{N_1}{N_2} \cdot \frac{N_3}{N_4}\left(J_3\ddot{\theta}_3 + b_3\dot{\theta}_3 + T_L\right) = T_m$$

Eliminating θ_2 and θ_3 from this last equation and writing the resulting equation in terms of θ_1 and its time derivatives, we can obtain as

$$J_1\ddot{\theta}_1 + b_1\dot{\theta}_1 + \frac{N_1}{N_2}\left(J_2\ddot{\theta}_1 \frac{N_1}{N_2} + b_2\dot{\theta}_1 \frac{N_1}{N_2}\right) + \frac{N_1}{N_2} \cdot \frac{N_3}{N_4}\left(J_3\ddot{\theta}_1 \frac{N_1}{N_2} \cdot \frac{N_3}{N_4} + b_3\dot{\theta}_1 \frac{N_1}{N_2} \cdot \frac{N_3}{N_4} + T_L\right) = T_m$$

or

$$J_1\ddot{\theta}_1 + b_1\dot{\theta}_1 + \left(\frac{N_1}{N_2}\right)^2 J_2\ddot{\theta}_1 + \left(\frac{N_1}{N_2}\right)^2 b_2\dot{\theta}_1$$

$$+\left(\frac{N_1}{N_2}\right)^2 \cdot \left(\frac{N_3}{N_4}\right)^2 J_3\ddot{\theta}_1 + \left(\frac{N_1}{N_2}\right)^2 \cdot \left(\frac{N_3}{N_4}\right)^2 b_3\dot{\theta}_1 + \left(\frac{N_1}{N_2}\right) \cdot \left(\frac{N_3}{N_4}\right) T_L = T_m$$

Thus,

$$\left[J_1+\left(\frac{N_1}{N_2}\right)^2 J_2+\left(\frac{N_1}{N_2}\right)^2\left(\frac{N_3}{N_4}\right)^2 J_3\right]\ddot{\theta}_1$$

$$+\left[b_1+\left(\frac{N_1}{N_2}\right)^2 b_2+\left(\frac{N_1}{N_2}\right)^2\left(\frac{N_3}{N_4}\right)^2 b_3\right]\dot{\theta}_1+\left(\frac{N_1}{N_2}\right)\cdot\left(\frac{N_3}{N_4}\right)T_L=T_m$$

Therefore,

$$\left[J_1+\left(\frac{N_1}{N_2}\right)^2 J_2+\left(\frac{N_1}{N_2}\right)^2\left(\frac{N_3}{N_4}\right)^2 J_3\right]\ddot{\theta}_1$$

$$+\left[b_1+\left(\frac{N_1}{N_2}\right)^2 b_2+\left(\frac{N_1}{N_2}\right)^2\left(\frac{N_3}{N_4}\right)^2 b_3\right]\dot{\theta}_1+\left(\frac{N_1}{N_2}\right)\cdot\left(\frac{N_3}{N_4}\right)T_L=T_m$$

Thus, the *equivalent moment of inertia* and *viscous-friction coefficient* of the gear train referred to shaft 1 are given, respectively, by these relations.

$$J_{1eq}=J_1+\left(\frac{N_1}{N_2}\right)^2 J_2+\left(\frac{N_1}{N_2}\right)^2\left(\frac{N_3}{N_4}\right)^2 J_3$$

$$b_{1eq}=b_1+\left(\frac{N_1}{N_2}\right)^2 b_2+\left(\frac{N_1}{N_2}\right)^2\left(\frac{N_3}{N_4}\right)^2 b_3$$

Similarly, the equivalent moment of inertia and viscous-friction coefficient of the gear train referred to as the load shaft (shaft 3) are given, respectively, by these equations.

$$J_{3eq}=J_3+\left(\frac{N_4}{N_3}\right)^2 J_2+\left(\frac{N_2}{N_1}\right)^2\left(\frac{N_4}{N_3}\right)^2 J_1$$

$$b_{3eq}=b_3+\left(\frac{N_4}{N_3}\right)^2 b_2+\left(\frac{N_2}{N_1}\right)^2\left(\frac{N_4}{N_3}\right)^2 b_1$$

The relationship between J_{1eq} and J_{3eq} can be said as

$$J_{1eq}=\left(\frac{N_1}{N_2}\right)^2\left(\frac{N_3}{N_4}\right)^2 J_{3eq}$$

The relationship between b_{1eq} and b_{3eq} can be said as

$$b_{1eq}=\left(\frac{N_1}{N_2}\right)^2\left(\frac{N_3}{N_4}\right)^2 b_{3eq}$$

- The effect of J_2 and J_3 on an equivalent moment of inertia is determined by the gear ratios $\frac{N_1}{N_2}$ and $\frac{N_3}{N_4}$;
- For speed-reducing gear trains, the ratios $\frac{N_1}{N_2}$ and $\frac{N_3}{N_4}$ are usually less than unity;
- If $\frac{N_1}{N_2} \ll 1$ and $\frac{N_3}{N_4} \ll 1$ then the effect of J_2 and J_3 on the equivalent moment of inertia J_{1eq} is negligible;
- Similar comments apply to the equivalent viscous-friction coefficient b_{1eq} of the gear train;
- In terms of the equivalent moment of inertia J_{1eq} and equivalent viscous-friction coefficient b_{1eq}, thus J_{1eq} and b_{1eq} can be simplified to give

$$J_{1eq}\ddot{\theta}_1 + b_{1eq}\dot{\theta}_1 + nT_L = T_m$$

where

$$n = \frac{N_1 N_3}{N_2 N_4}$$

2. *SSEs for disturbances*

Now, I am going to show about SSEs for disturbances. Please look at the feedback control system showing disturbance $D(s)$ as in the following.

Feedback control system shows disturbance in Figure 4.76. The transformation of the output is given by this relationship as

$$C(s) = E(s)G_1(s)G_2(s) + D(s)G_2(s)$$

From this control system,

$$C(s) = R(s) - E(s)$$

Then we can say as

$$E(s) = \frac{1}{1+G_1(s)G_2(s)}R(s) - \frac{G_2(s)}{1+G_1(s)G_2(s)}D(s)$$

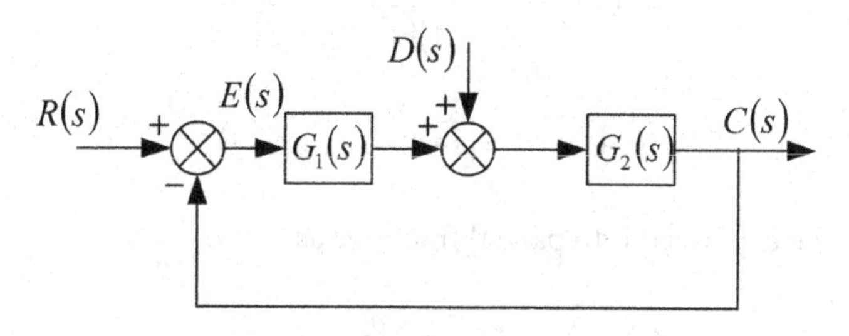

FIGURE 4.76 Steady-state errors for disturbances.

Now, apply the final value theorem, which we obtain as

$$e_{ss} = \lim_{s \to 0} sE(s) = \lim_{s \to 0} \frac{s}{1+G_1(s)G_2(s)} R(s) - \lim_{s \to 0} \frac{sG_2(s)}{1+G_1(s)G_2(s)} D(s)$$

where e_{ssR} and e_{ssD} can be said as

$$e_{ssR} = \lim_{s \to 0} \frac{s}{1+G_1(s)G_2(s)} R(s)$$

$$e_{ssD} = -\lim_{s \to 0} \frac{sG_2(s)}{1+G_1(s)G_2(s)} D(s)$$

Therefore, the SSEs for Disturbances in the control system may write as

$$e_{ss} = e_{ssR} + e_{ssD}$$

4.8 MATLAB AND MATLAB SIMULINK

Example 4.12

A closed-loop system transfer function is $G(s) = \frac{5(s^2+5s+6)}{s^3+6s^2+10s+8}$. Determine the time response when the system has unit-step input.

Solution

Taking the factorization of the closed-loop transfer function, we get

$$G(s) = \frac{5(s+2)(s+3)}{(s+4)(s^2+2s+2)}$$

As,

$$R(s) = \frac{1}{s}$$

Then,

$$C(s) = \frac{5(s+2)(s+3)}{s(s+4)(s^2+2s+2)}$$

Here,

$$s^2+2s+2 = (s+1-j)(s+1+j)$$

Thus, $C(s)$ can be expanded into partial fractions as

$$C(s) = \frac{a_0}{s} + \frac{a_1}{s+4} + \frac{a_2}{s+1+j} + \frac{a_3}{s+1-j}$$

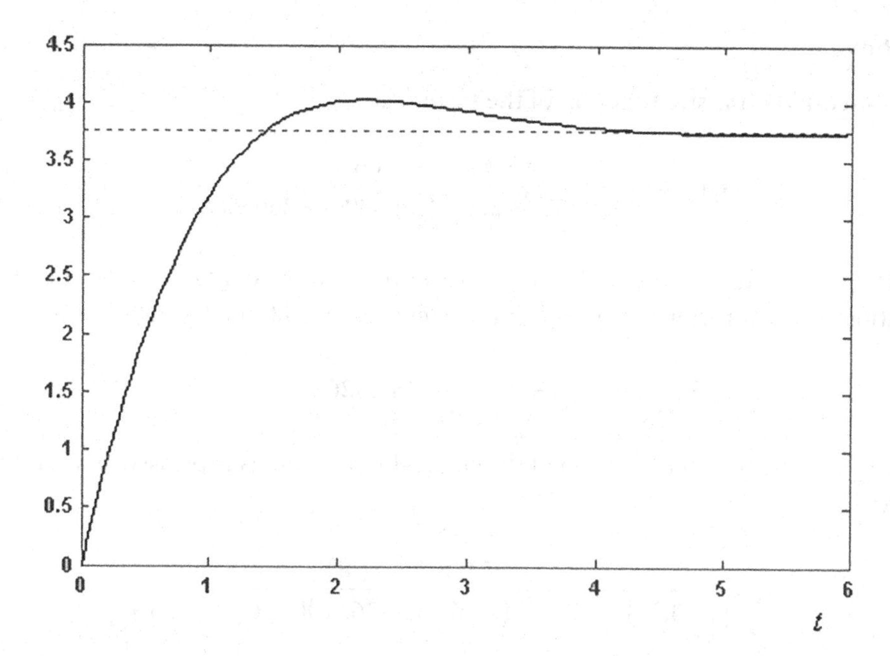

FIGURE 4.77 Step response curve.

And we get the coefficients

$$a_0 = \frac{15}{4}, \quad a_1 = -\frac{1}{4}, \quad a_2 = \frac{-7+j}{4}, \text{ and } a_3 = \frac{-7-j}{4}$$

Therefore, the output response $c(t)$ is

$$c(t) = \frac{15}{4} - \frac{1}{4}e^{-4t} + \frac{1}{4}e^{-t}(-14\cos t + 2\sin t)$$

$$= \frac{15}{4} - \frac{1}{4}e^{-4t} - \frac{\sqrt{200}}{4}e^{-t}\cos(t + 352^\circ)$$

The output response curve $c(t)$ is shown in Figure 4.77.

Example 4.13

An open-loop transfer function of a unity negative feedback control system is

$$G(s) = \frac{20(0.208s+1)}{s(0.1s+1)(0.05s+1)(0.015s+1)}$$

Determine the unit-step response of the system. If it can be equivalent to the low-order system, determine its performance index.

Solution

The closed-loop transfer function of the system is

$$G(s)=\frac{55466.7s+266666.7}{s^4+96.7s^3+2200s^2+68800s+266666.7} \tag{4.178}$$

By using the roots of MATLAB software to solve the roots of higher-order algebraic equations, the characteristic roots of the system can be obtained as follows:

$$-4.38,-79.33,-6.48\pm j26.95$$

Therefore, if the transfer function of the closed-loop system is expressed as a product, we get

$$G(s)=\frac{55466.7(s+4.8)}{(s+4.38)(s+79.33)(s+6.48-j26.95)(s+6.48+j26.95)}$$

The distribution of zero poles of the closed-loop system is shown in Figure 4.78, and the Laplace transform of the unit-step response of the closed-loop system is written as follows,

$$C(s)=\frac{55466.7(s+4.8)}{(s+4.38)(s+79.33)(s+6.48-j26.95)(s+6.48+j26.95)}\times\frac{1}{s}$$

or

$$C(s)=\frac{A}{s}+\frac{A_1}{s+4.38}+\frac{A_2}{s+79.33}+\frac{B_1}{s+6.48-j26.95}+\frac{B_2}{s+6.48+j26.95} \tag{4.179}$$

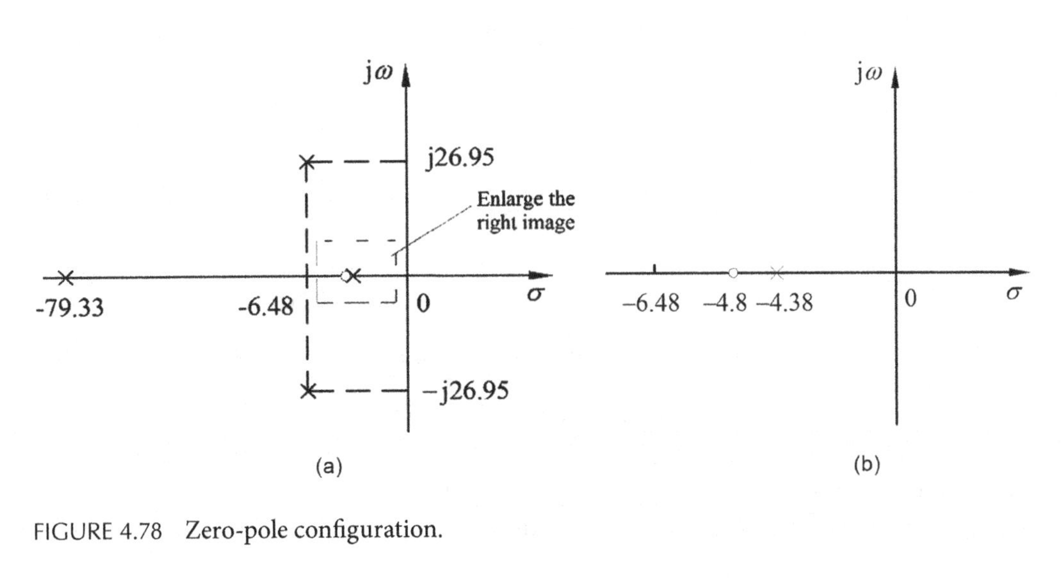

FIGURE 4.78 Zero-pole configuration.

Setting $A = G(0) = 1$, then we get

$$A_1 = \left.\frac{55466.7(s+4.8)}{s(s+79.33)(s+6.48-j26.95)(s+6.48+j26.95)}\right|_{s=-4.38} = -0.097$$

$$A_2 = \left.\frac{55466.7(s+4.8)}{s(s+4.38)(s+6.48-j26.95)(s+6.48+j26.95)}\right|_{s=-79.33} = -0.115$$

Similarly,

$$B_{1,2} = \left.\frac{55466.7(s+4.8)}{s(s+4.38)(s+79.33)}\right|_{s=-6.48\pm j26.95} = 0.477e^{\pm j2.53}$$

The unit-step response of the system is written by the Laplace inverse transformation by substituting each coefficient into Eq. (4.179), and then we get

$$c(t) = 1 - 0.097e^{-4.38t} - 0.115e^{-79.33t} + 0.954e^{-6.48t}\cos(26.95t + 2.53)$$

where $t \geq 0$. Figure 4.56 shows the pole $s_1 = -4.38$ is closest to the imaginary axis, but a zero $z = -4.8$ point is located close to this pole. The distance is s_1 between zero and the pole; thus, the distance to the imaginary axis is

$$\frac{|s_1 - z|}{|s_1|} = 10.42$$

Thus, s_1 forms with z dipole; the distance from s_2 to s_3 is the ratio of the imaginary axis as $79.33/6.48 = 12.24$; therefore, s_2 the impact of the transition process can be ignored. Thus, the system can be reduced to an equivalent second-order system. In order to make the equivalent second-order system, we have the same gain as the original system, and the items to be omitted in the original system should be simplified into a typical relationship, i.e.,

$$\begin{aligned} G(s) &= \frac{55466.7(s+4.8)}{(s+4.38)(s+79.33)(s+6.48-j26.95)(s+6.48+j26.95)} \\ &= \frac{766.24(0.208s+1)}{(0.228s+1)(0.0126s+1)(s+6.48-j26.95)(s+6.48+j26.95)} \\ &\approx \frac{766.24}{(s+6.48-j26.95)(s+6.48+j26.95)} \\ &= \frac{766.24}{s^2+12.96s+766.24} \\ &= \frac{\omega_n^2}{s^2+2\xi\omega_n s+\omega_n^2} \end{aligned}$$

We compare and we get

$$\omega_n = \sqrt{766.24}\,\text{rad/s} = 27.68\,\text{rad/s}, \ \xi = 0.234$$

The equivalent transient performance index of the second-order system is

The maximum overshoot: $M_p = e^{-\frac{\pi\xi}{\sqrt{1-\xi^2}}} = 46.9\%$

$$\text{The settling time: } t_s = \begin{cases} \dfrac{4}{\xi\omega_n} = \dfrac{4}{0.234 \times 27.68}\,\text{s} = 0.62\,\text{s} & ;(\Delta = \pm 2\%) \\ \dfrac{3}{\xi\omega_n} = \dfrac{3}{0.234 \times 27.68}\,\text{s} = 0.46\,\text{s} & ;(\Delta = \pm 5\%) \end{cases}$$

$$\text{The rise time: } t_r = \frac{\pi - \arccos\xi}{\omega_n\sqrt{1-\xi^2}} = \frac{3.14 - \arccos 0.234}{27.68\sqrt{1-0.234^2}}\,\text{s} = 0.067\,\text{s}$$

$$\text{The peak time: } t_p = \frac{\pi}{\omega_n\sqrt{1-\xi^2}} = \frac{3.14}{27.68\sqrt{1-0.234^2}}\,\text{s} = 0.117\,\text{s}$$

The unit-step response of the equivalent second-order system is

$$c(t) = 1 - 1.058e^{-6.48t}\sin(26.91t + 1.33)$$

The output response of the system is shown in Figure 4.79. The following MATLAB program is shown the Figure 4.57:

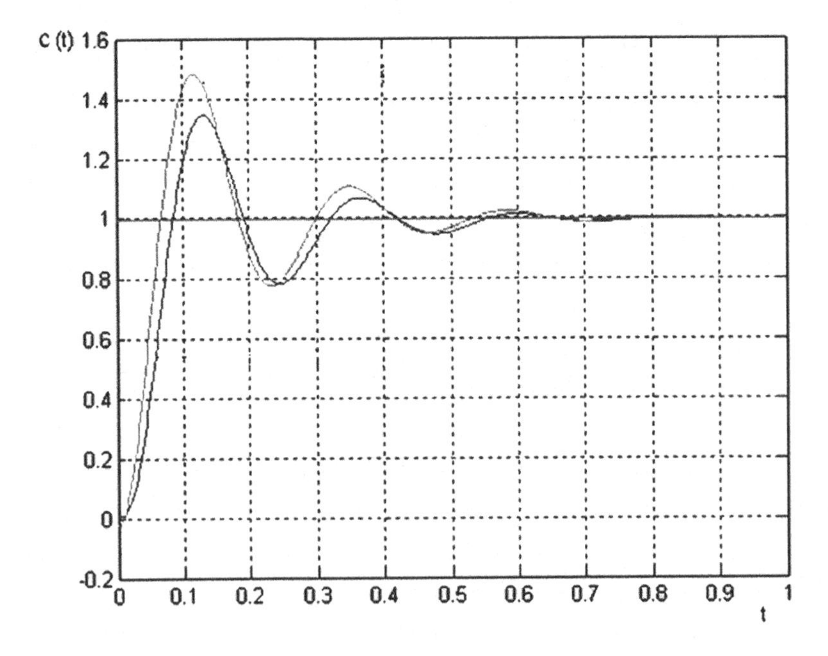

FIGURE 4.79 Response curve.

%The MATLAB program for Figure 4.79:

```
for i=1:500
tt(i)=(i-1)/500;
x11(i)=1-0.097.*exp(-4.38.*t(i))-0.115.*exp(-79.33.*t(i));
x22(i)=0.954.*exp(-6.48.*t(i)).*cos(26.95.*t(i)+2.53);
xo11(i)=x1(i)+x2(i);
xo22(i)=1-1.058.*exp(-6.48.*t(i)).*sin(26.91.*t(i)+1.33);
xoo(i)=1;
end
plot(tt, xo11,tt, xo22,tt, xoo)
grid
xlabel('Time, t')
ylabel('The output response, c(t)')
```

Similarly, the step response function step (num, den, t) in MATLAB can be used to execute, and the corresponding simplified step response curve can be obtained.

```
t=0:0.002:1;
num=[55466.7 266666.7];
den=[1 96.7 2200 68800 266666.7];
step(num, den, t)
grid
hold on
```

The unit-step response curve of the nonsimplified higher-order system and the unit step response curve of the simplified equivalent second-order system can be obtained, and the equivalent second-order system can be equivalent to the original higher-order system after 0.4 s; thus, proving that the simplification is feasible and effective.

Example 4.14

Let the transfer function of an aircraft autopilot system be

$$G(s)=\frac{4.5s^3+41.355s^2+104.904s+77.9803}{s^4+10.04s^3+62.689s^2+144.048s+77.9803}$$

Solution

The unit-step response is determined by MATLAB program as

```
t=0:0.01:8;
num=[4.5 41.355 104.904 77.9803];
den=[1 10.04 62.689 144.048 77.9803];
step(num, den, t);
```

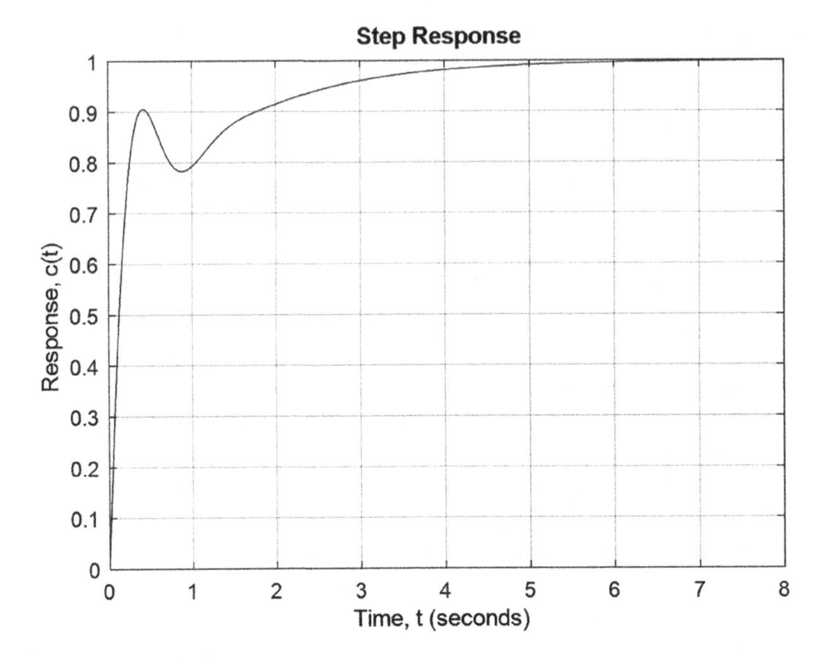

FIGURE 4.80 The step response output by the system.

```
grid
xlabel('Time, t')
ylabel ('Response, c(t)')
```

The solution result is shown in Figure 4.80.

Example 4.15

Find the unit-step response of the system. The transfer function of a system known as

$$G(s) = \frac{20}{s^4 + 8s^3 + 36s^2 + 40s + 20}$$

Solution

According to the transfer function, we get the system MATLAB program

```
num=[20];
den=[1 8 36 40 20];
step(num, den);
grid
xlabel('Time, t')
ylabel ('Response, c(t)')
```

The solved results are shown in Figure 4.81.

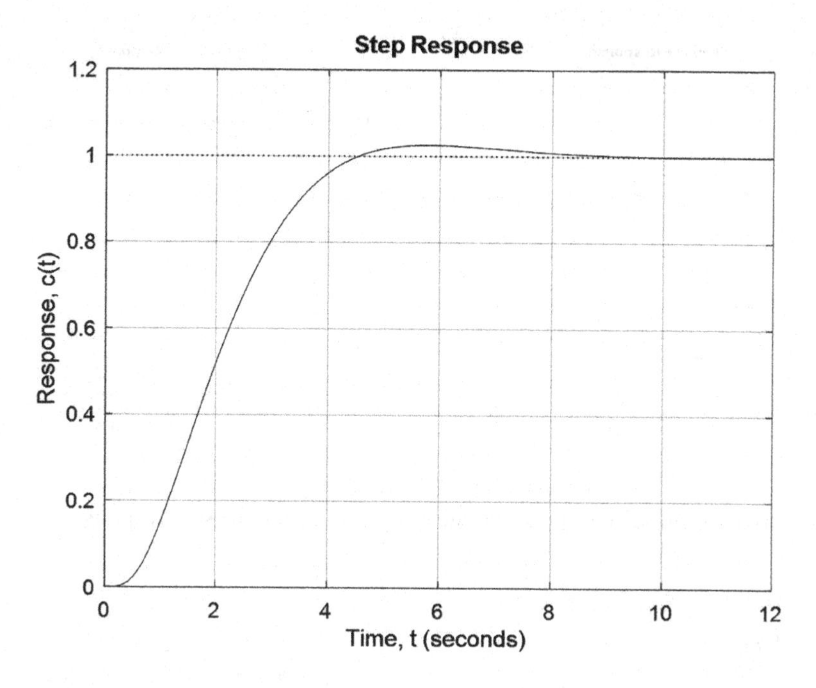

FIGURE 4.81 Unit-step response of system.

Example 4.16

An open-loop transfer function of a unit feedback system is

$$G(s)=\frac{2s+1}{s^2}$$

Find the unit-impulse response of the system and step response.

Solution

According to the transfer function of the system, we can write the MATLAB program for the system is

```
num= [2 1];
den= [1 0 0];
sys1= tf(num, den)
sys= feedback(sys1,1)
impulse(sys)
grid
figure
step(sys)
grid
xlabel('Time, t')
ylabel ('Response, c(t)')
```

The results are shown in Figure 4.82a and b.

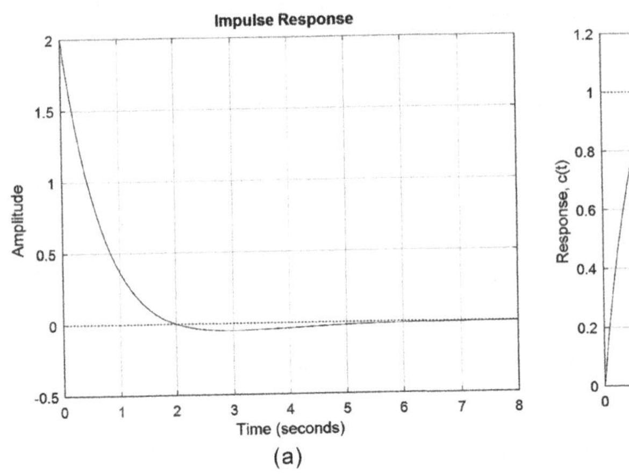

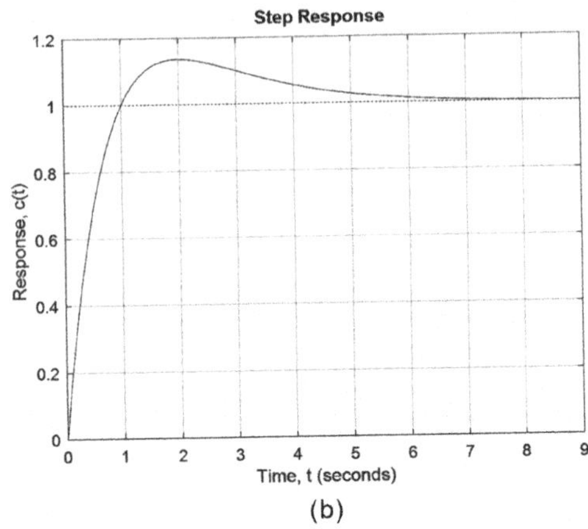

FIGURE 4.82 (a) Unit-impulse response of the system and (b) Unit-step response of the system.

Example 4.17

The closed-loop transfer function of a system is

$$G(s)=\frac{6s^3+26s^2+6s+20}{s^4+3s^3+4s^2+2s+2}$$

Determine unit step response of system and unit-impulse response.

Solution

According to the transfer function of the system, we can write the MATLAB program for the system as follows:

```
t=0:100;
num=[6 26 6 20]
den=[1 3 4 2 2 ]
[A, B, C, D]=tf2ss(num, den)
[y, x]= step(A, B, C, D, 1,t)
plot(t, y, t, x)
grid
xlabel('Time, t')
ylabel ('Response, c(t)')
figure
impulse(A, B, C, D, 1,t)
grid
xlabel('Time, t')
ylabel ('Response, c(t)')
```

The results are shown in Figure 4.83a and b.

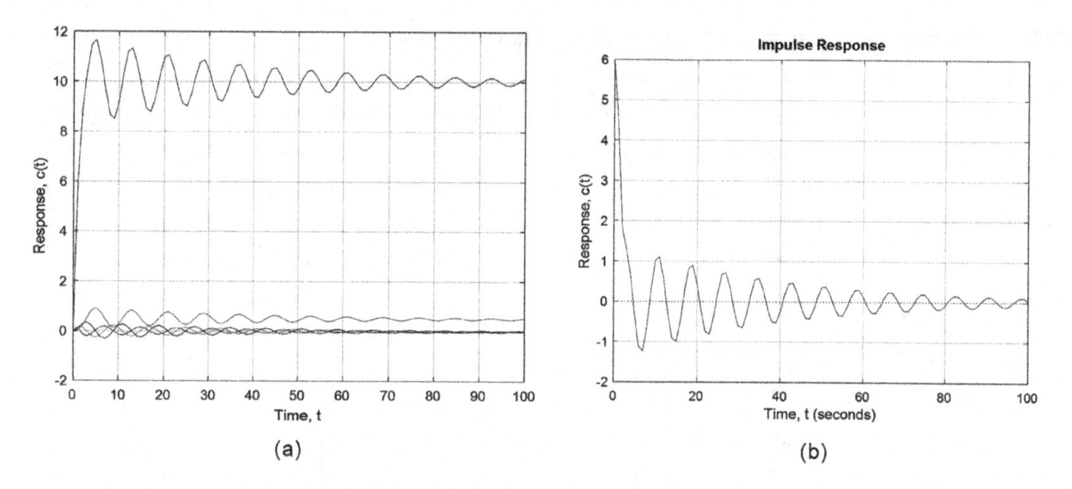

FIGURE 4.83 (a) Unit-step response of the system and (b) Unit-impulse response of the system.

Example 4.18

The state equation of a system is

$$\frac{dx}{dt}=\begin{bmatrix} -5 & 2 & 0 & 0 \\ 0 & -4 & 0 & 0 \\ -3 & 2 & -4 & -1 \\ -3 & 2 & 0 & 4 \end{bmatrix}x(t)+\begin{bmatrix} 1 \\ 2 \\ 0 \\ 1 \end{bmatrix}u(t) \quad y(t)=\begin{bmatrix} 1 & 2 & 3 & 4 \end{bmatrix}x(t)$$

Find out the unit step response, the unit-impulse response, and the zero-output response of the system.

Solution

The MATLAB program for the system is

```
A=[-5 2 0 0; 0-4 0 0; -3 2-4-1; -3 2 0 4];B=[1 2 0 1]; C=[1 2
3 4]; D=0;
t=0:0.0001:0.6;
x0=[0 0 0 1];
step(A, B, C, D, 1,t)
figure
impulse(A, B, C, D, 1,t)
figure
initial(A, B, C, D, x0,t)
```

The results are shown in Figure 4.84.

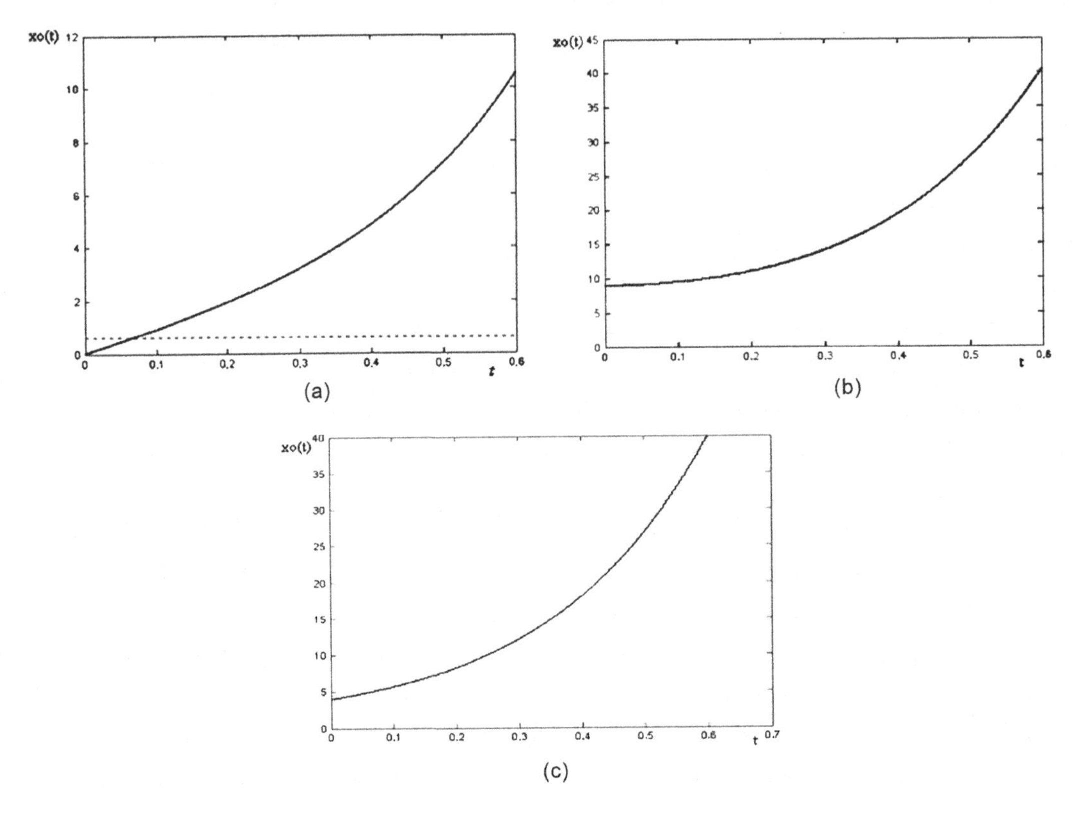

FIGURE 4.84 (a) Unit-step response of the system, (b) Unit-impulse response of the system, and (c) Zero-output response of the system.

Example 4.19

A known system with delay link transfer function

$$G(s)=\frac{1}{(s+1)^3}e^{-2s}$$

Obtain the unit-step response and the unit-impulse response of the system.

Solution

The system is divided into two parts drawing, with delay and no delay relationship.

```
num1=[1];
den1=poly([-1-1-1]);
sys1=tf(num1,den1)
step(sys1)
hold on
[num2,den2]= pade(2,4);
```

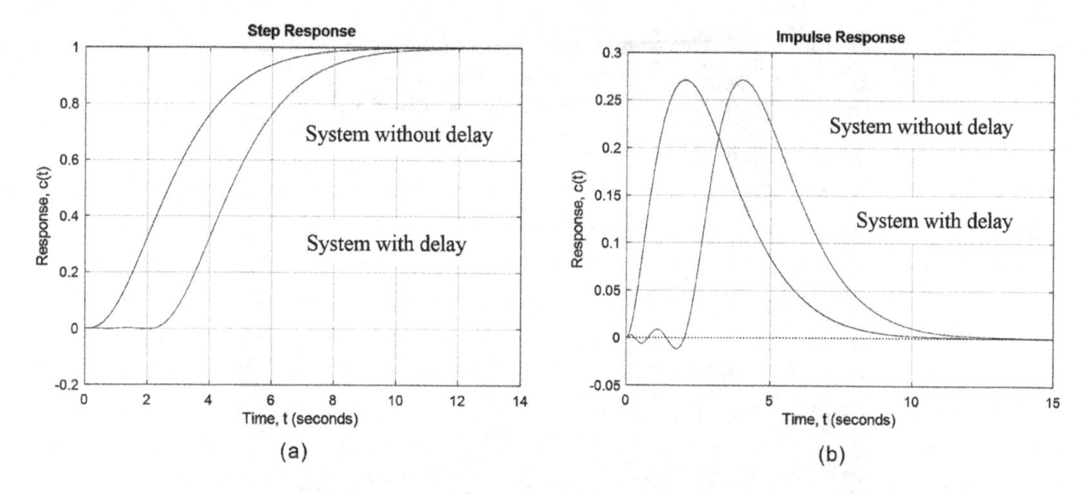

FIGURE 4.85 (a) Unit-step response of the system and (b) Unit-impulse response of the system.

```
sys2= tf(num2,den2)
step(sys1*sys2)
xlabel('Time, t')
ylabel ('Response, c(t)')
grid
figure
impulse(sys1);
hold on
impulse(sys1*sys2)
xlabel('Time, t')
ylabel ('Response, c(t)')
grid
```

The solution results are shown in Figure 4.85.

MATLAB SIMULINK

The Command Windows of MATLAB is shown in Figure 4.86a, after typing the SIMULINK command, the command window shown in 4.86b will appear in the system. From Figure 4.86, the control system of the MATLAB R2017b has toolbox such as Control System Toolbox, Fuzzy Logic Toolbox, Communication Blockset, NCD Blockset, and so on, which have been incorporated into the Simulink. To open the Simulink browser and then double-clicking the Continuous, there will be an icon as shown in Figure 4.86b and c.

According to Figure 4.86c, we can build the simulation model—click on the File to select New, then select model and double-click the mouse can get the untitled edit box. Use the mouse to drag the selected module of the control system into the untitled edit box or click the right mouse button to use the "add to Untitled" Edit bar. Now, for the use of the detailed process, please refer to the help file in MATLAB. The following is an example of the unit-step response of a control system (as shown in Figure 4.87a and b).

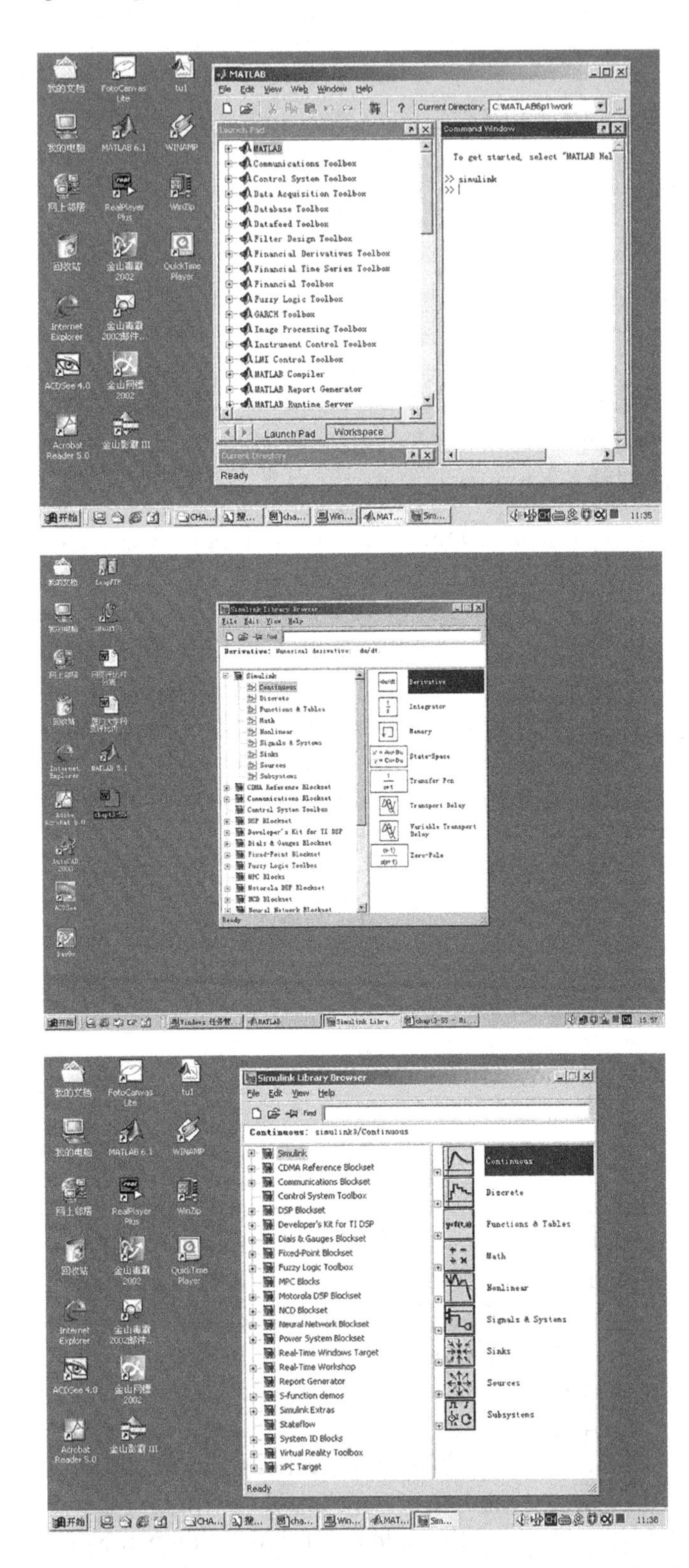

FIGURE 4.86 (a) Command Windows of MATLAB, (b) Icon that appears after double-clicking the Continuous, and (c) Command window of the SIMULINK Library Browser.

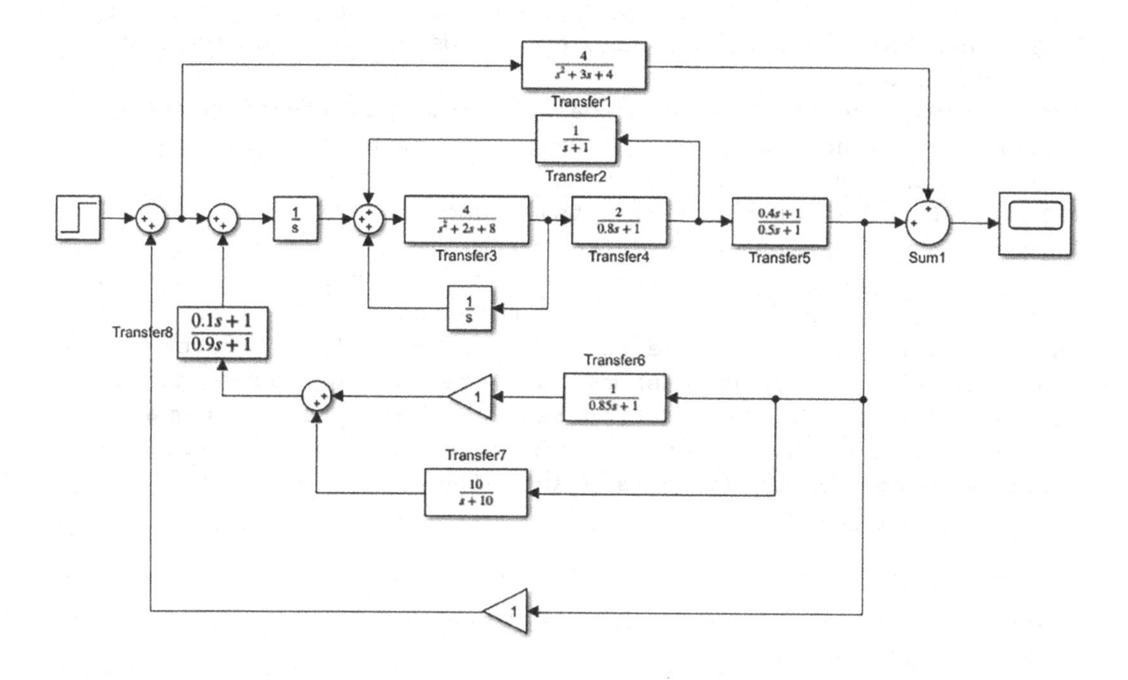

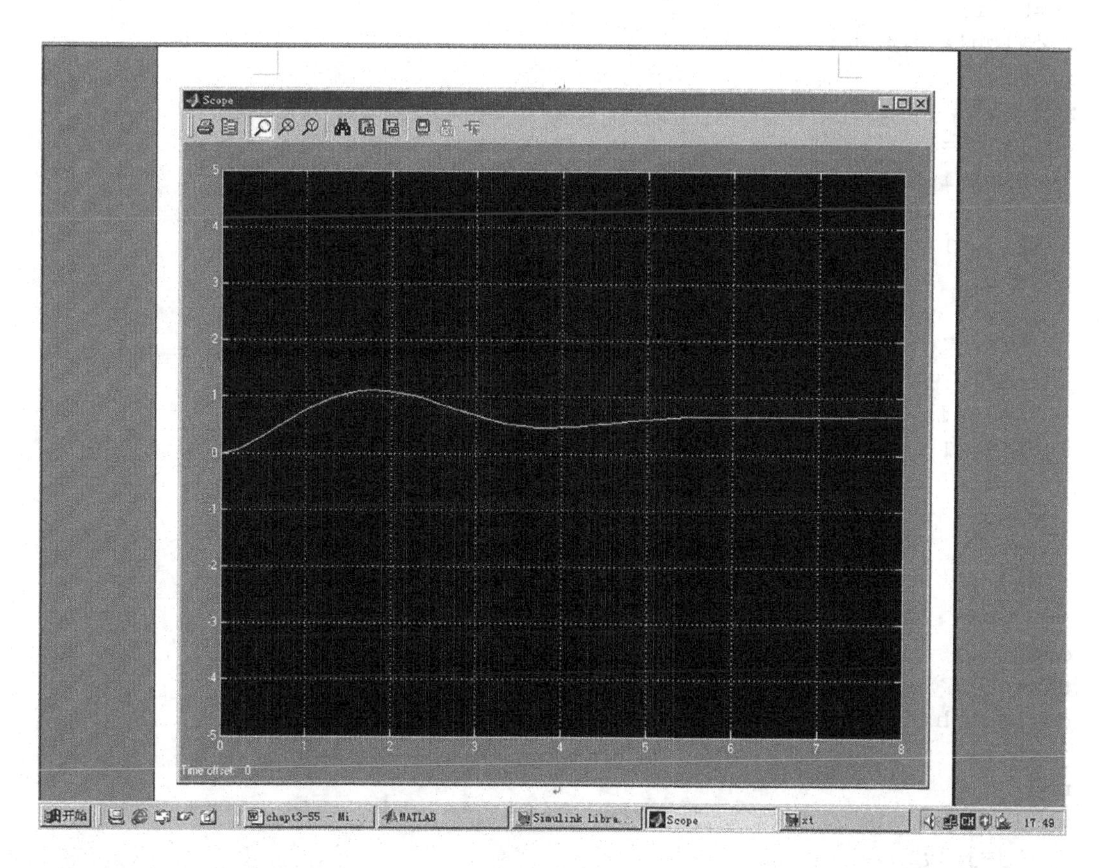

FIGURE 4.87 (a) Block diagram of a control system, and (b) Unit-step response of the block diagram of the control system.

```
% This program for solving the overlap, peak time, and modulation
time and so on.
% This procedure is a function called sigmatpts()that is used to
obtain performance index of the unit step response(overshoot
volume, peak time, regulation time).
% The call format of the sigmatpts() function:
% [sigma, tp, ts]=sigmatpts(delta, y, t)
% Among them, delta is used to select the 5% or 2% error band of
adjusting the time Ts, when delta=1 then choose the 5% error band
and when delta=2 then choose the 2% error band.[y, t] is the
function value of the corresponding system step response and its
corresponding time. The function returns the overshoot sigma, the
peak time TP and the adjustment time ts of the step response.
function [sigma,tp,ts,tr,td]=sigmatpts(delta,y,t)
% calculation the sigma and tp
[mp,tf]=max(y)
cs=length(t);
yss=y(cs);
sigma=100*(mp-yss)/yss
tp=t(tf)
% calculation ts
i=cs+1;
n=0;
while n==0
    i=i-1;
    if delta==1
  if i==1
   n=1;
  elseif y(i)>1.05*yss
   n=1;
  end
    elseif delta==2
  if i==1
   n=1;
  elseif y(i)>1.02*yss
   n=1;
  end
    end
end
t1=t(i);
cs=length(t);
j=cs+1;
n=0;
while n==0
    j=j-1;
    if delta==1
```

```
  if j==1
   n=1;
  elseif y(j)<0.95*yss
   n=1;
  end
    elseif delta==2
  if j==1
   n=1;
  elseif y(j)<0.98*yss
   n=1;
  end
    end
end
t2=t(j);
if t2<tp
    if t1>t2
  ts=t1
    end
elseif t2>tp
    if t2<t1
  ts=t2
    else
  ts=t1
    end
end

% calculation rising time tr
i=1;
while y(i)<yss
 i=i+1;
   end
 tr=t(i)
% calculation delay time td
i=1;
while y(i)<yss/2
 i=i+1;
   end
 td=t(i)
```

Example 4.20

Obtain the performance index of the step response of a system whose transfer function is $G(s)=\dfrac{s^2+2s+4}{s^2+3s^2+4s+5}$

Solution

Select the error band for delta=±2%, that is select delta=1 in the program, then the MATLAB program of the system is

```
% This Program for solving the Characteristic parameters of
step response.
global t, y;
num=[1 2 4];
den=[1 3 4 5];
sys=tf(num, den)
step(sys);
[y, t]=step(sys);
sgmatpts(1,y, t)
```

The calculation results are:

```
Transfer function:d
s^2 + 2 s + 4
--------------------------------
s^3 + 3 s^2 + 4 s + 5
Overshoot volume:sigma = 23.1715%
Peak time:tp = 2.0173s
Regulation time:ts = 2.9667s
Rise time:tr = 1.1867s
Delay time:td = 0.5933s
The solution results are shown in Figure 4.88.
```

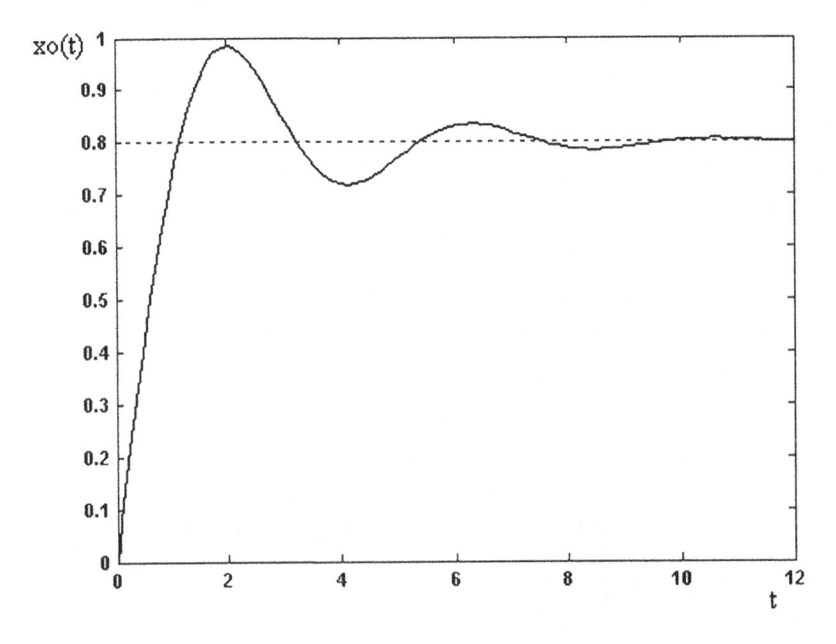

FIGURE 4.88 Step response of the system.

Example 4.21

Consider the closed-loop system defined by

$$\frac{C(s)}{R(s)} = \frac{1}{s^2 + 2\xi s + 1}$$

(The undamped natural frequency ω_n is normalized to 1.) Plot unit-step response curves $c(t)$ when ξ assumes the following values: $\xi = 0, 0.2, 0.6, 0.8, 1.0$; also plot a three-dimensional plot.

Solution

```
% ------- Two-dimensional plot and three-dimensional plot of
unit-step
% response curves for the standard second-order system with wn
= 1
% and zeta = 0, 0.2, 0.4, 0.6, 0.8, and 1. -------
t = 0:0.2:10;
zeta = [0 0.2 0.4 0.6 0.8 1];
for n = 1:6
num = [1];
den = [1 2*zeta(n) 1];
[yy(1:51,n), x, t] = step(num, den, t);
end
% To plot a two-dimensional diagram, enter the command plot(t,
y).
figure
plot(t, yy)
grid
title('2D Plot of Unit-Step Response Curves')
xlabel('Time, t (sec)')
ylabel('Response, c(t)')
text(4.1,1.86,' ξ = 0')
text(3.5,1.5,'0.2')
text(3,5,1.24,'0.4')
text(3.5,1.08,'0.6')
text(3.5,0.95,'0.8')
text(3.5,0.86,'1.0')
% To plot a three-dimensional diagram, enter the command
mesh(t, zeta, y').
figure
mesh(t, zeta, yy')
title('3D Plot of Unit-Step Response Curves')
xlabel('Time, t Sec')
ylabel(' ξ ')
zlabel('Response, c(t)')
```

The solution results are shown in Figure 4.89.

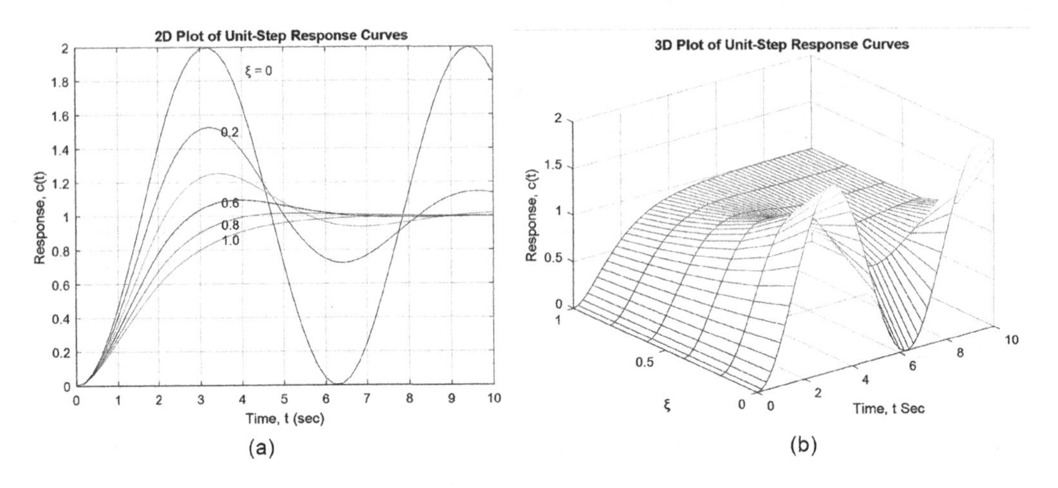

FIGURE 4.89 (a) 2D plot of unit-step response curve and (b) 3D plot of unit-step response curve.

Example 4.22

Obtain the unit-impulse response of the following system:

$$\frac{C(s)}{R(s)} = G(s) = \frac{1}{s^2 + 0.2s + 1}$$

Solution

The MATLAB program is as follows:

```
num = [1];
den = [1 0.2 1];
impulse(num, den);
grid
xlabel('Time, t (sec)')
ylabel('Response, c(t)')
title('Unit-Impulse Response of G(s) = 1/(s^2 + 0.2s + 1)')
```

The solution results are shown in Figure 4.90.

Example 4.23

Using the lsim command, obtain the unit-ramp response of the following system:

$$\frac{C(s)}{R(s)} = G(s) = \frac{2s + 1}{s^2 + s + 1}$$

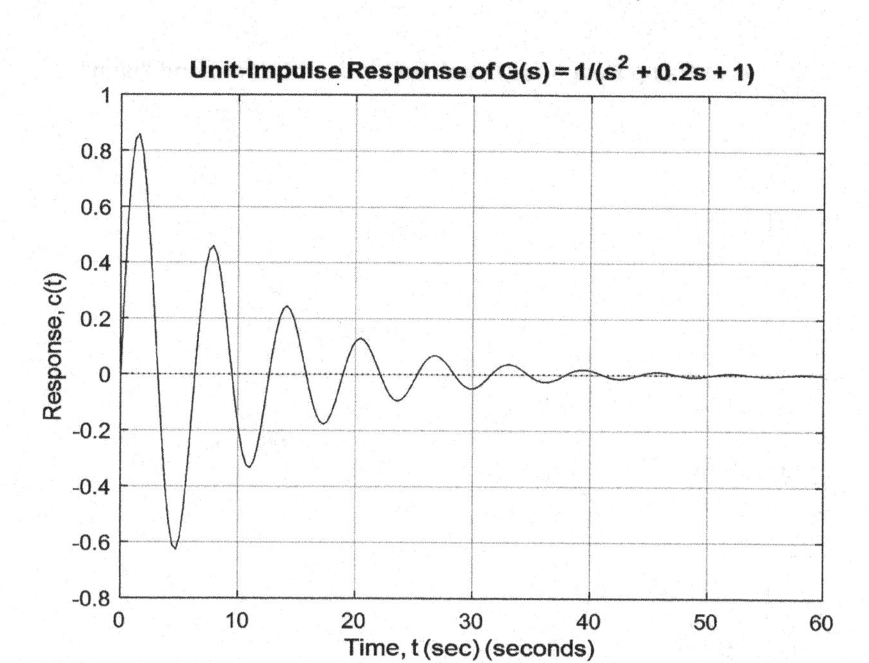

FIGURE 4.90 Unit-impulse response curve.

Solution

```
% ------- Ramp Response -------
num = [2 1];
den = [1 1 1];
t = 0:0.1:10;
r = t;
y = lsim(num, den, r, t);
plot(t, r, '-', t, y, 'o')
grid
title('Unit-Ramp Response Obtained by Use of Command "lsim"')
xlabel('t Sec')
ylabel('Unit-Ramp Input and System Output')
text(6.3,4.6,'Unit-Ramp Input')
text(4.75,9.0,'Output')
```

The solution results are shown in Figure 4.91.

Example 4.24

When the closed-loop system involves a numerator dynamic, the unit-step response curve may exhibit a large overshoot. Obtain the unit-step response of the following system with MATLAB:

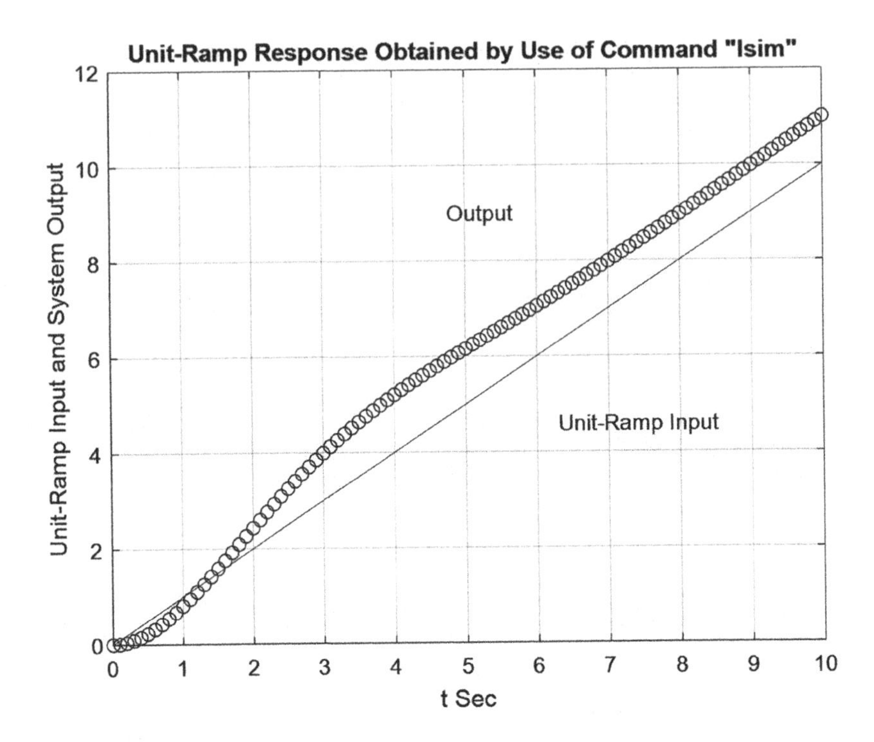

FIGURE 4.91 Unit-ramp response obtained by use of command "lsim."

$$\frac{C(s)}{R(s)} = G(s) = \frac{10s + 4}{s^2 + 4s + 4}$$

Obtain also the unit-ramp response with MATLAB.

Solution

The MATLAB program is as follows:

```
num = [10 4];
den = [1 4 4];
t = 0:0.02:10;
y = step(num, den, t);
figure
plot(t, y)
grid
title('Unit-Step Response')
xlabel('Time, t (sec)')
ylabel('Response, c(t)')
num1 = [10 4];
den1 = [1 4 4 0];
y1 = step(num1,den1,t);
figure
plot(t, t, '--', t, y1)
```

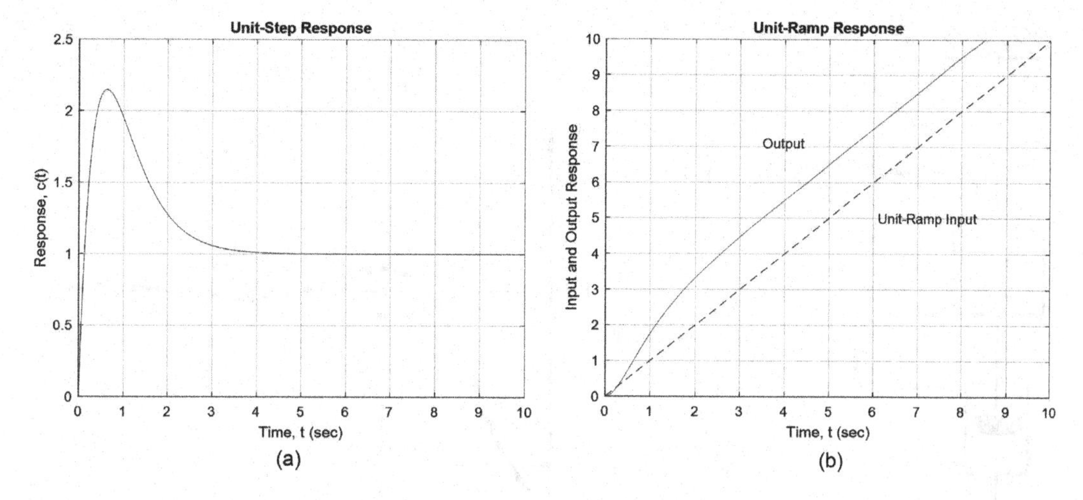

FIGURE 4.92 (a) Unit-step response curve; (b) unit-ramp response curve plotted with unit-ramp input.

```
v = [0 10 0 10]; axis(v);
grid
title('Unit-Ramp Response')
xlabel('Time, t (sec)')
ylabel('Input and Output Response')
text(6.1,5.0,'Unit-Ramp Input')
text(3.5,7.1,'Output')
```

The solution results are shown in Figure 4.92.

Example 4.25

Figure 4.93a shows a mechanical vibratory system. When 2 lb of force (step input) is applied to the system, the mass oscillates, as shown in Figure 4.93b. Determine m, b, and k of the system from this response curve. The displacement x is measured from the equilibrium position.

Solution

The transfer function of this system is

$$\frac{X(s)}{P(s)} = \frac{1}{ms^2 + bs + k}$$

Since

$$P(s) = \frac{2}{s}$$

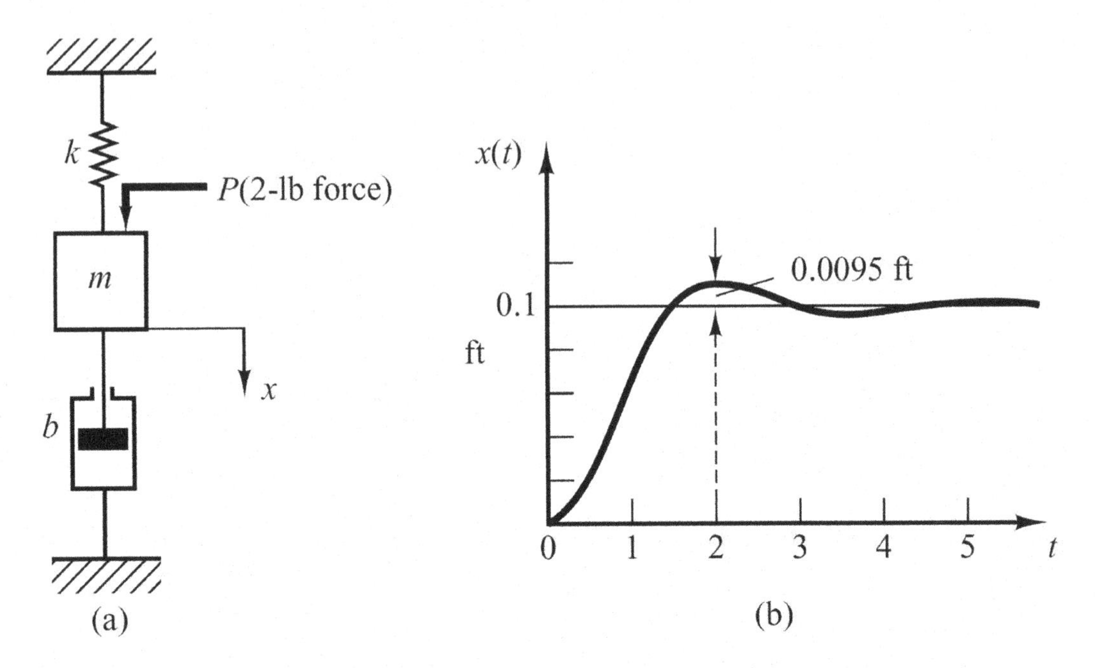

FIGURE 4.93 (a) Mechanical vibratory system; (b) step response curve.

we obtain

$$X(s)=\frac{2}{s\left(ms^2+bs+k\right)}$$

It follows that the steady-state value of x is

$$x(\infty)=\lim_{s\to 0} sX(s)=\frac{2}{k}=0.1 \text{ ft}$$

Hence,

$$k=20 \text{ lb}_f/\text{ft}$$

Note that $Mp=9.5\%$ corresponds to $\xi=0.6$.

The experimental curve shows that $t_p=2\,\text{s}$. Therefore,

$$t_p=\frac{\pi}{\omega_d}=\frac{\pi}{\omega_n\sqrt{1-\xi^2}}=\frac{\pi}{0.8\omega_n}$$

Since $\omega_n^2=\dfrac{k}{m}=20/m$, we obtain

$$m=\frac{20}{\omega_n^2}=\frac{20}{1.96^2}=5.2\,\text{slugs}=167\,\text{lb}$$

(Note that 1 slug = 1 lb_f-s/ft.) Then b is determined from

$$2\xi\omega_n = \frac{b}{m}$$

or

$$b = 2\xi\omega_n m = 2\times 0.6\times 1.96\times 5.2 = 12.2\,\text{lb}_f\text{-ft/s}$$

Example 4.26

Consider the unit-step response of the second-order system

$$\frac{C(s)}{R(s)} = \frac{\omega_n^2}{s^2 + 2\xi\omega_n s + \omega_n^2}$$

The amplitude of the exponentially damped sinusoid changes as a geometric series. At time $t = t_p = \dfrac{\pi}{\omega_d}$, the amplitude is equal to $e^{-\left(\frac{\sigma}{\omega_d}\right)\pi}$. After one oscillation, or at $t = t_p + \dfrac{2\pi}{\omega_d} = \dfrac{3\pi}{\omega_d}$, the amplitude is equal to $e^{-\left(\frac{\sigma}{\omega_d}\right)3\pi}$; after another cycle of oscillation, the amplitude is $e^{-\left(\frac{\sigma}{\omega_d}\right)5\pi}$. The logarithm of the ratio of successive amplitudes is called the *logarithmic decrement.* Determine the logarithmic decrement for this second-order system. Describe a method for experimental determination of the damping ratio from the rate of decay of the oscillation.

Solution

Let us define the amplitude of the output oscillation at $t = t_i$ to be x_i, where $t = t_p + (i-1)T$ (T = period of oscillation). The amplitude ratio per one period of damped oscillation is

$$\frac{x_1}{x_2} = \frac{e^{-\left(\frac{\sigma}{\omega_d}\right)\pi}}{e^{-\left(\frac{\sigma}{\omega_d}\right)3\pi}} = e^{2\left(\frac{\sigma}{\omega_d}\right)\pi} = e^{2\xi\pi/\sqrt{1-\xi^2}}$$

Thus, the logarithmic decrement δ is

$$\delta = \ln\frac{x_1}{x_2} = \frac{2\xi\pi}{\sqrt{1-\xi^2}}$$

It is a function only of the damping ratio ξ. Thus, the damping ratio ξ can be determined by the use of logarithmic decrement.

In the experimental determination of the damping ratio ξ from the rate of decay of the oscillation, we measure the amplitude x_1 at $t = t_p$ and amplitude x_n at

$t = t_p + (i-1)T$. Note that it is necessary to choose n large enough so that the ratio x_1 / x_n is not near unity. Then,

$$\frac{x_1}{x_2} = e^{(n-1)2\xi\pi/\sqrt{1-\xi^2}}$$

or

$$\ln\frac{x_1}{x_2} = (n-1)\frac{2\xi\pi}{\sqrt{1-\xi^2}}$$

Hence,

$$\xi = \frac{\frac{1}{n-1}\left(\ln\frac{x_1}{x_2}\right)}{\sqrt{4\pi^2 + \left[(n-1)\frac{2\xi\pi}{\sqrt{1-\xi^2}}\right]}}$$

Example 4.27

In the system shown in Figure 4.94, the numerical values of m, b, and k are given as $m = 1\,\text{kg}$, $b = 2\,\text{N-s/m}$, and $k = 100\,\text{N/m}$. The mass is displaced 0.05 m and released without initial velocity. Find the frequency observed in the vibration. In addition, find the amplitude four cycles later. The displacement x is measured from the equilibrium position.

Solution

The equation of motion for the system is

$$m\ddot{x} + b\dot{x} + kx = 0$$

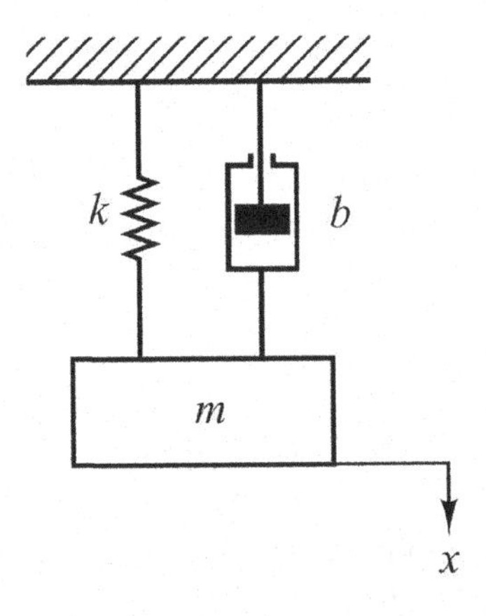

FIGURE 4.94 Spring-mass-damper system.

Substituting the numerical values for *m*, *b*, and *k* into this equation gives

$$\ddot{x}+2\dot{x}+100x=0$$

where the initial conditions are $x(0)=0.05$ and $x(0)=0$. From this last equation, the undamped natural frequency ω_n and the damping ratio ξ are found to be

$$\omega_n=10,\quad \xi=0.1$$

The frequency actually observed in the vibration is the damped natural frequency ω_d.

$$\omega_d=\omega_n\sqrt{1-\xi^2}=10\sqrt{1-0.01^2}=9.95\,\text{rad/s}$$

In the present analysis, $x(0)$ is zero. Thus, the solution $x(t)$ can be written as

$$x(t)=x(0)\,e^{\xi\omega_n t}\left(\cos\,\omega_d t+\frac{\xi}{\sqrt{1-\xi^2}}\sin\,\omega_d t\right)$$

It follows that at $t=nT$, where $T=2\pi/\omega_d$,

$$x(nT)=x(0)e^{-\xi\omega_n nT}$$

Consequently, the amplitude of four cycles later becomes

$$x(4T)=x(0)e^{-\xi\omega_n 4T}=x(nT)=x(0)e^{-(0.1)(10)(4)(0.6315)}$$

$$=0.05e^{-2.526}=0.05\times 0.07998=0.004\ m$$

EXAMPLES

This chapter requires students to learn the step response and impulse response of the first- and second-order systems; however, the transient response of other input signals and the higher-order system are only required for general understanding.

Example 4.28

For the system shown in Figure 4.95, obtain its natural frequency of nondamping ω_n, damping ratio ζ, overshoot M_p (%), peak time t_p, and adjustment time t_s (enter the error band of ±5%).

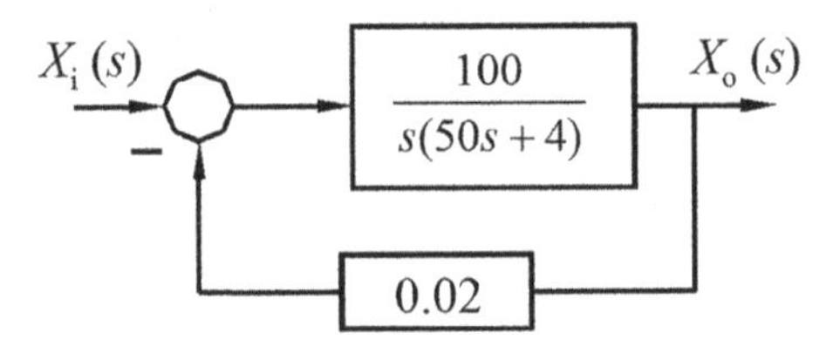

FIGURE 4.95 Closed-loop system with 0.02 feedback value.

Solution

For the system shown in Figure 4.95, first, obtain its transfer function, turn it into a standard form, and then use the formulas to calculate each characteristic quantity and the transient response index.

$$\frac{X_o(s)}{X_i(s)} = \frac{\frac{100}{s(50s+4)}}{1+\frac{100}{s(50s+4)}\times 0.02} = \frac{100}{s(50s+4)+2} = \frac{50}{5^2 s^2 + 2\times 0.2\times 5s + 1}$$

Thus, $\omega_n = 0.2\,\text{rad/s}\ldots\xi = 0.2$.

And

$$M_p = e^{-\frac{\pi\xi}{\sqrt{1-\xi^2}}} = e^{-\frac{\pi\times 0.2}{\sqrt{1-0.2^2}}} \approx 52.7\%$$

Consequently,

$$t_p = \frac{\pi}{\omega_n\sqrt{1-\xi^2}} = \frac{\pi}{0.2\sqrt{1-0.2^2}} \approx 16.03\,\text{s}$$

and

$$t_s \approx \frac{3}{\xi\omega_n} = \frac{3}{0.2\times 0.2} = 75\,\text{s}$$

Example 4.29

The unit-impulse response of a system is

$$x_o(t) = 10e^{-0.2t} + 5e^{-0.5t}$$

1. Obtain the transfer function of the system.
2. Obtain the time required for the unit-step response of the system to reach a steady-state value of 95%.

Solution

1. The transfer function of the system is the Laplace transform of the unit-impulse response of the system, that is

$$G(s)=\frac{X_o(s)}{X_i(s)}=X_o(s)=L\left[10e^{-0.2t}+5e^{-0.5t}\right]=\frac{10}{s+0.2}+\frac{5}{s+0.5}=\frac{15(s+0.4)}{s^2+0.7s+0.1}$$

2. According to the relationship between the impulse response and the step response:

$$x_1(t)=\int_0^t x_\delta(t)dt$$

The unit-step response of the system is

$$x_1(t)=\int_0^t x_\delta(\tau)d\tau=\int_0^t\left(10e^{-0.2\tau}+5e^{-0.5\tau}\right)d\tau=60-50e^{-0.2t}-10e^{-0.5t}$$

The time required for the unit-step response of the system reaches 95% of the steady-state value is t.

$$x_o(t)=60-50e^{-0.2t}-10e^{-0.5t}=60\times 95\%$$

Thus, $t=14.1$ s

Example 4.30

The open-loop transfer function of the unit feedback system is

$$G(s)=\frac{2s+1}{s^2}$$

Obtain the unit-step response and unit-impulse response of the system.

Solution

To get the system response, the closed-loop transfer function of the system can be obtained first, then get the image function of the output variable, the corresponding time-domain transient response can be obtained by the inverse Laplace transform.

$$\frac{X_o(s)}{X_i(s)}=\frac{\frac{2s+1}{s^2}}{1+\frac{2s+1}{s^2}}=\frac{2s+1}{(s+1)^2}$$

1. When inputting the unit step, $x_i(t)=1(t)$ then

$$X_i(s)=\frac{1}{s}$$

$$X_o(s)=\frac{X_o(s)}{X_i(s)}\cdot X_i(s)=\frac{2s+1}{(s+1)^2}\cdot\frac{1}{s}=\frac{1}{s}+\frac{1}{(s+1)^2}-\frac{1}{s+1}$$

Thus,

$$x_o(t)=\left[1+\left(te^{-t}-e^{-t}\right)\right]1(t)$$

2. The response of the linear time-invariant system to the input signal derivative can be obtained through the differential response to an input signal of the system, when imputing the unit impulse

$$x_i(t)=\delta(t)=\frac{d\left[1(t)\right]}{dt}$$

And then

$$x_o(t)=\frac{d\left[1+\left(te^{-t}-e^{-t}\right)\right]}{dt}\times 1(t)=\left(2e^{-t}-te^{-t}\right)\times 1(t)$$

Example 4.31

The unit-step response of the system shown in Figure 4.96a is shown in Figure 4.96b. Determine the numerical value of K_1, K_2 and a.

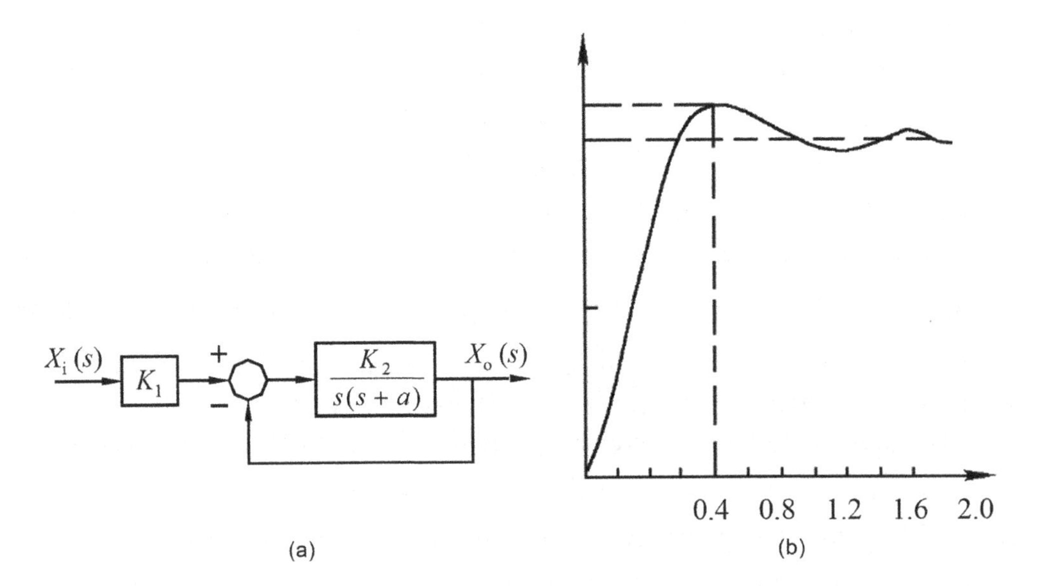

FIGURE 4.96 Control system and its unit-step response.

Solution

By Figure 4.96b, we can get

$$x_o(\infty)=2$$

$$\sigma\%=\frac{x_o(t_p)-x_o(\infty)}{x_o(\infty)}=\frac{2.18-2}{2}0.09$$

Thus, $t_p=0.8\,\text{s}$.

The closed-loop transfer function of the system is

$$\Phi(s)=\frac{K_1K_2}{s^2+as+K_2}$$

Thus,

$$X_o(s)=\Phi(s)\cdot X_i(s)=\frac{K_1K_2}{s(s^2+as+K_2)}$$

According to the final value theorem,

$$x_o(\infty)=\lim_{s\to 0} sX_o(s)=\Phi(0)=K_1$$

Thus, $K_1=2$. It can be seen the numerical value of the closed-loop transfer function in $s=0$ is the steady-state output value of the step response. According to the calculation formula of the overshoot and the peak time,

$$\begin{cases}\sigma\%=e^{-\frac{\pi\zeta}{\sqrt{1-\zeta^2}}}\\ t_p=\dfrac{\pi}{\omega_n\sqrt{1-\zeta^2}}\end{cases}$$

That we can get

$$\begin{cases}\zeta=\sqrt{\dfrac{(In\sigma)^2}{\pi^2+(In\,\sigma)^2}} & =0.608\\ \omega_n=\dfrac{\pi}{t_p\sqrt{1-\zeta^2}}=4.946\end{cases}$$

By the closed-loop transfer function,

$$\omega_n^2=K_2\quad 2\zeta\omega_n=a$$

Therefore, $K_2=24.46$ $a=6.01$.

Example 4.32

The open-loop transfer function of a unit feedback control system is

$$G(s)\frac{0.4s+1}{s(s+0.6)},$$

Find out the system response to the unit-step input and seek the rise time and the maximum overshoot.

Solution

The transfer function that is obtained is not a typical second-order oscillation element, its molecule has a differential effect. Therefore, using the formula to calculate the rise time and the maximum overshoot will cause large errors, so it is better to seek its numerical value according to the definition:

$$\frac{X_o(s)}{X_i(s)}=\frac{\frac{0.4s+1}{s(s+0.6)}}{1+\frac{0.4s+1}{s(s+0.6)}}=\frac{0.4s+1}{s^2+s+1}$$

When $x_i(t)=1(t)$, we can get $X_i(s)=\frac{1}{s}$.

$$X_o(s)=\frac{X_o(s)}{X_i(s)}\cdot X_i(s)=\frac{0.4s+1}{s^2+s+1}\cdot\frac{1}{s}=\frac{1}{s}-\frac{\left(s+\frac{1}{2}\right)+\frac{\sqrt{3}}{15}(\frac{\sqrt{3}}{2})}{(s+\frac{1}{2})^2+(\frac{\sqrt{3}}{2})^2}$$

By using the Laplace inverse transform, we can get

$$x_o(t)=1(t)-e^{-\frac{t}{2}}\left(\cos\frac{\sqrt{3}}{2}t+\frac{\sqrt{3}}{15}\sin\frac{\sqrt{3}}{2}t\right)\cdot 1(t)$$

We may find out the rise time, that is to find out the time that reach the steady-state value for the first time, then we can get

$$x_o(t_r)=1-e^{-\frac{t_r}{2}}\left(\cos\frac{\sqrt{3}}{2}t_r+\frac{\sqrt{3}}{15}\sin\frac{\sqrt{3}}{2}t_r\right)=1$$

We get

$$t_r\approx 1.946\,\text{s}$$

The maximum overshoot is the difference between the maximum peak value and the steady-state value, and the derivative of the peak value is zero.

Thus, we get

$$\left.\frac{dx_o(t)}{dt}\right|_{t_p} = 0$$

$$t_p \approx 3.156\,\text{s}$$

Therefore,

$$M_p = x_o(t_p) - 1 \approx 18\%$$

EXERCISES

1. The open-loop transfer function of the unit feedback system is

$$G(s) = \frac{4}{s(s+5)},$$

Find out the unit-step response and unit-impulse response of the system.

2. The open-loop transfer function of the unit feedback system is

$$G(s) = \frac{4}{s(s+2)},$$

Write out the expression of the unit-step response and the unit-ramp response of the system.

3. The differential equations of the system are as follows:

   Find out the unit-impulse response and the unit-step response of the system.

   a. $0.2\frac{dx_o(t)}{dt} = 2x_i(t)$

   b. $0.4\frac{d^2x_o(t)}{dt^2} + 0.24\frac{dx_o(t)}{dt} + x_o(t) = x_i(t)$

4. The impulse response of each system is as follows and find out the closed-loop transfer function $\Phi(s)$of the system.

   a. $x_o(t) = 0.0125e^{-1.25t}$

   b. $x_o(t) = 5t + 10\sin(4t + 45^\circ)$

   c. $x_o(t) = 0.1(1 - e^{-t/3})$

5. The block diagram of the spacecraft attitude control system is shown in Figure 4.97, suppose that the time constant T of the controller is equal to 3 s, and the ratio of moment and inertia is $K/J = 2/9\text{rad/s}^2$. Find out the System Damping Ratio.

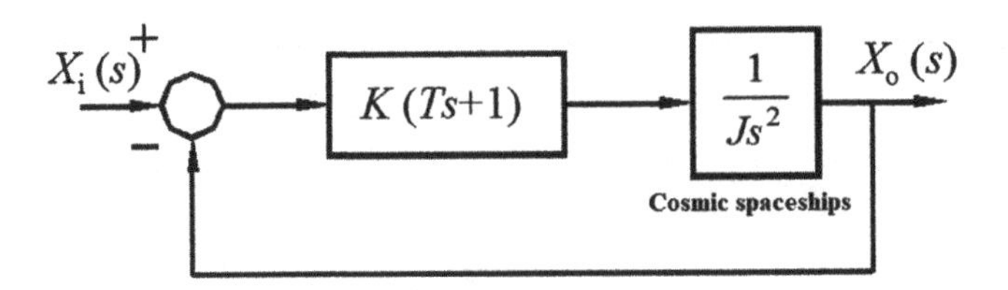

FIGURE 4.97 Block diagram of spacecraft attitude control system.

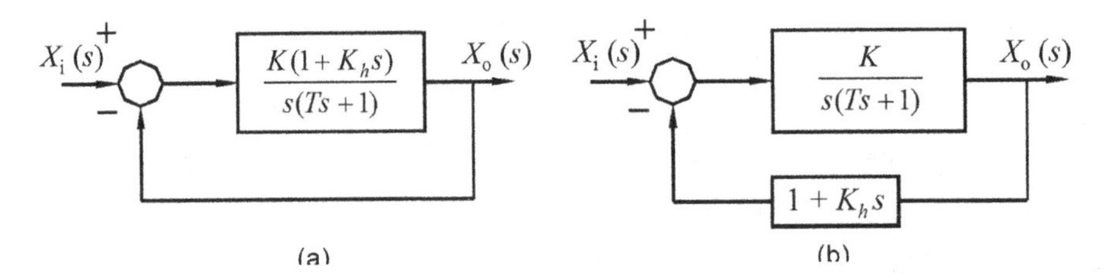

FIGURE 4.98 Block diagram of two unit-step response system.

6. The response of a servomechanism is measured in the case of unit-step input, and that is $c(t) = 1+0.2e^{-60t} - 1.2e^{-10t}$
   a. Obtain the closed-loop transfer function.
   b. Obtain the natural frequency without damping and damping ratio of the system.

7. The open-loop transfer function of the unit feedback control system is given

$$G(s)=\frac{25}{s(s+5)}$$

   a. Determining the damping coefficient ξ, natural frequency ω_n and damping natural frequency ω_d of the system.
   b. Find out the output response of the system to the unit-step function.
   c. Obtaining the performance index of the system.

8. Compare the unit-step response of the two systems that are shown in Figure 4.98a and Figure 4.98b.
9. A simplified flight control system structure diagram is shown in Figure 4.99; choose parameters that are K_1 and K_t, giving $\omega_{n=}6$, $\xi=1$.
10. The open-loop transfer function of the unit feedback system is

$$G(s)=\frac{1}{s(s+1)},$$

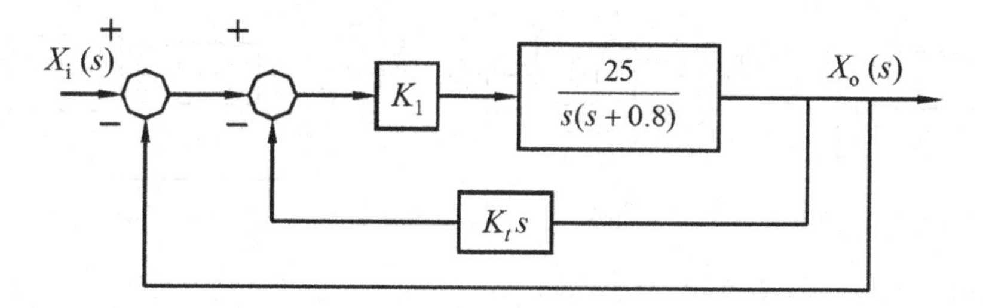

FIGURE 4.99 Block diargram of flight control system.

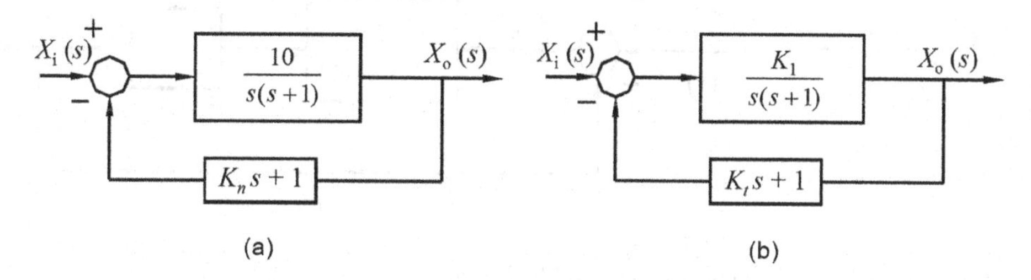

FIGURE 4.100 Two negative feedback system.

Determine the dynamic performance index of the system under the action of unit-step input.

When $G(s)=\frac{K}{s(s+1)}$, analyze the impact of amplification factor K on output dynamic process characteristic of unit-step input.

11. The transfer function of a system is

$$\frac{X_o(s)}{X_i(s)}=\frac{\omega_n^2}{s^2+2\zeta\omega_n s+\omega_n^2}$$

In order to make the system response to the step response has 5% overshoot and 2S adjustment time. Find out the numerical value of ξ *and* ω_n.

12. The open-loop transfer function of a unit feedback system is $\frac{10}{s(s+1)}$, the damping ratio of the system is 0.157, and the natural frequency without damping is 3.16 rad/s. Now the system is changed as shown in Figure 4.100(a), and the damping ratio is 0.5. Obtain the numerical value of K_n.

13. The open-loop transfer function of the unit feedback system is $G(s)=\frac{K}{s(Ts+1)}$, and $K>0$, $T>0$, so how much will the gain of the amplifier reduce that can make the maximum overshoot of the system step response from 75% down to 25%?

14. A system is shown in Figure 4.100(b). If the overshoot volume of the system is equal to 15%, peak time is equal to 0.8s. Obtain the gain K_1 and tachometer generator output slope K_t. At the same time, find out the delay time, rise time, and adjusting time of the system under K_1 and K_t values.

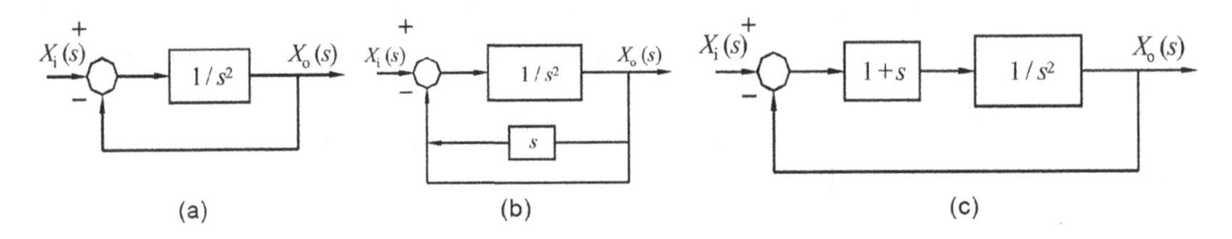

FIGURE 4.101 Three negative feedback system.

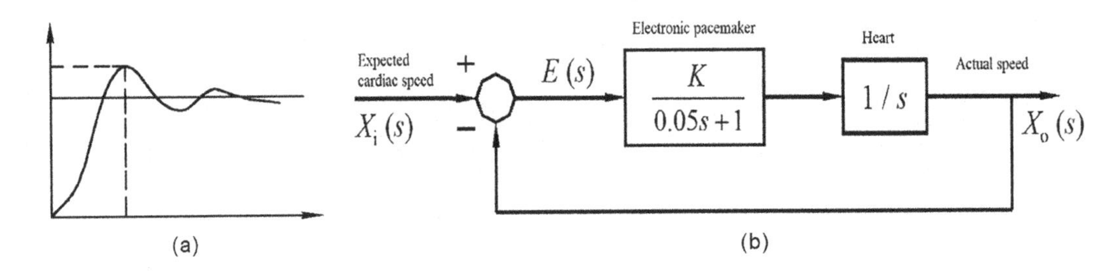

FIGURE 4.102 (a) Unit-step response and (b) block diagram.

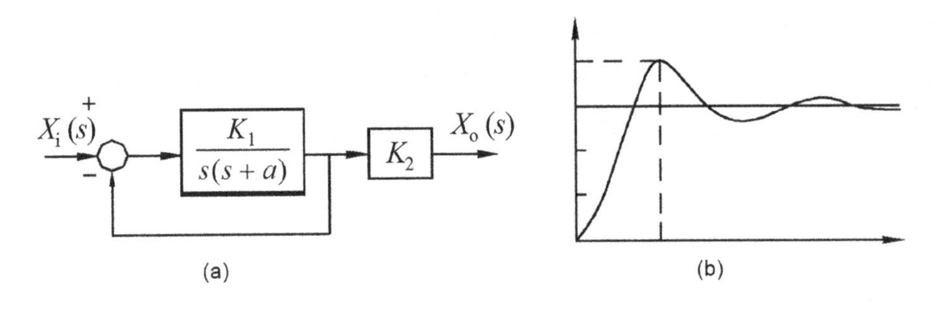

FIGURE 4.103 The control system and its unit-step response.

15. Obtain the value of ω_n, ξ and unit-step response expression of each system that is shown in Figure 4.101 and compare the dynamic performances of t_r, t_p, t_s, and σ %.

16. The unit-step response curve of the second-order system is shown in Figure 4.102a. If the system belongs to the unit feedback control system, obtain the open-loop transfer function of the system.

17. The electronic cardiac pacemaker system is shown in Figure 4.102b, and the transfer function be used to imitate the heart is equivalent to a pure integrator.

    Determine: 1) If $\zeta = 0.5$ corresponds to the best response, then how much the pacemaker gain K should be?

    2) If the expected heart rate is 60 beats per minute, and suddenly switched on the pacemaker, then how much is the actual heart rate after 1 second? And how much is the instantaneous maximum heart rate?

18. The unit-step response curve of the system shown in Figure 4.103a is shown in Figure 4.103b. Obtain the numerical value of system parameters K_1, K_2, and a.

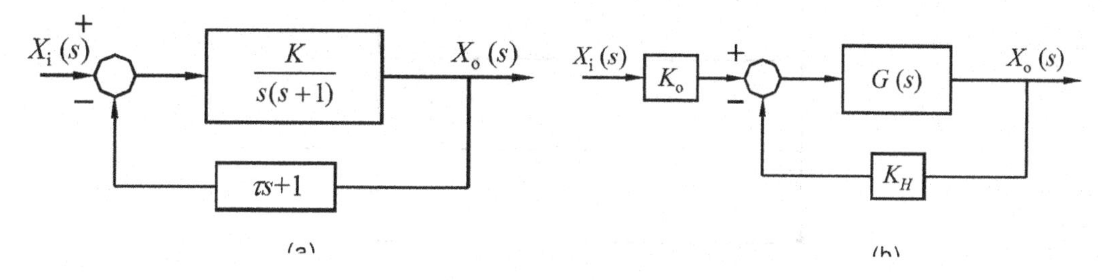

FIGURE 4.104 Negative feedback system with response.

19. The system structure diagram is shown in Figure 4.104a. If the system has performance indicators $Mp = 20\%$ and $t_{p=}1$ s. Obtain the system parameters K and τ and calculate the unit-step response eigenvalues t_r and t_s.

20. The transfer function of the first-order element is known as

$$G(s) = \frac{10}{0.2s+1}.$$

If the negative feedback method is used to make the adjust time t_s be reduced to 10% of the original (the structure diagram is shown in Figure 4.104(b)) and to ensure that the total amplification factor has not changed, try to select the numerical values of K_H and K_O.

21. A high-order system can be described by the following first-order differential equation:

$$T\frac{dx_o(t)}{dt} + x_o(t) = \tau\frac{dx_i(t)}{dt} + x_i(t),$$

And $0 < (T-\tau) < 1$. Prove the dynamic performance of the system is

$$t_d = [0.693 + In(T-t)]T$$

$$t_r = 2.2T$$

$$t_s = [3 + In(T-t)]T$$

22. The open-loop transfer function of a unit feedback system is

$$G(s) = \frac{1}{s(0.5s+1)(0.2s+1)}$$

Determine: (1) Using the Laplace Inverse transform method to obtain the unit-step response of the system. (2) Using the dominant pole method to obtain the unit-step response of the system.

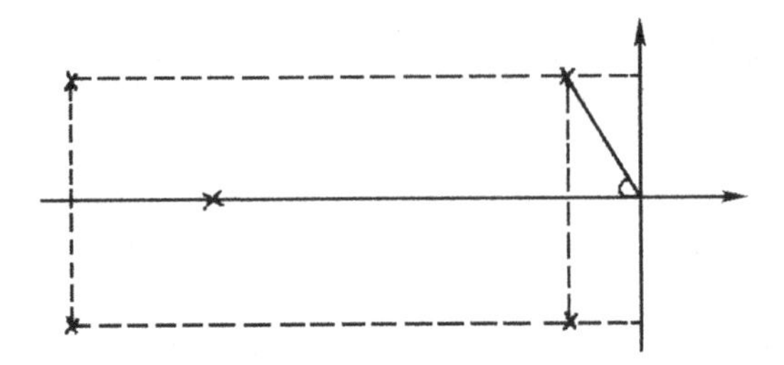

FIGURE 4.105 Closed-loop system of higher order system.

23. There is a high-order system, and the closed-loop poles are shown in Figure 4.105, and no zero point. Estimate its step response.

24. The closed-loop transfer function of the higher-order system is

$$\Phi(s) = \frac{16320(s+0.125)}{(s+0.12)(s+20)(s+50)(s+1-j4)(s+1+j4)}$$

MCQ AND TRUE/FALSE QUESTIONS

1. If the inputs to a control system are gradually changing the functions of time, then a function may be a good test input signal.

   a. ramp

   b. step

2. If sudden disturbances have occurred in the system, a function may be a good test signal.

   a. ramp

   b. step

3. For a system subjected to shock inputs, it may be best input signal to analyze the system.

   a. impulse function

   b. step function

4. The test input signal can be increased the Performance of the system, and it can be used to compare the performance of many systems.
   Answer: True/False

5. The most important characteristic of the dynamic behavior of a control system is absolute stability—that is, whether the system is stable or unstable.
   Answer: True/False

6. A control system is in equilibrium if, in the absence of any disturbance or input, the output stays in the:

   a. same state

   b. different state

7. A linear time-invariant (LTI) control system is stable if the output eventually comes back to its equilibrium state when the system is subjected to an initial condition.
   Answer: True/False

8. A linear time-invariant (LTI) control system is critically stable if oscillations of the output continue forever.
   Answer: True/False

9. A linear time-invariant (LTI) is unstable if the output diverges without bound from its equilibrium state when the system is subjected to an initial condition.
   Answer: True/False

10. What will be the unit-step response function of the above first-order system?

    a. $c(t)=1-e^{-\frac{t}{T}}$, for $t\geq 0$

    b. $c(t)=1+e^{-\frac{t}{T}}$, for $t\leq 0$

11. The response function of the first-order systems states that initially the output c(t) is zero and finally it becomes unity.
    Answer: True/False

12. One important characteristic of such an exponential response curve c(t) is that at $t=T$, the value of $c(t)$ is

    a. 0.632

    b. 0.630

13. The smaller the time constant T, the slower the system response.
    Answer: True/False

14. Another important characteristic of the exponential response curve is that the slope of the tangent line at t=0 is

    a. $1/T$

    b. T

15. The output would reach the final value at $t=T$ if it maintained its initial speed of response. However, we see that the slope of the response curve $c(t)$ decreases monotonically from

    a. $1/T$ at $t=0$ to zero at $t=\infty$

    b. 1/T at t=0 to zero at $t=-\infty$

16. The exponential response curve c(t) has gone from 0% to 63.2% of the final value.

    a. If t=0 to T

    b. If t=0 to ∞

17. The exponential response curve c(t) has gone from 0% to 86.5% of the final value.

    a. If t=2T

    b. If t=3T

18. For $t \geq 4T$, the response remains within 2% of the final value.
    Answer: True/False

19. What will be the transfer function between C(s) and T(s) in the above servo control system?

    a. $\frac{C(s)}{T(s)} = \frac{1}{s(Js+B)}$

    b. $\frac{C(s)}{T(s)} = \frac{1}{s(Js+B)+1}$

20. To determine the closed-loop transfer function of the given following system, we can get,

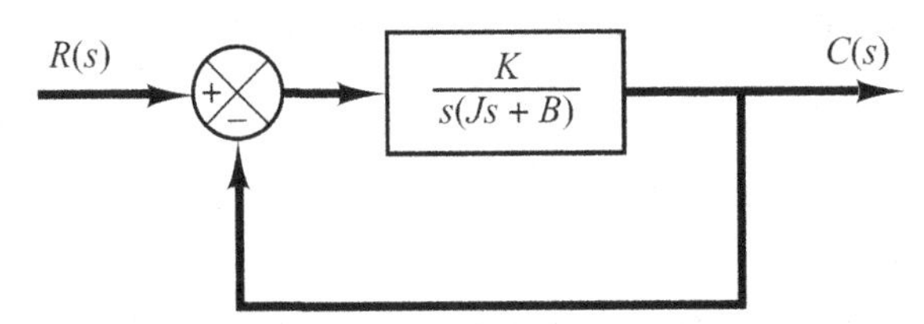

FIGURE 4.106 A second-order system.

    a. $G(s) = \frac{C(s)}{R(s)} = \frac{K}{s^2 + Bs + K}$

    b. $G(s) = \frac{C(s)}{R(s)} = \frac{K}{Js^2 + Bs + K}$

21. If the discriminant of the above system is $B_c^2 - 4JK = 0$, then we say,

    a. It is critical damping

    b. It is underdamping

22. The standard form of the second-order system closed-loop transfer function can be written as

    a. $s^2 + 2\zeta\omega_n s + \omega_n^2 = 0$

    b. $\frac{C(s)}{R(s)} = \frac{\omega_n^2}{s^2 + 2\zeta\omega_n s + \omega_n^2}$

23. The standard form of the second-order system is very important thus if $0<\zeta<1$, then the roots of the standard second-order system can be written as

$$S_{1,2} = -\zeta\omega_n \pm j\omega_n\sqrt{1-\zeta^2} = -\sigma_d \pm j\omega_d$$

Answer: True/False

24. If $\zeta=0$, then $S_{1,2} = \pm j\omega_n$; that means the two roots of the system are on the imaginary axis as shown in Figure 4.106.
Answer: True/False

25. After taking the inverse Laplace transform, the response of the undamped case can be determined as
   a. $c(t) = 1 - \cos(\omega_n t)$
   b. $c(t) = 1 - \cos(\omega_d t)$

26. If $\zeta=1$, then the two roots will be $S_1 = S_2 = -\omega_n$; the system is called critically damped; that means the two roots of the system are on the imaginary axis as shown in Figure 4.110.
Answer: True/False

27. If $\zeta=1$, then the two roots will be $S_1 = S_2 = -\omega_n$; the system is called critically damped; therefore, the system response will give as
   a. $c(t) = 1 - (1 - \omega_n t)e^{-\omega_n t}$
   b. $c(t) = 1 - (1 + \omega_n t)e^{-\omega_n t}$

28. For $0 < \zeta < 1$, the closed-loop poles are complex conjugates and lie in the left-half s-plane; underdamped, and the transient response is oscillatory; $\zeta = 0$, the transient response does not die out; $\zeta = 1$, system is called critically damped; and $\zeta > 1$, overdamped systems correspond. The above statement is
Answer: True/False

29. For the underdamped case, $0 < \zeta < 1$, from the transfer function, the damped natural frequency is
   a. $\omega_d = \omega_n\sqrt{1-2\zeta^2}$
   b. $\omega_d = \omega_n\sqrt{1-\zeta^2}$

30. If the damping ratio $\zeta=0$, the response becomes undamped and oscillations continue indefinitely. The response c(t) for the zero-damping case may be obtained by substituting $\zeta=0$ in the equation of this.

$$c(t)=1-e^{-\zeta\omega_n t}\left(\cos\omega_d t+\frac{\zeta}{\sqrt{1-\zeta^2}}\sin\omega_d t\right)$$

Then we can get

a. $c(t)=1-\cos\omega_d t, \text{for } t\geq 0$

b. $c(t)=1-\cos\omega_n t, \text{for } t\geq 0$

31. For the critically damped case, $\zeta=1$, that means if two poles or two roots are equal and the unit-step input is R(s)=1/s then from the transfer function by using the unit-step input, we can see as:

$$C(s)=\frac{1}{s}-\frac{\omega_n^2}{(s+\omega_n)^2}$$

Therefore, the critically damped response function can be written as

a. $c(t)=1-e^{-\omega_n t}(1-\omega_n t), \text{for } t\geq 0$

b. $c(t)=1-e^{-\omega_n t}(1+\omega_n t), \text{for } t\geq 0$

32. The performance characteristics of a control system are specified in terms of the transient response to a unit-step input, since it is easy to generate and is sufficiently drastic.
Answer: True/False

33. The transient response of a practical control system often exhibits damped oscillations before reaching steady state.
Answer: True/False

34. The delay time is the time required for the response to reach a peak of the final value the very first time.
Answer: True/False

35. The rise time is the time required for the response to rise from 10% to 90%, 5% to 95%, or 0% to 100% of its final value. For underdamped second-order systems, the 0%–100% rise time is normally used. For overdamped systems, the 10%–98% rise time is commonly used.
Answer: True/False

36. The peak time is the time required for the response to reach the first peak of the overshoot.
Answer: True/False

37. The maximum overshoot is the maximum peak value of the response curve measured from unity.
Answer: True/False

38. The settling time is the time required for the response curve to reach and stay within a range about the final value of size specified by an absolute percentage of the final value (usually 2% or 5%). The settling time is related to the largest time constant of the control system. Which percentage error criterion to use may be determined from the objectives of the system design in question.

    Answer: True/False

39. Except for certain applications where oscillations cannot be tolerated, it is desirable that the transient response be sufficiently fast and be sufficiently damped. Thus, for a desirable transient response of a second-order system, the damping ratio must be between 0.4 and 0.8. Small values of ζ (that is, $\zeta > 0.4$) yield excessive overshoot in the transient response, and a system with a large value of ζ (that is, $\zeta < 0.8$) responds sluggishly.

    Answer: True/False

40. We shall see later that the maximum overshoot (Mp) and the rise time (tr) conflict with each other. In other words, both the maximum overshoot (Mp) and the rise time (tr) cannot be made smaller simultaneously. If one of them is made smaller, the other necessarily becomes larger.

    Answer: True/False

41. The peak time t_p corresponds to one-half cycle of the frequency of damped oscillation.

    Answer: True/False

42. The maximum (percent) overshoot can be said as

    a. $M_p = e^{-\left(\zeta/\sqrt{1-\zeta^2}\right)\pi}$

    b. $M_p = e^{-\left(\zeta/\sqrt{1-\zeta^2}\right)\pi} \times 100\%$

43. The settling time corresponding to a $\pm$ 2% or $\pm$ 5% tolerance band may be measured in terms of the time constant $T = 1/\zeta\omega_n$ from the curves of the figure for different values of ζ.

    Answer: True/False

44. For $0 < \zeta < 0.9$, if the 2% criterion is used, the settling time t_s is approximately four times the time constant (T) of the system. If the 5% criterion is used, then the settling time t_s is approximately three times the time constant (T) of the system. Note that the settling time t_s reaches a minimum value around $\zeta = 0.76$ (for the 2% criterion) or $\zeta = 0.68$ (for the 5% criterion) and then increases almost linearly for large values of ζ.

    Answer: True/False

45. For convenience in comparing the responses of systems, we commonly define the settling time t_s to be, $t_s = 4T = \frac{4}{\sigma} = \frac{4}{\zeta\omega_n}$;(for 2% criterion), and $t_s = 3T = \frac{3}{\sigma} = \frac{3}{\zeta\omega_n}$ (for 5% criterion).

    Answer: True/False

46. The settling time is linearly proportional to the product of the damping ratio (ζ) and the undamped natural frequency (ω_n) of the system.
    Answer: True/False

47. For a rapid response, the undamped natural frequency (ω_n) must be large. To limit the maximum overshoot (Mp) and to make the settling time (t_s) small, the damping ratio (ζ) should not be too small.
    Answer: True/False

48. If the damping ratio (ζ) is between 0.4 and 0.7, then for step response c(t) the maximum percent overshoot (Mp) is between 25%, and 4% of the respective damping ratio (ζ).
    Answer: True/False

49. The derivative of the output signal can be used to improve system performance.
    Answer: True/False

50. The output of the potentiometer should be followed by a differentiating element.
    Answer: True/False

51. The output of a tachometer is proportional to the angular velocity of the motor. In any servo system, such a velocity signal can be easily generated by a tachometer.
    Answer: True/False

52. Potentiometers are commonly used to control electrical devices such as volume controls on audio equipment.
    Answer: True/False

53. The output of a potentiometer is a continuous voltage signal because, as the potentiometer brush is moving on the windings, voltages are induced in the switchover turns and thus generate transients.
    Answer: True/False

54. The effect of decreasing the damping ratio without affecting the undamped natural frequency of the system.
    Answer: True/False

55. The maximum overshoot Mp for the unit-step response can be found from the corresponding unit-impulse response. That is, the area under the unit-impulse response curve from t=1 to the time of the first zero.
    Answer: True/False

56. If there is a closed-loop zero close to a closed-loop pole, then the residue at this pole is small and the coefficient of the transient response term corresponding to this pole becomes larger.
    Answer: True/False

57. A pair of closely located poles and zeros will effectively cancel each other.
    Answer: True/False

58. If a pole is located very far from the origin, the residue at this pole may be large.
    Answer: True/False

59. The Characteristics of High-Order Response as the transients corresponding to such a remote pole are small and last a short time.
    Answer: True/False

60. The response curve of a stable higher-order system is the sum of a number of exponential curves and damped sinusoidal curves.
    Answer: True/False

61. If all closed-loop poles lie in the left-half s-plane, then the exponential terms and the damped exponential terms will approach zero as time t increases to infinity. The steady-state output is then $c(\infty)=0$.
    Answer: True/False

62. For a stable system, the closed-loop poles that are located far from the $j\omega$ axis have large negative real parts. The exponential terms that correspond to these poles decay very rapidly to zero.
    Answer: True/False

63. The horizontal distance from a closed-loop pole to the $j\omega$ axis determines the settling time (t_s) of transients due to that pole. The smaller the distance is, the shorter the settling time.
    Answer: True/False

64. The type of transient response is determined by the closed-loop poles, while the shape of the transient response is primarily determined by the closed-loop zeros.
    Answer: True/False

65. The relative dominance of closed-loop poles is determined by the ratio of the real parts of the closed-loop poles, as well as by the relative magnitudes of the residues evaluated at the closed-loop poles. The magnitudes of the residues depend on both the closed-loop poles and zeros.
    Answer: True/False

66. If the ratios of the real parts of the closed-loop poles exceed 6 and there are no zeros nearby, then the closed-loop poles nearest the $j\omega$ axis will dominate in the transient response behavior because these poles correspond to transient response terms that decay slowly.
    Answer: True/False

67. Those closed-loop zeros that have dominant effects on the transient response behavior are called dominant closed-loop poles.
    Answer: True/False

68. The problem of absolute stability can be solved readily by choosing no closed-loop poles in the right-half s-plane, including the $j\omega$ axis.
    Answer: True/False

69. The damping ratio, $\zeta > 0.4$ and settling time, $t_s < \frac{4}{\sigma}$, the relative stability and transient response performance of a closed-loop control system are directly related to the closed-loop pole-zero configuration in the s-plane.
    Answer: True/False

70. Routh's stability criterion, enables us to determine the number of closed-loop poles that lie in the right-half s-plane without having to factor the numerator polynomial.
    Answer: True/False

71. The Routh's stability criterion states that the number of roots of this equation $(a_0 s^n + a_1 s^{n-1} + \ldots + a_{n-1} s + a_n = 0)$ with positive real parts is equal to the number of changes in sign of the coefficients of the first column of the array. It should be noted that the exact values of the terms in the first column need not be known; instead, only the signs are needed.
    Answer: True/False

72. The necessary and sufficient condition that all roots of this equation $(a_0 s^n + a_1 s^{n-1} + \ldots + a_{n-1} s + a_n = 0)$ lie in the left-half s-plane is that all the coefficients of this equation $(a_0 s^n + a_1 s^{n-1} + \ldots + a_{n-1} s + a_n = 0)$ are positive and all terms in the first column of the array have positive signs.
    Answer: True/False

73. A nonzero actuating error signal at steady state means that there is an offset.
    Answer: True/False

74. Integral control action, while removing offset or SSE, may lead to an oscillatory response of slowly decreasing amplitude or even increasing amplitude, both of which are usually undesirable.
    Answer: True/False

75. The proportional control of a system without an integrator will result in a SSE with a ramp input.
    Answer: True/False

76. The proportional control action tends to stabilize the system, while the integral control action tends to eliminate or reduce SSE in response to various inputs. However, adding to a proportional controller provides high sensitivity.
    Answer: True/False

77. Although derivative control does not affect the SSE directly, it adds damping to the system and thus permits the use of a larger value of the gain K, which will result in an improvement in the steady-state accuracy.
    Answer: True/False

78. It is possible to make both the SSE (e_{ss}) for a ramp input and the maximum overshoot (M_p) for a step input small by making B small, Kp large, and Kd large enough so that ζ is between 0.4 and 0.7.
    Answer: True/False

79. A system may have no SSE (SSE) to a step input, but the same system may exhibit nonzero SSE to a ramp input. Please remember that the only way we may be able to eliminate this error is to modify the system structure.

    Answer: True/False

80. The magnitudes of the SSEs due to these individual inputs are indicative of the goodness of the system.

    Answer: True/False

81. As the type number is increased, the accuracy is improved; however, increasing the type number aggravates the stability problem. A compromise between steady-state accuracy and relative stability is always necessary.

    Answer: True/False

82. The static error constants defined the merit of control systems. The higher the constants, the larger the SSE.

    Answer: True/False

83. For a type 0 system, the static position error constant Kp is finite, while for a type 1 or higher system, Kp is infinite.

    Answer: True/False

84. The type 2 or higher system can follow a step input with zero error at steady state.

    Answer: True/False

85. If $\frac{N_1}{N_2} < 1$ or $\frac{N_2}{N_1} < 1$, the gear ratio reduces the speed as well as magnifies the torque.

    Answer: True/False

CHAPTER 5

Root-Locus Method

Control System Analysis and Design

In the *ROOT-LOCUS METHOD*, when a parameter K is $0 \rightarrow \infty$ (usually the open-loop gain is K), it moves toward the root-locus in the S-plane of the closed-loop system for the characteristic equation. It is noted that when $s=-1$ of the open-loop transfer function, it must meet the requirements of the system characteristic equation.

5.1 ROOT-LOCUS PLOTS OF NEGATIVE-FEEDBACK SYSTEMS

In 1948, W.R. Evans established a simple method of solving the root of the characteristic equation for the root-locus method. The root-locus method is the graphical method of analyzing and designing linear time-invariant (LTI) system. Especially, applying the root-locus method is easier than other methods for studying multiloop systems. Therefore, it is widely used in engineering practices such as control engineering, robotics, intelligent systems, and industrial automation system.

In this section, we will learn about root-locus method of negative-feedback systems and root-locus method: angle and magnitude conditions. Look at the closed-loop feedback system in Figure 5.1. From Figure 5.1, the closed-loop transfer function of the negative-feedback system is seen as

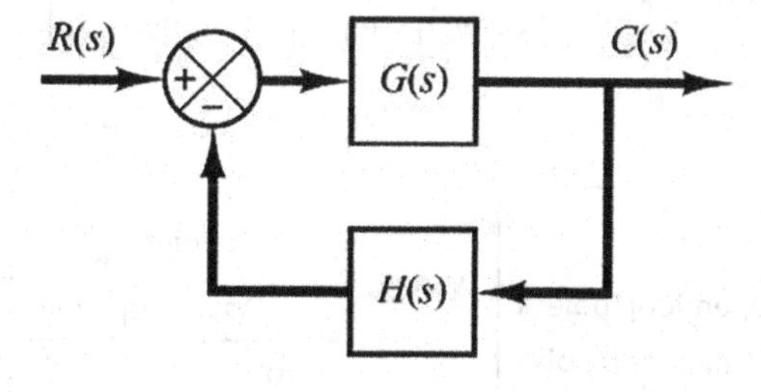

FIGURE 5.1 Block diagram of a closed-loop control system.

DOI: 10.1201/9781003293859-5

$$\frac{C(s)}{R(s)} = \frac{G(s)}{1+G(s)H(s)} \tag{5.1}$$

The characteristic equation for the closed-loop system is obtained by setting the denominator of the right-hand side to zero. That is,

$$1+G(s)H(s)=0 \tag{5.2}$$

or

$$G(s)H(s)=-1 \tag{5.3}$$

Here, we assume that $G(s)H(s)$ is a ratio of polynomials in s. It is noted that we can extend the analysis to the case when $G(s)H(s)$ involves the transport lag e^{-Ts}.

Since $G(s)H(s)$ is a complex quantity, it can be split into two equations by equating the angles and magnitudes of both sides; to obtain the following, the magnitude and angle condition are

$$|G(s)H(s)|=1 \tag{5.4}$$

and

$$\angle G(s)H(s)=\pm 180°(2k+1), \text{ for } k=0,1,2,\ldots \tag{5.5}$$

5.1.1 Locate Poles and Zeros on S-Plane

To begin sketching the root-loci of a system by the root-locus method, we must know the location of the poles and zeros of $G(s)H(s)$, as shown in Figure 5.2. Remember that the angles of the complex quantities originating from the open-loop poles and open-loop zeros to the test point s are measured in the counterclockwise direction, as shown in Figure 5.2.

To locate poles and zeros on S-plane, we can consider an example, if $G(s)H(s)$ is given by the following expression,

$$G(s)H(s)=\frac{K(s+z_1)}{(s+p_1)(s+p_2)(s+p_3)(s+p_4)}$$

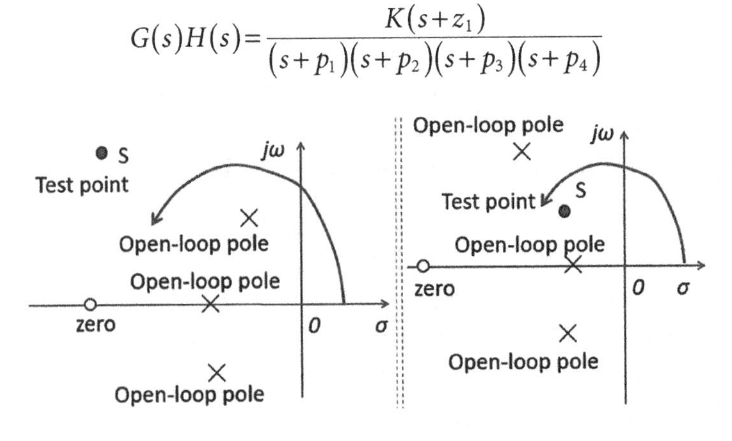

FIGURE 5.2 Location of the poles and zeros of $G(s)H(s)$.

Here, z_1 is called zero and p_1, p_2, p_3, and p_4 are called poles, whereas you can say p_2 and p_3 may be called complex-conjugate poles, as shown in Figure 5.3.

However, look at Figure 5.4 carefully. Diagrams show the angle measurements from open-loop poles and open-loop zero to test point s.

where $\varnothing_1$, θ_1, θ_2, θ_3, and θ_4 are measured counterclockwise as shown in Figure 5.4a and b. The magnitude of $G(s)H(s)$ and the angle condition are written as

$$|G(s)H(s)| = \frac{KB_1}{A_1A_2A_3A_4}$$

and

$$\angle G(s)H(s) = \varnothing_1 - \theta_1 - \theta_2 - \theta_3 - \theta_4$$

Here, A_1, A_2, A_3, A_4 and B_1 are the magnitudes of the complex quantities $s+p_1$, $s+p_2$, $s+p_3$, $s+p_4$, and $s+z_1$, respectively, as shown in Figure 5.4.

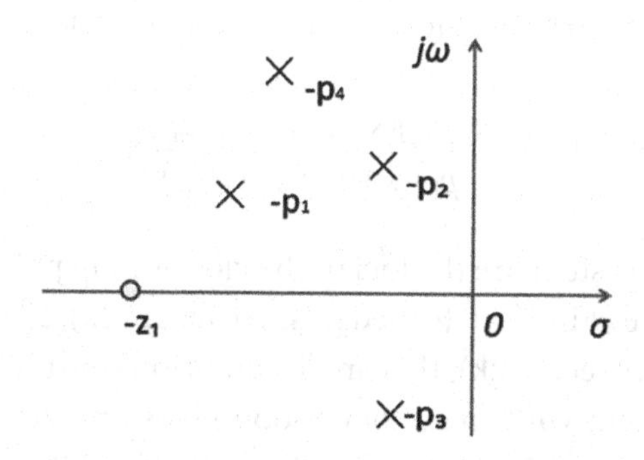

FIGURE 5.3 Locate poles and zeros with complex-conjugate poles.

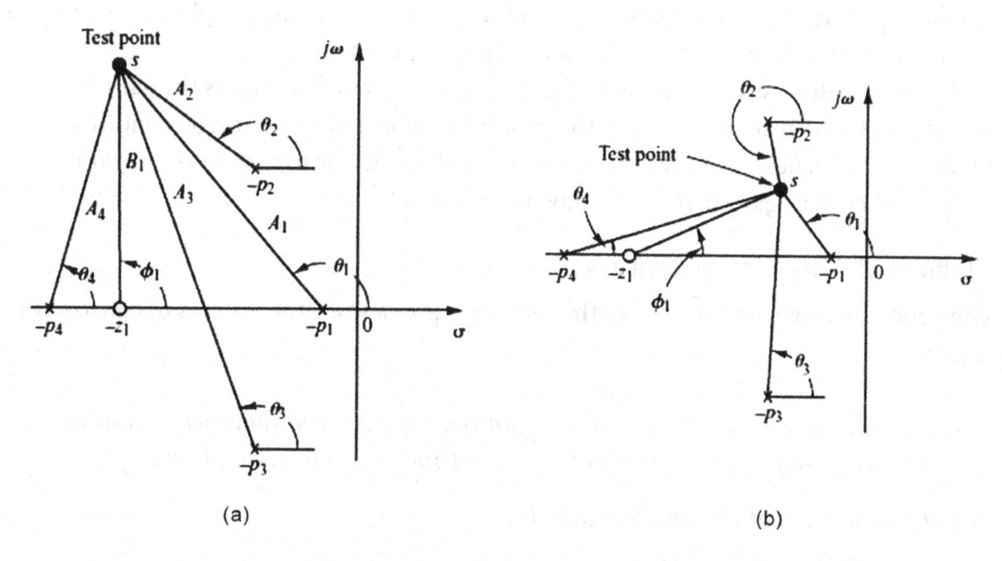

FIGURE 5.4 Locate poles and zeros with magnitudes and angles from test point s.

Note that because the open-loop complex-conjugate poles and complex-conjugate zeros, if any, are always located symmetrically about the real axis, the root-loci are always symmetrical with respect to this axis. Therefore, we only need to construct the upper half of the root-loci and draw the mirror image of the upper half in the lower-half of the S-plane.

5.2 ROOT-LOCUS METHOD

For constructing root-loci, first, we should obtain the characteristic equation as $1+G(s)H(s)=0$ and then if $H(s)=1$, we can say as $1+G(s)=0$.

5.2.1 General Rules for Constructing Root-Loci

The values of s that fulfill both the angle and magnitude conditions are the roots of the characteristic equation, or the closed-loop poles. A locus of the points in the complex plane satisfying the angle condition alone is the root-locus. The roots of the characteristic equation (the *closed-loop poles*) corresponding to a given value of the gain can be determined from the magnitude condition. The details of applying the angle and magnitude conditions to obtain the closed-loop poles are presented later. In many cases, $G(s)H(s)$ involves a gain parameter K, and the characteristic equation may be written as

$$1+\frac{K(s+z_1)(s+z_2)\ldots(s+z_m)}{(s+p_1)(s+p_2)\ldots(s+p_n)}=0 \tag{5.6}$$

Then the root-loci for the system are the loci of the closed-loop poles as the gain K is varied from zero to infinity. If the gain $K>0$ (called positive feedback), so we are interested in this, but if $K<0$ (called negative feedback), this angle condition must be modified.

For a complicated system with many open-loop poles and zeros, constructing a root-locus plot may seem complicated, but actually, it is not difficult if the rules for constructing the root-loci are applied. By locating particular points and asymptotes and by computing angles of departure from complex poles and angles of arrival at complex zeros, we can construct the general form of the root-loci without difficulty.

In the present discussions, we assume that the parameter of interest is the gain K, where $K>0$. (If $K<0$, which corresponds to the positive-feedback case, the angle condition must be modified). Note, however, that the method is still applicable to systems with parameters of interest other than gain that we already have seen before.

5.2.2 Preview Constructing Root-Loci

To construct the root-loci, we can see the following preview for the process of constructing root-loci:

1. *Locate the poles and zeros of G(s)H(s) on the S-plane. The root-locus branches start from open-loop poles and terminate at zeros (finite zeros or zeros at infinity);*
2. *Determine the root-loci on the real axis;*
3. *Determine the asymptotes of root-loci;*

4. *Find the breakaway and break-in points;*
5. *Determine the angle of departure (angle of arrival) of the root-locus from a complex pole (at a complex zero);*
6. *Find the points where the root-loci may cross the imaginary axis;*
7. *Taking a series of test points in the broad neighborhood of the origin of the S-plane, sketch the root-loci; and*
8. *Determine closed-loop poles.*

1. *Locate poles and zeros on S-plane*
   Locate the poles and zeros of $G(s)H(s)$ on the $S-$Plane. The root-locus branches start from open-loop poles and terminate at zeros (finite zeros or zeros at infinity):
   As we have seen in Figure 5.1, then in Eqs. (5.1) and (5.6), we can say that as $1+G(s)H(s)=0$, then the factored form of the open-loop transfer function locate the open-loop poles and zeros in the S-plane. Note that the open-loop zeros are the zeros of $G(s)H(s)$, while the closed-loop zeros consist of the zeros of $G(s)$ and the poles of $H(s)$. Note that the root-loci are symmetrical about the real axis of the S-plane, because the complex poles and complex zeros occur only in conjugate pairs. A root-locus plot will have just as many branches as there are roots of the characteristic equation. Since the number of open-loop poles generally exceeds that of zeros, the number of branches equals that of poles.
   If the number of closed-loop poles is the same as the number of open-loop poles, then the number of individual root-locus branches terminating at finite open-loop zeros is equal to the number m of the open-loop zeros. The remaining n-m branches terminate at infinity ($n-m$ implicit zeros at infinity) along asymptotes. If we include poles and zeros at infinity, the number of open-loop poles is equal to that of open-loop zeros. Hence, we can always state that the root-loci start at the poles of $G(s)H(s)$ and end at the zeros of $G(s)H(s)$, as K increases from zero to infinity, where the poles and zeros include both those in the finite S-plane and those at infinity.
2. *Determine root-loci on real axis*
   Root-loci on the real axis are determined by open-loop poles and zeros lying on it. The complex-conjugate poles and complex-conjugate zeros of the open-loop transfer function have no effect on the location of the root-loci on the real axis because the angle contribution of a pair of complex-conjugate poles or complex-conjugate zeros is 360° on the real axis. Each portion of the root-locus on the real axis extends over a range from a pole or zero to another pole or zero. In constructing the root-loci on the real axis, choose a test point S on it. *If the total number of real poles and real zeros to the right of this test point is odd, then this point lies on a root-locus.* If the open-loop poles and open-loop zeros are simple poles and simple zeros, then the root-locus and its complement form alternate segments along the real axis. Note that to begin sketching the root-loci of a system by the root-locus method, we must know the location of the poles and zeros of $G(s)H(s)$, as we can see in Figure 5.2.

Remember that the angles of the complex quantities originating from the open-loop poles and open-loop zeros to the test point s are measured in the counterclockwise direction, where $s=\sigma+j\omega$.

3. *Determine asymptotes of root-loci*
If the test point s is located far from the origin, then the angle of each complex quantity may be considered the same. The asymptotes of the root-loci as $s\to\infty$; if a test point s is very far from the origin. One open-loop zero and one open-loop pole then cancel the effects of the other. Therefore, the root-loci for very large values of s must be asymptotic to straight lines whose angles (slopes) are given by this relation.

$$\text{Angle of asymptotes}=\frac{\pm 180°(2k+1)}{n-m}\text{; for } k=0,1,2,... \tag{5.7}$$

where n number of finite poles of $G(s)H(s)$, m number of finite zeros of $G(s)H(s)$. Here, $k=0$ corresponds to the asymptotes with the smallest angle with the real axis. Although k assumes an infinite number of values, as k is increased the angle repeats itself, and the number of distinct asymptotes is $n-m$. All asymptotes intersect at a point on the real axis. The point can be obtained from the expanded form

$$G(s)H(s)=\frac{K\left[s^m+(z_1+z_2+...+z_m)s^{m-1}+...+z_1z_2...z_m)\right]}{s^n+(p_1+p_2+...+p_n)s^{n-1}+...+p_1p_2...p_n} \tag{5.8}$$

If a test point S is located very far from the origin, then dividing the denominator by the numerator gives

$$G(s)H(s)=\frac{K}{s^{n-m}+\left[(p_1+p_2+...+p_n)-(z_1+z_2+...+z_m)\right]s^{n-m-1}+...}$$

or

$$G(s)H(s)=\frac{K}{\left[s+\frac{(p_1+p_2+...+p_n)-(z_1+z_2+...+z_m)}{n-m}\right]^{n-m}}$$

The abscissa of the intersection of the asymptotes and the real axis is then obtained by setting the denominator of the right-hand side of the equation equal to zero and solving for s or

$$\left[s+\frac{(p_1+p_2+...+p_n)-(z_1+z_2+...+z_m)}{n-m}\right]^{n-m}=0$$

And solve it as

$$s=-\frac{(p_1+p_2+...+p_n)-(z_1+z_2+...+z_m)}{n-m} \tag{5.9}$$

If the intersection is determined, the asymptotes can be readily drawn in the complex plane; the asymptotes show the behavior of the root-loci for $|s| \gg 1$ and a root-locus branch may lie on one side of the corresponding asymptote or may cross the corresponding asymptote from one side to the other.

4. *Find breakaway and break-in points*
   Because of the conjugate symmetry of the root-loci, the breakaway points and break-in points either lie on the real axis or occur in complex-conjugate pairs.

   If a root-locus lies between two adjacent open-loop poles on the real axis, then there exists at least one breakaway point between the two poles. Similarly, if the root-locus lies between two adjacent zeros (one zero may be located at $-\infty$) on the real axis, then there always exists at least one break-in point between the two zeros.

   If the root-locus lies between an open-loop pole and a zero (finite or infinite) on the real axis, then there may exist no breakaway or break-in points or there may exist both breakaway and break-in points. To plot root-loci accurately, one must find the breakaway point, where the root-locus branches originating from the poles break away (as K is increased) from the real axis and move into the complex plane.

   The breakaway point corresponds to a point in the S-plane where multiple roots of the characteristic equation occur.

   If we say

$$f(s) = B(s) + KA(s) = 0 \tag{5.10}$$

Then,

$$K = -\frac{B(s)}{A(s)} \tag{5.11}$$

And after differentiating with respect to s, then we can say

$$\frac{dK}{ds} = -\frac{B'(s)A(s) - B(s)A'(s)}{A^2(s)} \tag{5.12}$$

Therefore, breakaway or break-in points can be determined from the roots if and only if

$$\frac{dK}{ds} = 0 \tag{5.13}$$

- If two roots $s = s_1$ and $s = -s_1$, a complex-conjugate pair;
- If it is not certain whether they are on root-loci, then check the corresponding K value;

- If $K>0$ or $K=0$ to a root $s=s_1$ of $dK/ds=0$ is positive, point $s=s_1$ is an actual breakaway or break-in point;
- If $K<0$ or a complex quantity, then point $s=s_1$ is neither a breakaway nor a break-in point.

5. *Determine the angle of departure (angle of arrival) of the root-locus from a complex*
   To sketch the root-loci with reasonable accuracy, we must find the directions of the root-loci near the complex poles and zeros. If a test point is chosen and moved in the very vicinity of a complex pole (or complex zero), the sum of the angular contributions from all other poles and zeros can be considered to remain the same.

   Therefore, the angle of departure (or angle of arrival) of the root-locus from a complex pole (or at a complex zero) can be found by subtracting from 180° the sum of all the angles of vectors from all other poles and zeros to the complex pole (or complex zero) in question, with appropriate signs included.

   Angle of departure:

   Angle of departure from a complex pole = 180° − (sum of the angles of vectors to a complex pole in question from other poles) ± (sum of the angles of vectors to a complex pole in question from zeros); For example, we get from Figure 5.4 as,

$$\theta_1 = 180° - (\theta_2 + \theta_3 + \theta_4) \pm \varnothing_1 \tag{5.14}$$

   or

$$\theta_2 = 180° - (\theta_1 + \theta_3 + \theta_4) \pm \varnothing_1 \tag{5.15}$$

   or we may say

$$\theta = 180° - (\theta_1 + \theta_2) + \varnothing_1 \tag{5.16}$$

   Angle of arrival at a complex zero = 180° − (sum of the angles of vectors to a complex zero in question from other zeros) ± (sum of the angles of vectors to a complex zero in question from poles) pole (at a complex zero). For example, from Figure 5.4, we get

$$\varnothing = 180° - \varnothing_1 \pm (\theta_1 + \theta_2) \tag{5.17}$$

6. *Cross points of root-loci on imaginary axis*
   The points where the root-loci intersect the $j\omega$ axis can be found easily by
   a. using Routh's stability criterion or
   b. letting $s=j\omega$ in the characteristic equation, equating both the real part and the imaginary part to zero, and solving for ω and K.

The values of ω thus found to give the frequencies at which root-loci cross the imaginary axis. The K value corresponding to each crossing frequency gives the gain at the K crossing point.

7. *Choose a test point in the neighborhood of $j\omega$ axis & origin*
   At first, we should determine the root-loci in the broad neighborhood of the $j\omega$ axis and the origin. The root-loci are on neither the real axis nor the asymptotes but are in the broad neighborhood of the $j\omega$ axis and the origin. The shape of the root-loci in this important region in the S-plane must be obtained with reasonable accuracy. If a test point is on the root-loci, then the angle condition, $\theta_1+\theta_2+\theta_3=180°$, must be satisfactory. If the test point does not satisfy the angle condition, select another test point until it satisfies the condition. The sum of the angles at the test point will indicate the direction in which the test point should be moved. We must continue this process and locate a sufficient number of points satisfying the angle condition (Figure 5.5).
8. *Determine closed-loop poles*
   A particular point on each root-locus branch will be a closed-loop pole if the value of K at that point satisfies the magnitude condition. Conversely, the magnitude condition enables us to determine the value of the gain K at any specific root location on the locus. (If necessary, the root-loci may be graduated in terms of K. The root-loci are continuous with K.)

   The value of K corresponding to any point s on a root-locus can be obtained using the magnitude condition, or

$$K=\frac{\text{product of lengths between points to poles}}{\text{product of lengths between points to zeros}} \tag{5.18}$$

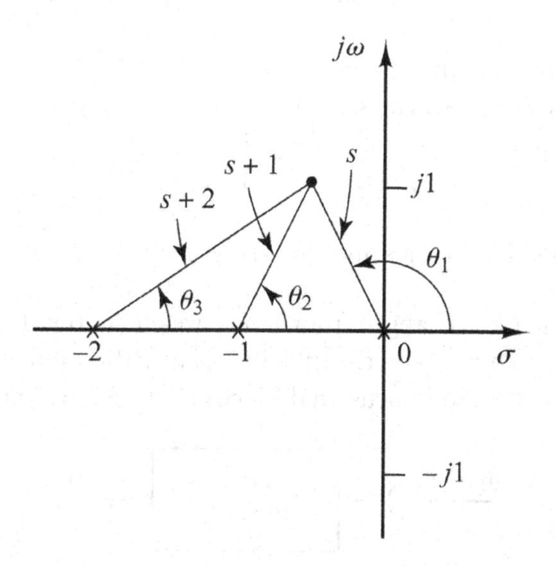

FIGURE 5.5 Choose a test point in the neighborhood of $j\omega$ axis and origin.

This value can be evaluated either graphically or analytically. MATLAB can be used for graduating the root-loci with K. If the gain K of the open-loop transfer function is given in the problem, then by applying the magnitude condition, we can find the correct locations of the closed-loop poles for a given K on each branch of the root-loci by a trial-and-error approach or by the use of MATLAB.

Example 5.1

Solution

As shown in Figure 5.6, the system open-loop transfer function is

$$G(s)=\frac{K}{s(0.5s+1)} \tag{5.19}$$

For the poles of the open-loop transfer function $p_1=0, p_2=-2$, the system has no zero, and K is the open-loop gain.

The close-loop transfer function is

$$G(s)=\frac{G(s)}{1+G(s)}=\frac{2K}{s^2+2s+2K} \tag{5.20}$$

The close-loop characteristic equation is expressed as $N(s)=s^2+2s+2K=0$

Thus, the roots of the characteristic equation are written as

$$s_1=-1+\sqrt{1-2K}\text{, and } s_2=-1-\sqrt{1-2K}$$

The root-locus is the locus of s_1 or s_2 along with the changing of K.

$$\begin{aligned}
&\text{When } K=0\text{, then } s_1=0\text{, and } s_2=-2;\\
&\text{When, } K=0.5\text{, then } s_1=-1\text{, and } s_2=-1;\\
&\text{When, } K=1\text{, then } s_1=-1+j\text{, and } s_2=-1-j;\\
&\text{When, } K=2\text{, then } s_1=-1+\sqrt{3}j\text{, and } s_2=-1-\sqrt{3}j;\\
&\quad\vdots \qquad\qquad \vdots \qquad\qquad \vdots\\
&\text{When, } K=\infty\text{, then } s_1=-1+j\infty\text{, and } s_2=-1-j\infty;
\end{aligned}$$

From Figure 5.7, when K is changed from $0\rightarrow\infty$, we obtain the characteristic root of the closed-loop system moving on the left half S-plane. This kind of movement of the root(s) is called the system root-locus. In the locus Figure 5.7, we know that "×" stands

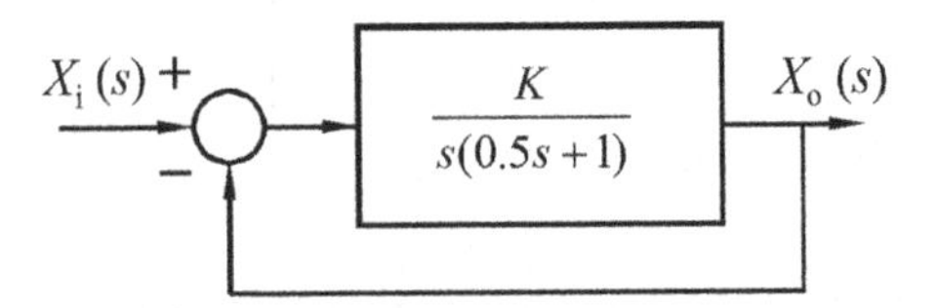

FIGURE 5.6 A control system.

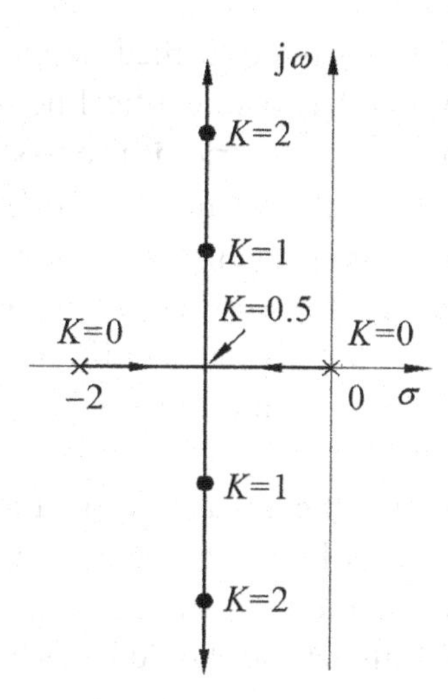

FIGURE 5.7 The root-locus of Example 5.1.

for pole(s) and "o" stands for zero(s) point. With the increasing of K, the arrow on the root-locus stands for root-locus variation. However, the numeric marked stands for closed-loop poles with the corresponding open-loop gain of K.

5.2.3 Root-Locus and the System Performance

The application of the root-locus system has the following characteristics:

Stability characteristic: When the open-loop gain $K=0\to\infty$ of the root-locus is in the left half S-plane (not in the right half S-plane) of the S-plane, the system is stable, as shown in Figure 5.7. For high-order system, root-locus may enter the right half S-plane of the S-plane, the root-locus and the imaginary axis intersection point of the K value is the critical open-loop gain.

9. *Steady-state characteristic*
   The open-loop system has one pole at the origin of coordinates, which is a type I system. If there are poles at the origin of coordinates, the system is a type 1 system. The K value of the system on the root-locus of $\lambda=1$ is the steady-state velocity error coefficient of the system. The K value of the system $\lambda=2$ on the root-locus is the steady-state acceleration error coefficient of the system. If a given system has steady-state error requirements, the allowable range of closed-loop pole position can be determined from the root-locus. In general, the parameter marked on the root-locus diagram is in the root-locus of the closed-loop gain, but it is not in the open-loop gain. The differences between the open-loop gain and the root-locus gain are only a proportionality constant. The root-locus of the parameter changes for the nonopen-loop gain which is a similar situation.

10. *Dynamic characteristics*
    (a) In the range of $0<K<0.5$ as shown in Figure 5.7, the characteristics of the closed-loop system is negative real roots; the roots for the system presents the overdamping

situation, and the step response is performed as an aperiodic process; (b) when $K = 0.5$, the closed-loop system has two identical negative real roots and it is critically damped; the step response is an aperiodic process, and the speed of the response is relatively faster than $0 < K < 0.5$; (c) when $K > 0.5$, the closed-loop characteristic roots are complex, and the system is in an underdamped situation; the step response is in an oscillating process, and the overshoot increases with the increase of gain K but the adjustment time does not change significantly. When $K = 1$, the roots of characteristic equation are $s_1 = -1 + j$ and $s_2 = -1 - j$, and the system has the damping ratio $\xi = 0.707$.

The above analysis shows that there is a closed relationship between the root-locus and the system performance. For higher-order system, it is very difficult to draw the system root-locus by analytical method, and it is not applicable in engineering. It is obvious that a simple graphical method should be used to quickly plot the root-locus of the closed-loop system according to the known open-loop transfer function.

5.2.4 Relationship between the Closed-Loop Zero-Pole and the Open-Loop Zero-Pole

Since the open-loop zero-pole is known, the relationship between open-loop zero-pole and closed-loop zero-pole is established, which is helpful for the plotting of root-locus of closed-loop system and leads to the root-locus equation.

Suppose a control system which is shown in Figure 5.1, and we know that its closed-loop transfer function is

$$\phi(s) = \frac{G(s)}{1 + G(s)H(s)} \tag{5.21}$$

The forward channel transfer function $G(s)$ and the feedback channel transfer function $H(S)$ are given by the following relationship:

$$\phi(s) = \frac{K_G(\tau_1 s + 1)(\tau_2^2 s^2 + 2\tau_2\xi_1 s + 1)\cdots}{s^\lambda (T_1 s + 1)(T_2^2 s^2 + 2T_2\xi_2 s + 1)\cdots} = K_G^* \frac{\prod_{l=1}^{M}(s - z_l)}{\prod_{i=1}^{N}(s - p_i)} \tag{5.22}$$

In Eq. (5.22), K_G is the forward channel gain and K_G^* is the feedback channel root-locus gain that satisfy the following relationship:

$$K_G^* = K_G \frac{\tau_1 \tau_2^2 \cdots}{T_1 T_2^2 \cdots} \tag{5.23}$$

Similarly,

$$H(s) = K_H^* \frac{\prod_{j=1}^{L}(s - z_j)}{\prod_{k=1}^{Q}(s - p_k)} \tag{5.24}$$

In Eq. (5.24), K_H^* is the feedback channel root-locus gain. Thus, the open-loop system transfer function is defined as

$$G(s)H(s)=K^*\frac{\prod_{l=1}^{M}(s-z_l)\prod_{j=1}^{L}(s-z_j)}{\prod_{i=1}^{N}(s-p_i)\prod_{k=1}^{Q}(s-p_k)}=K^*\frac{\prod_{i=1}^{m}(s-z_i)}{\prod_{j=1}^{n}(s-p_j)} \tag{5.25}$$

In Eq. (5.25), $K^*=K_G^*K_H^*$ is the root-locus gain of the open-loop system. However, the relationship between the gain K of the open-loop only differs from a proportional constant. Here, m is the number of open-loop zero and n is the number of open-loop poles of the system; thus, we can say $m=M+L$ and $n=N+Q$. Consequently, we let Eqs. (5.22) and (5.25) into Eq. (5.21), then the transfer function of the closed-loop system is written as

$$\phi(s)=\frac{K_G^*\prod_{l=1}^{M}(s-z_l)\prod_{k=1}^{Q}(s-p_k)}{\prod_{i=1}^{n}(s-p_i)+K^*\prod_{j=1}^{m}(s-z_j)} \tag{5.26}$$

Comparing Eqs. (5.25) and (5.26), we obtain

1. The root-locus gain of the closed-loop system is equal to the gain of the open-loop forward channel. For a unity-feedback system, the root-locus gain of the closed-loop system is the same as the open-loop system;
2. The zeros of the closed-loop system consist of the zero of the open-loop forward channel transfer function and the pole of the feedback channel transfer function; and
3. The poles of the closed-loop system are connected with the open-loop zeros, poles, and the root-locus gain K^*.

Therefore, to study the root-locus, the following contents should be mainly focused:

1. First, we have to find the open-loop zero; then we should distribute the pole and the root-locus gain, and then we must find the closed-loop poles graphically;
2. After knowing the pole(s) of the closed-loop system, we must find zero of the closed-loop systems from Eq. (5.22); however, we get the transfer function; and
3. If the transfer function of the closed-loop system is known, the time response of the closed-loop system should be solved by inverse Laplace transformation.

5.2.5 The Condition of the Root-Locus: Angle and Amplitude

The root-locus is the combination of all poles of the closed-loop transfer function; the purpose of the graphical method is to determine all the poles of the closed-loop system; the root-locus equation of the system is obtained by setting the open-loop transfer function equal to −1, i.e., we can say as we have seen in Figure 5.1 and then we recall Eq. (5.3); thus, we get

$$G(s)H(s) = -1$$

Thus, by rewriting Eq. (5.25), we get

$$K^* \frac{\prod_{j=1}^{m}(s - z_j)}{\prod_{i=1}^{n}(s - p_i)} = -1 \tag{5.27}$$

where we know that K^* is the root-locus gain of the open-loop system. K^* can be changed from 0 to ∞, z_j is the open-loop zero, and p_i is the open-loop pole. Therefore, Eq. (5.27) is called the root-locus equation.

According to Eq. (5.27), we can draw the root-locus with the continuous changing of the K^* from 0 to ∞. In general, as long as the closed-loop characteristic equation can be changed into Eq. (5.27), and then according to Eq. (5.27) to determine the root-locus gain of the closed-loop system, the other parameters of the system can also be determined. We must be determined zero and pole by using Eq. (5.27) in the S-plane; otherwise, the root-locus cannot be plotted. If more than one parameter changes in the system, to draw the root-locus plot is a set of curves.

In fact, Eq. (5.27) is a vector equation, because

$$-1 = 1e^{\pm j(2k+1)\pi} \tag{5.28}$$

where $k = 0,\ 1,\ 2,\ 3\ldots$, and thus, the root-locus equation can be expressed by the following equation:

$$\sum_{j=1}^{m} \angle(s - z_j) - \sum_{i=1}^{n} \angle(s - p_i) = \pm(2k+1)\,\pi \tag{5.29}$$

and

$$K^* = \frac{\prod_{i=1}^{n}|s - p_i|}{\prod_{j=1}^{m}|s - z_j|} \tag{5.30}$$

The points on the root-locus meet by Eqs. (5.29) and (5.30). Eq. (5.29) is called phase angle condition and Eq. (5.30) is called magnitude condition in the root-locus method. According to Eqs. (5.29) and (5.30), we can determine the root-locus of the system and the value K^* of the root-locus.

In particular, it is pointed out that the sufficient and necessary conditions satisfy Eq. (5.29) for root-locus on S-plane, that is, the phase angle condition is satisfied. Using the phase angle condition, we can draw the root-locus of the system, and K^* has nothing to do with the value of phase angle conditions, whereas the magnitude is used for only the special requirements of K^* value.

We suppose that we have an open-loop system, and the transfer function is written as

$$G(s)=\frac{K^*(s-z_1)}{s(s-p_2)(s-p_3)} \tag{5.31}$$

Thus, in the system, z_1 is the zero and 0, p_2, p_3 are the poles, as shown in Figure 5.8. We assume that there is a test point s_1 on the S-plane and we can draw the vector from open-loop zero(s) and the pole(s) to s_1. If s_1 is the test point on the root-locus, the s_1 must meet the following requirement:

$$\angle(s-z_1)-\angle s-\angle(s-p_2)-\angle(s-p_3)=\pm(2k+1)\pi$$

Thus,

$$\phi_1-(\theta_1+\theta_2+\theta_3)=\pm(2k+1)\pi \tag{5.32}$$

where $k=0,\ 1,\ 2,\ \cdots$, whereas ϕ_1 is the vector angle between line z_1s_1 and the real axis, θ_i is the vector angle from the open-loop zero and the poles to the real axis, as shown in Figure 5.8. If s_1 is the point on the root-locus, we can get the root-locus gain according to the magnitude condition

$$K^*=\frac{|s_1-0|\bullet|s_1-p_2|\bullet|s_1-p_3|}{|s_1-z_1|} \tag{5.33}$$

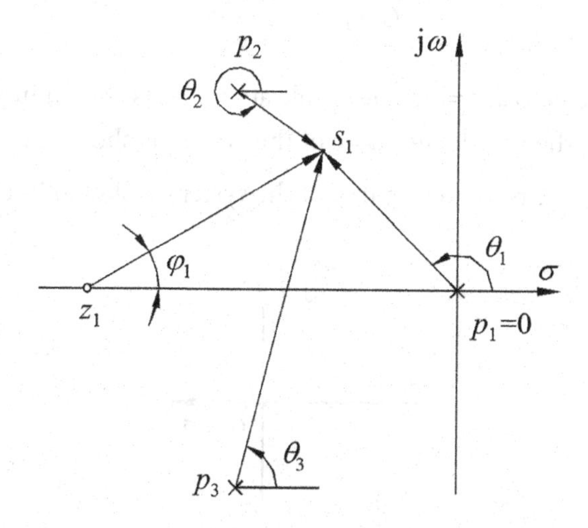

FIGURE 5.8 Distribution of zero-poles for an open-loop system.

According to the open-loop transfer function of the zero(s) and the pole(s), we can draw the root-locus of these following steps as

1. Find all the point s_i to meet the conditions of phase angle in the S-plane; connect all the points with a smooth curve, that is the root-locus gain K^* selections from $0 \rightarrow \infty$, then we obtain the root-locus curve of the closed-loop system.
2. With the given open-loop transfer function K^*, find the root-locus to satisfy the modulus or magnitude condition for all closed-loop poles at s_i, where $i = 0, 1, 2\ldots$.

5.3 ROOT-LOCUS CURVE OF A SIMPLE SYSTEM

1. *Integral factor*
   The transfer function of the integral factor is expressed as

$$G(s) = \frac{K}{s}$$

   The system has a pole $s = 0$ at the origin, as shown in Figure 5.9.

2. *Proportional-inertial factor*
   The transfer function of the proportional-inertial factor is written as

$$G(s) = \frac{K^*}{s+p}$$

   The system has a pole at $s = -p$, as shown in Figure 5.10.

3. *Differential inertial factor*
   The transfer function of the differential inertia factor is written as

$$G(s) = \frac{K^*(s+z)}{s+p}$$

   The system has a zero at $s = -z$ and a pole at $s = -p$, as shown in Figure 5.11.
   When $|p| < |z|$, the root-locus curve of the system is shown in Figure 5.11a.
   When $|p| > |z|$, the root-locus curve of the system is shown in Figure 5.11b.

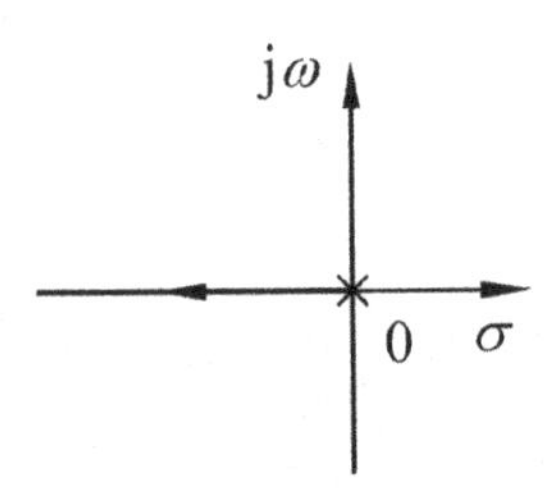

FIGURE 5.9 Integral factor of the root-locus.

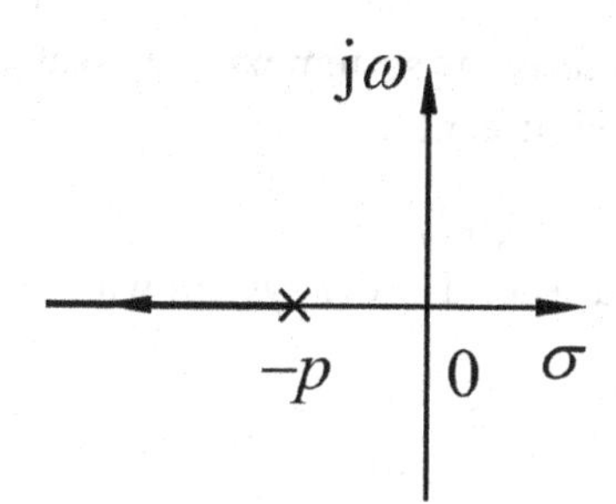

FIGURE 5.10 Root-locus plot of proportional-inertial factor.

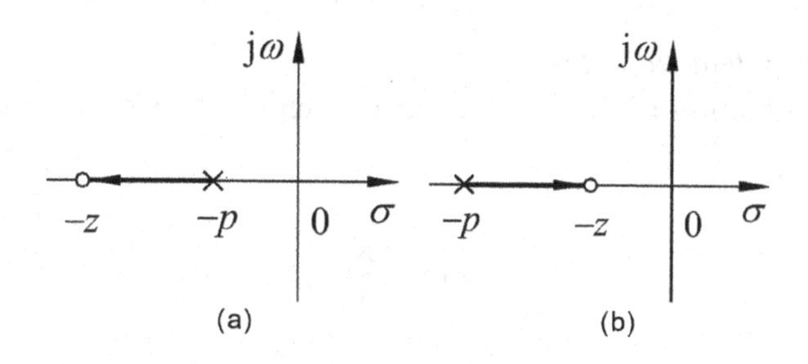

FIGURE 5.11 Root-locus plot of differential inertial factor.

4. *First-order system*

   The transfer function of the first-order system is written as,

$$G(s)=\frac{K^*(s+z)}{s(s+p)};(|z|<|p|)$$

   The system has a zero at $s=-z$ and two poles at $s=0$ and $s=-p$, as shown in Figure 5.12.

5. *Second-order system with differential factor*

   The transfer function of the second-order system with differential factors is written as

$$G(s)=\frac{K^*(s+z)}{(s+p_1)(s+p_2)}$$

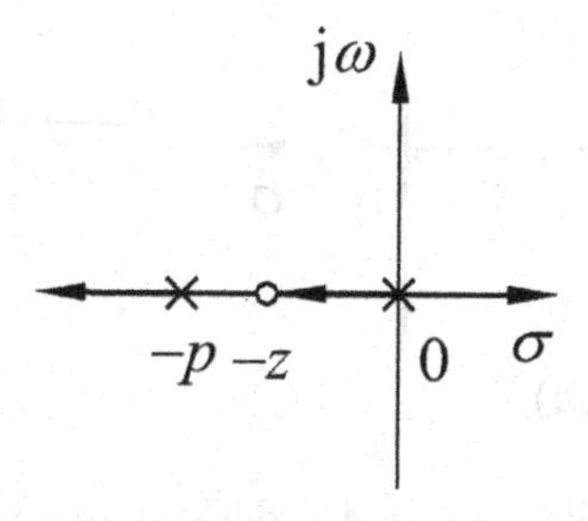

FIGURE 5.12 Root-locus system of the first-order system.

where $|z|<|p_1|<|p_2|$. Therefore, the system has a zero at $s=-z$, and two poles at $s=-p_1$ and $s=-p_2$, as shown in Figure 5.13a.

6. *Second-order system with integral factor*
   The transfer function of a second-order system with an integral factor is written as

$$G(s)=\frac{K^*}{s^2}$$

   The system has no zero, but it has two poles at $s=0$, as shown in Figure 5.13a.

7. *Second-order system with ($\xi=0$)*
   The transfer function of the second-order system with zero damping ratio ($\xi=0$) is written as

$$G(s)=\frac{K^*}{s^2+\omega^2}$$

   The system has no zero but it has two poles at $s=-j\omega$ and $s=j\omega$, as shown in Figure 5.14a.

8. *Second-order system with underdamped case ($0<\xi<1$)*
   The transfer function of the second-order system with underdamped case ($\xi=1$) is expressed as

$$G(s)=\frac{K^*}{(s+\sigma)^2+\omega^2}$$

   The system has no zero, but it has two poles at $s=-\sigma-j\omega$ and $s=-\sigma+j\omega$, as shown in Figure 5.14b.

9. *Second-order system with critically damped case ($\xi=1$)*
   The transfer function of the second-order system is defined as

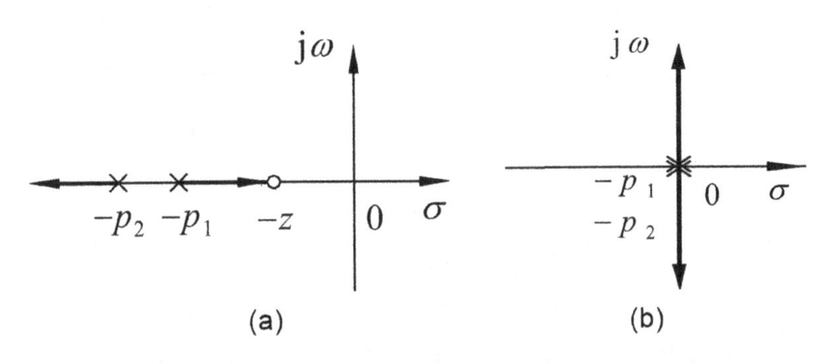

FIGURE 5.13 Root-locus system of the second-order system with differential factor, and (b) Root-locus system of the second-order system with integral factor.

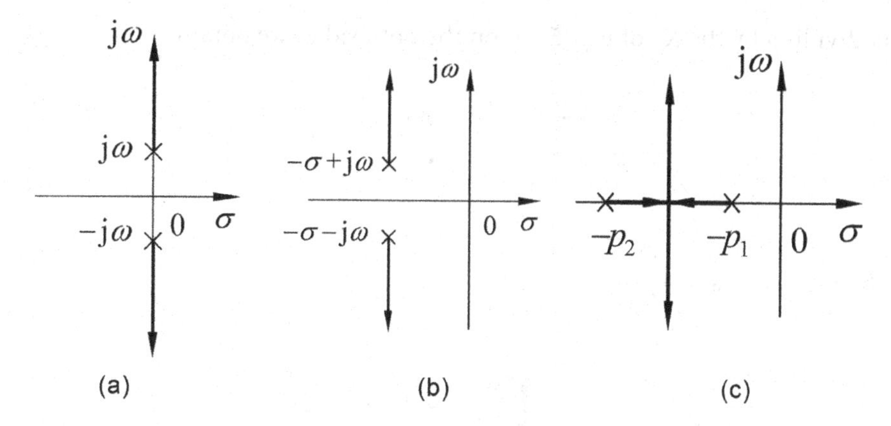

FIGURE 5.14 Root-locus system of the second-order system with (a) zero damping ($\xi = 0$) case, (b) underdamped case ($0 < \xi < 1$), and (c) critically damped ($\xi = 1$).

$$G(s) = \frac{K^*}{(s+p_1)(s+p_2)}$$

Therefore, the system has no zero but it has two poles at $s = -p_1$ and $s = -p_2$, as shown in Figure 5.14c.

5.4 BASIC RULES FOR PLOTTING OF THE ROOT-LOCUS DIAGRAM

In order to draw the root-locus diagram of the system, the rules of root-locus must be clearly understood. There are 12 rules for drawing root-locus plotting, which are summarized as follows:

1. *The starting and ending points of the root-locus*
   The root-locus begins at the system open-loop pole and ends at the system open-loop zero or infinite zero. If the number of opening zeros m is less than the number of opening poles n, the root-locus of $(n-m)$ tends to infinity.
   The starting point of the root-locus is the root-locus gain $K^* = 0$, and the ending point is $K^* \to \infty$. From Eq. (5.26), the closed-loop transfer function can be obtained from the closed-loop system characteristic by setting $K^* = 0$ as

$$\prod_{i=1}^{n}(s-p_i) = 0 \tag{5.34}$$

   where $s = p_i$;$(i = 1, \ 2, \ 3\cdots, \ n)$. Thus, when $K^* = 0$, the root of the closed-loop system characteristic equation is the pole of the open-loop characteristic equation. Thus, the root-locus must come from open-loop pole. From Eq. (5.26), we can obtain the closed-loop system characteristic equation as

$$\prod_{i=1}^{n}(s-p_i) + K^* \prod_{j=1}^{m}(s-z_j) = 0 \tag{5.35}$$

Let dividing by the K^* of Eq. (5.35) on the both sides, we obtain

$$\frac{1}{K^*}\prod_{i=1}^{n}(s-p_i)+\prod_{j=1}^{m}(s-z_j)=0 \tag{5.36}$$

When $K^* \to \infty$, and $\frac{1}{K^*} \to 0$, we get

$$\prod_{j=1}^{m}(s-z_j)=0 \tag{5.37}$$

Thus, we can say

$$s = z_j$$

where $j=1, 2\cdots, m$; thus, the root-locus must end in the open-loop zero. In the actual control system, if m is the degree of numerator polynomial of the open-loop transfer function and n is the degree of the denominator polynomial of the open-loop transfer function, the numerator polynomial and the denominator polynomial must satisfy the condition $n \geq m$. Thus, there is $(n-m)$ root-locus ending in infinity. When $s \to \infty$, then from Eq. (5.36) we obtain

$$\frac{1}{K^*}=\lim_{s\to\infty}\frac{\prod_{j=1}^{m}|s-z_j|}{\prod_{i=1}^{n}|s-p_i|}=\lim_{s\to\infty}\frac{1}{|s|^{n-m}}=0 \tag{5.38}$$

If the zero of a finite number is called the finite zero, and the zero of an infinite distance or number is called the infinite zero, then the root-locus must finally be open-loop zero. Under the definition that an infinite distance or number is regarded as an infinite zero, the open-loop transfer function has the same number of zeros and poles.

2. *The number of branches of the root-locus*

   The number of branches of the root-locus (called root-trajectory) in the S-plane is equal to the order n of the closed-loop characteristic equation, that is, the number of branches is the same as the number of closed-loop poles. Because the characteristic equation of order n has n characteristic roots, when $K^* \to \infty$, the roots n of the characteristic equation changes with K^* by changing the root-locus curve for n root-locus. While the closed-loop characteristic equation is the function of the root-locus gain of K^*, whereas when K^* changes from 0 to ∞, some parameters of the characteristic equation are also changed. Thus, the changes of the characteristic equation are necessarily continuous. Therefore, the root-locus has continuity.

3. *The symmetry of root-locus*

The open-loop poles, zeros, or closed-loop poles are real or a pair of conjugate complex number. They are distributed in the S-plane symmetrically to the real axis; thus, the root-locus are symmetrical to the real axis.

Since the closed-loop transfer function is a rational fraction function, the roots of the closed-loop characteristic equation are only real and complex numbers. The real roots are located on the real axis, whereas the complex roots must be conjugated in pairs, and the root-locus is the set of roots; therefore, the root-locus must be symmetric to the real axis.

According to the symmetry of the root-locus, only the root-locus of the upper half S-plane need to be obtained, and then the root-locus of the lower half S-plane can be drawn by using the symmetric relation.

4. *The root-locus on the real axis*

To the right side of the root-locus segment on the real axis, the summation of the number of the open-loop zero-poles is odd. That's, on the region of the real axis, if the summation of the open-loop zero-poles is odd, the region must be in the root-locus.

The transfer function of the open-loop zero-poles is shown in Figure 5.15. We assume that s_1 is a test point on the real axis, $\phi_i(i=1,\ 2,\ 3)$ is the phase angle from every open-loop zero to point s_1. However, $\theta_j(j=1,\ 2,\ 3,\ 4)$ is the phase angel from every open-loop pole to point s_1. From Figure 5.15, we obtain that the number of phase angles is odd for the complex-conjugate point to any point on the real axis and that is 2π.

It is same that the open-loop system has the complex-conjugate zero. Thus, when we make sure the root-locus of the real axis, we can't think about the influence of complex-conjugate zero-pole.

From Figure 5.15, we obtain that the phase angle from the s_1 point to the left side open-loop real zero and the vector phase angle from pole to s_1 is 0, while the phase angle of s_1 from right open-loop real zero-pole to s_1 is π.

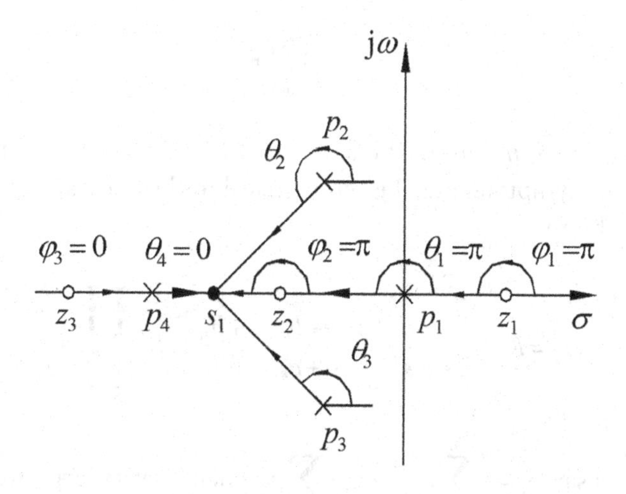

FIGURE 5.15 Root-locus on the real axis.

Let ϕ_j stand for the summation of the right segment vector phase angle of the open-loop real zeros and poles to the point s_1 and θ_i stand for the summation of the left segment vector phase angle of the open-loop real zeros and poles to the point s_1. According to the necessary and sufficient condition of the root-locus, we obtain

$$\sum \phi_j - \sum \theta_i = (2k+1)\pi \tag{5.39}$$

where $k = 0,1,2,..,n$, and $(2k+1)\pi$ is the odd number. Because every phase angle is π, π and $-\pi$ stand for the same vector angle. Thus, it's the same as poles π and $-\pi$. Therefore, the s_1 is on the root-locus; thus, we can rewrite Eq. (5.39) as

$$\sum \phi_j + \sum \theta_i = (2k+1)\pi \tag{5.40}$$

It is obvious that only when the summation of the right segment vector phase angle of the open-loop real zeros and poles to the point s_1 is odd, we can see the summation of the zeros and poles as we can see on the right side of Eq. (5.40). In Figure 5.15, the real axis is the part of the root-locus between z_1 and p_1, z_2 and p_4, z_3 and $-\infty$.

5. *The asymptote of the root-locus*
   When the finite number (n) of poles of the open-loop transfer function is greater than the finite number of zeros and the open-loop gain is greater than the finite number (m) of zeros as changes $K^* \to \infty$; then, the finite number $n-m$ root-locus branches intersect with the real axis at σ_α; and the angle φ_α is a set of asymptotes tending to ∞. Thus, we may write as

$$\sigma_\alpha = \frac{\sum_{i=1}^{n} p_i - \sum_{j=1}^{m} z_j}{n-m} \tag{5.41}$$

or

$$\varphi_\alpha = \frac{2k+1}{n-m}\pi \tag{5.42}$$

where $k = 0, 1, 2, \cdots, n-m-1$. The asymptote is the root-locus curve when s tends to ∞. Thus, the asymptote must be symmetrical to the real axis. From the open-loop function, we obtain

$$G(s)H(s) = K^* \frac{s^m + b_1 s^{m-1} + \cdots + b_{m-1}s + b_m}{s^n + a_1 s^{n-1} + \cdots + a_{n-1}s + b_n} = K^* \frac{\prod_{j=1}^{m}(s - z_j)}{\prod_{i=1}^{n}(s - p_i)} \tag{5.43}$$

In Eq. (5.43), let say $b_1 = -\sum_{j=1}^{m} z_j$, $a_1 = -\sum_{i=1}^{n} p_i$, when s is very large, we get the following from Eq. (5.43)

$$G(s)H(s)=\frac{K^*}{s^{n-m}+(a_1-b_1)s^{n-m-1}} \tag{5.44}$$

Let, $G(s)H(s)=-1$, and put it into Eq. (5.44); then we obtain the asymptote equation as in the following:

$$s^{n-m}\left(1+\frac{a_1-b_1}{s}\right)=-K^* \tag{5.45}$$

or

$$s\left(1+\frac{a_1-b_1}{s}\right)^{\frac{1}{n-m}}=(-K^*)^{\frac{1}{n-m}} \tag{5.46}$$

According to the binomial theorem, we can rewrite Eq. (5.46) as

$$\left(1+\frac{a_1-b_1}{s}\right)^{\frac{1}{n-m}}=1+\frac{a_1-b_1}{(n-m)s}+\frac{1}{2!}\frac{1}{n-m}\left(\frac{1}{n-m}-1\right)\left(\frac{a_1-b_1}{s}\right)^2+\cdots \tag{5.47}$$

When s is very large, then Eq. (5.47) can be simplified as

$$\left(1+\frac{a_1-b_1}{s}\right)^{\frac{1}{n-m}}=1+\frac{a_1-b_1}{(n-m)s} \tag{5.48}$$

If we substitute Eq. (5.48) into Eq. (5.46), then we get

$$s\left(1+\frac{1}{n-m}\frac{a_1-b_1}{s}\right)=(-K^*)^{\frac{1}{n-m}} \tag{5.49}$$

Substituting $s=\sigma+j\omega$ into Eq. (5.49), then according to De Moivre formula, we obtain

$$\left(\sigma+\frac{a_1-b_1}{n-m}\right)+j\omega=(K^*)^{\frac{1}{n-m}}\left(\cos\frac{2k+1}{n-m}\pi+j\sin\frac{2k+1}{n-m}\pi\right) \tag{5.50}$$

where $k=0, 1, 2, \cdots, n-m-1$, and we separate the real and imaginary parts from Eq. (5.50) and then we obtain

$$\sigma+\frac{a_1-b_1}{n-m}=(K^*)^{\frac{1}{n-m}}\left(\cos\frac{2k+1}{n-m}\pi\right) \tag{5.51}$$

and

$$\omega = (K^*)^{\frac{1}{n-m}}\left(\sin\frac{2k+1}{n-m}\pi\right) \tag{5.52}$$

Solving Eqs. (5.51) and (5.52), we get

$$(K^*)^{\frac{1}{n-m}} = \frac{\sigma - \sigma_\alpha}{\cos\varphi_\alpha} \tag{5.53}$$

and

$$\omega = (\sigma - \sigma_\alpha)\tan\varphi_a \tag{5.54}$$

Now, we may rewrite Eq. (5.53) as in the following:

$$\varphi_\alpha = \frac{2k+1}{n-m}\pi \tag{5.55}$$

or

$$\sigma_\alpha = \frac{b_1 - a_1}{n-m} = \frac{\sum_{i=1}^{n} p_i - \sum_{j=1}^{m} z_j}{n-m} \tag{5.56}$$

where $k=0,\ 1,\ 2,\ \cdots,\ n-m-1$. In the S-plane, Eq. (5.54) represents the line equation. The point of intersection is included the angle with the real axis. When k takes different values, we obtain $(n-m)\varphi_\alpha$ but σ_α cannot be changed. Thus, the number of n–m asymptotes of the root-locus are intersecting with the real axis at σ_α, including the angle with the real axis at φ_α.

Example 5.2

The open-loop transfer function of a feedback system is given by $G(s) = \dfrac{K^*}{s(s+1)(s+2)}$. Find the asymptotes of the root-locus.

Solution

The open-loop transfer function of the feedback system has three poles—0, −1, and −2, but it has no zero. Thus, there are $n-m=3-0=3$ root-locus curves tending to infinity. The intersection point (σ_α) in the coordinate by the asymptote(s) and the real axis is found at

$$\sigma_\alpha = \frac{p_1 + p_2 + p_3}{n-m} = \frac{0+(-1)+(-2)}{3-0} = -1$$

The angle between the asymptote and the real axis is written as

$$\varphi_{\alpha} = \frac{2k+1}{3}\pi$$

When, $k=0$, $\varphi_{\alpha} = \frac{1}{3}\pi$; when, $k=1$, $\varphi_{\alpha} = \pi$; and when, $k=2$, $\varphi_{\alpha} = \frac{5}{3}\pi$. Three asymptotes are shown in Figure 5.16a. Figure 5.16b–f shows the common transfer functions and their asymptotes.

6. *The starting (angle of arrival) and ending angles (angle of departure) of the root-locus*

   1. *The starting angle (angle of arrival)*: The starting angle is the angle between the tangent line of the open-loop root-locus and the positive real axis at the complex pole of the open-loop system, also called the angle of arrival or rising angle which is denoted by θ_{p_i}. The starting angle can be expressed as in the following:

$$\theta_{p_i} = \pi + \sum_{j=1}^{m} \phi_{z_j p_i} - \sum_{\substack{j=1 \\ j \neq i}}^{n} \theta_{p_j p_i} \tag{5.57}$$

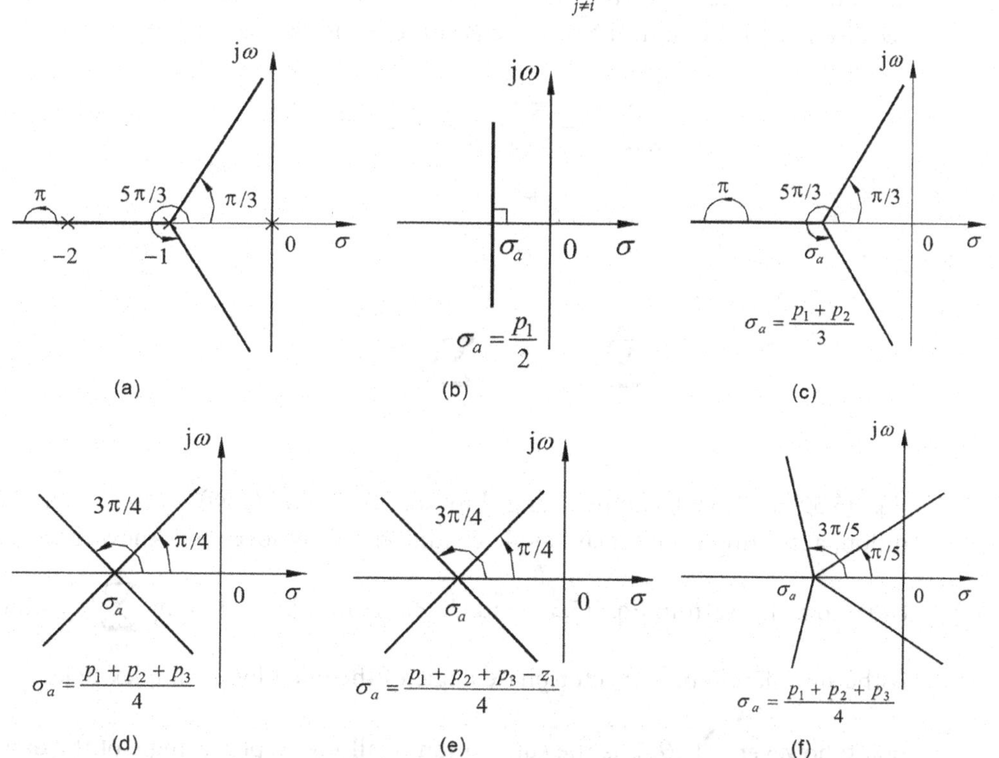

FIGURE 5.16 (a) Asymptotes and the angle, (b) asymptotes of $\frac{K^*}{s(s-p_1)}$, (c) asymptotes of $\frac{K^*}{s(s-p_1)(s-p_2)}$, (d) asymptotes of $\frac{K^*}{s(s-p_1)(s-p_2)(s-p_3)}$, (e) asymptotes of $\frac{K^*(s-z_1)}{s^2(s-p_1)(s-p_2)(s-p_3)}$, and (f) asymptotes of $\frac{K^*}{s^2(s-p_1)(s-p_2)(s-p_3)}$.

2. *The ending angle (angle of departure)*: The ending angle is the angle between the tangent and the open-loop root-locus curve with the real axis; it is also called incident angle or angle of departure which is denoted by ϕ_{z_i}. The ending angle can be expressed as follows:

$$\phi_{z_i} = \pi + \sum_{j=1}^{n} \theta_{p_j z_i} - \sum_{\substack{j=1 \\ j \neq i}}^{m} \phi_{z_j z_i} \tag{5.58}$$

In an open-loop system there are n finite poles and m finite zeros; at any point s_0 on the root-locus of the complex poles (or zeros) that are very close to the starting (or ending) angle. As s_0 is infinite which is closed to the complex pole of the starting angle p_i (Or the complex zero of the ending angle z_i), in addition to p_i (or z_i), all the vector phase angle of the open-loop zeros-poles to s_0 point can be replaced by the phase angle to $\phi_{z_j s_0}$ and $\theta_{p_j s_0}$. However, we can use as p_i (or z_i) such that we can say as $\phi_{z_j p_i}$ (or $\phi_{z_j z_i}$) and $\theta_{p_j p_i}$ (or $\theta_{p_j z_i}$). The vector phase angle of p_i (or z_i) is the starting angle θ_{p_i} (or the ending angle ϕ_{z_i}) must meet phase angle condition because s_0 is the point in the root-locus curve. Thus, we may write

$$\sum_{j=1}^{m} \phi_{z_j p_i} - \sum_{\substack{j=1 \\ j \neq i}}^{n} \theta_{p_j p_i} - \theta_{p_i} = -\pi \tag{5.59}$$

And,

$$\sum_{\substack{j=1 \\ j \neq i}}^{m} \phi_{z_j z_i} + \phi_{z_i} - \sum_{j=1}^{n} \theta_{p_j z_i} = \pi \tag{5.60}$$

Eqs. (5.59) and (5.60) can be changed to Eqs. (5.57) and (5.58). In the above condition, according to the phase angle condition of root-locus, we know that $-\pi$ is equal to π. Thus, from Eqs. (5.57) and (5.58), we may be written by $\sum_{j=1}^{m} \phi_{z_j p_i}$ which is the summation of all vector phase angle of the open-loop complex poles to p_i point; however, $\sum_{\substack{j=1 \\ j \neq i}}^{n} \theta_{p_j p_i}$ is the summation of all vector phase angle of the open-loop complex poles to p_i point, and $\sum_{\substack{j=1 \\ j \neq i}}^{m} \phi_{z_j z_i}$ is the summation of all vector phase angle of the open-loop complex poles to z_i. The root-locus curve starting angle is shown in Figure 5.17a, and the ending angle is shown in Figure 5.17b.

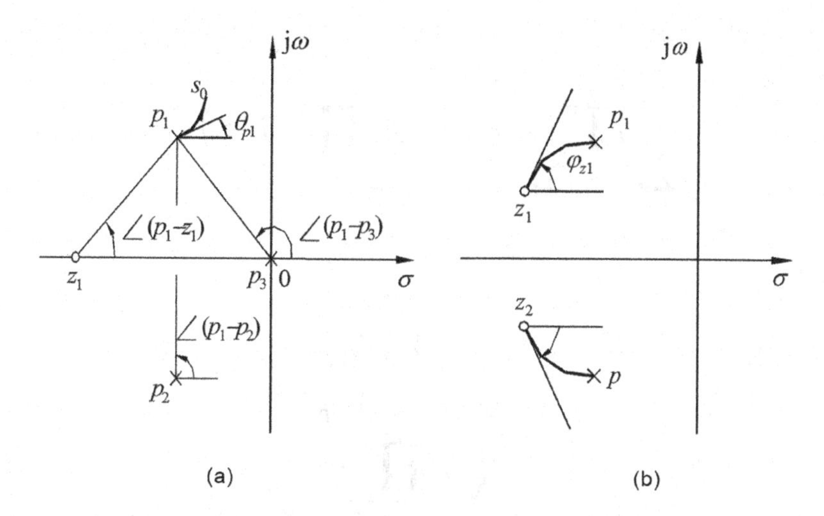

FIGURE 5.17 (a) The starting angle of the root-locus, and (b) The ending angle of the root-locus.

7. *Breakaway point in the coordinates and angle of departure*
   To understand the breakaway point (separation of a point) in the root-locus curve, when there are two or more branches of the root-locus curve, which meet in a common point, then we should separate it immediately in S-plane; it is called separation of point in the coordinate system. The separation of point d in coordinates can be solved in the following equation:

$$\sum_{j=1}^{m}\frac{1}{d-z_j}=\sum_{i=1}^{n}\frac{1}{d-p_i} \tag{5.61}$$

Because $p_i\left(i=1,2,\ldots,n\right)$ and $z_i\left(i=1,2,\ldots,m\right)$ are the open-loop poles and the open-loop zeros of the system, respectively; thus, the open-loop transfer function of the system is defined as

$$G(s)H(s)=K^*\frac{\prod_{i=1}^{m}(s-z_i)}{\sum_{i=1}^{n}(s-p_i)}$$

The system closed-loop characteristic equation is written as

$$D(s)=\prod_{i=1}^{n}(s-p_i)+K^*\prod_{i=1}^{m}(s-z_i)=0$$

We assume the separation point of the curve is d. So, d is the multiple roots of the closed-loop characteristic equation. We obtain

$$\left.\frac{dD(s)}{ds}\right|_{s=d}=0$$

Thus,

$$\sum_{i=1}^{n}\frac{\prod_{j=1}^{n}(d-p_j)}{d-p_i}+K^*\sum_{i=1}^{m}\frac{\prod_{j=1}^{m}(d-z_j)}{d-z_i}=0$$

Therefore,

$$K^*=-\frac{\sum_{i=1}^{n}\frac{\prod_{j=1}^{n}(d-p_j)}{d-p_i}}{\sum_{i=1}^{m}\frac{\prod_{j=1}^{m}(d-z_j)}{d-z_i}}$$

According to the root-locus condition, we obtain

$$K^*=-\frac{\prod_{j=1}^{n}(s-p_j)}{\prod_{j=1}^{m}(s-z_j)}$$

Hence, at the separation point $s=d$ we get,

$$\frac{\sum_{i=1}^{n}\frac{\prod_{j=1}^{n}(s-p_j)}{s-p_i}}{\sum_{i=1}^{m}\frac{\prod_{j=1}^{m}(s-z_j)}{s-z_i}}=\frac{\prod_{j=1}^{n}(s-p_j)}{\prod_{j=1}^{m}(s-z_j)}$$

Thus,

$$\sum_{i=1}^{n}\frac{1}{d-p_i}=\sum_{i=1}^{m}\frac{1}{d-z_i}$$

According to the principle of symmetry of the root-locus, we obtain the separation point of the root-locus, which must appear on the real axis or in the complex plane in the form of conjugate. Usually, separate points are in two root-locus branches of the real axis. If the root-locus is between two neighboring poles, one of them may be an infinity pole. Then there is at least one separate point. Similarly, if root-locus is between two neighboring open-loop poles, one of them may be an infinity pole, and there is also at least one separation point, as we can see in Figure 5.18.

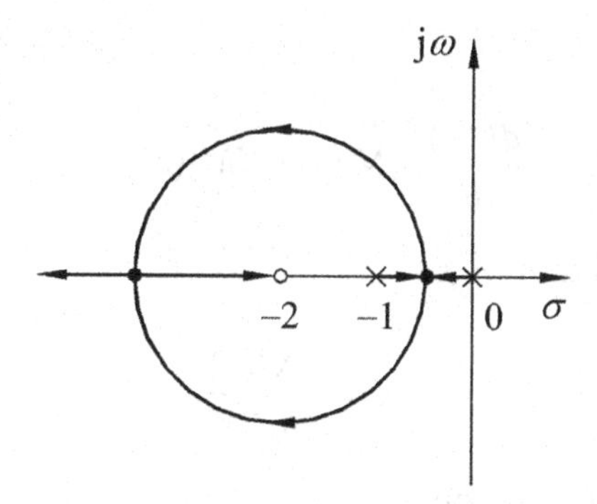

FIGURE 5.18 The separate point on the real axis.

The separation point on the real axis can be solved according to $\frac{dK}{ds}=0$. Here, K is the open-loop amplification gain constant. Because

$$G(s)H(s)=\frac{KM(s)}{N(s)}=\frac{K^*\prod_{j=1}^{m}(s-z_j)}{\prod_{i=1}^{n}(s-p_i)} \tag{5.62}$$

and

$$1+G(s)H(s)=0 \tag{5.63}$$

Now, we substitute Eq. (5.62) into Eq. (5.63), we obtain

$$1+K\frac{M(s)}{N(s)}=0 \tag{5.64}$$

or

$$N(s)+KM(s)=0$$

Let us consider

$$F(s)=N(s)+KM(s)=0 \tag{5.65}$$

Thus,

$$\frac{dF(s)}{ds}=0 \tag{5.66}$$

or

$$\frac{dF(s)}{ds}=N'(s)+KM'(s)=0 \tag{5.67}$$

From Eq. (5.67), we obtain

$$K=-\frac{N'(s)}{M'(s)}$$

Putting K into Eq. (5.65), we obtain

$$F(s)=N(s)-\frac{N'(s)}{M'(s)}M(s)=0 \tag{5.68}$$

Thus,

$$N(s)M'(s)-N'(s)M(s)=0 \tag{5.69}$$

From Eq. (5.65), we obtain

$$K=-\frac{N(s)}{M(s)}$$

Taking the derivative of K with respect to s and we put it into in Eq. (5.59), then we get,

$$\frac{dK}{ds}=-\frac{N'(s)M(s)-N(s)M'(s)}{M^2(s)}=0 \tag{5.70}$$

According to Eq. (5.70), we obtain separation point d.

Example 5.3

We assume a system open-loop transfer function is

$$G(s)=\frac{K^*(s+1.5)(s+2+j)(s+2-j)}{s(s+2.5)(s+0.5+j1.5)(s+0.5-j1.5)}$$

Draw the system's general root-locus.

Solution

1. Make sure the root-locus on the real axis. The area of the root-locus on the real axis is $[0.0,-1.5],[-\infty,-2.5]$.
2. Make sure the asymptote(s) of the root-locus. Here, $n=4, m=3$, then there is only one asymptote $(n-m=4-3=1)$.

and

$$\sigma_\alpha = \frac{\sum_{i=1}^{n} p_i - \sum_{j=1}^{m} z_j}{n-m} = 2.0$$

$$\varphi_\alpha = \frac{(2k+1)}{n-m}\pi = \pi$$

Hence, the asymptote is a straight line with the starting point of (2.0, 0.0). And the phase angle is π.

3. Make sure the starting point and the ending point. First, make sure of the starting point. Thus, we can draw the vector line from open-loop zero and poles to complex poles $(-0.5+j1.5)$; however, we can measure the angles, as shown in Figure 5.19. Hence, the starting point can be found as

$$\phi_1 = \tan^{-1}\left(\frac{1.5-0}{-0.5-(-1.5)}\right) = \tan^{-1}(1.5) = 56.3°$$

$$\phi_2 = \tan^{-1}\left(\frac{1.5-1}{-0.5-(-2)}\right) = 18.4°$$

$$\phi_3 = \tan^{-1}\left(\frac{1.5-(-1)}{-0.5-(-2)}\right) = 59°$$

And the ending point also can be calculated in a similar way as

$$\theta_1 = \tan^{-1}\left(\frac{1.5-0}{-0.5-0}\right) = 108.4°$$

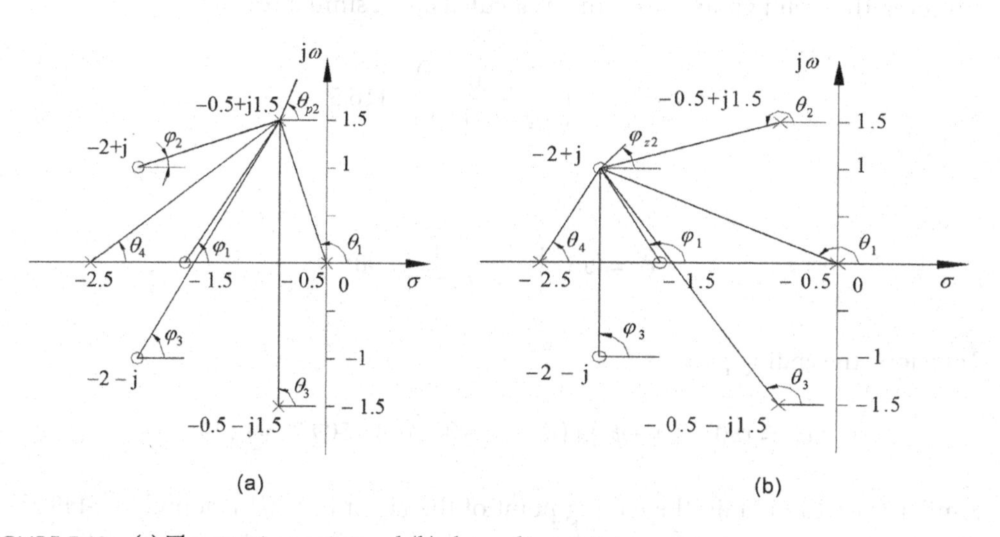

FIGURE 5.19 (a) The starting point and (b) the ending point.

$$\theta_3 = \tan^{-1}\left(\frac{1.5-(-1.5)}{-0.5-(-0.5)}\right) = 90°$$

$$\theta_4 = \tan^{-1}\left(\frac{1.5-0}{-0.5-(-2.5)}\right) = 36.9°$$

The starting point of $(-0.5+j1.5)$ is determined as

$$\theta_{p_2} = 180+(\phi_1+\phi_2+\phi_3)-(\theta_1+\theta_3+\theta_4) \approx 78.4°$$

According to the symmetry of root-locus, we obtain the symmetry of root-locus $(-0.5-1.5j)$ found as $-78.4°(281.6°)$. Then, we make sure of the ending point in the following:

$$\theta_1 = \tan^{-1}\left(\frac{1-0}{-2-0}\right) = 153.4°$$

$$\theta_2 = \tan^{-1}\left(\frac{1.5-1}{-0.5-(-2)}\right) = 198.4°$$

$$\theta_3 = \tan^{-1}\left(\frac{-1.5-1}{-0.5-(-2)}\right) = 121°$$

and

$$\theta_4 = \tan^{-1}\left(\frac{1-0}{-2-(-2.5)}\right) = 63.4°$$

However, the ending point also can be calculated in a similar way as

$$\phi_1 = \tan^{-1}\left(\frac{1-0}{-2-(-1.5)}\right) = 116.5°$$

and

$$\phi_3 = \tan^{-1}\left(\frac{1-(-1)}{-2-(-2)}\right) = 90°$$

Therefore, the ending point is

$$\phi_{z_2} = 180°-(\phi_1+\phi_3)+(\theta_1+\theta_2+\theta_3+\theta_4) = 509.7° = 149.7°$$

Similarly, we can obtain the ending point of the factor $(-2-j)$ is found as $-149.7°$. Thus, we can get the root-locus of the system that we can see in Figure 5.20.

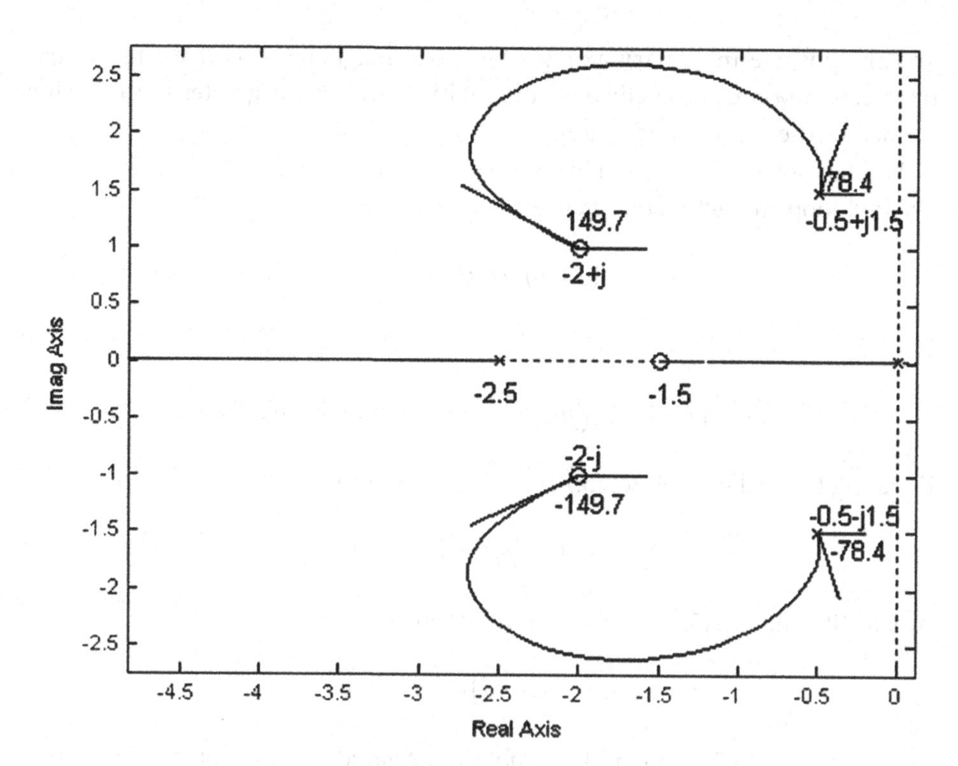

FIGURE 5.20 The root-locus of the system.

8. *The angle of departure and break-in point on the real axis is* $\pm\frac{\pi}{2}$
   The angle of departure (angle of separation) and break-in point (meeting angle) on the real axis are constant and it is $\pm\frac{\pi}{2}$. When the root-locus leaves the angle of departure point, the angle of inclination of the tangent of the root-locus is called breakaway angle. The angle of departure of the breakaway point on the real axis is always $\pm\frac{\pi}{2}$, the meeting angle of the break-in point is also always $\pm\frac{\pi}{2}$.
9. *The intersection or crossing point of the root-locus and the imaginary axis*
   The intersection of the root-locus and the imaginary axis can be determined by the value of K^* and ω from the characteristic equation of the closed-loop system based on the Routh or Hurwitz criteria. Since the root-locus path and the virtual axes intersection point, the system has a purely imaginary root, which means that the closed-loop system is critically stable with a value of gain K^*. Therefore, the first column of the Routh array contains a zero, and we can determine the value of gain K^* at the intersection point of the root-locus and the imaginary axis.

   Since a pair of purely imaginary roots is the same but the difference is in the numerical symbols of the root(s), the Routh array can be obtained from the purely imaginary roots from the coefficient of s^2 of an auxiliary equation. It is the value of ω at the intersection point of the root-locus and the imaginary axis. If the root-locus and the positive imaginary axis (or negative imaginary axis) have more than one intersection point, the coefficients of s^2 with power of s is greater than 2, which are formed by the auxiliary equation.

We may solve this problem by another method which is led to substitute $s = j$ into a closed-loop characteristic equation, i.e.,

$$1+G(j\omega)H(j\omega)=0 \tag{5.71}$$

or

$$\text{Re}\left(1+G(j\omega)H(j\omega)\right)+j\,\text{Im}\left(1+G(j\omega)H(j\omega)\right)=0 \tag{5.72}$$

Equating the real part of Eq. (5.72) to zero, we obtain

$$\text{Re}(1+G(j\omega)H(j\omega))=0 \tag{5.73}$$

By substituting Eq. (5.73) into Eq. (5.72), then we get

$$\text{Im}\left(1+G(j\omega)H(j\omega)\right)=0 \tag{5.74}$$

By solving Eqs. (5.73) and (5.74), we obtain the value of gain K^* and ω at the intersection point of the root-locus and the imaginary axis.

Example 5.4

Consider a unity-feedback system whose open-loop transfer function is given by

$$G(s)=\frac{K^*}{s(s+1)(s+2)}.$$

Determine the intersection point of the root-locus and the imaginary axis.

Solution

1. Using Routh's criterion, the characteristic equation of the closed-loop system is written as

$$N(s)=s(s+1)(s+2)+K^*=s^3+3s^2+2s+K^*$$

Thus, the Routh's array becomes

s^3	1	2
s^2	3	K^*
s^1	$\frac{6-K^*}{3}$	
s^0	K^*	

The value of K^* that makes the term s^1 in the first column is equal to zero; then the system critical stability is found

$$\frac{6-K^*}{3}=0 \Rightarrow K^*=6$$

The auxiliary equation is obtained from the coefficients of s^2 row, that is,

$$3s^2+K^*=0 \Rightarrow 3s^2+6=0$$

Thus,

$$s=\pm j\sqrt{2}$$

This is the intersection point of the root-locus and the imaginary axis.

2. By substituting $s=j$ into the equation of $N(s)$, we get

$$N(j\omega)=(j\omega)^3+3(j\omega)^2+2(j\omega)+K^*=0$$

Substituting both the real and imaginary parts of $N(j\omega)$ into Eqs. (5.73) and (5.74), respectively, we obtain

$$\begin{cases} -3\omega^2+K^*=0 \\ -\omega^3+2\omega=0 \end{cases}$$

By solving the above two equations we get

$$\begin{cases} \omega=\pm\sqrt{2},0 \\ K^*=6 \\ K=3 \end{cases}$$

Here, K is the open-loop gain and K^* is the root-locus gain.

10. *The direction of the root-locus at infinity*
    If the system has two or more root-locus approaches from zero to infinity, some root-locus must be laid in the left half S-plane and some root-locus must be laid to the right half S-plane.

11. *The summation of the closed-loop poles of the system is constant*
    Consider a system whose open-loop transfer function is given by

$$G(s)H(s)=\frac{K^*\prod_{i=1}^{m}(s-z_i)}{\prod_{j=1}^{n}(s-p_j)} \tag{5.75}$$

If the system is satisfied for $n-m \geq 2$, the characteristic equation is written as

$$\prod_{j=1}^{n}\left(s-p_j\right)+K^*\prod_{i=1}^{m}\left(s-z_i\right)=0 \tag{5.76}$$

It can be obtained from Vieta's theorem (1615), the summation of n roots of the n -order algebraic equation equals the coefficient of the $(n-1)$ term multiplication (-1). We obtain

The summation of the closed-loop poles

$$=\sum_{j=1}^{n} p_j \tag{5.77}$$

When $n-m \geq 2$, the summation of the closed-loop poles equals the summation of the closed-loop poles. Usually, $\left(\sum_{j=1}^{n} p_j\right)/n$ is called the pole center of gravity.

When the value of K^* changes, the pole center remains unchanged. This property can be used to estimate the changing trend of the root-locus curve, which is helpful to determine the pole position and the corresponding value of gain K^*.

12. *The product of the closed-loop poles of the system*

    Equation (5.76) is obtained according to Vieta's theorem (1615) (Table 5.1).

$$\text{The product of the closed-loop poles} = \prod_{j=1}^{n} p_j + K^*\prod_{i=1}^{m} z_i \tag{5.78}$$

TABLE 5.1 Rules for Constructing Root-Loci

No.	Content	Rules
1	Starting point and ending point	The root-locus starts in open-loop poles and ends in the open-loop zero (infinity).
2	The number of branches	The order of closed-loop characteristics equation of the root-locus $(m \leq n)$.
3	Symmetry	The root-locus is symmetric about the real axis.
4	Root-locus on the real axis	On the right side of the root-locus interval on the real axis, the sum of the open-loop poles and zeros is odd.
5	Asymptotes	The number of the open-loop pole is n, and the zeros is m, such that there is $n-m$ asymptote(s). The intersection points and the phase angle between the asymptote and real axis is $\sigma_\alpha = \dfrac{\sum_{i=1}^{n} p_i - \sum_{j=1}^{m} z_j}{n-m}$, and $\varphi_\alpha = \dfrac{2k+1}{n-m}\pi$, respectively. where, $k=0,\ 1,\ 2\cdots,\ n-m-1$.

(*Continued*)

TABLE 5.1 (*Continued*) Rules for Constructing Root-Loci

No.	Content	Rules
6	Starting angle (angle of arrival) and ending angle (Angle of Departure)	The starting angle (angle of arrival): The starting angle is the angle between the tangent line of the open-loop root-locus and the positive real axis at the complex pole of the open-loop system, also called the angle of arrival or rising angle which is denoted by θ_{p_i}, i.e., $$\theta_{p_i} = \pi + \sum_{j=1}^{m} \phi_{z_j p_i} - \sum_{\substack{j=1 \\ j \neq i}}^{n} \theta_{p_j p_i}$$ The ending angle (angle of departure): The ending angle is the angle between the tangent and the open-loop root-locus curve with the real axis; it is also called incident angle or angle of departure which is denoted by ϕ_{z_i}, i.e., $$\phi_{z_i} = \pi + \sum_{j=1}^{n} \theta_{p_j z_i} - \sum_{\substack{j=1 \\ j \neq i}}^{m} \phi_{z_j z_i}.$$
7	Breakaway point in the coordinates and angle of departure	The point that two or more than two of the root-locus met in the S-plane and separated in the S-plane. The breakaway point(s) in the coordinates are determined by the following condition such that $$\sum_{j=1}^{m} \frac{1}{d - z_j} = \sum_{i=1}^{n} \frac{1}{d - p_i} \Rightarrow \frac{dK}{ds} = 0$$ And the angle of departure $= \frac{2k+1}{l}\pi$; where, $k = 0, 1, 2, \cdots, l-1$; and l is the number of root-locus branches.
8	Angle of departure and break-in point (meeting angle) on the real axis	The angle of departure (angle of separation) and break-in point (meeting angle) on the real axis is constant and it is always $\frac{\pi}{2}$.
9	Intersection or crossing point of the root-locus and the imaginary axis	The intersection point of the root-locus and the imaginary axis can be obtained by Routh's criterion; and also, it can obtain by substituting $s = j$ into closed-loop characteristic equation, then we may use $\text{Re}(1 + G(j\omega)H(j\omega)) = 0$ and $\text{Im}(1 + G(j\omega)H(j\omega)) = 0$.
10	Direction of root-locus	If the system has two or more root-locus approaches from zero to infinity, then some root-locus must lie in the left half S-plane and some root-locus must lie in the right half S-plane.
11	Summation of the closed-loop poles	The summation of the closed-loop poles of the system is constant as $\sum_{j=1}^{n} p_j$.
12	Product of the closed-loop poles	The product of the closed-loop poles is a constant as $\prod_{j=1}^{n} p_j + K^* \prod_{i=1}^{m} z_i$.

Example 5.5

Consider a unity-feedback system whose open-loop transfer function is given by

$$G(s)H(s)=\frac{K^*(s+2)}{s(s+3)(s^2+2s+2)}$$

Plot the root-locus of the system.

Solution

According to the rules for constructing root-loci, we know that

1. The root-locus is symmetric about the real axis.
2. The root-locus is located in [0, –2] and [–, –3] on the real axis.
3. The number of asymptotes is $n-m=4-1=3$. The angle of asymptote is $\varphi_\alpha=\pm\frac{\pi}{3},-\pi$, and the intersection point with the real axis is $\sigma_\alpha=-1$.
4. The root-locus is between the poles and zeros on the real axis; thus, there is no angle of departure (separation angle) and break-in point (meeting angle) on the real axis.
5. The starting angle of complex-conjugate pole p_2 is expressed as

$$\theta_{p_2}=180°-(\theta_1+\theta_2+\theta_3)+\phi_1=180°-(135°+90°+22.6°)+45°=-22.6°$$

6. To determine the points where the root-loci cross the imaginary axis. The characteristic equation for the system is

$$s^4+5s^3+8s^2+(6+K^*)s+2K^*=0$$

Based on Routh's criterion, Routh's array becomes

s^4	1	8	$2K^*$
s^3	5	$6+K^*$	
s^2	$\frac{34-K^*}{5}$	$2K^*$	
s^1	$\frac{204-22K^*-K^{*2}}{34-K^*}$		
s^0	$2K^*$		

To find the equation of the intersection point on the imaginary axis, we get

$$\begin{cases}\frac{204-22K^*-K^{*2}}{34-K^*}=0\\2K^*=0\end{cases}\quad\text{or}\quad\begin{cases}K^*=7.03\\K^*=0\end{cases}$$

However, the auxiliary equation can be obtained third row of the array as

$$\frac{34-K^*}{5}s^2+2K^*=0$$

which gives

$$s^2=\frac{-10K^*}{34-K^*}=-2.61$$

Thus, we obtain

$$s=\pm j1.61 \text{ and } \omega=\pm 1.61$$

Figure 5.21 shows a root-locus plot for the system considerations.

Using the MATLAB program, we can draw the root-locus plot, as shown in Figure 5.22.

```
num=[1 2];
den=conv(poly([0-3]), [1 2 2]);
sys=tf(num, den)
% Transfer function:
%s+2
% -------------------------
% s^4+5 s^3+8 s^2+6 s
rlocus(sys)
```

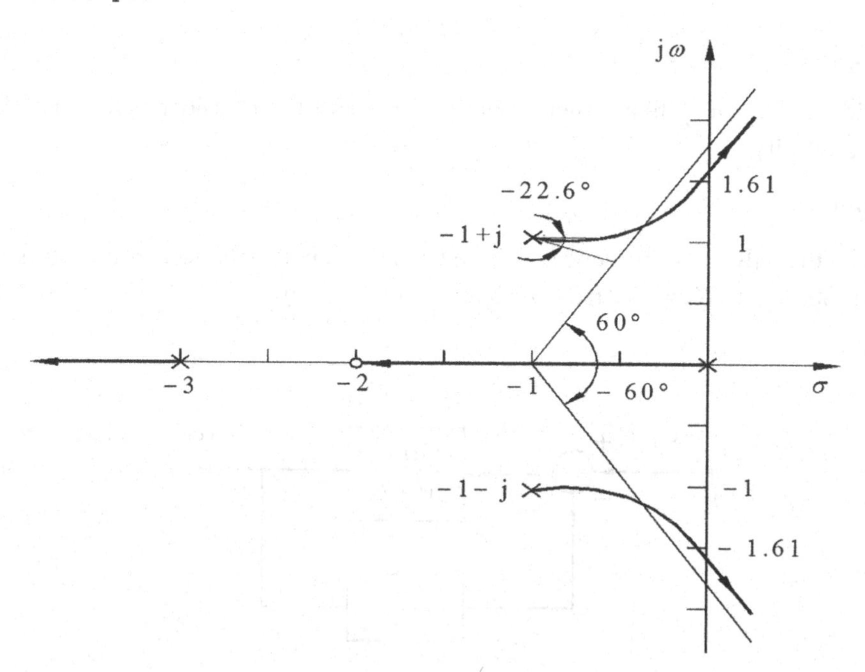

FIGURE 5.21 Root-locus plot.

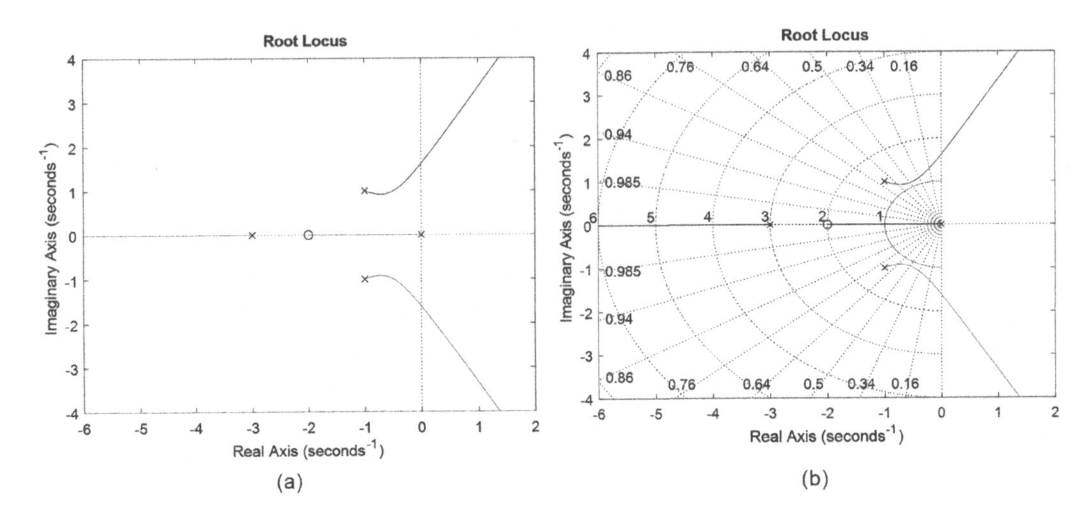

FIGURE 5.22 Root-locus plot using MATLAB software.

5.5 GENERALIZED ROOT-LOCUS (PARAMETER ROOT-LOCUS)

The abovementioned root-locus takes the open-loop amplification gain factor K or the open-loop gain factor K^* as the variable coefficient of the root-locus. If some parameters of the system vary from zero to infinity ($0 \sim \infty$), the root-locus of the closed-loop system will be also changed. Here, based on these parameters of the root-locus are called generalized root-locus or parameter root-locus.

Example 5.6

Consider a control system shown in Figure 5.23. Plot the root-loci as K_h varies from zero to infinity.

Solution

In order to analyze the influence of K_h on the system performance, we plot the generalized root-loci as K_h varies from zero to infinity ($0 \sim \infty$).

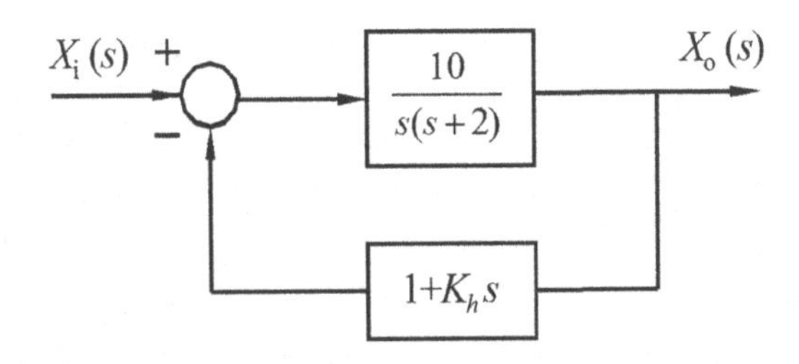

FIGURE 5.23 A kind of control system.

The transfer function of the closed-loop system is written as

$$G(s)H(s)=\frac{\frac{10}{s(s+2)}}{1+\frac{10}{s(s+2)}(1+K_h s)}=\frac{10}{s(s+2)+10(1+K_h s)}$$

Thus, the closed-loop characteristic equation for the system is written as

$$s^2+2s+10K_h s+10=0$$

In order to plot the parameters root-locus on K_h, we divide the characteristics of the equation on both sides by the characteristic equation not included K_h as

$$\frac{10K_h s}{s^2+2s+10}=-1$$

where

$$G'(s)H'(s)=\frac{10K_h s}{s^2+2s+10}$$

Due to $G'(s)H'(s)$, the equivalent open-loop transfer function has one zero and two poles; thus, the root-locus has two branches that started or originated from open-loop pole to finite zero z_1 and infinite zeros. Therefore, we can say that the left side of the entire real axis is a root-locus, as shown in Figure 5.24.

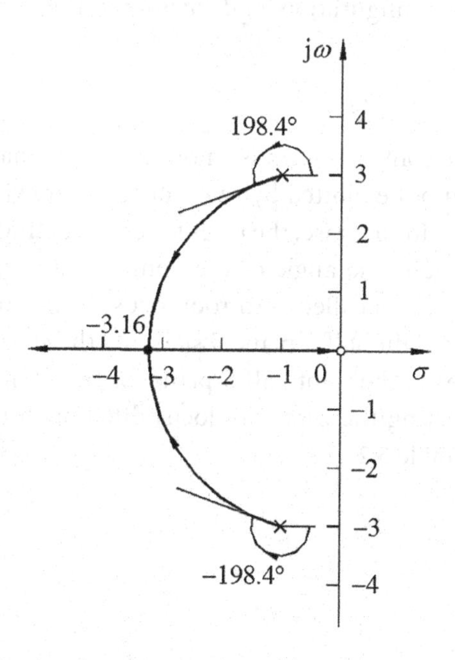

FIGURE 5.24 The root-locus for the control system.

The break-in point (meeting point) on the real axis can be determined by the following relationship as

$$\sum_{j=1}^{m} \frac{1}{d-z_j} = \sum_{j=1}^{n} \frac{1}{d-p_j}$$

or

$$\frac{1}{d+0} = \frac{1}{d+1-j3} + \frac{1}{d+1+j3}$$

Thus, we obtain

$$d = \pm\sqrt{10}$$

Since $\sqrt{10}$ is not in the root-locus, it can be omitted. Hence, the meeting point is at $-\sqrt{10}$, and the meeting angles are $\pm\frac{\pi}{2}$. Therefore, the departure angle of the complex-conjugate pole is found as

$$\theta_{p_1} = \pi + \left(\pi - \tan^{-1} 3\right) - \frac{\pi}{2} = 198.4°$$

Similarly, the departure angle of the complex-conjugate pole $-1-j3$ is $-198.4°$. The root-locus of the system is shown in Figure 5.24.

A typical pole-zero configuration and corresponding root-loci are shown in Figure 5.25.

5.6 THE ZERO ROOT-LOCUS

For a control system, if the control system is a nonminimum phase system, the root-locus of the control system cannot be plotted by the root-locus drawing rule of the minimum phase control system. Thus, in this case, the rules for constructing root-locus of the control system are unavailable. The phase angle of the nonminimum phase systems is $0° + 2k\pi$ instead of $(2k+1)\pi$. Hence, it is called zero root-locus. Nonminimum phase system has open-loop pole-zero in the right half S-plane. Especially the characteristic equation of the system has zeros and poles in the right half S-plane for the positive-feedback control system. The rules of constructing the zero root-locus differ slightly from the general root-locus, which is shown in Table 5.2.

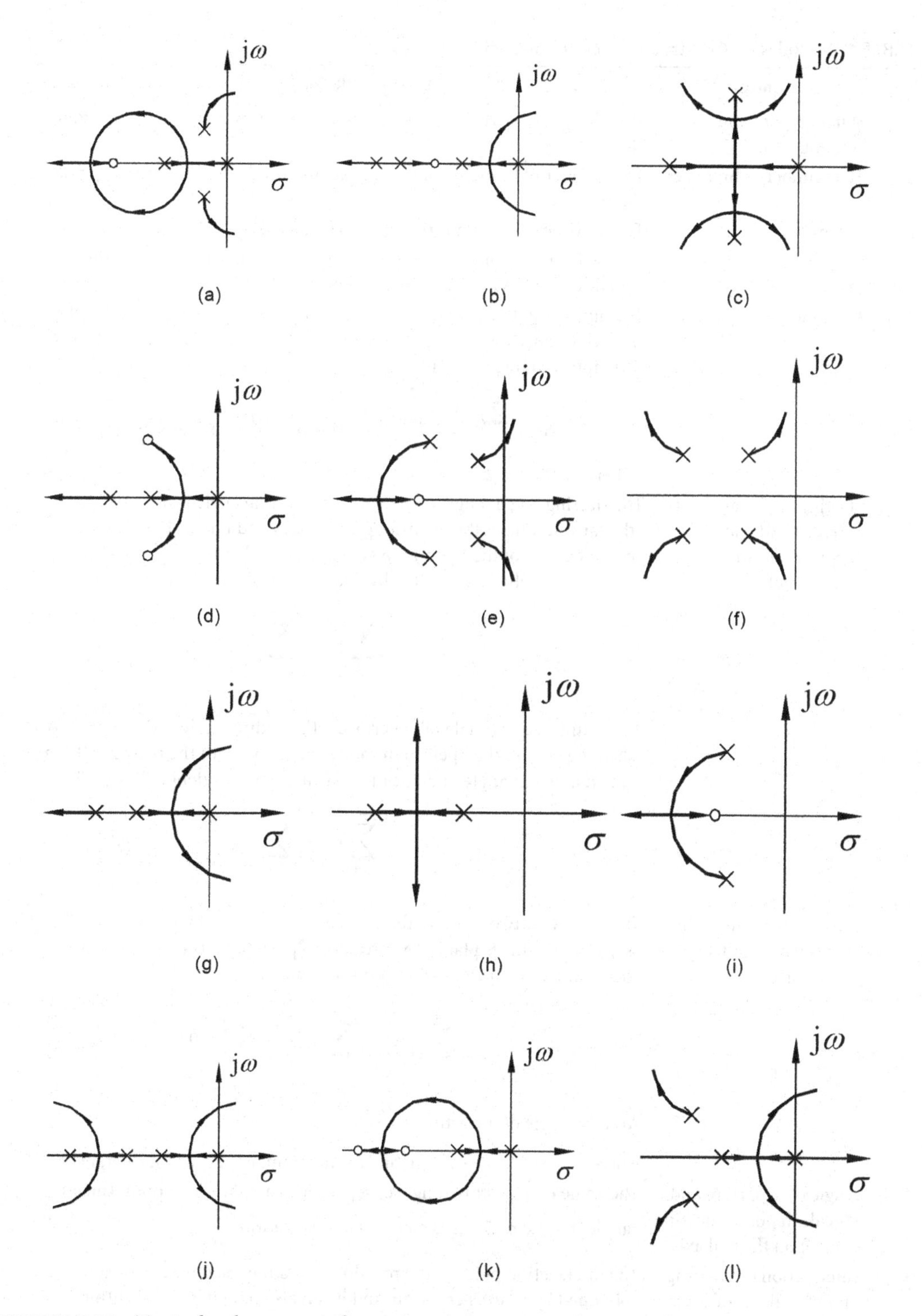

FIGURE 5.25 Typical pole-zero configuration and corresponding root-loci.

TABLE 5.2 Rules for Constructing Zero Root-Loci

No	Content	Rules
1	Starting point and ending point	The root-locus starts in open-loop poles and ends in the open-loop zero (infinity).
2	The number of branches	The number of branches of the root-locus is equal to the open-loop poles or zeros $(m \leq n)$.
3	Symmetry	The root-locus is symmetric about the real axis.
4	Root-locus on the real axis	On the right side of the root-locus interval on the real axis, the sum of the open-loop poles and zeros is an even.
5	Asymptotes	The number of the open-loop pole is n, and the zeros is m, such that there is $n-m$ asymptote(s). The intersection points and the phase angle between the asymptote and real axis is $$\sigma_\alpha = \frac{\sum_{i=1}^{n} p_i - \sum_{j=1}^{m} z_j}{n-m},\ and\ \varphi_\alpha = \frac{2k+1}{n-m}\pi, \text{ respectively.}$$ where, $k = 0,\ 1,\ 2\cdots,\ n-m-1$
6	Starting angle (angle of arrival) and ending angle (angle of departure)	The starting angle (angle of arrival): The starting angle is the angle between the tangent line of the open-loop root-locus and the positive real axis at the complex pole of the open-loop system; also called the angle of arrival or rising angle which is denoted by θ_{p_i}, i.e., $$\theta_{p_i} = \sum_{j=1}^{m} \phi_{z_j p_i} - \sum_{\substack{j=1 \\ j \neq i}}^{n} \theta_{p_j p_i}$$ The ending angle (angle of departure): The ending angle is the angle between the tangent and the open-loop root-locus curve with the real axis; it is also called incident angle or angle of departure which is denoted by ϕ_{z_i}. i.e. $$\phi_{z_i} = \sum_{j=1}^{n} \theta_{p_j z_i} - \sum_{\substack{j=1 \\ j \neq i}}^{m} \phi_{z_j z_i}$$
7	Breakaway point in the coordinates and angle of departure	The point that two or more than two of the root-locus met in the S-plane and separated in the S-plane. The breakaway point(s) in the coordinates are determined by the following condition such that $$\sum_{j=1}^{m} \frac{1}{d - z_j} = \sum_{i=1}^{n} \frac{1}{d - p_i} \Rightarrow \frac{dK}{ds} = 0$$ And the angle of departure $= \frac{2k+1}{l}\pi$; where, $k = 0,\ 1,\ 2,\ \cdots,\ l-1$; and l is the number of root-locus branches.
8	Angle of departure and break-in point (meeting angle) on the real axis	The angle of departure (angle of separation) and break-in point (meeting angle) on the real axis is constant and it is always $\frac{\pi}{2}$.
9	Intersection or crossing point of the root-locus and the imaginary axis	The intersection point of the root-locus and the imaginary axis can be obtained by Routh's criterion, and it can also obtain by substituting $s = j$ into closed-loop characteristic equation, then we may use $\text{Re}\big(1 + G(j\omega)H(j\omega)\big) = 0$ and $\text{Im}\big(1 + G(j\omega)H(j\omega)\big) = 0$.
10	Direction of root-locus	If the system has two or more root-locus approaches from zero to infinity, then some root-locus must lie in the left half S-plane and some root-locus must lie in the right half S-plane.

Example 5.7

Consider a unity positive-feedback system is shown in Figure 5.26. Plot the root-loci of the system.

Solution

Since this is a nonminimum phase system, we might plot the root-loci with the rules of zero root-locus. The zero-pole on the S-plane is

$$p_1 = -1 + j,\ p_2 = -1 - j,\ p_3 = -3,\ z_1 = -2$$

Based on the rules of constructing zero root-locus, we can say

1. Root-locus starts in open-loop poles and ends in open-loop zeros as K^* varies from 0 to ∞ (including infinity zeros).
2. The root-locus on the real axis is located in [–2,+∞] and [–∞,–3].
3. The number of asymptotes of the root-locus is $i = n - m = 2$. That is, there are two asymptotes and their phase angle is

$$\phi_\alpha = \frac{2k\pi}{n-m};\ \text{for}\ k = 0, 1$$

When $k = 0$, $\phi_\alpha = 0$, and when $k = 1$, $\phi_\alpha = \pi$. Thus, we can say that the asymptote of the root-locus is exactly on the real axis.
4. The breakaway point in the coordinates can be written as

$$\sum_{j=1}^{m} \frac{1}{d - z_j} = \sum_{i=1}^{n} \frac{1}{d - p_i}$$

or

$$\frac{1}{d+2} = \frac{1}{d+3} + \frac{1}{d+1-j} + \frac{1}{d+1+j} \Rightarrow d = -0.8$$

We obtain that the breakaway point in the coordinate on the real axis is $d = -0.8$.

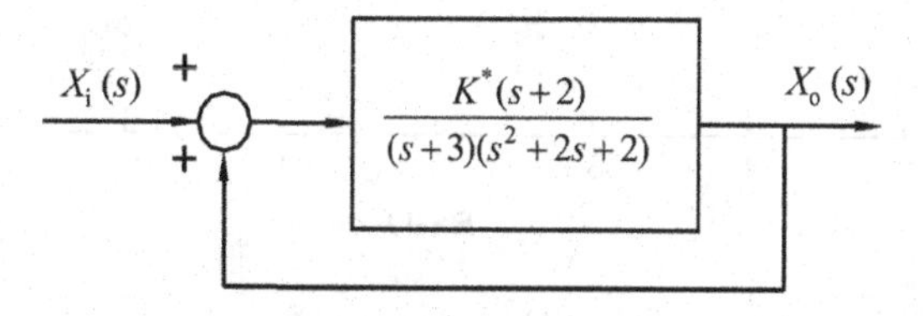

FIGURE 5.26 A unity positive-feedback system.

5. Angles of departure on the real axis are $\pm\frac{\pi}{2}$.
6. For the complex point $p_1 = -1 + j$, the angle of departure is

$$\theta_{p_1} = 45° - (90° + 26.6°) = -71.6°$$

   Based on the symmetry rules, the angle of departure of the complex pole $p_2 = -1 - j$ is 71.6°.
7. To obtain the critical root-locus gain, we may consider the root-locus of the system across the origin of coordinates according to the root-locus. Thus, it can be obtained under normal conditions, i.e.,

$$K^* = \frac{|0 - p_1| \cdot |0 - p_2| \cdot |0 - p_3|}{|0 - z_1|}$$

$$= \frac{|0+1-j| \cdot |0+1+j| \cdot |0+3|}{|0+2|} = 3$$

   We obtain $K = \frac{K^*}{3} = 1$ based on the root-locus of the open-loop transfer function. It is noted that in order to stabilize the positive-feedback control system, the open-loop gain must be less than 1. The root-locus plot is shown in Figure 5.27.

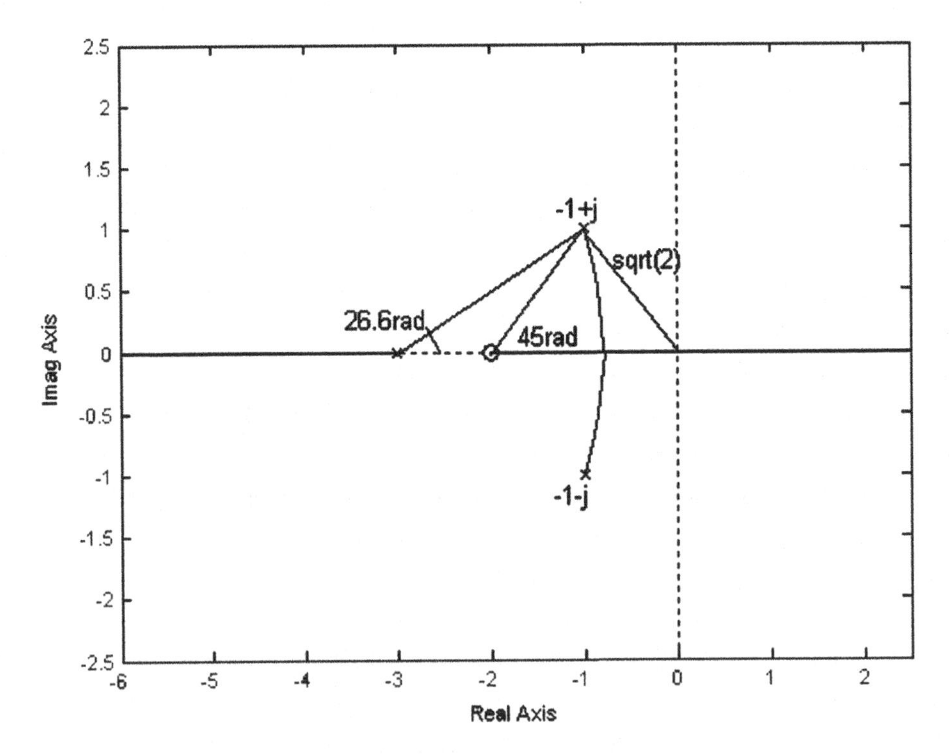

FIGURE 5.27 The root-locus of $\frac{K^*(s+2)}{(s+3)(s^2+2s+2)}$.

5.7 COMMENTS ON ROOTS-LOCUS PLOTS OF NEGATIVE-FEEDBACK SYSTEMS

In this section, we may learn comments on the root-locus plots, typical pole-zero configurations, constant root-loci and conditionally stable system, nonminimum phase system, and orthogonality of root-loci and constant-gain loci.

5.7.1 Comments on the Root-Locus Plots

Let's comment on the root-locus plots. Look at the closed-loop control system, as shown in Figure 5.28.

It is noted that the characteristic equation of the negative-feedback control system whose open-loop transfer function is known as

$$G(s)H(s)=\frac{K\left(s^m+b_1s^{m-1}+\ldots+b_m\right)}{s^n+a_1s^{n-1}+\ldots+a_n}; \text{ for } n\geq m$$

This is an nth-degree algebraic equation in s. If $m<n$, then the coefficient a_1 is the negative sum of the roots of the equation and is independent of K. In such a case, if some of the roots move on the locus toward the left as K is increased, then the other roots must move toward the right as K is increased. This information is helpful in finding the general shape of the root-loci. *What looks like the shape of root-loci?*

Since we know that if K increases, some of the roots move on locus toward the left, and other roots must move toward the right; thus, it is also noted that a slight change in the pole-zero configuration may cause significant changes in the root-locus configurations, as shown in Figure 5.29.

Therefore, we can say that Figure 5.29 demonstrates the fact that a slight change in the location of a zero or pole will make the root-locus configuration look quite different.

5.7.2 Typical Pole-Zero Configurations and Corresponding Root-Loci

Now, let's talk about typical pole-zero configurations and corresponding root-loci:

- The pattern of the root-loci depends only on the relative separation of the open-loop poles and zeros;
- *Unstable System:* If the number of open-loop poles exceeds, the number of finite zeros by three or more, for a gain K, *the root-loci enter the right-half S-plane*;

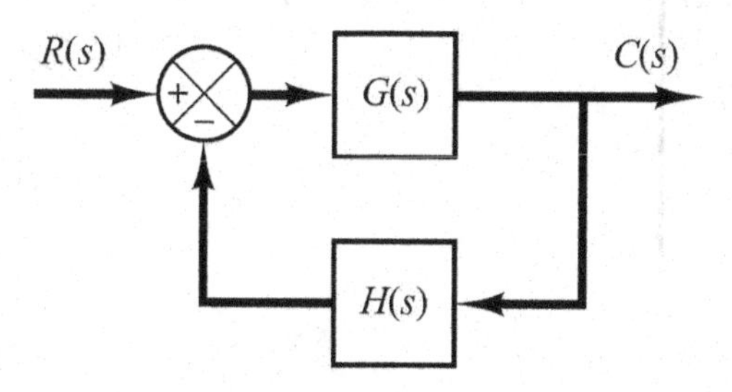

FIGURE 5.28 A closed-loop control system.

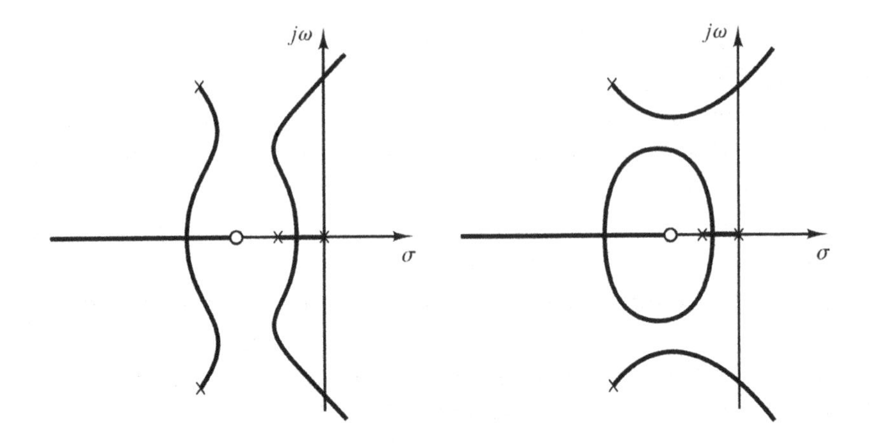

FIGURE 5.29 A slight change in the pole-zero configuration.

- *Stable System:* Closed-loop poles in *left-half S-plane*;
- The changes in the root-loci are due to the changes in the number and location of the open-loop poles and zeros by visualizing the root-locus plots resulting from various pole-zero configurations.

5.7.3 Constant ξ Loci and Constant ω_n Loci

In the complex plane, recalling the damping ratio ζ of a pair of complex-conjugate poles can be expressed in terms of the angle ∅, which is measured from the negative real axis with = cos, as shown in Figure 5.30a.

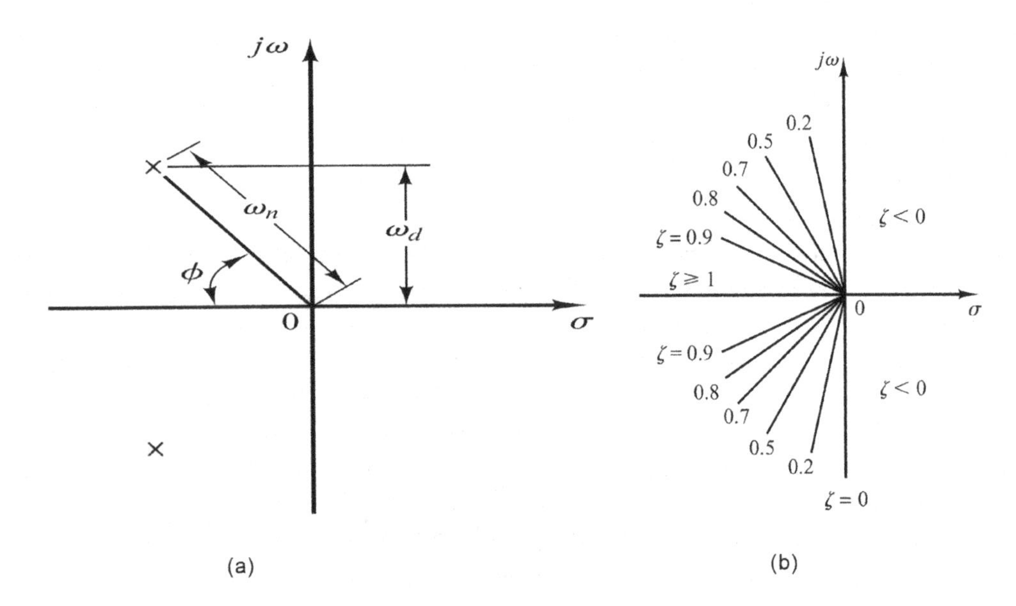

FIGURE 5.30 Constant ξ loci and constant ω_n loci.

In other words, the lines of constant damping ratio ξ are radial lines passing through the origin as shown in Figure 5.30b. For example, a damping ratio $\xi = 0.5$ requires that the complex-conjugate poles lie on the lines drawn through the origin making angles of $\pm 60°$ with the negative real axis.

- If the real part of a pair of complex-conjugate poles is positive, which means that the system is *unstable*, the corresponding ξ is negative.
- The damping ratio determines the angular location of the poles, while the distance of the pole from the origin is determined by the undamped natural frequency ω_n. The constant ω_n loci are circles.
- If $\xi = 0 \sim 1$, the complex-conjugate poles lie on the lines $\phi = 0 \sim \pm 90°$.
- If $\xi \geq 0$, poles are in the negative real axis (stable).
- If $\xi < 0$, poles are in the positive real axis (unstable).
- If $\xi = 0 \sim 1$, $\omega_n = 0.5, 1, 2$ circles, then root-locus plot of a negative-feedback system can be seen in this figure by using the MATLAB software.
- If $\xi = 0.5, 0.707$ lines and $\omega_n = 0.5, 1, 2$ circles with gain K on the root-locus plot as in Figure 5.31.
- The complex-conjugate poles lie on the lines drawn through the origin making angles.

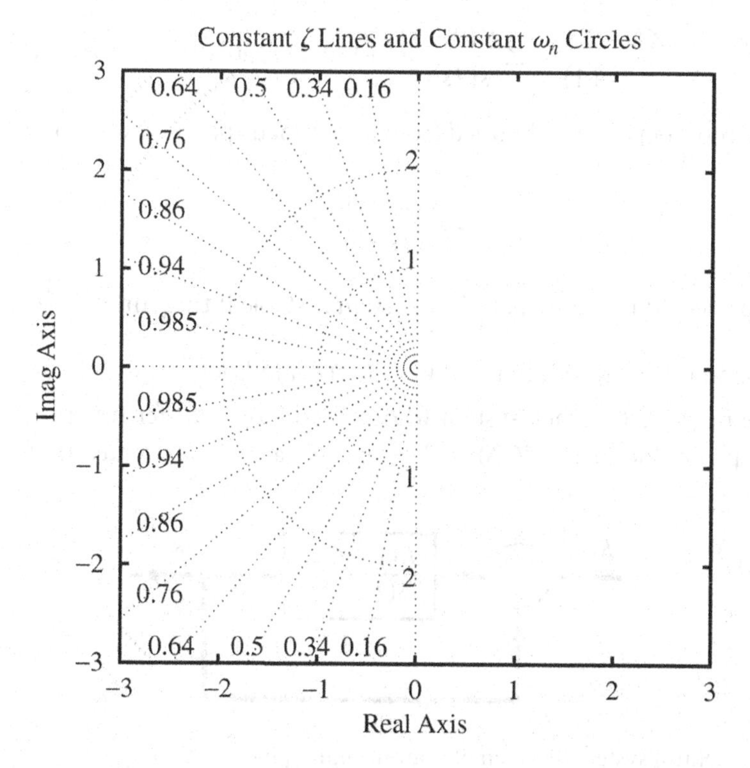

FIGURE 5.31 Constant ξ loci and constant ω_n loci (using MATLAB).

5.7.4 Conditionally Stable Systems

A system is conditionally stable if the unstable operation for certain K; however the system may break down, or be nonlinear due to a saturation nonlinearity. Conditionally stable systems are not desirable; it is dangerous, but it does occur in certain systems; and it exists unstable feedforward path, whereas the unstable feedforward path may occur if the system has a minor loop; thus, we must avoid such conditional stability. If the gain K drops for any critical value, the system is unstable. The conditional stability may be eliminated by adding proper compensation.

5.7.5 Nonminimum Phase Systems

If all the poles and zeros of a system lie in the left half S-plane, then the system is called minimum phase. If a system has at least one pole or zero in the right-half S-plane, then the system is called nonminimum phase. The term nonminimum phase comes from the phase-shift characteristics of such a system when subjected to sinusoidal inputs, as shown in Figure 5.32.

Here, we can see the transfer function $G(s)$ for $T_a > 0$ as

$$G(s)=\frac{K(1-T_a s)}{s(Ts+1)},\ H(s)=1;\ \text{for } T_a>0$$

There is one zero in the right-half S-plane; thus, it has nonminimum phase system. The angle condition as we can say is

$$\angle G(s)=\angle-\frac{K(1-T_a s)}{s(Ts+1)}=\angle\frac{K(T_a s-1)}{s(Ts+1)}+180°=\pm180°(2k+1),\ \text{for } k=0,1,2,\ldots$$

However, the root-loci can be obtained from a root-locus plot and we can say

$$\angle\frac{K(T_a s-1)}{s(Ts+1)}=0°$$

This can be presented in Figure 5.33. The system is stable if the gain $K<\frac{1}{T_a}$.

5.7.6 Orthogonality of Root-Loci and Constant-Gain Loci

Consider the negative-feedback system whose open-loop transfer function is $G(s)H(s)$. In the $G(s)H(s)$ plane, the loci of $|G(s)H(s)| = \text{constant}$ are circles centered at the origin, as

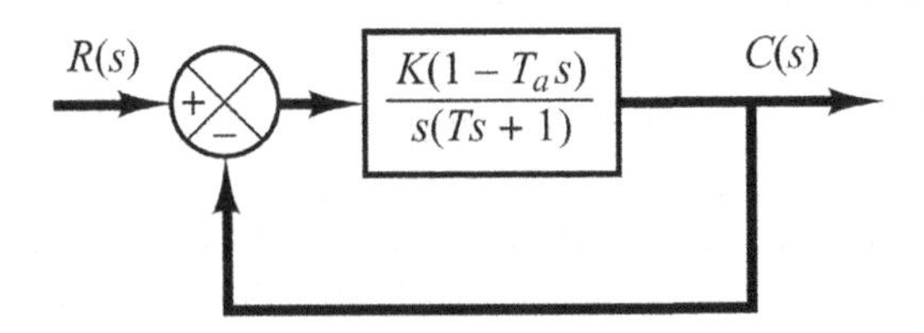

FIGURE 5.32 Control system diagram of nonminimum phase systems.

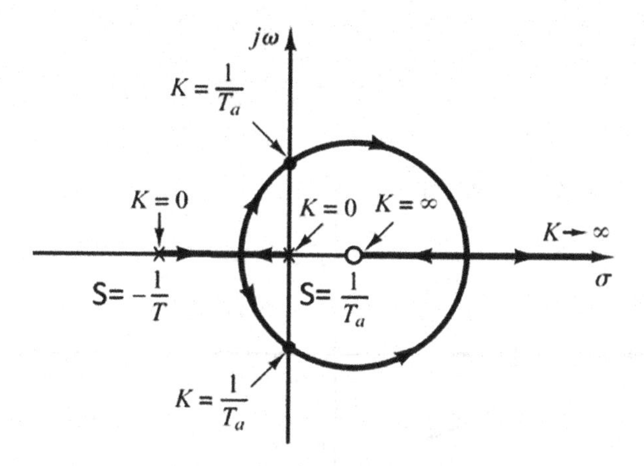

FIGURE 5.33 Root-locus plot of nonminimum phase systems.

shown in Figure 5.34a. The loci correspond to $\angle G(s)H(s) = \pm 180°(2k+1)$ (k = 0, 1, 2,..., n) lie on the negative real axis of $G(s)H(s)$ plane (which is not on S-plane), as shown in Figure 5.34b.

In the S-plane, it is conformal mapping (domain and an image in the complex plane). The constant-phase and constant-gain loci in the $G(s)H(s)$ plane are orthogonal, the root-loci and constant-gain loci in the S-plane are orthogonal. The root-loci and constant-gain loci are in the Figure 5.34 can be formed by

$$G(s) = \frac{K(s+2)}{S^2+2s+3}, H(s) = 1$$

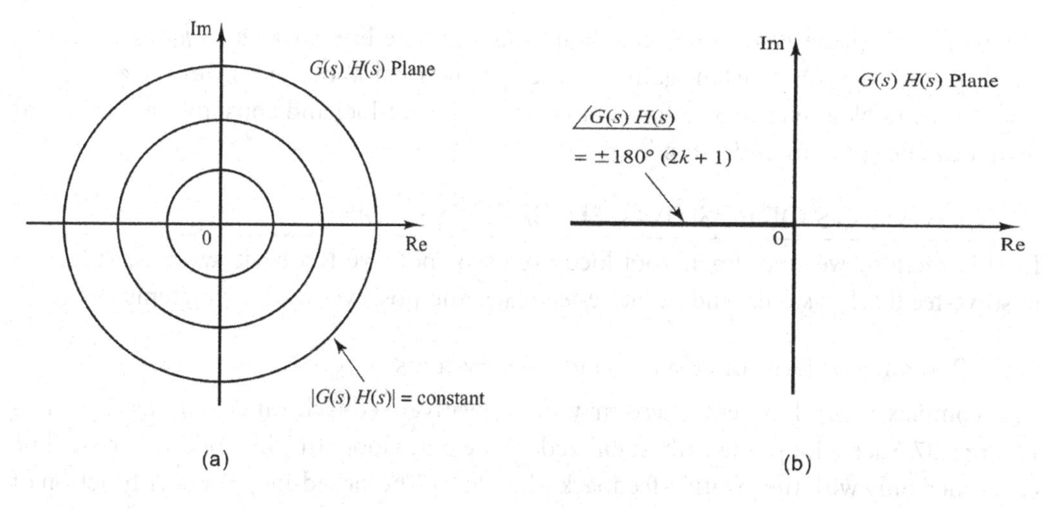

FIGURE 5.34 Orthogonality of root-loci and constant-gain loci.

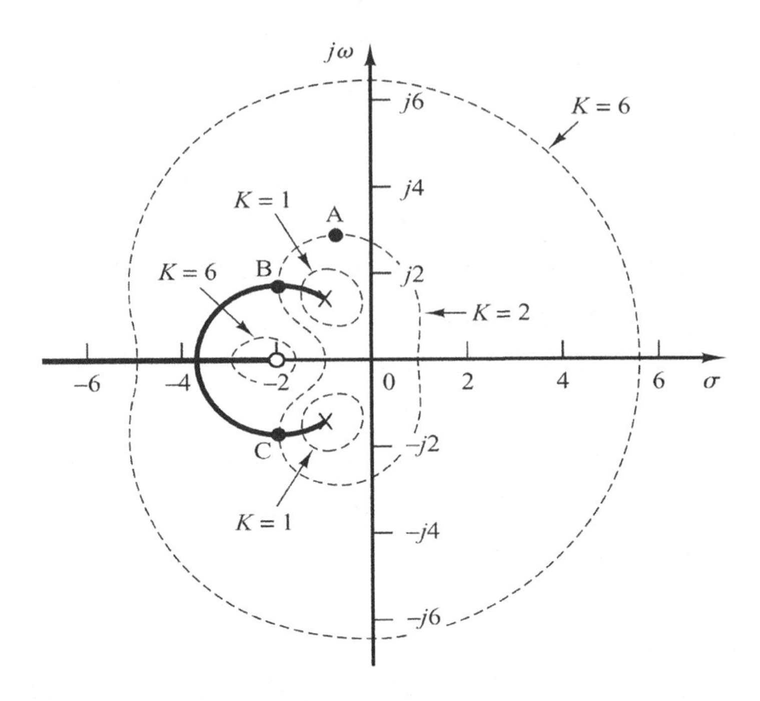

FIGURE 5.35 Symmetrical pole-zero and constant-gain loci.

The pole-zero and constant-gain loci are symmetrical about the real axis as we can see in Figure 5.35.

The root-loci and constant-gain loci are shown in Figure 5.35 from the following relations:

$$G(s)=\frac{K}{s(s+1)s(s+2)},H(s)=1$$

The poles in S-plane is symmetrical about real axis; the line parallel to $j\omega$ axis passing point ($\sigma = -1$, $\omega = 0$); constant-gain loci are symmetrical about $\omega = 0$ line (real axis) & $\sigma = -1$ line. Look at Figure 5.36 for orthogonality of root-loci and constant-gain loci, and then it can be easier to understand.

5.8 COMMENTS OF POSITIVE-FEEDBACK SYSTEMS

In this section, we may learn root-locus plots of positive-feedback systems, rules for positive-feedback systems, and negative-feedback and positive-feedback systems.

5.8.1 Roots-Locus Plots of Positive-Feedback Systems

In a complex control system, there may be a positive-feedback inner loop as shown in Figure 5.37. Such a loop is usually stabilized by the outer loop. In what follows, we shall be concerned only with the positive-feedback inner loop. The closed-loop transfer function of the inner loop is written as

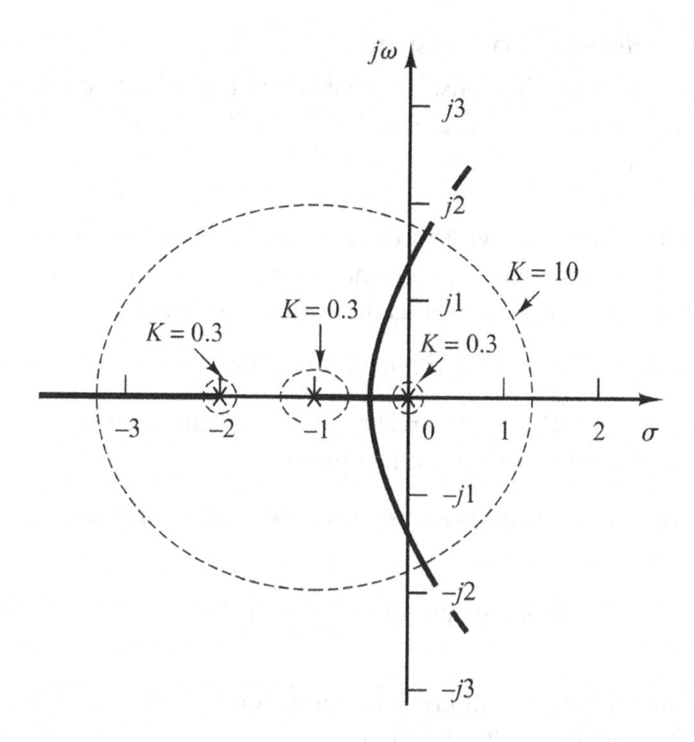

FIGURE 5.36 Pole-zero and constant-gain loci.

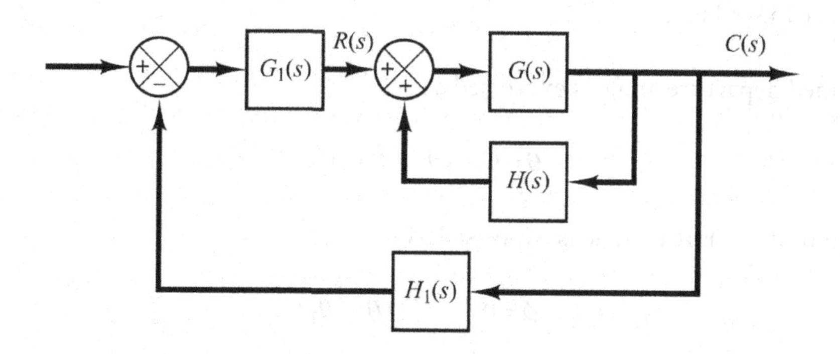

FIGURE 5.37 Positive-feedback systems.

$$\frac{C(s)}{R(s)} = \frac{G(s)}{1 - G(s)H(s)}$$

The characteristic equation is $1 - G(s)H(s) = 0$ or, $G(s)H(s) = 1$. Thus, the angle condition can be described as

$$\angle G(s)H(s) = 0° \pm k360°, \text{ for } k = 0,1,2\ldots \text{ and } |G(s)H(s)| = 1$$

5.8.2 Rules for Positive-Feedback Systems

To construct the Root-Loci of the positive-feedback system, we can construct the root-loci as like the drawing procedure of the negative-feedback system but some modified rules should be applied. Here,

Rule 1. Locate the poles and zeros of $G(s)H(s)$ on the S-plane. The root-locus branches start from open-loop poles and terminate at zeros (finite zeros or zeros at infinity), which will be the same procedure as the negative-feedback system.

Rule 2. The root-loci in the real axis must be modified.

If the total number of real poles and real zeros to the right of a test point on the real axis is even, then this test point lies on the root-locus;

Rule 3. The asymptotes of root-loci must be modified. The angle of asymptotes will be as,

$$\text{Angle of asymptotes} = \frac{\pm k360^\circ}{n-m}; \text{ for } k = 0,1,2,\ldots$$

Rule 4. Find the breakaway and break-in points for a positive-feedback system will be the same as a negative-feedback system.

Rule 5. Determine the angle of departure (angle of arrival) of the root-locus from a complex pole (at a complex zero) that needs to modify for a positive-feedback system as we can see here.

The modified departure angle is expressed as

$$\theta = 0^\circ - (\theta_1 + \theta_2) \pm \varnothing_1$$

And the modified arrival angle is expressed as

$$\varnothing = 0^\circ - \varnothing_1 \pm (\theta_1 + \theta_2)$$

Rule 6. Find the points where the root-loci may cross the imaginary axis;

Rule 7. Taking a series of test points in the broad neighborhood of the origin of the S-plane, sketch the root-loci; and

Rule 8. Determine closed-loop poles that remain the same as a negative-feedback system.

5.8.3 Negative-Feedback and Positive-Feedback Systems

Let's look at the negative-feedback and positive-feedback systems. The root-loci for negative (–ve) feedback systems are shown with heavy lines and curves and the root-loci for +ve feedback systems are shown with dashed lines and curves, as shown in Figure 5.38.

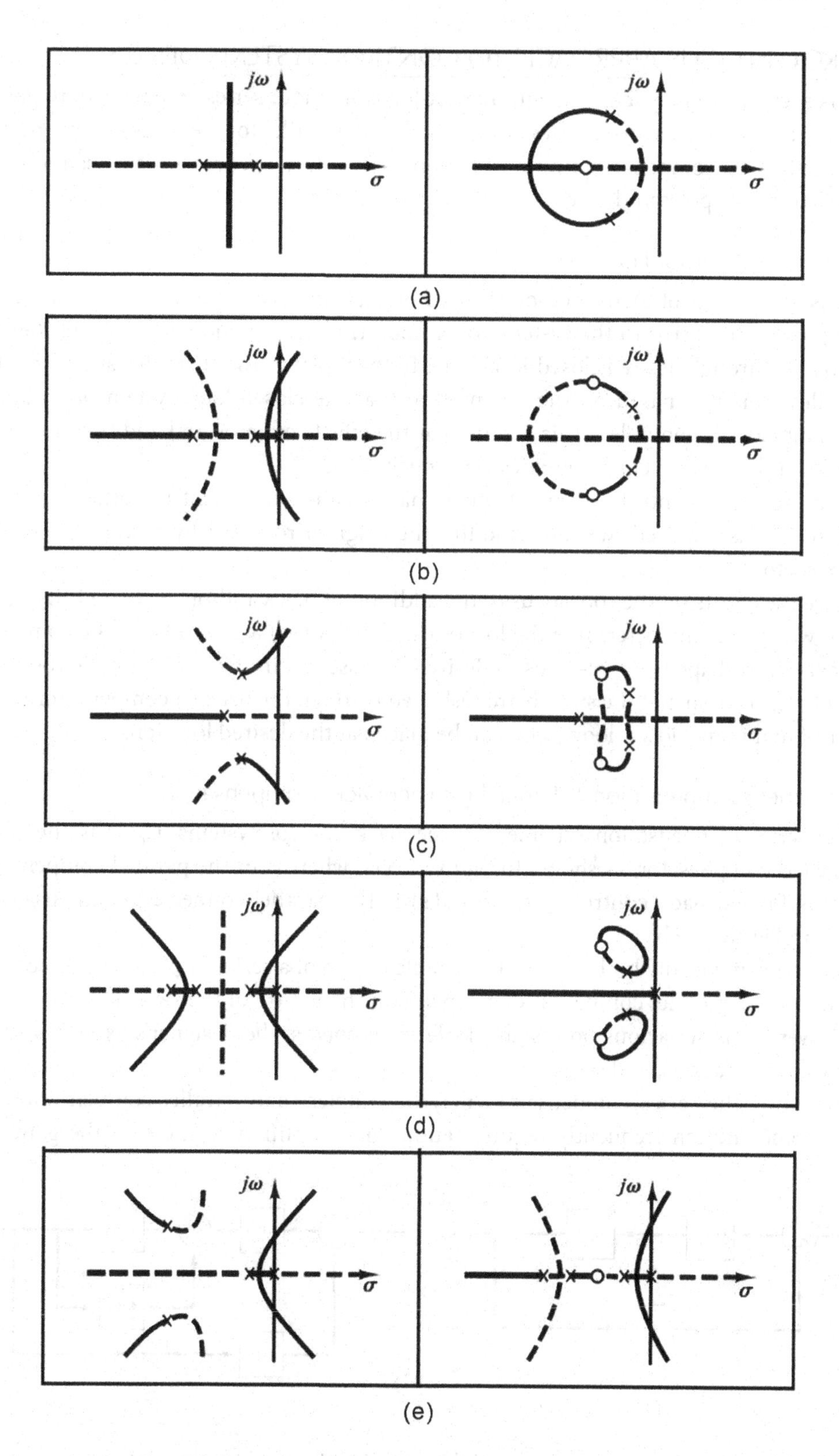

FIGURE 5.38 Negative-feedback and positive-feedback systems.

5.9 ROOT-LOCUS APPROACH TO CONTROL SYSTEMS DESIGN

In this section, we may learn design by root-locus method; series compensation; parallel (or feedback) compensation; series compensation & parallel (or feedback) compensation; commonly used compensators-output, region, types, device, control systems; and effects of the additional poles and zeros.

5.9.1 Design by Root-Locus Method

The design by the root-locus method is based on reshaping the root-locus of the system by adding poles and zeros to the system's open-loop transfer function and forcing the root-loci to pass through desired closed-loop poles in the S-plane. The characteristic of the root-locus design is it is based on the assumption that the closed-loop system has a pair of dominant closed-loop poles. This means that the effects of zeros and additional poles do not affect the response characteristics very much.

In designing a control system, if other than a gain adjustment (or other parameter adjustments) is required, we must modify the original root-loci by inserting a suitable compensator.

Once the effects on the root-locus of the addition of poles and/or zeros are fully understood, we can readily determine the locations of the pole(s) and zero(s) of the compensator that will reshape the root-locus as desired. In essence, in the design by the root-locus method, the root-loci of the system are reshaped through the use of a compensator so that a pair of dominant closed-loop poles can be placed at the desired location.

5.9.2 Series Compensation & Parallel (or Feedback) Compensation

In the series compensation schemes for feedback control systems, $G_c(s)$ is the series-connected compensator, as shown in Figure 5.39a, whereas in the parallel compensation schemes for feedback control systems, $G_c(s)$ is the parallel-connected compensator, as shown in Figure 5.39b.

The problem usually boils down to a suitable design of a series or parallel compensator. The series and parallel compensations depend on the nature of the signals in the system, the power levels at various points, available components, the designer's experience, economic considerations, and so on.

Therefore, the series compensation may be simpler than parallel compensation and series compensation frequently requires additional amplifiers to increase the gain and/

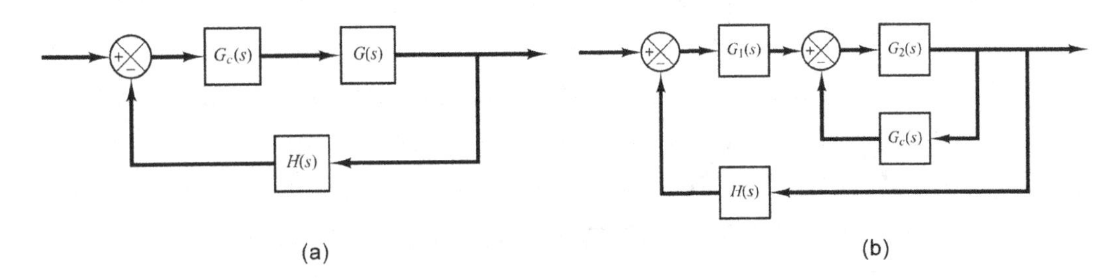

FIGURE 5.39 (a) Series compensation, and (b) parallel (or feedback) compensation.

or to provide isolation. To avoid power dissipation, the series compensator is inserted at the lowest energy point in the feedforward path. The number of components required in parallel compensation will be less than the number of components in series compensation; the energy transfer is from a higher power level to a lower level; this means that additional amplifiers may not be necessary. Thus, a parallel compensation technique uses a design of a velocity-feedback control system.

1. *Commonly used compensators (types)*
   The compensator is needed to meet the performance of a physical device of a system for physical construction. If a sinusoidal input is applied to the input of a network, and the steady-state output (which is also sinusoidal) has a phase lead, then the network is called a lead network. If the steady-state output has a phase lag, then the network is called a lag network. In a lag–lead network, both phase lag and phase lead occur in the output but in different frequency regions. The phase lag occurs in the low-frequency region but the phase lead occurs in the high-frequency region.
   A compensator has a characteristic of a lead network, lag network, or lag–lead network which is called a lead compensator, lag compensator, or lag–lead compensator. Many kinds of compensators are for the system considerations such as
   - Lead compensators,
   - Lag compensators,
   - Lag–lead compensators, and
   - Velocity-feedback (tachometer) compensators.
2. *Commonly used compensators-devices*
   Commonly used compensators-devices as lead, lag, and lag–lead compensators may be electronic devices using operational amplifiers circuits or RC networks such as electrical, mechanical, pneumatic, hydraulic, or combinations thereof and amplifiers.
3. *Commonly used compensators-control systems*
   The series compensators are used as lead, lag, and lag–lead compensators in the control systems. The PID controllers are used in industrial control systems. The final result is not unique to designing a control system by the root-locus or frequency-response methods; the best or optimal solution may not be precisely defined if the time domain specifications or frequency-domain specifications are given.

5.9.3 Effects of the Addition of Poles and Zeros

The effect(s) of pulling the root-locus to the right due to adding of a pole to the open-loop transfer function tends to lower the system's relative stability and to slow down the settling of the response. The system is less stable if add a pole at the origin of integral control. Figure 5.40a shows examples of root-loci illustrating the effects of the addition of a pole to a single-pole system; and two poles to a single-pole system as we can see in Figure 5.40b.

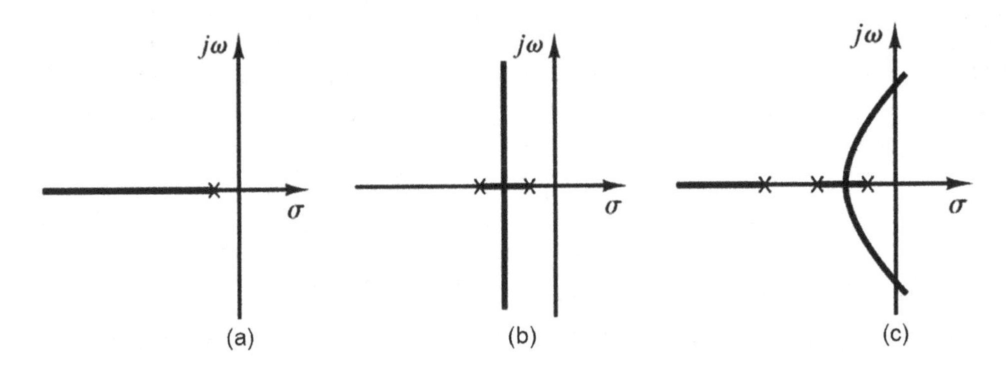

FIGURE 5.40 Effects of the addition of pole.

The effect(s) of pulling the root-locus to the left due to the addition of a zero to the open-loop transfer function tends to make the system more stable and speeds up the settling of the response. Physically, the addition of a zero in the feedforward transfer function means the addition of derivative control to the system. The effect(s) of such control is to introduce a degree of anticipation into the system and speed up the transient response. Figure 5.40c shows the root-loci for a system.

The system is stable for a small gain but unstable for a large gain as we can see in Figure 5.41. Figure 5.41 shows a root-locus plot for the system with zero in the open-loop transfer function. A zero is added to the first system to become stable for all values of gain.

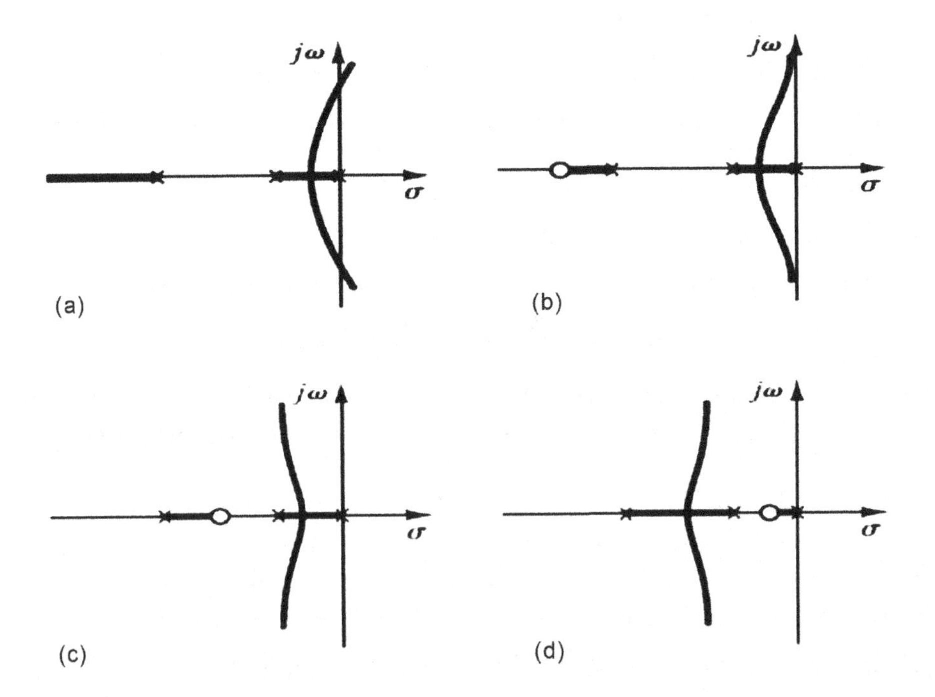

FIGURE 5.41 Effects of the addition of zero.

5.10 EXAMPLES OF ROOT-LOCUS APPLICATION BY MATLAB FUNCTION

5.10.1 Plot the Root-Loci

In MATLAB, the system provides a special function for root-locus such as **rlocous()** for the root-locus of the system; **rlocfind()** for the calculation of the root-locus gain as a given root; **pzmap()** for the plot of the zero-pole of the system and so on. These functions can be used conveniently, simply, and accurately to plot the system root-locus and calculate the parameters of the root-locus.

In the complex plane, by the function of MATLAB, we can plot the zeros and poles of the system at the same time. We use the notation "o" as zeros, while "×" as poles.

The instruction format of the root-locus in MATLAB function as in the following:

1. *Plot the pole-zero diagram by the system function pzmap()*

[*p*, *z*]=pzmap(A, B, C, D);	% **A, B, C, D** are the state matrixes representation by state space.
[*p*, *z*]=pzmap(sys);	% **Sys** is the transfer function of the system.
[*p*, *z*]=pzmap(num, den);	% **Num, den** is coefficient matrix function of numerator and denominator coefficients polynomial, respectively.
[*p*, *z*]=pzmap(*p*, *z*);	% ***p*, *z*** are the poles and zeros of the system.
pzmap(*p*, *z*);	% Plot the poles and zeros of the system directly.

2. *The function for plotting the poles and zeros of the system*

[*r*, *k*]=rlocus(A, B, C, D);	% **A, B, C, D** are the state matrixes representation by state space.
[*r*, *k*]=rlocus(sys);	% **Sys** is the transfer function of the system.
[*r*, *k*]=rlocus(num, den)	% ***r*, *k*** respectively is the complex root-locus matrix of the system and the corresponding gain vector.
rlocus()	% Plot the root-locus of the system directly.

3. *Determine the root-locus gain function as a given root of the system rlocfind()*

[*k*, *poles*]=rlcofind(sys)

[*k*, *poles*]=rlcofind(sys, p)

The **rlocfind()** function can be calculated with the root-locus corresponding gain (***k***) on the root-locus (location for the poles).

When **[*k*, *poles*]=rlcofind(sys)** performs, the cross cursor is displayed in the root-loci window. When the user selects any point on the root-locus, its corresponding gain is recorded as *k*, and all the associated poles with the gain are controlled by ***poles***.

4. *Determine the damping factor and natural frequency for continuous system root-locus and pole-zero diagrams function sgrid()*

sgrid(z, wn)

sgrid() command can be used in continuous-time system's root-locus or pole-zero diagram to plot grid lines. Grid lines are formed by the damping factor and natural frequency. Damping coefficient linear step is 0.1, ranging from $\xi=0$ to $\xi=1$, and the step of the natural frequency is 1 rad/s, ranging from $\omega_n=0$ to $\omega_n=10$. Before plotting grid lines, the window must have root-locus or pole-zero diagrams in a continuous-time system or it must use the function **pzmap()** or **rlocus()** together. However, **sgrid(z, wn)** can be specified as the damping coefficient z or ξ and the natural frequency ω_n.

Example 5.8

Consider one system whose open-loop transfer function is given by

$$G(s)=\frac{K(s+3)}{s(s+2)(s^2+10s+50)}$$

Plot the root-loci and the pole-zero diagram of the system.

Solution

According to the transfer function of the system, we obtain MATLAB programs of the root-locus as

```
num=[1 3];
den=conv(conv([1 0], [1 2]), [1 10 50]);
rlocus(num, den)
```

We obtain the root-locus as shown in Figure 5.42.

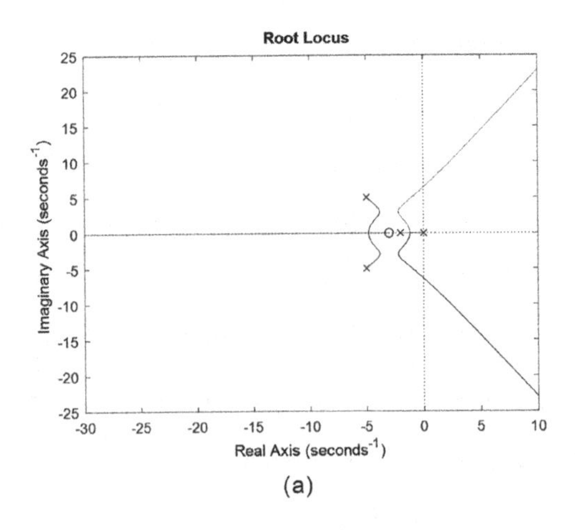

(a)

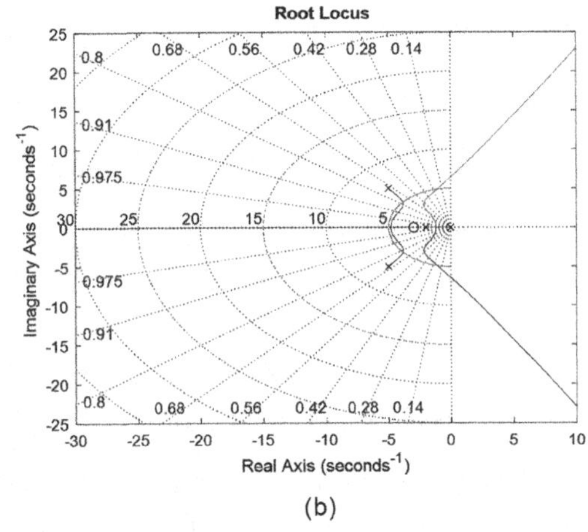

(b)

FIGURE 5.42 Root-locus of the system.

Example 5.9

Consider a system whose state-space model is given by

$$\frac{dx}{dt}=\begin{bmatrix} 5 & 2 & 0 & 0 \\ 0 & 4 & 0 & 0 \\ -3 & 2 & -4 & -1 \\ -3 & 2 & 0 & -4 \end{bmatrix}x(t)+\begin{bmatrix} 1 \\ 2 \\ 0 \\ 1 \end{bmatrix}u(t) \qquad y=[1,\ 2,\ 3,\ 4]x(t)$$

Determine the root-locus of the system.

Solution

According to the state of the system, we obtain MATLAB programs of the root-locus for

```
clear
A= [5 2 0 0; 0 4 0 0; -3 2 -4-1; -3 2 0 -4]
B= [1 2 0 1]'
C= [1 2 3 4]
D= [0]
rlocus(A, B, C, D)
[z, p]=rlocfind(A, B, C, D)
```

Select a point in the graphics window as,

```
selected_point= -8.7737+0.2531i
z = 0.9346
p = -8.7812
5.6640
-2.8687
-1.4254
```

And select a point in the graphics window as,

```
selected_point = -2.4290-0.4467i
z = 1.2936
p = -11.4357+0.0000i
5.6990+0.0000i
-2.4528+0.4347i
-2.4528-0.4347i
```

[*z*, *p*] on above is optional from the root-locus window. For the selected point = $-8.7737+0.2531i$, we obtain the root-locus as shown in Figure 5.43a, and for the selected point = $-2.4290-0.4467i$, we obtain the root-locus as shown in Figure 5.43b.

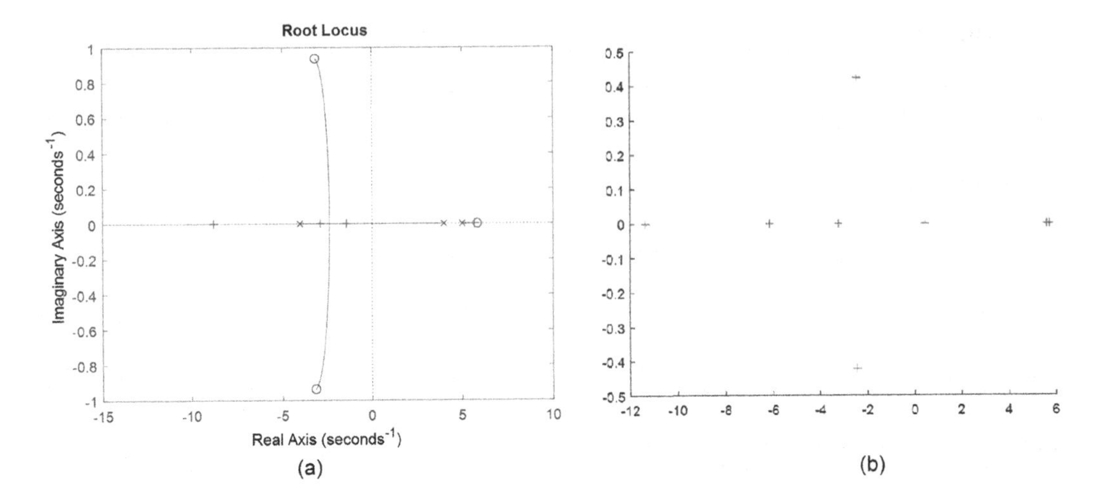

FIGURE 5.43 Root-locus of the state-space system.

Example 5.10

Consider a system whose delay transfer function is given by

$$G(s)=\frac{K(s+1)^2}{s^3\left(s^2+1.05s+12.25\right)}e^{-3s}$$

Determine the root-locus with delay factor and without delay factor of the system.

Solution

According to the transfer function of the system, we obtain MATLAB programs for

```
num1=conv([1 1], [1 1]);
den1=conv(conv(conv([1 0], [1 0]), [1 0]), [1 1.05 12.25]);
sys1=tf(num1, den1)
rlocus(sys1); sgrid
figure
[num2,den2]=pade(3,, 4);
sys2=tf(num2,den2)
rlocus(sys2)
figure
num=conv(num1,num2);
den=conv(den1,den2);
sys=tf(num, den)
rlocus(sys); sgrid
```

We obtain the root-locus as shown in Figure 5.44a–d. Figure 5.44a shows the root-locus of the system without a delay factor, Figure 5.44b shows the root-locus of the

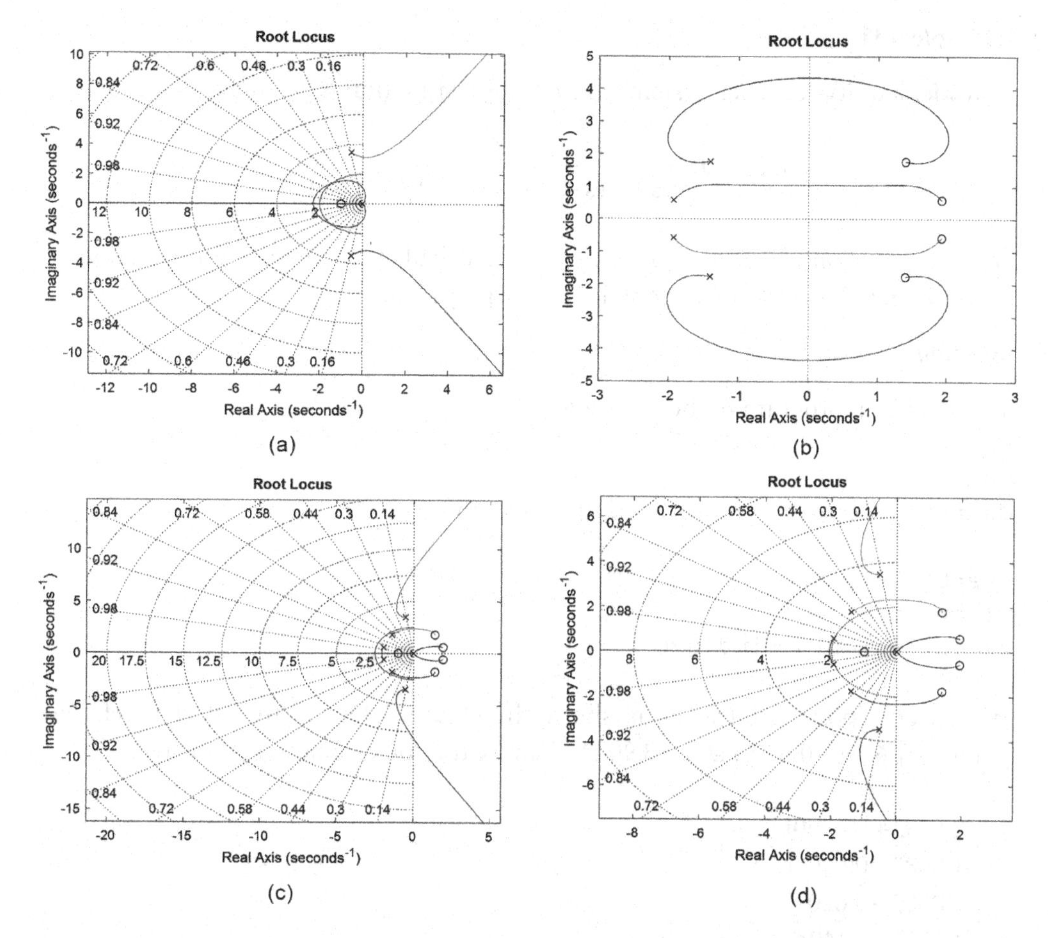

FIGURE 5.44 (a) The root-locus of the system without delay link, (b) the root-locus of the system with pure delay link, (c) the root-locus of the system with delay link, and (d) the root-locus of the system with local amplified delay factor.

system with a pure delay factor, Figure 5.44c shows the root-locus of the system with a delay factor and Figure 5.44d shows the root-locus of the system with local amplified delay factor.

5.10.2 Judge the Stability of the System Using the Root-Locus Methods

Firstly, we should plot root-locus of the system by the root-locus function **rlocus()**, and we should calculate all poles of the closed-loop system by the function of **rlocfind()**. The command of the called function is **[k, ploes] = rlocfind().** After the command is performed, the cross cursor is shown in the graphics window. When the user selects any point on the root-locus, its corresponding gain is controlled by k, all the poles with the gain are controlled by ***poles***, and their values are displayed on the screen, then we can judge the stability of the system.

Example 5.11

Consider a unity-feedback system whose transfer function is given by

$$G(s) = K\frac{s+2}{s(s+1)(s+3)(s^2+2s+2)(s^2+3s+3)}$$

Select a point on closed-loop root-locus and calculate the gain k and location of closed-loop poles, and judge the stability of the system.

Solution

The MATLAB program of the system is

```
num=[1 2];
den=conv(conv(conv(conv([1 0], [1 1]), [1 3]), [1 2 2]), [1 3
3]);
sys=tf(num, den);
rlocus(sys)
[k, poles]=rlocfind(sys)
```

We select a point arbitrarily as shown in Figure 5.45a; If we select a selected point = $-1.6231-0.0397i$ at $k = 3.9042$, then we the corresponding poles are

$$\begin{aligned}
&-2.9532+0.0000\text{i}\\
&-1.8087+0.6209\text{i}\\
&-1.8087-0.6209\text{i}\\
&-1.1272+1.3404\text{i}\\
&-1.1272-1.3404\text{i}\\
&-0.0875+0.4776\text{i}\\
&-0.0875-0.4776\text{i}
\end{aligned}$$

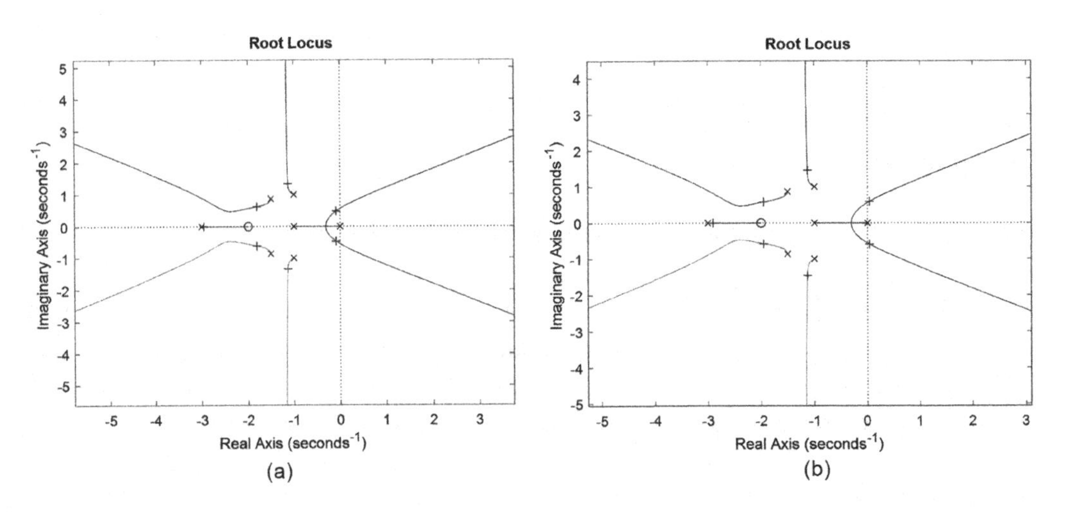

FIGURE 5.45 Root-loci of Example 5.10.

According to the output, we can judge that the system is stable as shown in Figure 5.45a. If we select another $K = 7.2723$ at the selected point $= -1.7292 + 0.0000i$, as shown in Figure 5.45b, the corresponding poles are

```
-2.9058+0.0000i
-1.9526+0.5815i
-1.9526-0.5815i
-1.1319+1.4583i
-1.1319-1.4583i
0.0375+0.5937i
0.0375-0.5937i
```

According to the output, we can judge that the system is unstable, as shown in Figure 5.45b.

EXAMPLES

This chapter requires students to learn the basic root-locus concept and the basic rules of plotting the root-locus and also learn how to plot root-loci of a simple system by using MATLAB, and analyze the system qualitatively based on root-loci.

Example 5.12

Prove that a part of root-loci is round as shown in Figure 5.46 and determine the center and radius of the circle.

Solution

Consider an open-loop system whose zero is at -4, and poles are at origin (0, 0 as two poles). Based on the rules of plotting root-loci, we can determine the root-loci of the system which is qualitatively shown in Figure 5.46.

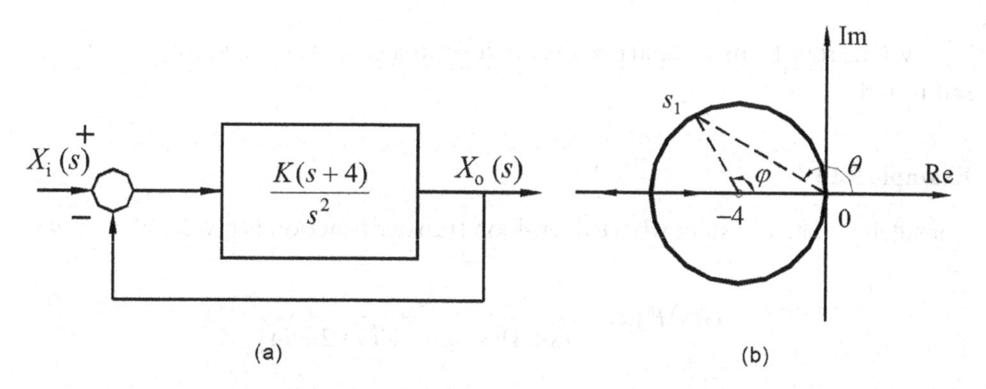

FIGURE 5.46 Negative feedback system and its root-loci.

Let's consider a point that is located in the curve section of the root-locus so that the point, $s_1 = - + j$.

From Figure 5.46b, the angle condition of root-locus, we obtain

$$\phi - 2\theta = \pm\pi$$

whereas

$$\tan\phi = -\frac{\omega}{\sigma - 4} \Rightarrow \phi = \tan^{-1}\left(-\frac{\omega}{\sigma - 4}\right)$$

However,

$$\tan\theta = -\omega/\sigma \Rightarrow \tan^{-1}(-\omega/\sigma)$$

Thus, we obtain

$$\tan^{-1}(-\frac{\omega}{\sigma - 4}) - 2\tan^{-1}(-\omega/\sigma) = \pm\pi$$

or

$$\tan^{-1}(-\frac{\omega}{\sigma - 4}) = \pm\pi + 2\tan^{-1}(-\omega/\sigma)$$

Taking tangents of both sides, we obtain

$$-\frac{\omega}{\sigma - 4} = \frac{2\left(-\frac{\omega}{\sigma}\right)}{1 - (-\frac{\omega}{\sigma})^2}$$

The simplified expression can be written as

$$(\sigma - 4)^2 + \omega^2 = 4^2$$

The system root-locus is a part of the circle with a center at the point (–4, 0) whose radius is 4.

Example 5.13

Consider a control system whose open-loop transfer function is given by

$$G(s)H(s) = \frac{K^*}{s(s+4)(s+2-j4)(s+2+j4)}$$

Sketch of the corresponding root-loci of the system.

Solution

1. Locate the open-loop poles $p1=0, p2=-4, p3=-2+j4, p4=-2-j4$ as shown in Figure 5.47.
2. Determine the root-locus on the real axis. We have only one real pole among the poles on real axis in the interval of [−4,0]; thus, the root-locus is on the real axis in the interval of [−4,0].
3. Determine the number of the root-locus branch(s) and the asymptote(s). Since $n=4, m=0$, there are $n-m=4$ root-locus which all approach infinity as K^* approaches to infinity. The asymptotes and the real axis intersection are

$$\sigma_a=\frac{\sum_{j=1}^{n}p_j-\sum_{i=1}^{m}z_i}{n-m}=-2$$

The angle between the root-locus and real axis is

$$\phi_a=\frac{(2k+1)\pi}{n-m}=\frac{(2k+1)\pi}{4};\text{ for } k=0,\pm1,\pm2$$

4. Determine the breakaway and break-in point. The equation for the break-in points is determined as

$$\frac{1}{d}+\frac{1}{d+4}+\frac{1}{d+2+j4}+\frac{1}{d+2-j4}=0$$

Thus, we get

$$d_1=-2, d_{2,3}=-2\pm j\sqrt{6}$$

When $d_1=-2$, then $K^*=64$, when $d_{2,3}=-2\pm j\sqrt{6}$, then $K^*=100$. Therefore, d_1, d_2, and d_3 all are the breakaway point of the root-locus.
5. To determine the departure angle (starting angle) of the root-locus, the complex open-loop poles are $p_{3,4}=-2\pm j4$, where

$$\theta_{p_3}=-90,\ \theta_{p_4}=90$$

6. Determine the crossing point or intersection point of the root-locus on the imaginary axis. The characteristic equation of the closed-loop the system is

$$s^4+8s^3+26s^2+80s+K^*=0$$

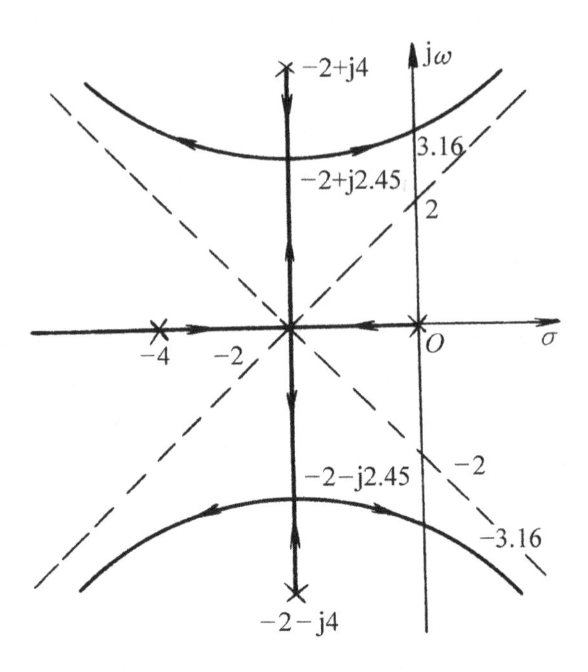

FIGURE 5.47 Intersection point of root-locus.

Thus, Routh's array becomes

s^4	1	36	K^*
s^3	8	80	
s^2	26	K^*	
s^1	$\dfrac{80\times 26-8K^*}{26}$		
s^0	K^*		

When $K^*=260$, Routh's array appears to all zero; thus, the auxiliary equation is written as

$$26s^2+260=0$$

We obtain the crossing point or intersection point of the root-locus and imaginary axis at $s_{1,2}=\pm j3.16$ from the above equation. We sketch the corresponding closed-loop root-loci plot as shown in Figure 5.47.

EXERCISES

1. Knowing the open-loop zero z and pole p, sketch the corresponding closed-loop root-locus. Considering the open-loop zero z, pole p, sketch the corresponding closed-loop root-loci plot.

1) $z = -2,-6, p=0,-3;$ 2) $z = -2,-4, p=0,-6;$
3) $p1 = -1, p_{2,3} = -2 \pm j1;$ 4) $z = -6,-8, p=0,-3;$
5) $p=0,-2, z=-4 \pm j4;$ 6) $p=0,-1,-5, z=-4,-6$

2. Consider unity-feedback systems whose transfer functions are given by

$$G(s)H(s) = \frac{K^*(s+z)}{s(s+p)};\ for\ z > p > 0$$

Plot the closed-loop root-loci as K^* varies from zero to infinity and prove that the root-locus is a circle (except the root-locus on the real axis), and calculate the center and radius.

3. Consider unity-feedback systems whose transfer functions are given in the following. Sketch the corresponding closed-loop root-loci plot. (It is required to determine the separation point coordinates d.)

1. $G(s) = \dfrac{K}{s(0.2s+1)(0.5s+1)}$; 2. $G(s) = \dfrac{K(s+1)}{s(2s+1)}$; 3. $G(s) = \dfrac{K^*(s+5)}{s(s+2)(s+3)}$

4. Consider unity-feedback systems whose transfer functions are given in the following. Sketch the corresponding closed-loop root-loci plot. (It is required to calculate the initial angle and termination angle.)

1. $G(s) = \dfrac{K^*(s+2)}{(s+1+j2)(s+1-j2)}$; 2. $G(s) = \dfrac{K^*(s+20)}{s(s+10+j10)(s+10-j10)}$;

3. $G(s) = \dfrac{K^*(s^2+2s+2)}{(s+5)(s^2+s+4)}$

5. Consider a feedback control system that is given by

$$G(s) = \frac{K^*}{s^2(s+2)(s+5)},\ H(s) = 1$$

Requirement:

1. Sketch the root-loci plot, and judge the stability of the closed-loop system.
2. If the transfer function of the feedback transfer function is changed to $H(s) = 1+2s$, judge the stability of the system after changing of $H(s)$, and study the effect due to change of $H(s)$.

6. Consider systems whose open-loop transfer functions are given as in the following. Sketch the root-loci plot of the system, and determine the range of system stability.

1. $G(s)H(s) = \dfrac{K^*(s^2+2s+4)}{s(s+3)(s^2+0.5)}$; 2. $G(s)H(s) = \dfrac{K^*(s^2+2s+4)}{s(s+4)(s+6)(s^2+1.4s+1)}$

7. Consider a feedback control system that is given by

   1. Sketch the root-loci plot for the closed-loop system with $H(s)=\frac{s+4}{s}$.
   2. Sketch the root-loci plot for the closed-loop system with $H(s)=\frac{s+1.05}{s}$.
   3. Compare the effects of the open-loop zero-point change on the shape of the root-locus.

8. Consider a unity negative-feedback system whose open-loop transfer function is given by

$$G(s)=\frac{K^*(s+2)}{s(s+1)(s+3)}$$

   1. Plot the closed-loop root-loci as K* varies from zero to infinity.
   2. Determine a pair of domain poles where $\xi = 0.707$ and calculate the value of the K.

9. Consider a unity-feedback system whose transfer function is given by

$$G(s)=\frac{K^*}{s(s+1)(s+2)(s+8)}$$

   Determine the value of K such that the system closed main poles have the damping ratio $\xi = 0.5$ (K is open-loop gain of the system).

10. Consider a control system whose open-loop transfer function is given by

$$G(s)=\frac{K^*(s+1)}{s^2(s+2)(s+4)}$$

   Plot root-locus for the positive-feedback system and negative-feedback system, respectively, and point out their difference in stability.

11. Consider the control system shown in Figure 5.48. What is the value of K_f that the system has the complex-conjugate pole $\xi = 0.7$? Are they the dominant poles of the system and why?

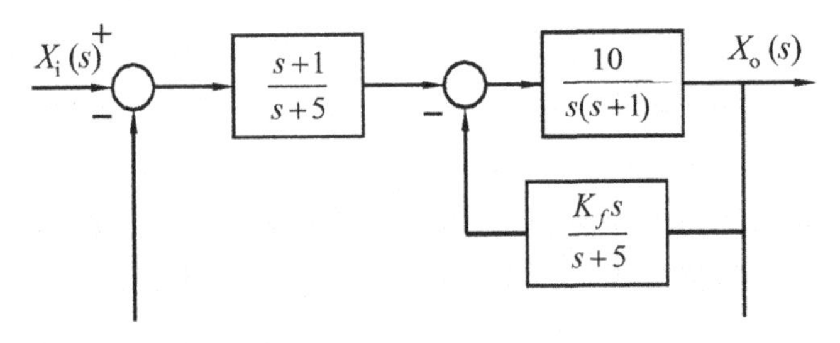

FIGURE 5.48 An example of control system.

12. Consider a unity-feedback system whose open-loop transfer function is given by

$$G(s)=\frac{K}{s(s+1)(0.5s+1)}$$

Determine the system closed-loop poles of a pair of conjugate complex-conjugate poles who's the damping ratio $\xi=0.5$. Determine open-loop gain K and analyze the time domain performance of the system approximately.

13. Consider the block diagrams of the system with local feedback as shown in Figure 5.49.
Requirements:
    1. Plot root-loci of the closed-loop system as K varies from zero to infinity.
    2. Determine the value of K and the value of the closed-loop poles using the root-locus method such that the system has the damping ratio $\xi=0.5$ (for a pair of complex-conjugate closed-loop poles).
    3. Determine the allowable accuracy value of the steady-state control system using the root-locus method in the action of unit-step signal.

14. Consider the system block diagrams shown in Figure 5.50.
Requirements:
    1. Sketch the root-loci plot of the system.
    2. Determine the range of system stability using the root-locus method.
    3. Determine the maximum value of the Kg of the system step response without overshoot using the root-locus method.

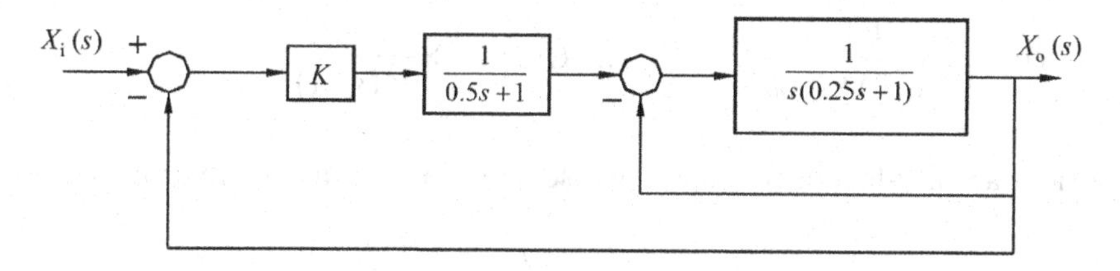

FIGURE 5.49 An example of control system for step response.

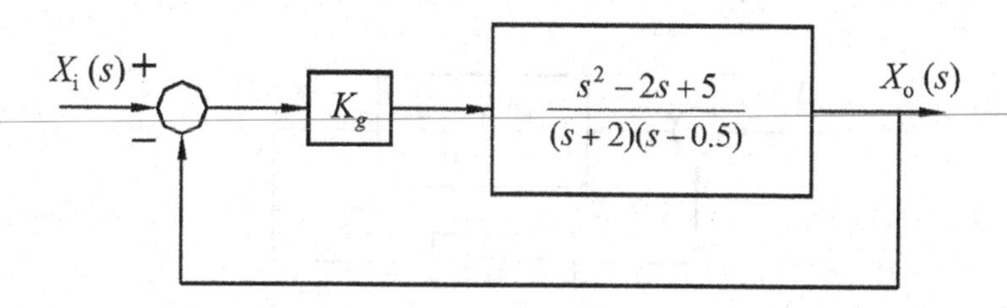

FIGURE 5.50 An example of control system with K_g.

15. Consider a unity-feedback system whose open-loop transfer function is given by

$$G(s)=\frac{K(0.5s-1)^2}{(0.5s+1)(2s-1)}$$

Requirements:

1. Plot the root-loci roughly of the closed-loop system as K varies from zero to infinity.
2. Determine the range of gain K for stability.
3. Determine the value of the absolute minimum $|e_{ss}|_{min}$ such that the steady-state error of the system could reach a minimum in the action of a unit-step input.

16. Consider the system block diagrams shown in Figure 5.51.

1. Plot the root-loci of the closed-loop system with $Kh=0$ as K varies from zero to infinity.
2. Determine the closed-loop poles of the system and the corresponding value for $Kh=0.5$, $K=10$.
3. Plot the Parameter root-loci as K varies from zero to infinity.
4. When $K=1$, determine the step response index *Mp, ts* for $Kh=0, 0.5, 4$, respectively, and discuss the impact of the size of Kh on the dynamic performance of the system.

17. Consider open-loop systems whose transfer functions are as follows, trying to plot the root-locus as b varies from zero to infinity.

1. $G(s)=\dfrac{20}{(s+4)(s+b)}$
2. $G(s)=\dfrac{30(s+b)}{s(s+10)}$
3. $G(s)=\dfrac{100}{s[(s+10)+100b]}$
4. $G(s)=\dfrac{1}{4}\times\dfrac{(s+b)}{s^2(s+1)}$

18. Consider a unity-feedback system whose open-loop transfer function is given by

$$G(s)=\frac{K^*}{s(s+5)(s^2+6s+25)}$$

Calculate and plot the system root-loci.

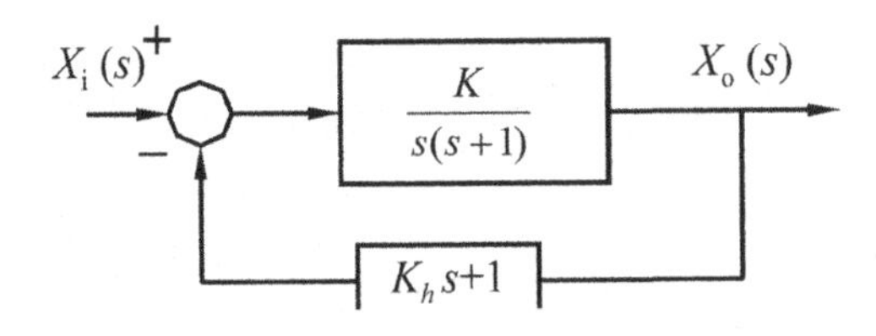

FIGURE 5.51 An example of feedback control system with K_h.

19. Consider a unity-feedback system whose open-loop transfer function is given by

$$G(s)=\frac{K^*(s+1)}{s(s+0.5)(s^2+10s+50)}$$

Using the procedures for calculation of the root-loci, calculate and plot the system root-loci; and also require selecting the appropriate value of K^* to enable the system to have better dynamic performance.

MCQ AND TRUE/FALSE QUESTIONS

1. The roots of the characteristic equation (the closed-loop poles) corresponding to a given value of the gain can be determined from the magnitude condition.
   Answer: True/False
2. The root-loci for the system are the loci of the closed-loop poles as the gain K is varied from zero to infinity. If the gain $K<0$ (called positive feedback), we are interested in this; if $K>0$ (called negative feedback), for this angle, condition must be modified.
   Answer: True/False
3. The complex quantities originating from the open-loop poles and open-loop zeros to the test point s are measured in the counterclockwise direction.
   Answer: True/False
4. The open-loop complex-conjugate poles and complex-conjugate zeros, if any, are always located symmetrically about the real axis, the root-loci are always symmetrical with respect to this axis. Therefore, we only need to construct the upper half of the root-loci and draw the mirror image of the upper half in the lower-half S-plane.
   Answer: True/False
5. The root-loci are symmetrical about the real axis of the S-plane, because the complex poles and complex zeros occur only in conjugate pairs.
   Answer: True/False
6. The number of open-loop poles generally exceeds that of zeros, the number of branches equals that of poles.
   Answer: True/False
7. If the number of closed-loop poles is the same as the number of open-loop poles, then the number of individual root-locus branches terminating at finite open-loop zeros is equal to the number m of the open-loop zeros.
   Answer: True/False
8. If we include poles and zeros at infinity, the number of open-loop poles is equal to that of open-loop zeros. Hence, we can always state that the root-loci start at the poles of $G(s)H(s)$ and end at the zeros of $G(s)H(s)$, as K increases from zero to

infinity, where the poles and zeros include both those in the finite S-plane and those at infinity.

Answer: True/False

9. The complex-conjugate poles and complex-conjugate zeros of the open-loop transfer function have no effect on the location of the root-loci on the real axis because the angle contribution of a pair of complex-conjugate poles or complex-conjugate zeros is 180° on the real axis.

   Answer: True/False

10. If the total number of real poles and real zeros to the right of this test point is odd, then this point lies on a root-locus.

    Answer: True/False

11. The asymptotes of the root-loci as $s \to 0$, if a test point s is very far from the origin.

    Answer: True/False

12. The root-loci for very large values of s must be asymptotic to straight lines whose angles (slopes) are given by this relation.

$$\text{Angle of asymptotes} = \frac{\pm 180°(2k+1)}{n-m}; \text{ for } k = 0,1,2,\ldots$$

    Answer: True/False

13. The conjugate symmetry of the root-loci, the breakaway points, and break-in points either lie on the real axis or occur in complex-conjugate pairs.

    Answer: True/False

14. If a root-locus lies between two adjacent open-loop poles on the real axis, then there exists at least one breakaway point between the two poles.

    Answer: True/False

15. If the root-locus lies between two adjacent zeros (one zero may be located at ∞) on the real axis, then there always exists at least one break-in point between the two zeros.

    Answer: True/False

16. If the root-locus lies between an open-loop pole and a zero (finite or infinite) on the real axis, then there may exist no breakaway or break-in points or there may exist both breakaway and break-in points.

    Answer: True/False

17. To plot root-loci accurately, we must find a breakaway point, where the root-locus branches originating from the poles at breakaway (as K is decreased) from the real axis and move into the complex plane.

    Answer: True/False

18. The breakaway point corresponds to a point in the S-plane where multiple roots of the characteristic equation occur.
    Answer: True/False

19. The angle of departure (or angle of arrival) of the root-locus from a complex pole (or at a complex zero) can be found by subtracting from 180° the sum of all the angles of vectors from all other poles and zeros to the complex pole (or complex zero) in question, with appropriate signs included.
    Answer: True/False

20. Angle of departure from a complex pole = 280° − (sum of the angles of vectors to a complex pole in question from other poles) ± (sum of the angles of vectors to a complex pole in question from zeros).
    Answer: True/False

21. If a test point is on the root-loci, then must satisfy the angle condition, $\theta_1+\theta_2+\theta_3=360°$.
    Answer: True/False

22. If K increases, some of the roots move on locus toward the left, and other roots must move toward the right. It is also noted that a slight change in the pole-zero configuration may cause significant changes in the root-locus configurations.
    Answer: True/False

23. The complex plane the damping ratio ξ of a pair of complex-conjugate poles can be expressed in terms of the angle $\varnothing$, which is measured from the negative real axis with $\xi=2\cos$.
    Answer: True/False

24. If the real part of a pair of complex-conjugate poles is positive, which means that the system is *unstable*, the corresponding ξ is negative.
    Answer: True/False

25. The damping ratio determines the angular location of the poles, while the distance of the pole from the origin is determined by the undamped natural frequency ω_n. The constant ω_n loci are lines.
    Answer: True/False

26. If $\xi=0\sim1$, the complex-conjugate poles lie on the lines $\phi=0\sim\pm190°$.
    Answer: True/False

27. If $\xi\geq0$, ϕ & poles are in the negative real axis, (stable).
    Answer: True/False

28. If $\xi<0$, ϕ & poles are in the positive real axis, (unstable).
    Answer: True/False

29. Conditional stability may be eliminated by adding proper compensation.
    Answer: True/False

30. The characteristic equation of a positive-feedback system is $1-G(s)H(s)=0$ or, $G(s)H(s)=1$. Thus, the angle condition can be described like this.

$$\angle G(s)H(s)=0^\circ \pm k360^\circ, \text{ for } k=0,1,2\ldots \text{ and } |G(s)H(s)|=1$$

Answer: True/False

31. Determine the asymptotes of root-loci must be modified for a positive-feedback system. The angle of asymptotes will be as follows:

$$\text{Angle of asymptotes} = \frac{\pm k360^\circ}{n-m}; \text{ for } k=0,1,2,\ldots$$

Answer: True/False

32. However, Rule 5. Determine the angle of departure (angle of arrival) of the root-locus from a complex pole (at a complex zero) that needs to modify for a positive-feedback system as we can see here.

The modified departure angle is expressed as

$$\theta = 0^o - (\theta_1 + \theta_2) \pm \varnothing_1$$

And the modified arrival angle is expressed as

$$\varnothing = 0^o - \varnothing_1 \pm (\theta_1 + \theta_2)$$

Answer: True/False

33. In designing a control system, if other than a gain adjustment (or other parameter adjustments) is required, we must modify the original root-loci by inserting a suitable compensator.
Answer: True/False

34. If a sinusoidal input is applied to the input of a network, and the steady-state output (which is also sinusoidal) has a phase lead, then the network is called a lead network.
Answer: True/False

35. A compensator having a characteristic of a lead network, lag network, or lag–lead network is called a lead compensator, lag compensator, or lag–lead compensator.
Answer: True/False

36. The effect of pulling the root-locus to the left due to adding of a pole to the open-loop transfer function.
Answer: True/False

37. The system is less stable if add a pole at the origin of integral control.
Answer: True/False

38. Physically, the addition of a pole in the feedforward transfer function means the addition of derivative control to the system.
    Answer: True/False

39. The effect of such control is to introduce a degree of anticipation into the system and speed up the transient response.
    Answer: True/False

CHAPTER 6

Control System Analysis and Design by Frequency-Response Analyses

6.1 CHARACTERISTICS OF FREQUENCY

In Chapter 4, the time domain method is introduced. However, without the aid of a computer, it is very difficult to analyze higher-order systems. Therefore, some other methods need to be developed to analyze the control system. Among them, frequency characteristic is an indirect method of analysis and synthesis system often used in engineering.

In mechatronics engineering, mechanical vibration is closely related to frequency characteristics. When the mechanical equipment is affected by a certain frequency, the system will produce downbeat vibrations, and the positive feedback will also cause the system to produce self-excited vibrations. The resonance frequency, spectral density, dynamic stiffness, and antivibration stability in mechanical vibration dynamics can be attributed to the frequency characteristics of electromechanical systems in the frequency domain. The frequency-domain method will establish these concepts simply and clearly.

6.1.1 Mathematical Basis of Frequency Method

When the system inputs are a nonsinusoidal periodic signal, its input can be expanded by the Fourier series into the superposition of a sine wave, and its output is still the superposition of a sine wave, as shown in Figure 6.1. When the input is an aperiodic signal, you can view it as a periodic signal with period T going to infinity. The mathematical basis of the frequency method is described by Fourier transform.

$$\sum_{n=1}^{N} a_n \sin(n\omega t+\theta_n) \qquad \sum_{n=1}^{N} A_n \sin\left[(n\omega t+\theta_n)+\varphi_m\right]$$

DOI: 10.1201/9781003293859-6

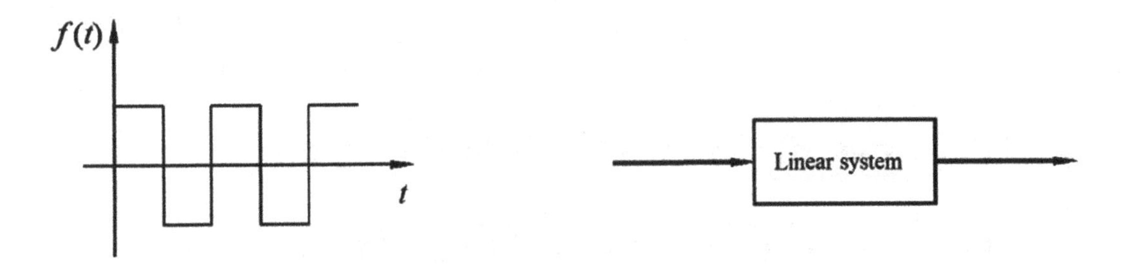

FIGURE 6.1 Steady-state response of periodic signal input of a linear system.

Suppose that the period of periodic function $f(t)$ is T, then the angular frequency is

$$\omega = 2\pi / T$$

The function $f(t)$ can be expanded by Fourier's law, and then we get

$$f(t) = A_0 + \sum_{n=1}^{\infty} (A_n \cos n\omega t + B_n \sin n\omega t) \tag{6.1}$$

Using the Euler method, we get

$$\cos n\omega t = \frac{e^{jn\omega t} + e^{-jn\omega t}}{2} \tag{6.2}$$

$$\sin n\omega t = \frac{e^{jn\omega t} - e^{-jn\omega t}}{2j} \tag{6.3}$$

Substituting Eqs. (6.2) and (6.3) into Eq. (6.1), we get

$$C_n = \frac{A_n - jB_n}{2}$$

$$D_n = \frac{A_n + jB_n}{2}$$

Therefore,

$$f(t) = A_0 + \sum_{n=1}^{\infty} \left(C_n e^{jn\omega t} + D_n e^{-jn\omega t} \right) \tag{6.4}$$

According to the Fourier series, we get the coefficient as

$$A_n = \frac{2}{T} \int_{\frac{-T}{2}}^{\frac{T}{2}} f(t) \cos n\omega t dt \quad (n = 0, 1, 2, 3 \ldots) \tag{6.5}$$

$$B_n = \frac{2}{T}\int_{\frac{-T}{2}}^{\frac{T}{2}} f(t)\sin n\omega t dt \quad (n=1,2,3\ldots) \tag{6.6}$$

and

$$C_n = \frac{A_n - jB_n}{2} = \frac{1}{T}\int_{-\frac{T}{2}}^{\frac{T}{2}} f(t)\, \mathrm{e}^{-jn\omega t} dt \quad (n=1,2,3,\ldots) \tag{6.7}$$

$$D_n = \frac{A_n + jB_n}{2} = \frac{1}{T}\int_{-\frac{T}{2}}^{\frac{T}{2}} f(t)\, \mathrm{e}^{jn\omega t} dt \quad (n=1,2,3,\ldots) \tag{6.8}$$

From Eqs. (6.7) and (6.8), we get

$$C_{-n} = D_n \tag{6.9}$$

and because of

$$C_0 = \frac{1}{T}\int_{-\frac{T}{2}}^{\frac{T}{2}} f(t)\, \mathrm{e}^0 dt = \frac{1}{T}\int_{-\frac{T}{2}}^{\frac{T}{2}} f(t)\, \mathrm{d}t = A_0/2$$

Eq. (6.4) can be expressed as

$$f(t) = \sum_{-\infty}^{+\infty} C_n e^{jn\omega t} \tag{6.10}$$

Consequently,

$$C_n = \frac{1}{T}\int_{-\frac{T}{2}}^{\frac{T}{2}} f(t)\, \mathrm{e}^{-jn\omega t} dt \quad (n=0,\pm1,\pm2\ldots) \tag{6.11}$$

Fourier transform is derived on the basis of the Fourier series. For the general nonperiodic function $f(t)$, it can be regarded as the periodic function when the period t approaches infinity, and then Eq. (6.10) can be written as

$$f(t) = \lim_{T\to\infty} \sum_{-\infty}^{\infty} \left[\frac{1}{T}\int_{-\frac{T}{2}}^{\frac{T}{2}} f(\tau)\, e^{-jn\omega\tau} d\tau \right] e^{jn\omega t}$$

$$= \lim_{T\to\infty} \sum_{-\infty}^{+\infty} \left[\frac{\Delta\omega}{2\pi} \int_{-\frac{T}{2}}^{\frac{T}{2}} f(\tau)\, e^{-jn\omega\tau} d\tau \right] e^{jn\omega t}$$

$$= \frac{1}{2\pi} \lim_{\substack{T\to\infty \\ \Delta\omega\to 0}} \sum_{-\infty}^{+\infty} \left[\int_{-\frac{T}{2}}^{\frac{T}{2}} f(\tau)\, e^{-jn\omega\tau} d\tau \right] e^{jn\omega t} \Delta\omega$$

$$= \frac{1}{2\pi} \int_{-\infty}^{+\infty} \left[\int_{-\infty}^{+\infty} f(\tau)\, e^{-j\omega\tau} d\tau \right] e^{j\omega t} d\omega \tag{6.12}$$

According to the definition of Fourier's law, we get

$$F(\omega) = \int_{-\infty}^{+\infty} f(\tau)\, e^{-j\omega\tau} d\tau$$

Thus, the integral property is expressed as

$$F(\omega) = \int_{-\infty}^{+\infty} f(t)\, e^{-j\omega t} dt \tag{6.13}$$

According to Eqs. (6.12) and (6.13), the following can be obtained

$$f(t) = \frac{1}{2\pi} \int_{-\infty}^{+\infty} F(\omega)\, e^{j\omega t} d\omega \tag{6.14}$$

Eq. (6.13) is called Fourier positive transform, which transforms the time domain function $f(t)$ into the frequency domain function $F(\omega)$, while Eq. (6.14) is called the inverse Fourier transform, which transforms the frequency-domain function $F(\omega)$ into the time domain function $f(t)$.

Sufficient conditions for the existence of Fourier transformations:

1. $F(t)$ and $f'(t)$ should be piecewise.
2. $f(t)$ is in the interval $(-\infty, \infty)$and absolutely integrable as $\int_{-\infty}^{+\infty} |f(t)|\, dt$.

Some basic properties of Fourier's law:

1. Additive law or asymmetry:
   Thus,

$$F[x_1(t)] = X_1(\omega)$$

$$F[x_2(t)] = X_2(\omega)$$

   Therefore,

$$F[a_1x_1(t) + a_2x_2(t)] = a_1X_1(\omega) + a_2X_2(\omega)$$

2. Differential theorem:
   If

$$\lim_{t \to \pm\infty} x^{(i)}(t) = 0 (i = 1, 2 \ldots n-1)$$

   Then

$$F\left[\frac{d^n x(t)}{dt^n}\right] = (j\omega)^n X(\omega)$$

3. Integral property:
   If

$$F[x(t)] = X(\omega)$$

   Then

$$F\left[\int_{-\infty}^{t} f(t)dt\right] = \frac{1}{j\omega} X(\omega)$$

4. Frequency shifting property:
   If

$$F[x(t)] = X(\omega)$$

   Then

$$F\left[e^{\pm j\omega_0 t} x(t)\right] = X(\omega \mp \omega_0)$$

5. Delay theorem:
   If

$$F[x(t)] = X(\omega)$$

   Then

$$F[x(t-\tau)] = e^{-j\omega t} X(\omega)$$

6. Scaling theorem:
   If

$$F[x(t)] = X(\omega)$$

   Then

$$F\left[x\left(\frac{t}{a}\right)\right] = aX(a\omega)$$

7. Convolution theorem:
   If

$$F[x_1(t)] = X_1(\omega)$$

$$F[x_2(t)] = X_2(\omega)$$

   Then

$$F[x_1(t)^* x_2(t)] = X_1(\omega)X_2(\omega)$$

According to the above seven properties of Fourier transform, the property of Fourier transforms and Laplace transform is a similar phenomenon. The *s* in the Laplace transform is changed to $j\omega$, which we call the Fourier transform formula. According to the definition of Fourier transform and Laplace transform, the Laplace transform can be used as a kind of unilateral generalized Fourier transform; in the interval $(0, +\infty)$, the Fourier transform is obviously stronger than the Laplace transform; however, the Laplace transform is applied to a wider range than Fourier transform. Therefore, to study the frequency characteristic of the system, the Laplace transform $G(s)$ can be obtained by replacing *s* in $j\omega$.

Frequency characteristic is defined as the ratio of Fourier transform of output to Fourier transform of input, namely

$$G(j\omega)=\frac{Y(j\omega)}{X(j\omega)}$$

6.1.2 Steady-State Output (O/P) to Sinusoidal Input (I/P)

We shall show that the steady-state output of a transfer function system can be obtained directly from the sinusoidal transfer function—that is, the transfer function in which s is replaced by $j\omega$, where ω is the frequency.

Consider the stable, linear, time-invariant (LTI) system shown in Figure 6.2.

The input and output of the system, whose transfer function is $G(s)$, are denoted by $x(t)$ and $y(t)$, respectively. If the input $x(t)$ is a sinusoidal signal, the steady-state output will also be a sinusoidal signal of the same frequency, but with possibly different magnitude and phase angle.

Let us assume that the input signal to the system is given by

$$x(t)=X\sin\omega t$$

Suppose that the transfer function $G(s)$ of the system can be written as a ratio of two polynomials in s, and we can say

$$G(s)=\frac{p(s)}{q(s)}=\frac{p(s)}{(s+s_1)(s+s_2)\ldots(s+s_n)}$$

The Laplace-transformed output $Y(s)$ of the system is expressed as

$$Y(s)=G(s)X(s)=\frac{p(s)}{q(s)}X(s)$$

where $X(s)$ is the Laplace transform of the input $x(t)$. It will be shown that after waiting until steady-state conditions are reached, the frequency response can be calculated by replacing s in the transfer function by $j\omega$. It will also be shown that the steady-state response can be given by the following relationship.

$$G(j\omega)X(s)=Me^{j\varphi}=M\angle\varphi$$

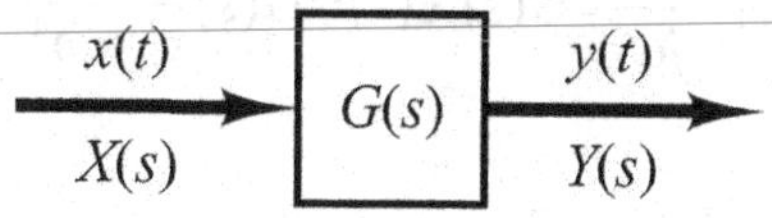

x(t)
X(s)
G(s)
y(t)
Y(s)

FIGURE 6.2 Block diagram of a stable, linear, time-invariant (LTI) system.

where M is the amplitude ratio of the output and input sinusoids and φ is the phase shift between the input sinusoid and the output sinusoid. In the frequency-response test, the input frequency ω is varied until the entire frequency range of interest is covered.

The steady-state response of a stable, linear, time-invariant system to a sinusoidal input does not depend on the initial conditions.

Thus, we can assume the zero-initial condition. If $Y(s)$ has only distinct poles, then the partial fraction expansion of $Y(s)=G(s)X(s)$; when $x(t)=X\sin\omega t$, we can say

$$Y(s)=G(s)X(s)=G(s)\frac{\omega X}{s^2+\omega^2}$$

or

$$Y(s)=\frac{a}{s+j\omega}+\frac{\bar{a}}{s-j\omega}+\frac{b_1}{s+s_1}+\frac{b_2}{s+s_2}+\cdots+\frac{b_n}{s+s_n}$$

where a and the b_i (where $i=1, 2, \ldots, n$) are constants and $\bar{a}$ is the complex conjugate of a. The inverse Laplace transform of $Y(s)$ gives

$$y(t)=ae^{-j\omega t}+\bar{a}e^{j\omega t}+b_1e^{-s_1t}+b_2e^{-s_2t}+\cdots+b_ne^{-s_nt},\text{for}\,t\geq 0$$

For a stable system, $-s_1, -s_2, \ldots, -s_n$ have negative real parts. Therefore, as t approaches infinity, the terms $e^{-s_1t}, e^{-s_2t}, \ldots,$ and e^{-s_nt} approach zero. Thus, all the terms on the right-hand side of $y(t)$, except the first two terms, drop out at a steady state. If $Y(s)$ involves multiple poles s_j of multiplicitym_j, then $y(t)$ will involve the terms such as $t^{h_j}e^{-s_jt}$ $(h_j=0,1,2,\ldots, m_j-1)$. For a stable system, the terms $t^{h_j}e^{-s_jt}$ approach zero as t approaches infinity. Thus, regardless of whether the system is of the distinct-pole type or multiple-pole type, the steady-state response becomes

$$y_{ss}(t)=ae^{-j\omega t}+\bar{a}e^{j\omega t}$$

where the constant a can be evaluated from

$$Y_{ss}(s)=\frac{a}{s+j\omega}+\frac{\bar{a}}{s-j\omega}$$

Note that, these two terms can be simplified as

$$\frac{a}{s+j\omega}=G(s)X(s)=G(s)\frac{\omega X}{s^2+\omega^2}$$

and

$$\frac{\bar{a}}{s-j\omega}=G(s)X(s)=G(s)\frac{\omega X}{s^2+\omega^2}$$

Then, the quantity a and $\bar{a}$ can be found as

$$a = G(s)\frac{\omega X}{s^2+\omega^2}(s+j\omega)\Big|_{s=-j\omega} = -\frac{XG(-j\omega)}{2j}$$

and

$$\bar{a} = G(s)\frac{\omega X}{s^2+\omega^2}(s-j\omega)\Big|_{s=j\omega} = \frac{XG(j\omega)}{2j}$$

Since $G(j\omega)$ is a complex quantity, the magnitude and the angle can be written in the following form:

$$G(\pm j\omega) = |G(\pm j\omega)|e^{\pm j\varphi} = |G(j\omega)|e^{\pm j\varphi}$$

and

$$\varphi = \angle G(\pm j\omega) = \tan^{-1}\left[\frac{\text{imaginary part of } G(\pm j\omega)}{\text{real part of } G(\pm j\omega)}\right]$$

Then, noting that the quantity a and $\bar{a}$ can be found as

$$a = -\frac{X|G(j\omega)|e^{-j\varphi}}{2j},\ \bar{a} = \frac{X|G(j\omega)|e^{j\varphi}}{2j}$$

Thus, $Y_{ss}(t)$ can be expressed as

$$y_{ss}(t) = -\frac{X|G(j\omega)|e^{-j\varphi}}{2j}e^{j\omega t} + \frac{X|G(j\omega)|e^{j\varphi}}{2j}e^{j\omega t}$$

or

$$y_{ss}(t) = X|G(j\omega)|\frac{e^{j(\omega t+\varphi)} - e^{-j(\omega t+\varphi)}}{2j}$$

$$= X|G(j\omega)|\sin(\omega t+\varphi) = Y\sin(\omega t+\varphi)$$

Thus,

$$y_{ss}(t) = Y\sin(\omega t+\varnothing)$$

Here, $Y = X|G(j\omega)|$. We see that a stable, linear, time-invariant system subjected to a sinusoidal input will, at steady state, have a sinusoidal output of the same frequency as the input,

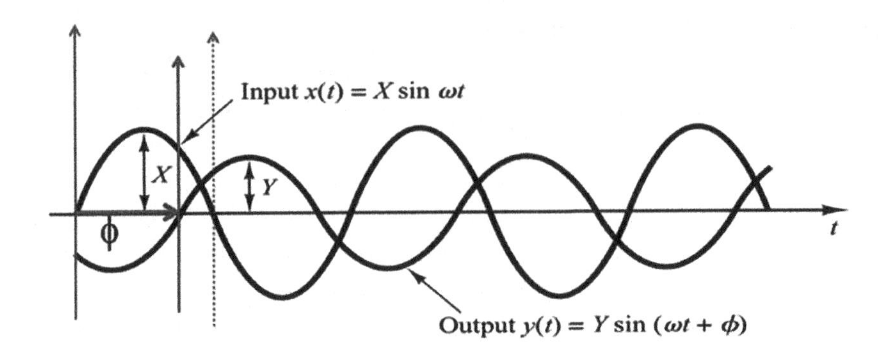

FIGURE 6.3 Input and output sinusoidal signal.

but the amplitude and phase of the output will, in general, be different from those of the input. In fact, the amplitude of the output is given by the product of that of the input and $|G(j\omega)|$, while the phase angle differs from that of the input by the amount $\varnothing = \angle G(j\omega)$. An example of input and output sinusoidal signals is shown in Figure 6.3.

On the basis of this, we obtain this important result: For sinusoidal inputs, the steady-state response characteristics of a system to a sinusoidal input can be obtained directly from

$$\frac{Y(j\omega)}{X(j\omega)} = G(j\omega)$$

The function $G(j\omega)$ is called the sinusoidal transfer function. It is the ratio of $Y(j\omega)$ to $X(j\omega)$, is a complex quantity, and can be represented by the magnitude and phase angle with frequency as a parameter.

$$|G(j\omega)| = \left|\frac{Y(j\omega)}{X(j\omega)}\right|$$

and

$$\angle G(j\omega) = \angle\left(\frac{Y(j\omega)}{X(j\omega)}\right)$$

The sinusoidal transfer function of any linear system is obtained by substituting $j\omega$ for s in the transfer function of the system where $s = j\omega$. As already mentioned previously, a positive phase angle is called phase lead, and a negative phase angle is called phase lag. A network that has phase-lead characteristics is called a lead network, while a network that has phase-lag characteristics is called a lag network.

6.1.3 The Basic Concept of Frequency Characteristics

Figure 6.4 shows the Resistance-Capacitance (RC) network. For the system shown in Figure 6.4, when the input is a sinusoidal signal, the output is also a sinusoidal signal when

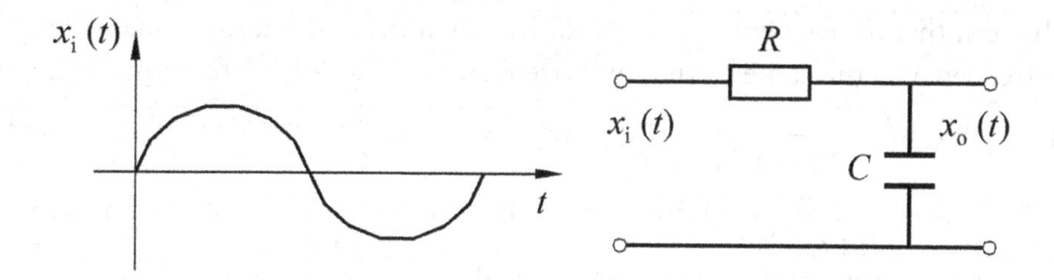

FIGURE 6.4 Steady-state response of sinusoidal input and electrical network.

it is stable. The frequency of the output sinusoidal signal is the same as that of the input signal, and the amplitude of the output is less than that of the input; thus, the output phase lags behind the input phase. However, when the amplitude of the input signal remains unchanged but the frequency changes, the output amplitude and phase generally change with the frequency of the input signal. The network can be described as

$$T\frac{dc(t)}{dt}+c(t)=r(t)$$

Then, the transfer function of RC network is expressed as

$$\frac{C(s)}{R(s)}=\frac{1}{Ts+1} \tag{6.15}$$

Suppose the input signal of the network is

$$x_i(t)=A\sin\omega t$$

Then, it can be obtained from Eq. (6.15)

$$C(s)=\frac{1}{Ts+1}R(s)=\frac{1}{Ts+1}\frac{A\omega}{s^2+\omega^2} \tag{6.16}$$

The inverse Laplace transform of Eq. (6.16) is written as

$$c(t)=\frac{A\omega T}{1+\omega^2T^2}e^{-\frac{t}{T}}+\frac{A}{\sqrt{1+\omega^2T^2}}\sin\left(\omega t-\tan^{-1}\omega T\right) \tag{6.17}$$

The first term in Eq. (6.17) is the transient component of the output voltage. The second term is the steady-state component. As $t\to\infty$, the first term goes to zero, and then we get

$$\lim_{t\to\infty}x_o(t)=\frac{A}{\sqrt{1+\omega^2T^2}}\sin\left(\omega t-\tan^{-1}\omega T\right) \tag{6.18}$$

It can be seen that the steady-state output of RC networks is still a sinusoidal voltage curve and its frequency is the same as the input frequency, the steady-state output amplitude of the input amplitude is $\frac{1}{\sqrt{1+\omega^2T^2}}$; the phase angle is led by the input voltage phase angle $\tan^{-1}\omega T$, whereas $\frac{1}{\sqrt{1+\omega^2T^2}}$ is a function of $-\tan^{-1}\omega T$; $\frac{1}{\sqrt{1+\omega^2T^2}}$ is called the amplitude-frequency characteristic of RC, $-\tan^{-1}\omega T$ is called the phase-frequency characteristic of RC.

As we can see, when $\omega \to \infty$, the amplitude-frequency characteristics $\frac{1}{\sqrt{1+\omega^2T^2}} \to 0$ and the phase-frequency characteristics $-\tan^{-1}\omega T \to -\frac{\pi}{2}$. In the RC network, x_i, x_o, and x_r are the input voltages, that are effective value at both ends of the output voltage and resistance; because the ω is very small, the tolerance is very large; the amplitude of x_i, and x_o are almost equal and the phase angles are also almost equal, as $\omega \to \infty$, then the capacitive reactance goes to zero; thus, the amplitude of x_o approaches to 0, and the effective value at both ends of phase-lag resistance, x_r is 90°, as shown in Figure 6.5.

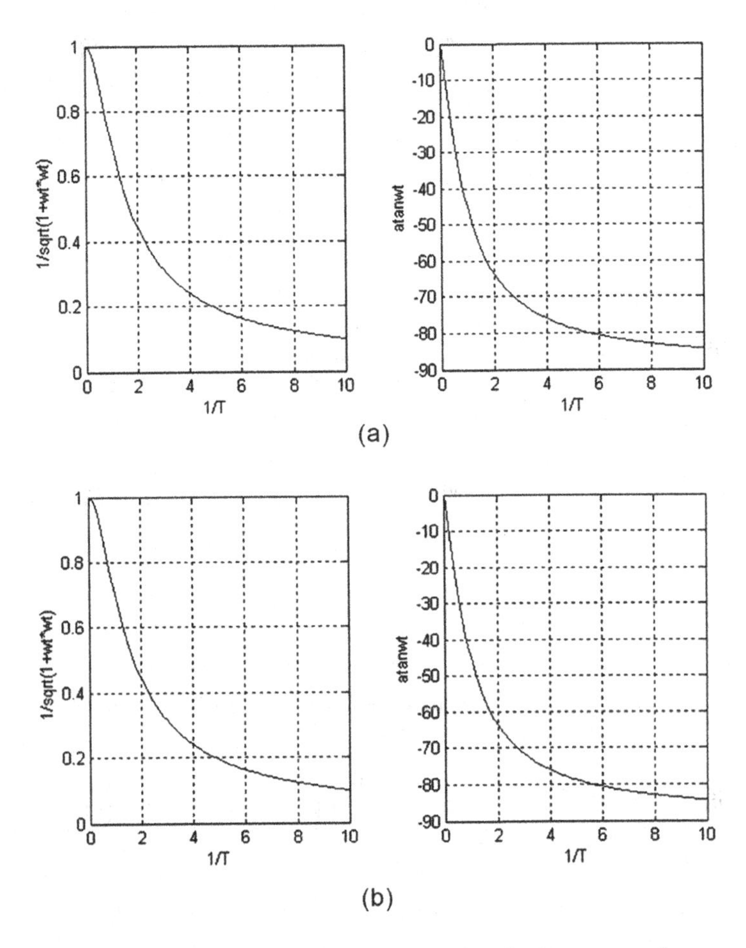

FIGURE 6.5 Phase-frequency characteristics: (a) amplitude-frequency characteristic, and (b) phase-frequency characteristic.

Figure 6.2 shows the transfer function of RC network as

$$G(s)=\frac{1}{RCs+1} \tag{6.19}$$

By substituting $s=j\omega$ Eq. (6.19), the amplitude-phase frequency characteristics of RC network can be obtained:

$$G(j\omega)=\frac{1}{RCj\omega+1} \tag{6.20}$$

This property of RC circuits is true for any stable linear network. The differential equation of general linear system as shown in Figure 6.6b is

$$\begin{aligned} &x_o^{(n)}(t)+a_1x_o^{(n-1)}(t)+\cdots+a_{n-1}\dot{x}_o(t)+a_nx_o(t) \\ &=b_0x_i^{(m)}(t)+b_1x_i^{(m-1)}(t)+\cdots+b_{m-1}\dot{x}_i(t)+b_mx_i(t) \end{aligned} \tag{6.21}$$

The transfer function of the system is obtained by Laplace transform from Eq. (6.21),

$$G(s)=\frac{b_0s^m+b_1s^{m-1}+\cdots+b_{m-1}s+b_m}{s^n+a_1s^{n-1}+\cdots+a_{n-1}s+a_n} \tag{6.22}$$

If a time harmonic function is added to the system as an input, then we may say

$$x_i(t)=x_o\cos(\omega t+\varphi) \tag{6.23}$$

In Eq. (6.23), x_o is the amplitude, ω is the angular frequency, and φ is the phase angle. For simplicity, let's say $\varphi=0$, and then

$$x_i(t)=x_o\cos(\omega t)$$

Hence,

$$x_i(t)=x_o\cos\omega t=\frac{x_o}{2}e^{j\omega t}+\frac{x_o}{2}e^{-j\omega t} \tag{6.24}$$

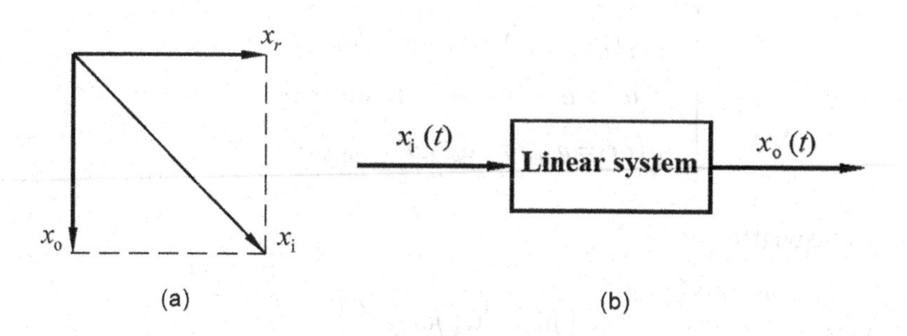

FIGURE 6.6 (a) Vector diagram of RC network, and (b) Block diagram of linear system.

Only the exponential function of the various changes is still exponential because of the derivatives and integrals; thus, the sines and cosines into the exponential function to study more convenient.

From Eq. (6.21), we get

$$x_i(t)=\frac{x_o}{2}e^{j\omega t} \tag{6.25}$$

The steady-state component of the particular solution of Eq. (6.21) is assumed to be

$$x_{o1}(t)=\frac{x_o}{2}G(j\omega)e^{j\omega t} \tag{6.26}$$

We obtain from Eq. (6.21),

$$\frac{x_o}{2}\left[(j\omega)^n+a_1(j\omega)^{n-1}+\quad+a_{n-1}j\omega+a_n\right]G(j\omega)e^{j\omega t}$$

$$=\frac{x_o}{2}\left[b_0(j\omega)^m+b_1(j\omega)^{m-1}+\quad+b_{m-1}j\omega+b_m\right]e^{j\omega t}$$

Hence,

$$G(j\omega)=\frac{b_0(j\omega)^m+b_1(j\omega)^{m-1}+\cdots+b_{m-1}j\omega+b_m}{(j\omega)^n+a_1(j\omega)^{n-1}+\cdots+a_{n-1}j\omega+a_n} \tag{6.27}$$

Recall $G(j\omega)$ as the frequency characteristic of the system and write as the real part and imaginary part of the numerator and denominator of Eq. (6.27), and then we get

$$G(j\omega)=\frac{a(\omega)+jb(\omega)}{c(\omega)+jd(\omega)} \tag{6.28}$$

In Eq. (6.28), we say

$$\begin{cases} a(\omega)=b_m-b_{m-2}\omega^2+b_{m-4}\omega^4-\cdots \\ b(\omega)=b_{m-1}\omega-b_{m-3}\omega^3+b_{m-5}\omega^5-\cdots \\ c(\omega)=a_n-a_{n-2}\omega^2+a_{n-4}\omega^4-\cdots \\ d(\omega)=a_{n-1}\omega-a_{n-3}\omega^3+a_{n-5}\omega^5-\cdots \end{cases} \tag{6.29}$$

Eq. (6.28) can be written as

$$G(j\omega)=\left|G(j\omega)\right|e^{j\angle G(j\omega)} \tag{6.30}$$

From Eq. (6.30), the magnitude and the phase angle can be found as

$$\begin{cases} \left|G(j\omega)\right| = \sqrt{\dfrac{a^2(\omega)+b^2(\omega)}{c^2(\omega)+d^2(\omega)}} \\ \angle G(j\omega) = \tan^{-1}\dfrac{b(\omega)c(\omega)-a(\omega)d(\omega)}{a(\omega)c(\omega)+b(\omega)d(\omega)} \end{cases} \tag{6.31}$$

Thus, the output of Eq. (6.26) is

$$x_{o1}(t) = \frac{x_o}{2}\left|G(j\omega)\right|e^{j(\omega t+\angle G(j\omega))} \tag{6.32}$$

If $x_i(t) = \frac{x_o}{2}e^{-j\omega t}$, then the output of Eq. (6.21) can be written as

$$x_{o2}(t) = \frac{x_o}{2}G(-j\omega)\ e^{-j\omega t} \tag{6.33}$$

Since in Eq. (6.29), $a(\omega)$ and $c(\omega)$ are the even function of ω, $a(\omega)$ and $c(\omega)$ are the odd function of ω; however, in Eq. (6.31), $\left|G(j\omega)\right|$ is the even function of ω, and $\angle G(j\omega)$ are the odd function of ω.

$$x_{o2}(t) = \frac{x_o}{2}\left|G(j\omega)\right|e^{-j[\omega t+\angle G(j\omega)]} \tag{6.34}$$

Thus, $x_i(t) = x_o\cos\omega t$ is the steady-state output component of the input signal, we get

$$x_o(t) = x_{o1}(t)+x_{o2}(t) = \frac{x_o}{2}\left|G(j\omega)\right|\left[e^{j(\omega t+\angle G(j\omega))}+e^{-j(\omega t+\angle G(j\omega))}\right]$$

or

$$x_o(t) = x_o\left|G(j\omega)\right|\cos\left(\omega t+\angle G(j\omega)\right) \tag{6.35}$$

Eq. (6.35) shows that the steady-state component caused is a harmonic function of time by harmonic action in a stable linear time-invariant (LTI) system. The amplitude and phase of the steady-state component are different from the amplitude and phase of the input signal, but the frequency is the same. The amplitude of the steady-state component is $\left|G(j\omega)\right|$, and the phase angle is $\angle G(j\omega)$, as shown in Eq. (6.31). The following relation exists between the frequency characteristics of the system and the transfer function of the system by comparing Eqs. (6.22) and (6.27).

$$G(j\omega) = G(s)\big|s = j\omega \tag{6.36}$$

The frequency characteristic function can be rewritten as follows:

$$G(j\omega)=U(\omega)+jV(\omega) \tag{6.37}$$

Here,

$U(\omega):G(j\omega)$ is real component which is called the real-frequency characteristics, and

$V(\omega):G(j\omega)$ is imaginary part which is called imaginary or virtual frequency characteristics.

The amplitude-frequency characteristic of $G(j\omega)$ is defined as

$$A(\omega)=|G(j\omega)|=\sqrt{U^2(\omega)+V^2(\omega)} \tag{6.38}$$

The phase-frequency characteristic of $G(j\omega)$ is defined as

$$\Phi(\omega)=\angle G(j\omega)=\tan^{-1}\left[\frac{V(\omega)}{U(\omega)}\right] \tag{6.39}$$

For every certain ω, there are certain phase angles, $\Phi(\omega)\in\left[-\frac{\pi}{2},\frac{\pi}{2}\right]$; to define the primary phase-frequency characteristics, use lowercase $\varphi(\omega)$, and then we get

$$\varphi(\omega)=\tan^{-1}\left[\frac{V(\omega)}{U(\omega)}\right] \tag{6.40}$$

The above vector can be expressed in the following vector: as

$$U(\omega)=A(\omega)\cos\varphi(\omega) \tag{6.41}$$

$$V(\omega)=A(\omega)\sin\varphi(\omega) \tag{6.42}$$

$$G(j\omega)=U(\omega)+jV(\omega)=A(\omega)[\cos\varphi(\omega)+j\sin\varphi(\omega)]=A(\omega)e^{j\varphi(\omega)} \tag{6.43}$$

The vector diagram is shown in Figure 6.7.

Example 6.1

The closed-loop transfer function of a control system is

$$G(s)=\frac{K}{s(T_1s+1)(T_2s+1)}$$

Find the frequency characteristic of the system.

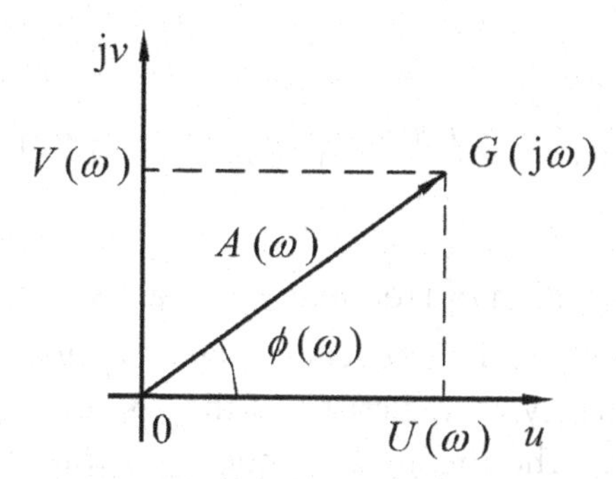

FIGURE 6.7 Vector diagram of frequency characteristics.

Solution

Before solving this example, briefly review the complex number properties as we can say

$$(x_1 + jy_1)(x_2 + jy_2)$$

$$= \sqrt{x_1^2 + y_1^2}\, e^{j\tan^{-1}\frac{y_1}{x_1}} \sqrt{x_2^2 + y_2^2}\, e^{j\tan^{-1}\frac{y_2}{x_2}}$$

$$= \sqrt{x_1^2 + y_1^2}\sqrt{x_2^2 + y_2^2}\, e^{j\left(\tan^{-1}\frac{y_1}{x_1} + \tan^{-1}\frac{y_2}{x_2}\right)}$$

Thus, the frequency characteristic of the system is

$$G(j\omega) = K\frac{1}{j\omega(T_1 j\omega + 1)(T_2 j\omega + 1)}$$

According to the complex property, we get

$$G(j\omega) = \frac{K}{\omega}e^{j\left(-\frac{\pi}{2}\right)} \cdot \frac{e^{j\left(-\tan^{-1}T_1\omega\right)}}{\sqrt{(T_1\omega)^2 + 1}} \cdot \frac{e^{j\left(-\tan^{-1}T_2\omega\right)}}{\sqrt{(T_2\omega)^2 + 1}}$$

$$= \frac{K}{\omega\sqrt{(T_1\omega)^2 + 1}\sqrt{(T_2\omega)^2 + 1}}e^{j\left[-\left(\frac{\pi}{2} + \tan^{-1}T_1\omega + \tan^{-1}T_2\omega\right)\right]}$$

Therefore, the amplitude-frequency characteristic and phase-frequency characteristic of the system are expressed, respectively, as

$$A(\omega) = \frac{K}{\omega\sqrt{(T_1\omega)^2 + 1}\sqrt{(T_2\omega)^2 + 1}}$$

and

$$\varphi(\omega) = -\left(\frac{\pi}{2} + \tan^{-1} T_1\omega + \tan^{-1} T_2\omega\right)$$

6.1.4 Geometric Representation of Frequency Characteristics

In control engineering analysis, the frequency characteristic is usually drawn into some curves, and then the frequency characteristic of the system is studied from the curve of the frequency characteristic. The amplitude-frequency characteristic curve and the phase-frequency characteristic curve are commonly used in RC network.

The amplitude-frequency characteristic curve and the phase-frequency characteristic curve are called amplitude-phase curve. They are an important set of curves in the frequency response; they are characterized by ω taking as an independent variable; the amplitude-phase curve of the frequency characteristic is plotted on the complex plane.

The amplitude and phase curves of the RC network are plotted on the s-plane, as shown in Figure 6.7. The direction of the positive real axis of the graph is the zero line of the phase angle, the angle of clockwise rotation is negative, and the angle of counterclockwise rotation is positive. For a certain ω, there must be a definite amplitude-frequency characteristic $A(\omega)$ and a certain phase-frequency characteristic $\phi(\omega)$.

For example,

$$\omega = 0,\ A(\omega) = 1,\ \phi(\omega) = 0$$

and $\omega = \frac{1}{T}$, $A(\omega) = 1/\sqrt{1+(\omega T)^2} = \sqrt{2}/2$, $\phi(\omega) = -\tan^{-1}\omega T = -\frac{\pi}{4}$;

However, if $\omega \to \infty$, $A(\omega) = 0, \varphi(\omega) = -\frac{\pi}{2}$.

Corresponding the points on s-plane, we say as $(1,0)$, $\left(\frac{\sqrt{2}}{2}, -\frac{\pi}{4}\right)$, and $\left(0, -\frac{\pi}{2}\right)$ which are both represented as a vector. The frequency ω changes from $0 \sim \infty$ as the amplitude-phase vector trajectory that is called the amplitude-phase curve, this set of phase curves is called a polar plot or diagram; also known as Nyquist plot. The mapping relationship between the imaginary axes on the s-plane and the Nyquist plots is shown in Figure 6.8c.

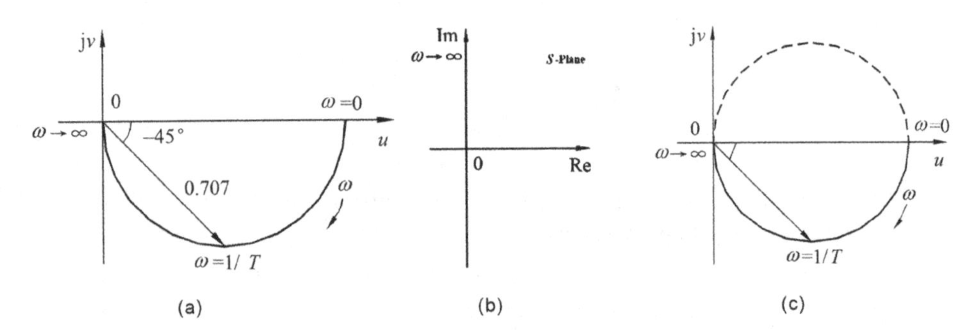

FIGURE 6.8 (a) Nyquist plots of RC networks, (b) Nyquist curve and the mapping, and (c) relationship between imaginary axis.

6.2 CHARACTERISTICS OF FREQUENCY RESPONSE

In this section, we can discuss presenting frequency-response characteristics in graphical forms, Bode diagram or logarithmic plot, basic factors of $G(j\omega)H(j\omega)$; gain K, and integral and derivative factors. For presenting frequency-response characteristics in graphical forms, the sinusoidal transfer function, a complex function of the frequency ω, is characterized by its magnitude and phase angle, with frequency as the parameter. There are three commonly used representations of sinusoidal transfer functions as

1. Bode diagram or logarithmic plot;
2. Frequency response of the Nyquist plot or polar plot; and
3. Log-magnitude-versus-phase plot (that is also called Nichols plots).

6.2.1 Bode Diagram or Logarithmic Plot

A Bode diagram consists of two graphs:

a. plot of the logarithm of the magnitude of a sinusoidal transfer function and
b. plot of the phase angle; both are plotted against the frequency on a logarithmic scale.

The standard representation of the logarithmic magnitude of $G(j\omega)$ as

$$20\log_{10}\left|G(j\omega)\right| = 20\log\left|G(j\omega)\right| \; dB$$

The curves are drawn on semilog paper, using the log scale for frequency and the linear scale for either magnitude (but in decibels) or phase angle (in degrees).

Advantages: The main advantage of using the Bode diagram is summarized as

- Multiplication of magnitudes can be converted into addition;
- Log-magnitude curve is based on asymptotic approximations;
- Straight-line asymptotes are sufficient if rough information on the frequency-response characteristics is needed;
- Expanding the low-frequency range by use of a logarithmic scale for the frequency;
- It is not possible to plot the curves right down to zero frequency as $\log 0 = -\infty$;
- The transfer function may be simple if frequency-response data are presented in the form of a Bode diagram; and
- Using the logarithmic plot is the relative ease of plotting frequency-response curves.

Note that the experimental determination of a transfer function can be made simple if frequency-response data are presented in the form of a Bode diagram.

6.2.2 Basic Factors of $G(j\omega)H(j\omega)$

Now, let's discuss about basic factors of $G(j\omega)H(j\omega)$. As stated earlier, the main advantage of using the logarithmic plot is the relative ease of plotting frequency-response curves. The basic factors that very frequently occur in an arbitrary transfer function $G(j\omega)H(j\omega)$ are

1. Gain K,
2. Integral and derivative factors $(j\omega)^{\pm 1}$,
3. First-order factors $(1+j\omega)^{\pm 1}$, and
4. Quadratic factors $\left[1+2\xi(\omega/\omega_n)+(j\omega/\omega_n)^2\right]^{\pm 1}$.

6.2.2.1 Gain K

Once we become familiar with the logarithmic plots of these basic factors, it is possible to utilize them in constructing a composite logarithmic plot for any general form of $G(j\omega)H(j\omega)$ by sketching the curves for each factor and adding individual curves graphically, because adding the logarithms of the gains corresponds to multiplying them together.

A number greater than unity has a positive value in decibels, while a number smaller than unity has a negative value. The log-magnitude curve for a constant gain K is a horizontal straight line at the magnitude of $20\log K$ decibels. The phase angle of the gain K is zero. The effect of varying the gain K in the transfer function is that it raises or lowers the log-magnitude curve of the transfer function by the corresponding constant amount, but it does not affect the phase curve. A number-decibel conversion line is given in Figure 6.9.

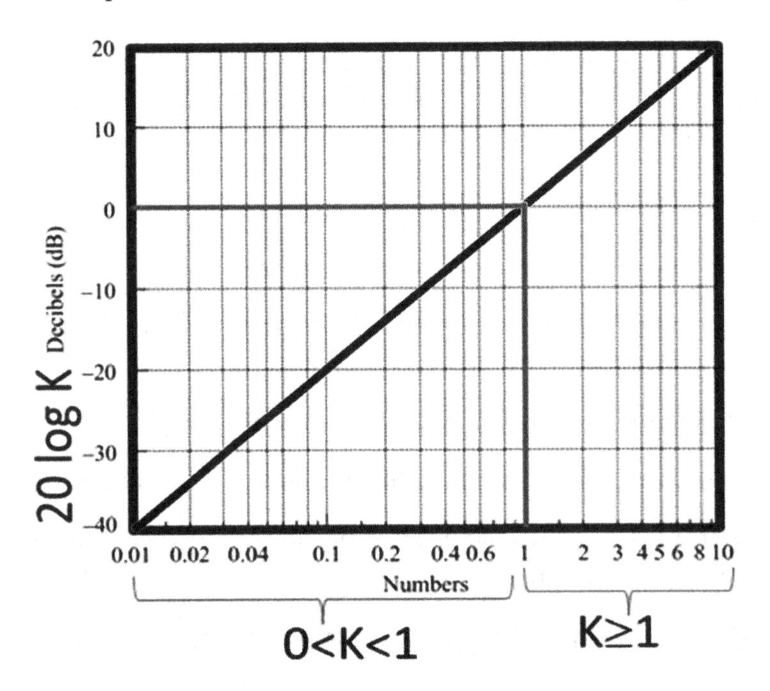

FIGURE 6.9 A number-decibel conversion line.

The decibel value of any number can be obtained from this line. As a number increases by a factor of 10, the corresponding decibel value increases by a factor of 20. This may be seen in the following:

$$20\log(K\times 10)=20\log K+20$$

$$20\log\left(K\times 10^n\right)=20\log K+20n$$

Note that when expressed in decibels, the reciprocal of a number differs from its value only in sign, that is, for the number K, we can say

$$20\log K=-20\log\left(\frac{1}{K}\right)$$

6.2.2.2 Integral and Derivative Factors, $(j\omega)^{\pm 1}$

For integral and derivative factors, the logarithmic magnitude of $1/j\omega$ can be expressed as

$$20\log\left|\frac{1}{j\omega}\right|=-20\log\omega\,\text{dB}$$

The phase angle of $1/j\omega$ is constant and equal to –90°.

In Bode diagrams, frequency ratios are expressed in terms of octaves or decades. An octave is a frequency band from ω_1 to $2\omega_1$, where ω_1 is any frequency value. A decade is a frequency band from ω_1 to 10 ω_1, where again ω1 is any frequency. (On the logarithmic scale of semilog paper, any given frequency ratio can be represented by the same horizontal distance. For example, the horizontal distance from $\omega=1$ to $\omega=10$ is equal to that from $\omega=3$ to $\omega=30$.)

If the log-magnitude –20 logω dB is plotted against ω on a logarithmic scale, it is a straight line. To draw this straight line, we need to locate one point (0 dB, $\omega=1$) on it. Since

$$\left(-20\log 10\omega\right)dB=\left(-20\log\omega-20\right)\text{dB}$$

The slope of the line is –20 dB/decade (or –6 dB/octave). Similarly, the log magnitude of $j\omega$ in decibels is

$$20\log\left|j\omega\right|=20\log\omega\,\text{dB}$$

The phase angle of $j\omega$ is constant and equal to 90°.

The log-magnitude curve is a straight line with a slope of 20 dB/decade, as shown in Figure 6.10.

Figure 6.10 shows frequency-response curves for $1/j\omega$ and $j\omega$, respectively. We can clearly see that the differences in the frequency responses of the factors $1/j\omega$ and $j\omega$ lie in

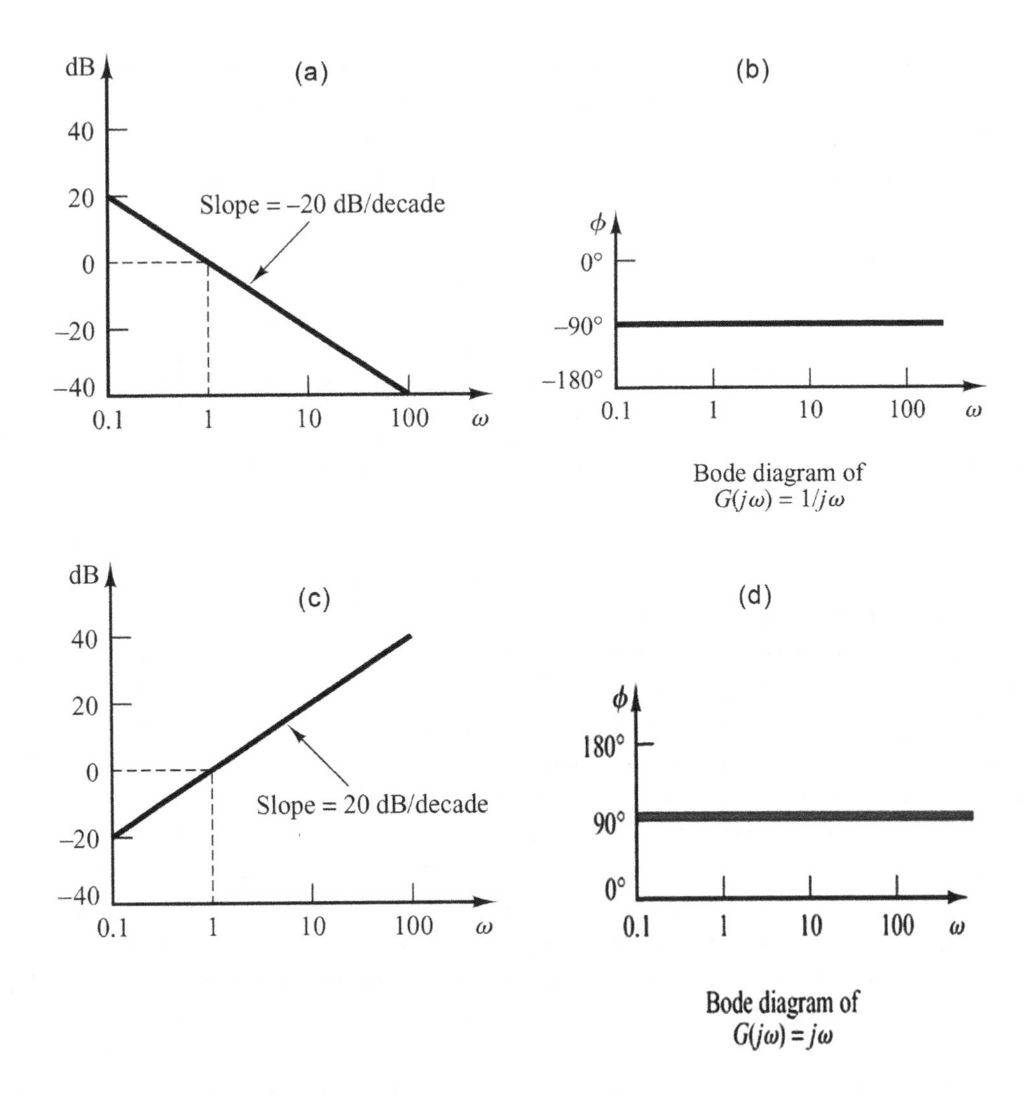

FIGURE 6.10 (a) and (c) Log-magnitude, and (b) and (d) phase curve.

the signs of the slopes of the log-magnitude curves and in the signs of the phase angles. Both log magnitudes become equal to 0 dB at $\omega = 1$.

If the transfer function contains the factor $(1/j\omega)^n$ or $(j\omega)^n$, the log magnitude becomes, –20n dB/decade and 20n dB/decade respectively. The slopes of the log-magnitude curves for the factors $(1/j\omega)^n$ and $(j\omega)^n$ are thus –20n dB/decade and 20n dB/decade, respectively. The phase angle of $(1/j\omega)^n$ is equal to $-90° \times n$ over the entire frequency range, while that of $(j\omega)^n$ is equal to $90° \times n$ over the entire frequency range. The magnitude curves will pass through the point (0 dB, $\omega = 1$).

6.2.2.3 *First-Order Factors* $(1 + j\omega T)^{-1}$

The log magnitude of the first-order factor $1/(1 + j\omega T)$ is expressed as

$$20\log\left|\frac{1}{1+j\omega T}\right| = -20\log\sqrt{1+\omega^2T^2}\,\text{dB}$$

For low frequencies, such that $\omega \ll 01/T$, the log magnitude may be approximated by the following expression:

$$-20\log\sqrt{1+\omega^2T^2} \approx -20\log 1 = 0 \text{ dB}$$

Thus, the log-magnitude curve at low frequencies is the constant 0-dB line. For high frequencies, such that $\omega \gg 1/T$,

$$-20\log\sqrt{1+\omega^2T^2} \approx -20\log T\omega \text{ dB}$$

This is an approximate expression for the high-frequency range. At $\omega = 1/T$, the log magnitude equals 0 dB; at $\omega = 10/T$, the log magnitude is –20 dB. Thus, the value of $-20 \log \omega\, T$ dB decreases by 20 dB for every decade of ω. For $\omega \gg 1/T$, the log-magnitude curve is thus a straight line with a slope of –20 dB/decade (or –6 dB/octave), as shown in Table 6.1.

Our analysis shows that the logarithmic representation of the frequency-response curve of the factor $1/(1+j\omega T)$ can be approximated by two straight-line asymptotes, one is a straight line at 0 dB for the frequency range $0<\omega<1/T$ and the other a straight line with slope –20 dB/decade (or –6 dB/octave) for the frequency range $1/T<\omega<\infty$. The exact log-magnitude curve, the asymptotes, and the exact phase-angle curve are shown in Figure 6.11.

TABLE 6.1 Frequency vs Log-Magnitude

Frequency, ω	Log-Magnitude
$1/T$	0
$10/T$	–20 dB

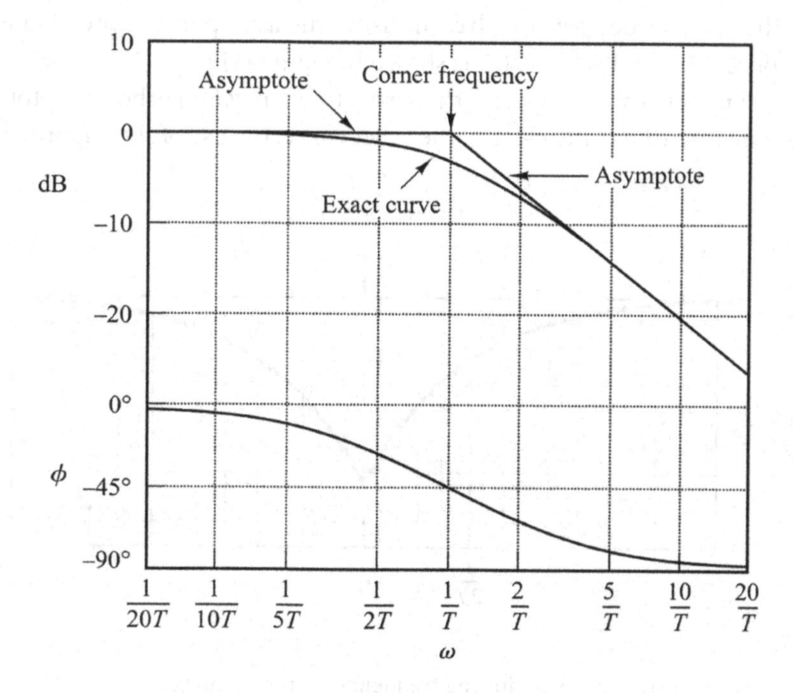

FIGURE 6.11 The log-magnitude curve, asymptotes, and phase-angle curve.

What is meant by corner frequency or break frequency? The frequency at which the two asymptotes meet is called the corner frequency or break frequency. For the factor $1/(1+j\omega T)$, the frequency $\omega = 1/T$ is the corner frequency, since at $\omega = 1/T$ the two asymptotes have the same value. The low-frequency asymptotic expression at $\omega = 1/T$ is $20\log 1$ dB $= 0$ dB, and the high-frequency asymptotic expression at $\omega = 1/T$ is also $20\log 1$ dB $= 0$ dB. The corner frequency divides the frequency-response curve into two regions: a curve for the low-frequency region and a curve for the high-frequency region. The corner frequency is very important in sketching logarithmic frequency-response curves.

The exact phase angle φ of the factor $1/(1+j\omega T)$ is

$$\varphi = -\tan^{-1}\omega T$$

At zero frequency, the phase angle is 0°. At the corner frequency, the phase angle is

$$\varphi = -\tan^{-1}\frac{T}{T} = -45^\circ$$

At infinity, the phase angle becomes –90°. Since the phase angle is given by an inverse-tangent function, the phase angle is skew symmetric about the inflection point at $\varphi = -45^\circ$. The error in the magnitude curve caused by the use of asymptotes can be calculated. The maximum error occurs at the corner frequency $\omega = 1/T$ and is approximately equal to –3 dB; the Error at one octave below the corner frequency $\omega = 1/2T$ is approximated to –1 dB; and the error at one octave above the corner frequency $\omega = 2/T$ is approximated to –1 dB.

The error at one decade below the corner frequency $\omega = 1/10T$ is approximated to –0.043 dB and the error at one decade above the corner frequency $\omega = 10/T$ is approximated to –0.043 dB. The error in decibels involved in using the asymptotic expression for the frequency-response curve of $1/(1+j\omega T)$ is shown in Figure 6.12.

The error is symmetric with respect to the corner frequency. Since the asymptotes are quite easy to draw and are sufficiently close to the exact curve, the use of such approximations in

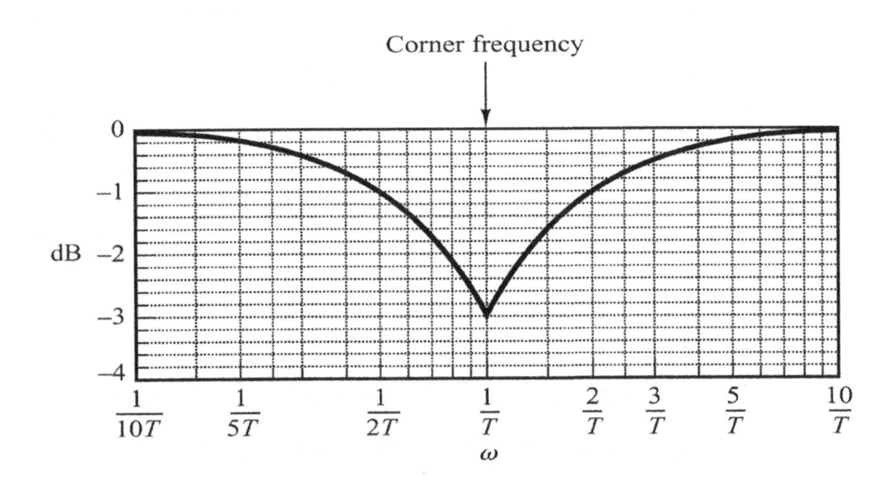

FIGURE 6.12 Asymptotic expression for the frequency-response curve.

drawing Bode diagrams is convenient in establishing the general nature of the frequency-response characteristics quickly with a minimum amount of calculation and may be used for most preliminary design work. If accurate frequency-response curves are desired, corrections may easily be made by referring to the curve given in this Figure 6.12. In practice, an accurate frequency-response curve can be drawn by introducing a correction of 3 dB at the corner frequency and a correction of 1 dB at points one octave below and above the corner frequency and then connecting these points by a smooth curve.

Note that varying the time constant T shifts the corner frequency to the left or to the right, but the shapes of the log-magnitude and the phase-angle curves remain the same.

The transfer function $1/(1+j\omega T)$ has the characteristics of a low-pass filter (LPF). For frequencies above $\omega = 1/T$, the log magnitude falls off rapidly toward $-\infty$. This is essentially due to the presence of the time constant. In the LPF, the output can follow a sinusoidal input faithfully at low frequencies. But as the input frequency is increased, the output cannot follow the input because a certain amount of time is required for the system to build up in magnitude.

Thus, at high frequencies, the amplitude of the output approaches zero and the phase angle of the output approaches −90°. Therefore, if the input function contains many harmonics, then the low-frequency components are reproduced faithfully at the output, while the high-frequency components are attenuated in amplitude and shifted in phase. Thus, a first-order element yields exact, or almost exact, duplication only for constant or slowly varying phenomena.

6.2.2.4 *First-Order Factors* $(1+j\omega T)^1$

An advantage of the Bode diagram is that for reciprocal factors—for example, the factor $1+j\omega T$ —the log-magnitude and the phase-angle curves need only be changed in sign as we can get from the following relationship :

$$20\log|1+j\omega T| = -20\log\left|\frac{1}{1+j\omega T}\right|$$

$$\angle(1+j\omega T) = \tan^{-1}\omega T = -\angle\left(\frac{1}{1+j\omega T}\right)$$

The corner frequency is the same for both cases. The slope of the high-frequency asymptote $1+j\omega T$ is 20 dB/decade, and the phase angle varies from 0° to 90° as the frequency ω is increased from zero to infinity. The log-magnitude curve, together with the asymptotes, and the phase-angle curve for the factor $1+j\omega T$ are shown in Figure 6.13.

To draw a phase curve accurately, we have to locate several points on the curve. The phase angles of $(1+j\omega T)^{\mp 1}$ are shown in Table 6.2.

6.2.2.5 *First-Order Factors* $(1+j\omega T)^{\pm n}$

For the case where a given transfer function involves terms like $(1+j\omega T)^{\mp n}$, a similar asymptotic construction may be made. The corner frequency is still at $\omega = 1/T$, and the

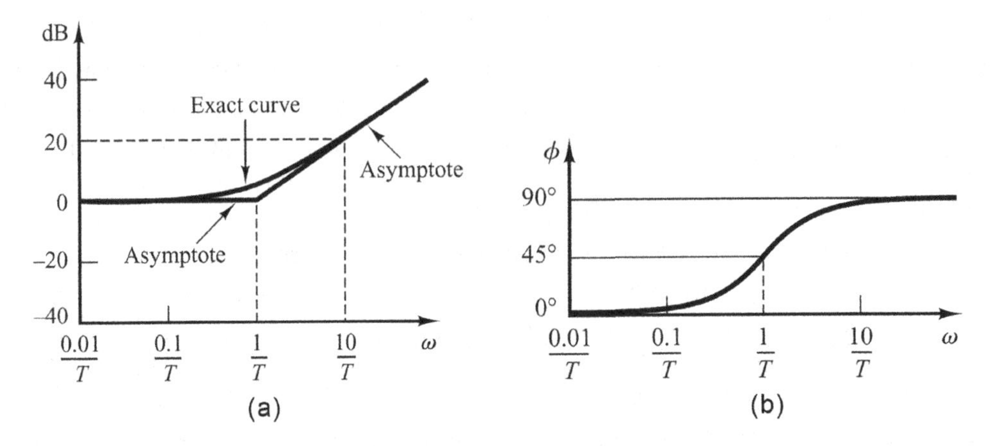

FIGURE 6.13 The asymptotes, and the phase-angle curve for the factor $1+j\omega T$.

TABLE 6.2 Frequency and Phase Angle

ω	**Phase Angles**
$1/T$	∓45°
$1/2T$	∓26.6°
$1/10T$	∓5.7°
$2/T$	∓63.4°
$10/T$	∓84.3°

asymptotes are straight lines. The low-frequency asymptote is a horizontal straight line at 0 dB, while the high-frequency asymptote has a slope of –20 or 20n dB/decade.

The error involved in the asymptotic expressions is n times that for $(1+j\omega T)^{\mp 1}$. The phase angle is n times that of $(1+j\omega T)^{\mp 1}$ at each frequency point.

6.2.2.6 Quadratic Factors & Resonant Frequency

1. *Quadratic factors* $\left[1+2\xi(\omega/\omega_n)+(j\omega/\omega_n)^2\right]^{\pm 1}$

   Control systems often possess quadratic factors of the form as

$$G(j\omega)=\frac{1}{1+2\xi\left(j\frac{\omega}{\omega_n}\right)+\left(j\frac{\omega}{\omega_n}\right)^2}$$

- If $\xi > 1$, this quadratic factor can be expressed as a product of two first-order factors with real poles.
- If $0 < \xi < 1$, this quadratic factor is the product of two complex-conjugate factors.
- Asymptotic approximations to the frequency-response curves are not accurate for a factor with low values of ξ.

- This is because the magnitude and phase of the quadratic factor depend on both the corner frequency and the damping ratio ξ.

The asymptotic frequency-response curve may be obtained as follows:

$$20\log\left|\frac{1}{1+2\xi\left(j\frac{\omega}{\omega_n}\right)+\left(j\frac{\omega}{\omega_n}\right)^2}\right|=-20\log\sqrt{\left(1-\frac{\omega^2}{\omega_n^2}\right)^2+\left(2\xi\frac{\omega}{\omega_n}\right)^2}$$

For low frequencies such that $\omega \ll \omega_n$, the log magnitude becomes

$$-20\log 1=0\,\text{dB}$$

The low-frequency asymptote is thus a horizontal line at 0 dB. For high frequencies such that $\omega \gg \omega_n$, the log magnitude becomes as

$$-20\log\left(\frac{\omega^2}{\omega_n^2}\right)=-40\log\left(\frac{\omega}{\omega_n}\right)$$

The equation for the high-frequency asymptote is a straight line having the slope −40 dB/decade. So we can say $-40\log\left(\frac{10\omega}{\omega_n}\right)=-40-40\log\left(\frac{\omega}{\omega_n}\right)$

The high-frequency asymptote intersects the low-frequency one at $\omega=\omega_n$, since at this frequency we can say

$$-40\log\left(\frac{\omega}{\omega_n}\right)=-40\log 1=0\,\text{dB}$$

This frequency, ω_n, is the corner frequency for the quadratic factor considered. The two asymptotes just derived are independent of the value of ξ. Near the frequency $\omega=\omega_n$, *a resonant peak occurs* as may be expected from

$$G(j\omega)=\frac{1}{1+2\xi\left(j\frac{\omega}{\omega_n}\right)+\left(j\frac{\omega}{\omega_n}\right)^2}$$

- The damping ratio ξ determines the magnitude of this resonant peak;
- Errors obviously exist in the approximation by straight-line asymptotes;
- The magnitude of the error depends on the value of ξ; and
- It is large for small values of ξ.

Figure 6.14 shows the exact log-magnitude curves, together with the straight-line asymptotes and the exact phase-angle curves for the quadratic factor with several values of ξ. If corrections are desired in the asymptotic curves, the necessary amounts of correction at a sufficient number of frequency points may be obtained from Figure 6.14.

The phase angle of the quadratic factor is described by the following relationship:

$$\varphi = \angle\left(\frac{1}{1+2\xi\left(j\frac{\omega}{\omega_n}\right)+\left(j\frac{\omega}{\omega_n}\right)^2}\right) = -\tan^{-1}\left[\frac{2\xi\frac{\omega}{\omega_n}}{1-\left(\frac{\omega}{\omega_n}\right)^2}\right]$$

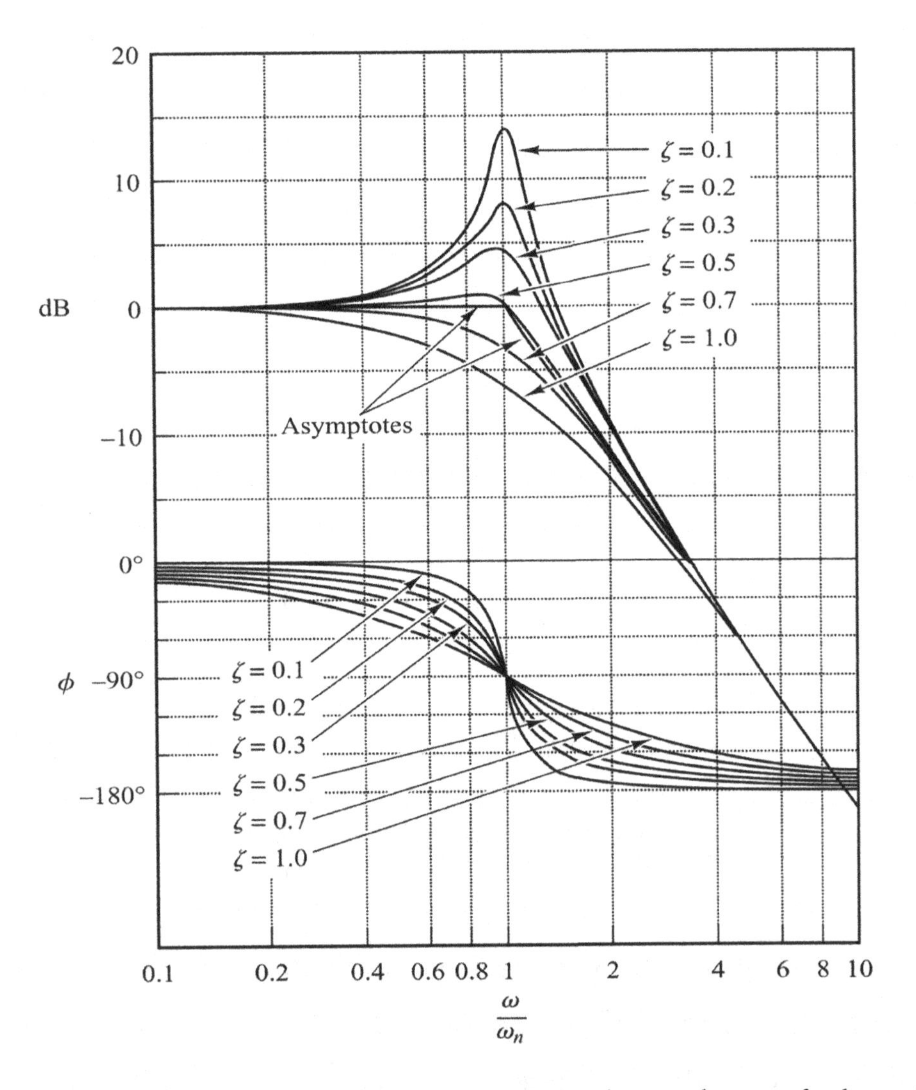

FIGURE 6.14 The log-magnitude curves, asymptotes, and the phase-angle curves for the quadratic factor.

- The phase angle is a function of both ω and ξ. At $\omega = 0$, the phase angle equals 0°;
- At the corner frequency $\omega = \omega_n$, the phase angle is –90° regardless of ξ, as we can see here;
- At $\omega = \infty$, the phase angle becomes –180°;
- The phase-angle curve is skew symmetric about the inflection point—the point where $\varnothing = -90°$; and
- There are no simple ways to sketch such phase curves. We need to refer to the phase-angle curves shown in Figure 6.14.

The frequency-response curves for the factor can be obtained by merely reversing the sign of the log magnitude and that of the phase angle of the factor as

$$\frac{1}{1+2\xi\left(j\frac{\omega}{\omega_n}\right)+\left(j\frac{\omega}{\omega_n}\right)^2}$$

To obtain the frequency-response curves of a given quadratic transfer function, we must first determine the value of the corner frequency ω_n and that of the damping ratio ξ. Then, by using the family of curves (Figure 6.14), the frequency-response curves can be plotted as shown in Figure 6.14.

6.2.2.7 Resonant frequency ω_r and the Resonant Peak Value M_r

To discuss the resonant frequency ω_r and the resonant peak value M_r, the magnitude of $G(j\omega)$ can be seen from the following expression:

$$\left|G(j\omega)\right| = \frac{1}{\sqrt{\left(1-\frac{\omega^2}{\omega_n^2}\right)^2+\left(2\xi\frac{\omega}{\omega_n}\right)^2}}$$

If $\left|G(j\omega)\right|$ has a peak value at some frequency, this frequency is called the resonant frequency. Since the numerator of $\left|G(j\omega)\right|$ is constant, a peak value of $\left|G(j\omega)\right|$ will occur when it will be

$$g(\omega) = \left(1-\frac{\omega^2}{\omega_n^2}\right)^2+\left(2\xi\frac{\omega}{\omega_n}\right)^2$$

It is a minimum when it will be as

$$g(\omega) = \left[\frac{\omega^2-\omega_n^2\left(1-2\xi^2\right)}{\omega_n^2}\right]+4\xi^2\left(1-\xi^2\right)$$

TABLE 6.3 Comments on Damping Ratio ξ.

Parameter	Characteristics/Comments
$\xi > 0.707$	No resonant peak
Magnitude, $\lvert G(j\omega)\rvert$; $\lvert G(j\omega)\rvert < 0$	Decreases monotonically with increasing frequency ω;
$0 < \xi < 1$	step response is well damped oscillatory & hardly perceptible;

The minimum value of $g(\omega)$ occurs at $\omega = \omega_n\sqrt{1-2\xi^2}$. Thus, the resonant frequency ω_r is

$$\omega_r = \omega_n\sqrt{1-2\xi^2} \qquad ;\text{for } 0 \le \xi \le 0.707$$

As the damping ratio ξ approaches zero, the resonant frequency approaches ω_n. For $0 < \xi \le 0.707$, the resonant frequency ω_r is less than the damped natural frequency $\omega_d = \omega_n\sqrt{1-\xi^2}$, which is exhibited in the transient response. From $\omega = \omega_n\sqrt{1-2\xi^2}$, it can be derived that for $\xi > 0.707$, there is no resonant peak. The magnitude decreases monotonically with increasing frequency ξ. The magnitude is less than 0 dB for all values of $\omega > 0$. Recall that for $0.7 < \xi < 1$, the step response is oscillatory, but the oscillations are well damped and are hardly perceptible, as summarized in Table 6.3.

For $0 \le \xi \le 0.707$, the magnitude of the resonant peak, $M_r = \lvert G(j\omega_r)\rvert$, then we can find

$$M_r = \lvert G(j\omega)\rvert_{\max} = \lvert G(j\omega_r)\rvert = \frac{1}{2\xi\sqrt{1-\xi^2}}$$

For, ξ >0.707, and $M_r = 1$ as well as if $\xi \to 0$, $M_r \to \infty$ that we can observe in Figure 6.15.

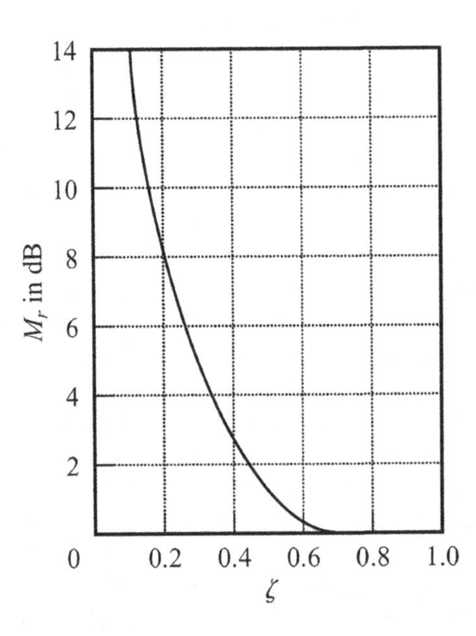

FIGURE 6.15 The resonant peak value M_r and ξ.

The phase angle of $G(j\omega)$ at the frequency where the resonant peak occurs can be obtained from this expression:

$$\varphi = \angle\left(\frac{1}{1+2\xi\left(j\frac{\omega}{\omega_n}\right)+\left(j\frac{\omega}{\omega_n}\right)^2}\right) = -\tan^{-1}\left[\frac{2\xi\frac{\omega}{\omega_n}}{1-\left(\frac{\omega}{\omega_n}\right)^2}\right]$$

Thus, at the resonant frequency ω_r, the phase angle can be seen like this.

$$\angle G(j\omega_r) = -\tan^{-1}\frac{\sqrt{1-2\xi^2}}{\xi} = -90° + \sin^{-1}\frac{\xi}{\sqrt{1-\xi^2}}$$

6.3 FREQUENCY RESPONSE OF THE NYQUIST PLOT OR POLAR PLOT

6.3.1 Polar Plots and General Shapes of Polar Plots

In this section, we will discuss polar plots, factors in polar plots, and natural and resonant frequency in a polar plot. The polar plot of a sinusoidal transfer function $G(j\omega)$ is a plot of the magnitude of $G(j\omega)$ versus the phase angle of $G(j\omega)$ on polar coordinates as ω is varied from zero to infinity. Thus, the polar plot is the locus of vectors $|G(j\omega)|\angle G(j\omega)$ as ω is varied from zero to infinity. Note that in polar plots, a positive (negative) phase angle is measured counterclockwise (clockwise) from the positive real axis. The polar plot is often called the Nyquist plot. An example of such a plot is shown in Figure 6.16.

Each point on the polar plot of $G(j\omega)$ represents the terminal point of a vector at a particular value of ω. In the polar plot, it is important to show the frequency graduation of the locus. The projections of $G(j\omega)$ on the real and imaginary axes are its real and imaginary components. MATLAB may be used to obtain a polar plot $G(j\omega)$ or to obtain $|G(j\omega)|$ and $\angle G(j\omega)$ accurately for various values of ω in the frequency range of interest.

Advantage & Disadvantage of Polar Plots: An advantage of using a polar plot is that it depicts the frequency-response characteristics of a system over the entire frequency range in a single plot. One disadvantage is that the plot does not clearly indicate the contributions of each individual factor of the open-loop transfer function.

Factors in Polar Plots: To understand the polar plots and their shape, we will study factors in polar plots such as

1. Integral and Derivative Factors;
2. First-Order Factors; and
3. Quadratic Factors.

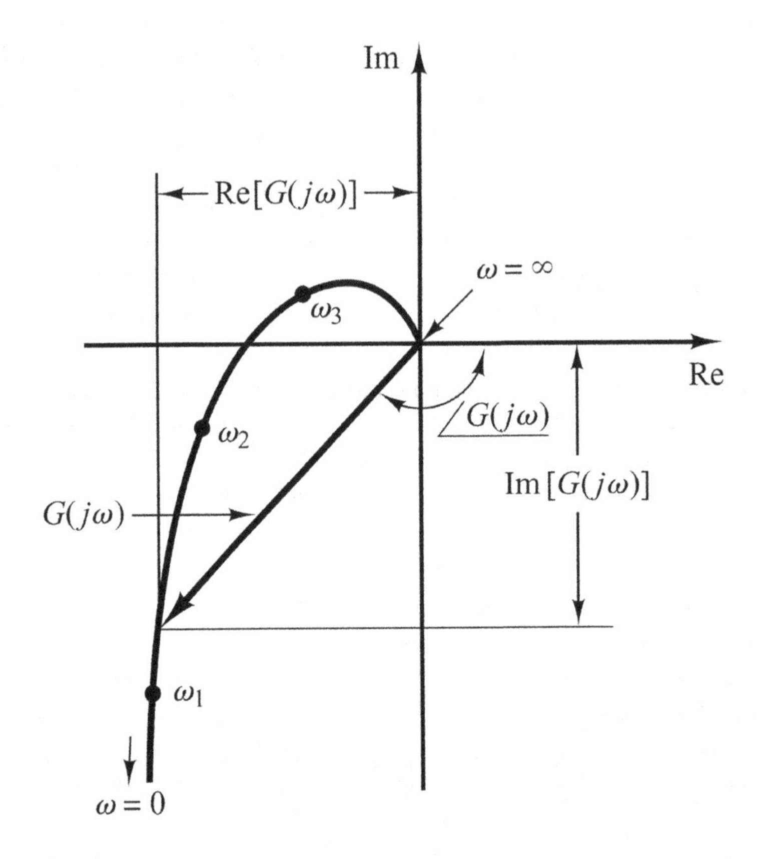

FIGURE 6.16 Polar plot.

6.3.1.1 *Integral and Derivative Factors* $(j\omega)^{\pm 1}$

For the Integral and Derivative Factors $(j\omega)^{\mp 1}$, the polar form can be seen from the polar plot of $G(j\omega)=1/j\omega$, which is the negative imaginary axis. Since $G(j\omega)=1/j\omega$, then we can say

$$G(j\omega)=-j\frac{1}{\omega}=\frac{1}{\omega}\angle-90^\circ$$

The polar plot of $G(j\omega)=j\omega$ is a positive imaginary axis.

6.3.1.2 *First-Order Factors* $(1+j\omega T)^{\mp 1}$

For the First-Order Factors $(1+j\omega T)^{\mp 1}$, the polar form can be seen from the polar plot of $G(j\omega)=1/(1+j\omega T)$, which is the negative imaginary axis. Since $G(j\omega)=1/(1+j\omega T)$, we can say

$$G(j\omega)=\frac{1}{1+j\omega T}=\frac{1}{\sqrt{1+\omega^2T^2}}\angle-\tan^{-1}\omega T$$

The values of $G(j\omega)$ at $\omega=0$ and $\omega=1/T$ are,

$$G(j0)=1\angle 0^\circ$$

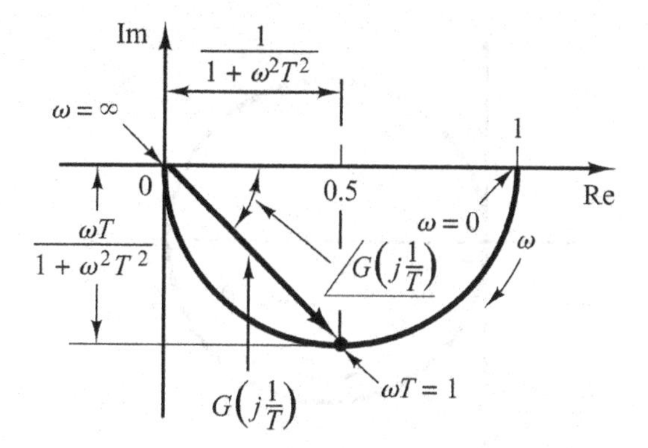

FIGURE 6.17 The polar plot of transfer function as a semicircle.

and

$$G\left(j\frac{1}{T}\right) = \frac{1}{\sqrt{2}}\angle -45^\circ$$

If ω approaches infinity, the magnitude of $G(j\omega)$ approaches zero and the phase angle approaches −90°. The polar plot of this transfer function is a semicircle as the frequency ω is varied from zero to infinity, as shown in Figure 6.17.

The center is located at 0.5 on the real axis, and the radius is equal to 0.5. To prove that the polar plot of the first-order factor $G(j\omega) = 1/(1 + j\omega T)$ is a semicircle, it can be defined as

$$G(j\omega) = X + jY$$

where

$$X = \frac{1}{1+\omega^2T^2} = \text{Real part of } G(j\omega)$$

$$Y = \frac{-\omega T}{1+\omega^2T^2} = \text{Imaginary part of } G(j\omega)$$

Then we obtain the Circle equation at (0.5,0; 0.5) as

$$\left(X - \frac{1}{2}\right)^2 + Y^2 = \left[\frac{1}{2}\left(\frac{1-\omega^2T^2}{1+\omega^2T^2}\right)\right]^2 + \left(\frac{-\omega^2T^2}{1+\omega^2T^2}\right)^2 = \left(\frac{1}{2}\right)^2$$

Thus, in the X–Y plane $G(j\omega)$ is a circle with a center at and with radius as shown in Figure 6.18.

The lower semicircle corresponds to $0 \le \omega \le \infty$, and the upper semicircle corresponds to $-\infty \le \omega \le 0$. The polar plot of the transfer function $1 + j\omega T$ is simply the upper half of the

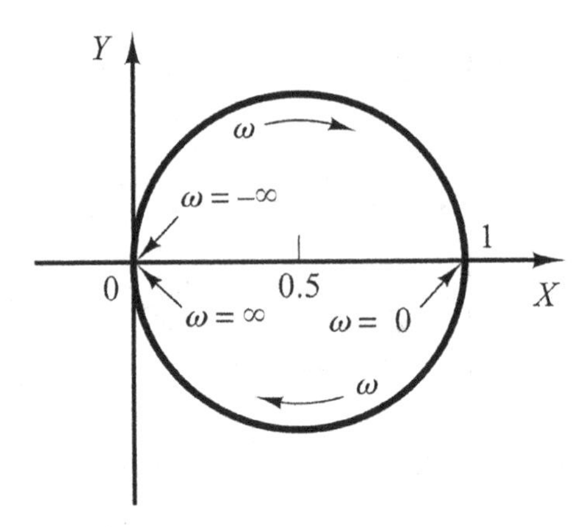

FIGURE 6.18 The polar plot of the first-order factor.

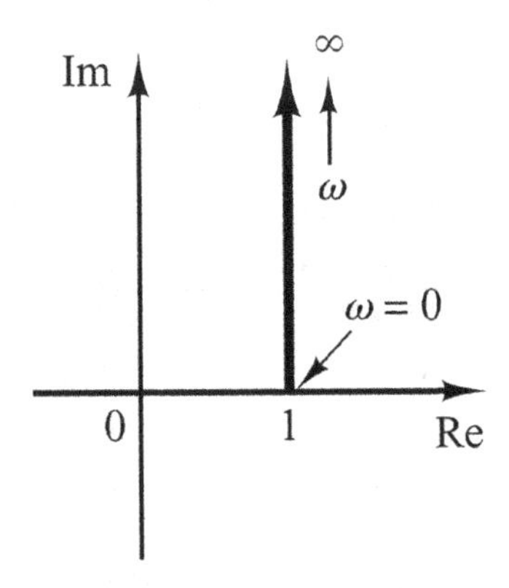

FIGURE 6.19 The polar plot of the transfer function $1 + j\omega T$.

straight line passing through point (1,0) in the complex plane and parallel to the imaginary axis, as shown in Figure 6.19.

The polar plot of $1 + j\omega T$ has an appearance completely different from that of $1/(1 + j\omega T)$.

6.3.1.3 Quadratic Factors $\left[1+2\xi(j\omega/\omega_n)+(j\omega/\omega_n)^2\right]^{\pm 1}$

The low- and high-frequency portions of the polar plot of the following sinusoidal transfer function as we can see here.

$$G(j\omega)=\frac{1}{1+2\xi\left(j\frac{\omega}{\omega_n}\right)+\left(j\frac{\omega}{\omega_n}\right)^2};\text{for } \xi>0$$

which are given, respectively, by this relationship. If ω increases→0 to ∞, then the low-frequency portions of polar plot:

$$\lim_{\omega\to 0} G(j\omega)=1\angle 0^\circ$$

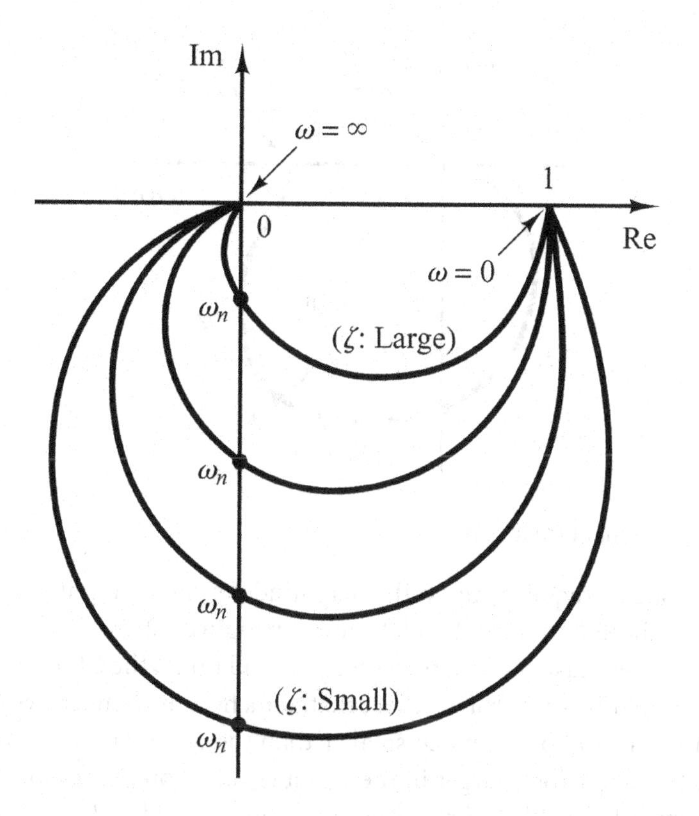

FIGURE 6.20 Polar plots of the transfer function.

and the high-frequency portions of polar:

$$\lim_{\omega \to \infty} G(j\omega) = 0\angle -180^\circ$$

The polar plot of this sinusoidal transfer function starts at $1\angle 0^\circ$ and ends at $0\angle -180^\circ$ as ω increases from zero to infinity. Thus, the high-frequency portion of $G(j\omega)$ is tangent to the negative real axis.

6.3.2 Natural and Resonant Frequency in Polar Plot

To understand natural and resonant frequency in polar plot, the following examples of polar plots of the transfer function just considered are shown in Figure 6.20.

The exact shape of a polar plot depends on the value of the damping ratio ξ, but the general shape of the plot is the same for both the underdamped case ($1 > \xi > 0$) and overdamped case ($\xi > 1$).

For the underdamped case at $\omega = \omega_n$, we have $G(j\omega_n) = 1/(j2\xi)$, and the phase angle at $\omega = \omega_n$ is -90°. Therefore, it can be seen that the frequency at which the $G(j\omega)$ locus intersects the imaginary axis is the undamped natural frequency ω_n. In the polar plot, the frequency point(s) whose distance from the origin is maximum corresponds to the resonant frequency ω_r. The peak value of $G(j\omega)$ is obtained as the ratio of the magnitude of the

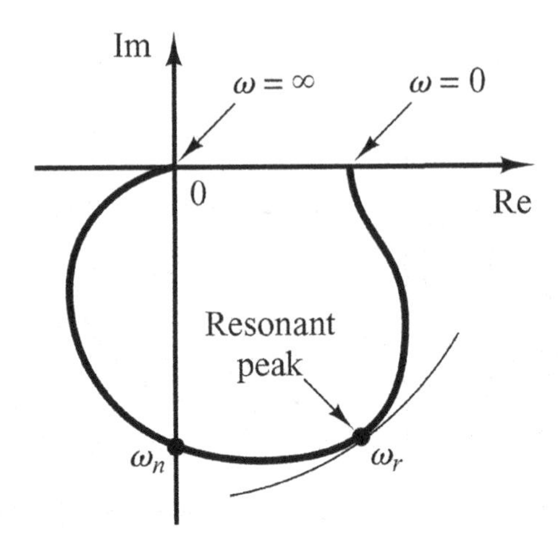

FIGURE 6.21 The resonant frequency.

vector at the resonant frequency ω_r to the magnitude of the vector at $\omega = 0$. The resonant frequency ω_r is indicated in the polar plot shown in Figure 6.21.

For the overdamped case, as ξ increases well beyond unity, the $G(j\omega)$ locus approaches a semicircle. This may be seen from the fact that, for a heavily damped system, the characteristic roots are real, and one is much smaller than the other. Since, for sufficiently large ξ, the effect of the larger root (larger in the absolute value) on the response becomes very small, the system behaves like a first-order one. Next, consider the following sinusoidal transfer function as

$$G(j\omega) = 1 + 2\xi\left(j\frac{\omega}{\omega_n}\right) + \left(j\frac{\omega}{\omega_n}\right)^2$$

or

$$G(j\omega) = \left(1 - \frac{\omega^2}{\omega_n^2}\right)^2 + 2j\left(\frac{2\xi\omega}{\omega_n}\right)$$

The low-frequency portion of the curve is

$$\lim_{\omega \to 0} G(j\omega) = 1\angle 0^\circ$$

and the high-frequency portion is

$$\lim_{\omega \to \infty} G(j\omega) = \infty\angle -180^\circ$$

Since the imaginary part of $G(j\omega)$ is positive for $\omega > 0$ and is monotonically increasing, and the real part of $G(j\omega)$ is monotonically decreasing from unity, the general shape of the polar plot of $G(j\omega)$ is as shown in Figure 6.22. The phase angle is between 0° and 180°.

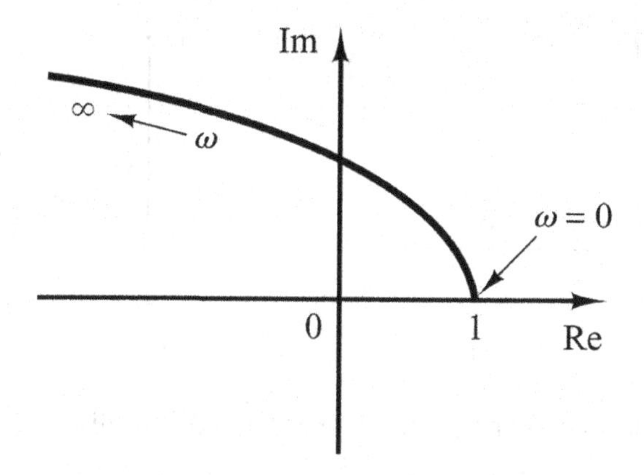

FIGURE 6.22 The general shape of the polar plot of $G(j\omega)$.

6.3.3 Nyquist Diagram of a Typical Part

Typical relations are commonly used in engineering control system that are as follows:

1. *The proportional amplification gain factor*: $G(j\omega)=K$
2. *Integral part*: $G(j\omega)=\dfrac{1}{j\omega}$
3. *Differential relation*: $G(j\omega)=j\omega$
4. *First-order differential relation*: $G(j\omega)=Tj\omega+1$
5. *First-order reciprocal relation:* $G(j\omega)=\dfrac{1}{Tj\omega+1}$
6. *Second-order oscillation relation*: $G(j\omega)=\dfrac{1}{T^2(j\omega)^2+2\xi T(j\omega)+1}$
7. *Second-order differential relation*: $G(j\omega)=T^2(j\omega)^2+2\xi Tj\omega+1$
8. *Delay relation:* $G(j\omega)=e^{-j\omega T}$

The Nyquist plots of these eight cases are discussed in below.

6.3.3.1 The Proportional Amplification Gain Factor, $G(j\omega)=K$

The amplitude-phase characteristics of the proportional relation are as follows:

$$A(\omega)=|G(j\omega)|=K$$

$$\varphi(\omega)=\angle G(j\omega)=0°$$

The Nyquist plots of proportional relations are shown in Figure 6.23a.

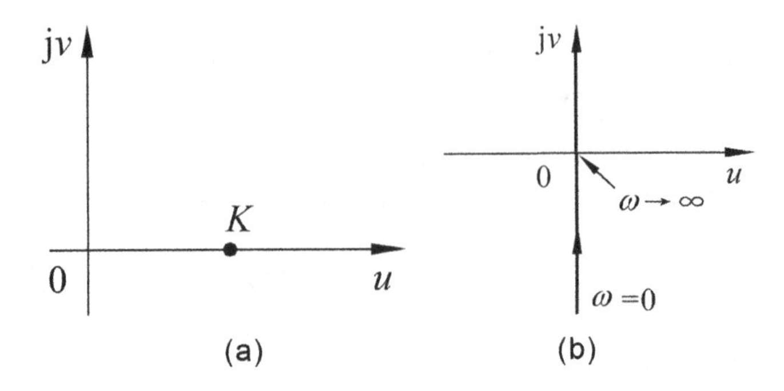

FIGURE 6.23 (a) Nyquist plots of proportional factor, and (b) Nyquist plots of integral factor.

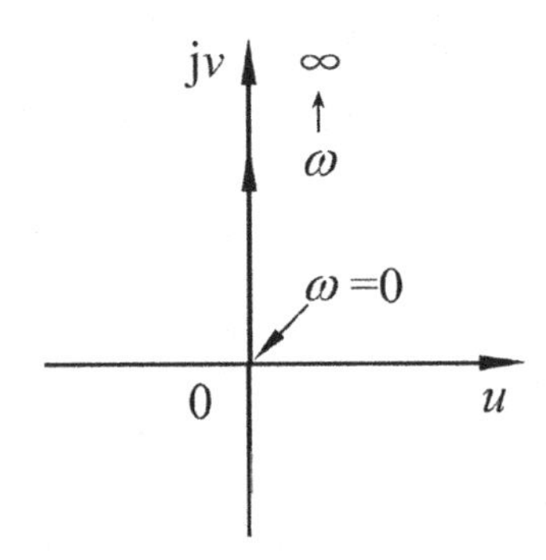

FIGURE 6.24 Nyquist plots of differential factor.

6.3.3.2 *Integral Factor,* $G(j\omega)=\frac{1}{j\omega}$

The amplitude-phase characteristics of the integral relation are as follows:

$$A(\omega)=|G(j\omega)|=\frac{1}{\omega}$$

$$\varphi(\omega)=\angle G(j\omega)=-90^\circ$$

The Nyquist plots of integral elements are shown in Figure 6.23b.

6.3.3.3 *Differential Factor,* $G(j\omega)=j\omega$

The amplitude-phase characteristics of the differential element are as follows:

$$A(\omega)=|G(j\omega)|=\omega$$

$$\varphi(\omega)=\angle G(j\omega)=90^\circ$$

The Nyquist plots of differential elements or factor are shown in Figure 6.24.

6.3.3.4 *First-Order Differential Factor,* $G(j\omega)=Tj\omega+1$

The amplitude-phase characteristics of the first-order differential segment are as follows:

$$A(\omega)=|G(j\omega)|=\sqrt{1+(\omega T)^2}$$

$$\varphi(\omega)=\angle G(j\omega)=\tan^{-1}\omega T$$

For $\omega=0$ then $\begin{cases} A(\omega)=1 \\ \varphi(\omega)=0 \end{cases}$,

and

For $\omega=\infty$ then $\begin{cases} A(\omega)=\infty \\ \varphi(\omega)=\dfrac{\pi}{2} \end{cases}$

The Nyquist plots of first-order differential elements or factor are shown in Figure 6.25.

6.3.3.5 *First-Order Interial Factor,* $G(j\omega)=\dfrac{1}{Tj\omega+1}$

The amplitude and phase characteristics of the first-order inertial element are as follows:

$$A(\omega)=|G(j\omega)|=\frac{1}{\sqrt{\omega^2T^2+1}}$$

$$\varphi(\omega)=\angle G(j\omega)=-\tan^{-1}(\omega T)$$

For $\omega=0$, then $\begin{cases} A(\omega)=1 \\ \varphi(\omega)=0° \end{cases}$

and

For $\omega=\infty$, then $\begin{cases} A(\omega)=0 \\ \varphi(\omega)=-90° \end{cases}$

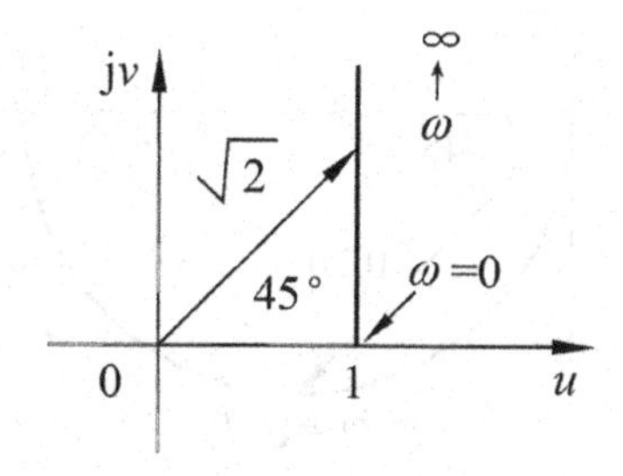

FIGURE 6.25 Nyquist plots of first-order differential elements.

The Nyquist plots of first-order inertial elements are shown in Figure 6.26. The Nyquist plot of the first-order inertial elements is a semicircle. Thus, we get

$$G(j\omega)=\frac{1}{j\omega T+1}=\frac{1}{1+\omega^2T^2}+j\frac{-\omega T}{1+\omega^2T^2}$$

$$U(\omega)=\frac{1}{T^2\omega^2+1},\quad V(\omega)=\frac{-\omega T}{\omega^2T^2+1} \tag{6.44}$$

$$\tan\varphi(\omega)=\frac{V(\omega)}{U(\omega)}=-\omega T \tag{6.45}$$

Substituting Eq. (6.45) into Eq. (6.44), we get

$$U(\omega)=\frac{1}{\left(\frac{V(\omega)}{U(\omega)}\right)^2+1}=\frac{U^2(\omega)}{U^2(\omega)+V^2(\omega)}$$

$$U^2(\omega)+V^2(\omega)=U(\omega)$$

$$\left(U(\omega)-\frac{1}{2}\right)^2+V^2(\omega)=\left(\frac{1}{2}\right)^2 \tag{6.46}$$

In Eq. (6.46), the point $\left(\frac{1}{2},0\right)$ is the center of the semicircle in the Nyquist plot and the radius of the circle is $\frac{1}{2}$ of the semicircle, as shown in Figure 6.26.

6.3.3.6 Second-Order Oscillation Relation

$$G(j\omega)=\frac{1}{(j\omega)^2T^2+j2\xi\omega T+1};(T>0,\ 0<\xi<1)$$

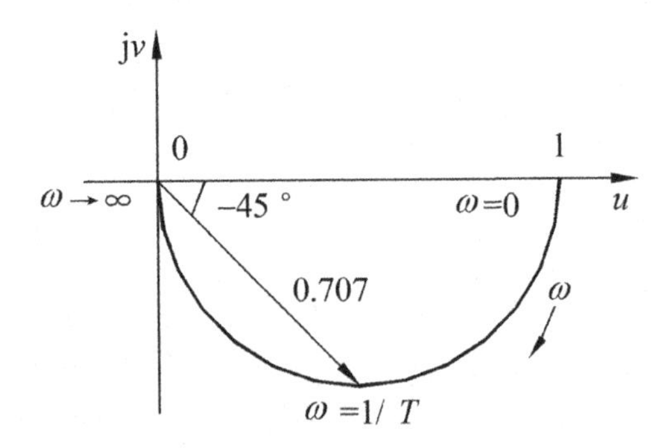

FIGURE 6.26 Nyquist plots of first-order inertial elements.

The amplitude-phase characteristics of the second-order oscillating relation are as follows:

$$A(\omega)=\left|G(j\omega)\right|=\frac{1}{\sqrt{\left(1-\omega^2T^2\right)^2+\left(2\xi\omega T\right)^2}}$$

$$\varphi(\omega)=\angle G(j\omega)=\begin{cases} -\tan^{-1}\dfrac{2\xi\omega T}{1-\omega^2T^2}\left(\omega \quad \dfrac{1}{T}\right) \\ -\pi-\tan^{-1}\dfrac{2\xi\omega T}{1-\omega^2T^2}\left(\omega>\dfrac{1}{T}\right) \end{cases}$$

For $\omega=0$ then $\begin{cases} A(\omega)=1 \\ \varphi(\omega)=0° \end{cases}$

For $\omega=\infty$ then $\begin{cases} A(\omega)=0 \\ \varphi(\omega)=-180° \end{cases}$

The Nyquist plots of the second-order oscillating elements are shown in Figure 6.27.

Frequency characteristics of the second-order oscillation relation can also be expressed as

$$G(j\omega)=\frac{1}{\sqrt{\left(1-\dfrac{\omega^2}{\omega_n^2}\right)^2+4\xi^2\dfrac{\omega^2}{\omega_n^2}}}e^{-j\tan^{-1}\left(\frac{2\xi\frac{\omega}{\omega_n}}{1-\frac{\omega^2}{\omega_n^2}}\right)} \tag{6.47}$$

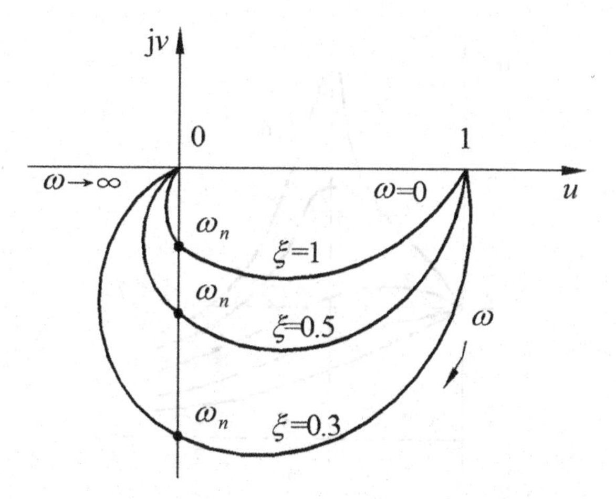

FIGURE 6.27 Nyquist plots of second-order oscillating elements.

If $u=\dfrac{\omega}{\omega_n}$, then the amplitude-phase characteristic of the second-order oscillating element is

$$A(\omega)=\left|G(j\omega)\right|=\frac{1}{\sqrt{\left(1-\dfrac{\omega^2}{\omega_n{}^2}\right)^2+4\xi^2\dfrac{\omega^2}{\omega_n{}^2}}}=\frac{1}{\sqrt{\left(1-u^2\right)^2+4\xi^2u^2}} \tag{6.48}$$

According to Eq. (6.48), it can be obtained

$$A(u)=\frac{1}{\sqrt{(1-u^2)^2+4\xi^2u^2}} \tag{6.49}$$

$$\varphi(\omega)=\angle G(j\omega)=-\tan^{-1}\frac{2\xi\dfrac{\omega}{\omega_n}}{1-\dfrac{\omega^2}{\omega_n^2}}=-\tan^{-1}\left(\frac{2\xi u}{1-u^2}\right) \tag{6.50}$$

According to Eq. (6.50), it can be obtained

$$\varphi(u)=-\tan^{-1}\left(\frac{2\xi u}{1-u^2}\right) \tag{6.51}$$

For any value of ξ, we say $u=1$, then we obtain $\dfrac{\omega}{\omega_n}=1$, thus, we get $\varphi(u)=-\dfrac{\pi}{2}$. According to Eq. (6.49), the maximum value of the amplitude-frequency characteristic increases with the decrease of ξ; the curve of $A(u)$ can be obtained by changing with u, as shown in Figure 6.28. Figure 6.28 shows that when the damping ratio ξ decreases to a certain value of u; the amplitude-frequency characteristic shows the resonance peak; the frequency

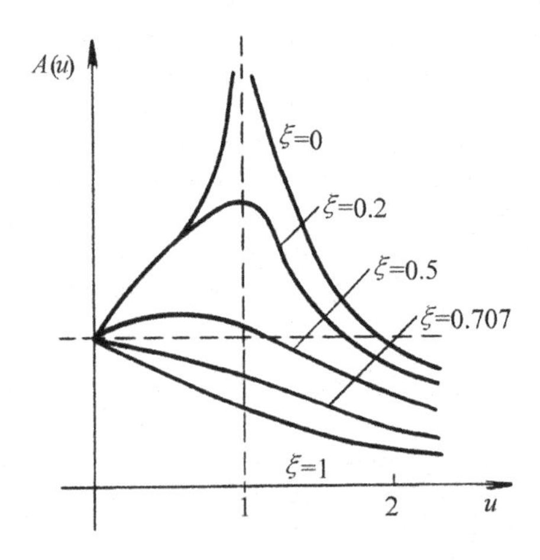

FIGURE 6.28 The $A(u)\sim u$ curve of the second-order oscillation.

corresponding to the resonant peak is called the resonant frequency which is ω_r. The dimensionless resonant frequency is defined as $u_r = \frac{\omega_r}{\omega_n}$. We can see that the u_r decreases with the increasing of ξ. Finally, setting its derivative to zero to the derivative Eq. (6.49), we get

$$u_r = \sqrt{1-2\xi^2} \quad ;(\xi \le 0.707) \tag{6.52}$$

and

$$\omega_r = \omega_n\sqrt{1-2\xi^2} \quad ;(\xi \le 0.707) \tag{6.53}$$

Substituting Eq. (6.52) into Eq. (6.49), the resonance peak value of the frequency characteristic of Eq. (6.49) is

$$M_r = A(\omega)\Big|\max\Big|\frac{1}{2\xi\sqrt{1-\xi^2}} \quad ;\left(\xi \le \frac{\sqrt{2}}{2}\right) \tag{6.54}$$

According to Eqs. (6.52) and (6.54), the resonance peaks make physical sense. Since $\xi > \frac{\sqrt{2}}{2}$, the slope of the amplitude-frequency characteristic curve of Eq. (6.49) is always negative; therefore, there is no resonance peak. The $M_r \sim \xi$ curve is shown in Figure 6.29a. In Chapter 4, we have discussed the unit-step response of the peak $c(t_p) \sim \xi$ relationships; thus, we say

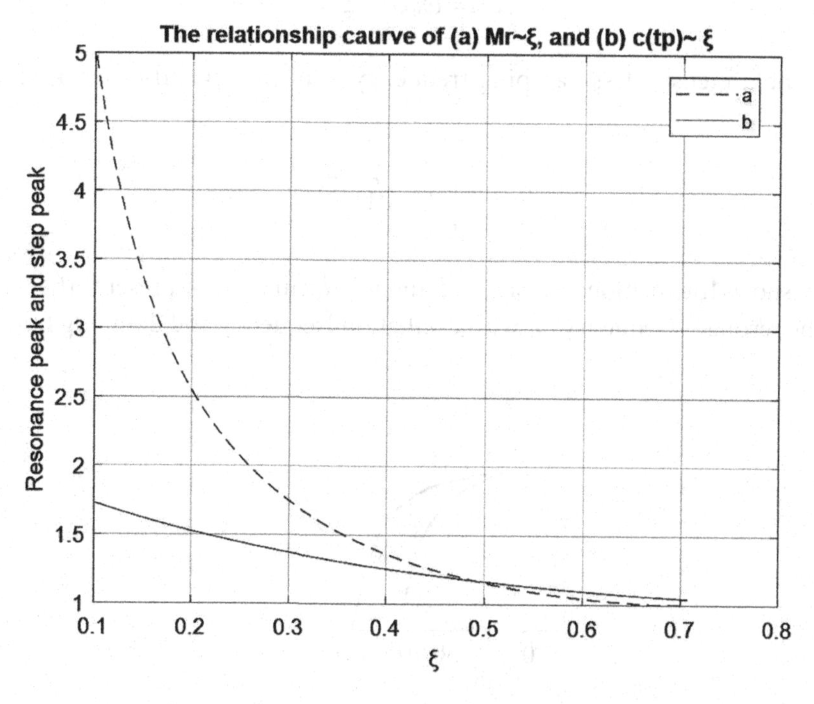

FIGURE 6.29 The relationship caurve of (a) $M_r \sim \xi$ and (b) $c(t_p) \sim \xi$.

$$c(t_p)=x_o(t_p)=1+e^{-\frac{\pi\xi}{\sqrt{1-\xi^2}}} \tag{6.55}$$

The $c(t_p) \sim \xi$ relationship curve is shown in Figure 6.16b.

MATLAB program of resonance peak and step peak change with damping ratio is as follows:

```
for i=1:607
Z(i)=(100+i)/1000;
mr(i)=1/(2*Z(i)*sqrt(1-Z(i)*Z(i)));
xotp(i)=1+exp(-pi*Z(i)/sqrt(1-Z(i)*Z(i)));
end
plot(Z,mr,'--k',ct,xotp,'k')
legend('a','b')
xlabel('ξ')
ylabel('Resonance peak and step peak')
title('The relationship caurve of (a) Mr~ξ, and (b) c(tp)~ξ ')
grid
```

The relation between damping frequency ω_d and damping ratio ξ of the second-order oscillation relationship is

$$\omega_d=\omega_n\sqrt{1-\xi^2}$$

Therefore, the dimensionless damping frequency u_d of the second-order oscillation system is

$$u_d=\frac{\omega_d}{\omega_n}=\sqrt{1-\xi^2} \tag{6.56}$$

Figure 6.30 shows the relationship of $u_r \sim \xi$ and $u_d \sim \xi$ curves. It represents the relationship between the resonant frequency ω_r with oscillation frequency and damping frequency ω_d.

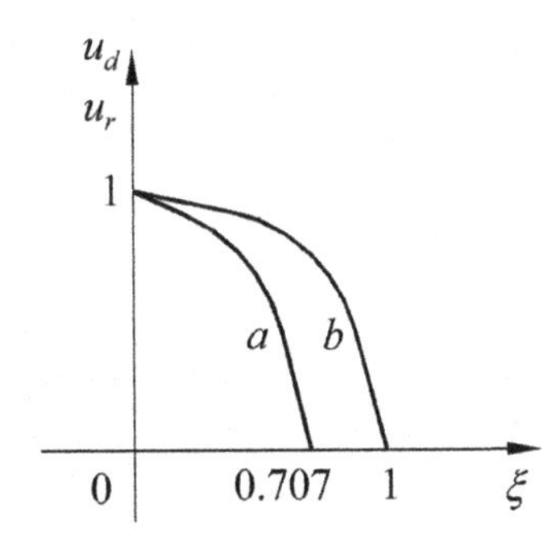

FIGURE 6.30 Curves of (a) $u_r \sim \xi$ and (b) $u_d \sim \xi$.

Figures 6.29 and 6.30 are related in the time domain performance and frequency-domain indicators of the system. Figure 6.29 shows that the resonance peak M_r of the frequency-domain index is closely correlated with the peak $x_o(t_p)$ of the time domain index. If M_r is larger, $x_o(t_p)$ is also larger, whereas if M_r is smaller, $x_o(t_p)$ is also smaller. The size of the visibility of M_r directly presents the amount of the overshoot; therefore, it is called oscillation performance metrics or indicators.

If the oscillation performance indicator M_r is larger, then the amount of the system overshoot is also larger.

The step input can be thought of as the sum of harmonics whose frequency varies from $0 \sim \infty$, M_r is larger which means that the frequency $\omega_r = 0$, and the peak value of the nearby ω_r component increases after the oscillation relation. Therefore, the unit-step response has a large overshoot.

6.3.3.7 Second-Order Differential Relation, $G(\omega) = T^2(j\omega)^2 + 2\xi j\omega T + 1; \ (T > 0, 0 < \xi < 1)$

The amplitude-phase frequency characteristics of second-order differential elements are as follows:

$$A(\omega) = \sqrt{\left[1-(\omega T)^2\right]^2 + (2\omega\xi T)^2}$$

and

$$\varphi(\omega) = \begin{cases} \tan^{-1}\left(\dfrac{2\xi\omega T}{1-(\omega T)^2}\right) & ;(\omega T<1) \\ \pi + \tan^{-1}\left(\dfrac{2\xi\omega T}{1-(\omega T)^2}\right) & ;(\omega T>1) \end{cases}$$

For $\omega = 0$ then, $A(\omega) = 1$, $\phi(\omega) = 0°$.

And for $\omega = \infty$ then, $A(\omega) = \infty$, $\phi(\omega) = 180°$.

The Nyquist plots of second-order differential elements are shown in Figure 6.31.

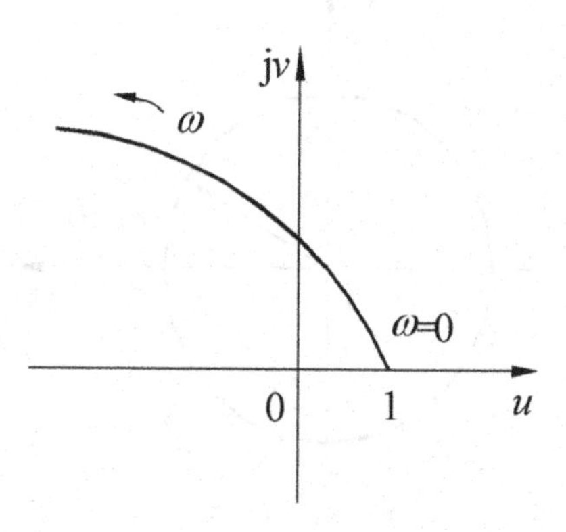

FIGURE 6.31 Nyquist plots of second-order differential elements.

6.3.3.8 Delay Relation, $G(j\omega)=e^{-j\omega T}$

The amplitude-phase frequency characteristics of the delay relation are as follows:

$$A(\omega)=1\varphi(\omega)=-\omega T$$

For, $\omega=0$ then $\begin{cases} A(\omega)=1 \\ \varphi(\omega)=0^\circ \end{cases}$.

For $\omega=\infty$ then $\begin{cases} A(\omega)=1 \\ \varphi(\omega)=-\infty \end{cases}$.

The Nyquist plots of delay relation are shown in Figure 6.32.

6.3.4 Drawing of Nyquist Plot

The typical Nyquist plots are drawn as it can be summarized in the following steps:

a. Find the expression of $A(\omega)$ and $\varphi(\omega)$.
b. Find $G(j\omega)$ when $\omega=0$ and $\omega=\infty$.
c. Find the intersection point of the Nyquist plot in the axes; the point of intersection with the real axis is $\text{Im}[G(j\omega)]=0$ and $\varphi(\omega)=n$ (where n is an integer); the point of intersection with the imaginary axis $\text{Re}[G(j\omega)]=0$ and $\varphi(\omega)=n\frac{\pi}{2}$ (where n is an integer).
d. Put a couple of intermediate points in the middle of the plot; and then
e. Draw the Nyquist plot.

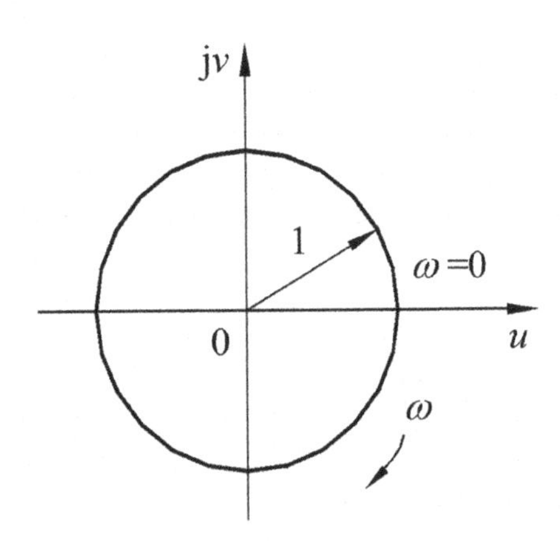

FIGURE 6.32 Nyquist plots of delay relation.

Example 6.2

We know that the transfer function of a system is $G(j\omega)=\dfrac{e^{-j\omega\tau}}{j\omega T+1}$. Plot the Nyquist plots of the system.

Solution

The amplitude-phase frequency characteristics of the solution for the given system are as follows:

$$A(\omega)=\left|G(j\omega)\right|=\frac{1}{\sqrt{1+\omega^2T^2}}$$

$$\varphi(\omega)=-\tau\omega-\tan^{-1}(\omega T)$$

For $\omega=0$ then $\begin{cases} A(\omega)=1 \\ \varphi(\omega)=0 \end{cases}$.

And for $\omega=\infty$ then $\begin{cases} A(\omega)=0 \\ \varphi(\omega)=-\infty^\circ \end{cases}$.

According to the above analysis, the Nyquist plots of $G(j\omega)$ have infinitely many intersection points with real and imaginary axes. With the increase of ω, the curve is getting closer and closer to the origin and the amplitude approaches zero but phase angle approaches to $-\infty^\circ$, as shown in Figure 6.33.

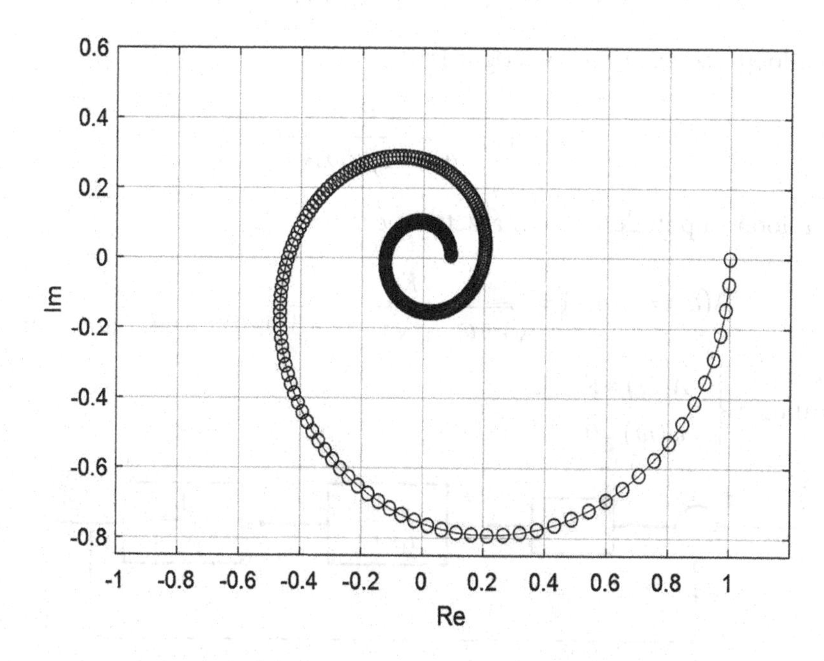

FIGURE 6.33 Nyquist plot of $\dfrac{e^{-j\omega\tau}}{j\omega T+1}$.

The MATLAB program as follows:

```
T=1;
w = linspace(0,7*pi/2,300);
A=exp(-complex(0,w));
B=1+complex(0,w);
Gw=A./B;
plot(Gw, 'o')
hold on
plot(Gw, '-')
axis equal;
axis([-1 1.2-0.85 0.6]);
xlabel('Re')
ylabel('Im')
grid
```

Example 6.3

The structure of a feedback control system of a certain unit is shown in Figure 6.34, where $T_1 > T_2$. Draw the open-loop Nyquist plots of the system.

Solution

The open-loop transfer function of the system is

$$G(s)=\frac{K}{(T_1s+1)(T_2s+1)}$$

The open-loop frequency characteristic is

$$G(j\omega)=\frac{K}{(j\omega T_1+1)(j\omega T_2+1)}$$

The open-loop amplitude-phase characteristic is

$$A(\omega)=\left|G(j\omega)\right|=\frac{K}{\sqrt{1+\omega^2T_1^2}\sqrt{1+\omega^2T_2^2}}\ \varphi(\omega)=-\tan^{-1}\omega T_1-\tan^{-1}\omega T_2$$

For $\omega=0$ then $\begin{cases} A(\omega)=K \\ \varphi(\omega)=0 \end{cases}$.

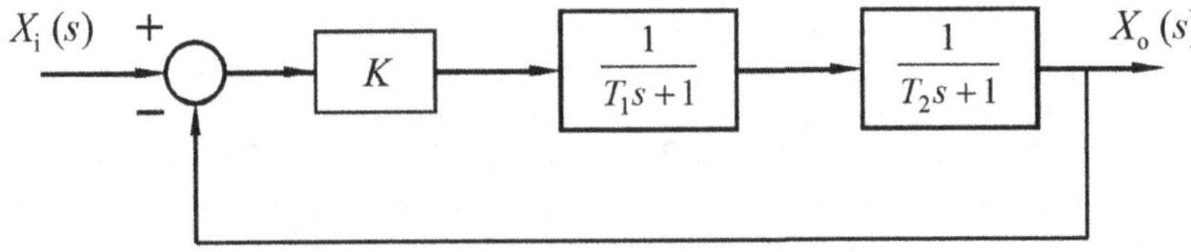

FIGURE 6.34 A feedback control system.

And for $\omega = \infty$ then $\begin{cases} A(\omega)=0 \\ \varphi(\omega)=-\pi \end{cases}$.

If $T_1 = 1$, $T_2 = 1/2$, then we can see the Nyquist plot in Figure 6.35. The MATLAB program is as follows:

```
T1=1; T2=1/2; K=10;
for i=1:1000
w(i)=(i-1)/100;
Aw(i)=10/sqrt(w(i)*w(i)*(T1*T1)+1)/sqrt(w(i)*w(i)*(T2*T2)+1);
Phi(i)=-atan(w(i)*T1)-atan(w(i)*T2);
end
polar(Phi, Aw, '.k')
ylabel('Polar coordinates')
```

In this case, there are two first-order inertia elements; when $\omega \to \infty$, the Nyquist curve approaches to $0\angle -180°$. The Nyquist plots of the first-order inertial elements are shown in Figure 6.13, and the Nyquist plots of the two first-order inertial elements are shown in Figure 6.35. If the system consists of nth first-order inertial elements, then the system approaches to $0\angle -n\times\frac{\pi}{2}$. Figure 6.36a shows the approximated shape of the Nyquist plots when $n=1 \sim 4$.

If the zero-type system contains inertia and proportional elements, it also contains first-order differential equation, when $\omega = 0 \sim \infty$. In the first-order differential

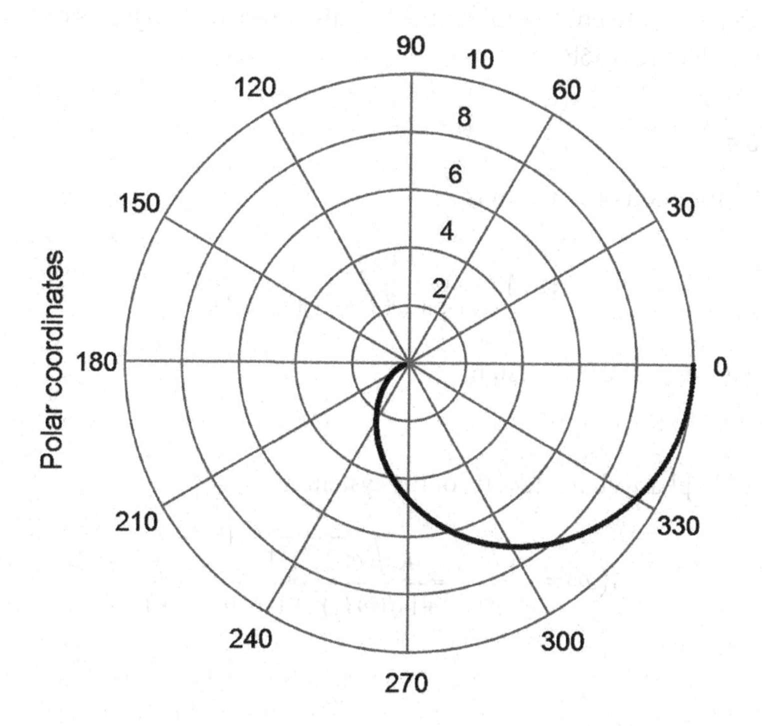

FIGURE 6.35 Nyquist plot of Example 6.3.

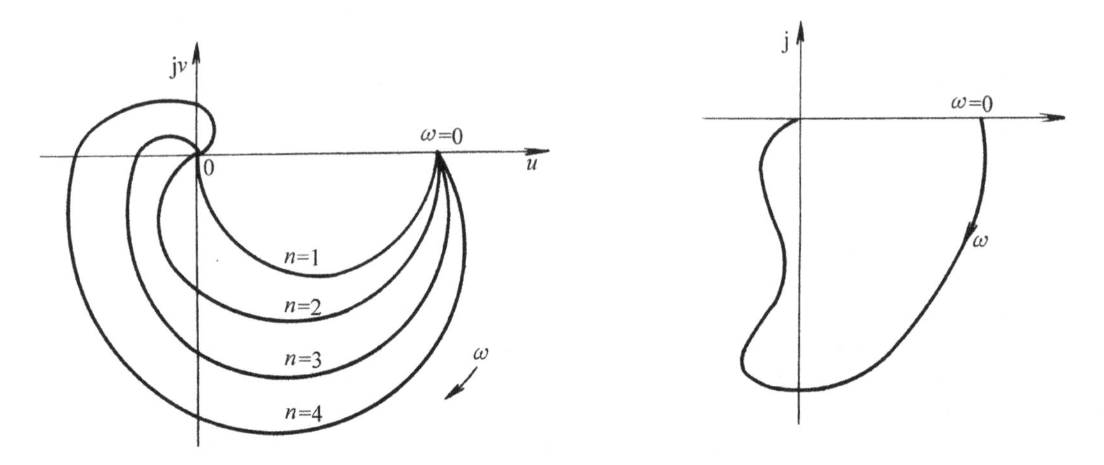

FIGURE 6.36 (a) Nyquist plots of the zero-type system with n inertial elements and (b) concave and convex Nyquist plots.

equation, the phase frequency varies from 0 to $\frac{\pi}{2}$. The total phase-frequency characteristic of the system must satisfy the following relation:

$$\varphi(\omega)=(m-n)\frac{\pi}{2} \tag{6.57}$$

In Eq. (6.57), m and n are the first-order differential equation and the number of first-order inertia equation, respectively. Therefore, the Nyquist curve is changed as concave and convex, as shown in Figure 6.36b.

Example 6.4

The transfer function of a system is

$$G(s)=\frac{K(T_1s+1)}{(T_2s+1)(T_3s+1)(T_4s+1)}$$

Plot the Nyquist plots of the system.

Solution

The amplitude-phase characteristic of the system is

$$A(\omega)=\frac{K\sqrt{(\omega T_1)^2+1}}{\sqrt{(\omega T_2)^2+1}\sqrt{(\omega T_3)^2+1}\sqrt{(\omega T_4)^2+1}}$$

and

$$\varphi(\omega)=\tan^{-1}\omega T_1-\tan^{-1}\omega T_2-\tan^{-1}\omega T_3-\tan^{-1}\omega T_4$$

When $\omega \to 0$, $\begin{cases} A(\omega)=K \\ \varphi(\omega)=0 \end{cases}$.

When $\omega \to \infty$, $\begin{cases} A(\omega)=0 \\ \varphi(\omega)=(1-3)\frac{\pi}{2}=-\pi \end{cases}$.

If the time constant $T_2 > T_1, T_3 > T_1, T_1 > T_4$, the Nyquist plots of the system are shown in Figure 6.36b.

Example 6.5

The open-loop transfer function of a feedback system is

$$G(s)=\frac{K}{s(T_1s+1)(T_2s+1)(T_3s+1)}$$

The characteristics of the open-loop Nyquist plots were analyzed. Determine the frequency corresponding to the intersection of the Nyquist plots and the negative real axes at ω_x. Calculate $|G(j\omega_x)|$ and roughly draw the open-loop Nyquist curve.

Solution

The frequency characteristic of the open-loop system is

$$G(j\omega)=\frac{K}{j\omega(j\omega T_1+1)(j\omega T_2+1)(j\omega T_3+1)}$$

The open-loop amplitude-phase characteristic of the system is

$$A(\omega)=\frac{K}{\omega\sqrt{(\omega T_1)^2+1}\sqrt{(\omega T_2)^2+1}\sqrt{(\omega T_3)^2+1}}$$

and

$$\varphi(\omega)=-\frac{\pi}{2}-\tan^{-1}\omega T_1-\tan^{-1}\omega T_2-\tan^{-1}\omega T_3$$

If $T_1 > T_2 > T_3$, then

$$G(j0)=\infty\angle-\frac{\pi}{2}\; G(j\infty)=0\angle-2\pi$$

When $\omega \to 0$, the asymptote of the Nyquist curve is not an imaginary axis but the abscissa for V_x is parallel to the imaginary axis of a straight line. The solution of the V_x is the real part of $G(j\omega)$ which is separated from the imaginary part, i.e.

$$G(j\omega)=\frac{-K\left[(T_1+T_2+T_3)-\omega^2T_1T_2T_3\right]}{\left(1+\omega^2T_1^2\right)\left(1+\omega^2T_2^2\right)\left(1+\omega^2T_3^2\right)}+j\frac{-K\left[1-\omega^2(T_1T_2+T_2T_3+T_3T_1)\right]}{\omega\left(1+\omega^2T_1^2\right)\left(1+\omega^2T_2^2\right)\left(1+\omega^2T_3^2\right)}$$

If $\omega\to0$, the real part of $G(j\omega)$ is

$$V_x=-K(T_1+T_2+T_3)$$

There are two ways to solve the problem of ω_x, which are as follows:

1. According to $\varphi(\omega)=-\frac{\pi}{2}-\tan^{-1}\omega T_1-\tan^{-1}\omega T_2-\tan^{-1}\omega T_3$

   Now, $\varphi(\omega)=-\pi$, then calculate ω as ω_x; thus

$$\pi=\frac{\pi}{2}+\tan^{-1}\omega T_1+\tan^{-1}\omega T_2+\tan^{-1}\omega T_3$$

$$\frac{\pi}{2}-\tan^{-1}\omega T_1=\tan^{-1}\omega T_2+\tan^{-1}\omega T_3$$

   If we take the tangent of both sides, then we get

$$\frac{1}{\omega_xT_1}=\frac{\omega_xT_2+\omega_xT_3}{1-\omega_xT_2\omega_xT_3}$$

   or

$$1-\omega_x{}^2T_2T_3=\omega_x^2(T_2+T_3)T_1$$

   Thus, $\omega_x=\frac{1}{\sqrt{T_1T_2+T_2T_3+T_3T_1}}$

2. ω_x can also be used for $\text{Im}\left[G(j\omega)\right]=0$, and thus

$$\frac{-K\left[1-\omega_x^2(T_1T_2+T_2T_3+T_3T_1)\right]}{\omega_x\left(1+\omega_x^2T_1^2\right)\left(1+\omega_x^2T_2^2\right)\left(1+\omega_x^2T_3^2\right)}=0$$

   After solving the above equation, we get

$$\omega_x=\frac{1}{\sqrt{T_1T_2+T_2T_3+T_3T_1}}$$

   Thus, we get

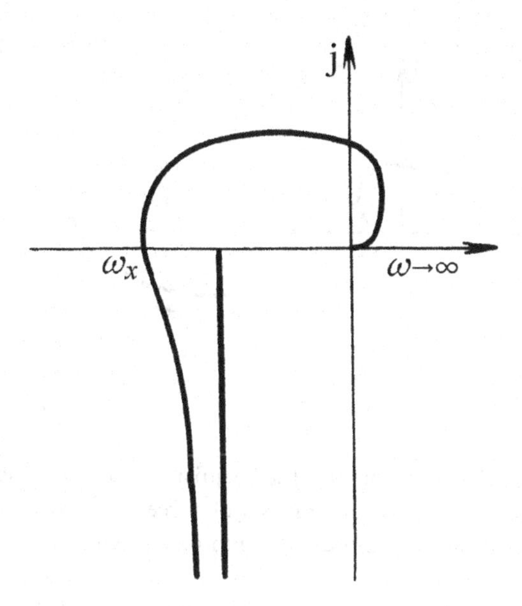

FIGURE 6.37 Nyquist plots of a type I system.

$$|G(j\omega_x)| = K\frac{(T_1+T_2+T_3)-\omega_x^2 T_1 T_2 T_3}{(1+\omega_x^2 T_1^2)(1+\omega_x^2 T_2^2)(1+\omega_x^2 T_3^2)}$$

The Nyquist plots of the system are shown in Figure 6.37.

Type II system contains two first-order inertia equations. Its open-loop Nyquist plots must start at infinity which is parallel to the negative real axis. For example, an open-loop transfer function is

$$G(s) = \frac{K}{s^2(T_1 s+1)(T_2 s+1)}$$

The Nyquist plots of the system are shown in Figure 6.38a.

Type II system consists of three first-order inertial elements and one first-order differential element and the Nyquist plots of the system may have concave and convex shapes. For example, an open-loop transfer function is $G(s) = \frac{K(T_1 s+1)}{s^2(T_2 s+1)(T_3 s+1)(T_4 s+1)}$ and the characteristics of the Nyquist curve is $G(j0^+) = \infty\angle -\pi, G(j\infty) = 0\angle -2\pi$. If the time constant $T_1 > T_2 > T_3 > T_4$, the Nyquist plots of the system are shown in Figure 6.38b. The frequency corresponding to the intersection of the Nyquist plots and the negative real axis is obtained by the same method as the previously described.

For the Nyquist plots of higher-order systems, the LTI system for a general case, its transfer function is

$$G(s) = \frac{b_0 s^m + b_1 s^{m-1} + \cdots + b_{m-1}s + b_m}{a_0 s^n + a_1 s^{n-1} + \cdots + a_{n-1}s + a_n} = \frac{K(\tau_1 s+1)(\tau_2 s+1)\cdots(\tau_M s+1)}{s^\lambda (T_1 s+1)(T_2 s+1)\cdots(T_N s+1)} \tag{6.58}$$

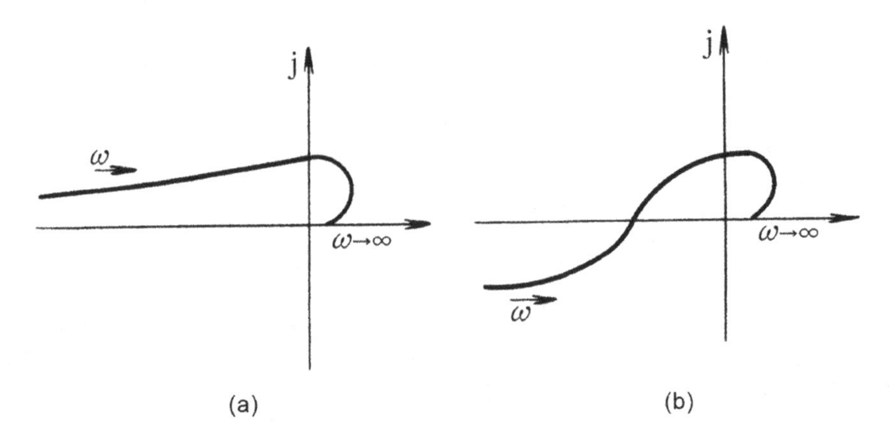

FIGURE 6.38 (a) The type II system contains the Nyquist plots of two first-order inertial elements, and (b) the type II system consists of three first-order inertial elements and the Nyquist plots of one first-order differential element.

Then the frequency characteristic of the system is

$$G(j\omega)=\frac{K(j\omega\tau_1+1)(j\omega\tau_2+1)\cdots(j\omega\tau_M+1)}{(j\omega)^{\lambda}(j\omega T_1+1)(j\omega T_2+1)\cdots(j\omega T_N+1)} \tag{6.59}$$

It is obtained from Eq. (6.59). When $\lambda=0$, the system is type zero; when $\lambda=1$, the system is type I; and when $\lambda=1$, the system is type II.

1. When $\omega=0$, the characteristics of the Nyquist plots were determined by the K of the proportional equation of the system and the type of the system.
2. When $\omega\to 0$, then $G(j\omega)\approx\frac{K}{\omega^{\lambda}}\angle\lambda\left(-\frac{\pi}{2}\right)$, when $K>0$, the Nyquist plots of zero-type systems begin with finite values of the positive real axis at point (K, j_0); the asymptotes of the Nyquist plots of type I systems are the lines that are parallel to the imaginary axis, and their coordinates are determined by the following equation.

$$V_x=\lim_{\omega\to 0^+}\text{Re}\left[G(j\omega)\right] \tag{6.60}$$

3. For the general control systems, in Eq. (6.59), we say $N<\lambda+M$, *when* $\omega\to\infty$. Thus, the amplitude-phase characteristic of the system is

$$G(j\infty)=0\angle[M-N-\lambda]\frac{\pi}{2} \tag{6.61}$$

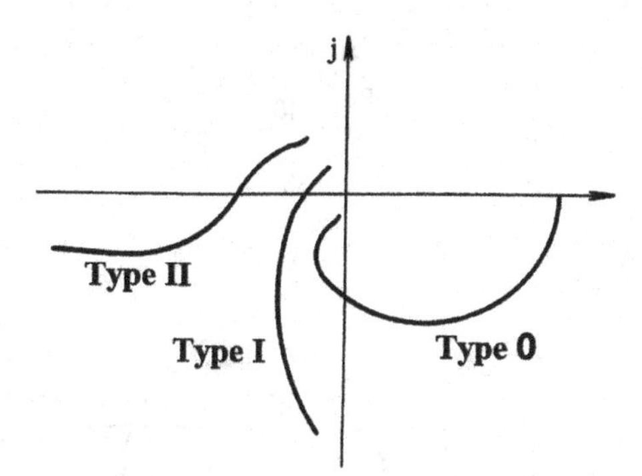

FIGURE 6.39 Various types of low-frequency characteristics of Nyquist plots.

The Nyquist curve is measured by $[M-N-\lambda]\frac{\pi}{2}$ phase angle of the tangent and the origin.

4. The Nyquist curve with negative real axis intersection of frequency is ω_x and the corresponding amplitude is obtained as $\text{Im}[(G(j\omega))]=0$, when $\varphi(\omega)=-\pi$.
5. When there is no first-order differential equation, the phase angles of the Nyquist plots decreased continuously. On the other hand, the phase angle of the Nyquist plot may not necessarily decrease continuously. Therefore, the Nyquist plots may have concave and convex shapes.

The low-frequency characteristics of the Nyquist plots of various types are shown in Figure 6.39.

Example 6.6

The transfer function of the feedback control system of a certain system is

$$G(s)=\frac{10(0.1s+1)(100s+1)}{(20s+1)(10s+1)(5s+1)(0.01s+1)(0.005s+1)(0.0006s+1)}$$

Plot the Nyquist plots of the system.

Solution

For the transfer function of this example, it is difficult to determine its Nyquist plots, thus we need to solve it by simulation. We may solve this by using MATLAB program, the Nyquist plot we can see is shown in Figure 6.40. When $\omega\to0$, the Nyquist curve tends to 0°; when $\omega\to\infty$, the Nyquist curve tends to 360°.

The MATLAB program is as follows:

```
for j=1:2000
w(j)=(j-1)/1000;
```

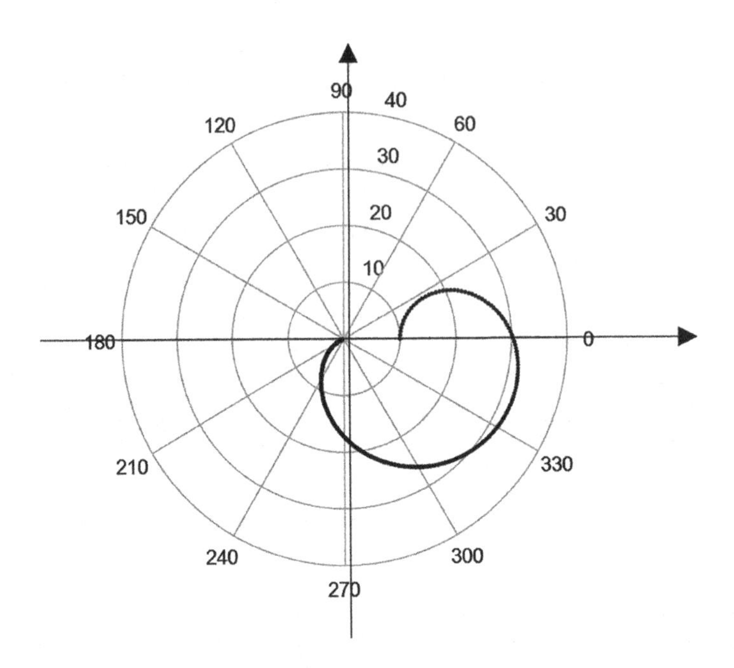

FIGURE 6.40 Nyquist curve of the Example 6.6.

```
a1(j)=10*sqrt((0.1*w(j))^2+1);
a2(j)=sqrt((100*w(j))^2+1);
a3(j)=1/sqrt((20*w(j))^2+1);
a4(j)=1/sqrt((10*w(j))^2+1);
a5(j)=1/sqrt((5*w(j))^2+1);
a6(j)=1/sqrt((0.01*w(j))^2+1);
a7(j)=1/sqrt((0.005*w(j))^2+1);
a8(j)=1/sqrt((0.0006*w(j))^2+1);
a(j)=a1(j)*a2(j)*a3(j)*a4(j)*a5(j)*a6(j)*a7(j)*a8(j);
fa1(j)=atan(0.1*w(j))+atan(100*w(j));
fa2(j)=-atan(20*w(j))-atan(10*w(j))-atan(5*w(j));
fa3(j)=-atan(0.01*w(j))-atan(0.005*w(j))-atan(0.0006*w(j));
fa(j)=fa1(j)+fa2(j)+fa3(j);
end
polar(fa, a, '.')
```

6.4 BODE DIAGRAM OR LOGARITHMIC PLOT OF FREQUENCY RESPONSE

A Bode diagram or graph is a curve drawn on two semilog axes by plotting the relationship between amplitude and phase angle in frequency characteristics with the frequency ω. *Bode diagram* is widely used in frequency response; its amplitude and phase angle are divided by uniform scale, that is, linear scale. The frequency is divided by logarithm. The coordinates of amplitude-frequency characteristics are shown in Figure 6.41.

The abscissa of the logarithmic frequency characteristic is represented in the frequency ω which is divided logarithmically; the units are (rad/s), and the logarithmic

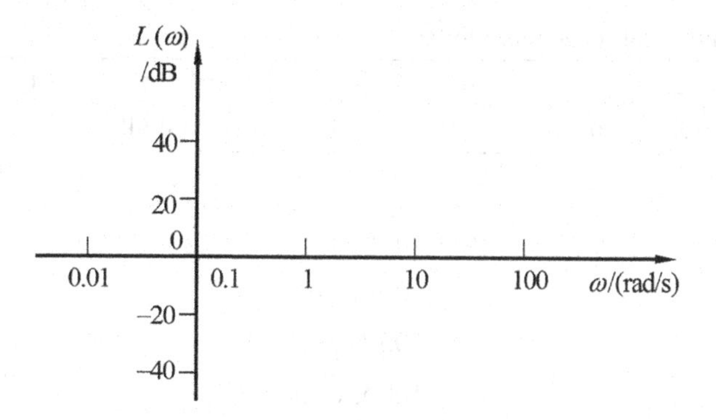

FIGURE 6.41 Bode diagram of frequency characteristics.

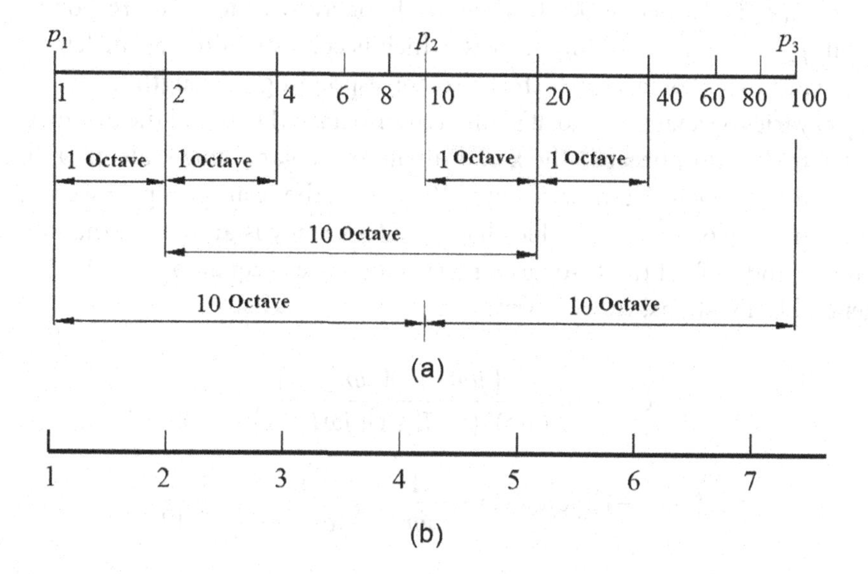

FIGURE 6.42 (a) Logarithmic scale and (b) linear scale.

amplitude-frequency curve represents the function value of the log-frequency characteristic, its unit is decibels (dB); the logarithmic phase-frequency characteristics is along with vertical axis which is the function of phase-frequency characteristics, its unit is the degrees. Suppose that there are two signal power such as N_1 and N_2, then we say, $\log N_2 - \log N_1 = \log(N_2 / N_1) = 1$; they differ by 1 dB, and because dB unit is larger, the decibel (dB) is used as the unit in engineering. When $\log N_2 / N_1 = 0.1$, then $10\lg N_2 / N_1 = 1$, and thus, $N_2 > N_1 > 1$ dB. The square of some physical quantity, such as voltage, is proportional to the power, that is,

$$10\log\frac{N_2}{N_1} = 10\log\left(\frac{V_2}{V_1}\right)^2 = 20\log\frac{V_2}{V_1}$$

Thus, V_2 and V_1 differ by 1 dB. In mechatronic and control engineering, any number N can be expressed by the decibel value n, defined as n (dB) = 20log$N$. The difference between logarithmic scale and linear scale is shown in Figure 6.42.

TABLE 6.4 Logarithmic Scale of ω from 1 to 10.

ω	1	2	3	4	5	6	7	8	9	10
$\log\omega$	0	0.303	0.477	0.602	0.699	0.778	0.845	0.903	0.954	1

$$\omega = 1, \log\omega = 0$$
$$\omega = 10, \log\omega = 1$$
$$\omega = 100, \log\omega = 2$$
$$\omega = 1000, \log\omega = 3$$
$$\cdots\cdots\ \cdots\cdots\ \cdots\cdots$$

In Figure 6.42, p_1 is defined as $\omega=1$, when $\omega=10$ octaves, which is corresponding to the length from p_2 to p_1; when $\omega=100$ octaves, which is corresponding to the length from p_3 to p_1; and when $\omega=10$ octaves, which is also corresponding the length from p_3 to p_2. The frequency ω varies by a factor of 10, it's called a ten-octave (Dec), and the distance between the abscissa is one unit, because the x-coordinate is a logarithmic scale of ω; if ω is not uneven or uniform, log ω is uniform. Every 10 octaves, the relationship between ω and log ω is shown in Table 6.4. Every double change of frequency is an octave, the interval distance is 0.301, and each of the 10-octave interval distance is equal to $1/0.301=3.32$ times.

For general LTI systems, we may write

$$G(j\omega)=\frac{K(j\omega\tau_1+1)(j\omega\tau_2+1)\cdots}{(j\omega)^{\lambda}(j\omega T_1+1)(j\omega T_2+1)\cdots}$$

$$=K(j\omega\tau_1+1)(j\omega\tau_2+1)\cdots\frac{1}{(j\omega)^{\lambda}}\cdots\frac{1}{j\omega T_1+1}\frac{1}{j\omega T_2+1}\cdots$$

$$=A_1(\omega)e^{j\varphi_1(\omega)}\cdot A_2(\omega)e^{j\varphi_2(\omega)}\cdots A_n(\omega)e^{\varphi_n(\omega)}$$

$$=A_1(\omega)A_2(\omega)\cdots A_n(\omega)e^{j[\varphi_1(\omega)+\varphi_2(\omega)\cdots\varphi_n(\omega)]}$$

$$L(\omega)=20\log\left|G(j\omega)\right|=20\log\left(A_1(\omega)A_2(\omega)\cdots A_n(\omega)\right)$$

$$=20\log A_1(\omega)+20\log A_2(\omega)+\cdots+20\log A_n(\omega)$$

According to the above analysis, the Bode diagram or plot has the following advantages:

1. Multiplication and division are simplified to addition and subtraction, and the square root operation is simplified to multiplication and division.
2. Logarithmic amplitude-frequency diagram or graph can be drawn by a simple and convenient method. The amplitude-frequency characteristic can be approximated by broken lines. The amplitude-frequency characteristic of a system can be combined with the broken lines of the amplitude-frequency characteristic of each relation of the system.

3. The expression of frequency characteristic and the transfer function of the system can be easily determined by drawing the numerical value of frequency characteristic and which is obtained by experiment into the logarithmic frequency characteristic curve.
4. Semilog paper uses limited paper space to express a wide range of frequencies.

6.4.1 Bode Diagram or Logarithmic Plot of Typical Relations

The Bode diagram of eight typical relations is commonly used to discuss here. A Bode diagram of a complex system can be constructed by superimposing these eight typical relations. The eight typical relations are as follows:

1. *Proportional gain factor (Amplification gain) K.*
2. *Integral factor 1/s (Including second-order factor* $1/s^2$).
3. *Derivative factor* s.
4. *First-order inertia factor* $1/(T_s+1)$.
5. *First-order differential factor* (T_s+1).
6. *Second-order inertia factor* $1/\left(s^2/\omega_n^2+2\xi s/\omega_n+1\right)$.
7. *Second-order differential factor* $s^2/\omega_n^2+2\xi s/\omega_n+1$.
8. *Delay factor* $e^{-j\omega\tau}$.

The Bode diagram of these eight typical factors is shown below.

6.4.1.1 *Proportional Gain Factor* $G(j\omega)=K$

The logarithmic amplitude-phase characteristics of the amplification factors are as follows:

$$L(\omega)=20\log K$$

$$\varphi(\omega)=0°$$

The Bode diagram of the amplification gain factors is shown in Figure 6.43.

6.4.1.2 *Integral Factors* $G(j\omega)=\dfrac{1}{j\omega}$

The logarithmic amplitude-phase characteristics of the integral factor are as follows:

$$L(\omega)=20\log\left|\frac{1}{j\omega}\right|=-20\log\omega$$

$$\varphi(\omega)=-\frac{\pi}{2}$$

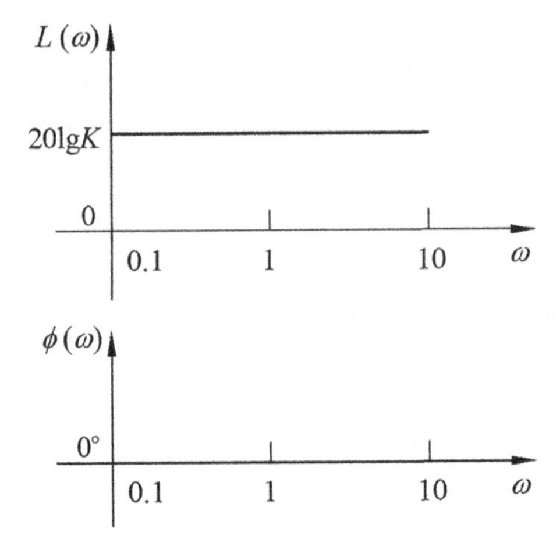

FIGURE 6.43 Bode diagram of proportional gain factor.

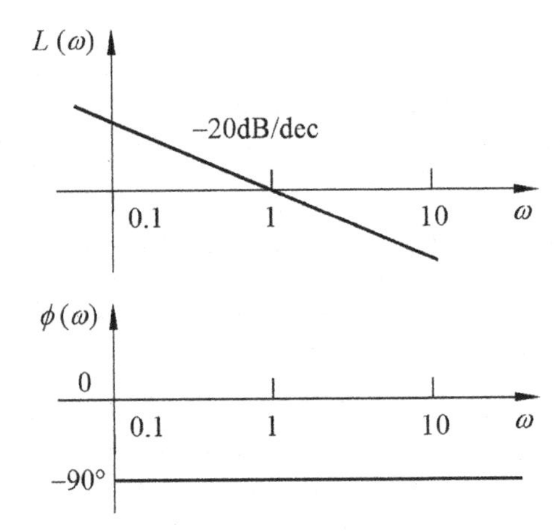

FIGURE 6.44 Bode diagram of integral factor.

The Bode diagram of integral factors is shown in Figure 6.44.

6.4.1.3 *Second-Order Integral Factors* $G(j\omega)=\dfrac{1}{(j\omega)^2}$

The logarithmic amplitude and phase characteristics of the second-order integral factors are as follows:

$$L(\omega)=20\log\left|G(j\omega)\right|=20\log\left|\frac{1}{(j\omega)^2}\right|=-40\log\omega$$

$$\varphi(\omega)=-\pi$$

The Bode diagram of second-order integral factors is shown in Figure 6.45a.

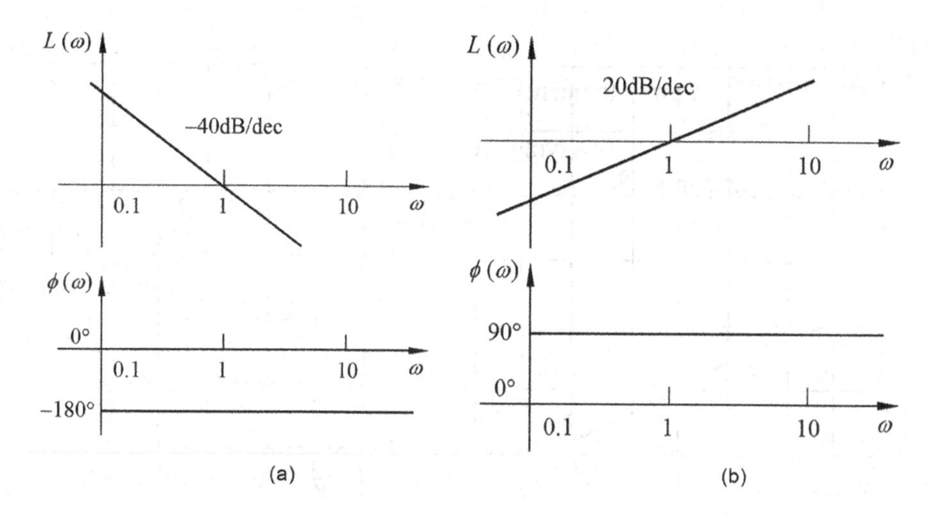

FIGURE 6.45 Bode diagram of (a) second-order integral factor and (b) derivative factors.

6.4.1.4 Derivative Factors $G(j\omega)=j\omega$

The logarithmic amplitude-phase characteristics of the derivative factors are as follows:

$$L(\omega)=20\log\left|G(j\omega)\right|=20\log\omega$$

$$\varphi(\omega)=\frac{\pi}{2}$$

The Bode diagram of the differential factors is shown in Figure 6.45b.

6.4.1.5 First-Order Inertia Factor $G(j\omega)=\dfrac{1}{j\omega T+1}$

The logarithmic amplitude-phase characteristics of the first-order inertial elements are as follows:

$$L(\omega)=20\left|G(j\omega)\right|=-20\log\sqrt{\omega^2T^2+1},\text{and}$$

$$\varphi(\omega)=-\tan^{-1}\omega T$$

At low frequency, when ω *is* very small, then $\omega T \ll 1$; thus $L(\omega)=0$ dB.

At high frequency, when ω is very larger, then $\omega T \gg 1$; thus $L(\omega)=$ 20 $\log T$.

The Bode diagram or plot of its amplitude-frequency characteristics can be approximated by the broken lines of the two asymptotes of the low-frequency and high-frequency bands, as shown in Figure 6.46a; when $\omega T=1$, $\omega_T=1/T$, which is called turning or transition or corner frequency. The logarithmic amplitude-frequency error at the corner frequency is expressed as

$$L(\omega)=-20\log\sqrt{1+1}=-3\text{dB}$$

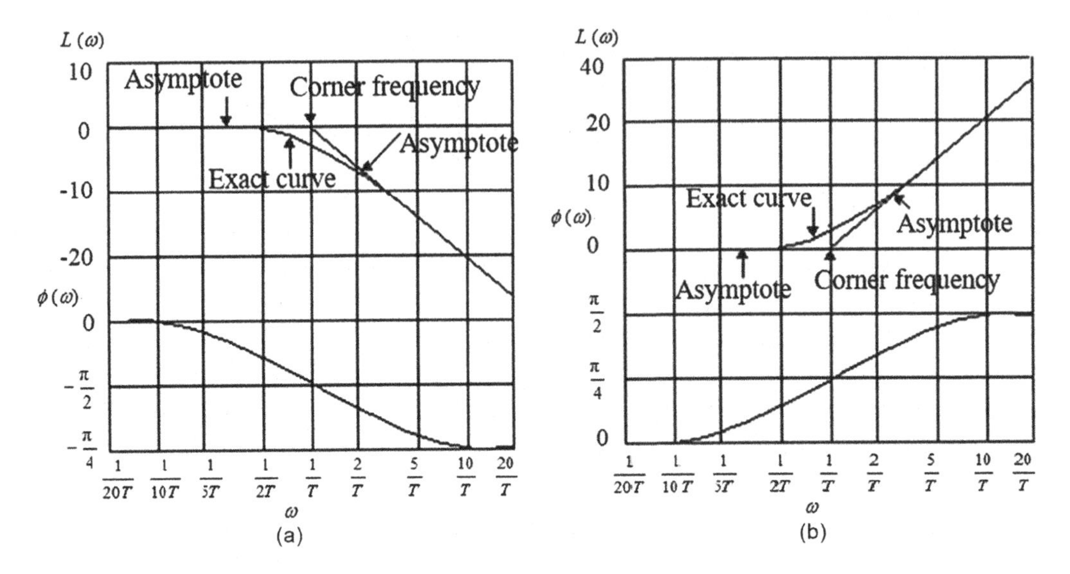

FIGURE 6.46 Bode diagram of (a) first-order inertial factor and (b) first-order differential factor.

TABLE 6.5 Modified Amplitude-Frequency for Bode Diagram of First-Order Inertial Factor

ωT	1/10	2/10	5/10	1	2	5	10
Correction/dB	−0.04	−0.17	−0.97	−3.01	−0.97	−0.17	−0.04

In the Bode diagram, ω_T is the amplitude-frequency characteristics of the maximum error, it differs with the accurate values about −3 dB. The accurate values can be modified on the basis of the approximated Bode diagram.

By using $L(\omega) = -20\log\sqrt{\omega^2 T^2 + 1}$, each frequency can be determined by corresponding amplitude-frequency values. The error at the frequency one octave below the corner frequency (ω_T) that is at $\omega_T = 1/2T$; thus, the error is defined as

$$\Delta L(\omega)_{1/2T} = L(\omega)_{\text{Exact}} - L(\omega)_{\text{Asymptote}} = -20\log\sqrt{\frac{1}{4}+1} + 20\lg 1 = -0.97\text{dB}$$

The error at the frequency one octave above the corner frequency (ω_T) that is at $\omega_T = 2/T$; thus, the error is defined as

$$\Delta L(\omega)_{2/T} = L(\omega)_{\text{Exact}} - L(\omega)_{\text{Asymptote}} = -20\log\sqrt{2^2+1} + 20\log 2 = -0.97\text{dB}$$

Therefore, when the frequency range is lower or higher than one octave of the corner frequency, the error is 0.97 dB. The same method can be used to calculate the error of any point in the frequency range 10 octaves of the corner frequency, as shown in Table 6.5. It may be plotted as the error into a graph and corrected according to the graph, but using computer numerical simulation, it is very convenient to calculate the exact value of the system, and there is no need to make the graph of error correction.

TABLE 6.6 Phase Angle-Frequency for Bode Diagram of First-Order Inertial Factor

ωT	0	1/10	2/10	3/10	5/10	1	2	3	5	10	∞
ϕ/rad	0	−0.099	−0.197	−0.291	−0.464	−0.785	−1.107	−1.249	−1.373	−1.471	−1.571

Since $\varphi(\omega)=-\tan^{-1}\omega T$; at low frequency, ω is very small, then $\varphi(\omega)=0$. However, at high frequency, ω approaches to ∞, then we can say $\varphi(\omega)=-\frac{\pi}{2}$. At the corner frequency, the phase angle is defined as

$$\varphi(\omega_T)=-\tan^{-1}(\omega_T T)=-\tan^{-1}1=-\frac{\pi}{4}$$

The Bode diagram of the first-order inertial factor is shown in Figure 6.46a, and the phase-frequency characteristics are shown in Table 6.6.

6.4.1.6 First-Order Differential Factor $G(j\omega)=j\omega\tau+1$

The amplitude-phase characteristic of the first-order differential factor is

$$L(\omega)=20\log\sqrt{\tau^2\omega^2+1}, \text{And}$$

$$\varphi(\omega)=\tan^{-1}(\omega\tau)$$

The analysis method of first-order differential factor is similar to that of first-order inertial factor, and the result is exactly opposite to that of first-order inertial factor. Its Bode diagram is shown in Figure 6.46b.

6.4.1.7 Second-Order Inertia Factor

$$G(j\omega)=\frac{1}{(j\omega/\omega_n)^2+2\xi(j\omega/\omega_n)+1};\ (\omega_n=1/T>0,\ 0<\xi<1)$$

The logarithmic amplitude-frequency characteristic of the second-order inertia factor is expressed as $L(\omega)=20\log|G(j\omega)|$

$$=-20\log\sqrt{(1-(\omega/\omega_n)^2)^2+(2\xi\omega/\omega_n)^2}$$

$$=-20\log\sqrt{(1-(\omega T)^2)^2+(2\xi\omega T)^2}$$

At low frequency, when ω is very small, $\omega T \ll 1$;

Thus, $L(\omega)=0$ dB; and

At high frequency, when ω is larger, $\omega T \gg 1$;

Thus, $L(\omega)=-20\log\omega^2T^2=-40\log\omega T$.

The Bode diagram of amplitude-frequency characteristics of the second-order inertia factor can be approximated by the asymptotes composed of two straight lines in the low-frequency band and the high-frequency band, the asymptotes are shown in Figure 6.46.

TABLE 6.7 Amplitude-Frequency for Bode Diagram of Second-Order Inertia Factor

	ωT										
ξ	**0.1**	**0.2**	**0.4**	**0.6**	**0.8**	**1**	**1.25**	**1.66**	**2.5**	**5**	**10**
0.1	0.086	0.348	1.48	3.728	8.094	13.98	8.094	3.728	1.48	0.348	0.086
0.2	0.08	0.325	1.36	3.305	6.345	7.96	6.345	3.305	1.36	0.325	0.08
0.3	0.071	0.292	1.179	2.681	4.439	4.439	4.439	2.681	1.179	0.292	0.071
0.5	0.044	0.17	0.627	1.137	1.137	0.00	1.137	1.137	0.627	0.17	0.044
0.7	0.001	0.00	−0.08	−0.472	−1.41	−2.92	−1.41	−0.472	−0.08	0.00	0.001
1.00	−0.086	−0.34	−1.29	−2.76	−4.296	−6.20	−4.296	−2.76	−1.29	−0.34	−0.086

TABLE 6.8 Phase Angle Frequency for Bode Diagram of Second-Order Inertia Factor

	ωT							
ξ	**0.1**	**0.2**	**0.5**	**1**	**2**	**5**	**10**	**20**
0.1	0.021	−0.042	−0.133	−1.571	−3.009	−3.100	−3.121	−3.131
0.2	−0.040	−0.084	−0.260	−1.571	−2.829	−3.058	−3.101	−3.121
0.3	−0.061	−0.124	−0.380	−1.571	−2.761	−3.018	−3.081	−3.112
0.5	−0.101	−0.206	−0.588	−1.571	−2.553	−2.936	−3.040	−3.091
0.7	−0.141	−0.284	−0.750	−1.571	−2.391	−2.857	−3.000	−3.072
1.00	−0.199	−0.394	−0.927	−1.571	−2.215	−2.747	−2.943	−3.037

Two asymptotes are intersected at the undamped natural frequency ω_n. If the abscissa is ω/ω_n, then two asymptotes are intersected at $\omega/\omega_n = 1$. In fact, the Bode diagram varies with the damping ratio ξ, as shown in Figure 6.46, and the modification of its amplitude is shown in Table 6.7, and the phase angle is shown in Table 6.8.

The phase-frequency characteristic is expressed as

$$\varphi(\omega) = \begin{cases} -\tan^{-1}\left(\dfrac{2\xi T\omega}{1-\omega^2T^2}\right) & ;\left(\omega \le \dfrac{1}{T}\right) \\ -\pi - \tan^{-1}\left(\dfrac{2\xi\omega T}{1-\omega^2T^2}\right) & ;\left(\omega > \dfrac{1}{T}\right) \end{cases}$$

At low frequency, when ω is very small, $\varphi(\omega) = 0$;

At high frequency, when ω is larger, $\varphi(\omega) = -\pi$; and

At corner frequency, $\omega = \omega_n = \dfrac{1}{T}$, $\varphi(\omega_n) = -\dfrac{\pi}{2}$.

The exact curve varies with the damping ratio ξ, as we can see that Figure 6.47 can be drawn according to the data in Table 6.7. The Bode diagram of the second-order inertia

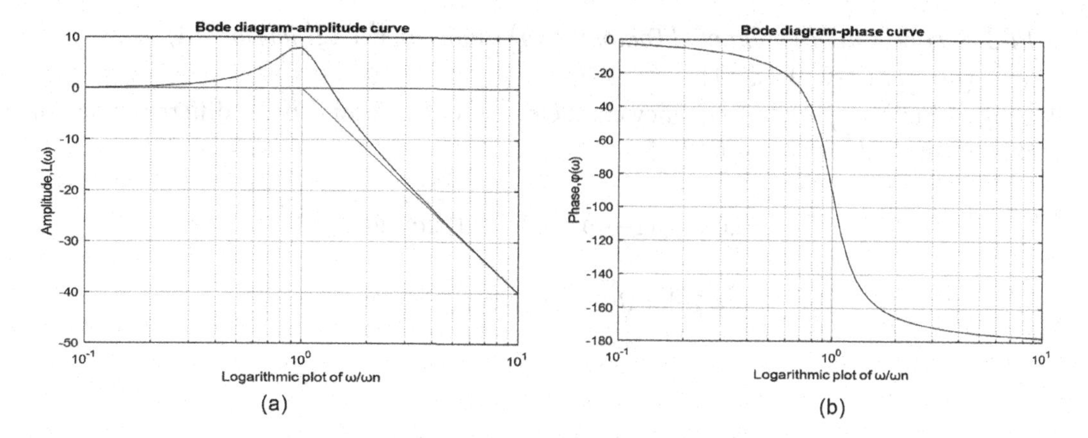

FIGURE 6.47 Bode diagram of second-order inertia factor.

factor is drawn by MATLAB procedures in the following. When using this program, we just change the damping ratio ξ, then we obtain Figure 6.47.

MATLAB program for Figure 6.47 is as follows:

```
ct=0.2;
for j=1:100
wt(j)=j/9.99999999999;
L(j)=-20*log10(sqrt((1-wt(j)*wt(j))^2+(2*ct*wt(j))^2));
l0(j)=0;
fa2(j)=-180;
if wt(j)<1
      l1(j)=0;
fa(j)=-180/pi*atan((2*ct*wt(j))/(1-wt(j)^2));
         else
        l1(j)=-40*log10(wt(j));
          fa(j)=-180/pi*(pi+atan((2*ct*wt(j))/(1-wt(j)^2)));
        end
end
subplot(2,1,1)
semilogx(wt,L,wt,l0,wt,l1)
title('Bode diagram-amplitude curve')
xlabel('Logarithmic plot of ω/ωn')
ylabel('Amplitude, L(ω) ')
grid
subplot(2,1,2)
semilogx(wt,fa,wt,fa2)
title('Bode diagram-phase curve')
xlabel('Logarithmic plot of ω/ωn')
ylabel('Phase,φ(ω)')
grid
```

6.4.1.8 Second-Order Differential Factor $G(j\omega)=(j\omega/\omega_n)^2+2\xi(j\omega/\omega_n)+1$; $(\omega_n=1/T>0,\ 0<\xi<1)$

The logarithmic amplitude-frequency characteristic of the second-order differential factor is expressed as $L(\omega)=20\log|G(j\omega)|$

$$=20\log\sqrt{(1-(\omega/\omega_n)^2)^2+(2\xi\omega/\omega_n)^2}$$

$$=20\log\sqrt{(1-(\omega T)^2)^2+(2\xi\omega T)^2}$$

and

$$\varphi(\omega)=\begin{cases}\tan^{-1}\dfrac{2\xi T\omega}{1-\omega^2T^2} & ;\left(\omega\le\dfrac{1}{T}\right)\\ \pi+\tan^{-1}\dfrac{2\xi\omega T}{1-\omega^2T^2} & ;\left(\omega>\dfrac{1}{T}\right)\end{cases}$$

The exact curve varies with the damping ratio ξ, as we can see that Figure 6.48 can be drawn. The Bode diagram of the second-order differential factor is drawn by MATLAB procedures in the following. When using this program, we just change the damping ratio ξ, and then we obtain Figure 6.48.

MATLAB program for Figure 6.48:

```
ct=0.2;
for j=1:100
wt(j)=j/9.99999999999;
L(j)=20*log10(sqrt((1-wt(j)*wt(j))^2+(2*ct*wt(j))^2));
l0(j)=0;
fa2(j)=200;
if wt(j)<1
```

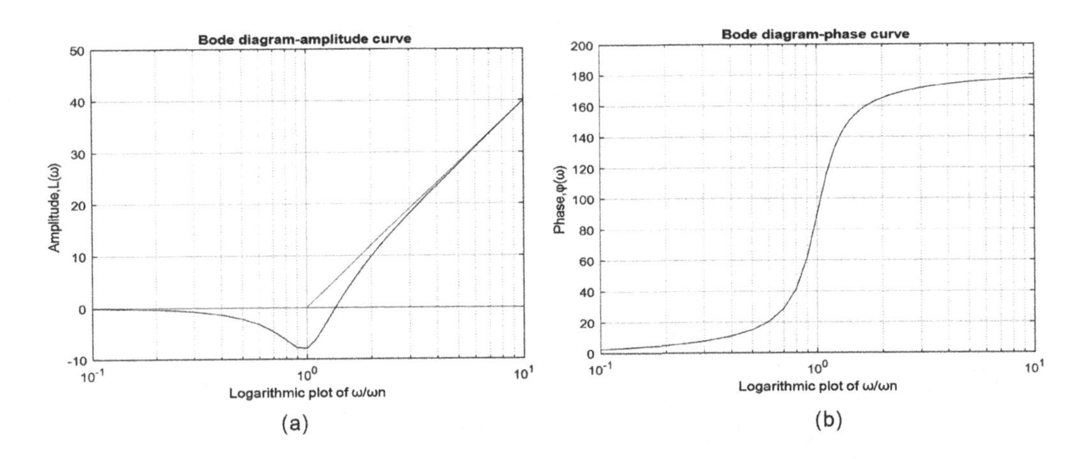

FIGURE 6.48 Bode diagram of second-order differential factor.

```
        ll(j)=0;
    fa(j)=180/pi*atan((2*ct*wt(j))/(1-wt(j)^2));
            else
           ll(j)=40*log10(wt(j));
             fa(j)=180/pi*(pi+atan((2*ct*wt(j))/(1-wt(j)^2)));
           end
end
subplot(2,1,1)
semilogx(wt,L,wt,l0,wt,ll)
title('Bode diagram-amplitude curve')
xlabel('Logarithmic plot of  ω/ωn')
ylabel('Amplitude,L(ω) ')
grid
subplot(2,1,2)
semilogx(wt,fa,wt,fa2)
title('Bode diagram-phase curve')
xlabel('Logarithmic plot of  ω/ωn')
ylabel('Phase,φ(ω)')
grid
```

6.4.1.9 Delay Factor $G(j\omega)=e^{-j\omega\tau}$

The logarithmic frequency characteristic of the delay factor is expressed as

$$L(\omega)=20\log\left|G(j\omega)\right|=20\log 1=0\text{dB};\text{and}\,\varphi(\omega)=-\omega\tau$$

The Bode diagram of the delay factor is shown in Figure 6.49a. The phase-frequency characteristic curve of the delay factor in Figure 6.49b in the linear coordinate system is compared with Figure 6.49a in the logarithmic coordinate system. If you take linear coordinates, the phase-frequency characteristic of the delay factor is a straight line, as shown in

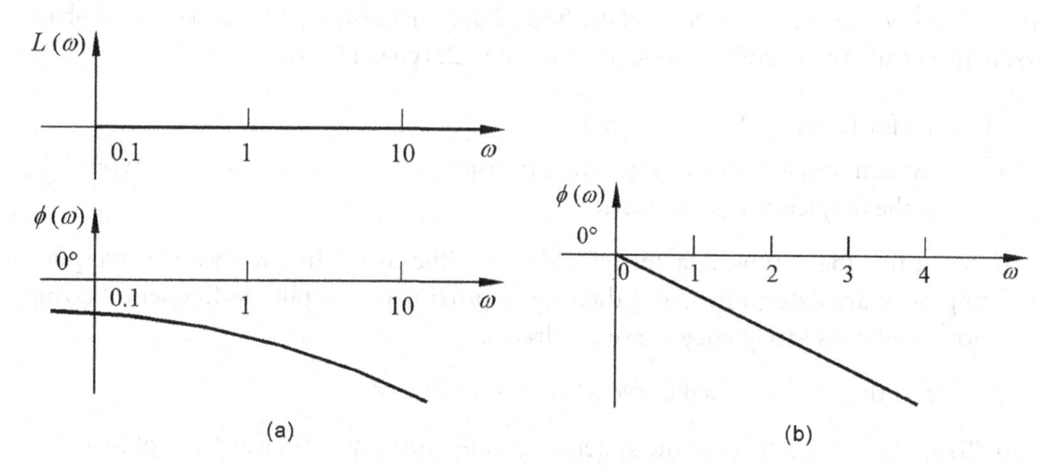

FIGURE 6.49 Bode diagram of (a) the delay factor, and (b) the phase-frequency characteristic curve of the delay factor in linear coordinates.

Figure 6.49b. In fact, if the phase-frequency characteristic of an element is a straight line in linear coordinates, the factor is the delayed factor.

6.4.2 General Method of Bode Diagram Drawing

For a general system,

$$G(j\omega)=\frac{K\prod_{i=1}^{M}(\tau_i j\omega+1)\prod_{l=1}^{N}(\tau_l^2(j\omega)^2+2\xi\tau_l j\omega+1)}{(j\omega)^{\lambda}\prod_{m=1}^{Q}(T_m j\omega+1)\prod_{n=1}^{P}(T_n^2(j\omega)^2+2\xi T_n j\omega+1)}$$

The logarithmic amplitude-frequency characteristic of the system is expressed as

$$L(\omega)=20\log K+\sum_{i=1}^{M}20\log\sqrt{(\tau_i\omega)^2+1}+\sum_{l=1}^{N}20\log\sqrt{(1-\omega^2\tau_l^2)^2+(2\xi_l\tau_l\omega)^2}$$

$$-20\lambda\log\omega-\sum_{m=1}^{Q}20\log\sqrt{1+(\omega T_m)^2}-\sum_{n=1}^{P}20\log\sqrt{(1-\omega^2T_n^2)^2+(2\xi_n\omega T_n)^2}$$

The Bode diagrams of the amplitude-frequency characteristic of the typical factors are superimposed by the Bode diagram of the amplitude-frequency characteristic of each typical factor. The logarithmic phase-frequency characteristic of the system is expressed as

$$\varphi(\omega)=\sum_{i=1}^{M}\tan^{-1}\omega\tau_i+\sum_{l=1}^{N}\tan^{-1}\frac{2\xi_l\tau_l\omega}{1-\omega^2\tau_l^2}+\lambda\left(-\frac{\pi}{2}\right)$$

$$+\sum_{m=1}^{Q}\left(-\tan^{-1}\omega T_m\right)+\sum_{n=1}^{P}\left[-\tan^{-1}\frac{2\xi_n\omega T_n}{1-\omega^2T_n^2}\right]$$

The phase-frequency characteristic of the Bode diagram is also superimposed by the phase-frequency characteristic of the Bode diagram of each typical factor.

6.4.3 Steps for Drawing Bode Diagram

a. The system frequency characteristic is transformed into the product of a typical factor in the frequency characteristic.

b. According to the typical factors of the system, the corner frequency and corresponding slope are determined, and then the approximated amplitude-frequency asymptotes and phase-frequency curve are drawn.

c. Correct the approximated curve when it is necessary.

d. To analyze the features of the system accurately, the powerful functions of MATLAB language can quickly make a MATLAB program.

Example 6.7

Given the frequency characteristic of a system is

$$G(j\omega)=\frac{10(j\omega+3)}{j\omega(j\omega+2)\left[(j\omega)^2+j\omega+2\right]}$$

Draw the Bode diagram of the system.

Solution

The frequency characteristic of the system can be written as

$$G(j\omega)=\frac{7.5\left(\frac{1}{3}j\omega+1\right)}{j\omega\left(\frac{1}{2}j\omega+1\right)\left[\left(\frac{1}{\sqrt{2}}\right)^2(j\omega)^2+2\frac{1}{2\sqrt{2}}\frac{1}{\sqrt{2}}j\omega+1\right]}$$

Therefore, the system can be divided into five typical links as

$$G_1(j\omega)=7.5;$$

$$G_2(j\omega)=\frac{1}{j\omega};$$

$$G_3(j\omega)=\frac{1}{\left(\frac{1}{\sqrt{2}}\right)^2(j\omega)^2+2\frac{1}{2\sqrt{2}}\frac{1}{\sqrt{2}}j\omega+1}$$

$$G_4(j\omega)=\frac{1}{\frac{1}{2}j\omega+1};$$

and

$$G_5(j\omega)=\frac{1}{3}j\omega+1$$

The logarithmic amplitude-frequency characteristics of these five typical factors are written as

$$L_1(\omega)=20\log 7.5$$

$$L_2(\omega)=-20\log\omega$$

$$L_3(\omega)=-20\log\sqrt{\left[1-\left(\frac{1}{\sqrt{2}}\omega\right)^2\right]^2+\left(\frac{1}{2}\omega\right)^2}$$

$$=-20\log\sqrt{\left(1-\frac{1}{2}\omega^2\right)^2+\frac{\omega^2}{4}}$$

$$L_4(\omega)=-20\log\sqrt{1+\left(\frac{1}{2}\omega\right)^2}$$

and

$$L_5(\omega)=20\log\sqrt{1+\left(\frac{1}{3}\omega\right)^2}$$

The total logarithmic amplitude-phase characteristic of the system is written as

$$L(\omega)=L_1(\omega)+L_2(\omega)+L_3(\omega)+L_4(\omega)+L_5(\omega)$$

and

$$\varphi(\omega)=\tan^{-1}\left(\frac{1}{3}\omega\right)-\frac{\pi}{2}-\tan^{-1}\left(\frac{1}{2}\omega\right)-\tan^{-1}\left(\frac{\omega/2}{1-\frac{1}{2}\omega^2}\right)$$

The transition or corner frequency of the system are $\sqrt{2}$, 2, and 3 rad/s). The system Bode diagram is shown in Figure 6.50. This example is relatively simple for drawing multiple lines to practice. MATLAB is given below to draw the Bode diagrams.

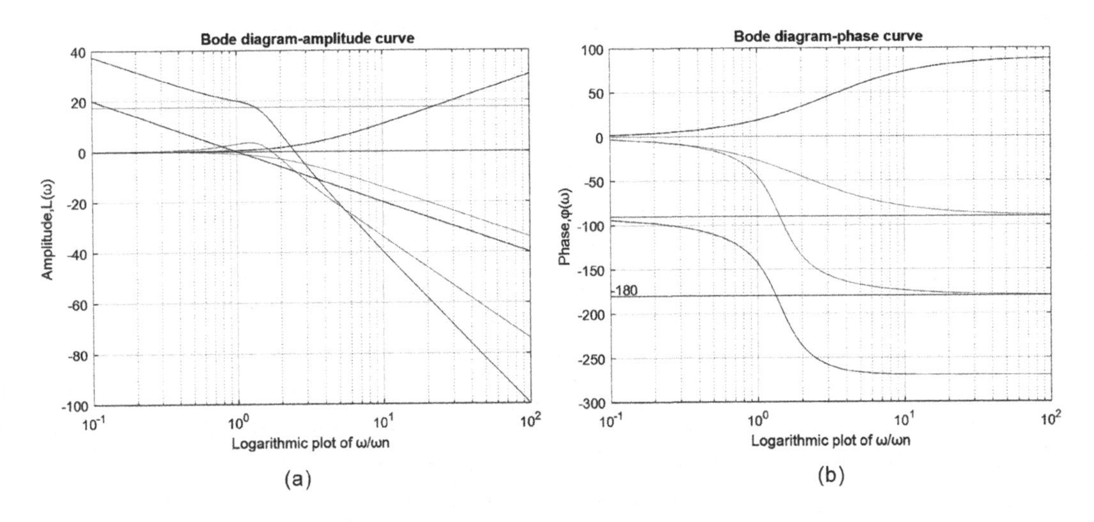

FIGURE 6.50 Bode diagram of a system.

MATLAB program for Figure 6.37:

```
for i=1:1000
w(i)=i/10; lw0(i)=0;
faw0(i)=-180;
a=sqrt(w(i)*w(i)/9+1);
b=sqrt(w(i)*w(i)/4+1);
c=sqrt((1-w(i)*w(i)/2)*(1-w(i)*w(i)/2)+w(i)*w(i)/4);
lw(i)=20*log10(7.5*a/(w(i)*b*c));
lw1(i)=20*log10(7.5);
lw2(i)=20*log10(1/w(i));
lw3(i)=20*log10(1/c);
lw4(i)=20*log10(1/b);
lw5(i)=20*log10(a);
faw1(i)=0; faw2(i)=-90;
faw4(i)=-atan(w(i)/2)*180/pi;
faw5(i)=atan(w(i)/3)*180/pi;
if w(i)<=1.414
faw(i)=atan(w(i)/3)-pi/2-atan(w(i)/2)-atan((w(i)/2)/
(1-w(i)*w(i)/2));
faw(i)=faw(i)*180/pi;
faw3(i)=-atan((w(i)/2)/(1-w(i)*w(i)/2))*180/pi;
else
faw3(i)=-atan((w(i)/2)/(1-w(i)*w(i)/2))*180/pi-180;
faw(i)=atan(w(i)/3)-3*pi/2-atan(w(i)/2)-atan((w(i)/2)/
(1-w(i)*w(i)/2));
faw(i)=faw(i)*180/pi;
end
end
subplot(2,1,1)
semilogx(w,lw,w,lw0,w,lw1,w,lw2,w,lw3,w,lw4,w,lw5)
title('Bode diagram-amplitude curve')
xlabel(' Logarithmic plot of  ω /ωn')
ylabel(' Amplitude, L(ω)')
grid
subplot(2,1,2)
semilogx(w,faw,w,faw0,w,faw1,w,faw2,w,faw3,w,faw4,w,faw5)
title('Bode diagram-phase curve')
xlabel(' Logarithmic plot of  ω /ωn')
ylabel(' Phase,φ(ω)')
text(0.1,-172,'-180')
grid
```

6.4.4 Minimum-Phase System

The transfer function which has neither pole nor zero in the right half s-plane is called the minimum-phase transfer function; otherwise, it is called non-minimum-phase transfer function. A system with a minimum-phase transfer function is called a minimum-phase system.

TABLE 6.9 Minimum-Phase System Correspondence between Amplitude-Frequency and Phase-Frequency

Factors	Amplitude Frequency	Phase Frequency
$j\omega$	20 ~ 20dB/dec	$\frac{\pi}{2} \sim \frac{\pi}{2}$
$\frac{1}{j\omega}$	−20 ~ −20dB/dec	$-\frac{\pi}{2} \sim -\frac{\pi}{2}$
$\tau\ j\omega + 1$	0 ~ 20dB/dec	$0 \sim \pi/2$
$1/\left(T^2(j\omega)^2 + 2\xi Tj\omega + 1\right)$	0 ~ −40dB/dec	$0 \sim -\pi$
$\tau^2(j\omega)^2 + 2\xi\tau\ j\omega + 1$	0 ~ +40dB/dec	$0 \sim +\pi$
……	……	……
$\frac{1}{\prod_{i=1}^{n}(T_i j\omega + 1)}$	0 ~ n(−20)dB/dec	$0 \sim n\left(-\frac{\pi}{2}\right)$
$\prod_{i=1}^{m}(\tau_i j\omega + 1)$	0 ~ m(20)dB/dec	$0 \sim m\left(+\frac{\pi}{2}\right)$

For the same order of the basic factor, when the frequency $\omega = 0 - \infty$ in the continuous range, the smallest change causes the smallest phase shift. For the minimum-phase system, the amplitude-frequency characteristic is known and the phase-frequency characteristic can be uniquely determined. The non-minimum-phase system cannot be uniquely determined. Most minimum-phase systems are practical in engineering; in order to simplify the system, only the amplitude-frequency characteristics are shown for the Bode diagram of the minimum-phase system (Table 6.9).

Example 6.8

The open-loop transfer functions of a feedback control system by two units are given below:

$$G_1(s) = \frac{T_1 s + 1}{T_2 s + 1}, \text{ and } G_2(s) = \frac{-T_1 s + 1}{T_2 s + 1}, \text{ Where } 0 < T_1 < T_2$$

Analyze the Bode diagram of the system.

Solution

According to the transfer function, the frequency characteristic of the system is described as shown in Figure 6.51: $G_1(j\omega) = \frac{T_1 j\omega + 1}{T_2 j\omega + 1}$, $G_2(j\omega) = \frac{-T_1 j\omega + 1}{T_2 j\omega + 1}$

Due to system 1, there are neither poles nor zeros in the right half s-plane; thus, $G_1(j\omega)$ has the minimum-phase system, whereas System 2 has a zero in the right half s-plane; thus, System 2 has non-minimum-phase system. The Bode diagrams of amplitude-frequency characteristics of the two systems are the same, but the Bode

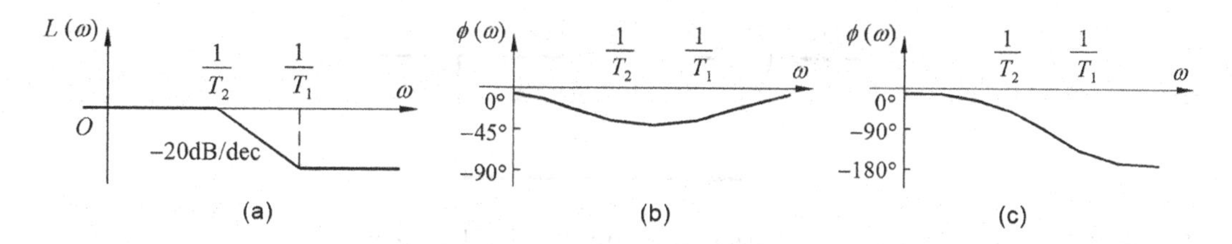

FIGURE 6.51 Bode diagram of $G_1(s)$ and $G_2(s)$: (a) amplitude-frequency curves of $G_1(s)$ and $G_2(s)$, (b) phase-frequency curves of $G_1(s)$, and (c) phase-frequency curves of $G_2(s)$.

diagrams of phase-frequency characteristics are different. Its phase-frequency characteristic of the two systems is described as

$$\varphi_1(\omega) = \tan^{-1}(\omega T_1) - \tan^{-1}(\omega T_2)$$

$$\text{And,}\quad \varphi_2(\omega) = -\tan^{-1}(\omega T_1) - \tan^{-1}(\omega T_2)$$

The Bode diagram of the two systems is shown in Figure 6.38. In order to determine whether it is a minimum-phase system, it is necessary to check the slope of the high-frequency asymptote of the logarithmic amplitude-frequency characteristic curve; however, we also have to check the angle as $\omega \to \infty$. If $\omega \to \infty$, the slope and phase angle of the amplitude-frequency characteristic curve are $-20\times(n-m)$ dB/dec and $-90\times(n-m)$, respectively, where n and m are the order of the denominator and the numerator polynomial in the transfer function, respectively, and then the system is the minimum-phase system, whereas for the non-minimum-phase system, the phase angle is not equal to $-90\times(n-m)$.

Non-minimum-phase system is mostly caused by the system containing delay factors, transmission-lag, small closed-loop instability, and other factors; thus, the starting performance is poor and the response is slow. Therefore, non-minimum-phase system is always avoided in the servo system that requires a faster response.

6.4.5 Obtain System Transfer Function from the Frequency-Response Curve

In control engineering applications, many physical models are difficult to abstract very precisely mathematical expressions, and the transfer function is difficult to be analyzed and obtained by pure mathematical methods. For this kind of system, the frequency characteristic curve of the system can be measured by the experimental method, and then the transfer function of the system can be obtained easily.

6.5 RELATIONSHIP BETWEEN SYSTEM TYPE AND LOG-MAGNITUDE CURVE

Consider the unity-feedback control system as shown in Figure 6.52. The static position, velocity, and acceleration error constants describe the low-frequency behavior of type 0, type 1, and type 2 systems, respectively. For a given system, only one of the static error constants is finite and significant. (The larger the value of the finite static error constant, the higher the loop gain, as ω approaches zero.)

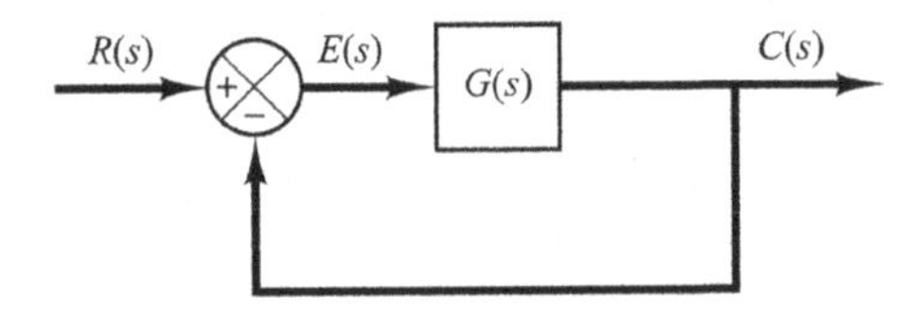

FIGURE 6.52 Closed-loop control system with unity feedback system.

The type of the system determines the slope of the log-magnitude curve at low frequencies. Thus, information concerning the existence and magnitude of the steady-state error of a control system to a given input can be determined from the observation of the low-frequency region of the log-magnitude curve.

1. *Type 0 System and Determination of Static Position Error Constants*
   Consider the unity-feedback control system shown in Figure 6.52. Assume that the open-loop transfer function is given by

$$G_0(s) = K\frac{(s)^m + b_1(s)^{m-1} + \cdots + b_{m-1}(s) + b_m}{(s)^n + a_1(s)^{n-1} + \cdots + a_{n-1}(s) + a_n} = \frac{K_p \prod_{i=1}^{m}(\tau_i s + 1)}{\prod_{l=1}^{n}(T_p s + 1)}$$

or

$$G_0(j\omega) = K\frac{(j\omega)^m + b_1(j\omega)^{m-1} + \cdots + b_{m-1}(j\omega) + b_m}{(j\omega)^n + a_1(j\omega)^{n-1} + \cdots + a_{n-1}(j\omega) + a_n} = \frac{K_p \prod_{i=1}^{m}(\tau_i j\omega + 1)}{\prod_{l=1}^{n}(T_p j\omega + 1)}$$

or

$$G_0(j\omega) = K\frac{(T_a s + 1)(T_b s + 1)\ldots(T_m s + 1)}{s^n (T_1 s + 1)(T_2 s + 1)\ldots(T_p s + 1)} = \frac{K_p \prod_{i=1}^{m}(\tau_i s + 1)}{\prod_{l=1}^{n}(T_p s + 1)}$$

Thus,

$$G_0(j\omega) = K\frac{(T_a j\omega + 1)(T_b j\omega + 1)\ldots(T_m j\omega + 1)}{(j\omega)^n (T_1 j\omega + 1)(T_2 j\omega + 1)\ldots(T_p j\omega + 1)} = \frac{K_p \prod_{i=1}^{m}(\tau_i j\omega + 1)}{\prod_{l=1}^{n}(T_p j\omega + 1)}$$

Figure 6.53 shows an example of the log-magnitude plot of a type 0 system. In such a system, the magnitude of $G_0(j\omega)$ equals K_p at low frequencies; in a type 0 system, $n = 0$, then

$$\lim_{\omega \to 0} G_0(j\omega) = K = K_p$$

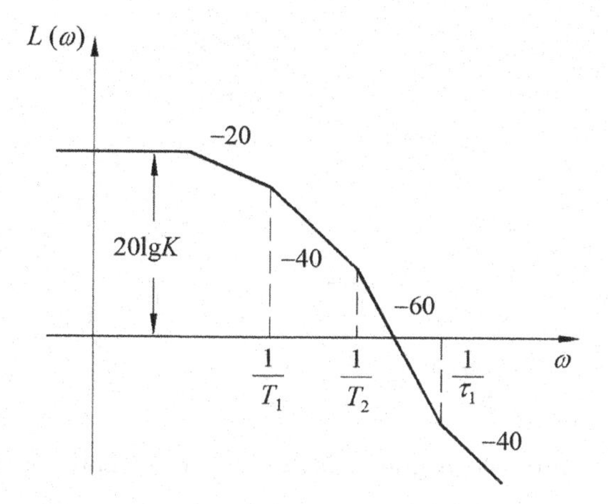

FIGURE 6.53 Bode diagram of type 0 system.

At the low-frequency band, when ω is very small, $G_0(j\omega)=K_p$, $K_p=\left|G_0(j0)\right|$, $L(0)=20\log\left|G_0(j0)\right|=20\log K_p$. The amplitude-frequency characteristics Bode diagram of type 0 system is $20\log K_p$ in the low frequency, as shown in Figure 6.53. It follows that the low-frequency asymptote is a horizontal line at $20\log K_p$ dB.

2. *Type 1 system and Determination of Static Velocity Error Constants*
Consider the unity-feedback control system shown in Figure 6.52. Figure 6.53 shows an example of the log-magnitude plot of a type 1 system. The intersection of the initial −20 dB/decade segment (or its extension) with the line $\omega=1$ has the magnitude 20 log K_v. This may be seen as follows:

$$G_1(j\omega)=K\frac{(j\omega)^m+b_1(j\omega)^{m-1}+\cdots+b_{m-1}(j\omega)+b_m}{(j\omega)^n+a_1(j\omega)^{n-1}+\cdots+a_{n-1}(j\omega)+a_n}=\frac{K_v\prod_{i=1}^{m}(\tau_i j\omega+1)}{j\omega\prod_{v=1}^{n}(T_v j\omega+1)}$$

or

$$G_1(j\omega)=K\frac{(T_a s+1)(T_b s+1)\ldots(T_m s+1)}{s^n(T_1 s+1)(T_2 s+1)\ldots(T_v s+1)}=\frac{K_v\prod_{i=1}^{m}(\tau_i s+1)}{\prod_{v=1}^{n}(T_v s+1)}$$

Thus,

$$G_1(j\omega)=K\frac{(T_a j\omega+1)(T_b j\omega+1)\ldots(T_m j\omega+1)}{(j\omega)^n(T_1 j\omega+1)(T_2 j\omega+1)\ldots(T_v j\omega+1)}=\frac{K_v\prod_{i=1}^{m}(\tau_i j\omega+1)}{\prod_{l=1}^{n}(T_v j\omega+1)}$$

In a type 1 system, $n=1$, then

$$G_1(j\omega)=\frac{K_v}{j\omega},\text{for}\,\omega\ll 1$$

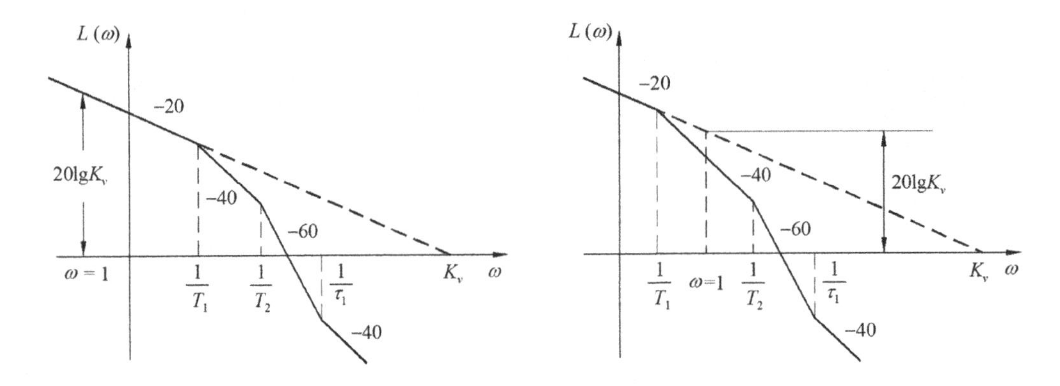

FIGURE 6.54 Type I system Bode diagram to determine the magnification of the system.

Thus,

$$20 \log\left|\frac{K_v}{j\omega}\right|_{\omega=1} = 20 \log K_v$$

The intersection of the initial −20 dB/decade segment (or its extension) with the 0 dB line has a frequency numerically equal to K_v. To see this, define the frequency at this intersection to be ω_1; then

$$\left|\frac{K_v}{j\omega_1}\right| = 1 \text{ or } K_v = \omega_1,$$

At the low-frequency band, when ω is very small, $G_1(j\omega) = \frac{K_v}{j\omega}, \left|G_1(j1)\right| = K_v$; as shown in Figure 6.54, each corner frequency of the system is greater than $\omega = 1$; at $\omega = 1$, the amplitude-frequency characteristic of type I system is $20 \log K_v$ if the system has an angular frequency less than $\omega = 1$.

3. *Type 2 System and Determination of Static Acceleration Error Constants*
   Consider the unity-feedback control system shown in Figure 6.52. Figure 6.55 shows an example of the log-magnitude plot of a type 2 system. The intersection of the initial −40 dB/decade segment (or its extension) with the $\omega = 1$ line has a magnitude of $20 \log K_a$. Since at low frequencies

$$G_2(j\omega) = K\frac{(j\omega)^m + b_1(j\omega)^{m-1} + \cdots + b_{m-1}(j\omega) + b_m}{(j\omega)^n + a_1(j\omega)^{n-1} + \cdots + a_{n-1}(j\omega) + a_n} = \frac{K_a\prod_{i=1}^{m}(\tau_i j\omega + 1)}{(j\omega)^n\prod_{a=1}^{n}(T_a j\omega + 1)}$$

or

$$G_2(j\omega) = K\frac{(T_a s+1)(T_b s+1)\ldots(T_m s+1)}{s^n(T_1 s+1)(T_2 s+1)\ldots(T_a s+1)} = \frac{K_a\prod_{i=1}^{m}(\tau_i s+1)}{\prod_{a=1}^{n}(T_a s+1)}$$

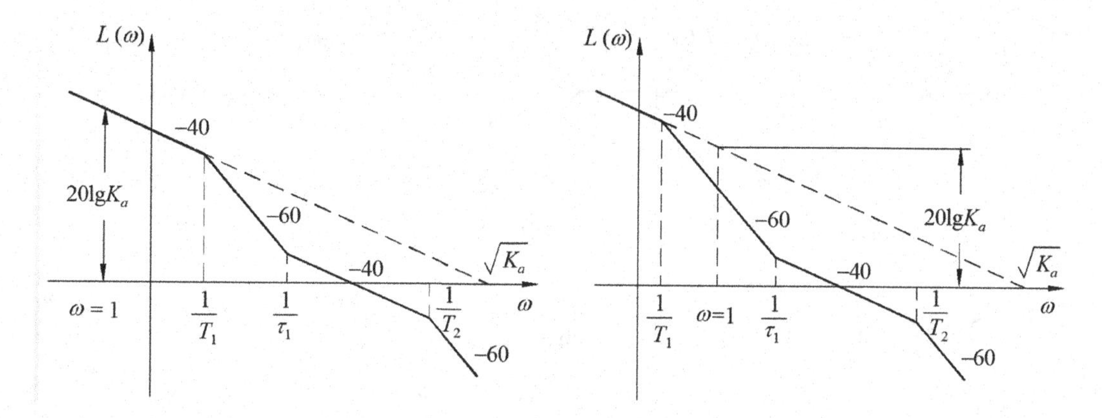

FIGURE 6.55 Log-magnitude curve of a type 2 system.

Thus,

$$G_2(j\omega)=K\frac{(T_a j\omega+1)(T_b j\omega+1)\ldots(T_m j\omega+1)}{(j\omega)^n(T_1 j\omega+1)(T_2 j\omega+1)\ldots(T_a j\omega+1)}=\frac{K_a\prod_{i=1}^{m}(\tau_i j\omega+1)}{\prod_{a=1}^{n}(T_a j\omega+1)}$$

In a type 2 system, $n=2$, then

$$G_1(j\omega)=\frac{K_a}{(j\omega)^2},\text{for}\,\omega\ll 1$$

Thus,

$$20\log\left|\frac{K_a}{(j\omega)^2}\right|_{\omega=1}=20\log K_a$$

The frequency ω_a at the intersection of the initial −40 dB/decade segment (or its extension) with the 0 dB line gives the square root of K_a numerically. This can be seen from the following:

$$20\log\left|\frac{K_a}{(j\omega)^2}\right|_{\omega=1}=20\log 1=0$$

or

$$\omega_a=\sqrt{K_a}$$

The open-loop gain values for eight common cases are listed below for reference in use, as shown in Figure 6.56.

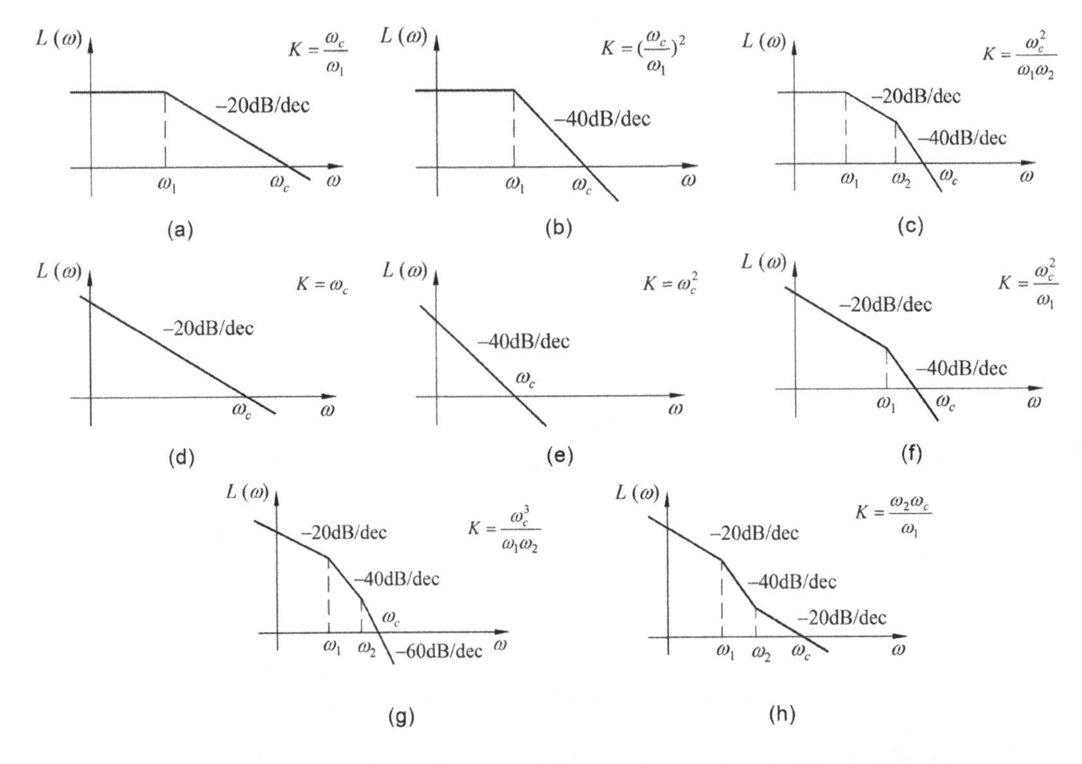

FIGURE 6.56 Determination of open-loop amplification factor of typical Bode diagrams.

Example 6.9

The Bode diagram obtained from the test of a control system is shown in Figure 6.57. Determine the transfer function of the system.

Solution

The low-frequency Bode diagram of amplitude-frequency characteristic shows that this is a type 0 system; thus, 20log $K = 0$, then $K = 1$. Using –20 dB/decade line of amplitude-frequency characteristics of asymptote, the middle band is –20 dB/dec.

If high frequency is –40 dB/dec, two angle frequencies are $\omega_1 = 1\ (\text{rad/s})$ and $\omega_2 = 2.4\ (\text{rad/s})$; therefore, this is a second-order system composed of two first-order systems, whose periods are, respectively,

$$T_1 = \frac{1}{\omega_1} = 1\ s$$

$$\text{and } T_2 = \frac{1}{\omega_2} = 0.417\ s$$

$$\varphi(1) = -85° = \left(-\tau_1 \times 1 - \tan^{-1} 1 - \tan^{-1} 0.417\right) \times \frac{180°}{\pi} = -85°$$

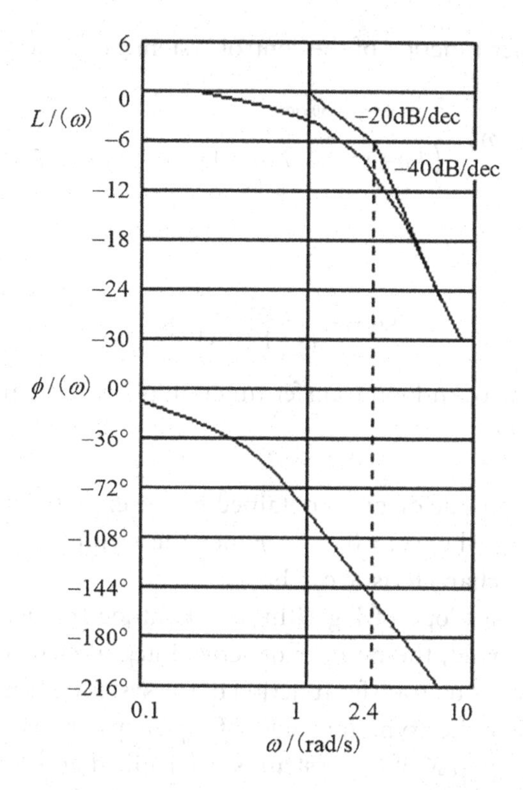

FIGURE 6.57 Tested Bode diagram of a control system.

Therefore, $\tau_1 = 0.303$, as shown in Figure 6.57,

$$\varphi(2.4) = -155° = \left(-\tau_2 \times 2.4 - \tan^{-1} 2.4 - \tan^{-1} 1\right) \times \frac{180°}{\pi} = -155°$$

Thus,

$$\tau_2 = 0.310$$

Similarly, we get

$$\varphi(4) = -205° = \left(-\tau_3 \times 4 - \tan^{-1} 4 - \tan^{-1}(0.417 \times 4)\right) \times \frac{180°}{\pi} = -205°$$

Thus,

$$\tau_3 = 0.305$$

The average value of τ

$$\tau = \frac{\tau_1 + \tau_2 + \tau_3}{3} = \frac{0.303 + 0.310 + 0.305}{3} = 0.306$$

Therefore, the transfer function of the control system can be written as

$$G(j\omega)=\frac{e^{-0.306j\omega}}{(j\omega+1)(0.417j\omega+1)}=\frac{e^{-0.306s}}{(s+1)(0.417s+1)}$$

or

$$G(s)=\frac{e^{-0.306s}}{(s+1)(0.417s+1)}$$

Thus, the steps of the open-loop transfer function of the system can be obtained by experimental method:

a. The slope of the Bode diagram obtained by the experimental method which is $\pm 20\lambda$ dB/decade, where ($\lambda = 0, 1, 2\ldots$); then the asymptote of logarithmic amplitude-frequency characteristic can be obtained.
b. According to the slope of logarithmic amplitude-frequency characteristic of low-frequency band, the number of series integral elements is determined in the system open-loop transfer function of the system. If there is an integral factor λ, the slope of the asymptote of low-frequency band is $\pm 20\lambda$ dB/decade.
c. The open-loop gain K of the system is determined according to the dB value, and the frequency value of the logarithmic amplitude-frequency characteristic is the shape of the part above the 0 dB.
d. According to the slope change of logarithmic amplitude-frequency asymptote at the corner frequency, the number of series factors is determined in the system.
e. Furthermore, according to the single value correspondence between the slope of log-amplitude-frequency characteristic of the minimum-phase system and the phase-frequency characteristic, the system is tested whether there is a series delay factor or the asymptote is modified.

6.6 THE CLOSED-LOOP FREQUENCY RESPONSE OF THE CONTROL SYSTEM

6.6.1 Estimation of Closed-Loop Frequency Characteristics

The closed-loop frequency characteristics are estimated from open-loop frequency-response characteristics. As shown in Figure 6.58, the open-loop frequency characteristic that refers to the open-loop system of the closed-loop transfer function $H(j\omega)$, and its open-loop frequency characteristic is $G(j\omega)H(j\omega)$.

The closed-loop characteristic of the system is

$$\frac{X_o(j\omega)}{X_i(j\omega)}=\frac{G(j\omega)}{1+G(j\omega)H(j\omega)} \tag{6.62}$$

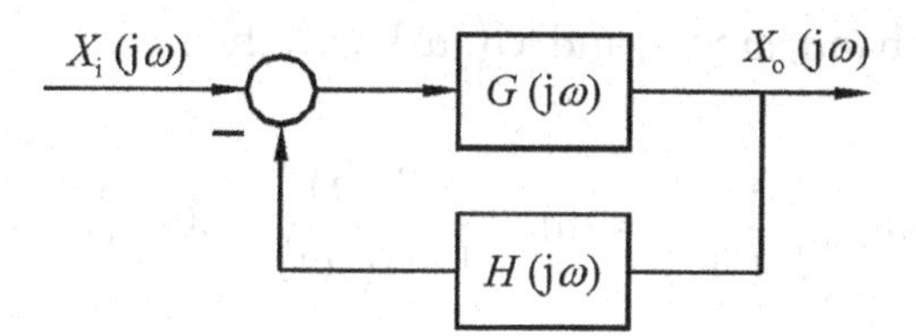

FIGURE 6.58 A feedback control system with feedback transfer function $H(j\omega)$.

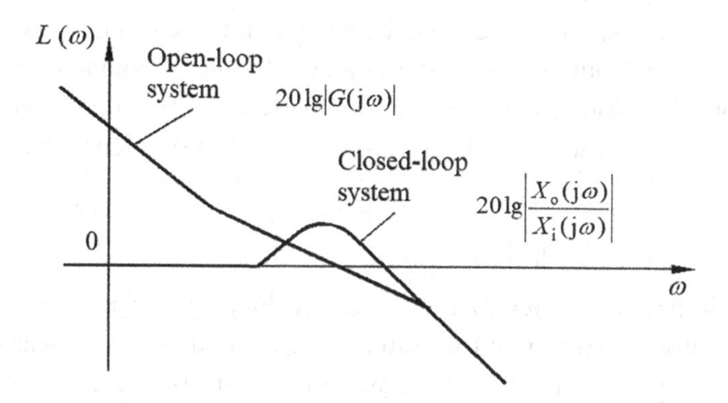

FIGURE 6.59 Bode diagram of open-loop system and closed-loop system.

Therefore, the frequency characteristics of the open-loop system are given, the frequency characteristics of the closed-loop system can be estimated, and the Bode diagram of the closed-loop system can be roughly drawn. Now, especially with the rapid development of computer and software technology, any complex and tedious calculation can be simulated through the computer; thus, it shows the computer in the control system of the strong vitality. However, in some field tests, in order to roughly estimate the characteristics of the system, it is necessary to have a clear understanding of the basics of how to use the open-loop frequency characteristics to estimate the closed-loop frequency characteristics.

Assume that the system is unity feedback, i.e.

$$H(j\omega)=1, \text{ and}$$

$$\frac{X_o(j\omega)}{X_i(j\omega)}=\frac{G(j\omega)}{1+G(j\omega)} \tag{6.63}$$

In control engineering, the frequency characteristics of all the general open-loop system have the characteristics of LPF system, as shown in Figure 6.59.

At the low-frequency band, $\omega \langle\langle 1$, and $|G(j\omega)|\rangle\rangle 1$, then

$$20\log\left|\frac{X_o(j\omega)}{X_i(j\omega)}\right|=20\log\left|\frac{G(j\omega)}{1+G(j\omega)}\right|\approx 0 \tag{6.64}$$

At the high-frequency band, $\omega >> 1$, and $|G(j\omega)| << 1$, then

$$20\log\left|\frac{X_o(j\omega)}{X_i(j\omega)}\right| = 20\log\left|\frac{G(j\omega)}{1+G(j\omega)}\right| = 20\log|G(j\omega)| \tag{6.65}$$

In the spectrum diagram, nearby the corner frequency ω_c, the Bode diagram of the closed-loop system can be drawn by calculating tracing points, as shown in Figure 6.46. Thus, for any minimum-phase system, in the low-frequency band, the input signal of the system is basically the same as the output signal of the system. In high frequency, the high-frequency characteristic of closed-loop system is basically the same as that of open-loop system. The open-loop frequency characteristic is used to estimate the frequency characteristic of the closed-loop system.

6.6.2 The Frequency-Domain Index of the System

The frequency-domain characteristics of the closed-loop system are shown in Figure 6.60. Here, ω_b is the cutoff frequency of the system, it is defined as the decrease of logarithmic amplitude-frequency characteristic of the system at –3 dB or the amplitude is reduced to $A(0)/\sqrt{2}$ corresponding to the frequency.

Bandwidth (BW) means that the system's frequency ω starts at 0; it is defined as the decrease of logarithmic amplitude-frequency characteristic of the system at –3dB or the amplitude is reduced to $A(0)/\sqrt{2}$ corresponding to the frequency. In general, the desire of cutoff frequency ω_b is the BW of the system.

Resonance frequency ω_r refers to the system to produce a corresponding frequency peak. Resonance peak M_r refers to the resonance frequency ω_r corresponding amplitude.

For a second-order system, the transfer function is expressed as

$$G(j\omega) = \frac{\omega_n^2}{(j\omega)^2 + 2\xi\omega_n(j\omega) + \omega_n^2} \tag{6.66}$$

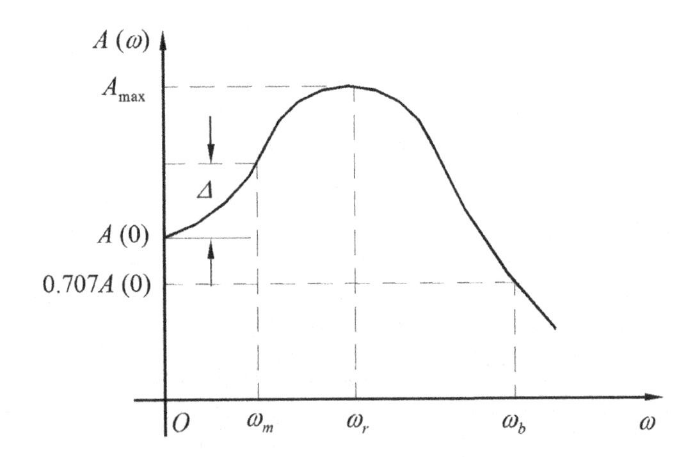

FIGURE 6.60 Frequency-domain characteristics of the system.

Thus,

$$G(j\omega)=\frac{1}{\sqrt{\left[1-\left(\frac{\omega}{\omega_n}\right)^2\right]^2+4\xi^2\left(\frac{\omega}{\omega_n}\right)^2}}e^{-j\tan^{-1}\left(\frac{2\xi\frac{\omega}{\omega_n}}{1-\frac{\omega^2}{\omega_n^2}}\right)} \tag{6.67}$$

Let $u=\frac{\omega}{\omega_n}$, then we get the amplitude-frequency characteristic from Eq. (6.67) as

$$A(u)=\frac{1}{\sqrt{(1-u^2)^2+(2\xi u)^2}} \tag{6.68}$$

In order to find the maximum value of $A(u)$, the derivative of Eq. (6.68) with respect to u is set to zero, and then simplified as

$$u^3-u+2u\xi^2=0 \tag{6.69}$$

The solution of Eq. (6.69) is obtained as follows:

$$\begin{cases} u_r=0 \\ u_r=\sqrt{1-2\xi^2} \end{cases} \tag{6.70}$$

When $u_r=0$, $\omega=0$ is the starting point of the system. It is obviously seen that it is not the maximum value of the system, so the dimensionless resonant frequency of the system is

$$u_r=\sqrt{1-2\xi^2} \qquad ;\left(0\le\xi\le 1/\sqrt{2}\right) \tag{6.71}$$

Therefore, the resonant frequency of the system is

$$\omega_r=\omega_n\sqrt{1-2\xi^2} \qquad ;\left(0\le\xi\le 1/\sqrt{2}\right) \tag{6.72}$$

The resonance peak value of the system can be obtained by substituting into Eq. (6.68), and then we get

$$M_r=\frac{1}{2\xi\sqrt{1-\xi^2}} \tag{6.73}$$

It can be seen that the resonance peak M_r is only related to the damping ratio ξ of the system, as shown in Figure 6.61a, whereas the resonance frequency ω_r is not directly related with damping ratio ξ and the undamped natural frequency ω_n. The relationship

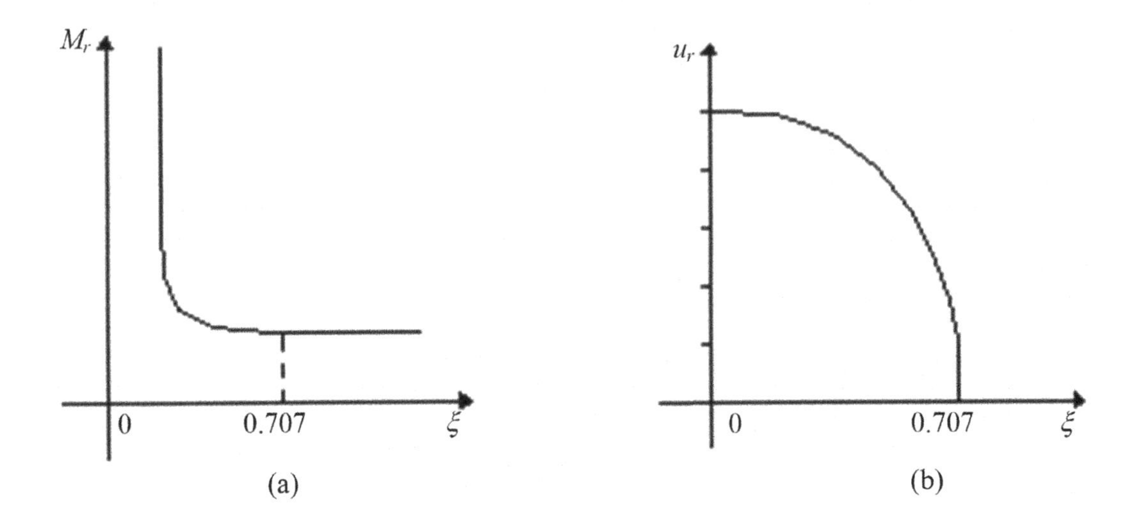

FIGURE 6.61 The relationship between (a) $M_r \sim \xi$, and (b) $u_r \sim \xi$.

curve between the dimensionless resonance frequency u_r with damping ratio ξ is shown in Figure 6.61b.

By definition of BW, $\omega = 0$ is the starting point of the system, and $A(\omega)$ is reduced to the initial value of $1/\sqrt{2}$ times to the corresponding frequency, that is,

$$A(u) = \frac{1}{\sqrt{[1-u^2)^2 + (2\xi u)^2}} = \frac{1}{\sqrt{2}} \tag{6.74}$$

Thus,

$$u^2 = \left(1-2\xi^2\right) + \sqrt{4\xi^4 - 4\xi^2 + 2} \tag{6.75}$$

The solution of Eq. (6.75) shows that the BW of the system is defined as

$$\omega_b = \omega_n \sqrt{\left(1-2\xi^2\right) + \sqrt{4\xi^4 - 4\xi^2 + 2}} \tag{6.76}$$

Example 6.10

An open-loop transfer function of a unity-feedback control system is $G(s) = \dfrac{K}{s(s+a)}$, resonance peak of the system, $M_r = 1.04$, and the resonant frequency, $\omega_r = 11.55$ rad/s. Determine K, *the* value of σ and the undamped natural frequency of the system ω_n, BW ω_b, damping ratio ξ and maximum overshoot M_p in percentage.

Solution

The closed-loop transfer function of the system is

$$G(s)=\frac{\frac{K}{s(s+a)}}{1+\frac{K}{s(s+a)}}=\frac{K}{s^2+as+K}=\frac{\omega_n^2}{s^2+2\xi\omega_n s+\omega_n^2}$$

And resonance peak of the system is $M_r=\frac{1}{2\xi\sqrt{1-\xi^2}}$.

Damping ratio is $\xi=\sqrt{\frac{1-\sqrt{1-1/M_r^2}}{2}}=0.6$.

The resonant frequency is $\omega_r=\omega_n\sqrt{1-2\xi^2}$.

The undamped natural frequency is $\omega_n=\frac{\omega_r}{\sqrt{1-2\xi^2}}=\frac{11.55}{\sqrt{1-2\times0.6^2}}=21.83$ rad/s.

Thus, $K=\omega_n^2=476.55$.

and $\sigma=2\xi\omega_n=2\times0.6\times21.83=26.2$.

The BW of the system is $\omega_b=\omega_n\sqrt{(1-2\xi^2)+\sqrt{4\xi^4-4\xi^2+2}}=25.1$ rad/s.

The maximum overshoot is $M_p(\%)=e^{-\frac{\xi\pi}{\sqrt{1-\xi^2}}}=9.5\%$.

6.7 NYQUIST STABILITY AND MAPPING THEOREM

In this section, we will learn the Nyquist stability criterion, a summary of the Nyquist stability criterion, the remarks on the Nyquist stability criterion, and the mapping theorem.

6.7.1 Nyquist Stability Criterion

The Nyquist stability criterion determines the stability of a closed-loop system from its open-loop frequency response and open-loop poles. This criterion, derived by *H. Nyquist*, is useful in control engineering because the absolute stability of the closed-loop system can be determined graphically from open-loop frequency-response curves, and there is no need for actually determining the closed-loop poles. Analytically obtained open-loop frequency-response curves, as well as those experimentally obtained, can be used for the stability analysis. This is convenient because, in designing a control system, it often happens that mathematical expressions for some of the components are not known; only their frequency-response data are available. The system is stable if all poles are in the left-half s-plane that we know. Therefore, the Nyquist stability criterion relates the open-loop frequency response $G(j\omega)H(j\omega)$ to the number of zeros and poles of $1+G(s)H(s)$ that lie in the right-half s-plane.

Nyquist stability criterion

- Useful for absolute stability;
- Determined graphically from open-loop frequency-response curves;
- No need for actually determining closed-loop poles;
- It is used for stability analysis;
- Unknown components;
- Based on theory of complex variables;
- Need mappings of contours in complex plane;
- $G(s)H(s)$ represents as a ratio of polynomials in s;
- For a physically realizable system, degree of denominator polynomial≥numerator polynomial;
- If $s \to \infty$, Limit of $G(s)H(s)=0$ or constant, for any physically realizable system.

The characteristic equation of the system can be expressed in the Nyquist stability criterion as (Figure 6.62)

$$F(s)=1+G(s)H(s)=0$$

6.7.2 Summary of Nyquist Stability Criterion

The encirclement of $-1+j0$ point is described by $G(j\omega)H(j\omega)$ as the Nyquist stability criterion for a special case when $G(j\omega)H(j\omega)$ has neither poles nor zeros on the $j\omega$ axis such that (a) if open-loop transfer function $G(j\omega)H(j\omega)$ has k poles in the right half S-plane and $\lim_{s\to\infty} G(j\omega)H(j\omega)=$ constant; and (b) for stability, $G(j\omega)H(j\omega)$ locus, as $\omega=-\infty$ *to* ∞, encircle $-1+j0$ point k times in the counterclockwise direction.

6.7.3 Remarks on Nyquist Stability Criterion

1. This criterion can be expressed as the Nyquist stability criterion, which can be expressed as

$$Z=N+P$$

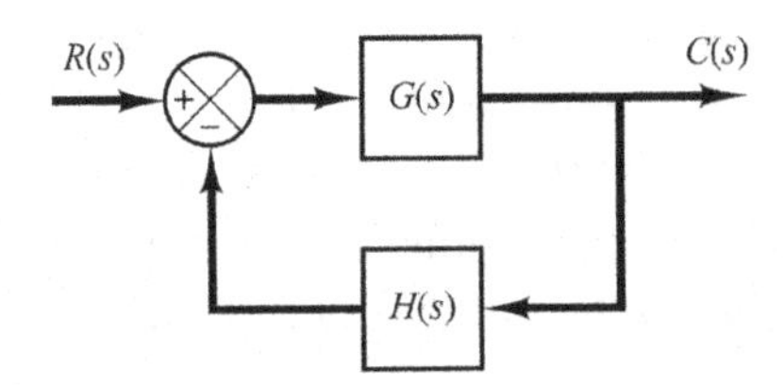

FIGURE 6.62 A feedback control system with feedback transfer function $H(s)$.

where Z is the number of zeros of $1+G(s)H(s)$ in the right-half s-plane, N is the number of clockwise encirclements of $-1+j0$ point, and P is the number of poles of $G(s)H(s)$ in right-half s-plane;

Here,

- If $P \neq 0$, for a stable control system, $Z=0$, or $N=-P$;
- P counterclockwise encirclements of $-1+j0$ point;
- If $G(s)H(s)$ does not have any poles in right-half s-plane, then $Z=N$;
- For stability no encirclement of $-1+j0$ point by $G(j\omega)H(j\omega)$ locus;
- Not necessary to consider locus for entire $j\omega$ axis for positive-frequency portion;
- Stability can be determined by $-1+j0$ point is enclosed by Nyquist plot of $G(j\omega)H(j\omega)$;
- Enclosed region by Nyquist plot is in Figure 6.63;
- For stability, $-1+j0$ point must lie outside the shaded region that we can see in Figure 6.63.

2. *Carefully test the stability of multiple-loop systems*: To test the stability of multiple-loop systems we may look at the following consideration as

   - Include poles in right-half s-plane;
   - Inner loop may be unstable;

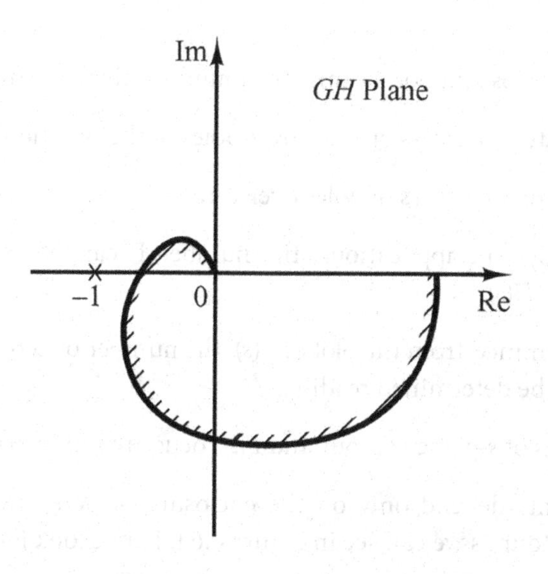

FIGURE 6.63 The $-1+j0$ point enclosed by Nyquist plot of $G(j\omega)H(j\omega)$.

- Entire closed-loop system can be stable by proper design;
- Simple inspection of encirclements of $-1+j0$ point by the $G(j\omega)H(j\omega)$ locus is not sufficient to detect instability in multiple-loop systems;
- Any pole of $1+G(s)H(s)$ is in right-half s-plane can be determined easily by applying the **Routh stability criterion**;

3. If the locus of $G(j\omega)H(j\omega)$ passes through $-1+$j0 point, then zeros are located on the $j\omega$ axis. If the locus of $G(j\omega)H(j\omega)$ passes through $-1+j0$ point, then zeros are located on the $j\omega$ axis; the following is what we can see
   - This is not desirable for practical control systems;
   - For a well-designed closed-loop system, none of the roots of the characteristic equation should lie on $j\omega$ axis.
4. *Mapping theorem*: Now let's discuss about mapping theorem. To discuss the mapping theorem briefly,
   - Let, $F(s)$ be a ratio of two polynomials in s;
   - Let, P=no. of poles & Z=no. of zeros of $F(s)$ in the Nyquist plot that lies inside some closed contour in s-plane;
   - Let, contour does not pass through any poles or zeros of $F(s)$;
   - Closed contour in s-plane is mapped into $F(s)$ plane as a closed curve;
   - Total no. N of clockwise encirclements of origin of $F(s)$ plane;
   - As a representative point s traces out the entire contour in a clockwise direction, $N=Z-P$;
   - Number of zeros and poles cannot be found—only their difference;
   - $N>0$ indicates an excess of zeros over poles of the function $F(s)$;
   - $N<0$ indicates an excess of poles over zeros;
   - In control system applications, the number P can be readily determined for $F(s)=1+G(s)H(s)$;
   - If N is determined from the plot of $F(s)$, the number of zeros in closed contour in s-plane can be determined readily;
   - Exact shapes of s-plane contour and $F(s)$ locus are immaterial (unimportant);
   - Encirclements depend only on the enclosure of poles and/or zeros of $F(s)$ by s-plane contour as we can see in Figure 6.64. Please look it carefully.

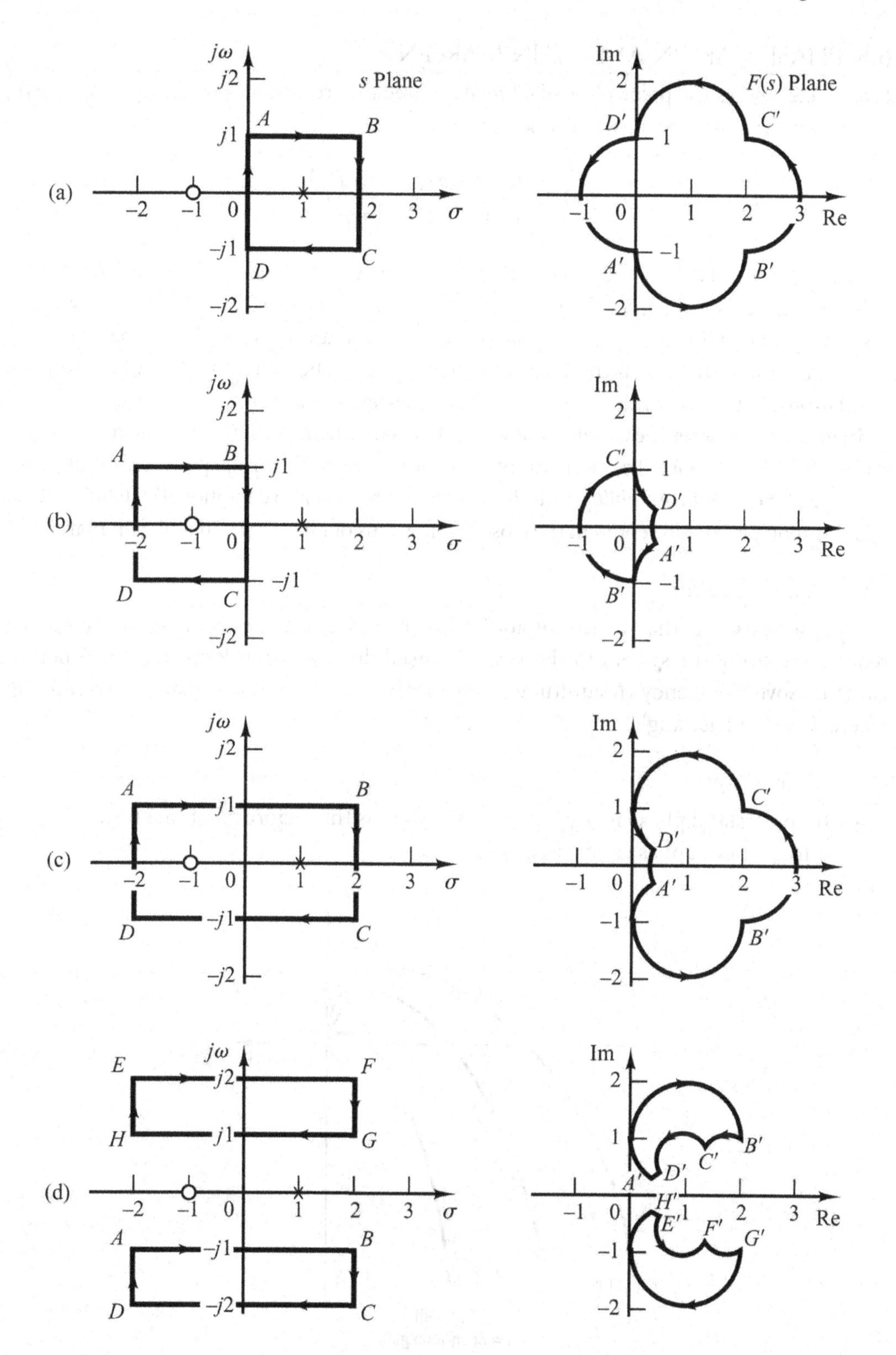

FIGURE 6.64 Enclosure of poles and/or zeros of $F(s)$ by s-plane contour.

6.8 PHASE MARGIN AND GAIN MARGIN

Figure 6.65 shows the polar plots of $G(j\omega)$for three different values of the open-loop gain K. We can derive this K from the following system:

$$G(j\omega)=\frac{K(1+j\omega T_a)(1+j\omega T_b)\ldots}{(j\omega)(1+j\omega T_1)(1+j\omega T_2)\ldots}$$

For a large value of the gain K, the system is *unstable*. As the gain K is decreased to a certain value, the $G(j\omega)$locus passes through the $-1+j0$ point. At (–1,0), gain K is on the verge of instability, and the system will exhibit sustained oscillations. This means that with this gain value, the system is on the verge of instability, and the system will exhibit sustained oscillations. For a small value of the gain K, the system is *stable*.

In general, the closer the $G(j\omega)$ locus comes to encircling the $-1+j0$ point, the more oscillatory the system response. The closeness of the $G(j\omega)$ locus to the $-1+j0$ point can be used as a measure of the *margin of stability*. This does not apply, however, to conditionally stable systems. It is common practice to represent the closeness in terms of phase margin and gain margin.

6.8.1 Phase Margin

The phase margin is the amount of additional phase lag at the gain crossover frequency required to bring the system to the verge of instability; for open-loop transfer function, gain crossover frequency (Magnitude), $|G(j\omega)|=1$; thus in the phase margin, $\gamma=180°+\phi$, where ϕ is the phase angle.

6.8.2 Gain Margin

- To understand phase margin, the gain margin is the reciprocal of magnitude $|G(j\omega)|$ at frequency at phase angle is, $\phi=-180°$.

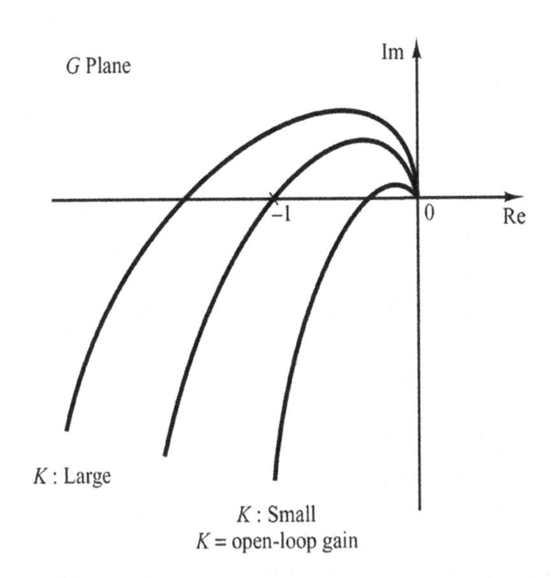

FIGURE 6.65 Polar plots of $G(j\omega)$ for three different values of the open-loop gain K.

- Defining phase crossover frequency ω_1 to be the frequency at a phase angle of an open-loop transfer function, $\phi = -180°$, *the gain margin is*

$$K_g = \frac{1}{\left|G(j\omega_1)\right|}$$

Here,

$$K_g\text{dB} = 20\log K_g = -20\log\left|G(j\omega_1)\right|$$

Therefore, we can say

$$K_g\ \text{dB} = \begin{cases} K_g > 0, & \text{for } K_g > 1 \\ K_g < 0, & \text{for } K_g < 1 \end{cases}$$

- Positive gain margin (in dB): *system is stable.*
- Negative gain margin (in dB): *system is unstable.*

Look at Figure 6.66 and observe carefully to understand phase margin and gain margin.

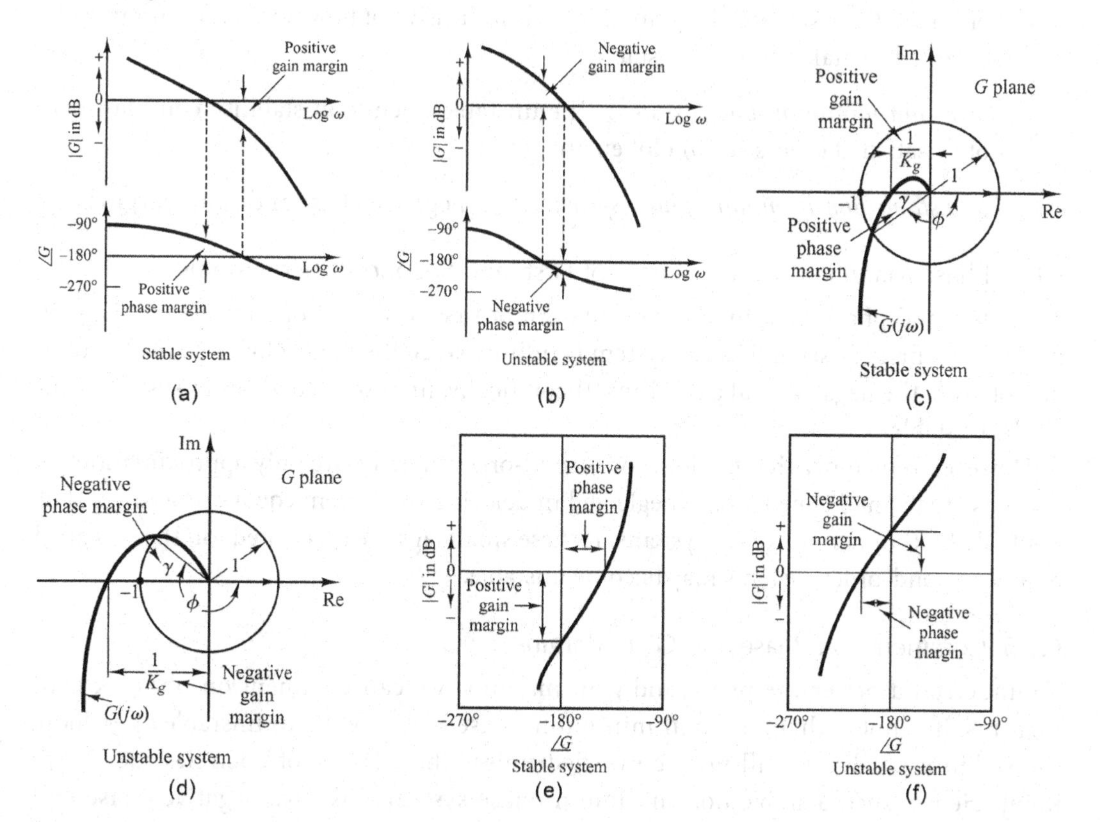

FIGURE 6.66 Phase margin and gain margin.

Figure 6.66 illustrates the phase margin of both *a stable system and an unstable system in (a) Bode diagrams, (b) polar plots, and (c) log-magnitude-versus-phase plots.* In the polar plot, a line may be drawn from the origin to the point at which the unit circle crosses the $G(j\omega)$ locus.

- If this line lies below (above) the negative real axis, then the angle γ is positive (negative).
- The angle from the negative real axis to this line is the phase margin. The phase margin is positive for $\gamma > 0$ and negative for $\gamma < 0$.
- For a minimum-phase system to be stable, the phase margin must be positive.
- In the logarithmic plots, the critical point in the complex plane corresponds to the 0 dB and –180° lines.

6.8.3 Minimum, Non-minimum-Phase Margin and Gain Margin

Now, let's discuss about minimum, non-minimum-phase margin and gain margin.

- For a *stable minimum-phase system*: The gain margin indicates how much the gain can be increased before the system becomes unstable.
- For an *unstable system*: The gain margin is indicative of how much the gain must be decreased to make the system stable.
- *For a non-minimum-phase system*: The unstable open-loop stability condition will not be satisfied unless $G(j\omega)$ plot encircles the $-1+j0$ point.
- *For a stable non-minimum-phase system*: Have negative phase and gain margins.

6.8.4 Phase Margin and Gain Margin of First- and Second-Order Systems

To understand phase margin and gain margin of first & second-order systems, the gain margin of a first- or second-order system is infinite since the polar plots for such systems do not cross the negative real axis. Thus, theoretically, first- or second-order systems cannot be unstable.

However, remember that the first- or second-order systems are only approximations in the sense that small time lags are neglected in deriving the system equations and are thus not truly first- or second-order systems. If these small lags are accounted for, the so-called first- or second-order systems may become unstable.

6.8.5 Comments on Phase and Gain Margins

To understand better the phase and gain margins, we can comment on phase & gain margins. It is noted that for a non-minimum-phase system with an unstable open loop, the stability condition will not be satisfied unless the $G(j\omega)$ plot encircles the $-1+j0$ point. Hence, such a stable non-minimum-phase system will have negative phase and gain margins.

It is also important to point out that conditionally stable systems will have two or more phase crossover frequencies, and some higher-order systems with complicated numerator dynamics may also have two or more gain crossover frequencies, as shown in Figure 6.67.

For stable systems having two or more gain crossover frequencies, the phase margin is measured at the highest gain crossover frequency. The phase and gain margins of a control system are a measure of the closeness of the polar plot to the $-1+j0$ point. Therefore, these margins may be used as design criteria. It should be noted that either the gain margin alone or the phase margin alone does not give a sufficient indication of the relative stability. Both should be given in the determination of relative stability. For a minimum-phase system, both the phase and gain margins must be positive for the system to be stable.

6.8.6 Negative Margins Indicate Instability

Proper phase and gain margins ensure us against variations in the system components and are specified for definite positive values. The two values bound the behavior of the closed-loop system near the resonant frequency ω_r. For satisfactory performance, the phase margin (γ) should be between 30° and 60°, and the gain margin (kg) should be greater than 6 dB.

With these values, a minimum-phase system has guaranteed stability, even if the open-loop gain and time constants of the components vary to a certain extent. Although the phase and gain margins give only rough estimates of the effective damping ratio of the closed-loop system, they do offer a convenient means for designing control systems or adjusting the gain constants of systems.

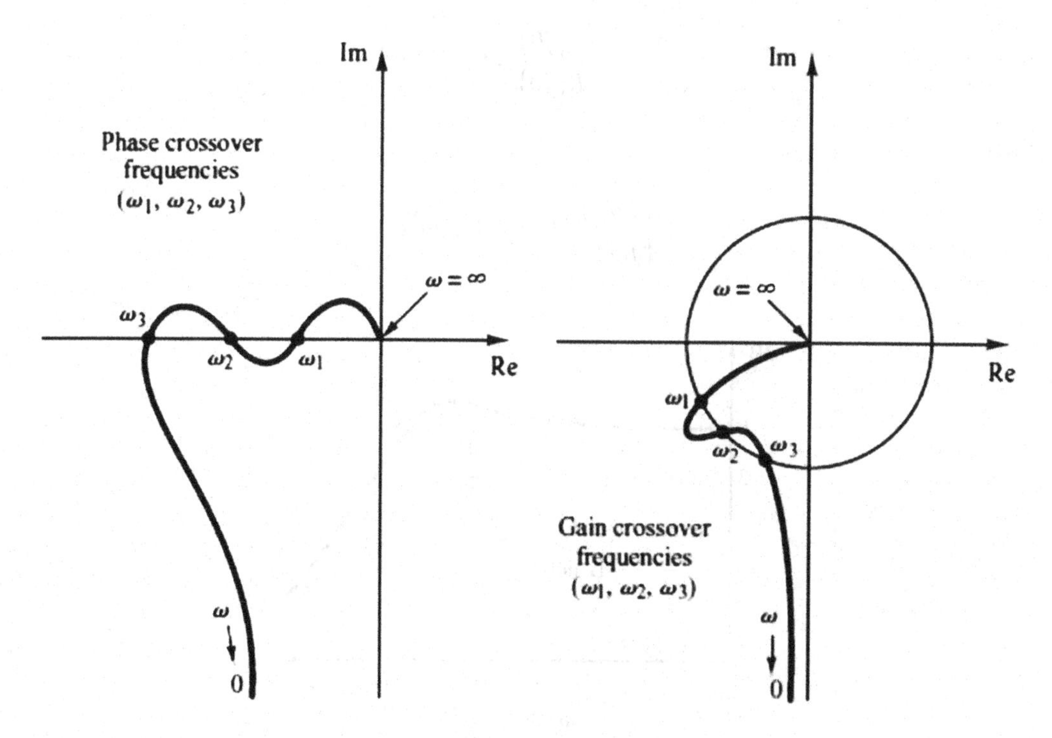

FIGURE 6.67 Higher-order systems gain crossover frequencies.

For minimum-phase systems, the magnitude and phase characteristics of the open-loop transfer function are definitely related. The requirement that the phase margin be between 30° and 60° means that in a Bode diagram, the slope of the log-magnitude curve at the gain crossover frequency should be more gradual than −40 dB/decade. In most practical cases, a slope of −20 dB/decade is desirable at the gain crossover frequency for stability.

If the slope is −40 dB/decade, the system could be either stable or unstable. Even if the system is stable, the phase margin is small. If the slope at the gain crossover frequency is −60 dB/decade or steeper, the system is most likely unstable.

For non-minimum-phase systems, the correct interpretation of stability margins requires careful study. The best way to determine the stability of non-minimum-phase systems is to use the Nyquist diagram approach rather than Bode diagram approach.

6.9 CUTOFF FREQUENCY, BANDWIDTH (BW), AND CUTOFF RATE

6.9.1 Cutoff Frequency

Referring to this Figure 6.68, the *frequency* ω_b at which the magnitude of the closed-loop frequency response is 3 dB below its *zero-frequency value* is called the *cutoff frequency.* Thus, it can be described by this expression as like this.

$$\left|\frac{C(j\omega)}{R(j\omega)}\right| \left\langle \left|\frac{C(j0)}{R(j0)}\right| - 3\text{dB} \quad ;\text{for } \omega \right\rangle \omega_b$$

If

$$\left|\frac{C(j0)}{R(j0)}\right| = 0$$

Then

$$\left|\frac{C(j\omega)}{R(j\omega)}\right| \langle -3\text{dB} \quad ;\text{for } \omega \rangle \omega_b$$

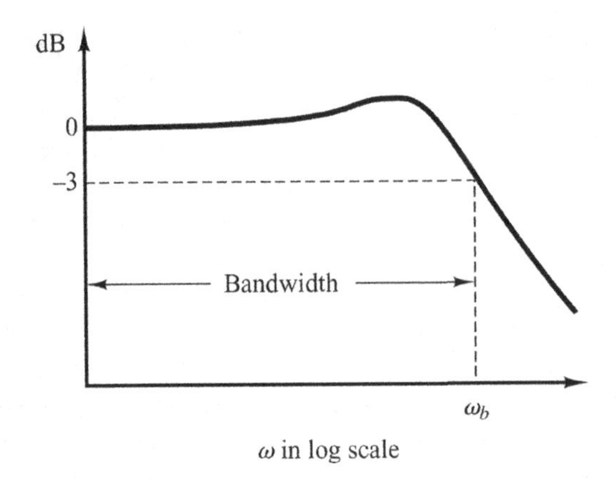

FIGURE 6.68 Cutoff frequency, bandwidth (BW), and cutoff rate.

The closed-loop system filters out the signal components whose frequencies are greater than the cutoff frequency and transmits those signal components with frequencies lower than the cutoff frequency.

6.9.2 Bandwidth (BW)

The frequency range $0 \le \omega \le \omega_b$ in which the magnitude of $C(j\omega)/R(j\omega)$ is greater than −3 dB is called the BW of the system. The BW indicates the frequency where the gain starts to fall off from its low-frequency value.

Thus, the BW indicates how well the system will track an input sinusoid. Note that for a given ω_n, the rise time increases with increasing damping ratio ξ. On the other hand, the BW decreases with the increase in ξ. Therefore, the rise time and the BW are inversely proportional to each other.

Specification factors of BW: Now, the specification of the BW may be determined by the following factors:

First, the ability to reproduce the input signal. A large BW corresponds to a small rise time (t_r), or fast response. Roughly speaking, we can say that the BW is proportional to the speed of response. For example, to decrease the rise time in the step response by a factor of 2, the BW must be increased by approximately a factor of 2.

Second, the necessary filtering characteristics for high-frequency noise. For the system to follow arbitrary inputs accurately, it must have a large BW. From the viewpoint of noise, however, the BW should not be too large. Thus, there are conflicting requirements on the BW, and a compromise is usually necessary for good design. Note that a system with large BW requires high-performance components, so the cost of components usually increases with the BW.

6.9.3 Cutoff Rate

The cutoff rate is the slope of the log-magnitude curve near the cutoff frequency. The *cutoff rate indicates* the ability of a system to distinguish the *signal from noise.* It is noted that a closed-loop frequency-response curve with a steep cutoff characteristic may have a large resonant peak magnitude, which implies that the system has a relatively small stability margin.

6.10 STEP TRANSIENT RESPONSE AND FREQUENCY RESPONSE, *M* AND *N* CIRCLES

In this section, we learn closed-loop frequency response of unity-feedback system; constant-magnitude loci (*M* circles); constant-phase-angle loci (*N* Circles); and the use of *M* and *N* circles.

6.10.1 Closed-Loop Frequency Response of Unity-Feedback System

To understand closed-loop frequency response of a unity-feedback system, for a stable, unity-feedback closed-loop system, the closed-loop frequency response can be obtained easily from that of the open-loop frequency response. Consider the unity-feedback system shown in Figure 6.69.

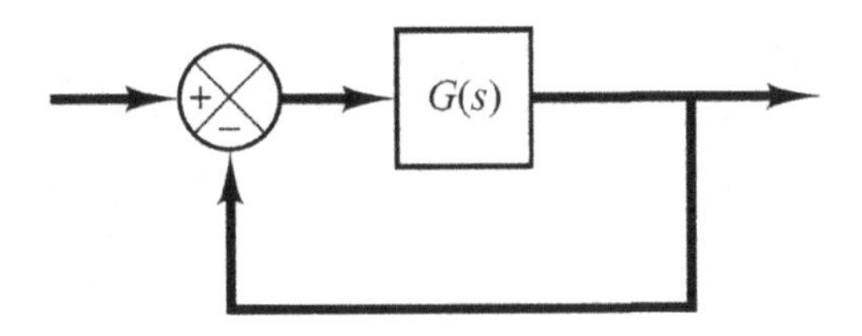

FIGURE 6.69 A unity-feedback system.

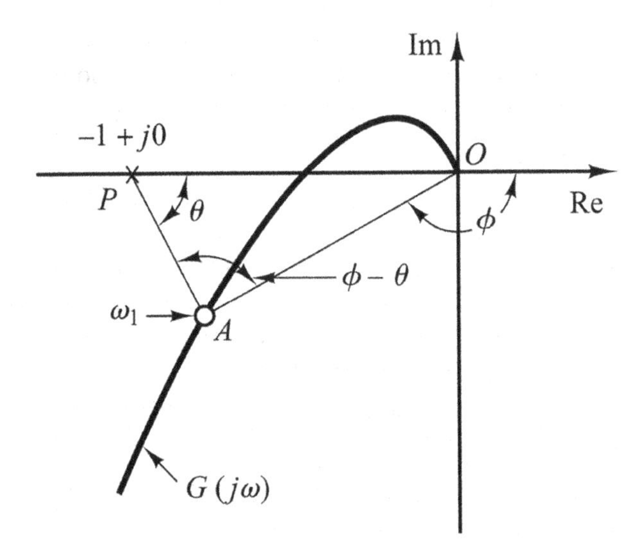

FIGURE 6.70 Nyquist or polar plot of unity-feedback system.

The closed-loop transfer function of unity-feedback system is described as

$$\frac{C(s)}{R(s)} = \frac{G(s)}{1+G(s)}$$

This can be represented in the Nyquist or polar plot which is shown in Figure 6.70.

From this figure, the vector $\overrightarrow{OA}$ represents $G(j\omega_1)$, where ω_1 is the frequency at point A. The length of the vector $\overrightarrow{OA}$ is $|G(j\omega_1)|$ and the angle of the vector $\overrightarrow{OA}$ is $\angle G(j\omega_1)$. The vector $\overrightarrow{PA}$, the vector from the $-1+j0$ point to the Nyquist locus, represents $1+G(j\omega_1)$. Therefore, the ratio of $\overrightarrow{OA}$, to $\overrightarrow{PA}$ represents the closed-loop frequency response, or we can say as

$$\frac{C(j\omega_1)}{R(j\omega_1)} = \frac{\overrightarrow{OA}}{\overrightarrow{PA}} = \frac{G(j\omega_1)}{1+G(j\omega_1)}$$

The magnitude of the closed-loop transfer function at $\omega = \omega_1$ is the ratio of the magnitudes of $\overrightarrow{OA}$ to $\overrightarrow{PA}$. The phase angle of the closed-loop transfer function at $\omega = \omega_1$ is the angle formed by the vectors $\overrightarrow{OA}$ to $\overrightarrow{PA}$ —that is $\varnothing - \theta$, shown in Figure 6.70.

By measuring the magnitude and phase angle at different frequency points, the closed-loop frequency-response curve can be obtained. Let us define the magnitude of the closed-loop frequency response as M and the phase angle as α, or it can be expressed as

$$\left|\frac{\overrightarrow{OA}}{\overrightarrow{PA}}\right| = M \text{ and } \angle\left(\overrightarrow{OA},\overrightarrow{PA}\right) = \alpha$$

$$\textit{Magnitude} \leftrightarrow \textit{Phase}:$$

$$\frac{C(s)}{R(s)} = Me^{j\alpha}$$

In the following, we shall find the constant-magnitude loci and constant-phase-angle loci. Such loci are convenient in determining the closed-loop frequency response from the polar plot or Nyquist plot.

6.10.2 Constant-Magnitude Loci (M Circles)

To understand constant-magnitude loci (M circles), we need to obtain the constant-magnitude loci, let us first note that $G(j\omega)$ is a complex quantity and can be written as

$$G(j\omega) = X + jY$$

where X and Y are real quantities. From

$$\frac{C(j\omega)}{R(j\omega)} = \frac{G(j\omega)}{1+G(j\omega)}$$

M is given by the expression as

$$M = \frac{|X+jY|}{|1+X+jY|}$$

And M^2 is obtained as

$$M^2 = \frac{X^2+Y^2}{(1+X)^2+Y^2}$$

Hence, we get the following equation: $X^2\left(1-M^2\right) - 2M^2X - M^2 + \left(1-M^2\right)Y^2 = 0$

If $M=1$, the equation of a straight line is parallel to the Y axis and passing through the point (–1/2,0); thus, we can simplify and express as

$$X^2 + \frac{2M^2}{M^2-1}X + \frac{M^2}{M^2-1} + Y^2 = 0$$

or

$$\left(X+\frac{M^2}{M^2-1}\right)^2+(Y-0)^2=\left(\frac{M}{M^2-1}\right)^2$$

Therefore, we can say that this is an equation of circle and its center is $\left(-\frac{M^2}{M^2-1},0\right)$ and radius is $r=\left|\frac{M}{M^2-1}\right|$.

If $M\neq 1$, the equation is a family of circles at the center and with a radius. Constant M loci on $G(s)$ plane are a family of circles, as shown in Figure 6.71. Different values of M that we can see in Table 6.7. We can find the center and radius of the circles; then we can draw the family of circles that we can see in Figure 6.71 (Table 6.10).

It is seen that as M becomes larger compared with 1, the M circles become smaller and converge to the $-1+j0$ point. For $M>1$, the centers of the M circles lie to the left of the $-1+j0$ point. Similarly, as M becomes smaller compared with 1, the M circle becomes smaller and converges to the origin. For $0<M<1$, the centers of the M circles lie to the right of the origin. $M=1$ corresponds to the locus of points equidistant from the origin and from the $-1+j0$ point. As stated earlier, it is a straight line passing through the point and parallel to the imaginary axis. The constant M circles corresponding to $M>1$ lie to the left of the $M=1$ line, and those corresponding to $0<M<1$ lie to the right of the $M=1$ line. The M circles are symmetrical with respect to the straight line corresponding to $M=1$and with respect to the real axis.

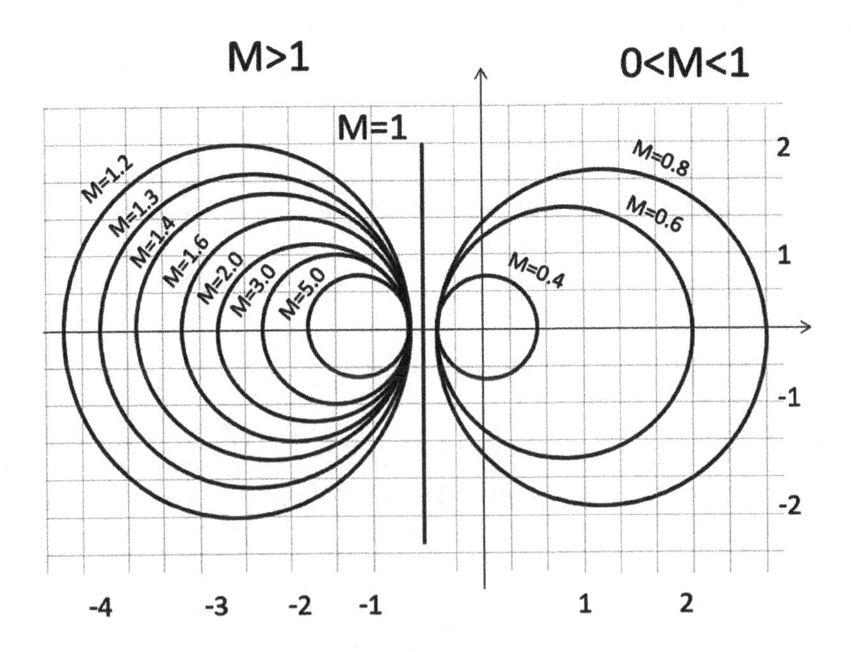

FIGURE 6.71 Constant-magnitude loci (M circles).

TABLE 6.10 Center and Radius for Different Values of *M*

M	*(a,b)*	*r*
0.4	(0.19,0)	0.48
0.6	(0.56,0)	0.94
0.8	(1.78,0)	2.22
1.0	Straight line	
1.2	(−3.27,0)	2.73
1.3	(−2.45,0)	1.88
1.4	(−2.04,0)	1.46
1.6	(−1.64,0)	1.03
2.0	(−1.33,0)	0.67
3.0	(−1.13,0)	0.38
5.0	(−1.04,0)	0.21

6.10.3 Constant-Phase-Angle Loci (N Circles)

To understand the constant-phase-angle loci (N circles), we can consider it similar to the constant-magnitude loci (M circles) function $G(j\omega) = X + jY$. The phase angle can be expressed as

$$\angle e^{j\alpha} = \angle\left(\frac{X + jY}{1 + X + jY}\right)$$

and

$$\alpha = \tan^{-1}\frac{Y}{X} - \tan^{-1}\frac{Y}{1+X}$$

Let, $N = \tan\alpha$, thus we can say N as

$$\Rightarrow N = \tan\left(\tan^{-1}\frac{Y}{X} - \tan^{-1}\frac{Y}{1+X}\right)$$

or

$$N = \frac{\dfrac{Y}{X} - \dfrac{Y}{1+X}}{1 + \dfrac{Y}{X}\left(\dfrac{Y}{1+X}\right)} = \frac{Y}{X^2 + X + Y^2}$$

Then, the simplified equation can be formed as

$$\left(X + \frac{1}{2}\right)^2 + \left(Y - \frac{1}{2N}\right)^2 = \left(\sqrt{\frac{1}{4} + \left(\frac{1}{2N}\right)^2}\right)^2$$

Therefore, we can say that this is an equation of a circle and its center is $\left(-\frac{1}{2}, \frac{1}{2N}\right)$ and radius is $r = \sqrt{\frac{1}{4} + \left(\frac{1}{2N}\right)^2}$. Different values of α in degree we can get different values of N that we can see in Table 6.8. Then we can find the center and radius of the circles, and we can draw the family of circles that we can see in Figure 6.72 (Table 6.11).

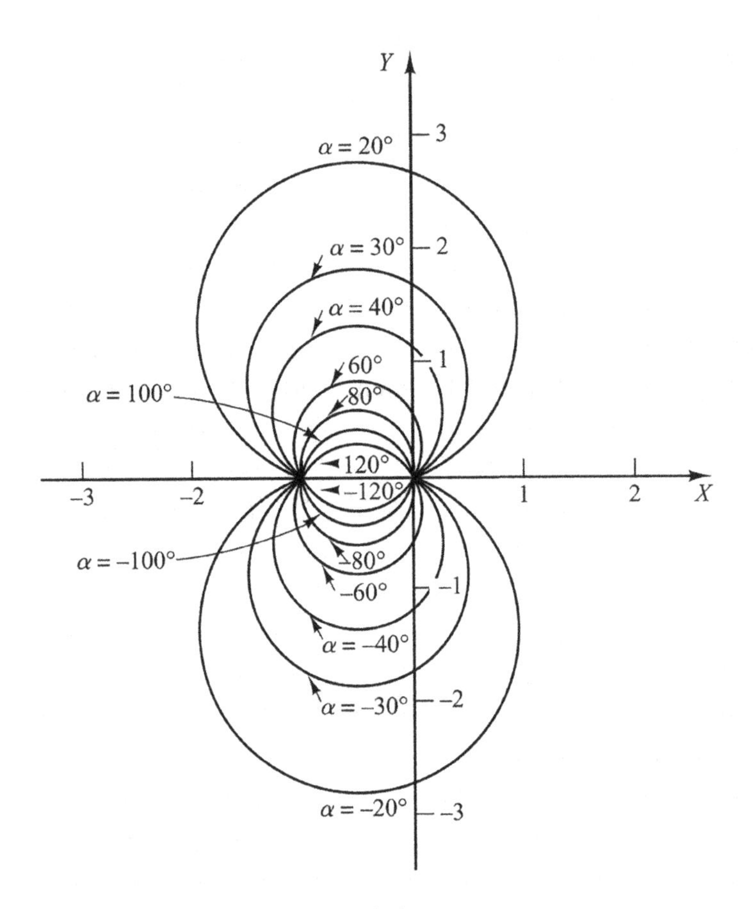

FIGURE 6.72 Constant-phase-angle loci (N circles).

TABLE 6.11 Different Values of α in Degree We Can Get Different Values of N

α	N	(a,b)	r
±20°	±0.364	(−0.5,±1.37)	1.46
±30°	±0.577	(−0.5,±0.87)	1.0
±40°	±0.840	(−0.5,±0.60)	0.78
±60°	±1.732	(−0.5, ±0.29)	0.58
±80°	±5.671	(−0.5,±0.08)	0.52
±100°	±5.672	(−0.5, ±0.09)	0.51
±120°	±1.732	(−0.5,±0.29)	0.50

Constant N locus for a given value of α is actually not the entire circle, but only an arc. The $\alpha = 30°$ and $\alpha = -150°$ arcs are parts of the same circle because tangent of an angle remains same if the angle ±180° (or multiples) is added to original angle.

6.10.4 Use of M and N Circles

The use of the M and N circles enables us to find the entire closed-loop frequency response from the open-loop frequency response $G(j\omega)$ without calculating the magnitude and phase of the closed-loop transfer function at each frequency. The intersections of the $G(j\omega)$ locus and the M circles and N circles give the values of M and N at frequency points on the $G(j\omega)$ locus.

The N circles are multivalued in the sense that the circle for $\alpha = \alpha_1$ and that for $\alpha = \alpha_1 \pm 180° \times n$; ($n = 1, 2, \ldots$) are the same. Combined M and N circles can be seen in Figure 6.73.

In using the N circles for the determination of the phase angle of closed-loop systems, we must interpret the proper value of α. To avoid any error, start at zero frequency, which corresponds to $\alpha = 0$ and proceed to higher frequencies. The phase-angle curve must be continuous.

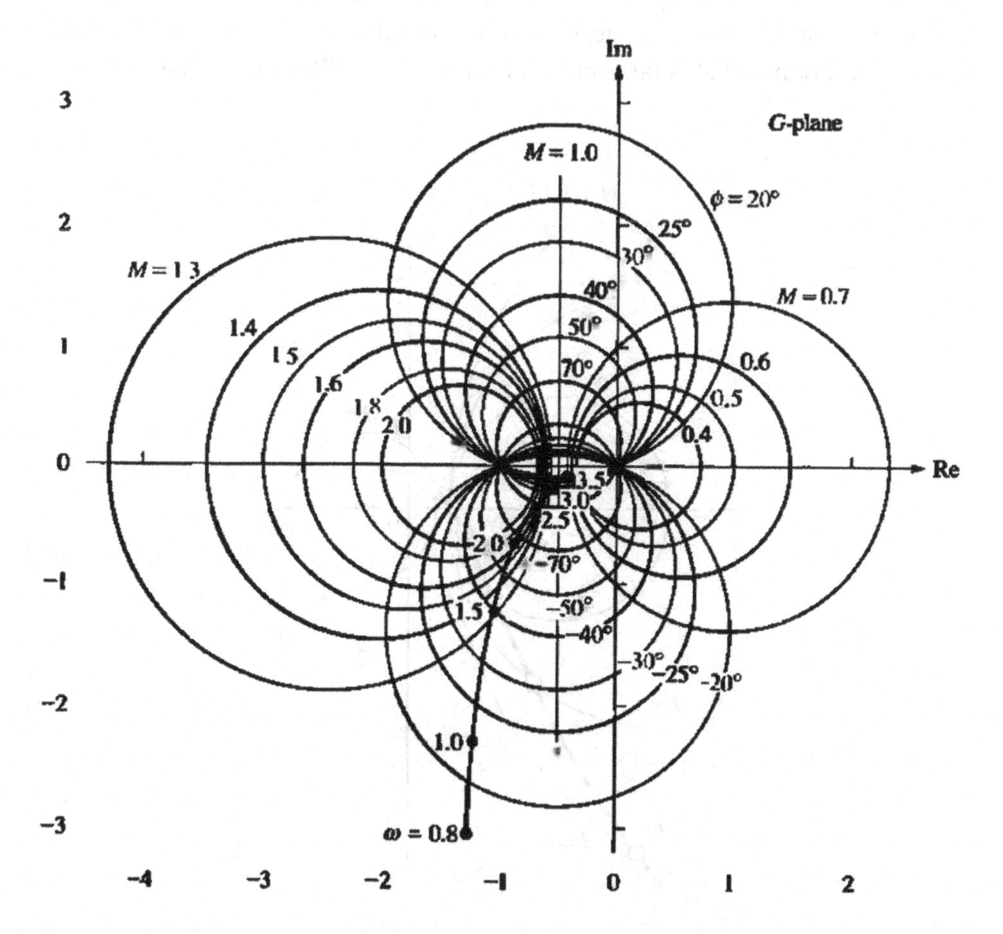

FIGURE 6.73 Constant-magnitude loci (M circles) and constant-phase-angle loci (N circles).

Graphically, the intersections of the $G(j\omega)$ locus and M circles give the values of M at the frequencies denoted on the $G(j\omega)$ locus. Thus, the constant M circle with the smallest radius that is tangent to the $G(j\omega)$ locus gives the value of the resonant peak magnitude M_r. If it is desired to keep the resonant peak value less than a certain value, then the system should not enclose the critical point ($-1+j0$ point) and, at the same time, there should be no intersections with the particular M circle and the $G(j\omega)$ locus, as shown in Figure 6.74.

Figure 6.75 shows the $G(j\omega)$ locus superimposed on a family of N circles, as shown in Figure 6.75.

Figure 6.75 shows the $G(j\omega)$ locus superimposed on a family of N circles. From these two plots, it is possible to obtain the closed-loop frequency response by inspection. It is seen that the $M=1.1$ circle intersects the $G(j\omega)$ locus at the frequency point $\omega=\omega_1$. This means that at this frequency, the magnitude of the closed-loop transfer function is 1.1. In this Figure, the $M=2$ circle is just tangent to the $G(j\omega)$ locus. Thus, there is only one point on the $G(j\omega)$ locus for which $\left|\dfrac{C(j\omega)}{R(j\omega)}\right|$ is equal to 2.

Figure 6.76 shows the closed-loop frequency-response curve for the system. The upper curve is the M-versus frequency ω curve, and the lower curve is the phase angle α-versus-frequency ω curve. The resonant peak value is the value of M corresponding to the M circle of smallest radius that is tangent to the $G(j\omega)$ locus. Thus, in the Nyquist diagram,

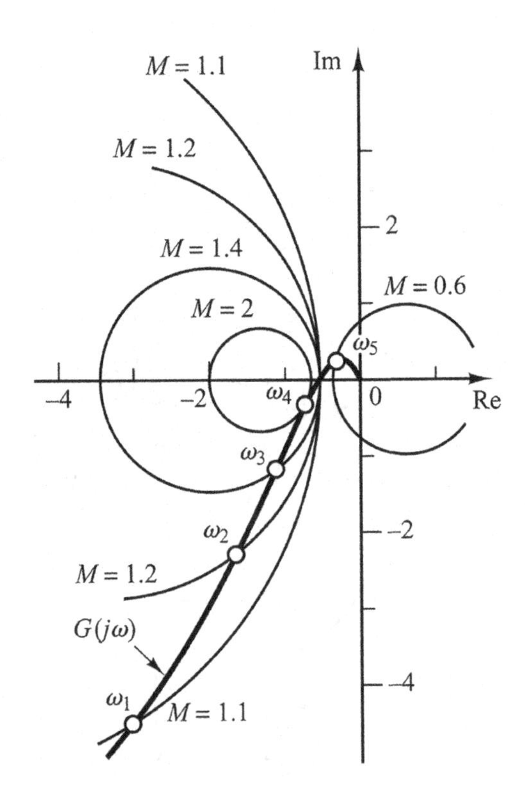

FIGURE 6.74 Particular M circle and the $G(j\omega)$ locus.

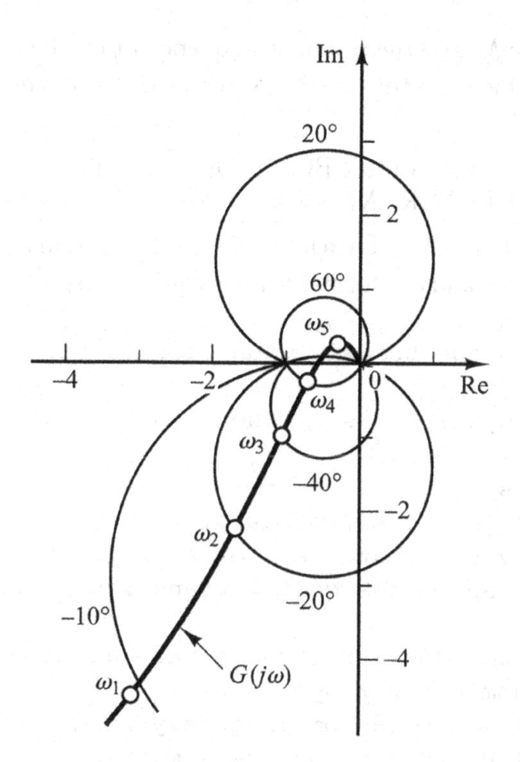

FIGURE 6.75 Particular N circle and the $G(j\omega)$ locus.

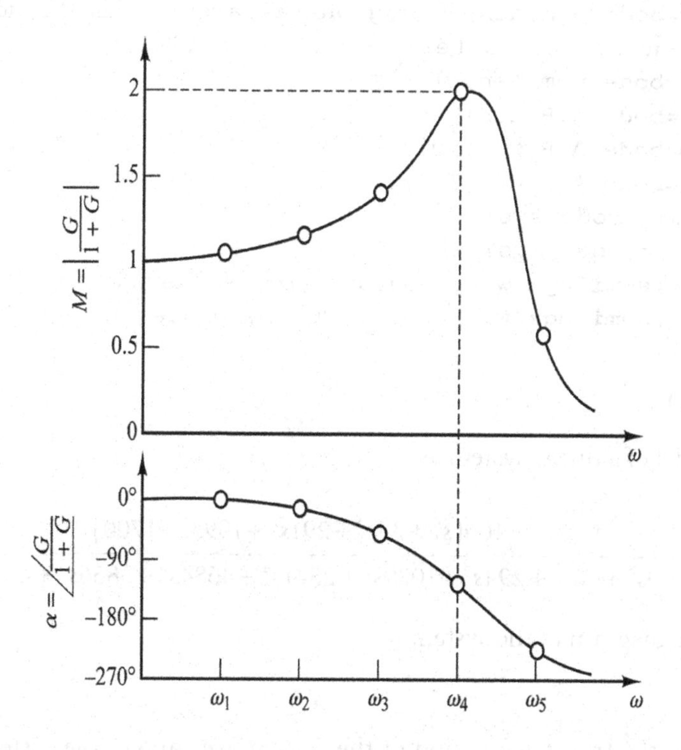

FIGURE 6.76 The closed-loop frequency-response curve.

the resonant peak value M_r and the resonant frequency ω_l can be found from the M circle tangency to the $G(j\omega)$ locus. In the present example, $M_r = 2$ and $\omega_r = \omega_4$.

6.11 MATLAB FUNCTIONS FOR BODE DIAGRAM AND NYQUIST DIAGRAM

The frequency response of the basic format for MATLAB function is shown in the following. The function bode() command call format of the frequency response of Bode diagram as

```
bode sys) % sys is the transfer function of the system.
bode sys,w)
bode sys1,sys2,…,sysn) % sys is the transfer function of the
subsystem.
bode sys1,sys2,…,sysn,w)
bode sys1,'plotstyle1',sys2,'plotstyle2'…,sysn,'plotstylen') %
plotstyle identify the characters or strings for the various
properties supported by the MATLAB standard function by plot ()
command.
bode num,den) %[num, den] are the polynomial coefficients of
numerator and denominator respectively.
bode num,den,w) % w for custom frequency range.
bode A,B,C,D) % A,B,C,D are the state matrix.
bode A,B,C,D,iu)
bode A,B,C,D,iu,w) % iu is the number of input variables.
[mag,phase]=bode num,den) %[mag,phase] are the amplitude and the
phase of frequency characteristics, respectively.
[mag,phase]=bode num,den,w)
[mag,phase]=bode A,B,C,D)
[mag,phase]=bode A,B,C,D,iu)
[mag,phase]=bode A,B,C,D,iu,w)
[mag,phase,w]= bode sys)
[mag,phase,w]= bode sys1,w)
subplot(211);semilogx w,20*log10 (mag)) % Custom amplitude curve
subplot(212);semilogx w,phase) % Custom phase curve
```

Example 6.11

Given a model of a linear system as

$$G(s)=\frac{100\left(s^4+35s^3+291s^2+1093s+1700\right)}{s^9+9s^8+66s^7+294s^6+1029s^5+2541s^4+4684s^3+5856s^2+4629s+1700}$$

Draw a Bode diagram of the system.

Solution

According to the transfer function of the system, we can get the system MATLAB program:

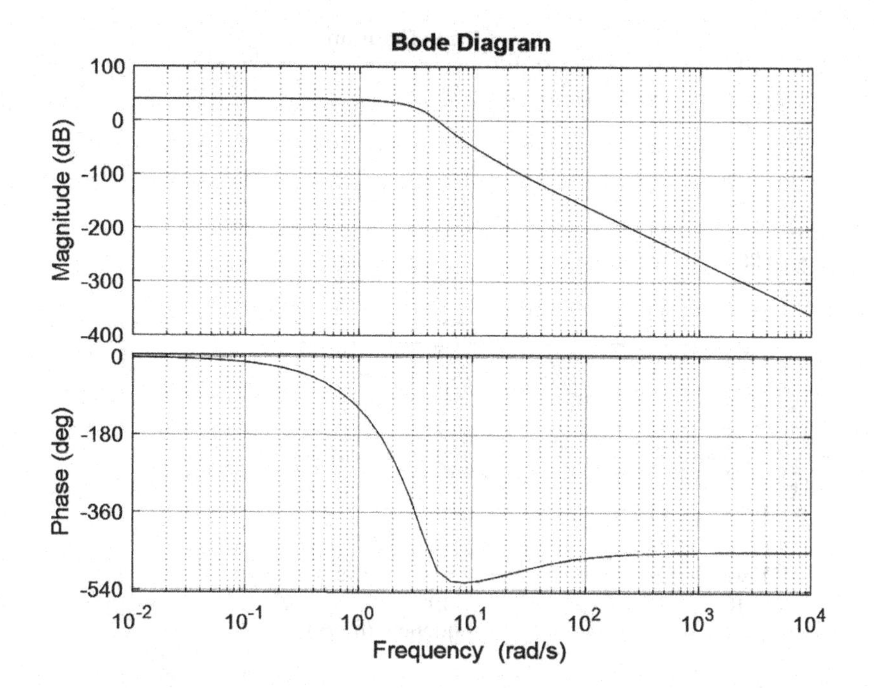

FIGURE 6.77 Bode diagram of the linear control system.

```
clear
w=logspace(-2,4);
num=100*[1 35 291 1093 1700];
den=[1 9 66 294 1029 2541 4684 5856 4629 1700];
sys=tf(num,den);
bode(sys,w,'k')
```

The Bode diagram of the system is shown in Figure 6.77.

Example 6.12

Given a model of a linear system as

$$G(s)=\frac{s(s+2)^3(s^2+3s+2)^2}{(s+1)(s-1)\left(s^3+3s^2+5s+2\right)(s^2+2s+4)^2}$$

Draw a Bode diagram of the system.

Solution

According to the transfer function of the system, we can get the system MATLAB program. The Bode diagram of the system is shown in Figure 6.78.

```
w=logspace(-2,7);
num1=conv(1,[1 0]);
```

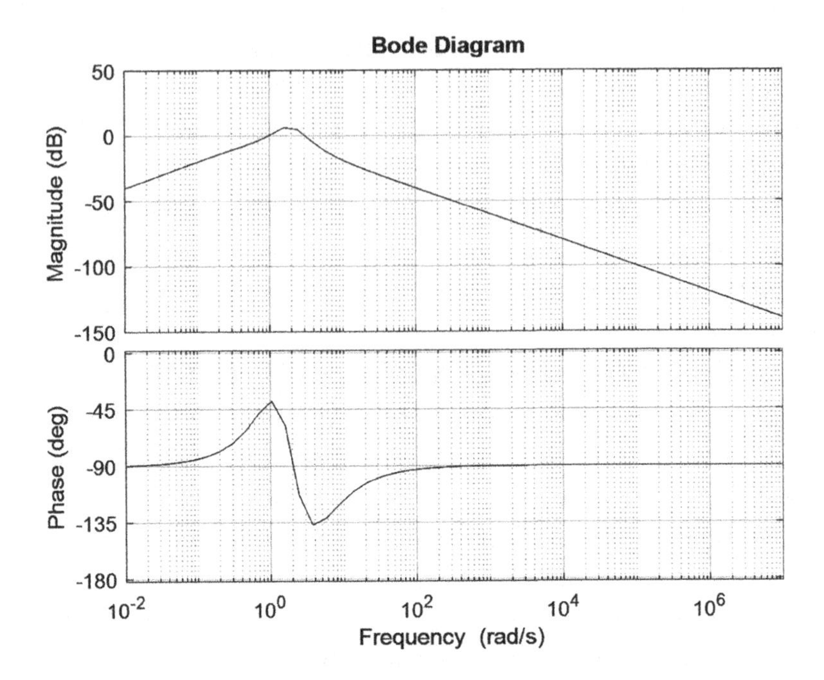

FIGURE 6.78 Bode diagram of the control system.

```
num2=conv(conv(conv(num1,[1 2]), [1 2]), [1 2]);
num=conv(conv(num2,[1 3 2]), [1 3 2]);
den1=conv([1 1], [1-1]);
den2=conv(den1,[1 3 5 2]);
den=conv(conv(den2,[1 2 4]), [1 2 4]);
sys=tf(num, den);
% Transfer function:
% s^8+12s^7+61s^6+170s^5+280s^4+272s^3+144s^2+32s
%-----------------------------------------------------------------------
% s^9+7s^8+28s^7+67s^6+103s^5+78 s^4-20s^3-120s^2-112s-32
bode(sys, w, 'k')
grid
```

Example 6.13

Given a model of an open-loop transfer function for a linear system as

$$G(s)=\frac{s(s+5)(s+7)}{(s+1)(s+2)(s+3)(s+4)(s+10)(s+12)(s+15)(s+20)}$$

Draw a Bode diagram of the system.

Solution

According to the transfer function of the system, we can get the system MATLAB program. The Bode diagram of the system is shown in Figure 6.79.

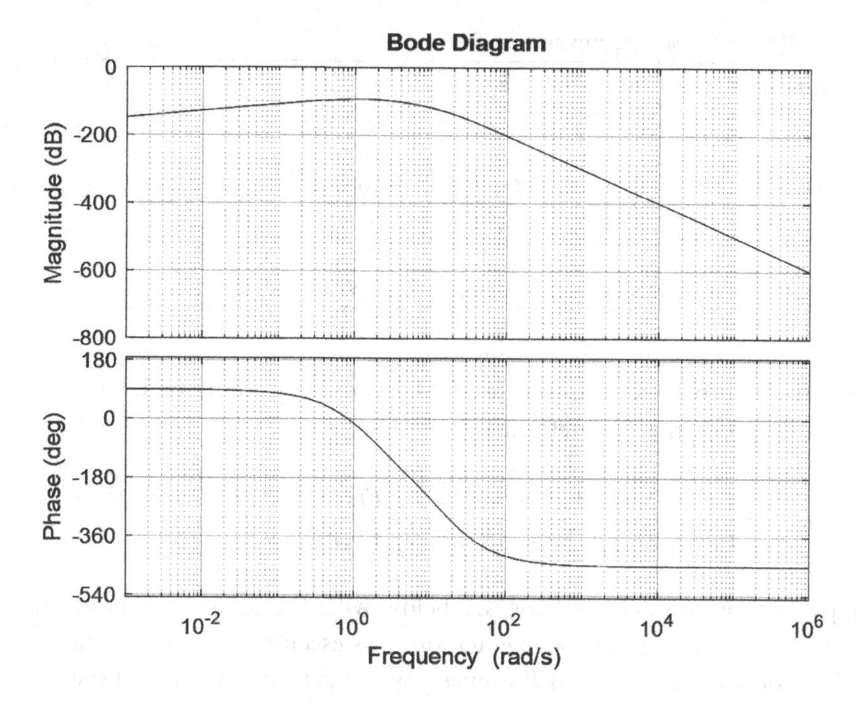

FIGURE 6.79 Bode diagram of the a open-loop control system.

```
w=logspace(-3,6);
num=poly([0,-5,-7]);
den=poly([-1-2-3-4-10-12-15-20]);
sys=tf(num, den);
% Transfer function:
% s^3 + 12 s^2 + 35 s
%-----------------------------------------------------------------------------
% s^8+67s^7+1795s^6+24745s^5+188524s^4+798868s^3+1.829e006s^2+
2.059e006s+864000
bode(sys, w, 'k');
grid
```

Example 6.14

Given a unity-feedback transfer function with a delayed open loop is

$$G(s) = \frac{(s+5)(s+7)e^{-1.5s}}{(s+1)(s+2)(s+3)(s+4)}$$

Draw a Bode diagram of the system.

Solution

About the delay factor e^{-sT} can use the rational approximation function class (i.e., the transfer function of the system denominator polynomial). In 1982, Pade, a French

TABLE 6.12 Pade Approximation Table

Iteration, n	p_1	p_2	p_3	p_4	p_5
1	0	0	0	0	0
2	1/12	0	0	0	0
3	1/10	1/120	0	0	0
4	3/28	1/84	1/1680	0	0
5	1/9	1/72	1/1008	1/30,240	0
6	5/44	1/66	1/792	1/15,840	1/66,528

mathematician, proposed a famous rational approximation method, which was later named Pade approximation method as

$$e^{-sT} = \frac{1 - Ts/2 + p_1(Ts)^2 - p_2(Ts)^3 + p_3(Ts)^4 - p_4(Ts)^5 + p_5(Ts)^6 - \cdots}{1 + Ts/2 + p_1(Ts)^2 + p_2(Ts)^3 + p_3(Ts)^4 + p_4(Ts)^5 + p_5(Ts)^6 + \cdots}$$

The approximate numbers of Pade are below with the six iterations in ascending order in Table 6.8 using the above equation. It is essential to observe that if you set $T=1$, then you use the Pade(1, n); however, you can get the figure out of the Pade coefficients. In the MATLAB control system toolbox, the call format is as follows:

```
[num,den]=pade(T,n) or, [A,B,C,D]=pade(T,n)
```

where, T is the delay time constant and n is the iteration order to be fitted; return the equivalent transfer function model of Pade approximation after the call [num, den] or equivalent equation of state model [A, B, C, D].

Let the delay factor be a fourth-order approximation: The system of the MATLAB program with e^{-sT} uses the Pade approximation equivalent function; its amplitude is at any T, $s=1$. The phase curve shown in Figure 6.52 is well fitted in the low-frequency band. Of course, the higher the order of fitting, the higher is the fitting accuracy. In general, a satisfactory accuracy can be obtained by taking the order of the Pade approximation to be 3 or 4. Obviously, the fitting accuracy of the one with a large delay time constant is often lower than that of the one with a small delay time constant (Table 6.12).

```
% This program for delay system
clear
w=logspace(-1,2);
num1=poly([-5-7]);
den1=poly([-1-2-3-4]);
[num2,den2]=pade(1.5,4); % The delay factor is a fourth-order
approximation
num=conv(num1,num2);
den=conv(den1,den2);
sys1=tf(num1,den1)
```

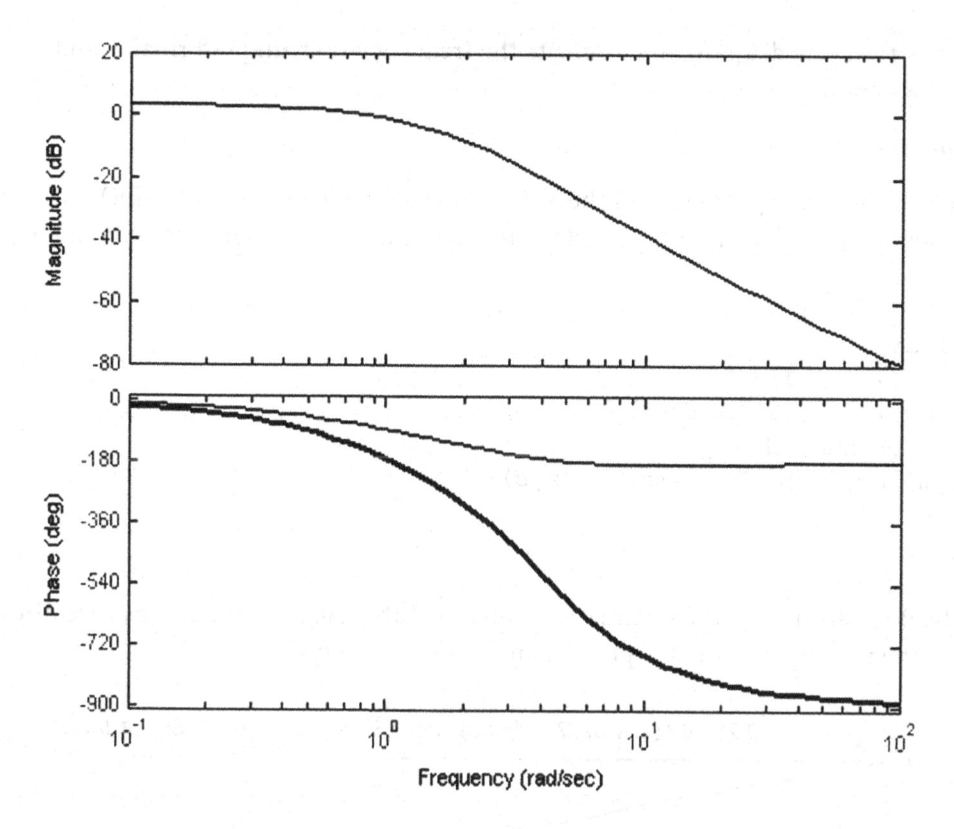

FIGURE 6.80 Bode curve with delay factor.

```
sys=tf(num, den)
bode(sys1,w);hold on
grid
% The bode curve was drawn without delay factor. The amplitude
curve in the figure was identical, and the phase angle curve
was fine and solid.
bode(sys, w)
% The bode curve was drawn with delay factor. The amplitude
curve in the figure is exactly the same, and the phase angle
curve is thick and solid.
```

The Bode diagram of the system is shown in Figure 6.80.

Example 6.15

Given the forward channel of a unity-feedback system, the transfer function of the channel is

$$G(s)=\frac{4s^3+6s^2+4s+1}{s^6+5s^5+10s^4+10s^3+5s^2+s}$$

Draw the Bode diagram and calculate the frequency-domain performance index of the system.

Solution

According to the problem, call the MATLAB function command margin() that can draw the Bode diagram of the system and calculate the frequency-domain index of the system.

```
clear
num=[4 6 4 1];
den=[1 5 10 10 5 1 0];
sys=tf(num, den)
[gm, pm, wcp, wcg]=margin(sys)
margin(sys)
grid
```

The Bode diagram of the system can be obtained after running the above procedures, as shown in Figure 6.81. The performance index of the system is

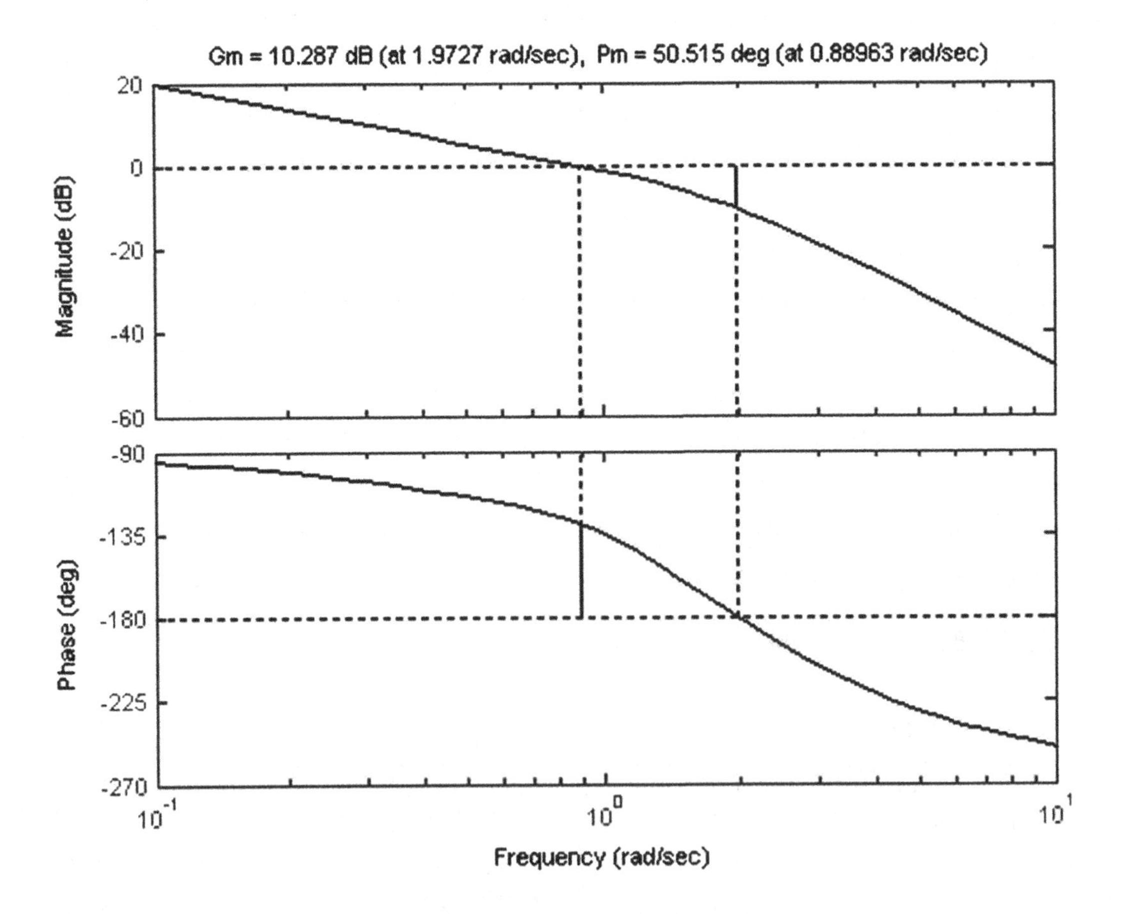

FIGURE 6.81 Bode diagram of the unity-feedback system.

gm=3.2685, pm=50.5153°, wcp =1.9727 rad/s, wcg=0.8896 rad/s
i.e., the performance index of the system is as follows:
Corner frequency, $\omega_c = 0.8896 \quad 1/s$
The phase margin, $\gamma = 50.5153°$
$-\pi$ cross over frequency, $\omega_g = 1.9727 \quad 1/s$
The amplitude margin, $L_h = 20\log(3.2685) = 10.287 \quad \text{dB}$

EXAMPLES

In this chapter, students are required to learn the concept of frequency characteristics, be familiar with the Nyquist plots and Bode plots of typical factors, and be able to draw Nyquist plots and Bode diagrams or plots based on the known system transfer functions.

Example 6.16

The transfer function of a system is $\frac{7}{3s+2}$, when the input is $\frac{1}{7}\sin\left(\frac{2}{3}t+45°\right)$. Find its steady-state output.

Solution

When a linear system is given as a sinusoidal signal, its system output will be a sinusoidal function of the same frequency as the input.

Thus, the transfer function is $G(s)=\frac{7}{3s+2}$.

Then, $G(j\omega)=\frac{7}{3(j\omega)+2}$.

The amplitude is

$$A(\omega)=\frac{7}{\sqrt{9\omega^2+4}}$$

And the phase is

$$\varphi(\omega)=-\tan^{-1}\left(\frac{3\omega}{2}\right)$$

$$\text{Therefore,} r(t)=\frac{1}{7}\sin\left(\frac{2}{3}t+45°\right)$$

$$\text{Then,} \frac{1}{7}A\left(\frac{2}{3}\right)=\frac{1}{7}\frac{7}{\sqrt{9\left(\frac{2}{3}\right)^2+4}}=\frac{\sqrt{2}}{4}$$

and

$$\varphi\left(\frac{2}{3}\right)+45°=-\tan^{-1}\left(\frac{3}{2}\times\frac{2}{3}\right)+45°=0°$$

Thus, the output response of the system is $c(t)=\frac{\sqrt{2}}{4}\sin\left(\frac{2}{3}t\right)$.

Example 6.17

The asymptotic logarithmic amplitude-frequency characteristic of the minimum-phase system is shown in Figure 6.82. Determine its transfer function and draw the logarithmic phase-frequency characteristic diagram.

Solution

According to the system transfer function, we get

$$G(s)=\frac{K\left(\frac{1}{0.8}s+1\right)}{s^2\left(\frac{1}{30}s+1\right)\left(\frac{1}{50}s+1\right)}$$

Its logarithmic amplitude-frequency characteristic is

$$L(\omega)=20\log K+20\log\sqrt{\left(\frac{\omega}{0.8}\right)^2+1}-20\log\omega^2$$

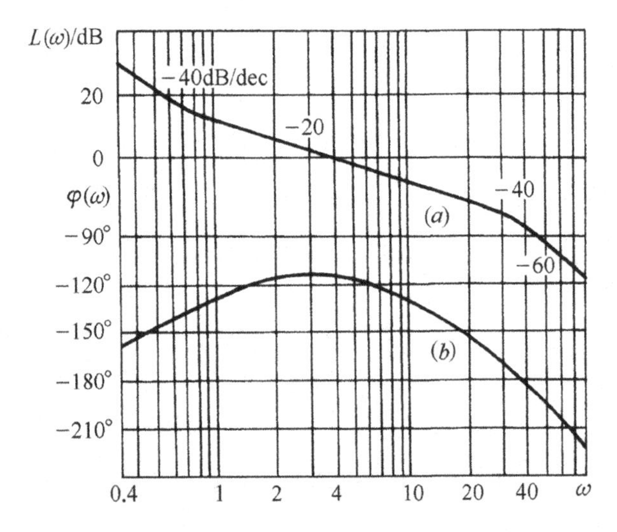

FIGURE 6.82 The asymptotic logarithmic amplitude-frequency characteristics.

$$-20\log\sqrt{\left(\frac{\omega}{30}\right)^2+1}-20\log\sqrt{\left(\frac{\omega}{50}\right)^2+1}$$

The asymptotic logarithmic amplitude frequency passes through the horizontal axis at $\omega=4\,\text{rad/s}$. Considering this relation, it can be obtained from the above equation as

$$L(4)=20\log K+20\log\frac{4}{0.8}-20\log 4^2=0\text{dB}$$

According to the above equation, $K=3.2$; then the system transfer function is rewritten as

$$G(s)=\frac{3.2\left(\frac{1}{0.8}s+1\right)}{s^2\left(\frac{1}{30}s+1\right)\left(\frac{1}{50}s+1\right)}$$

EXERCISES

1. Find the amplitude-frequency characteristics of the following functions $A(\omega)$, phase-frequency characteristics $\omega=\omega_1$, real-frequency characteristics $U(\omega)$, and imaginary-frequency characteristics $V(\omega)$.

   1. $G_1(j\omega)=\dfrac{5}{30j\omega+1}$

   2. $G_2(j\omega)=\dfrac{1}{j\omega(0.1j\omega+1)}$

2. The open-loop transfer function of a unity-feedback system is

   $$G(s)=\frac{10}{(s+1)}$$

   Find the expression of steady-state response output $x_o(t)$ with the following input.

   1. $x_i(t)=\sin(t+30°)$
   2. $x_i(t)=2\cos(2t-45°)$

3. The open-loop transfer function of the unity-feedback system is

$$G(s)=\frac{K}{s(Ts+1)}$$

Here, we have the sinusoidal signal $x_i(t) = \sin 10t$. The steady-state output of a closed-loop system is $x_o(t) = \sin(10t - 90°)$. Calculate the values of parameters K and T.

4. Can you find the Nyquist plots of the following transfer functions shown in Figure 6.83?

1. $G_1(s) = \dfrac{0.2(4s+1)}{s^2(0.4s+1)}$

2. $G_2(s) = \dfrac{0.14(9s^2+5s+1)}{s^3(0.3s+1)}$

3. $G_3(s) = \dfrac{K(0.1s+1)}{s(s+1)}$

4. $G_4(s) = \dfrac{K}{(s+1)(s+2)(s+3)}$

5. $G_5(s) = \dfrac{K}{s(s+1)(0.5s+1)}$

6. $G_5(s) = \dfrac{K}{(s+1)(s+2)}$

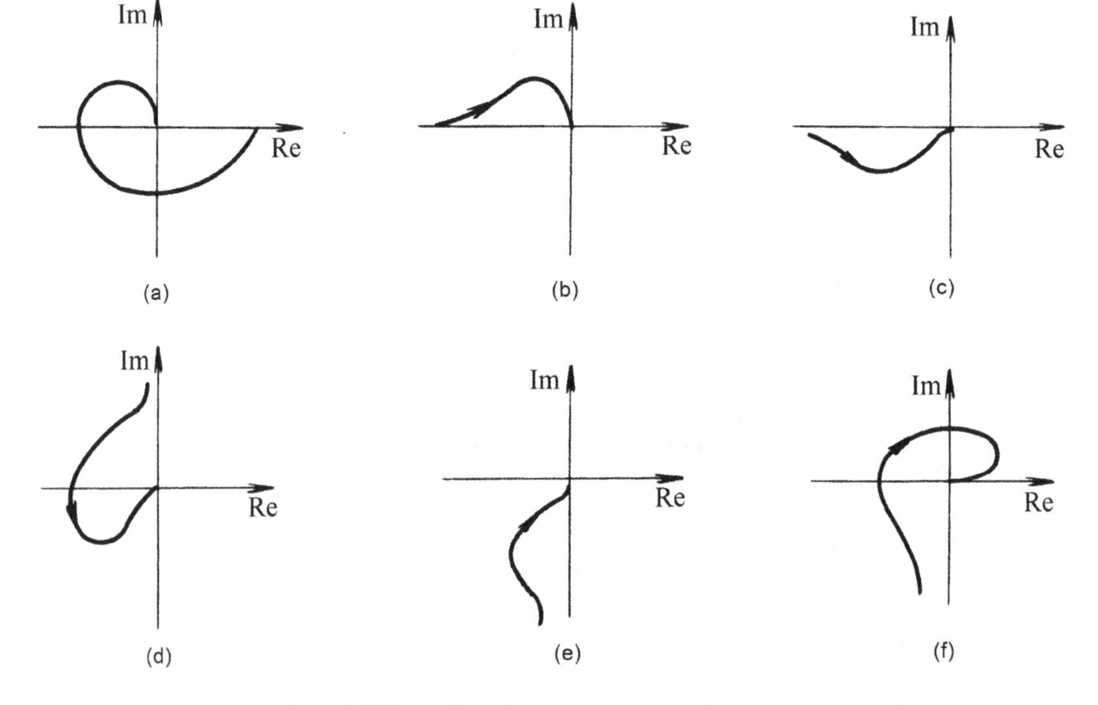

FIGURE 6.83 Nyquist plot of different functions.

5. Given the open-loop transfer function of the system in the following, draw the Nyquist plot of the system.

1. $G(s)=\dfrac{1}{(s+1)(2s+1)}$

2. $G(s)=\dfrac{1}{s^2(s+1)(2s+1)}$

3. $G(s)=\dfrac{(0.2s+1)(0.025s+1)}{s^2(0.005s+1)(0.001s+1)}$

4. $G(s)=\dfrac{1}{s(s+1)(0.5s+1)}$

5. $G(s)=\dfrac{s^3}{(s+0.31)(s+5.06)(s+0.64)}$

6. $G(s)=\dfrac{10}{(5s+1)(10s+1)}$

6. The open-loop transfer function of the system is

$$G(s)H(s)=\frac{K(T_a s+1)(T_b s+1)}{s^2(T_1 s+1)};(K>0)$$

Draw the Nyquist plots of the following two scenarios.

1. $T_a>T_1>0,\ T_b>T_1>0$
2. $T_1>T_a>0,\ T_1>T_b>0$

7. The open-loop transfer function of the unity-feedback system is given

$$G(s)=\frac{K(T_3 s+1)}{s^2(T_1 s+1)(T_2 s+1)};(K,T_1,T_2,T_3>0)$$

Draw the Nyquist plots.

8. Draw the Bode diagram of the following transfer function.

1. $G(s)=\dfrac{20}{s(0.5s+1)(0.1s+1)}$

2. $G(s)=\dfrac{2s^2}{(0.4s+1)(0.04s+1)}$

3. $G(s)=\dfrac{7.5(0.2s+1)(s+1)}{s\left(s^2+16s+100\right)}$

4. $G(s)=\dfrac{50(0.6s+1)}{s^2(45s+1)}$

5. $G(s)=\dfrac{8(s+0.1)}{s\left(s^2+s+1\right)\left(s^2+4s+25\right)}$

6. $G(s)=\dfrac{10(s+0.2)}{s^2(s+0.1)}$

7. $G(s)=\dfrac{2}{(2s+1)(8s+1)}$

8. $G(s)=\dfrac{200}{s^2(s+1)(10s+1)}$

9. $G(s)=\dfrac{50}{s^2\left(s^2+s+1\right)(6s+1)}$

10. $G(s)=\dfrac{40(s+0.5)}{s(s+0.2)\left(s^2+s+1\right)}$

11. $G(s)=\dfrac{50}{(0.2s+1)(s+2)(s+0.5)}$

12. $G(s)=\dfrac{100}{s(0.2s+1)}$

13. $G(s)=\dfrac{100(s+1)}{s(0.1s+1)(0.5s+1)(0.8s+1)}$

14. $G(s)=\dfrac{-10}{2s(1-20s)}$

9. The logarithmic amplitude-frequency asymptotic characteristics of the minimum-phase system are shown in Figure 6.84. Determine the transfer function of the system.

10. Asymptotes of open-loop logarithmic amplitude-frequency characteristics of the minimum-phase system are shown in Figure 6.85. Determine the open-loop transfer function of the system and draw the general graph of the phase-frequency characteristic of the system.

11. The open-loop transfer function of the non-minimum-phase system is

$$G(s)H(s)=\frac{10}{s(s-1)(0.2s+1)}$$

Draw a Bode diagram of the system.

12. Given the open-loop transfer function of the system is

$$G(s)H(s)=\frac{K(5s+1)(6s+1)(0.2s+1)}{s^2\left(s^2+s+4\right)(s+1)(3s+1)(4s+1)}.$$

When $\omega=0.3$, the logarithmic amplitude-frequency asymptotic characteristic of open-loop system is $L_a(0.3)=10\,\text{dB}$. Determine the system parameter K.

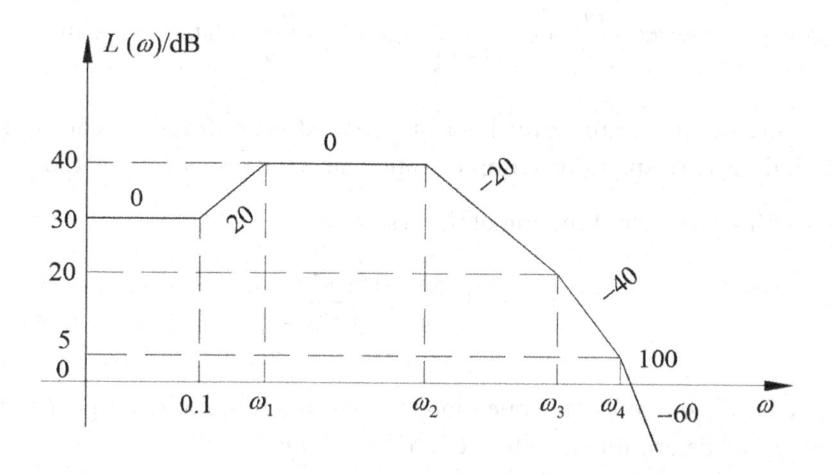

FIGURE 6.84 Logarithm amplitude-frequency curve of minimum-phase system.

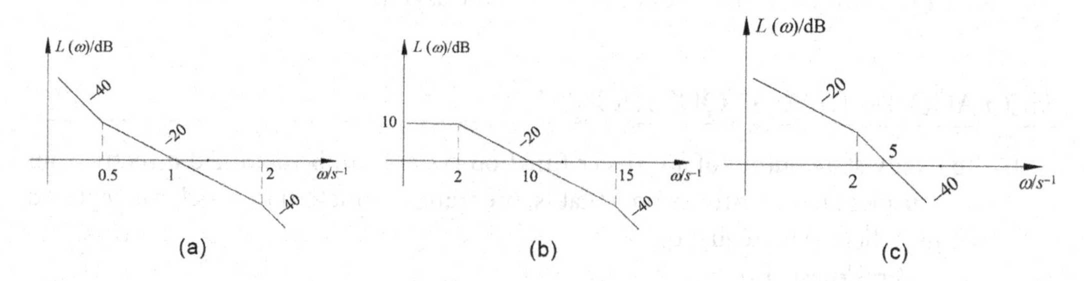

FIGURE 6.85 Aysmptotes of open-loop logarithm amplitude-frequency curve of minimum-phase system.

13. Determine the resonance peak, resonance frequency and BW of the following system.

$$\frac{C(j\omega)}{R(j\omega)} = \frac{5}{(j\omega)^2 + 2j\omega + 5}$$

14. If the unit-step response of the system is

$$x_o(t) = 1 - 1.8e^{-4t} + 0.8e^{-9t}; (t \geq 0)$$

Find the frequency characteristics of the system.

15. Determine the transfer function of the minimum-phase system by the following expression.

1. $\phi(\omega) = -\frac{\pi}{2} - \tan^{-1}\omega + \arctan\frac{\omega}{3} - \tan^{-1}10\omega \; ; A(5) = 2$

2. $\phi(\omega) = -\pi + \arctan\frac{\omega}{5} - \tan^{-1}\frac{\omega}{1-\omega^2} + \tan^{-1}\frac{\omega}{1-3\omega^2} - \tan^{-1}\frac{\omega}{10} \; ; A(10) = 1$

16. Given that the maximum overshoot of a second-order feedback control system is 25%, find the corresponding damping ratio ξ and the resonance peak M_r.

17. The open-loop transfer function of the system is

$$G(s)H(s) = \frac{0.5s + 1}{s(s+1)(0.2s+1)}$$

1. Draw an asymptote of the open-loop logarithmic amplitude-frequency characteristics and determine the intersection point at ω;
2. To modify asymptote, determine the system the accurate curve of corner frequency ω_c; and
3. Discuss the circumstances under which the asymptote must be modified.

MCQ AND TRUE/FALSE QUESTIONS

1. The steady-state output of a transfer function system can be obtained directly from the sinusoidal transfer function—that is, the transfer function in which s is replaced by $-j\omega$, where ω is frequency.

   Answer: True/False

2. If the input $x(t)$ is a sinusoidal signal, the steady-state output will also be a sinusoidal signal of the same frequency, but with possibly different magnitude only.
   Answer: True/False

3. The frequency response can be calculated by replacing s in the transfer function by $j\omega$. It will also be shown that the steady-state response can be given by this relationship.

$$G(j\omega)X(s) = Me^{j\varphi} = M\angle\varphi$$

   Answer: True/False

4. The steady-state response of an unstable, linear, time-invariant system to a sinusoidal input does not depend on the initial conditions.
   Answer: True/False

5. A network that has phase-lead characteristics is called a lead network, while a network that has phase-lag characteristics is called a lag network.
   Answer: True/False

6. For presenting frequency-response characteristics in graphical forms, the sinusoidal transfer function, a complex function of the frequency ω, is characterized by its magnitude and phase angle, with frequency as the parameter.
   Answer: True/False

7. The standard representation of the logarithmic magnitude of G(jω) as

$$20\log_{10}|G(j\omega)| = 20\log|G(j\omega)|\ \text{dB}$$

   Answer: True/False

8. It is not possible to plot the curves right down to zero frequency as $\log 0 = -\infty$.
   Answer: True/False

9. The experimental determination of a transfer function can be made simple if frequency-response data are presented in the form of a Bode diagram.
   Answer: True/False

10. The phase angle of the gain K is zero. The effect of varying the gain K in the transfer function is that it raises or lowers the log-magnitude curve of the transfer function by the corresponding constant amount, but it does not affect the phase curve.
    Answer: True/False

11. The exact phase angle f of the factor $1/(1+j\omega\ T)$ is, $\varphi = -\tan^{-1}\omega T$. At zero frequency, the phase angle is 90°. At the corner frequency, the phase angle is $\varphi = -\tan^{-1}\frac{T}{T} = -45°$
    Answer: True/False

12. At infinity, the phase angle becomes –90°. Since the phase angle is given by an inverse-tangent function, the phase angle is skew symmetric about the inflection point at $\varnothing = -45°$.
   Answer: True/False

13. If $0.7 < \xi < 1$, the step response is oscillatory, but the oscillations are well damped and are hardly perceptible.
   Answer: True/False

14. For $0 < \xi \le 0.707$, the resonant frequency ω_r is less than the damped natural frequency $\omega_d = \omega_n \sqrt{1-\xi^2}$, which is exhibited in the transient response.
   Answer: True/False

15. From $\omega = \omega_n \sqrt{1-2\xi^2}$, it can be seen that for $\xi > 0.707$, there is no resonant peak. The magnitude decreases monotonically with increasing frequency ξ. The magnitude is greater than 0 dB for all values of $\omega > 0$.
   Answer: True/False

16. The polar plot of a sinusoidal transfer function $G(j\omega)$ is a plot of the magnitude of $G(j\omega)$ versus the phase angle of $G(j\omega)$ on polar coordinates as ω is varied from zero to infinity.
   Answer: True/False

17. The polar plot is often called the Nyquist plot.
   Answer: True/False

18. An advantage in using a polar plot is that it depicts the frequency-response characteristics of a system over the entire frequency range in a single plot.
   Answer: True/False

19. One disadvantage is that the plot does not clearly indicate the contributions of each individual factor of the open-loop transfer function.
   Answer: True/False

20. The polar plot of this transfer function is a semicircle as the frequency ω is varied from zero to infinity,
   Answer: True/False

21. The lower semicircle corresponds to $0 \le \omega \le -\infty$, and the upper semicircle corresponds to $\infty \le \omega \le 0$.
   Answer: True/False

22. The exact shape of a polar plot depends on the value of the damping ratio ξ, but the general shape of the plot is the same for both the underdamped case ($1 > \xi > 0$) and the overdamped case ($\xi < 1$).
   Answer: True/False

23. The Nyquist stability criterion determines the stability of a closed-loop system from its open-loop frequency response and open-loop poles.
   Answer: True/False

24. The Nyquist stability criterion relates the open-loop frequency response $G(j\omega)H(j\omega)$ to the number of zeros & poles of $1+G(s)H(s)$ that lie in right-half s-plane.
    Answer: True/False

25. If $s\to\infty$, Limit of $G(s)H(s)=1$ or constant, for any physically realizable system.
    Answer: True/False

26. The Nyquist stability criterion can be expressed as, $Z=N+P$ Here, Z is the No. of zeros of $1+G(s)H(s)$ in right-half s-plane; N is the No. of clockwise encirclements of $-1+j0$ point; and P is the No. of poles of $G(s)H(s)$ in right-half s-plane.
    Answer: True/False

27. For stability, there is no encirclement of $-1+j0$ point by $G(j\omega)H(j\omega)$ locus; and it is not necessary to consider locus for entire $j\omega$ axis for positive-frequency portion.
    Answer: True/False

28. Stability can be determined by $-1+j0$ point is enclosed by Nyquist plot of $G(j\omega)H(j\omega)$; For stability, $-1+j0$ point must lie outside shaded region.
    Answer: True/False

29. If the locus of $G(j\omega)H(j\omega)$ passes through $-1+j0$ point, then zeros are located on the $j\omega$ axis; then we can see as, this is not desirable for practical control systems; For a well-designed closed-loop system, none of the roots of the characteristic equation should lie on $j\omega$ axis.
    Answer: True/False

30. For a large value of the gain K, the system is unstable. As the gain K is decreased to a certain value, the $G(j\omega)$ locus passes through the $-1+j0$ point. This means that with this gain value, the system is on the verge of instability, and the system will exhibit sustained oscillations. For a small value of the gain K, the system is stable.
    Answer: True/False

31. The phase margin, $\gamma=360°+\phi$; where ϕ is called phase angle.
    Answer: True/False

32. To understand phase margin, Gain margin is the reciprocal of magnitude $|G(j\omega)|$ at a frequency at phase angle is, $\phi=-360°$.
    Answer: True/False

33. For a non-minimum-phase system, unstable open-loop stability condition will not be satisfied unless $G(j\omega)$ plot encircles the $-1+j0$ point.
    Answer: True/False

34. For a stable non-minimum-phase system: Have negative phase and gain margins.
    Answer: True/False

35. To understand phase margin and gain margin of first- and second-order systems, the gain margin of a first- or second-order system is infinite since the polar plots for such

systems do not cross the negative real axis. Thus, theoretically, first- or second-order systems cannot be unstable.

Answer: True/False

36. The first- or second-order systems are only approximations in the sense that small time lags are neglected in deriving the system equations and are thus not truly first- or second-order systems. If these small lags are accounted for, the so-called first- or second-order systems may become unstable.

    Answer: True/False

37. Proper phase and gain margins ensure us against variations in the system components and are specified for definite positive values. The two values bound the behavior of the closed-loop system near the resonant frequency ω_r. For satisfactory performance, the phase margin (γ) should be between 30° and 90°, and the gain margin (kg) should be greater than 6 dB.

    Answer: True/False

38. If the slope is −40 dB/decade, the system could be either stable or unstable. Even if the system is stable, however, the phase margin is small.

    Answer: True/False

39. If the slope at the gain crossover frequency is −40 dB/decade or steeper, the system is most likely unstable.

    Answer: True/False

40. Referring to this Figure, the frequency ω_r at which the magnitude of the closed-loop frequency response is 3 dB below its zero-frequency value is called the cutoff frequency.

    Answer: True/False

41. The closed-loop system filters out the signal components whose frequencies are greater than the cutoff frequency and transmits those signal components with frequencies lower than the cutoff frequency.

    Answer: True/False

42. The frequency range $0 \leq \omega \leq \omega_b$ in which the magnitude $C(j\omega)/R(j\omega)$ is greater than -2 dB is called the bandwidth (BW) of the system. The bandwidth (BW) indicates the frequency where the gain starts to fall off from its low-frequency value.

    Answer: True/False

43. The ability to reproduce the input signal. A large bandwidth corresponds to a small rise time (t_r), or fast response. Roughly speaking, we can say that the bandwidth is proportional to the speed of response. For example, to decrease the rise time in the step response by a factor of 10, the bandwidth must be increased by approximately a factor of 2.

    Answer: True/False

44. It is seen that as M becomes larger compared with 1, the M circles become smaller and converge to the $-1+j0$ point. For $M>1$, the centers of the M circles lie to the left of the $-1+j0$ point.

    Answer: True/False

45. Similarly, as M becomes smaller compared with 1, the M circle becomes smaller and converges to the origin. For $0<M<1$, the centers of the M circles lie to the left of the origin.

    Answer: True/False

46. $M=1$ corresponds to the locus of points equidistant from the origin and the $-1+j0$ point. As stated earlier, it is a straight line passing through the point and parallel to the imaginary axis.

    Answer: True/False

47. The constant M circles corresponding to $M<1$ lie to the left of the $M=1$ line, and those corresponding to $0<M<1$ lie to the right of the $M=1$ line.

    Answer: True/False

48. The M circles are symmetrical with respect to the straight line corresponding to $M=1$ and with respect to the real axis.

    Answer: True/False

49. Constant N locu for a given value of α is actually not an entire circle, but it is only an arc; the $\alpha=30°$ & $\alpha=-150°$ arcs are part(s) of the same circle; because the tangent of an angle remains same if $\pm180°$ (or multiples) is added to original angle.

    Answer: True/False

50. The use of the M and N circles enables us to find the entire closed-loop frequency response from the open-loop frequency response $G(j\omega)$ without calculating the magnitude and phase of the closed-loop transfer function at each frequency. The intersections of the $G(j\omega)$ locus and the M circles and N circles give the values of M and N at frequency points on the $G(j\omega)$ locus.

    Answer: True/False

51. The N circles are multivalued in the sense that the circle for $\alpha=\alpha_1$ and that for $\alpha=\alpha_1\pm180°\times n$; ($n=1, 2, \ldots$) are the same.

    Answer: True/False

52. In using the N circles for the determination of the phase angle of closed-loop systems, we must interpret the proper value of α. To avoid any error, start at zero frequency, which corresponds to $\alpha = 0°$ and proceed to higher frequencies. The phase-angle curve must be continuous.

    Answer: True/False

CHAPTER 7

Nonlinearity in Control Systems

NONLINEAR AND LINEAR SYSTEMS have essential differences as follows:

1. Mathematical models describe the motional variation state of a system. Linear systems are described by linear differential equations, whereas nonlinear systems are described by nonlinear differential equations.
2. Linear systems can be used according to the superposition principle, whereas nonlinear systems cannot be used according to the superposition principle.
3. The stability of a linear system depends only on the structure and parameters of the system; it is independent of the initial conditions. The stability of a nonlinear system depends not only on the structure and parameters of the system but also on the initial conditions.

7.1 TYPICAL NONLINEAR SYSTEM

Typical nonlinear properties are as follows: insensitive zone (dead zone), saturation, clearance, relay, etc.

7.1.1 Insensitive Zone (Dead Zone)

When the input $|x_i|$ change between − and +, the system has no output. When the input $|x_i| \geq \Delta$, the output changes linearly, as shown in Figure 7.1. The dead zone of the system consists of the dead zone of the measuring element, the dead zone of the amplifier, the dead zone of the actuator, etc.

For example, as shown in Figure 7.2, The negative-opening valve-controlled hydraulic cylinder has a dead-zone that has output characteristic because the port of the valve has a positive amount of overlap Δ, only when the input x_i is more than the positive amount of overlap Δ. The hydraulic cylinder starts to move, and its output characteristic is expressed as

$$x_o = \begin{cases} 0 & ;|x_i| \leq \Delta \\ \tan \alpha(x_i - \Delta \sin x_i) & ;|x_i| > \Delta \end{cases} \tag{7.1}$$

DOI: 10.1201/9781003293859-7

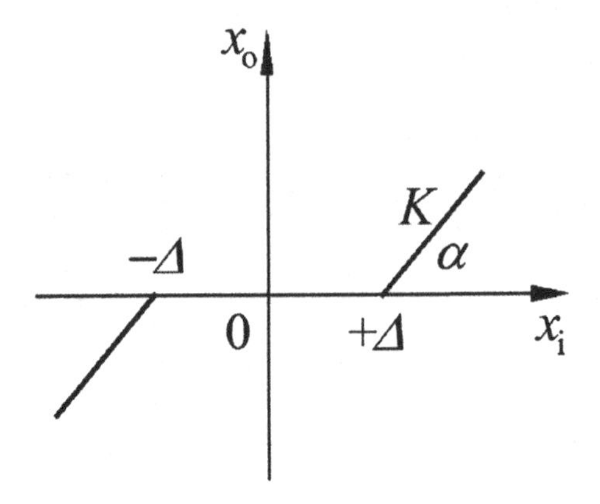

FIGURE 7.1 Insensitive zone (dead zone).

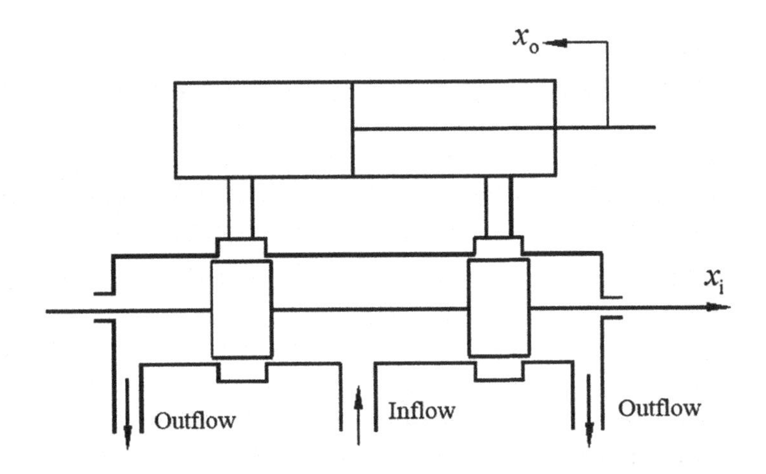

FIGURE 7.2 Hydraulic cylinder system.

7.1.2 Saturation Characteristic

When the input $|x_i|$ changes between−and +, the output arises linearly. When the input $|x_i| \geq \Delta$, the output doesn't change, as shown in Figure 7.3a. For example, in the operational amplifier, as shown in Figure 7.3b, we assume that its amplification factor is 10. The amplifier provides the power of −15 V to +15 V. Thus, when the input is more than ±1.5 V, the output can only be up to ±15 V, which is shown in the saturated state.

However, as shown in Figure 7.2, when the displacement of the spool x_i is greater than the width of the valve port Δ and the spool is moved again, the output of the hydraulic cylinder will not change. The input and output characteristics of such systems are

$$x_i = \begin{cases} K_a & ;x_i > \Delta \\ Kx_i & ;|x_i| \leq \Delta \\ -K_a & ;x_i < -\Delta \end{cases} \tag{7.2}$$

where $K = \tan\alpha$.

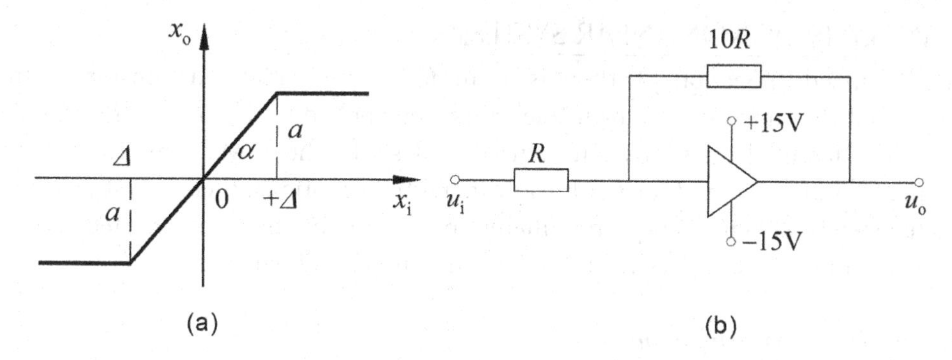

FIGURE 7.3 (a) Saturation characteristics and (b) operational amplifier.

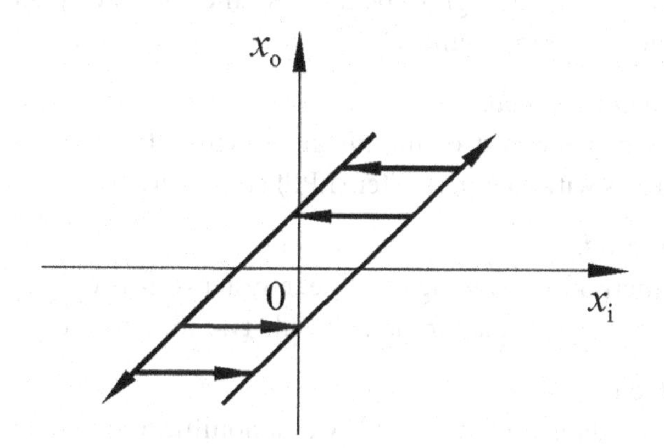

FIGURE 7.4 The gap or clearance characteristics.

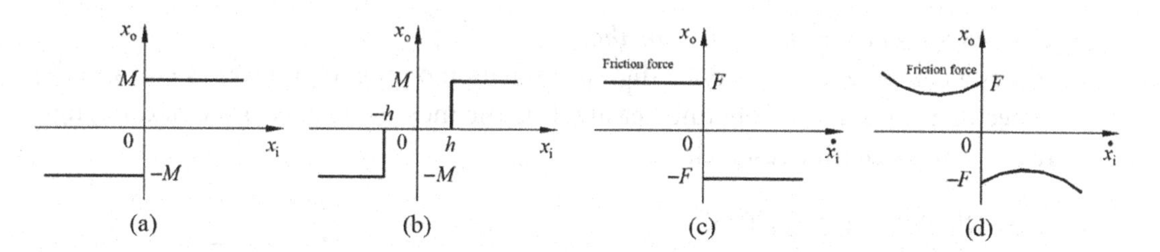

FIGURE 7.5 Relay characteristics.

7.1.3 The Gap or Clearance Characteristics

As shown in Figure 7.4, the gap or clearance characteristics is the common factor of mechanical systems, which mainly consists of thread screw transmission pair, gear transmission pair, and other factors.

7.1.4 Relay Characteristics

The relay characteristics are shown in Figure 7.5, which can be divided into two and three positions, as shown in Figure 7.5a and b. It is composed of two-position and three-position relays. In addition, the friction force can be approximated as the coulomb friction force in mechanical motion, which is a function of the velocity of motion, as shown in Figure 7.5c. The actual friction of the moving pair is shown in Figure 7.5d.

7.2 ANALYSIS OF NONLINEAR SYSTEMS

As we discussed in Section 7.1, there is no uniform differential equation for nonlinear systems, and the superposition principle cannot be applied to linear systems; thus, only nonlinear differential equations can be used to describe them. Mathematically, there is no unified solution method for nonlinear differential equations; therefore, it needs to be perfected step by step in theory, and different processing methods are selected according to the needs of engineering. Here are some common methods such as

1. *Linearized approximation:*
   This method has been discussed in the previous chapter. It requires that the nonlinear factors are small enough to be ignored and that the system must make small changes around a working point.
2. *Describing function method:*
   It is the frequency method of nonlinear systems. It is applicable to all kinds of nonlinear systems with low-pass filter (LPF) characteristics.
3. *Phase plane method*:
   It is a planar method for solving nonlinear systems. It is only applicable to second-order systems because its planar plane is only two dimensional.
4. *Liapunov method*:
   It is a method to determine the stability of a nonlinear system by using the generalized energy concept. In theory, it can be applied to all nonlinear systems, but in practice, it is difficult to find the Liapunov function.
5. *Computer numerical simulation method*:
   It uses a computer to carry out a digital simulation on the nonlinear system, and the expected results can be obtained easily. It is the most widely used method in engineering and scientific research.

7.3 DESCRIBING FUNCTION

Assuming that the input function of the nonlinear element is a sinusoidal signal, generally speaking, the output of the nonlinear element is not a sinusoidal signal, but often changes periodically, and the period of the input and output signals is the same (the output signal contains the higher-order harmonic component in addition to the first harmonic component). In the analysis of descriptive functions for nonlinear systems, it is assumed that only the first harmonic component of the output is significant. Because the amplitude of the higher harmonics in the output of the nonlinear element is usually smaller than that of the first harmonic component and because most control systems are low-pass filters (LPF), the higher harmonic component decodes much faster than the first harmonic component. The description function of the nonlinear element is defined as the ratio of the first harmonic component of the output to the complex number of the input, i.e.,

$$N = \frac{Y_1}{X} \angle \varnothing_1 \tag{7.3}$$

where N is the describing function, X is the amplitude of input sinusoidal signal, Y is the amplitude of the first harmonic component of the output, and $\varnothing_1$ is the phase shift of the first harmonic component of the output.

It has been stated previously that the differential equations of each nonlinear element are different, so there is no uniform description function. For each nonlinear element, a description function is derived. To derive the description function, we first need to find the first harmonic component of the output. When the input signal of the nonlinear element is sinusoidal input or $x(t) = X\sin t$, the output $y(t)$ can be expanded into Fourier series as

$$y(t) = A_0 + \sum_{n=1}^{\infty}(A_n \cos n\omega t + B_n \sin n\omega t)$$

$$= A_0 + \sum_{n=1}^{\infty} Y_n \sin(n\omega t + \phi_n) \tag{7.4}$$

From Eq. (7.4), we get

$$A_n = \frac{1}{\pi}\int_0^{2\pi} y(t)\cos\omega t\, d\omega t$$

$$B_n = \frac{1}{\pi}\int_0^{2\pi} y(t)\sin\omega t\, d\omega t$$

$$Y_n = \sqrt{A_n^2 + B_n^2}$$

$$\varphi_n = \tan^{-1}\frac{A_n}{B_n}$$

If the nonlinear function is an odd function, $A_0 = 0$, and the output of the first harmonic component is expressed as

$$y_1(t) = A_1 \cos\omega t + B_1 \sin\omega t = Y_1 \sin(\omega t + \phi_1) \tag{7.5}$$

Then we attain the describing function of output is

$$N = \frac{Y_1}{X}\angle\varphi_1 = \frac{\sqrt{A_1^2 + B_1^2}}{X}\angle\tan^{-1}\frac{A_1}{B_1} \tag{7.6}$$

It is obvious that when $\varnothing_1 \neq 0$, the description function N is a complex number.

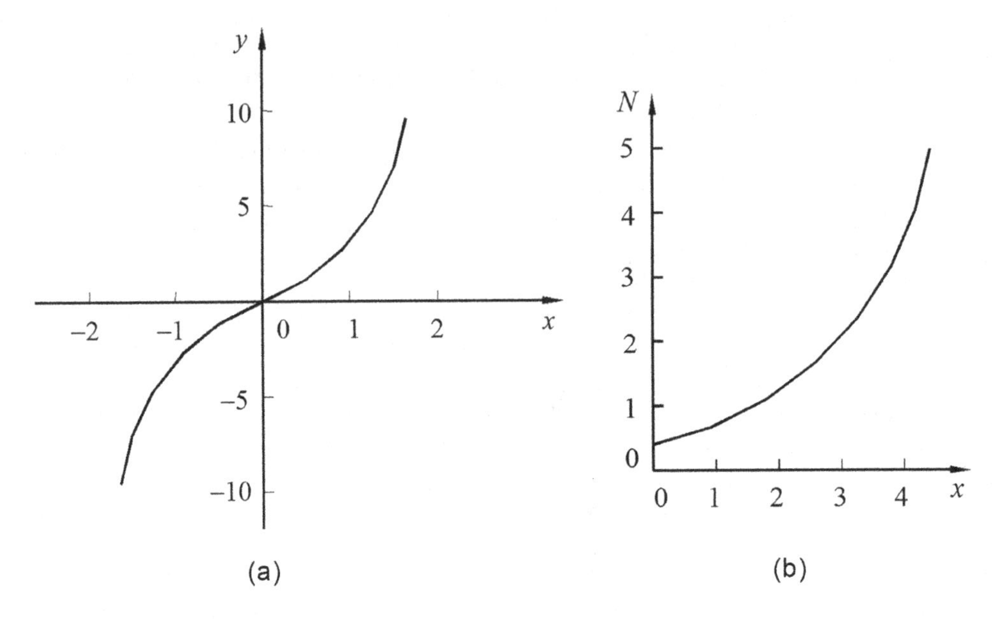

FIGURE 7.6 Input and output characteristics of Example 7.1.

Example 7.1

We assume that there is a nonlinear component characteristic in Figure 7.6 as

$$y(t) = \frac{1}{2}x + \frac{1}{4}x^3 \tag{7.7}$$

Attain the describing function of the nonlinear components.

Solution

According to the nonlinear characteristic of the components, it is a single-valued function. Thus, $A_1 = 0$, $\phi_1 = 0$

$$B_1 = \frac{1}{\pi}\int_0^{2\pi} y(t)\sin\omega t\, d\omega t = \frac{1}{\pi}\int_0^{2\pi}\left(\frac{1}{2}x + \frac{1}{4}x^3\right)\sin \omega t d\omega t$$

And substituting $x = X\sin t$, we obtain

$$B_1 = \frac{1}{\pi}\int_0^{2\pi}\left(\frac{1}{2}X\sin\omega t + \frac{1}{4}X^3\sin^3\omega t\right)\sin\omega t d\,\omega t$$

$$= \frac{1}{2\pi}\times 2X\int_0^{\pi}\left(\sin^2\omega t + \frac{X^2}{2}\sin^4\omega t\right)d\omega t$$

$$= \frac{1}{2}X + \frac{3}{16}X^3$$

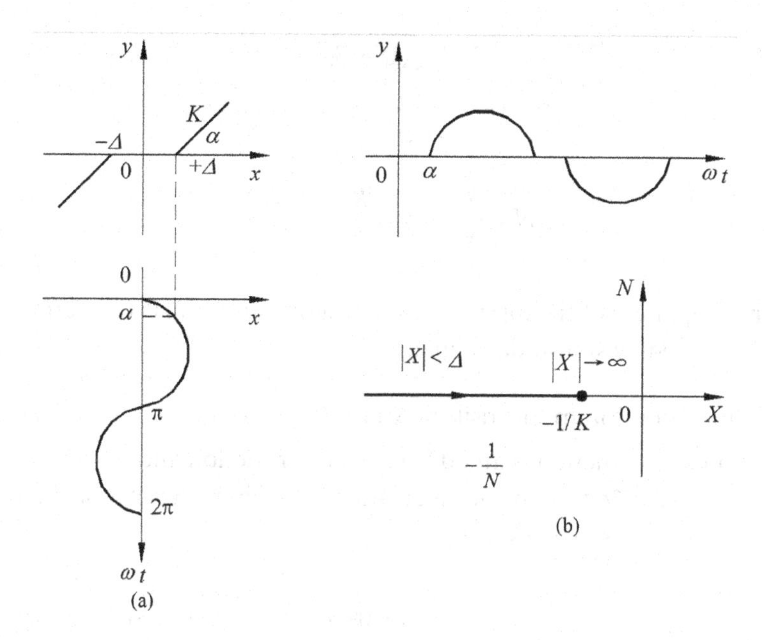

FIGURE 7.7 a–b The input and output of dead zone and the tracking of $-\frac{1}{N}$.

Thus, the describing function of the system is

$$N = \frac{B_1}{X} = \frac{1}{2} + \frac{3}{16}X^2$$

And we can obtain the first harmonic component of output is

$$y_1(t) = \left(\frac{1}{2} + \frac{3}{16}X^2\right) X \sin \omega t \tag{7.8}$$

According to Eq. (7.8), for the first harmonic of the output of the nonlinear element, the nonlinear factor is equivalent to an amplification factor whose gain is determined by the input amplitude. The nonlinear function expressed by the descriptive function is to replace the original nonlinear characteristic with a family of straight lines whose slope changes with the input amplitude in the sense of the first harmonic. For the nonlinear characteristics of Eq. (7.7), the relation between the slope and input amplitude of this family of lines is shown in Figure 7.7.

7.4 COMMON DESCRIBING FUNCTION OF NONLINEAR ELEMENT

There is some commonly used equation in deducing common describing function of a nonlinear element

$$A_1 = \frac{1}{\pi}\int_0^{2\pi} y(t)\cos \omega t \, d\omega t \tag{7.9}$$

$$B_1 = \frac{1}{\pi}\int_0^{2\pi} y(t)\sin \omega t \, d\omega t \tag{7.10}$$

$$Y_1 = \sqrt{A_1^2 + B_1^2} \tag{7.11}$$

$$\varphi_1 = \tan^{-1}\frac{A_1}{B_1} \tag{7.12}$$

$$N = \frac{Y_1}{X_1} < \phi_1 = \frac{B_1}{X} + \frac{A_1}{X} j \tag{7.13}$$

where X is the amplitude of the input sinusoidal signal and $y(t)$ is the output of nonlinear characteristic affected by sinusoidal signal.

7.4.1 Describing Function in Insensitive Area (Dead Zone)

Because the dead zone function is an odd symmetric function and $A_0=0$, $A_1=0$, $\phi_1=0$, we should only consider $0 \le \omega t \le \pi$ in the calculation. The characteristic of the input and the output in the dead zone is expressed as

$$y(t) = \begin{cases} 0 & ;\text{for } 0 < \omega t < \alpha \quad \text{and } \pi - \alpha < \omega t \le \pi \\ K(X \sin\omega t - \Delta) & ;\text{for } \alpha \le \omega t \le \pi - \alpha \end{cases} \tag{7.14}$$

Here, $\alpha = \sin^{-1}\frac{\Delta}{X}$.

Thus,

$$B_1 = \frac{1}{\pi}\int_0^{2\pi} y(t)\sin\omega t d\omega t = \frac{4}{\pi}\int_0^{\frac{\pi}{2}} y(t)\sin\omega t d\omega t = \frac{4}{\pi}\int_\alpha^{\frac{\pi}{2}} K(X\sin\omega t - \Delta)\sin\omega t d\omega t$$

$$= \frac{2KX}{\pi}\left[\frac{\pi}{2} - \sin^{-1}\left(\frac{\Delta}{X} - \frac{\Delta}{X}\sqrt{1-\left(\frac{\Delta}{X}\right)^2}\right)\right] \quad ;\text{ for } (X \ge \Delta)$$

Here, $X \ge \Delta$; and

$$N = \frac{Y_1}{X}\angle\varphi_1 = \frac{B_1}{X}\angle 0° = K - \frac{2K}{\pi}\left[\sin^{-1}\left(\frac{\Delta}{X} + \frac{\Delta}{X}\sqrt{1-\left(\frac{\Delta}{X}\right)^2}\right)\right] \tag{7.15}$$

Consequently, the describing function in the insensitive area (dead zone) is expressed as

$$N = \begin{cases} K - \frac{2K}{\pi}\left[\sin^{-1}\left(\frac{\Delta}{X} + \frac{\Delta}{X}\sqrt{1-\left(\frac{\Delta}{X}\right)^2}\right)\right] & ;\text{for } X \ge \Delta \\ 0 & ;\text{ for } X < \Delta \end{cases} \tag{7.16}$$

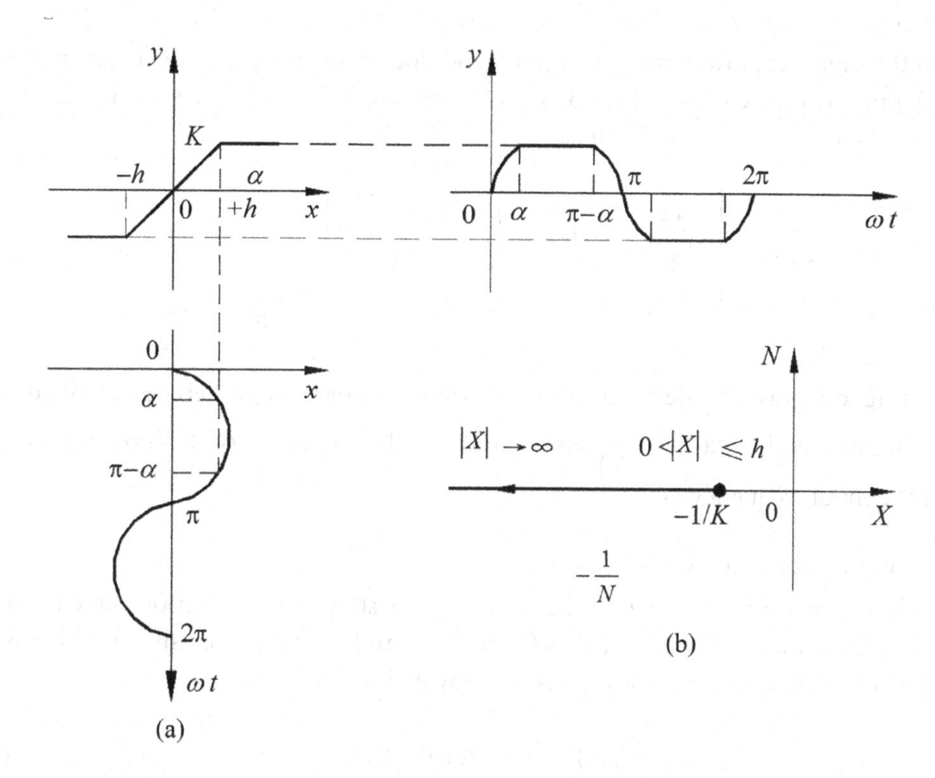

FIGURE 7.8 a–b Input and output waveform of saturation characteristics and $-\frac{1}{N}$ locus.

7.4.2 Describing Function of Saturation Characteristics

As shown in Figure 7.8, it is the saturation characteristics and its output waveform under the influence of a sinusoidal signal. According to the curve, when $X<h$, the system is totally linear. When $X\geq h$, the system is nonlinear.

Since the saturation characteristic is an odd function, then we can say $A_0=0,\ A_1=0,\ \varphi_1=0,\ \alpha=\sin^{-1}\frac{h}{x}$ in $0\leq\omega t\leq\pi$, the relationship between input and output of saturation characteristic is defined as

$$y(t)=\begin{cases} KX\sin\omega t & ;\text{for } 0\leq\omega t<\alpha \text{ and } -\alpha<\omega t\leq\pi \\ Kh & ;\text{for } \alpha\leq\omega t\leq\pi-\alpha \end{cases} \tag{7.17}$$

Thus,

$$B_1=\frac{4}{\pi}\int_0^{\frac{\pi}{2}} y(t)\sin\omega t d\omega t=\frac{4}{\pi}\left[\int_0^{\alpha} KX\sin^2\omega t d\omega t+\int_{\alpha}^{\frac{\pi}{2}} Kh\sin\omega t d\omega t\right]$$

$$=\frac{2KX}{\pi}\left[\sin^{-1}\left(\frac{h}{X}+\frac{h}{X}\sqrt{1-\left(\frac{h}{X}\right)^2}\right)\right] \qquad ;\text{ for } X>h$$

When the input amplitude is small enough and does not exceed the linear region, the factor is a proportional coefficient factor with the ratio of K. Thus, the describing function of saturation characteristic is defined as

$$N(X)=\begin{cases} \dfrac{2K}{\pi}\left[\sin^{-1}\left(\dfrac{h}{X}+\dfrac{h}{X}\sqrt{1-\left(\dfrac{h}{X}\right)^2}\right)\right] & ;\text{ for } X>h \\ K & ;\text{ for } X\leq h \end{cases} \tag{7.18}$$

According to Figure 7.8, the description function of saturation property is obtained $N(X)$ as the negative reciprocal $-\dfrac{1}{N(X)}$ associated with the amplitude of the input signal, but it is not the input frequency ω.

7.4.3 Two-Position Relay Characteristic

As is shown in Figure 7.9, the nonlinear characteristics of two-position relay is an odd function; thus, we can say $A_1=0$, $\varphi_1=0$, the input and output are satisfied in the range of $0\leq\omega t\leq\pi$ with its characteristics. Thus, we may define it as

$$y(t)=M \quad ;\text{for } 0\leq\omega t\leq\pi \tag{7.19}$$

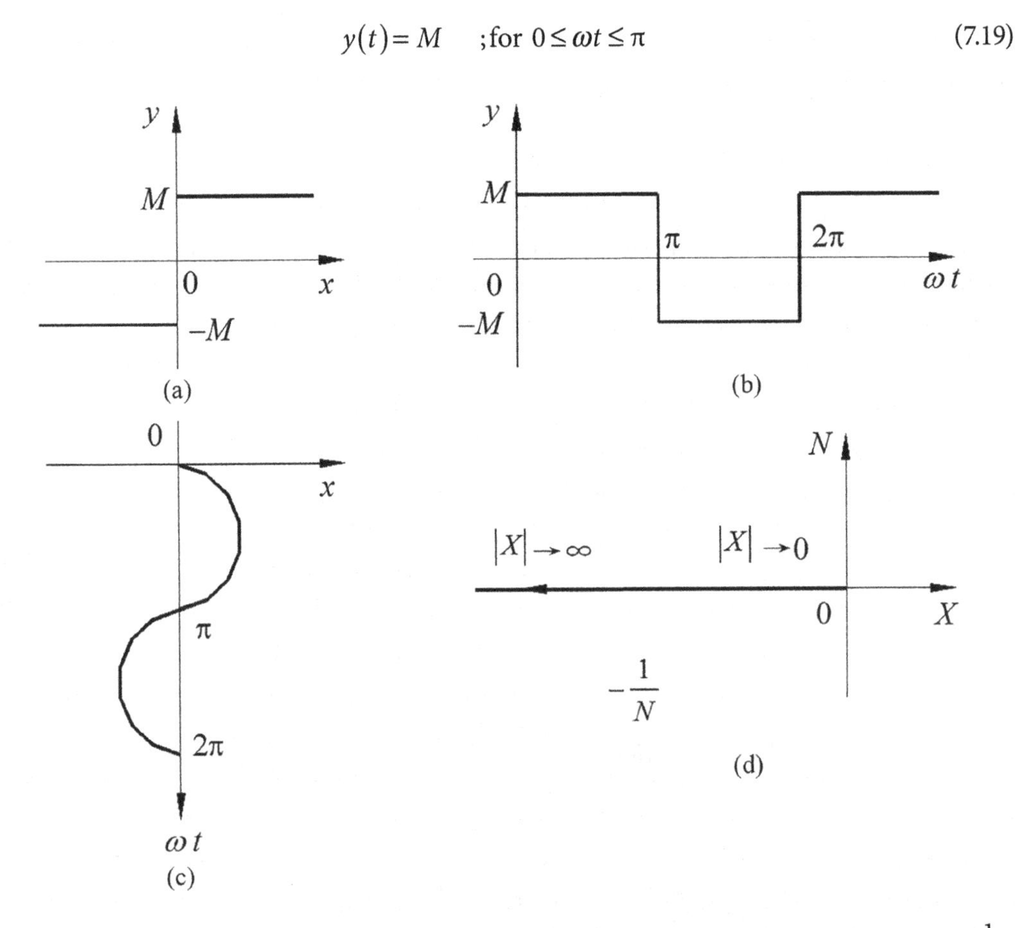

FIGURE 7.9 The input and output characteristics of the two-position relay and the locus of $-\dfrac{1}{N}$.

Therefore, we can say

$$B_1 = \frac{1}{\pi}\int_0^{2\pi} y(t)\sin \omega t d\omega t = \frac{2}{\pi}\int_0^{\pi} M \sin \omega t \, d\omega t = \frac{4M}{\pi}$$

Thus, the describing function of the two-position relay is defined as

$$N = \frac{4M}{\pi X} \tag{7.20}$$

According to Eq. (7.20), we obtain that the describing function of the two-position relay is only the function of input amplitude, but it is not associated with input frequency ω.

7.4.4 Three-Position Relay Characteristic

The input and output characteristics of the three-position relay characteristics are shown in Figure 7.10. According to Figure 7.10, we can get that it is an odd function; thus, when $A_1 = 0$, $\varphi_1 = 0$ is in the range of $0 \le \omega t \le \pi$, and then we can define as

$$y(t) = \begin{cases} M & ;\text{for } t < -\alpha \\ 0 & ;\text{for } t < \alpha \text{ and } -\alpha \le \omega t \le \pi \end{cases} \tag{7.21}$$

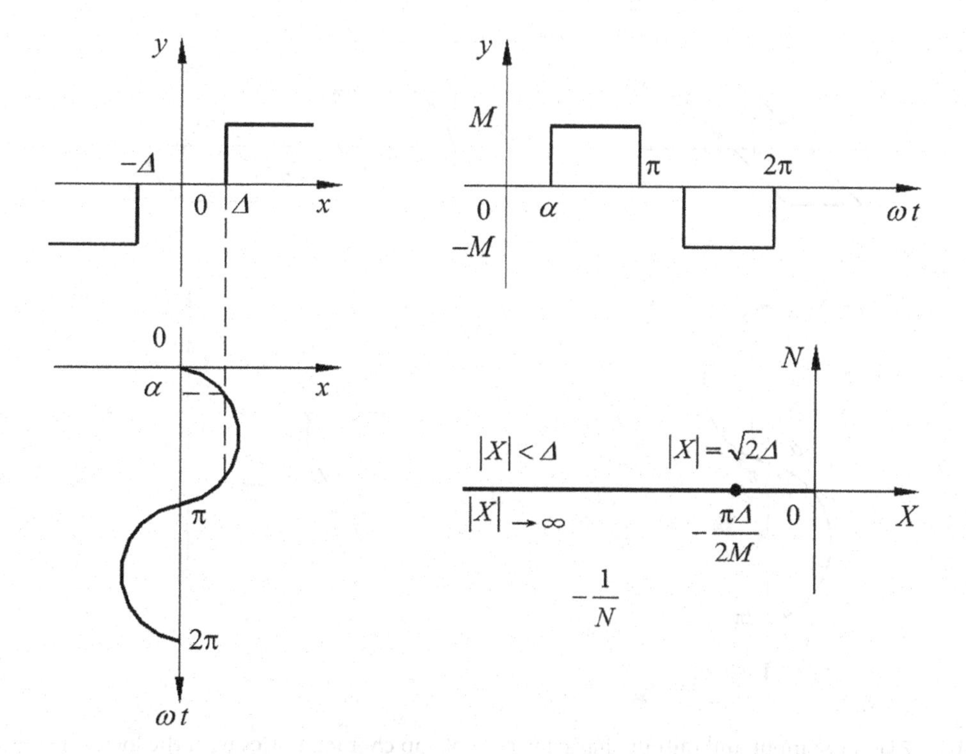

FIGURE 7.10 Input and output of three-position relay characteristics with the locus of $-\frac{1}{N}$.

Thus, we can say

$$B_1 = \frac{1}{\pi}\int_0^{2\pi} y(t)\sin\omega t\, d\omega t = \frac{4}{\pi}\int_\alpha^{\frac{\pi}{2}} M \sin\omega t\, d\omega t = \frac{4M}{\pi}\sqrt{1-\left(\frac{\Delta}{X}\right)^2}$$

where $\alpha = \sin^{-1}\frac{\Delta}{X}$

Therefore, we can obtain that the describing function of the system as

$$N(X) = \begin{cases} \frac{4M}{\pi X}\sqrt{1-\left(\frac{\Delta}{X}\right)^2} & ;\text{for } X \geq \Delta \\ 0 & ;\text{for } X < \Delta \end{cases} \quad (7.22)$$

It is obvious that the describing function is independent of frequency ω, and it is only the function of input amplitude associated with the negative reciprocal of N as $-\frac{1}{N}$ of the describing function which is shown in Figure 7.10.

7.4.5 Describing Function of Gap or Clearance Characteristics

Figure 7.11 shows the input and output characteristic of gap or clearance characteristics in the range of $0 \leq t \leq 2$. It can be defined as

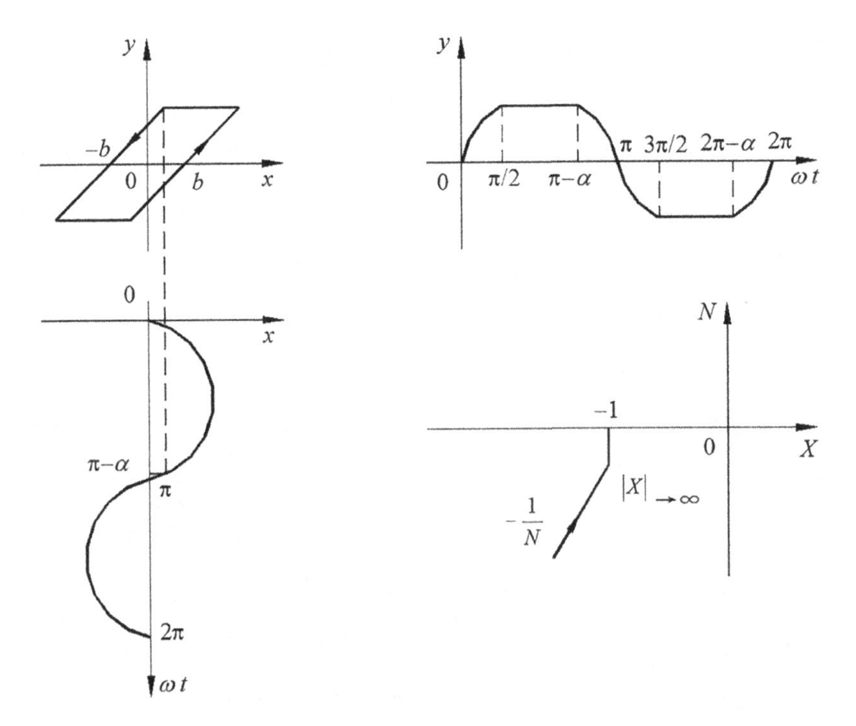

FIGURE 7.11 The input and output characteristics of gap characteristics with the locus of $-\frac{1}{N}$.

$$y(t)=\begin{cases} K(X\sin\omega t-b) & ;0\leq \omega t<\dfrac{\pi}{2} \text{ and } 2\pi-\alpha<\omega t\leq 2\pi \\ K(X-b) & ;\dfrac{\pi}{2}\leq \omega t<\pi-\alpha \\ K(X\sin\omega t+b) & ;\pi-\alpha\leq \omega t<\dfrac{3}{2}\pi \\ -K(X-b) & ;\dfrac{3}{2}\pi\leq \omega t\leq 2\pi-\alpha \end{cases} \tag{7.23}$$

The gap characteristic is a multi-valued function and it is not an odd function nor an even function to the effect of a sinusoidal signal. Therefore, we calculate A_1, and B_1, and it is obvious that $A_0=0$. Thus,

$$A_1=\frac{1}{\pi}\int_0^{2\pi} y(t)\cos\omega t\, d\omega t$$

$$=\frac{1}{\pi}\int_0^{\frac{\pi}{2}} K(X\sin\omega t-b)\cos\omega t d\omega t+\frac{1}{\pi}\int_{\frac{\pi}{2}}^{\pi-\alpha} K(X-b)\cos\omega t d\omega t$$

$$+\frac{1}{\pi}\int_{\pi-\alpha}^{\frac{3}{2}} K(X\sin\omega t+b)\cos\omega t\, d\omega t+\frac{1}{\pi}\int_{\frac{3}{2}\pi}^{2\pi-\alpha} -K(X-b)\cos\omega t d\omega t$$

$$+\frac{1}{\pi}\int_{2\pi-\alpha}^{2\pi} K(X\sin\omega t-b)\cos\omega t\, d\omega t=\frac{4Kb}{\pi}\left(\frac{b}{X}-1\right) \quad ;\text{for } X\geq b$$

and

$$B_1=\frac{1}{\pi}\int_0^{2\pi} y(t)\sin\omega t d\omega t$$

$$=\frac{1}{\pi}\left[\begin{array}{l} \displaystyle\int_0^{\frac{\pi}{2}} K(X\sin\omega t-b)\sin\omega t\, d\omega t+\int_{\frac{\pi}{2}}^{\pi-\alpha} K(X-b)\sin\omega t\, d\omega t \\ \displaystyle=\frac{1}{\pi}\int_0^{\frac{\pi}{2}} K(X\sin\omega t-b)\sin\omega t\, d\omega t+\int_{\frac{\pi}{2}}^{\pi-\alpha} K(X-b)\sin\omega t\, d\omega t \\ \displaystyle+\int_{2\pi-\alpha}^{2\pi} K(X\sin\omega t-b)\sin\omega t\, d\omega t \end{array}\right]$$

$$=\frac{KX}{\pi}\left[\frac{\pi}{2}+\sin^{-1}\left(\frac{X-2b}{X}\right)+2\left(\frac{X-2b}{X}\right)\sqrt{\frac{b}{X^2}(X-b)}\right] \qquad ;X\geq b$$

Therefore, we can obtain that the describing function of the gap characteristic is

$$N(X)=\frac{B_1}{X}+j\frac{A_1}{X}$$
$$=\frac{K}{\pi}\left[\frac{\pi}{2}+\sin^{-1}\left(\frac{X-2b}{X}\right)+2\left(\frac{X-2b}{X}\right)\sqrt{\frac{b}{X^2}(X-b)}\right]+j\frac{4Kb}{\pi X}\left(\frac{b}{X}-1\right)\quad ;X\geq b \tag{7.24}$$

or

$$|N|=\sqrt{\left\{\frac{K}{\pi}\left[\frac{\pi}{2}+\sin^{-1}\left(\frac{X-2b}{X}\right)+2\left(\frac{X-2b}{X}\right)\sqrt{\frac{b}{X^2}(X-b)}\right]\right\}^2\left[\frac{4Kb}{\pi X}\left(\frac{b}{X}-1\right)\right]^2} \tag{7.25}$$

and

$$\varphi_1=\tan^{-1}\left[\frac{4b(b-X)}{X^2\left[\frac{\pi}{2}+\sin^{-1}\left(\frac{X-2b}{X}\right)+2\left(\frac{X-2b}{X}\right)\sqrt{\frac{b}{X^2}(X-b)}\right]}\right] \tag{7.26}$$

According to the describing function in Eqs. (7.24)–(7.26), although the describing function is complicated, we obtain that it changes following its input amplitude, but it is not followed by its frequency ω. The negative reciprocal of the describing function is shown in Figure 7.11.

7.4.6 Describing Function of Relay Characteristics in Insensitive Region (Dead Zone)

Figure 7.12 shows the relay characteristics in the insensitive region (dead zone) with the effect of sinusoidal signal; its output characteristics ($X\geq h$) are defined as

$$y(t)=\begin{cases}0 & ;0\leq\omega t\leq\alpha_1;\ \alpha_2\leq\omega t\leq\pi+\alpha_1;\ 2\pi-\alpha_2\leq\omega t\leq 2\pi;\\ M & \alpha_1\leq\omega t\leq\pi-\alpha_2\\ -M & ;\pi+\alpha_1\leq\omega t\leq 2\pi-\alpha_2\end{cases} \tag{7.27}$$

where

$$\alpha_1=\sin^{-1}\frac{h}{X},\ \alpha_2=\sin^{-1}\frac{mh}{X}$$

According to Figure 7.12, it is obvious that $A_0=0$, and the relay characteristic in an insensitive area with hysteresis is a nonodd and noneven function. Thus, we obtain

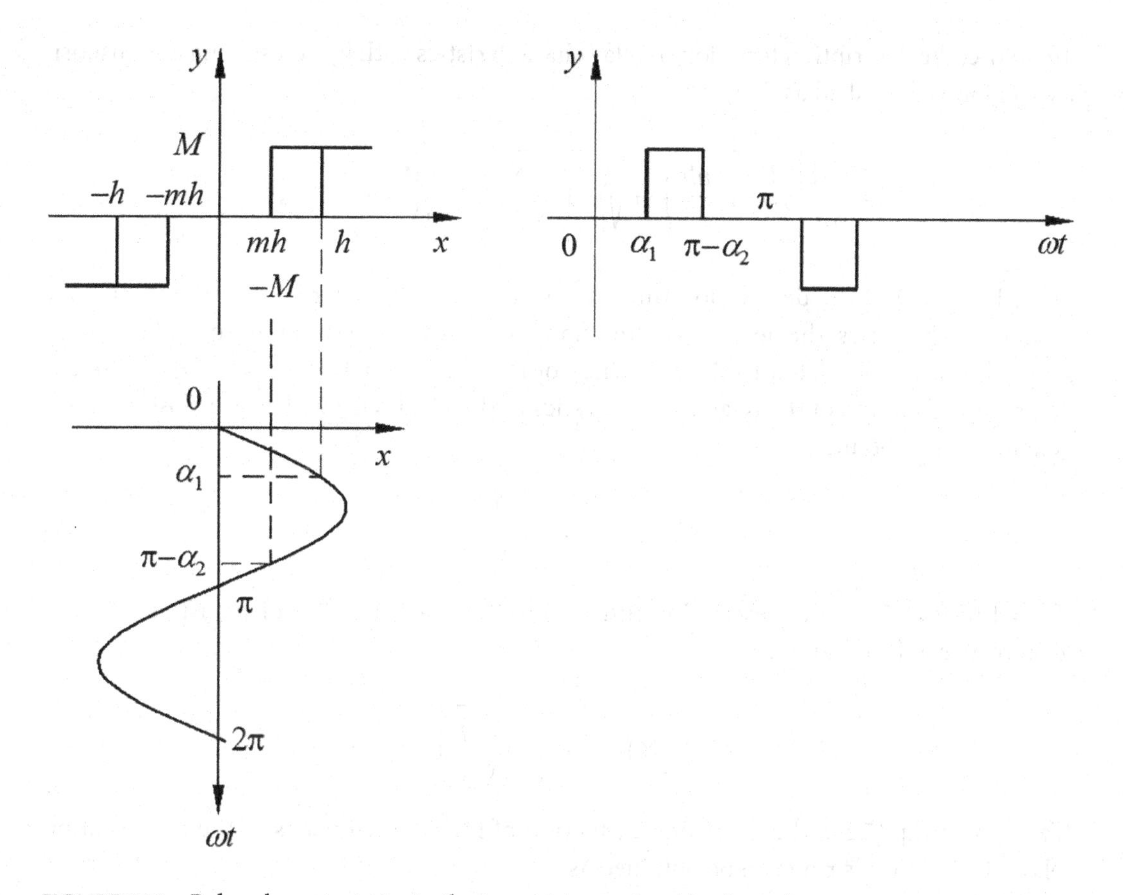

FIGURE 7.12 Relay characteristics in the insensitive region (dead zone).

$$A_1 = \frac{1}{\pi}\int_0^{2\pi} y(t)\cos\omega t\, d\omega t$$

$$= \frac{1}{\pi}\left[\int_{\alpha_1}^{\pi-\alpha_2} M\cos\omega t\, d\omega t - \int_{\pi+\alpha_1}^{2\pi-\alpha_2} M\cos\omega t\, d\omega t\right] = \frac{2Mh}{\pi X}(m-1)$$

and

$$B_1 = \frac{1}{\pi}\int_0^{2\pi} y(t)\sin\omega t d\omega t$$

$$= \frac{1}{\pi}\left[\int_{\alpha_1}^{\pi-\alpha_2} M\sin\omega t d\omega t - \int_{\pi+\alpha_1}^{2\pi-\alpha_2} M\sin\omega t d\omega t\right]$$

$$= \frac{2M}{\pi}\left[\sqrt{1-\left(\frac{mh}{X}\right)^2} + \sqrt{1-\left(\frac{h}{X}\right)^2}\right] \qquad ;\text{for } X \geq h$$

Therefore, the description function of relay characteristics with hysteresis ring and insensitive region is described as

$$N(X)=\frac{2M}{\pi X}\left[\sqrt{1-\left(\frac{mh}{X}\right)^2}+\sqrt{1-\left(\frac{h}{X}\right)^2}\right]+j\frac{2Mh}{\pi X^2}(m-1) \quad ;\text{for } X\geq h \tag{7.28}$$

From Eq. (7.28) of the description function of the relay characteristics in the insensitive region with hysterics, the description function is only related to the amplitude of the input signal, but it is not related to the frequency of the input signal. If $h=0$ in Eq. (7.28), the description function of the relay characteristics of the ideal relay with a two-position relay system can be obtained as

$$N(X)=\frac{4M}{\pi X} \tag{7.29}$$

If $m=1$ in Eq. (7.28), the description function of the characteristics of the three-position ideal relay can be obtained as

$$N(X)=\frac{4M}{\pi X}\sqrt{1-\left(\frac{h}{X}\right)^2} \tag{7.30}$$

If $m=-1$ in Eq. (7.28), the description function of the characteristics of the two-position relay with hysteresis ring can be obtained as

$$N(X)=\frac{4M}{\pi X}\sqrt{1-\left(\frac{h}{X}\right)^2}-j\frac{4Mh}{\pi X} \tag{7.31}$$

From the derivation of the describing functions of the above various nonlinear elements, it is found that if the nonlinear characteristics are single-valued functions, such as ideal relay, insensitive region, and so on, their description functions are only real variable functions of the amplitude of the input function. The nonlinear characteristics are multivalued functions, such as clearance, relay with the insensitive region, two-position relay with hysteresis, etc. Their description functions are complex functions of the amplitude of the input function. Table 7.1 shows the description functions of common nonlinear properties for easy searching.

7.5 STABILITY OF NONLINEAR SYSTEMS

A nonlinear system consists of two parts: (a) a linear part whose transfer function is expressed by $G(s)$ and (b) the nonlinear part. To describe the nonlinear function, it represents by $N(X)$ with its nonlinear characteristics, as shown in Figure 7.13. The function descriptive method is an approximated method, and its practical condition is that the linear part $G(j\omega)$ has the characteristics of low-pass filtering (LPF), and the high-order harmonics generated by the output of the nonlinear part can be fully attenuated. It is very

TABLE 7.1 Nonlinear Characteristics and Its Describing Functions

Nonlinear Characteristic	Describing Function
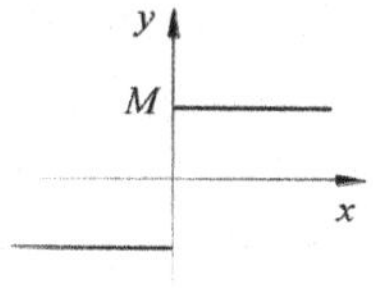 y, M, x	$\dfrac{4M}{\pi X}$
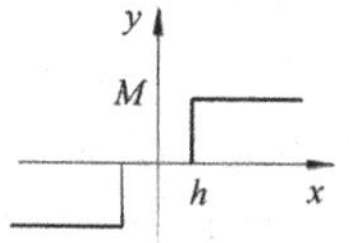 y, M, h, x	$\dfrac{4M}{\pi X}\sqrt{1-\left(\dfrac{h}{X}\right)^2} \quad ; X \ge h$
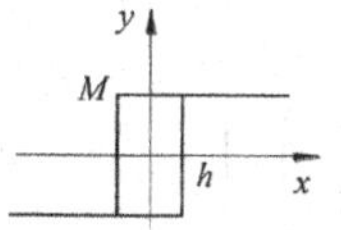 y, M, h, x	$\dfrac{4M}{\pi X}\sqrt{1-\left(\dfrac{h}{X}\right)^2} - j\dfrac{4Mh}{\pi X} \quad ; X \ge h$
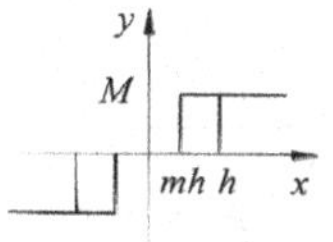 y, M, mh h, x	$\dfrac{2M}{\pi X}\left[\sqrt{1-\left(\dfrac{mh}{X}\right)^2}+\sqrt{1-\left(\dfrac{h}{X}\right)^2}\right] + j\dfrac{2mh}{\pi X^2}(m-1) \ ;X \ge h$
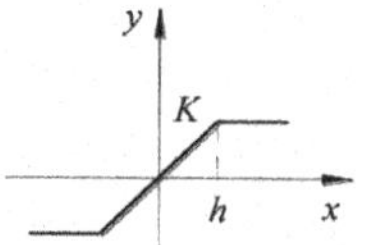 y, K, h, x	$\begin{cases} \dfrac{2K}{\pi}\left[\sin^{-1}\left(\dfrac{h}{X}+\dfrac{h}{X}\sqrt{1-\left(\dfrac{h}{X}\right)^2}\right)\right] & ;X > h \\ K & ;X \le h \end{cases}$
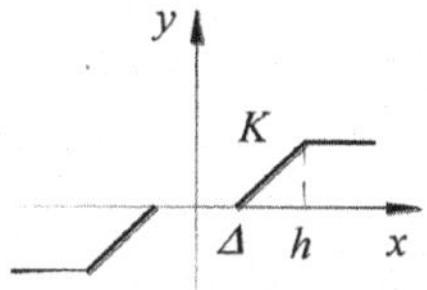 y, K, Δ h, x	$\dfrac{2K}{\pi}\left[\begin{matrix}\sin^{-1}\dfrac{h}{X}-\sin^{-1}\dfrac{\Delta}{X}+\dfrac{h}{X}\sqrt{1-\left(\dfrac{h}{X}\right)^2} \\ -\dfrac{\Delta}{X}\sqrt{1-\left(\dfrac{\Delta}{X}\right)^2}\end{matrix}\right] 1\,X \ge h$
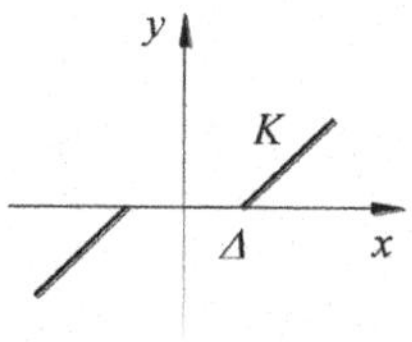 y, K, Δ, x	$\dfrac{2K}{\pi}\left[\begin{matrix}\dfrac{\pi}{2}-\sin^{-1}\dfrac{\Delta}{X}- \\ \dfrac{\Delta}{X}\sqrt{1-\left(\dfrac{\Delta}{X}\right)^2}\end{matrix}\right] ;X \ge \Delta$
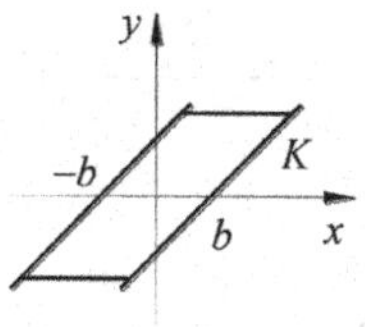 y, −b, K, b, x	$\dfrac{K}{\pi}\left[\begin{matrix}\dfrac{\pi}{2}+\sin^{-1}\left(\dfrac{X-2b}{X}\right)+ \\ 2\left(\dfrac{X-2b}{X}\right)\sqrt{\dfrac{b}{X^2}(X-b)}\end{matrix}\right] + j\dfrac{4Kb}{\pi X}\left(\dfrac{b}{X}-1\right);X \ge b$

(*Continued*)

TABLE 7.1 (*Continued*) Nonlinear Characteristics and Its Describing Functions

Nonlinear Characteristic	Describing Function
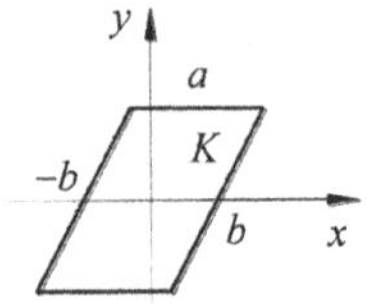 Hysteresis loop with a, $-b$, b, slope K	$\frac{K}{\pi}\left[\sin^{-1}\frac{a+Kb}{KX}+\sin^{-1}\frac{a-Kb}{KX}+\frac{a+Kb}{KX}\sqrt{1-\left(\frac{a+Kb}{KX}\right)^2}+\frac{a-Kb}{KX}\sqrt{1-\left(\frac{a-Kb}{KX}\right)^2}\right]-j\frac{4ab}{\pi X^2};X\geq\frac{a+Kb}{K}$
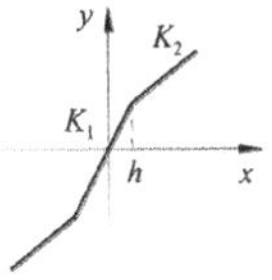 Piecewise linear with slopes K_1, K_2, breakpoint h	$K_2+\frac{2(K_1-K_2)}{\pi}\left[\sin^{-1}\frac{h}{X}+\frac{h}{X}\sqrt{1-\left(\frac{h}{X}\right)^2}\right];\quad X\geq h$
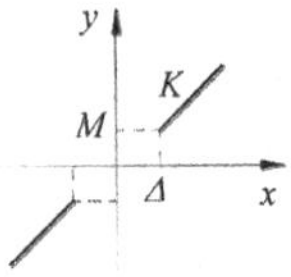 Slope K with M, Δ	$K-\frac{2K}{\pi}\sin^{-1}\frac{\Delta}{X}+\frac{4M-2K\Delta}{\pi X}\sqrt{1-\left(\frac{\Delta}{X}\right)^2};X\geq\Delta$
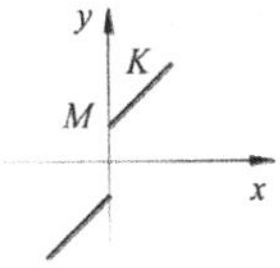 Slope K with offset M	$K+\frac{4M}{\pi X}$
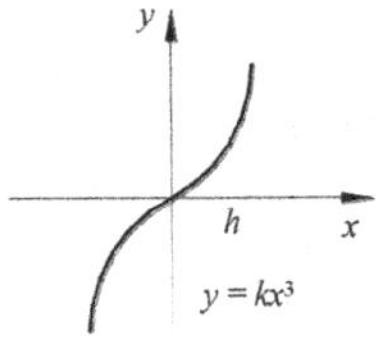 $y=kx^3$	$\frac{3K}{4}X^2$
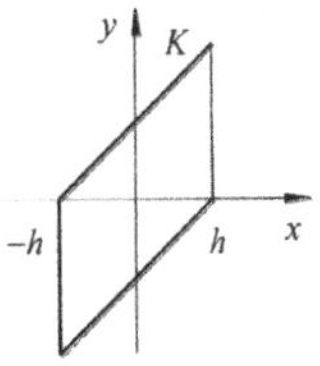 Backlash with slope K, $-h$, h	$\frac{K}{\pi}\sqrt{\pi^2+8\pi\frac{h}{X}\sqrt{1-\left(\frac{h}{X}\right)^2}+16\left(\frac{h}{X}\right)^2}\ \angle-\tan^{-1}\frac{4\left(\frac{h}{X}\right)^2}{\pi+4\frac{h}{X}\sqrt{1-\left(\frac{h}{X}\right)^2}}$

simple and efficient to use descriptive functions, especially for higher-order systems. The stability of the nonlinear system and its special self-oscillation problem is analyzed with descriptive function.

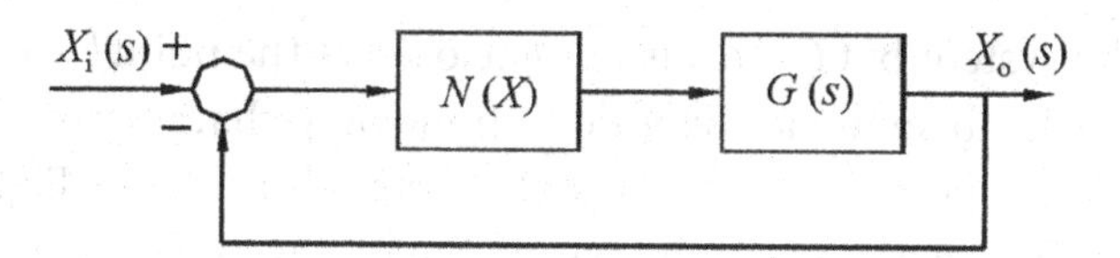

FIGURE 7.13 A nonlinear control system.

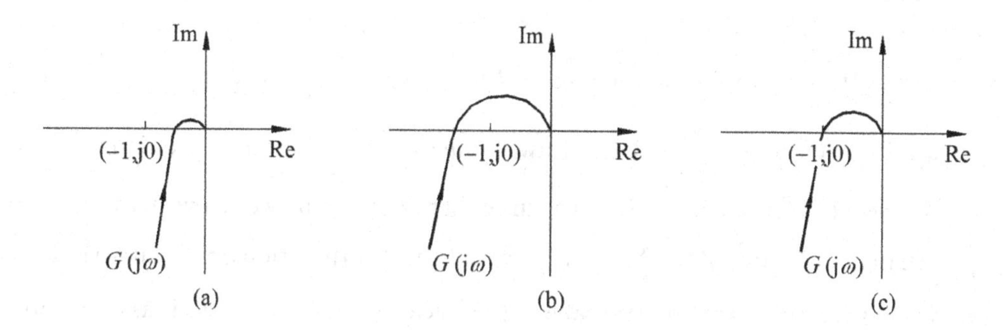

FIGURE 7.14 Open-loop linear systems by Nyquist criterion (a) stable system, (b) unstable system, and (c) critically stable.

7.5.1 Stability Analysis of Nonlinear Systems

Figure 7.14 shows the nonlinear systems, $G(j\omega)$ represents the frequency characteristic of the linear part, assuming that $N(X)$ is a nonlinear element after the higher harmonic produced by the linear part which is fully attenuated. Thus, the output of the nonlinear components has only the fundamental component, and it is influenced by the performance of the system, whereas it is used the describing function method to analyze the stability of the system. Especially, when the performance of the low-pass filter (LPF) of the system is excellent and the nonlinear distortion is not very serious, the stability of the system is more accurate. The descriptive function $N(X)$ can be treated as a gain for a variable (complex or real). When the stability of the system is studied, the closed-loop frequency characteristic of the system is defined as

$$\varnothing(j\omega)=\frac{X_o(j\omega)}{X_i(j\omega)}=\frac{N(X)G(j\omega)}{1+N(X)G(j\omega)} \tag{7.32}$$

The characteristic equation of Eq. (7.32) can be written as

$$1+N(X)G(j\omega)=0 \tag{7.33}$$

When

$$G(j\omega)=-\frac{1}{N(X)} \tag{7.34}$$

The critical oscillation system appears for the oscillation in the nonlinear systems. For the critical stable system, this is equivalent to the open-loop frequency characteristic of a linear system as $G(j\omega)=-1$ which is constant amplitude oscillations at the system for the critical stable system. For the linear systems, we may take advantage of the Nyquist criterion to judge the stability of linear system; Figure 7.14a shows an open-loop frequency characteristic of the system but $G(j\omega)$ is not enclosed at the point $(-1, j0)$; thus, the system is stable. Figure 7.14b shows an open-loop frequency characteristic of the system but $G(j\omega)$ is enclosed at the point $(-1, j0)$; thus, the system is unstable. Figure 7.14c shows an open-loop frequency characteristic of the system but $G(j\omega)$ is passed at the point $(-1, j0)$; thus, the system is critically stable and persistent oscillation system.

For the nonlinear control systems, the frequency characteristics of linear systems can be analyzed for critically stable system; the point $(-1, j0)$ is replaced by a nonlinear system that describes the root-locus enclosed by the function $-\frac{1}{N(X)}$. The critical point $(-1, j0)$ makes a critical curve $-\frac{1}{N(X)}$. Let's draw $G(j\omega) = -\frac{1}{N(X)}$ separately on the complex plane, as shown in Figure 7.15. $G(j\omega)$ is a curve that varies with frequency ω, the curve for $-\frac{1}{N(X)}$ changes with amplitude X. The arrows indicate the direction on the curve in which ω and X are increasing. In stability analysis, the linear part of the system is assumed to be an open-loop stability. At this point, the Nyquist criterion for system stability is defined as

1. The curve $G(j\omega)$ is not enclosed by $-\frac{1}{N(X)}$, that when $\omega = 0 \to \infty$ and $X \to \infty$, $-\frac{1}{N(X)}$ root-locus is always located on the left side of the curve $G(j\omega)$, as shown in Figure 7.15a; thus, the nonlinear system is stable. The other two parts of the system are stable.
2. If the curve $G(j\omega)$ is not enclosed by $-\frac{1}{N(X)}$ as shown in Figure 7.15b, the nonlinear system is unstable. The output of such a system will increase indefinitely, regardless of any interference signal, until a fault occurs, or it may be increased to a limit of other safety devices or mechanical performance limits are increased until the energy supply is disconnected.

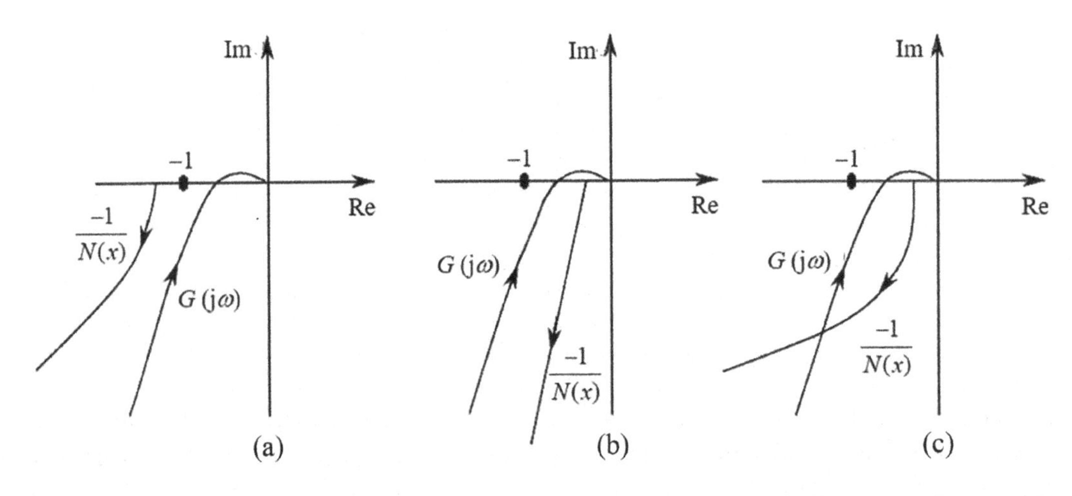

FIGURE 7.15 An open-loop Nyquist criterion of a nonlinear system (a) stable system, (b) unstable system, and (c) self-oscillating system or critically stable system.

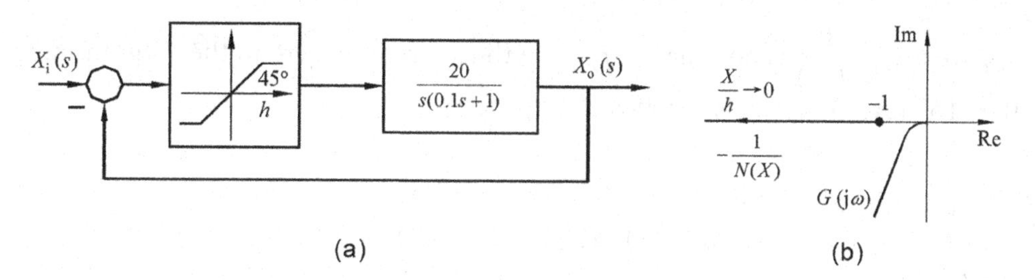

FIGURE 7.16 (a) The nonlinear control system and (b) the curve $G(j\omega)$ and $1/N(X)$ with saturation characteristics.

3. If the curve $G(j\omega)$ is not enclosed by $-\frac{1}{N(X)}$ as shown in Figure 7.15c, the output of such a system may be continuously self-oscillating. Generally speaking, this oscillation is not sinusoidal but it can be approximated by a sinusoidal oscillation. The frequency and amplitude of self-oscillation can be used at the intersection of the linear curve $G(j\omega)$ corresponding to ω_0 of $-\frac{1}{N(X)}$, and it is characterized by x_0; however, not all the intersections can produce self-oscillation, at this point.

Example 7.2

A nonlinear system with saturation properties is shown in Figure 7.16. Analyze the system stability.

Solution

According to Figures 7.15 and 7.16a, the saturation characteristic of the descriptive function of the negative expression is written as

$$-\frac{1}{N(X)}=\frac{-\pi}{2K\left[\sin^{-1}(h/X)+(h/X)\sqrt{1-(h/X)^2}\right.}$$

To facilitate the use of computer drawing the curve $G(j\omega)$ and the curve $-\frac{1}{N(X)}$ in the same coordinate system, $G(j\omega)$ will be divided into real and imaginary part, namely

$$G(j\omega)=\frac{20}{j\omega(0.1j\omega+1)}=-\frac{2}{\left(1+0.01\omega^2\right)}-j\frac{20}{\omega\left(1+0.01\omega^2\right)}$$

Let

$$\begin{cases} u=-\dfrac{2}{\left(1+0.01\omega^2\right)} \\ v=-\dfrac{20}{\omega\left(1+0.01\omega^2\right)} \end{cases}$$

Apparently $-\frac{1}{N(x)}$ is not imaginary part; thus, it is set to zero in the programming. The MATLAB program is as follows:

```
for j=1:12
nx(j)=-pi/(2*(asin(1/j)+1/j*sqrt(1-(1/j)^2)));
vx(j)=0;
u(j)=-2/(1+0.01*j*j);
v(j)=-20/(j*(1+0.01*j*j));
end
plot(u, v, nx, vx)
grid
```

Figure 7.16b shows the $-\frac{1}{N(X)}$ as $\frac{X}{h}=1\ \infty$ curve. The frequency characteristic curve of the linear part is shown in Figure 7.16b as the curve of $G(j\omega)$. The curve of $G(j\omega)$ does not enclose by the point $(1, j0)$; according to the Nyquist criterion, the linear part of the system itself is stable. However, the curve of $G(j\omega)$ is not enclosed by the curve of $-\frac{1}{N(X)}$; according to the Nyquist criterion of nonlinear system, the whole nonlinear system is stable.

7.5.2 Determination of Self-Oscillation

Self-oscillations in nonlinear systems correspond to periodic motions in stable systems. The previous analysis only shows the possible periodic motion of the system and its parameters. To judge whether there is self-oscillation in the system, the stability of periodic motion must also be studied. As shown in Figure 7.17, we can see two points of intersection $A(X_1,\omega_1)$ and $B(X_2,\omega_2)$ with the curve $-\frac{1}{N(X)}$ and the frequency curve of $G(j\omega)$, and let's say that $B(X_2,\omega_2)$ corresponds to track the curve $-\frac{1}{N(X)}$ on the maximum amplitude. The curve and the arrow on the curve increase in the direction of the maximum amplitude and frequency. Now, the functions of $A(X_1,\omega_1)$ and $B(X_2,\omega_2)$ are analyzed and studied, respectively.

Let's consider that the point $A(X_1,\omega_1)$ is the initial working point. If the point $A(X_1,\omega_1)$ is affected by a tiny disturbance signal, the input amplitude of the nonlinear element will increase. For example, the working point $A(X_1,\omega_1)$ will move from point $A(X_1,\omega_1)$ to point C; as the point C is enclosed by the curve $G(j\omega)$, the system is unstable; therefore, the oscillation will intensify and the amplitude will continue to increase, leaving the working point further away from the point C to the point B. Conversely, the input amplitude of the nonlinear element is decreased if the system is disturbed, and the working point $A(X_1,\omega_1)$ will move from point $A(X_1,\omega_1)$ to point D, whereas the point D is not enclosed by $G(j\omega)$; thus, the system is in stable state. However, when the system is in a stable state, the oscillation will weaken and the amplitude will continue to decrease; and then it shows that the

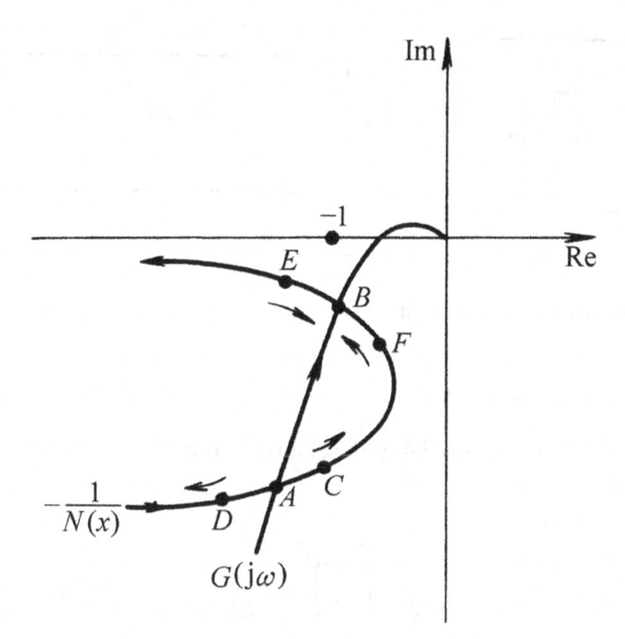

FIGURE 7.17 System oscillation stability analysis.

working point moves further downward from point D. Therefore, the point $A(X_1, \omega_1)$ is corresponded a periodic motion that is indicated as unstable limit cycles, and the point $A(X_1, \omega_1)$ is indicated as the characteristics of divergence; thus, the point $A(X_1, \omega_1)$ is unstable working point in this stage.

When the working point $B(X_2, \omega_2)$, if a tiny disturbance signal acts on the system, the input amplitude of the nonlinear element increases slightly, the working point $B(X_2, \omega_2)$ will move from point $B(X_2, \omega_2)$ to point E but the point E is not enclosed by the curve of $G(j\omega)$; thus, the system is in stable state. However, when the system is in a stable state, the oscillation is weakened, and the input amplitude of the nonlinear element is decreased; thus, the working point move from point E to point $B(X_2, \omega_2)$ until returning to the point $B(X_2, \omega_2)$. On the contrary, if the external interference or disturbance signal is decreased the nonlinear element with input amplitude, and the working point $B(X_2, \omega_2)$ is moved to be enclosed by the curve of $G(j\omega)$ at the point F, the system becomes unstable. However, it will show volatility and it will increase the amplitude, then the working point will return from the point F to the point $B(X_2, \omega_2)$. Therefore, the point $B(X_2, \omega_2)$ is limit cycle with convergence properties, and the point $B(X_2, \omega_2)$ is acted as stable working point.

Usually, the control system does not want to appear as the natural phenomenon of self-oscillation, but the self-oscillation with a small amplitude is allowed in some engineering applications.

Example 7.3

A nonlinear control system, as shown in Figure 7.18, where $M=10$, $h=1$. Judge the system whether there is natural frequency; if there is natural frequency, find the vibration amplitude and frequency.

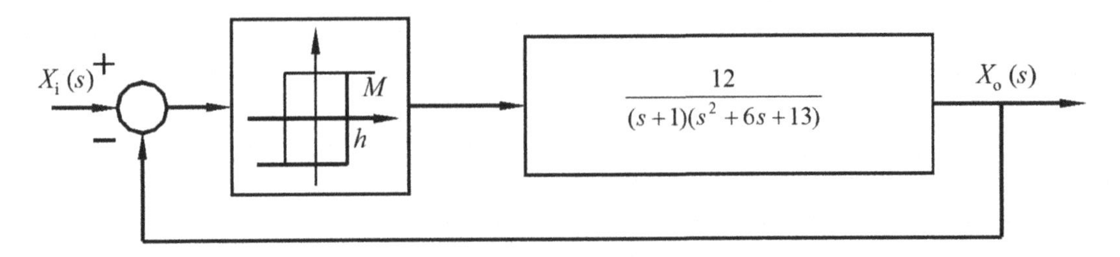

FIGURE 7.18 A nonlinear control system.

Solution

According to Table 7.1, the available describing function for nonlinear components can be written as

$$N(X)=\frac{40}{\pi X}\sqrt{1-\left(\frac{1}{X}\right)^2}-j\frac{40}{\pi X^2} \qquad ;\text{for } X\geq 1$$

The negative inverse of the description function $-\frac{1}{N(X)}$ is written as

$$-\frac{1}{N(X)}=\frac{\pi}{40}\left(\sqrt{X^2-1}+j\right)$$

For the linear part of the frequency characteristics is written as

$$G(j\omega)=\frac{12}{(j\omega+1)(j\omega)^2+6j\omega+13}$$

Thus, $G(j\omega)$ and $-\frac{1}{N(X)}$ is divided into real and imaginary part, respectively, we get

$$G(j\omega)=\frac{12\left(13-7\omega^2\right)-12\omega\left(19-\omega^2\right)j}{\left(\left(1+\omega^2\right)\left(13-\omega^2\right)^2+36\omega^2\right)}$$

Therefore,

$$\begin{cases} u=\dfrac{12\left(13-7\omega^2\right)}{\left(1+\omega^2\right)\left(\left(13-\omega^2\right)^2+36\omega^2\right)} \\ v=-\dfrac{12\omega\left(19-\omega^2\right)}{\left(1+\omega^2\right)\left(\left(13-\omega^2\right)^2+36\omega^2\right)} \end{cases}$$

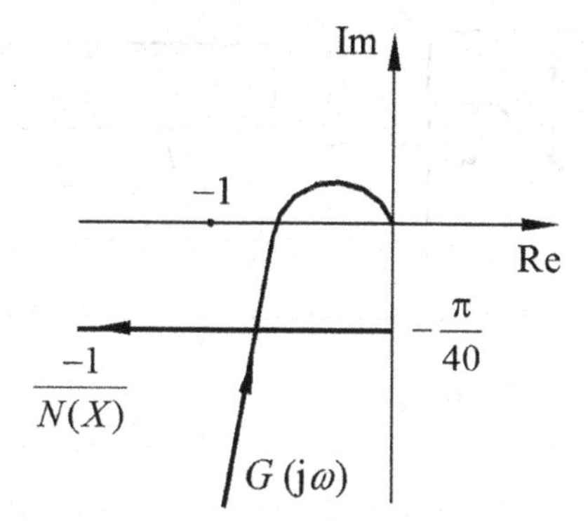

FIGURE 7.19 Curve of $G(j\omega)$ and $-1/N(X)$ as shown in Figure 7.18.

And from

$$-\frac{1}{X(X)}=\frac{\pi}{40}\left(\sqrt{X^2-1}+j\right)$$

We may write as

$$n_u=-\frac{\pi}{40}\left(\sqrt{X^2-1}\right)$$

$$n_v=-\frac{\pi}{40}$$

Figure 7.19 shows the curve of $-\frac{1}{N(X)}$ and $G(j\omega)$. Note, as a result of calculating on a computer, ω is impossible to infinity. Therefore, when using MATLAB program, there is a very small segment near the origin that is not connected with the origin. The two curves of $-\frac{1}{N(X)}$ and $G(j\omega)$ on the complex plane have an intersection point at $X_o=2.3$ and $\omega_o=3.2$, respectively. According to the stability criterion of nonlinear system, the system is in the function of signal $x=2.3\sin 3.2t$. No matter what kind of disturbance signal there is in this system, the system will finally show the state of natural vibration.

The MATLAB program is as follows:

```
for j=1:6
jw(j)=j+2;
uu(j)=(1+jw(j)^2)*((13-jw(j)^2)^2+36*jw(j)^2);
u(j)=12*(13-7*jw(j)^2)/uu(j);
v(j)=-12*(19-jw(j)^2)*jw(j)/uu(j);
v0(j)=0;
nxu(j)=-pi/40*sqrt(j*j-1);
nxv(j)=-pi/40;
end
plot(u, v, nxu, v0,nxu, nxv)
grid
```

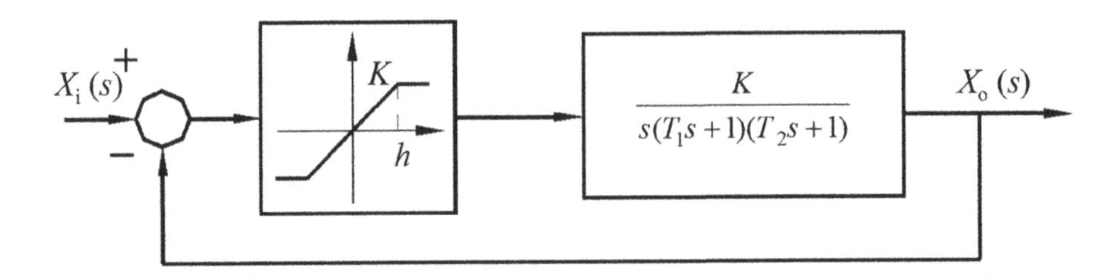

FIGURE 7.20 A nonlinear control system.

Example 7.4

The nonlinear control system is shown in Figure 7.20. Analyze the stability of the system.

Solution

According to Eq. (7.18) for the description of the nonlinear system function, we may rewrite as

$$N(X)=\begin{cases} \frac{2K}{\pi}\left[\sin^{-1}\left(\frac{h}{X}+\frac{h}{X}\sqrt{1-\left(\frac{h}{X}\right)^2}\right)\right] & ;\text{for } X>h \\ K & ;\text{for } X\le h \end{cases}$$

when $X\le h, -\frac{1}{N(X)}=-\frac{1}{K}$, and when $X\to\infty, -\frac{1}{N(X)}\to\infty$.

The partial frequency characteristics of the linear system are expressed as

$$G(j\omega)=\frac{K}{j\omega(T_1j\omega+1)(T_2j\omega+1)}$$

when $\omega\to 0, G(j\omega)=\infty\angle-\frac{\pi}{2}$, and when $\omega\to\infty, G(j\omega)=0\angle-\frac{3\pi}{2}$.

The Nyquist curve of linear part and negative real axis have an intersection point; the intersection point of the coordinates at $\left(-\frac{KT_1T_2}{T_1+T_2}, j0\right)$, the crossover frequency is $-\frac{1}{\sqrt{T_1T_2}}$. Figure 7.20 shows the nonlinear control system, the nonlinear element describes the function of negative reciprocal characteristic curve and linear part of the Nyquist frequency characteristic curve, as shown in Figure 7.21. The linear partial amplification factor K is sufficiently large to make $\frac{KT_1T_2}{T_1+T_2}>\frac{1}{K}$, the intersecting curves $G(j\omega)$ and $-\frac{1}{N(X)}$ generate a limit cycle. When the amplitude of the disturbance X is turned on left of the real axis to the intersection point A and then to the left point B, the curve of $G(j\omega)$ is not enclosed by the point B; thus, the system is stable.

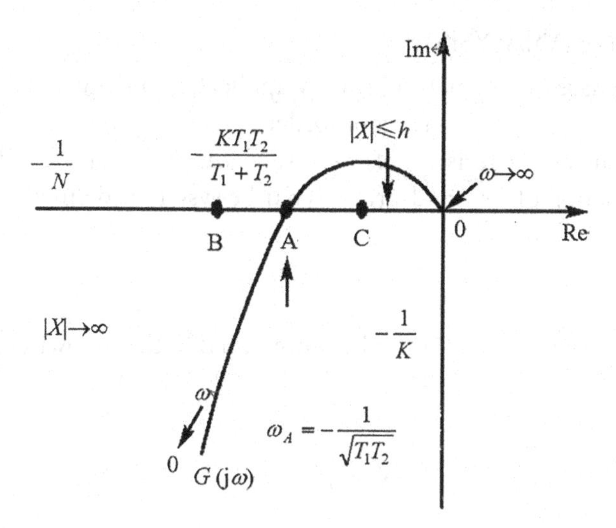

FIGURE 7.21 The Nyquist curve of $G(j\omega)$ and $-\frac{1}{N(X)}$.

However, the amplitude decreases gradually, and it backs to the point A. When the amplitude of the disturbance X is turned a little or half of the amplitude, the point A move to the right point C; the point C is enclosed by the curve of $G(j\omega)$; thus, the system is not stable. Whereas the amplitude increases gradually, and similarly it return to the point A. Therefore, we can get the point A is stable limit cycle, its amplitude is equal to the amplitude of the $-\frac{1}{N(X)}$ at point A, and its frequency is equal to $\omega_A = \frac{1}{\sqrt{T_1T_2}}$ of the curve of $G(j\omega)$ at point A. Of course, it will not exist the limit cycle in the system, as long as reduce part of the linear amplification coefficient K to make $\frac{KT_1T_2}{T_1+T_2} < \frac{1}{K}$, the system has no intersection curve of $G(j\omega)$ and the curve of $-\frac{1}{N(X)}$; thus, there are no limit cycle.

By the above discussion, we can summarize the stability of the system by describing the function analysis of the following steps:

1. Describe the function according to the nonlinear element;
2. Make the curve of $-\frac{1}{N(X)}$ and $G(j\omega)$ on the complex plane;
3. Determine whether a system is stable or unstable and the existence of a limit cycle; and
4. If there is a limit cycle, then judge its stability, and determine its frequency and amplitude.

To design the nonlinear control system with the describing function method, it is important to avoid the intersecting point of the curve of the linear part and nonlinear part. This can be corrected through implementation.

7.6 PHASE PLANE ANALYSIS

The phase plane (phase locus) method is only applicable to the geometric solution of second-order nonlinear systems. For a second-order dynamic system, if two state variables are known, the dynamic characteristics of the system can be completely described. Suppose the differential equation of a second-order dynamic system is defined as

$$\ddot{x} = f(x,\dot{x}) \tag{7.35}$$

If $x = x_1$, and $\dot{x} = x_2$, then Eq. (7.35) can be represented in the following equations as

$$\dot{x}_1 = x_2 \tag{7.36}$$

$$\dot{x}_2 = f(x_1,x_2) \tag{7.37}$$

By using Eq. (7.36) with Eq. (7.37), we can get

$$\frac{dx_2}{dx_1} = \frac{f(x_1,x_2)}{x_2} \tag{7.38}$$

In Eq. (7.38), x_1 is the independent variables and x_2 is the dependent variable of the first order differential equation. If Eq. (7.38) can be solved, then Eq. (7.36) with Eq. (7.37) can be solved by the function of x with respect to change over time t. Thus, we may substitute the solution of Eq. (7.35) into the solution of Eq. (7.38). In essence, Eq. (7.35) can be regarded as the equation of particle motion. Here, x_1 represents the position of the particle and x_2 represents on behalf of the particle velocity (it can be also on behalf of the other physical characteristics). In control theory, the coordinates of time variables are not shown in x_1,x_2, but it consists of a plane that is called the phase plane. Since there are only two state variables (e.g. x_1,x_2), the curve of state variables over time can be expressed on the phase plane. Each state of the system corresponds to a point on the phase plane, and the change of the state variable with time corresponds to the movement of the point on the phase plane. The curve of the points in the phase plane over time is called the phase trajectory curve or locus. The following has a discussion of plotting phase curves.

7.6.1 Plotting of Phase Curves

7.6.1.1 Analytical Method

In physics, an object at a unit particle is assumed to be in free-falling body. Its fundamental equation is written as

$$\ddot{x} = g \tag{7.39}$$

Let

$$x_1 = x,\ x_2 = \dot{x}$$

From which

$$\ddot{x}=\frac{dx_2}{dt}=\frac{dx_2}{dx_1}\times\frac{dx_1}{dt}=x_2\frac{dx_2}{dx_1}$$

Thus,

$$x_2\frac{dx_2}{dx_1}=g$$

$$x_2dx_2=gdx_1 \tag{7.40}$$

From Eq. (7.40), we can write as

$$x_2^2=2gx_1+C \tag{7.41}$$

where C is constant. For x_1 is the abscissa and x_2 is the coordinate phase track diagram, as shown in Figure 7.22.

From the phase track diagram as shown Figure 7.22, it is a family of parabola. The velocity is positive in the upper half plane; thus, the displacement goes up, the arrow goes to the right, whereas, in the lower half of the plane, since the velocity is negative, the displacement goes down, the arrow goes to the left. Suppose the particle is thrown vertically from the ground, the displacement is zero, the velocity is positive, and the initial point is point *A*. The motion of the phase curve motion of the particle starting from point *A* increases with the displacement of the particle. The velocity is getting smaller and smaller, and the particle displacement is maximized at point *B*. At this point, the velocity of the particle is zero, then the particle falls freely along the *BC* curve and reaches at point *C*, where the displacement is zero and the velocity becomes negative again. The curve of motion is different at different starting points, as shown in Figure 7.22 and the initial values of the three points *A*,*F*,*G* are different and it is drawn as three different curves.

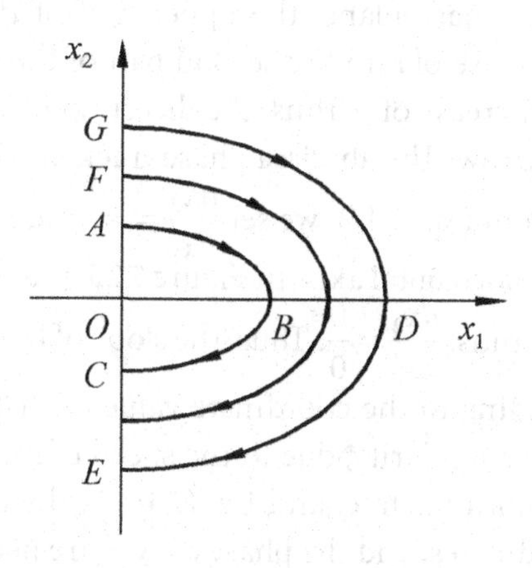

FIGURE 7.22 Ideal phase track for the free-falling body.

A linear system of differential equations can be written as

$$\ddot{x}+2\xi\omega_n\dot{x}+\omega_n^2x=0 \tag{7.42}$$

In phase coordinates $x_1=x$, $x_2=\dot{x}$, and Eq. (7.42) is simplified to

$$\frac{dx_2}{dx_1}=-\frac{2\xi\omega_n x_2+\omega_n^2x_1}{x_2} \tag{7.43}$$

We have already discussed before, Eq. (7.42) is called free-falling body equation; to study the characteristics, the distribution characteristics of the root can be divided into the following ways:

i. *Undamped natural frequency ($\xi=0$)*:
For the undamped natural frequency, Eq. (7.43) becomes

$$\frac{dx_2}{dx_1}=-\frac{\omega_n^2x_1}{x_2} \tag{7.44}$$

From Eq. (7.44), we yield

$$x_1^2+\frac{x_2^2}{\omega_n^2}=A^2 \tag{7.45}$$

In Eq. (7.45), $A=\sqrt{\frac{x_{20}^2}{\omega_n^2}+x_{10}^2}$, and x_{10},x_{20} are initial conditions; x_{10},x_{20} are different values; Eq. (7.45) is a family of concentric ellipse in the plane, as shown in Figure 7.23. Every ellipse corresponds to a simple harmonic oscillation with different initial conditions. In the phase plane, the upper half of the plane, $\dot{x}(x_2)>0$, and x increases with the increase of t; in the second half of the phase plane, $\dot{x}(x_2)<0$ and x decreases with the increase of t. Thus, the direction of the phase track is shown in Figure 7.23 with the arrow. The abscissa phase track and intersections can be found when $x_2=0, x_1\neq 0$. From Eq. (7.44), we set $\frac{dx_2}{dx_1}=\infty$, and then the phase curve passes vertically through the horizontal axis. In Figure 7.23, the coordinates are at the origin because of $x_1=x_2=0$; thus, $\frac{dx_2}{dx_1}=\frac{0}{0}$. Thus, the slope of the phase trajectory cannot be one-to-one corresponding to the coordinate value of that point, this point is called the singularity. At the singularity, due to the speed of x_2, the acceleration $\dot{x}_2$ is zero, the system is in equilibrium state, and Eq. (7.44) is the only isolated singularity in the origin of the coordinates, and the phase curves are near the singularity that are a family of closed curves. Such a singularity is called the center.

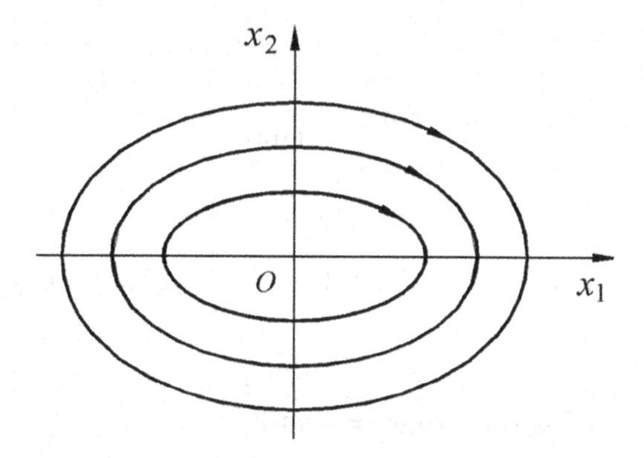

FIGURE 7.23 Undamped phase curve of the movement.

ii. *Underdamped natural frequency* ($0<\xi<1$):
The solutions of Eq. (7.42) will give

$$x_1 = x(t) = Ae^{-\xi\omega_n t}\cos(\omega_d t+\varphi) \tag{7.46}$$

where

$$\omega_d = \omega_n\sqrt{1-\zeta^2}$$

$$A = \frac{1}{\omega_d}\sqrt{x_{20}^2 + 2\zeta\omega_n x_{10}x_{20} + \omega_n^2 x_{10}^2}$$

$$\phi = -\arctan\frac{x_{20}+\zeta\omega_n x_{10}}{\omega_d x_{10}}$$

Thus, the derivative of Eq. (7.46) can be found as

$$\dot{x}(t) = x_2 = -A\xi\omega_n e^{-\xi\omega_n t}\cos(\omega_d t+\varphi) - A\omega_d e^{-\xi\omega_n t}\sin(\omega_d t+\varphi) \tag{7.47}$$

Now, $\xi\omega_n$ multiply in Eq. (7.46) and add with Eq. (7.47), and we obtain

$$x_2 + \xi\omega_n x_1 = -A\omega_d e^{-\xi\omega_n t}\sin(\omega_d t+\varphi) \tag{7.48}$$

and ω_d multiply in Eq. (7.46), and we obtain

$$\omega_d x_1 = A\omega_d e^{-\xi\omega_n t}\cos(\omega_d t+\varphi) \tag{7.49}$$

From Eqs. (7.48) and (7.49), we yield

$$(x_2+\xi\omega_n x_1)^2 + \omega_d^2 x_1^2 = A^2\omega_d^2 e^{-2\xi\omega_n t} \tag{7.50}$$

and

$$-\frac{x_2+\xi\omega_n x_1}{\omega_d x_1}=\tan(\omega_d t+\varphi) \tag{7.51}$$

By Eq. (7.51) to work out the t and substitute in Eq. (7.50), we get

$$(x_2+\xi\omega_n x_1)^2+\omega_d^2 x_1^2=A^2\omega_d^2 e^{\frac{2\xi\omega_n}{\omega_d}\left(\varphi+\tan^{-1}\left(\frac{x_2+\xi\omega_n x_1}{\omega_d}\right)\right)} \tag{7.52}$$

Eq. (7.52) may be considered the underdamped movement of the phase curve equation, which represents the family of curves around the phase plane as a spiral line on the origin of coordinates. In the polar coordinates, it is clear, and we can define

$$r\cos\theta=\omega_d x_1 \text{ and } r\sin\theta=-\left(\xi\omega_n x_1+x_2\right)$$

Thus, Eq. (7.52) can be simplified, and it gives

$$r^2=A^2\omega_d^2\exp\left[\frac{2\zeta\omega_n}{\omega_d}(\phi-\theta)\right]$$

or

$$r=A\omega_d\exp\left[\frac{\xi\omega_n}{\omega_d}(\varphi-\theta)\right] \tag{7.53}$$

Eq. (7.53) is the corresponding logarithmic spiral line equation in polar coordinates because

$$\tan\theta=-\frac{\xi\omega_n x_1+x_2}{\omega_d x_1}=\tan(\omega_d t+\varphi)$$

Thus, we get

$$\theta=\omega_d t+\varphi \tag{7.54}$$

According to Eqs. (7.53) and (7.54), θ increases with the increase of t, while r decreases with the increase of t, that is, the movement of the phase curve approaches the origin from the outside. The phase curve of the underdamped motion of the system is shown in Figure 7.24. It can be seen that the initial conditions do not matter, the oscillation eventually approaches to the equilibrium state after attenuating. Thus, the origin of coordinates is one of its singularities.

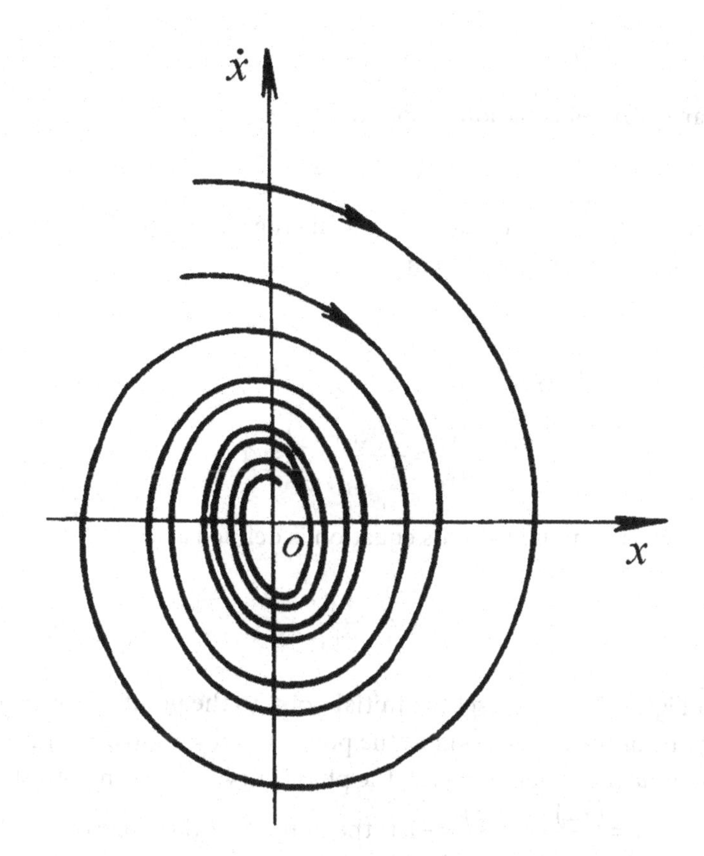

FIGURE 7.24 System damping movement owed to the phase trajectory.

7.6.1.2 Isoclinic Line Method

The isoclinic line is suitable for the general form of second-order differential equation as it can be written as

$$\ddot{x} = f(x,\dot{x})$$

It can be turned into the following:

$$\frac{dx_2}{dx_1} = \frac{f(x_1,x_2)}{x_2}$$

The isoclinic line refers to the phase plane within the corresponding phase curve with a slope which is equal to the line. Setting up the slope to k, we get

$$k = \frac{dx_2}{dx_1} = \frac{f(x_1,x_2)}{x_2} \tag{7.55}$$

The direction field of the tangent line of the phase curve can be obtained by drawing different isocline lines corresponding to different k values. From the sequence of the initial short slope lines, a phase curve diagram is formed by connecting adjacent short slope lines. Obviously, the denser the interval of the isocline, the more accurate the phase trajectory.

Example 7.5

The nonlinear system of equations is defined as

$$\ddot{x}+\dot{x}+x=0 \tag{7.56}$$

The phase curve of the system is drawn by using the oblique or tilt line method. Draw the phase track diagram of the system.

Solution

According to Eq. (7.56), we get

$$\frac{dx_2}{dx_1}=-\frac{(x_1+x_2)}{x_2}=\alpha \tag{7.57}$$

The isoclinic line is straight; thus, its equation is defined as

$$x_2=-\frac{1}{1+\alpha}x_1 \tag{7.58}$$

As shown in Figure 7.25, if we set the initial point to the point A, we get the slope of the isoclinic line of the initial point to the point A as $\alpha=-1$ to $\alpha=-1.2$, respectively. Under an isoclinic line slope $\alpha=-1.2$, the phase curves between both slopes, it used average slope as $\alpha=\dfrac{(-1)+(-1.2)}{2}=-1.1$; the point A of the slope $\alpha=-1.1$ of the line AB, and it intersects the second isocline at point B, which is the phase curve; because of the slope of the third isocline $\alpha=-1.4$; and also uses the average slope; thus the slope is $\alpha=\dfrac{(-1.2)+(-1.4)}{2}=-1.3$ of the point B as a straight line, whereas the line intersects the third isocline at point C, and the BC line is another phase curve, and so on. If it keeps continuing, we can figure out each point of A, B, C, D, E,... on the curve to get phase curve, as shown in Figure 7.25.

Example 7.6

One for the control system of nonlinear equations

$$\ddot{x}+0.2\left(x^2-1\right)\dot{x}+x=0 \tag{7.59}$$

Draw a phase track diagram of the system.

Solution

Define $x_1=x$, $x_2=\dot{x}$

The system of equations can be expressed as

$$\frac{dx_1}{dt}=x_2$$

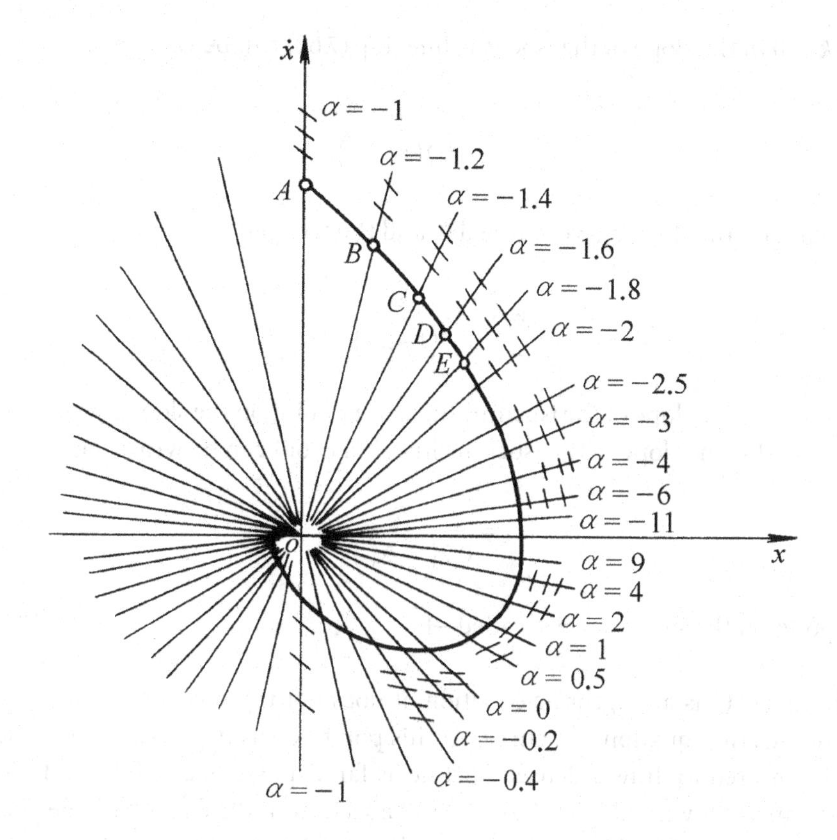

FIGURE 7.25 Phase curve diagram of $\ddot{x}+\dot{x}+x=0$.

$$\frac{dx_2}{dt}=-0.0\left(x_1^2-1\right)x_2-x_1$$

Thus,

$$\frac{dx_2}{dx_1}=\frac{-0.2\left(x_1^2-1\right)x_2-x_1}{x_2}=-0.2\left(x_1^2-1\right)-\frac{x_1}{x_2}$$

Note that

$$\frac{dx_2}{dx_1}=k$$

Thus,

$$k=-0.2\left(x_1^2-1\right)-\frac{x_1}{x_2}$$

Therefore,

$$x_2=\frac{x_1}{0.2\left(1-x_1^2\right)-k} \tag{7.60}$$

When $k = 0$ in the slope of the isoclinic line, Eq. (7.60) can be written as

$$x_2 = \frac{x_1}{0.2\left(1-x_1^2\right)}$$

Every point on the slope curve is called 0, and thus we get

$$x_2 = \frac{x_1}{0.2\left(1-x_1^2\right)-1}$$

When $k = 1$ on the slope of the isoclinic line, every point on the slope curve is called 1. When $k = -1$ is the slope of the isoclinic line, Eq. (7.60) can be written as

$$x_2 = \frac{x_1}{0.2\left(1-x_1^2\right)+1}$$

Every point on the slope curve is called –1.

Although Eq. (7.60) is an algebraic equation, it does not represent the equation of a line, but it represents the equation of a curve. At this point, to ensure the accuracy of the phase curve, it is required to draw a denser isocline as far as possible and finally, draw a phase curve diagram as shown in Figure 7.26. It can be seen from the figure that the phase curve eventually tends to a closed limit cycle regardless of whether the starting point is near or far from the origin.

7.6.1.3 The δ Method

When the isoclines are straight lines, it is more convenient to apply the isoclines method, but when the isoclines are curves, it is more convenient to use the δ method to draw phase curves through the known points. In the δ method, the phase trajectory is the continuous line of a series of arcs whose center slides along the X-axis. The δ method is applied to the following equation as

$$\ddot{x} = f(x,\dot{x})$$

where $f(x,\dot{x})$ is linear or nonlinear and it also can be time-varying, but it must be continuous with a single value. When we add $\omega^2 x$ to both sides of the above equation, we get

$$\ddot{x}+\omega^2 x = f(x,\dot{x})+\omega^2 x \tag{7.61}$$

In Eq. (7.61), we should choose an appropriate value of ω. In order to make the following definition of the δ function, the range of values is discussed by x and $\dot{x}$ as neither too big nor too small. Thus, the δ function is defined as

$$\delta(x,\dot{x}) = \frac{f(x,\dot{x})+\omega^2 x}{\omega^2} \tag{7.62}$$

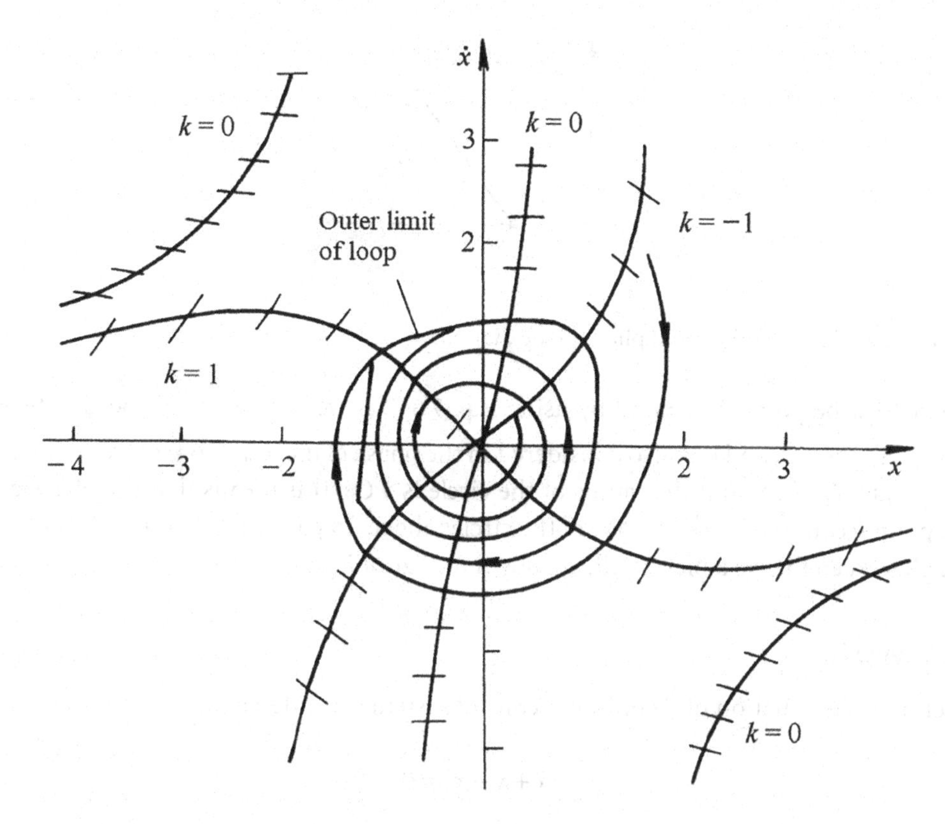

FIGURE 7.26 Phase curve diagram of $\ddot{x}+0.2\left(x^2-1\right)\dot{x}+x=0$.

The value of δ function changes with the value of x and $\dot{x}$. However, when x and $\dot{x}$ changes very much, $\delta(x,\dot{x})$ can be treated as a constant. For example, near a certain state as $x=x_1$ and $\dot{x}=\dot{x}_1$ which may be considered as $x=x_1 \pm \Delta x_1$ and $\dot{x}=\dot{x}_1 \pm \Delta \dot{x}_1$; if $\Delta x_1 \to 0$ and $\Delta \dot{x} \to 0$, it can be assumed that δ values can be in a constant δ_1 in Eq. (7.62) and then into Eq. (7.61) which gives

$$\ddot{x}+\omega^2\left(x-\delta_1\right)=0 \tag{7.63}$$

or

$$\frac{d\dot{x}}{dx}+\frac{\omega^2\left(x-\delta_1\right)dx}{\dot{x}}=0$$

By integrating, we get

$$\left(\frac{\dot{x}}{\omega}\right)^2+(x-\delta_1)^2=\left(\sqrt{\left(\frac{\dot{x}_1}{\omega}\right)^2+(x_1-\delta_1)^2}\right)^2 \tag{7.64}$$

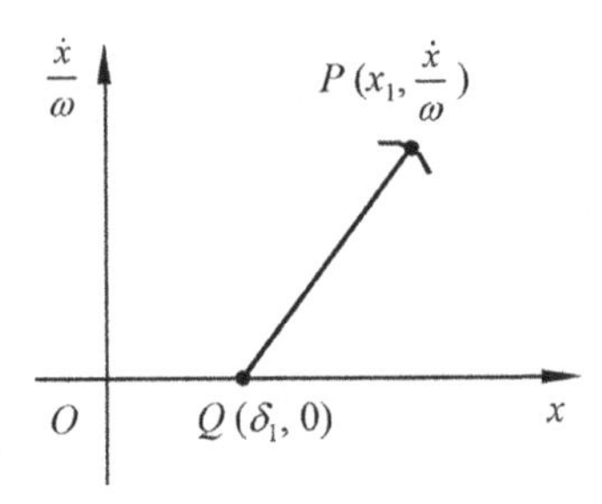

FIGURE 7.27 The δ method and phase plane diagram.

Apparently, the point P_1 is found by using Eq. (7.64). If we choose the Y-coordinate as $\frac{\dot{x}}{\omega}$ with the abscissa x, and it is also considered in the phase plane, Eq. (7.64) represents a circle at the center $Q(\delta_1, 0)$ and the radius of the circle is $|PQ|$. This means that the phase curve near point P can be replaced by one of the circles above. In particular, the arc must be short enough to keep the variables small, as shown in Figure 7.27.

Example 7.7

Setup of the equation of a nonlinear control system is written as

$$\ddot{x} + \dot{x} + x^3 = 0$$

Known initial point is $x_o = 1$ and $\dot{x}_o = 0$; draw the initial point of phase curve using the δ method.

Solution

According to ω, δ is selected value in principle, taking $\omega = 1$ in the both sides of the known equation combined with x at the same time, we get

$$\ddot{x} + x = -\dot{x} - x^3 + x$$

Hence,

$$\delta = -\dot{x} - x^3 + x$$

Since the value δ depends on x and $\dot{x}$, we must carry on the continuous test in the drawing. In Figure 7.28, the phase curve from the known points is $A(x = 1, \dot{x} = 0)$; thus, near point A, we can find $\delta_1 = -0 - 1 + 1 = 0$. The radius of the initial point is $\sqrt{\left(\frac{x_1}{\omega}\right)^2 + (x_1 - \delta_1)^2} = 1$ and the center is located at point (0,0). Therefore, we can make a point at (0, 0) as the center of the circle, the radius of 1 with a short arc. In order to make the drawing more accurate, we can use the average short arc x and $\dot{x}$; however, to reach more accuracy, we can find δ values.

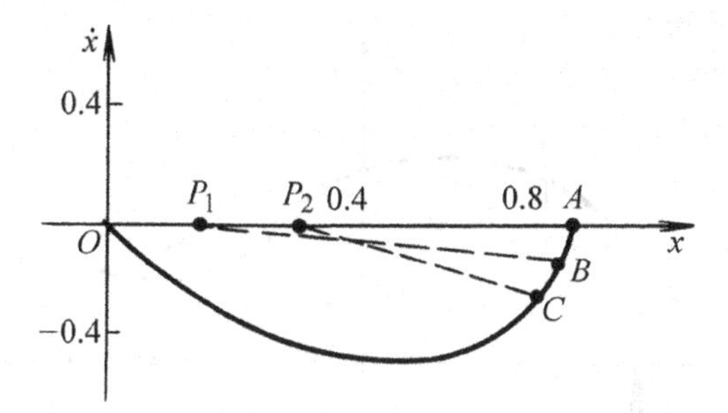

FIGURE 7.28 Phase curve diagram.

After several approaches, we can get a more accurate δ value. For this example, the first period of circular arc approach AB is as follows: first, as the center of the circle in the origin, the arc radius is 1, it is as $\dot{x} = -0.2$ in a straight line to point B', the coordinate of the point B' is $\dot{x} = -0.2$.

Thus, from Eq. (7.64), we get

$$x = \delta_1 + \sqrt{\left(\frac{\dot{x}_1}{\omega}\right)^2 + (x_1 - \delta_1)^2 - \left(\frac{\dot{x}}{\omega}\right)^2} = 0 + \sqrt{1 - 0.2^2} = 0.98$$

By substituting points A and B', we get the average of $\dot{x}_A = -0.1, x_A = (1 + 0.98)/2 = 0.99$ into the expression of $\delta(\dot{x}, x)$, and then we can calculate an approximated value of δ_1

$$\delta_1 = 0.1 - 0.99^3 + 0.99 \approx 0.12$$

In this way, the first section of arc $\widehat{AB}$ center is located at point $P_1\,(0.12, 0)$. In the same way, we can make the second period of a circular arc $\widehat{BC}$, while the center is located in $P_2\,(\delta = 0.37, 0)$. Therefore, it can be continuously carried out, and we can make the phase curve as shown in Figure 7.28.

7.6.2 Determine Time Information by Phase Plane

On the phase plane, the phase curve is drawn by eliminating $x_1 - x_2$ of time variable t. Although it can be intuitively described by the system state points, it is implicit in the time function. Therefore, the information of the state variable changing with time cannot be represented intuitively. Sometimes, we also have to know the time information, and that means we have to figure out the time information from the phase curve diagram.

1. *Time information according to* $t = \int \frac{1}{\dot{x}} dx$

   Since $\dot{x} = \frac{dx}{dt}$, we can use the time interval $t_B - t_A$ and it is said as

$$t_B - t_A = \int_{x_A}^{x_B} \frac{1}{\dot{x}} dx \tag{7.65}$$

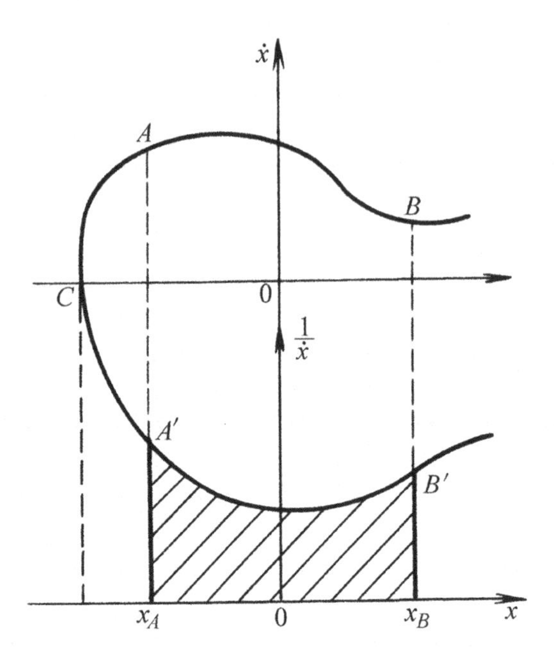

FIGURE 7.29 Timeline curve of $\frac{1}{\dot{x}}$.

For the time information, $\frac{1}{\dot{x}}$ can be in the vertical coordinates, x is along with the horizontal axis, we can draw $\frac{1}{\dot{x}} \sim x$ phase curve. The integral area of the $\frac{1}{\dot{x}}$ curve represents the corresponding time interval. As shown in Figure 7.29, the time interval between requirements AB show only the $\frac{1}{\dot{x}} \sim x$ phase curve diagram of the area $A'B'x_Bx_A$. It can be used in the calculation of actual analytical method or graphical method to determine the area. Usually, if $\frac{1}{\dot{x}}$ is a function of integrable, it uses the analytical method; if $\frac{1}{\dot{x}}$ is different, it uses the graphical method.

2. *Time information according to the arc approximation*
   In Figure 7.30, it is very difficult to find point C; hence, knowing the time information around the point is also difficult. Sometimes we can't even find a solution for it. At point C, finding time information is more convenient by using the arc approximation to determine the point C. The phase curves are approximated by a series of arcs with the center of the circle on the real axis. As shown in Figure 7.30, we can see the curve AD, it can use the variable x on the X-axis at the point P_1, P_2, P_3 as the center of the circle with the radius $|P_3A|$, $|P_2B|$ and $|P_1C|$ of the small circular arc approximation. Thus, we can get

$$t_{AD} = t_{AB} + t_{BC} + t_{CD} \approx t_{\widehat{AB}} + t_{\widehat{BC}} + t_{\widehat{CD}}$$

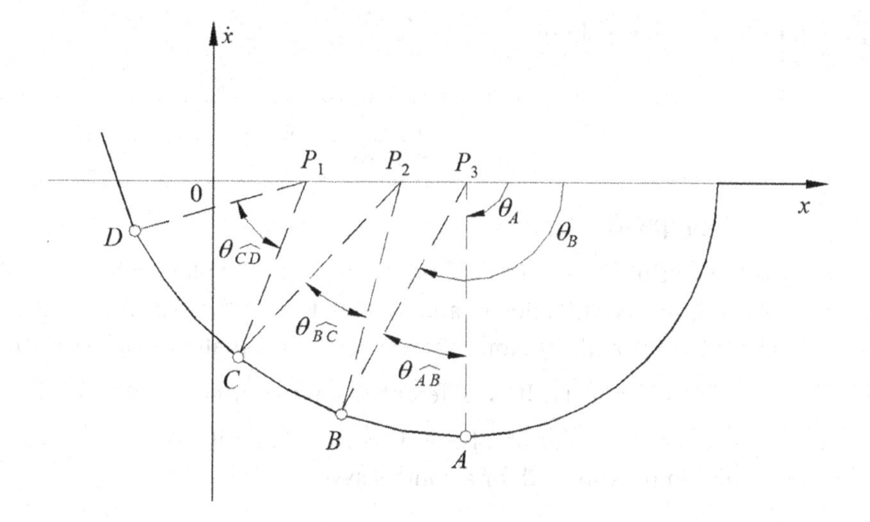

FIGURE 7.30 With a small circular arc approximation method to calculate time.

And each little arc segment has its due time, for example, $t_{\widehat{AB}}$ can be easily calculated. In fact, we say

$$\dot{x} = |P_3A|\sin\theta,\ x = |OP_3| + |P_3A|\cos\theta$$

By using Eq. (7.65), we can say

$$t_{\widehat{AB}} = \int_{\theta_A}^{\theta_B} \frac{-|P_3A|\sin\theta}{|P_3A|\sin\theta} d\theta = \theta_A - \theta_B = \theta_{\widehat{AB}} \tag{7.66}$$

By using Eq. (7.66), $t_{\widehat{AB}}$ is equal to the central angle of the $\widehat{AB}$. However, we should pay attention to angle $\theta_{\widehat{AB}}$ in the unit of radian to measure the angle.

3. *Time information using* $t = \int \frac{1}{f(x,\dot{x})} d\dot{x}$

   Since,

$$\ddot{x} = f(x,\dot{x})$$

Thus,

$$\frac{d\dot{x}}{dt} = f(x,\dot{x})$$

We can rearrange it and obtain

$$dt = \frac{1}{f(x,\dot{x})} d\dot{x}$$

Through the integral, we obtain

$$t_{AB} = t_B - t_A = \int_A^B \frac{1}{f(x, \dot{x})} d\dot{x} \tag{7.67}$$

7.6.3 Singularity of the Phase Plane

The singularity is the equilibrium point, the point on the phase plane where the system is in equilibrium. At singularity, the velocity and acceleration of the system are zero. Taking x_1 along the horizontal axis and x_2 along the vertical coordinate, the slope of the phase curve is $\frac{dx_2}{dx_1} = \frac{0}{0}$ at the singularity. It is different from a common point, but it does not apply to the uniqueness of the solution of the singularity, and it has a phase curve in the singularity. The equation is expressed for a control system as

$$\ddot{x} + 2\zeta\omega_n\dot{x} + \omega_n^2 = 0$$

let, $x = x_1$, and $\dot{x} = x_2$, we obtain

$$x_2 \frac{dx_2}{dx_1} + 2\zeta\omega_n x_2 + \omega_n^2 x_1 = 0$$

Thus,

$$\frac{dx_2}{dx_1} = -\frac{2\zeta\omega_n x_2 + \omega_n^2 x_1}{x_2}$$

The singularity of the system satisfies the condition

$$\frac{dx_2}{dx_1} = \frac{0}{0}$$

Here, namely

$$\begin{cases} -2\zeta\omega_n x_2 - \omega_n^2 x_1 = 0 \\ x_2 = 0 \end{cases}$$

From which we can say

$$\begin{cases} x_1 = 0 \\ x_2 = 0 \end{cases}$$

Therefore, these is the singularity point for the second-order system.

In the system, the damping ratio ξ is different and the corresponding phase plane is also different, as shown in Figure 7.31.

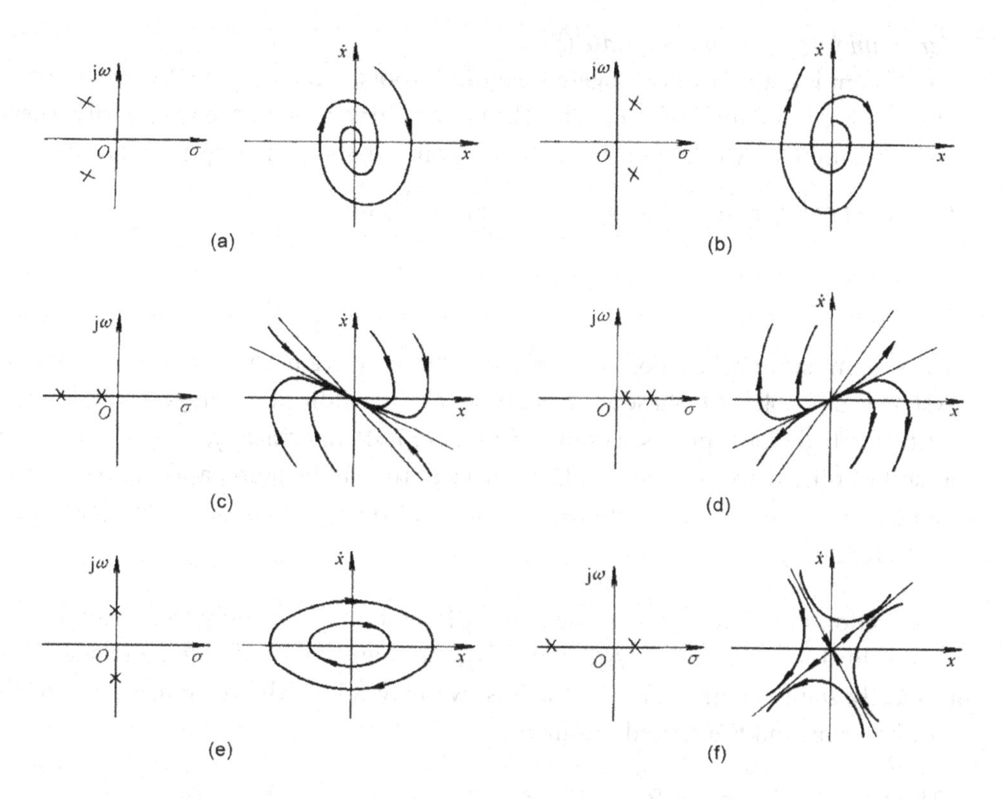

FIGURE 7.31 The different singularities of the second-order system at (a) stable state, (b) unstable state, (c) stable node, (d) unstable node, (e) center point, and (f) saddle point.

1. *Underdamped state* ($0 < \xi < 1$)

   The system has a pair of conjugate complex roots in the left half S-plane (i.e., the negative real part and complex conjugate root); the system shows stability, the phase curve is the spiral line, the phase curves converge to the singularity; therefore, the singularity is called stability in this state, as shown in Figure 7.31a.

2. *Negative damping state* ($-1 < \xi < 0$)

   The system has a pair of conjugate complex roots in the right half S-plane (that is, the positive real part and complex conjugate root); the system is not stable, it is also a spiral linear phase curve but the phase curve began emanating singularity; therefore, the singularity is called unstable state, as shown in Figure 7.31b.

3. *Overdamping* ($\xi > 1$)

   The system has two negative real roots; thus, the system shows stability and the phase curves converge to the singularity without oscillation; therefore, the singularity is called a stable node, as shown in Figure 7.31c.

4. *Negative overdamping* ($\xi < -1$)

   The system has two positive real roots, the phase curves emanating from the singularity; thus, the system is not stable; therefore, the singularity is called unstable node, as shown in Figure 7.31d.

5. *Zero damping or undamped state ($\xi = 0$)*
   The system has a pair of conjugate imaginary roots and a persistent oscillation; the phase curve is a family of curves for the closed curve around the singularity; therefore, the singularity is known as the center point, as shown in Figure 7.31e.

6. *If the item of linear* system $\ddot{x}$ *and* x *have different values, such as*

$$\ddot{x} - 2\zeta\omega_n\dot{x} - \varpi_n^2 = 0$$

   The system has to have a positive real root, but there is a negative real root; thus, the system is unstable. The phase curve assumes as the saddle point; therefore, the singularity is called saddle points, as shown in Figure 7.31f. Specifically, if the singularity is a saddle point, there is a special phase curve in the saddle point, and the error phase plane is divided into different casual areas, and then, the phase curve is called the separation line.

The classification of singularities in the above discussion applies only to a second-order system. For third- or high-order system, usually, the singularity is divided into stable and asymptotically stable or unstable. For details, we may follow the reference book of the Liapunov criterion and the related resources.

7.6.4 The Limit Cycle on the Phase Plane

A limit cycle is an isolated closed curve in the phase plane, and a system may have one or more limit cycles. A limit cycle represents a steady-state oscillation in which all phase curves near the limit cycle are rolled to or from the limit cycle. Thus, the limit cycle separates the phase plane into two planes such as internal and external. The phase curves cannot pass through the limit cycle outside the limit cycle and enter the limit cycle. Similarly, the phase curve cannot pass through the limit cycle inside the limit cycle to get out of the limit cycle. There are three basic forms of limit cycles which are as follows:

1. *Stable limit cycle*
   The phase curves converge to the limit cycle on the inside or outside of the limit cycle, and the system oscillates with equal amplitude, as shown in Figure 7.32a. In the sense that the phase curves diverge to the limit cycle on the inside of the limit cycle, the inside of the limit cycle is the unstable region. In the sense that the phase curves converge to the limit cycle of the outside the limit cycle, the outside of the limit cycle is the stable region. In this control system, the size of the limit cycle should be reduced as far as possible to meet the requirement of accuracy.

2. *Unstable limit cycle*
   The unstable limit cycles are near the limit cycle of the phase curves that are going to move away from the limit cycle. At this point, any small vibration will cause the limit cycle instability, as shown in Figure 7.32b. The inside limit cycle of the unstable region exists stable region while the outside limit cycle exists in the unstable region.

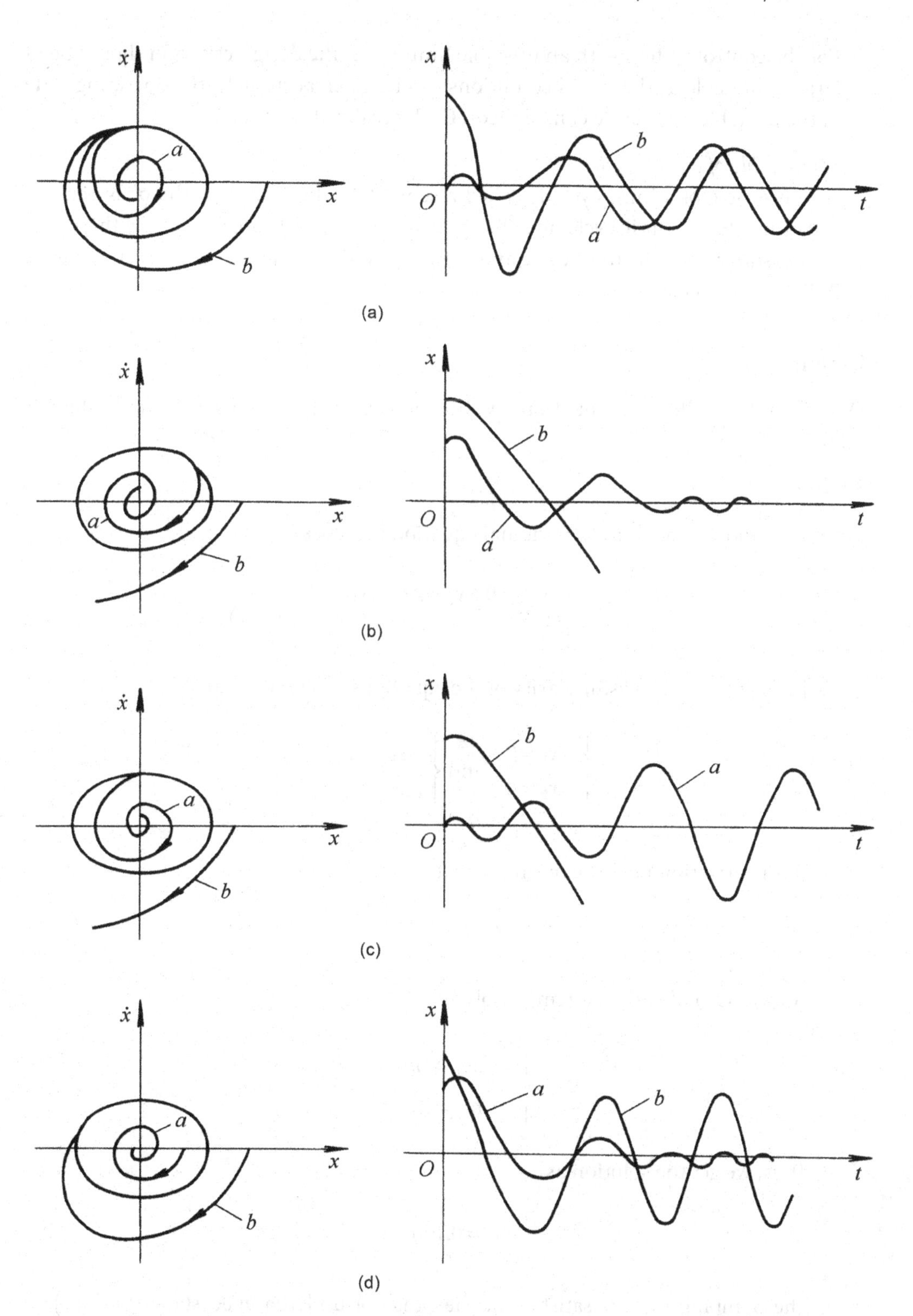

FIGURE 7.32 The limit cycle and the corresponding curve of $x\ (t)$: (a) stable limit cycle, (b) unstable limit cycle, (c) semi-stable limit cycle, and (d) stable and semi-stable limit cycle.

For the control system with an unstable limit cycle, the design criterion should be as large as possible. If the initial conditions can be given accurately, the operating state of the unstable limit cycle can be theoretically realized.

3. *Semi-stable of a limit cycle*
   A semi-stable of a limit cycle is called a semi-stable limit cycle if the phase curves start outside the limit cycle and dissipate from the limit loop; however, the phase curves start inside the limit cycle and converge to the limit cycle, as shown in Figure 7.32c, or vice versa.

Example 7.8

Described the following equation by the nonlinear phase plan of the system $\ddot{x}+0.5\dot{x}+2x+x^2=0$

Solution

Let $x_1=x$, and $x_2=\dot{x}$; thus, the original equation becomes

$$\frac{dx_2}{dx_1}=\frac{-0.5x_2-2x_1-x_1^2}{x_2}$$

1. The solution for the singularity of the equations can be written as

$$\begin{cases} x_1=0 \\ x_2=0 \end{cases} \text{and} \begin{cases} x_1=-2 \\ x_2=0 \end{cases}$$

   To linearization near the origin, we get

$$\ddot{x}+0.5\dot{x}+2x=0$$

   For the second-order system, we obtain

$$\begin{cases} 2\zeta\omega_n=0.5 \\ \omega_n^2=2 \end{cases}$$

   Thus, we get the solution as

$$\zeta=0.178$$

   The damping ratio is satisfied in the range of underdamped state $(0<\xi<1)$; thus, the singular point is a stable point. The root of the characteristic equation is $(-0.25\pm j1.39)$.

2. Near the point (–2,0), let $x^* = x + 2$; the system equation is

$$\ddot{x}^* + 0.5\dot{x}^* + 2(x^* - 2) + (x^* - 2)^2 = 0$$

The linearization is done in the vicinity of x^*; thus,

$$\ddot{x}^* + 0.5\dot{x}^* + 2x^*$$

Since $\ddot{x}^*$ and x^* are different symbols, the singularity is the saddle point. The root of the characteristic equation is 1.19 or –1.69. As shown in Figure 7.33, the phase curve uses the isoclinic line into the saddle point (–2, 0) of the two phase that divided the line. In the system, the line separating the plane will divide it into two different areas. The shaded part is the stable region. Any initial point in this region converges to the origin, and the system reaches the equilibrium state. The area under the shadow line is not enclosed by the curve; any initial point in the area of the phase curve will tend to infinity. With the increase of time, the displacement of the x increases indefinitely on the negative side.

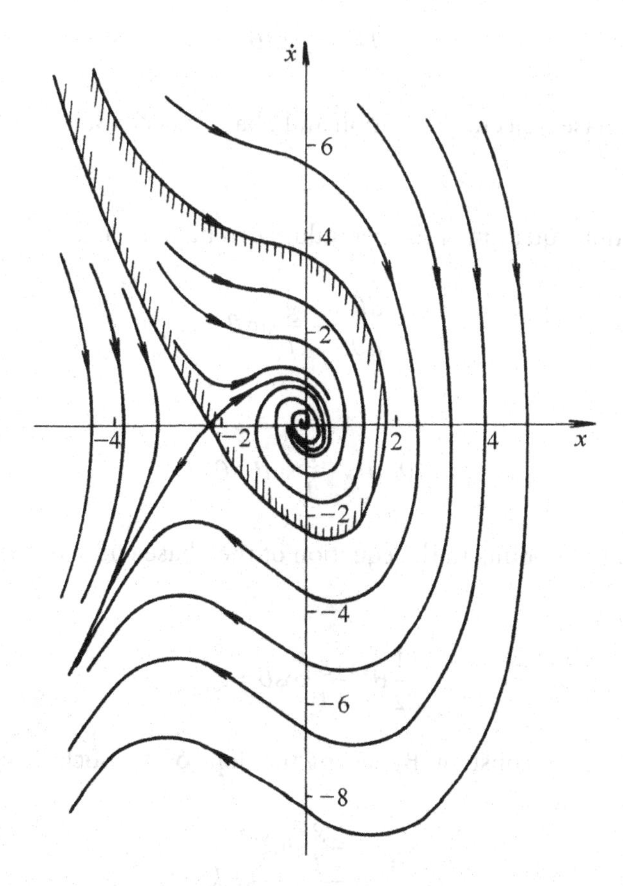

FIGURE 7.33 Phase curve diagram of $\ddot{x} + 0.5\dot{x} + 2x + x^2 = 0$.

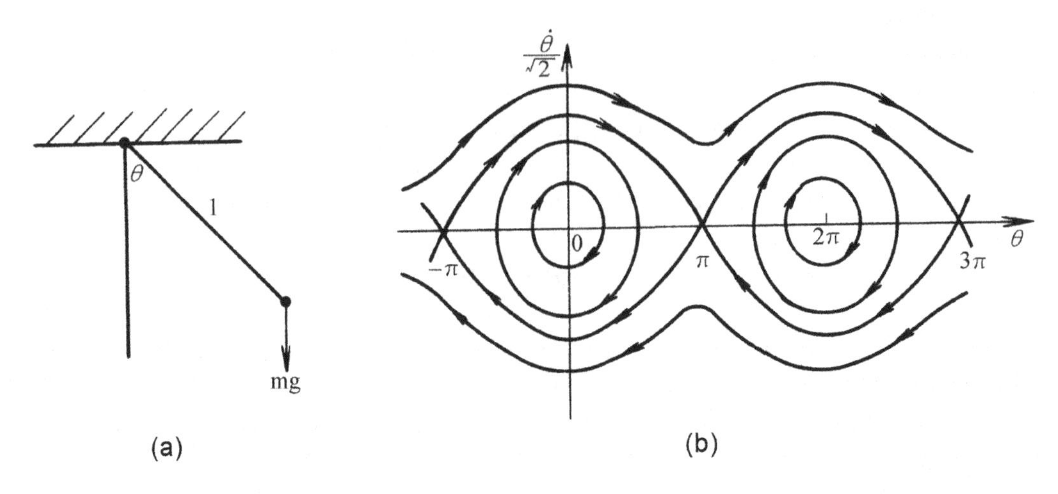

FIGURE 7.34 (a) Single pendulum, and (b) phase curve of a single pendulum system.

Example 7.9

As shown in Figure 7.34a, the equation of the single pendulum is

$$\ddot{\theta} = -\frac{g}{l}\sin\theta$$

Determine the phase trajectory equation and phase trajectory.

Solution

The transformation equation of the pendulum can be written as

$$\dot{\theta}\frac{d\dot{\theta}}{d\theta} = -\frac{g}{l}\sin\theta$$

Thus,

$$\dot{\theta}d\dot{\theta} = -\frac{g}{l}\sin\theta d\theta$$

For the two terminal points to the equation of the phase curve of the system can be written as

$$\frac{1}{2}\dot{\theta}^2 - \frac{g}{l}\cos\theta = c$$

where C is the integral constant. By setting the slope of the isoclinic line for k, we get

$$\frac{d\dot{\theta}}{d\theta} = \frac{-\frac{g}{l}\sin\theta}{\dot{\theta}} = k$$

which gives

$$\dot{\theta} = -\frac{g}{kl}\sin\theta$$

Therefore, the isoclinic line for the sine curve has a singularity, which gives $k\pi$, and the phase curves are shown in Figure 7.34b.

7.7 PHASE PLANE ANALYSIS OF NONLINEAR SYSTEMS

A nonlinear second-order system can be approximated by a linear system of several segments. The phase plane is then divided into regions, and each region is considered with a singularity. If the singularity is in the partition, it is called a real singularity. Outside of the partition, the phase curve can never reach this point in the region, which is called virtual singularity. There is only one real singularity in the second-order systems. The properties and positions of each singularity depend on the differential equations governing the region. The location of the singularity can also depend on the amount of input. The phase plane represents the phase plane of the linear system in each region. The phase curves are analyzed and synthesized by connecting the phase curves at the boundary of each working partition, which gives the transient response of the nonlinear system.

7.7.1 Nonlinear Phase Plane at Zero Input

Example 7.10

Describe the mechanical system is shown in Figure 7.34a. The mechanical model of the system is expressed as

$$\begin{cases} m\ddot{x} = -k\dot{x} - F & ;\forall \dot{x} > 0 \\ m\ddot{x} = -k\dot{x} + F & ;\forall \dot{x} < 0 \end{cases}$$

The Coulomb friction force is F, where m is the quality of the mechanical system, and K is the spring stiffness.

Solution

Let $x_1 = x$, and $x_2 = \dot{x}$; thus, the system equation can be rewritten as

$$\begin{cases} mx_2\dfrac{dx_2}{dx_1} = -kx_1 - F & ;\forall x_2 > 0 \\ mx_2\dfrac{dx_2}{dx_1} = -kx_1 + F & ;\forall x_2 < 0 \end{cases}$$

Thus,

$$\begin{cases} x_2 dx_2 = -\dfrac{k}{m}\left(x_1 + \dfrac{F}{k}\right)dx_1 & ;\forall x_2 > 0 \\ x_2 dx_2 = -\dfrac{k}{m}\left(x_1 - \dfrac{F}{k}\right)dx_1 & ;\forall x_2 < 0 \end{cases}$$

Integrating both sides, we obtain

$$\begin{cases} \dfrac{x_2^2}{C^2} + \dfrac{\left(x_1 + \dfrac{F}{k}\right)^2}{\left(C\sqrt{\dfrac{m}{k}}\right)^2} = 1 & ;\forall x_2 > 0 \quad \Leftarrow \text{The upper half phase plane} \\ \dfrac{x_2^2}{C^2} + \dfrac{\left(x_1 - \dfrac{F}{k}\right)^2}{\left(C\sqrt{\dfrac{m}{k}}\right)^2} = 1 & ;\forall x_2 < 0 \quad \Leftarrow \text{The upper half phase plane} \end{cases}$$

where C is the integral constant.

Therefore, when $x_2 > 0$, the phase curve has a family of ellipse curve at the center point $\left(-\dfrac{F}{k}, 0\right)$. Thus, when $x_2 < 0$, the phase curve has a family of ellipse curves at the center point $\left(\dfrac{F}{k}, 0\right)$. As we can see in Figure 7.35, when a particle moves along the phase curve to x_1, the shaft will stop movement between $\left(-\dfrac{F}{k}, 0\right)$ and $\left(\dfrac{F}{k}, 0\right)$; this creates the Coulomb friction by the dead zone. The x_1 shaft from $\left(-\dfrac{F}{k}, 0\right)$ to $\left(\dfrac{F}{k}, 0\right)$ parts is performed as singularity.

Example 7.11

A relay control system is shown in Figure 7.36; the features of mathematical expressions for nonlinear system are given as

$$z = \begin{cases} M & ;\forall y > h, \dot{y} < 0 \\ 0 & ;\forall |y| \langle h, |\dot{y}| \rangle 0 \\ -M & ;\forall y \langle -h, \dot{y} \rangle 0 \end{cases} \tag{7.68}$$

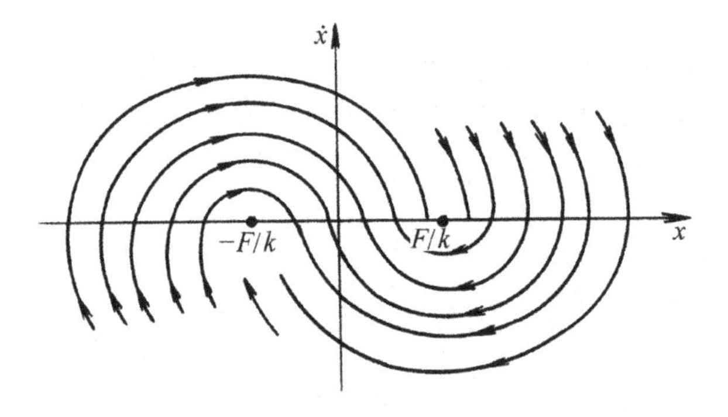

FIGURE 7.35 The phase curve of a mechanical system of Figure 7.34a.

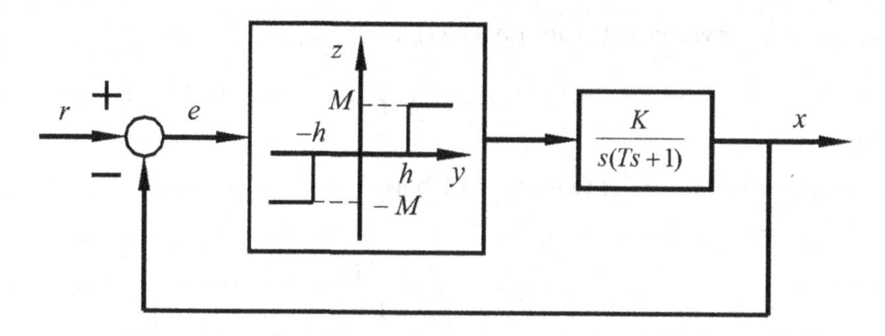

FIGURE 7.36 Block diagram of a relay system.

Draw a phase track diagram of the system.

Solution

For the system as shown in Figure 7.36, we assume that the input amount is $r = 0$, $e = x$, and then the system of equations can be described as

$$T\ddot{x}(t)+\dot{x}(t)=\begin{cases} kM & ;x>h \\ 0 & ;|x|<h \\ -kM & ;x<-h \end{cases} \tag{7.69}$$

If we consider $x(t)$ in the phase plane of the X-axis for the abscissa, the coordinate of $\dot{x}(t)$ is also in the phase plane. It is clear that the phase plane line $x=\pm h$ is divided into three different regions. The curve of the system is determined by a linear differential equation in the area of each phase. This partition is discussed below:

1. When $x>h$ region, the phase curve equation of the system is written as

$$T\ddot{x}(t)+\dot{x}(t)=-kM \tag{7.70}$$

Assume that the initial conditions $x(0)=x_0$ and $\dot{x}(0)=\dot{x}_0$; then the solution of Eq. (7.70) can be found as

$$x(t)=x_0+(\dot{x}_0+kM)T-(\dot{x}_0+kM)Te^{-\frac{1}{T}t}-kMt \tag{7.71}$$

Now we do differentiation in both sides of Eq. (7.71), then we get the derivative as

$$\dot{x}(t)=(\dot{x}_0+kM)e^{-\frac{1}{T}t}-kM \tag{7.72}$$

When $\dot{x}_0 = -kM$, we get from Eq. (7.72) as

$$\dot{x}(t) = -kM \tag{7.73}$$

When $\dot{x}_0 \neq -kM$, we get from Eq. (7.72) as

$$t = -T\ln\left|\frac{\dot{x}+kM}{\dot{x}_0+kM}\right| \tag{7.74}$$

Substituting Eq. (7.74) into Eq. (7.72), then we get

$$x(t) = x_0 + (\dot{x}_0 - \dot{x}(t))T + kMT\ln\left|\frac{\dot{x}(t)+kM}{\dot{x}_0+kM}\right| \tag{7.75}$$

Therefore, we get the phase curve of the system in the region of $x > h$ from Eqs. (7.73) and (7.75).

2. When $|x| < h$ region, the phase curve equation of the system is written as

$$T\ddot{x}(t) + \dot{x}(t) = 0 \tag{7.76}$$

Integrating Eq. (7.76) to the phase curve in the area of $|x| < h$ region, we get the slope as $-\frac{1}{T}$ line; the equation is expressed as

$$\dot{x} = \dot{x}_0 - \frac{1}{T}(x - x_0) \tag{7.77}$$

3. When $x < -h$ region, the phase curve equation of the system is written according to condition (1) as

$$\dot{x} = kM(\forall \dot{x}_0 = kM) \tag{7.78}$$

and

$$x = x_0 + (\dot{x}_0 - \dot{x})T - kMT\ln\left|\frac{\dot{x}-kM}{\dot{x}_0-kM}\right| \quad ;(\dot{x}_0 \neq kM) \tag{7.79}$$

The phase curve of the above three regions ($x > h$, $|x| < h$, $x < -h$) put them on the line to connect; then, we get the whole system phase curve, as shown in Figure 7.37.

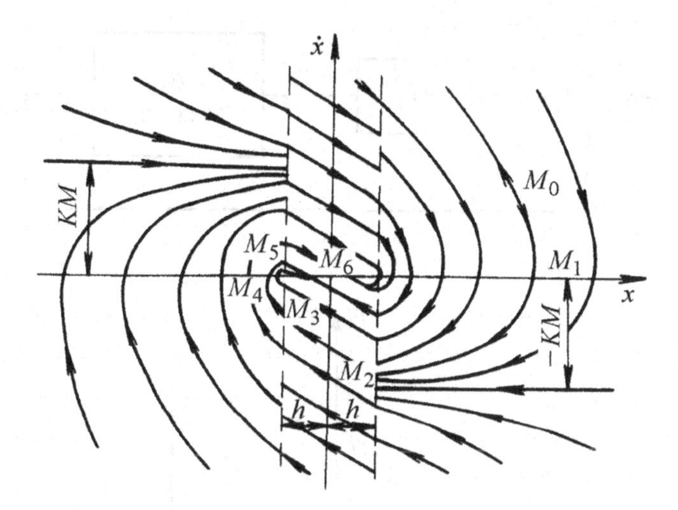

FIGURE 7.37 Phase curve diagram.

From Figure 7.37, the curve is converted in the line $x=\pm h$; this line in the control system is called switch line. It represents a kind of working transition state to another state. It starts from the point M_0 of a phase in the track or curve; after that it goes to the point M_1, M_2, M_3, M_4, and M_5, and finally it will arrive at the point M_6. At the point M_2, M_3, and M_5, the working state of the relay is converted, whereas the maximum value is x at the point M_1 and the minimum is x at the point M_4. As shown in Figure 7.37, $\dot{x}=0, |x|<h$ are the terminating line segment of a phase curve; it is called an equilibrium segment where each point corresponds to an equilibrium state.

7.7.2 Nonlinear Phase Curve of Step and Ramp Input

As mentioned above, we have discussed the phase plane analysis of a nonlinear control system with zero input. If the system input is a typical signal input such as step input, ramp input, etc., what is the phase curve of the system? The phase plane analysis of nonlinear control systems with step input and ramp input is discussed below.

Example 7.12

As we can see in Figure 7.38a, it shows the nonlinear control system. The nonlinear characteristics of it are shown in Figure 7.38b. Draw the step input and the ramp input in order to input nonlinearity, namely the system error of $\dot{e}$ and e as the phase curve diagram in the phase coordinates system.

Solution

The system equation of Figure 7.38a can be expressed as

$$T\ddot{x}+\dot{x}=Ku \tag{7.80}$$

N Functional diagram of nonlinear device.

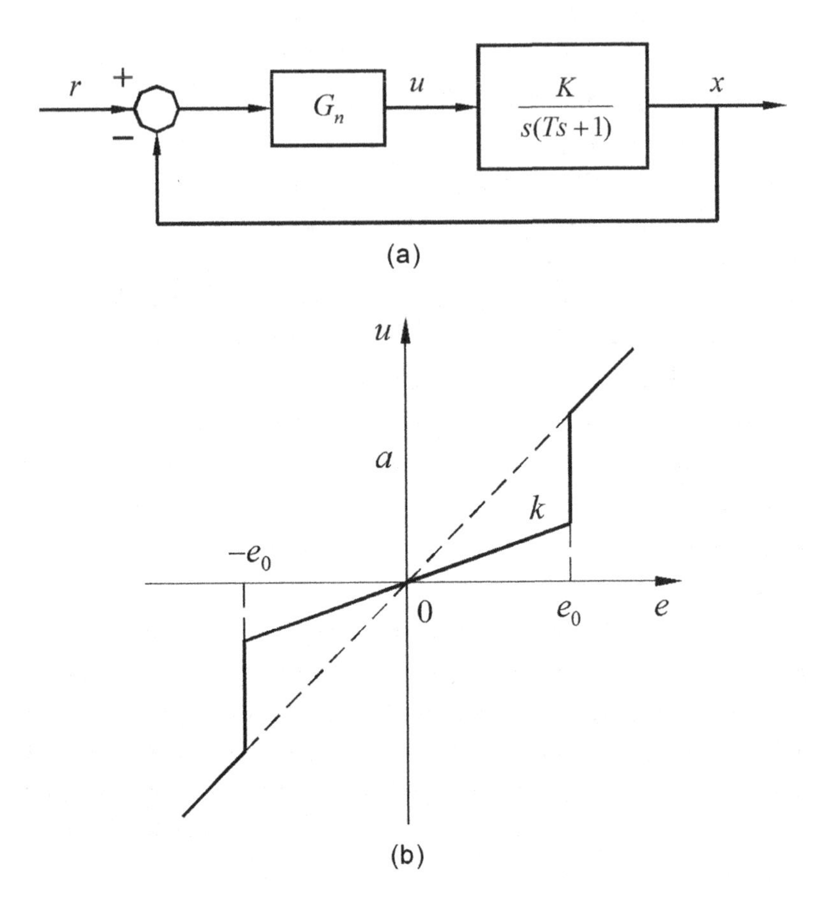

FIGURE 7.38 (a) A nonlinear control system and (b) input/output characteristics of an amplifier.

and the nonlinear characteristic of Figure 7.38b is expressed as

$$u=\begin{cases} ke & |e|<e_0 \\ e & |e|>e_0 \end{cases} \tag{7.81}$$

The error characteristics is

$$e=r-x \tag{7.82}$$

Substituting Eq. (7.82) into Eq. (7.80), then we get

$$T\ddot{e}+\dot{e}+Ku=T\ddot{r}+\dot{r} \tag{7.83}$$

It is assumed that the system parameters satisfy the following relationships as

$$\frac{1}{2\sqrt{KT}}<1<\frac{1}{2\sqrt{kKT}} \tag{7.84}$$

Therefore, the system phase curve can be divided into the following two cases for the discussion as

1. *Step input* $r(t) = R$
   when $t > 0$, then $\ddot{r} = \dot{r} = 0$, and then substituting Eq. (7.81) into Eq. (7.83), the differential equation of the system is written as

$$T\ddot{e} + \dot{e} + Ke = 0 \ \left(|e| > e_0\right) \tag{7.85}$$

$$T\ddot{e} + \dot{e} + kKe = 0 \ \left(|e| < e_0\right) \tag{7.86}$$

Let, e is in the phase plane, $\dot{e}$ is in the vertical coordinates the switch line $e = \pm e_0$ is divided into three regions of the phase plane. By using Eq. (7.85) and (7.86), it can be known in the region $e > e_0$ and $e < -e_0$. By using Eq. (7.85), the system is in an underdamped state $\left(\xi = \dfrac{1}{2\sqrt{KT}} < 1\right)$; when $R > e_0$, the initial point is assumed to be at $A(R,0)$. The phase curve starts from the point A, and its movement toward the origin as the spiral line shape is shown in Figure 7.39a as a AB segment. When the motion fall into the region of $|e| < e_0$, by using Eq. (7.85), the system is shown in overdamped condition $\left(\xi = \dfrac{1}{2\sqrt{kKT}} > 1\right)$. At phase curve by searching point B, the Helix variable is a parabola, which is trend to the origin, as shown in Figure 7.39a as BC segment, whereas the system exists at the underdamped case from the point C to D; the system is in damping state from D to E,

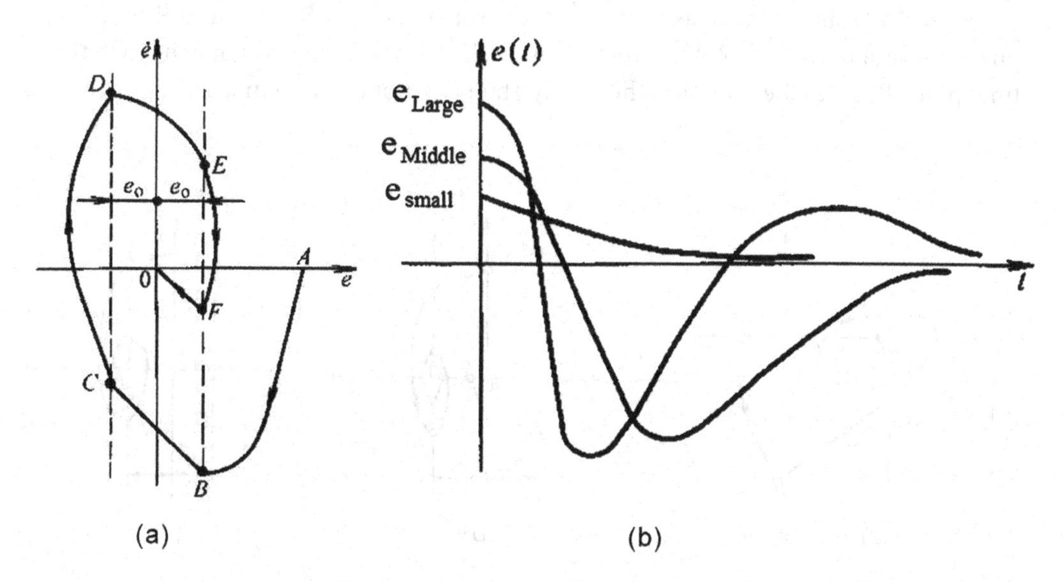

FIGURE 7.39 (a) The phase curve of the system under step signal input and (b) the transition curves of the system under different step signal inputs.

and so on. We can draw a given nonlinear control system of the phase curve as shown in Figure 7.39a as the curve is indicated as *ABCDEFO*. It is noted that the system does not exist as steady-state error for the unit step input.

From this type of nonlinear control system for phase track or curve, the response is a pure linear system (including overdamped state and underdamped state) with faster response. At the same time under the different sizes of step signal input, the transient response of the system is also different, as shown in Figure 7.39b.

2. *The input signal is $r(t) = Vt + R$ of the ramp signal*
   When $t > 0$, $\ddot{r} = 0$ and $\dot{r} = V$, $r_0 = R$, and then substituting Eq. (7.81) into Eq. (7.83), the differential equation of the system is written as

$$T\ddot{e} + \dot{e} + Ke = V \left(|e| > e_0\right) \tag{7.87}$$

$$T\ddot{e} + \dot{e} + kKe = V \left(|e| > e_0\right) \tag{7.88}$$

Compared with the step input signal, the switch line of the phase curve does not change, but the singularity position is different, and Eq. (7.86) corresponds to the singularity at the stable point $P_1\left(\dfrac{V}{K}, 0\right)$; Eq. (7.88) corresponds to the singularity at the point $P_2\left(\dfrac{V}{kK}, 0\right)$ which is called the stable node; when $k < 1$, $P_2\left(\dfrac{V}{kK}, 0\right)$ is always on the right side of $P_1\left(\dfrac{V}{K}, 0\right)$.

Figure 7.40a shows the phase curve of the error signal for $kKe_o > V$ and $R > e_0$. The phase curve started at $A(R, V)$, which then reached point B; however, it can reach the final point P_2. It can be seen that the steady-state error of the system is OP_2.

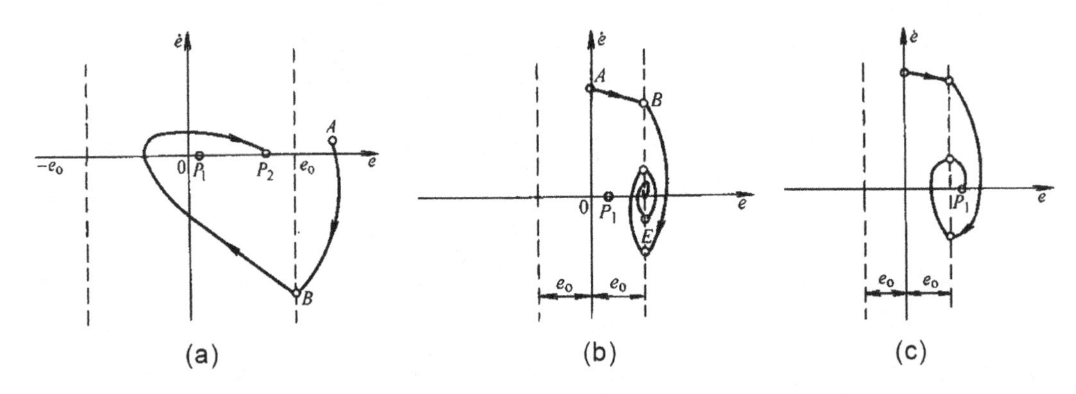

FIGURE 7.40 The phase curve of the error signal: (a) for $kKe_o > V$ and $R > e_0$, (b) when $Ke_o > V > kKe_o$ and $R = 0$, and (c) $R = 0$.

Figure 7.40b shows the phase curve of the error signal when $Ke_o > V > kKe_o$ and $R = 0$. The phase curve starts from the point $A(0, V)$, then the phase curve reaches at the point B; however, it can be reached as $ABCD$ where it ended at the final point $D(e_o, 0)$. The error signals exhibit oscillation characteristics and steady-state error e_o.

Figure 7.40c shows the phase curve of the error signal $R = 0$. The phase curve starting point is at $A(0, V)$, and then the phase curve reaches the final point P_1; the steady-state error $V > Ke_o$ difference is equal to OP_2.

$$Ke_o > V > kKe_o \text{ and } R = 0; \text{and } (c) \text{ for } R = 0$$

The above analyses show that an important characteristic of nonlinear systems is discussed, that is, the response curve can be in either aperiodic or oscillatory form if the input signal is of a different form or size.

7.8 LIAPUNOV METHOD

The Liapunov method is very effective for judging the stability of nonlinear systems and is suitable for linear systems. This book mainly discusses its application to the nonlinear stability criterion. The basic idea of the Liapunov direct method is to analyze the stability of the system from the point of view of energy change. If the energy stored in the system decreases with the passage of time, the system is stable. On the contrary, if the system constantly absorbs energy from outside in the process of motion, the system is in an unstable state. For example, we can consider a pendulum. Due to the resistance of the air, if the oscillation decreases, the system is considered to be in a stable state, whereas if a continuous force is applied to the pendulum and the oscillation of the simple pendulum increases, it can be referred to as an unstable state.

The scalar function $V(x)$ is used to express the energy of the system. According to the definition of the physical system, the energy function is always greater than 0, that is,

$$V(x) > 0 \tag{7.89}$$

where $\dot{V}(x)$ is the rate of change of energy with respect to time; thus, we get

$$\dot{V}(x) < 0 \tag{7.90}$$

That is, if the energy decreases with time, then the system is in a stable state, when

$$\dot{V}(x) > 0 \tag{7.91}$$

If the energy increases over time, the system is in an unstable state. The Liapunov is a direct method on $\dot{V}(x)$ to determine the stability of the system. From Eqs. (7.90) and (7.91), it can be seen that the Liapunov criterion is to use the energy function directly instead of solving the differential equation to determine the stability of the system; thus, it is also called the

Liapunov direct method. The scalar function $V(x)$ is called the Liapunov function; it can be a real energy function but it is also the energy function of fiction. The simplest form of the Liapunov function is considered the quadratic type of function. But being a quadratic type of function is not necessary, it may be any function. As long as it meets the Liapunov criteria, it can be used as a Liapunov function. However, the Liapunov function is not unique; thus, for determining the Liapunov function accurately, the stability of the system, by using the Liapunov direct method, is the key that can be used for judgment.

The stability criterion of the Liapunov direct method is as follows:

Theorem 7.1

The assumption of the system is subject to initial perturbations, then $x \neq 0$, $V(x) > 0$.

1. Such that $\dot{V}(x) < 0$, namely $\dot{V}(x)$ is definitely negative, then the system is gradually stable (if $x \to \infty$, $V(x) \to \infty$, the system is in a wide range of asymptotic stability)
2. If $\dot{V}(x) > 0$, namely $\dot{V}(x)$ is definitely positive, the system in the unstable state.
3. Such as $\dot{V}(x) \leq 0$ or less, or $\dot{V}(x)$ is semidefinitely negative, but $\dot{V}(x)$ is not identical to 0 (except $\dot{V}(0) = 0$), then the system is asymptotically stable.

Example 7.13

The governing equation of a control system is defined as

$$\dot{x}_1 = x_2 - x_1\left(x_1^2 + x_2^2\right)$$

$$\dot{x}_2 = -x_1 - x_2\left(x_1^2 + x_2^2\right)$$

Determine the stability of the system based on the Liapunov stability criterion.

Solution

According to the Liapunov criterion, the energy function can be expressed as

$$V(x) = x_1^2 + x_2^2$$

where the derivative of $V(x)$ can be performed by using the differentiation with respect to time *t*, thus we get

$$\dot{V}(x) = 2x_1\dot{x}_1 + 2x_2\dot{x}_2 = -2\left(x_1^2 + x_2^2\right)$$

It is obvious that $\dot{V}(x) < 0$ is negative. The system increases asymptotically stable with time t, which proves that $V(x)$ is a Liapunov function.

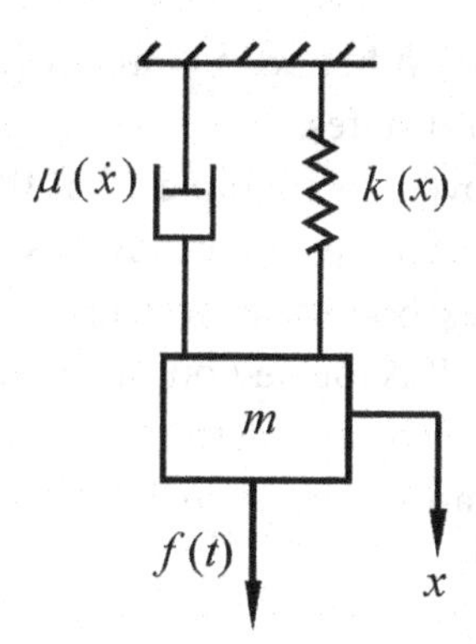

FIGURE 7.41 Mass-spring-damping system.

Example 7.14

A mass-spring-damping system is shown in Figure 7.41. Evaluate the stability of the system using the Liapunov stability criterion.

Solution

As shown in Figure 7.41 of the nonlinear system, the equation of motion for

$$m\ddot{x}+\mu(\dot{x})\dot{x}+k(x)x=f(t)$$

where μ is the damping coefficient, its value is changed along with the change of speed; K is the stiffness of the spring, and its value is changed with the change of displacement.

When $f(t)=0$ and $t\geq 0$, the equation of state is as follows:

$$\dot{x}_1=x_2$$

$$\dot{x}_2=-\frac{k(x_1)}{m}x_1-\frac{\mu(x_2)}{m}x_2$$

The total energy of the system is defined as the sum of the kinetic energy and the potential energy, and the Liapunov function is expressed as

$$V(x)=\frac{1}{2}mx_2^2+\int_0^{x_1}k(x_x)x_1dx_1$$

Since the stiffness of the spring is always greater than 0, such that $k(x_1)>0$ and $V(x)$ is a positive real number, we can say

$$\dot{V}(x)=\frac{d}{dt}\left(\frac{1}{2}mx_2^2\right)+\frac{d}{dx_1}\left[\int_0^{x_2}k(x_1)x_1\,dx_1\right]\frac{dx_1}{dt}=-\mu\left(x_2x_2^2\right)$$

Since $\mu>0$, $\dot{V}(x)$ is semidefinitely negative. In addition to the origin point $x_1=0$ and $x_2=0$, $\dot{V}(x)$ is not identical to zero. Thus, the system is asymptotically stable in an equilibrium state at the origin. When $x \to \infty$ and $V(x) \to -\infty$, the system is stable over a wide range of equilibrium states.

It can be seen from the above two examples that the Liapunov stability criterion of linear and nonlinear are very convenient but it is very difficult to construct the Liapunov function for complex nonlinear systems. Thus, the Liapunov criterion is applied by certain restrictions. It is formed on the basis of the Liapunov method and several methods such as the Krasovskii method, Schultz–Gibson method as variable gradient method, the Aizerman method, Lurie method, etc. To learn more, please refer to the relevant documents.

EXAMPLES

This chapter requires understanding the superposition principle that cannot be applied to nonlinear systems, that there are abnormal characteristics, and that there is no unified analysis method; emphatically, it is about describing function and phase plane method. The definition of the describing function N and the calculation methods will use $1/N$ curve and $G(j)$ curve analysis of system stability; however, it will determine the existence of the limit cycle system and its frequency as well as amplitude. It is clear that the basic concepts of the phase plane method, the analytic method, and the isoclinic line diagram method can be understood by the classification of the singularity and limit cycle; it will be simple to analyze with the phase plane method of the second-order nonlinear system.

Example 7.15

The structural diagram of a known nonlinear control system is shown in Figure 7.42; in order to make the system does not produce self-vibration, determine the relay characteristic parameter value a, b using the descriptive function method.

Solution

According to Table 7.1, we can get the descriptive function of the nonlinear element as

$$N(X) = \frac{4b}{\pi X}\sqrt{1-\left(\frac{a}{X}\right)^2} \quad ;\text{for } X \geq a$$

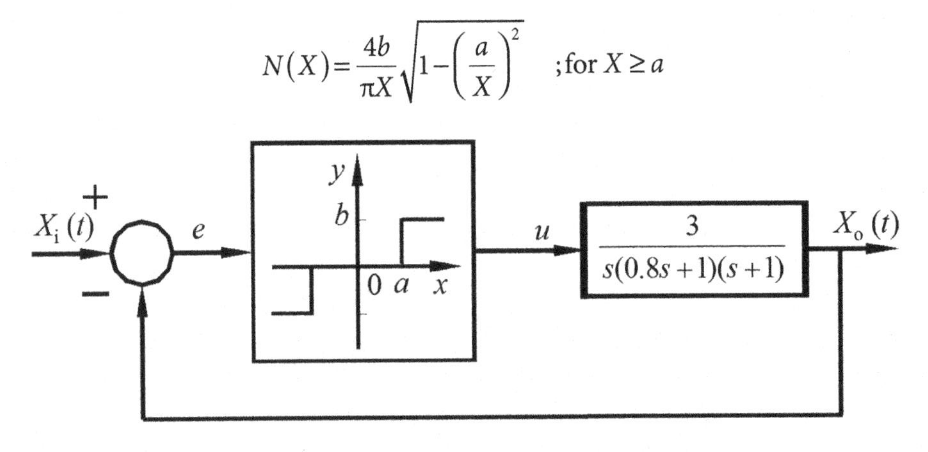

FIGURE 7.42 Diagram of a nonlinear control system.

The negative reciprocal function $\dfrac{-1}{N(X)}$ of the function can be described as

$$\frac{-1}{N(X)} = \frac{-\pi X}{4b\sqrt{1-\left(\frac{a}{X}\right)^2}}$$

When $X \to a, \dfrac{-1}{N(X)} \to \infty$.

When $X \to \infty, \dfrac{-1}{N(X)} \to -\infty$.

Thus, there is a maximum value, we get

$$\frac{d\left[-\frac{1}{N(X)}\right]}{dX} = -\frac{\pi}{4b} \cdot \frac{X^3 - 2Xa^2}{\left(X^2 - a^2\right)\sqrt{X^2 - a^2}} = 0$$

and, we get

$$X = \sqrt{2a}$$

Hence,

$$-\frac{1}{N(X)}\bigg|_{x=\sqrt{2a}} = -\frac{\pi a}{2b}$$

The transfer function of the system can be written as

$$G(j\omega) = \frac{3}{j\omega(0.8j\omega+1)(j\omega+1)}$$

Then, the intersection of the transfer function $G(j\omega)$ and the real axis is obtained. Let

$$\angle G(j\omega) = -\pi$$

Thus, we get

$$-\frac{\pi}{2} - \tan^{-1}(0.8\omega) - \tan^{-1}\omega = -\pi$$

or we obtain

$$1 - 0.8\omega^2 = 0 \;\Rightarrow \omega = \frac{\sqrt{5}}{2}$$

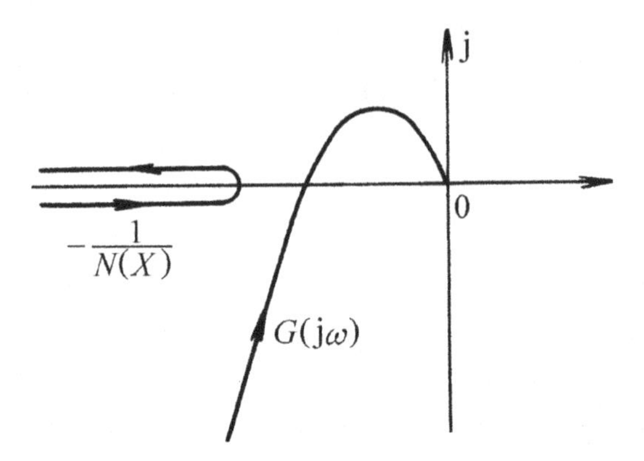

FIGURE 7.43 Curve of a known nonlinear control system.

Therefore,

$$\left|G(j\omega)\right|_{\omega}=\frac{\sqrt{5}}{2}=\frac{1}{\omega\sqrt{(0.8\omega)^2+1}.\sqrt{\omega^2+1}}\Bigg|_{\omega=\frac{\sqrt{5}}{2}}=\frac{4}{3}$$

The intersection point of $G(j)$ and the real axis is at $(-4/3, 0)$; $G(s)$ has a positive pole at $p=0$. Thus, the system does not generate natural vibration. In order to stop the system from vibrating, it should have two intersecting curves at the $-1/N(X)$ and $G(j)$, as shown in Figure 7.43.

Thus, we should have

$$-\frac{\pi a}{2b}<-\frac{4}{3}$$

and we also have

$$a>\frac{8}{3\pi}b$$

Example 7.16

The nonlinear systems with saturation properties are shown in Figure 7.44. If the system parameter $K = 2$, $T = 1$, $M = 0.2, e_o = 0.2$, draw the phase curve of the system under the condition of the unit-step signal.

Solution

The nonlinear properties of Figure 7.44 are written as

$$\begin{cases} m=e & ;(|e|<e_0) \\ m=M & ;(e\geq e_0) \\ m=-M & ;(e\leq -e_0) \end{cases}$$

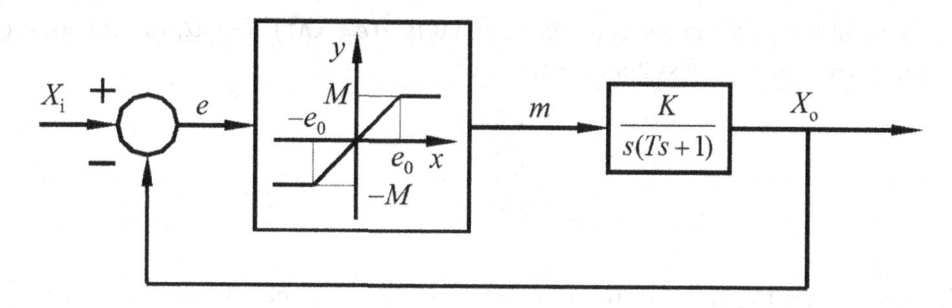

FIGURE 7.44 Nonlinear control systems.

The linear differential equations of the system are written as

$$T\ddot{x}_o + \dot{x}_o = Km$$

The system will generate the error $e = x_i - x_o$. If we substitute this error into the above equation, we get

$$T\dot{e} + \dot{e} + Km = T\ddot{x}_i + \dot{x}_i$$

For unit-step input, we get

$$x_i(t) = 1(t), \ddot{x}_i = \dot{x}_i = 0, (t > 0)$$

Thus, we may rewrite as

$$T\dot{e} + \dot{e} + Ke = 0$$

or

$$\frac{d\dot{e}}{de} = \frac{\ddot{e}}{\dot{e}} = -\frac{1}{T} \cdot \frac{\dot{e} + Ke}{\dot{e}}$$

When $e = 0, \dot{e} = 0$, $d\dot{e}/de = 0/0$; thus, the $\dot{e} - e$ plane is at origin (0, 0) and it is a singular point. With $K = 2$, $T = 1$, the roots of the characteristic equation is $\lambda_{1,2} = -\frac{1}{2} \pm j\frac{\sqrt{7}}{2}$; thus, the singularity is in a stable state. Therefore, the phase curve of the system in the linear region is a spiral line.

If the system work is in the saturated zone, the differential equations of motion can be expressed as

$$T\ddot{e} + \dot{e} + KM = 0 \qquad (e > e_0)$$

$$T\ddot{e} + \dot{e} - KM = 0 \qquad (e < -e_0)$$

The slope of the phase curve is $d\dot{e}/de=\alpha$, that is $-(\dot{e}+KM)/T\dot{e}=\alpha$, the equation of the saturated zone of the system is expressed as

$$\dot{e}=\frac{-KM}{\alpha T+1}\quad (e>e_0)$$

$$\dot{e}=\frac{-KM}{\alpha T+1}\quad (e<-e_0)$$

This shows that the waiting line is a cluster of parallel lines. When $e>e_0$, if $\alpha=0$, then all the phase curves are asymptotically in a straight line $\dot{e}=KM$.

According to the above analysis, we can know that the entire phase plane according to $|e|<e_0$, $e>e_0$, and $e<-e_0$ are divided into I, II, and III regions. When the system is under the unit-step input, phase track or curve starts at the point $A(1, 0)$. The phase curve of the system is plotted into a segment and the segment is in the direction of the isocline, as shown in Figure 7.45. The phase curve finally converges to the stable state, and there is no steady-state error. From the phase curve, the maximum overshoot of the damped oscillation, that is, the maximum length of the abscissa can also be obtained.

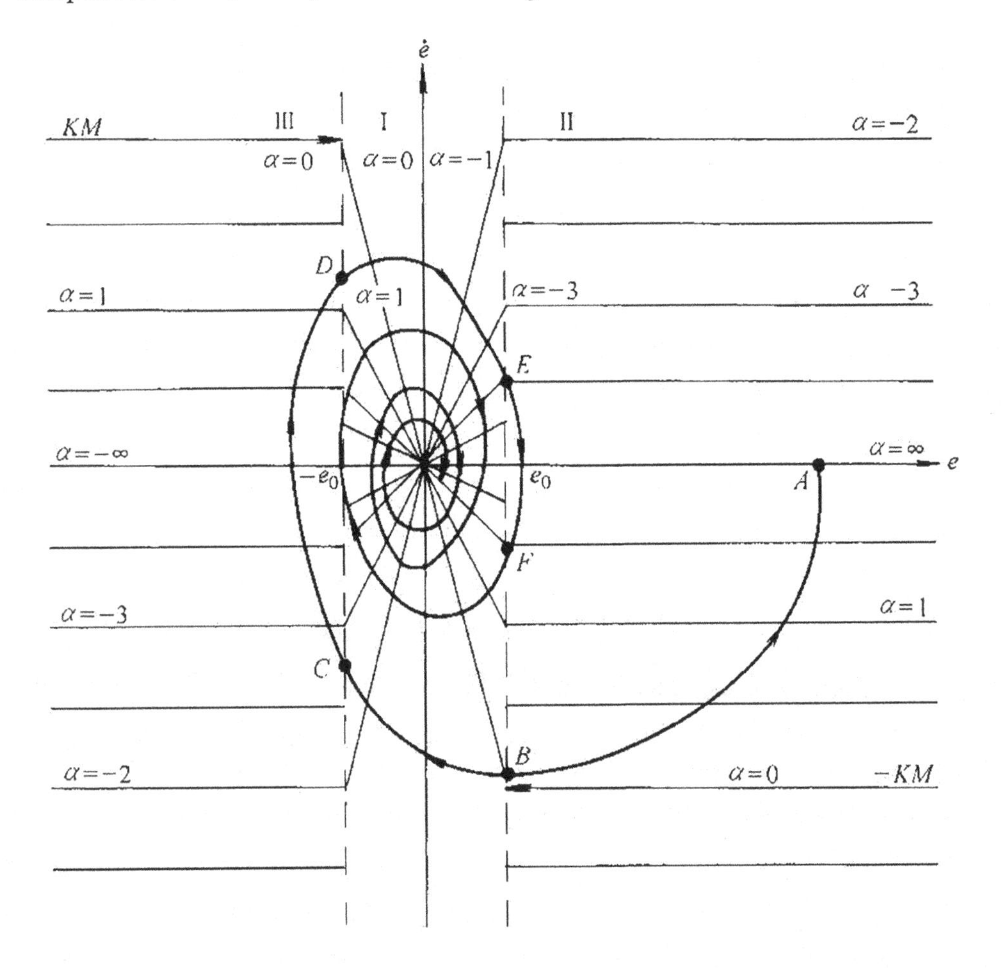

FIGURE 7.45 System phase curve.

EXERCISES

1. Find the descriptive function of the nonlinear device as shown in Figure 7.46. Draw $-1/N$ curve and point out that $X = 0$, $X = 1$ and $X = \infty$ for the value of $-1/N$.
2. Find the descriptive function N of the nonlinear system as shown in Figure 7.46b.
3. The control system with saturation nonlinearity is shown in Figure 7.47. Find the free motion of the system at $K=15$. And what is the critical stability of K in order for the system to work stably without self-oscillation?
4. Given the structure of the servomechanism system with a spring shaft, as shown in Figure 7.48, determine whether the system is stable of the following transfer functions using the descriptive function method. Is there a natural vibration? If so, what are the parameters?

   1. $G(s)=\dfrac{4000}{s(20s+1)(10s+1)}$

   2. $G(s)=\dfrac{20}{s(10s+1)}$

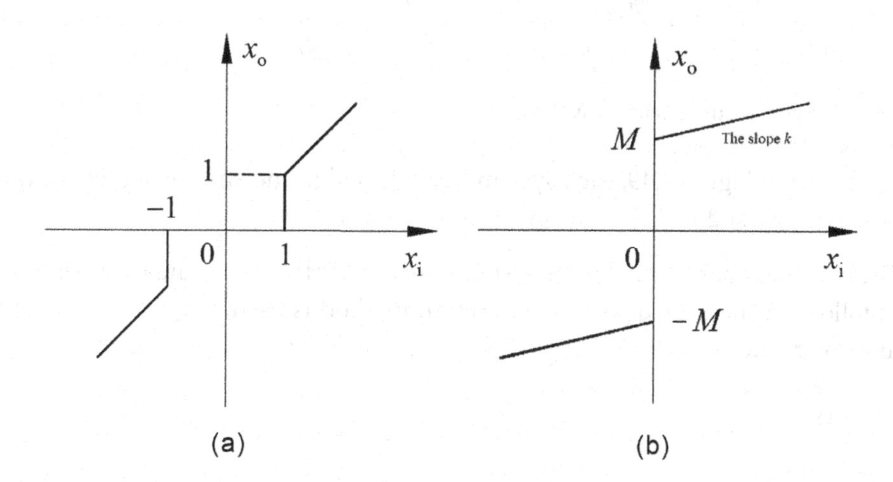

FIGURE 7.46 Functional diagram of nonlinear device.

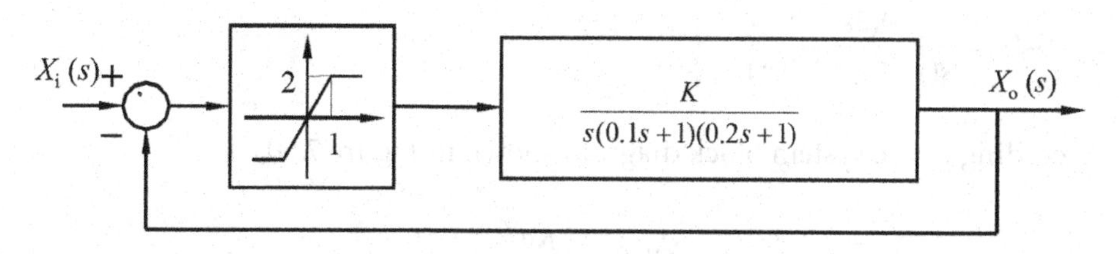

FIGURE 7.47 N functional diagram of nonlinear device.

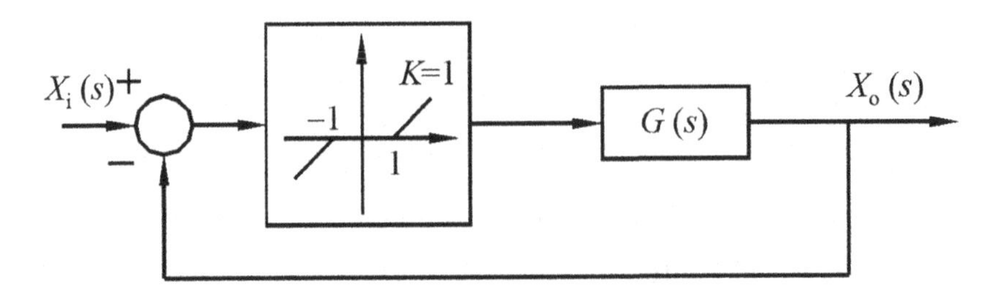

FIGURE 7.48 Structure of servomechanism system with a spring shaft.

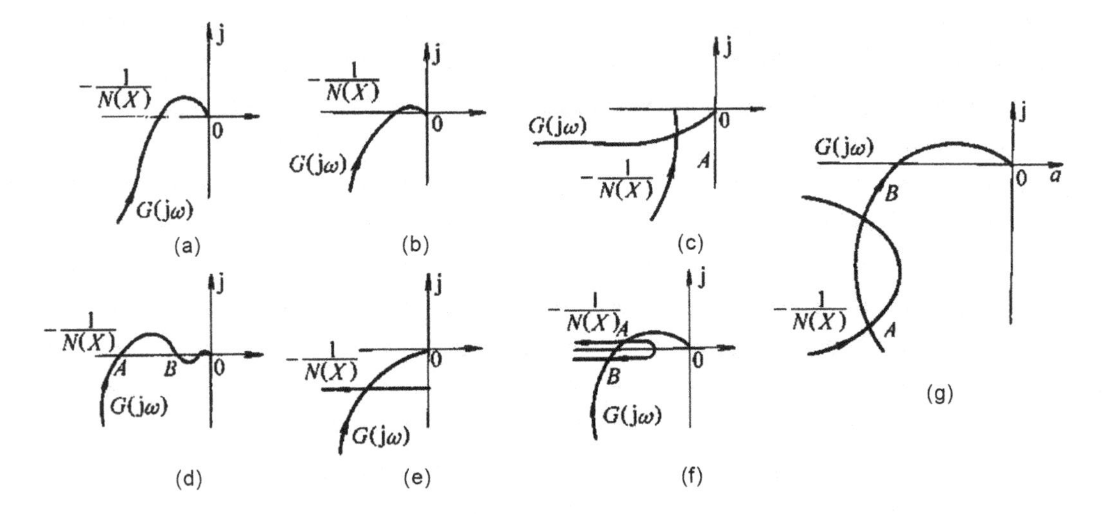

FIGURE 7.49 System curve with -1/$N(X)$ and $G(j\omega)$.

5. As shown in Figure 7.49, each system is stable; judge and show whether the intersection $-1/N(X)$ and $G(j\omega)$ is due to natural vibration.

6. The nonlinear parts of the three systems are exactly the same, and the linear parts are as follows. When the descriptive function method is used, which system analysis is more accurate?

   1. $G_1(s)=\dfrac{10}{s(s+1)}$

   2. $G_2(s)=\dfrac{10}{s(0.1s+1)}$

   3. $G_3(s)=\dfrac{10(3s+1)}{s(s+1)(0.1s+1)}$

7. According to the system block diagram shown in Figure 7.50,

$$G(s)=\frac{Ke^{-0.1s}}{s(0.1s+1)}$$

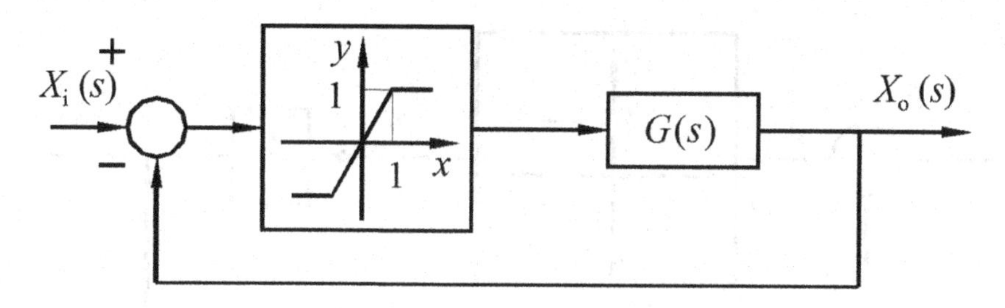

FIGURE 7.50 Block diagram of $G(s)=(K_e^{(-0.1s)})/s(0.1s+1)$.

Determine the stability of the system when $K=1$. What range should K be limited to so that the system will not self-oscillate?

8. Prove the isocline method of

$$\ddot{x}+2\zeta\omega_n\dot{x}+\omega_n^2x=0(\zeta>1)$$

There are two lines passing through the origin in the phase curve, and their slopes are the two characteristic roots of the differential equation.

9. Draw

$$T\ddot{x}+\dot{x}=M$$

When the curve satisfy as (T > 0, M > 0).

10. Determine the nature and the position of the singularity of the equations and draw the approximate figure of the phase curve.

    1. $\ddot{x}+\dot{x}+2x=0$
    2. $\ddot{x}+\dot{x}+2x=1$
    3. $\ddot{x}+3\dot{x}+x=0$
    4. $\ddot{x}+3\dot{x}+x+1=0$

11. Find the singularities of the following equations and determine the type of singularities.

    1. $2\ddot{x}+\dot{x}^2+x=0$
    2. $\ddot{x}-(1-\dot{x}^2)\dot{x}+x=0$

12. The characteristics of the phase curve of the system are shown in Figure 7.51. Analyze the system by using the phase plane method under the following conditions: $b=0$, $b<0$ and $b>0$.

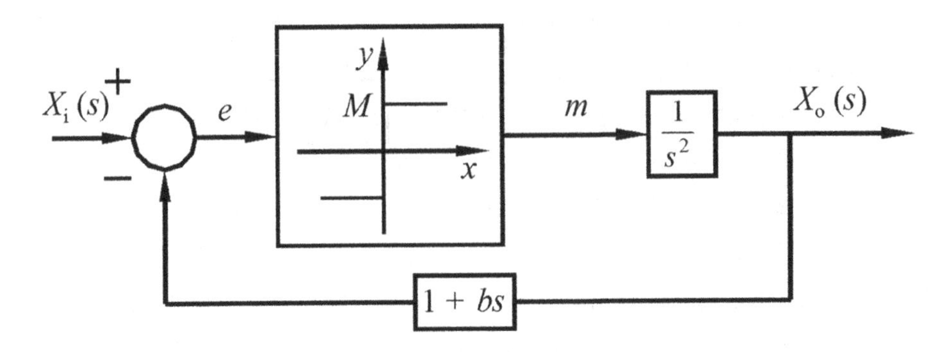

FIGURE 7.51 Block diagram of a phase curve.

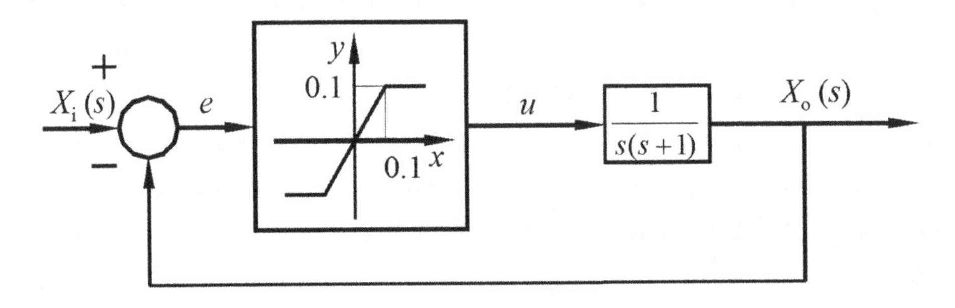

FIGURE 7.52 Structural block diagram of a nonlinear system.

13. Let the equation of the nonlinear system be

$$2\ddot{x}+\dot{x}^2+x=0$$

Draw the phase plane diagram of the system.

14. The structural diagram of the nonlinear system is shown in Figure 7.52. Describe the phase plan of the system. Set the input $x_i = 0$.

CHAPTER 8

Distributed Control System

8.1 THE Z-TRANSFORM

The effect(s) or role of the Z-transform on discrete systems is similar to the effect of the Laplace transform on continuous systems. After sampling the continuous signal, the discrete function is generated to analyze the distributed system as Z-transform. Therefore, we must discuss the sampler and the holder first.

8.1.1 Sampler and Holder

8.1.1.1 Sampling

The sampler is one of the most basic elements of a discrete system. An ordinary sampler throws the switch every T second to make the input signals that pass through the switch or sampler that is called sampling or sampling process, as shown in Figure 8.1a–c. Figure 8.1a shows the continuous signal and Figure 8.1b shows the sampler system. A sampler is used to change the continuous signals into a series of pulse signals that happened in 0, T, $2T$, $3T$, …, and so on. Figure 8.1c shows a series of pulse signals that is called discrete signals. Note that T is the sampling period. No information transmits through the sampler between two instants sampling. But at the instant sampling time, the two signals share the same value.

8.1.1.2 Holding

The holder can transfer sampled signals or discrete signals into continuous signals that can be approximately reproduced signals in the sampler. The simplest zero-order holder can change the sampling signals into signals between two continuous sampling at the same time with remain constant, as shown in Figure 8.1d and e.

When the incoming or input signal $x(t)$ is sampled at the instant discrete time, the sampling signals pass through the holder. In fact, this kind of holder corresponds to a low-pass filter (LPF), which can smoothen the signal $x^*(t)$ after being sampled, and then, it will produce the output signal $x_o(t)$. The output $x_o(t)$ will remain the last value before the next sampling, that is, $x_o(t)$ is a constant. Hence, the output signal $x_o(t)$ can be represented as

$$x_o(kT+t)=x_o(kT)\,;\ \text{for } 0\le t<T \tag{8.1}$$

DOI: 10.1201/9781003293859-8

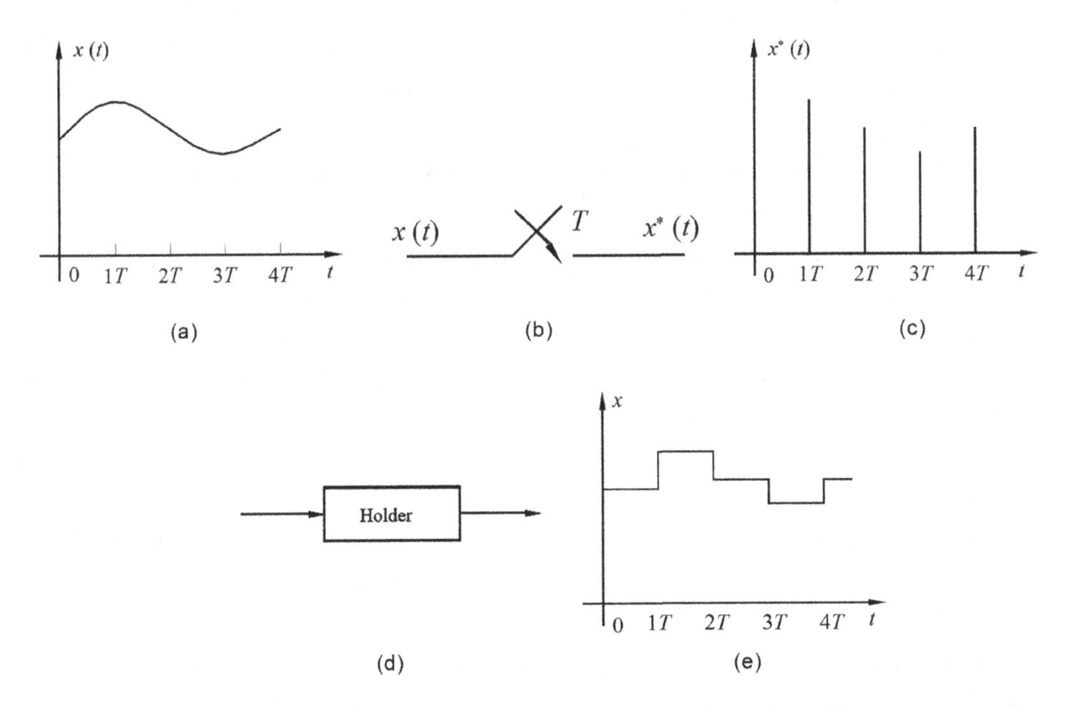

FIGURE 8.1 Sampling and holding: (a) continuous signals before sampling, (b) sampler, (c) discrete signals after sampling, (d) holder system, and (e) after holder.

Suppose that the sampler outputs are a series of weighted pulses so that continuous signals $x(t)$ can be sent through the sampler, and then we get the output signal $x^*(t)$ from the sampler with the pulse signal. By connecting an impulse signal $\delta_T(t)$, we get

$$x^*(t) = \delta_T(t)x(t) \tag{8.2}$$

It can be generated from a pulse train from Eq. (8.2), as shown in Figure 8.2a. Therefore, the sampler can be seen as a modulator, as shown in Figure 8.2b.

Let $x(t)$ be a modulating signal and $\delta_T(t)$ be the unit pulse train that can be used as a carrier signal in the modulator system; however, $x^*(t)$ is the output of the modulator. The mathematical expression of impulse signal $\delta_T(t)$ is as follows:

$$\delta_T(t) = \sum_{k=-\infty}^{\infty} \delta(t - kT) \tag{8.3}$$

After inputting the carrier signal $\delta_T(t)$ to the continuous signal $x(t)$, it will be modulated and then the output $x^*(t)$will be produced. The modulation processes are as follows:

$$x^*(t) = x(t)\delta_T(t) = x(t)\sum_{k=-\infty}^{\infty} \delta(t - kT) \tag{8.4}$$

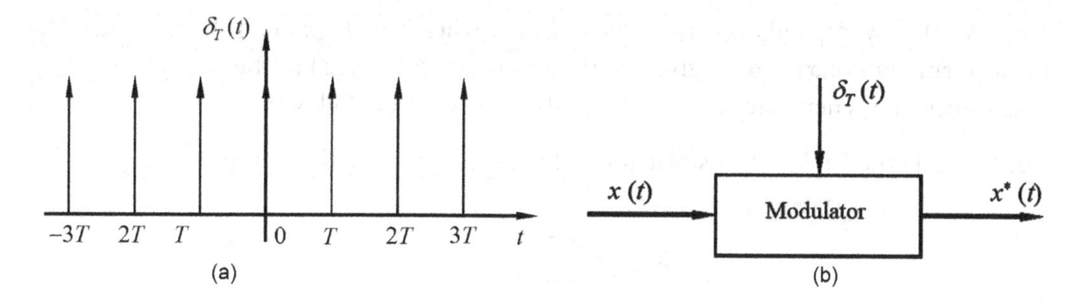

FIGURE 8.2 Sampling: (a) unit pulse train and (b) sampling from carrier modulator $\delta_T(t)$.

In control systems and engineering applications, when $t<0$, all the signals are zero. That is,

$$x(t)=0;(t<0) \tag{8.5}$$

Hence,

$$\begin{aligned} x^*(t) &= x(t)\sum_{k=0}^{+\infty}\delta(t-kT) \\ &= x(0)\delta(t)+x(T)\delta(t-T)+x(2T)\delta(t-2T)+\cdots+x(kT)\delta(t-kT)+\cdots \\ &= \sum_{k=0}^{\infty}x(kT)\delta(t-kT) \end{aligned} \tag{8.6}$$

8.1.2 Evaluate the Z-Transformations by Definition

8.1.2.1 Definition of Z-Transform

Given Laplace transformation in Eq. (8.6), we get

$$X^*(s)=\mathcal{L}\left[x^*(t)\right]=\sum_{k=0}^{\infty}x(kT)e^{-kTs} \tag{8.7}$$

Let $z=e^{sT}$ and substitute it into Eq. (8.7), and then change $X^*(s)$ into $X(z)$; we get

$$X(z)=X^*(s)\Big|_{s=\frac{1}{T}\ln z}=X^*\left(\frac{1}{T}\ln z\right)=\sum_{k=0}^{\infty}x(kT)z^{-k} \tag{8.8}$$

We can define that $X(z)$ is the *Z-transformation* of $x^*(t)$ which can be expressed by $Z\left[x^*(t)\right]$. Since we only consider instantaneous signals in Z-transform, the Z-transformation of $x(t)$ is equal to $x^*(t)$. That is,

$$Z\left[x(t)\right]=Z\left[x^*(t)\right]=X(z)=\sum_{k=0}^{\infty}x(kT)z^{-k} \tag{8.9}$$

where $X(z)$ only depends on the values of $x(t)$ when $t=kT$ (k=0, 1, 2, …); thus, the Z-transformation of $x(t)$ only gives us the information of $X(z)$ at the instant sampling time. Hence, the general steps to evaluate Z-transform are as follows:

1. Evaluate the Laplace transformation of the sampling signal $x^*(t)$, and then get

$$X^*(s)=\sum_{k=0}^{\infty}x(kT)e^{-kTs}$$

2. Put $z=e^{Ts}$ into $X^*(s)$ and then obtain $Z[\,x(t)]$, that is $X(z)$.

If the inputs are the switching signals, then we can evaluate the Z-transformations according to the above two steps. If it is unsatisfactory, we should use Eq. (8.9).

Example 8.1

Let the input $x(t)=\delta(t)$. Evaluate $Z[\delta(t)]$.

Solution

Since the input is the switching signal, first, we evaluate the Laplace transformation of $X^*(s)=1$ as

$$Z[\delta(t)]=X(z)=X^*(s)\Big|_{s=\frac{1}{T}\ln z}=1 \tag{8.10}$$

Example 8.2

Let the input $x(t)=\delta_T(t)=\sum_{k=0}^{\infty}\delta(t-kT)$. Note that T is the sampling period. Evaluate $Z[\delta_T(t)]$.

Solution

We have T is the sampling period. Thus, taking the Laplace transformation to the incoming signals, we get

$$X^*(s)=\sum_{k=0}^{\infty}e^{-kTs}$$

Hence,

$$X(z)=X^*(s)\Big|_{s=\frac{1}{T}\ln z}=\sum_{k=0}^{\infty}z^{-k}=1+z^{-1}+z^{-2}+\cdots$$

Thus,

$$Z[\delta_T(t)]=\frac{1}{1-z^{-1}}=\frac{z}{z-1}\qquad ;\left(\left|z^{-1}\right|<1\right) \tag{8.11}$$

Example 8.3

Let $x(t)=1(t)$. Evaluate $Z[1(t)]$.

Solution

According to Eq. (8.9), we get

$$Z[1(t)]=\sum_{k=0}^{\infty}1(kT)z^{-k}$$
$$=1+z^{-1}+z^{-2}+\cdots=\frac{z}{z-1} \quad (8.12)$$

From Eqs. (8.11) and (8.12), we can see that the different incoming signals may have the same output. Thus, this is a question that we must especially pay attention to in Z-transformation.

Example 8.4

Evaluate the Z-transformation of the incoming signal $x(t)=e^{-at}$ $(t\geq 0)$.

Solution

$$Z[e^{-at}]=\sum_{k=0}^{\infty}e^{-akT}z^{-k}=1+e^{-aT}z^{-1}+e^{-2aT}z^{-2}+\cdots=\frac{z}{z-e^{-aT}}.$$

Example 8.5

Evaluate the Z-transformation of $x(t)=\sin\omega t$ $(t\geq 0)$.

Solution

Since $\sin\omega t=\dfrac{e^{j\omega t}-e^{-j\omega t}}{2j}$

$$Z\left[\sin\omega t\right]=Z\left[\frac{e^{j\omega t}-e^{-j\omega t}}{2j}\right]=\frac{1}{2j}\left(\frac{z}{z-e^{j\omega T}}-\frac{z}{z-e^{-j\omega T}}\right)$$

$$=\frac{1}{2j}\frac{z\left(e^{j\omega T}-e^{-j\omega T}\right)}{z^2-z\left(e^{j\omega T}+e^{-j\omega T}\right)+1}$$

$$=\frac{z\sin\omega T}{z^2-2z\cos\omega T+1}$$

Example 8.6

Evaluate the Z-transformation of $X(s)=\dfrac{1}{s(s+1)}$.

Solution

We evaluate the partial fractions of $X(s)$, we get

$$X(s)=\frac{1}{s(s+1)}=\frac{1}{s}-\frac{1}{s+1}$$

And then evaluate the Z-transformation of each partial fraction. However, taking Z-transformation of $X(s)$, we obtain

$$Z[X(s)]=Z\left[\frac{1}{s}-\frac{1}{s+1}\right]$$

Since the Z-transformation of $\frac{1}{s}$ (or $1(t)$) is $\frac{z}{z-1}$, the Z-transformation of $\frac{1}{s+1}$ (or e^{-t}) is $\frac{z}{z-e^{-T}}$.

Hence, we get

$$Z[x(t)]=\frac{z}{z-1}-\frac{z}{z-e^{-T}}=\frac{z(1-e^{-T})}{(z-1)(z-e^{-T})}$$

8.1.3 Evaluate Z-Transformations by Residue Theorem

Let $X(s)$ is the ratio of the two polynomials of s, and then all the poles of $X(s)$ are in the left half S-plane, in which the time of the denominator is more than the time of the numerator. That is, $\lim_{s\to\infty} X(s)=0$. Then, the Z-transformation of $X(s)$ is expressed as

$$X(z)=\sum_{i=1}^{m}\left\{\frac{1}{(n_i-1)!}\frac{d^{n_i-1}}{ds^{n_i-1}}\left[\frac{(s-s_i)^{n_i}X(s)z}{z-e^{Ts}}\right]\right\}\Bigg|_{s=s_i} \tag{8.13}$$

$$=\sum\frac{X(s)z}{z-e^{Ts}}\ [X(s)\text{ is the residues of the poles}]$$

where s_i is the pole of $X(s)$, n_i is the multiplicity of the multiple pole s_i, and m is the number of the poles that are different from each other.

Example 8.7

Given $X(s)=\frac{s+3}{(s+1)(s+2)}$, evaluate $X(z)$.

Solution

Since $\lim_{s\to\infty} X(s)=0$, according to the residue theorem, we get

$$X(z)=\left[\frac{(s+3)z}{(s+2)(z-e^{Ts})}\right]\Bigg|_{s=-1}+\left[\frac{(s+3)z}{(s+1)(z-e^{Ts})}\right]\Bigg|_{s=-2}$$

Thus,

$$Z(z)=\frac{2z}{z-e^{-T}}-\frac{z}{z-e^{-2T}}$$

Example 8.8

Given $X(s)=\frac{1}{(s+a)^2}$, evaluate $X(z)$.

Solution

Since $\lim_{s\to\infty} X(s)=0$, according to the residue theorem, we get

$$X(z)=\frac{X(s)z}{z-e^{Ts}} \text{ ; where } s=-a \text{ is the residue}$$

$$=\frac{d}{ds}\left[\frac{z}{z-e^{Ts}}\right]\Bigg|_{s=-a}=\frac{Tze^{Ts}}{(z-e^{Ts})^2}\Bigg|_{s=-a}=\frac{Tze^{-aT}}{(z-e^{-aT})^2}$$

8.1.4 Properties of Z-Transform

8.1.4.1 Linear Theorem

Let,

$$Z[x_1(t)]=X_1(z), Z[x_2(t)]=X_2(z) \tag{8.14}$$

Then,

$$Z[a_1x_1(t)\pm a_2x_2(t)]=a_1X_1(z)\pm a_2X_2(z) \tag{8.15}$$

Proof:

$$Z[a_1x_1(t)\pm a_2x_2(t)]=\sum_{k=0}^{\infty}[a_1x_1(kT)\pm a_2x_2(kT)]z^{-k}$$

$$=a_1\sum_{k=0}^{\infty}x_1(kT)z^{-k}\pm a_2\sum_{k=0}^{\infty}x_2(kT)z^{-k}$$

$$=a_1X_1(z)\pm a_2X_2(z)$$

8.1.4.2 Initial Value Theorem

$$x(0)=\lim_{z\to\infty}X(z) \tag{8.16}$$

Proof: Since

$$X(z)=\sum_{k=0}^{\infty}x(kT)z^{-k}=x(0)+x(T)z^{-1}+x(2T)z^{-2}+\cdots+x(kT)z^{-k}+\cdots$$

When $z \to \infty$, we get

$$x(0) = \lim_{z\to\infty} X(z)$$

Example 8.9

Evaluate the initial value of $X(z) = \dfrac{z}{z - e^{-T}}$.

Solution

$$x(0) = \lim_{z\to\infty} X(z) = \lim_{z\to\infty} \frac{z}{z - e^{-T}} = 1$$

8.1.4.3 Final Value Theorem

$$x(\infty) = \lim_{z\to 1}\left[X(z)(z-1)\right] \tag{8.17}$$

Proof: Since $Z\left[x(kT)\right] = \sum_{k=0}^{\infty} x(kT) z^{-k} = X(z)$.

Thus,

$$Z\left\{x\left[(k+1)\right]T\right\} = \sum_{k=0}^{\infty} x\left[(k+1)T\right] z^{-k}$$

$$= \left\{\sum_{k=0}^{\infty} x\left[(k+1)T\right] z^{-(k+1)}\right\} z = zX(z) - zx(0)$$

Therefore,

$$\sum_{k=0}^{\infty} x\left[(k+1)T\right] z^{-k} - \sum_{k=0}^{\infty} x(kT) z^{-k} = zX(z) - zx(0) - X(z) \tag{8.18}$$

Thus,

$$\sum_{k=0}^{\infty} x\left[(k+1)T\right] z^{-k} - \sum_{k=0}^{\infty} x(kT) z^{-k} = (z-1)X(z) - zx(0) \tag{8.19}$$

We evaluate Eq. (8.19) by taking the limit on both sides of Eq. (8.19) and get

$$\lim_{z\to 1}\left[\sum_{k=0}^{\infty} x\left[(k+1)T\right] z^{-k} - \sum_{k=0}^{\infty} x(kT) z^{-k}\right] = \lim_{z\to 1}\left[(z-1)X(z) - zx(0)\right] \tag{8.20}$$

From Eq. (8.20), we can obtain

$$x(\infty) - x(0) = \lim_{z\to 1}\left[(z-1)X(z) - zx(0)\right] \tag{8.21}$$

Hence,

$$x(\infty)=\lim_{z\to 1}\left[X(z)(z-1)\right] \tag{8.22}$$

8.1.4.4 Real Shifting Theorem

Given the Z-transformation of $x(t)$ is $X(z)$, then

$$Z\left[x(t-kT)\right]=z^{-k}X(z) \tag{8.23}$$

and

$$Z\left[x(t+kT)\right]=z^{k}\left[X(z)-\sum_{j=0}^{k-1}x(jT)z^{-j}\right] \tag{8.24}$$

Proof: According to the definition of Z-transform, we get

$$Z\left[x(t-kT)\right]=\sum_{j=0}^{\infty}x(jT-kT)z^{-j} \tag{8.25}$$

Eq. (8.25) can be written as

$$Z\left[x(t-kT)\right]=z^{-k}\sum_{j=0}^{\infty}x(jT-kT)z^{-(j-k)} \tag{8.26}$$

Since $t<0$, and $x(t)=0$, Eq. (8.26) can be written as

$$Z\left[x(t-kT)\right]=z^{-k}\sum_{j=k}^{\infty}x(jT-kT)z^{-(j-k)}=z^{-k}X(z)$$

And from Eq. (8.24), we can proof that

$$Z\left[x(t+kT)\right]=\sum_{j=0}^{\infty}x(jT+kT)z^{-j}$$

$$=z^{k}\sum_{j=0}^{\infty}x(jT+kT)z^{-(j+k)}$$

$$=z^{k}\left[\sum_{k=0}^{\infty}x(kT)z^{-k}-\sum_{j=0}^{k-1}x(jT)z^{-j}\right]$$

$$=z^{k}\left[X(z)-\sum_{j=0}^{k-1}x(jT)z^{-j}\right]$$

When $k=1$, we get from Eq. (8.24) as

$$Z\left[x(t+T)\right]=zX(z)-zx(0) \tag{8.27}$$

When $k=2$, we get

$$Z\left[x(t+2T)\right]=z^2X(z)-z^2x(0)-zx(T) \tag{8.28}$$

When $k=m$, we get

$$\begin{aligned} Z\left[x(t+mT)\right]&=z^mX(z)-z^mx(0)-z^{m-1}x(T) \\ &-z^{m-2}x(2T)-\cdots-zx\left[(m-1)T\right] \end{aligned} \tag{8.29}$$

Therefore, the real shifting theorem is frequently used in differential equations.

Example 8.10

Evaluate the Z-transformation of $1(t-T)$.

Solution

According to Eq. (8.23), we obtain

$$Z\left[1(t-T)\right]=z^{-1}Z\left[1(t)\right]=z^{-1}\frac{z}{z-1}=\frac{1}{z-1}$$

Example 8.11

Evaluate the Z-transformation of $e^{-a(t-T)}$.

Solution

According to Eq. (8.23), we obtain

$$Z\left[e^{-a(t-T)}\right]=z^{-1}Z\left[e^{-at}\right]=z^{-1}\frac{z}{z-e^{-aT}}=\frac{1}{z-e^{-aT}}$$

8.1.4.5 Complex Shifting Theorem

Given the Z-transformation of $x(t)$ is $X(z)$, then

$$Z\left[x(t)e^{\mp at}\right]=X\left(ze^{\pm aT}\right) \tag{8.30}$$

Proof: According to the definition of Z-transform, we get

$$Z\left[x(t)e^{\mp at}\right]=\sum_{k=0}^{\infty}x(kT)e^{\mp akT}z^{-k}=\sum_{k=0}^{\infty}x(kT)(ze^{\pm aT})^{-k}$$

Let $z_1 = ze^{\pm aT}$, and then the above equation can be written as

$$Z\left[x(t)e^{\pm at}\right] = \sum_{k=0}^{\infty} x(kT) z_1^{-k} = X(z_1)$$

Therefore,

$$Z\left[x(t)e^{\mp at}\right] = X\left(ze^{\pm aT}\right)$$

Example 8.12

Evaluate the Z-transformation of $x(t) = e^{-at}\sin\omega t$.

Solution

From Form 10–1, we can obtain the Z-transformation of sin ωt is

$$Z[\sin\omega t] = \frac{z\sin\omega T}{z^2 - 2z\cos\omega T + 1}$$

According to the complex shifting theorem, we get

$$Z\left[e^{-at}\sin\omega t\right] = \frac{ze^{aT}\sin\omega T}{z^2e^{2aT} - 2ze^{aT}\cos\omega T + 1} = \frac{ze^{-aT}\sin\omega T}{z^2 - 2ze^{-aT}\cos\omega T + e^{-2aT}}$$

8.1.4.6 Translation Theorem of Impulse Train (Delay Theorem)

$$Z\left[x(k+m)\right] = z^m X(z) - z^m x(0) - z^{m-1}x(1) - \cdots - zx(m-1) \tag{8.31}$$

and

$$Z\left[x(k-m)\right] = z^{-m}X(z) \tag{8.32}$$

Proof: Since $Z\left[x(k+m)\right] = \sum_{k=0}^{\infty} x(k+m)z^{-k}$, then

Let, $k + m = j$
Thus, we get

$$Z\left[x(k+m)\right] = \sum_{j=m}^{\infty} x(j)z^{-(j-m)}$$

$$= z^m\left[\sum_{j=0}^{\infty} x(j)z^{-j} - \sum_{j=0}^{m-1} x(j)z^{-j}\right]$$

$$=z^m[X(z)-\sum_{j=0}^{m-1}x(j)z^{-j}$$

$$=z^m X(z)-z^m x(0)-z^{m-1}x(1)-\cdots-zx(m-1)$$

Since $k<m$, and $x(k-m)=0$.

Hence,

$$Z[x(k-m)]=z^{-m}X(z).$$

Example 8.13

Given $Z[\delta(k)]=1$, evaluate the Z-transformation of $Z[\delta(k-1)]$.

Solution

According to Eq. (8.32), we get

$$Z[\delta(k-1)]=z^{-1}$$

8.1.4.7 Shifting Theorem of Impulse Train

Given

$$Z[x(k)]=X(z)$$

Then, we get

$$Z[e^{\mp ak}x(k)]=X(ze^{\pm a}) \tag{8.33}$$

Proof: Since $Z[e^{-ak}x(k)]=\sum_{k=0}^{\infty}e^{-ak}x(k)z^{-k}=\sum_{k=0}^{\infty}x(k)(ze^{a})^{-k}=X(ze^{a})$.

Similarly,

$$Z[e^{ak}x(k)]=X(ze^{-a})$$

Therefore,

$$Z[e^{\mp ak}x(k)]=X(ze^{\pm a})$$

Example 8.14

If a pulse train is $x(k)=e^{-2k}\cos\omega k$, evaluate $Z[x(k)]$.

Solution

From Table 8.1, we get

$$Z[\cos\omega k] = \frac{z(z - \cos\omega)}{z^2 - 2z\cos\omega + 1}$$

TABLE 8.1 The Z-Transform Pairs

Order	$X(s)$	$X(t)$ or $X(k)$	$X(z)$
1	1	$\delta(t)$	1
2	e^{-KTs}	$\delta(t\text{-}KT)$	z^{-k}
3	$\frac{1}{s}$	$1(t)$	$\frac{z}{z-1}$
4	$\frac{1}{s^2}$	t	$\frac{Tz}{(z-1)^2}$
5	$\frac{1}{s+a}$	e^{-at}	$\frac{z}{z-e^{-aT}}$
6	$\frac{a}{s(s+a)}$	$1-e^{-at}$	$\frac{(1-e^{-aT})z}{(z-1)(z-e^{-aT})}$
7	$\frac{\omega}{s^2+\omega^2}$	$\sin\omega t$	$\frac{z\sin\omega T}{z^2-2z\cos\omega T+1}$
8	$\frac{s}{s^2+\omega^2}$	$\cos\omega t$	$\frac{z(z-\cos\omega T)}{z^2-2z\cos\omega T+1}$
9	$\frac{1}{(s+a)^2}$	te^{-at}	$\frac{Tze^{-aT}}{(z-e^{-aT})^2}$
10	$\frac{\omega}{(s+a)^2+\omega^2}$	$e^{-at}\sin\omega t$	$\frac{ze^{-aT}\sin\omega T}{z^2-2ze^{-aT}\cos\omega T+e^{-2aT}}$
11	$\frac{s+a}{(s+a)^2+\omega^2}$	$e^{-at}\cos\omega t$	$\frac{z^2-ze^{-aT}\cos\omega T}{z^2-2ze^{-aT}\cos\omega T+e^{-2aT}}$
12	$\frac{1}{s^3}$	$\frac{1}{2}t^2$	$\frac{T^2z(z+1)}{2(z-1)^3}$
13	$\frac{1}{s^{n+1}}$	$\frac{t^n}{n!}$	$\lim_{a\to 0}\frac{(-1)^n}{n!}\times\frac{\partial^n}{\partial a^n}\left(\frac{z}{z-e^{-aT}}\right)$
14	$\frac{1}{1-e^{-Ts}}$	$\delta_T(t)=\sum_{k=0}^{\infty}\delta(t-kT)$	$\frac{z}{z-1}$
15	$\frac{1}{(s+a)(s+b)}$	$\frac{e^{-at}-e^{-bt}}{b-a}$	$\frac{1}{b-a}\left(\frac{z}{z-e^{-aT}}-\frac{z}{z-e^{-bT}}\right)$
16		a^k	$\frac{z}{z-a}$
17		$a^k\cos k\pi$	$\frac{z}{z+a}$

Hence,

$$Z\left[x(k)\right]=Z\left[e^{-2k}\cos\omega k\right]=\frac{ze^2\left(ze^2-\cos\omega\right)}{(ze^2)^2-2ze^2\cos\omega+1}=\frac{ze^2\left(ze^2-\cos\omega\right)}{z^2e^4-2ze^2\cos\omega+1}$$

8.1.4.8 Z-Transformation of Weighted Impulse Train (Scaling Factor)

Let

$$Z\left[x(k)\right]=X(z)$$

Then, we get

$$Z\left[a^k x(k)\right]=X\left(\frac{z}{a}\right) \tag{8.34}$$

Proof: $Z\left[a^k x(k)\right]=\sum_{k=0}^{\infty}a^k x(k)z^{-k}=\sum_{k=0}^{\infty}x(k)\left(\frac{z}{a}\right)^{-k}=X\left(\frac{z}{a}\right)$.

Example 8.15

Evaluate the Z-transformation of $x(k)=4^k\cdot k$.

Solution

Since $Z(k)=\dfrac{z}{(z-1)^2}$

Hence,

$$Z\left[4^k x(k)\right]=\frac{\frac{z}{4}}{\left(\frac{z}{4}-1\right)^2}=\frac{4z}{(z-4)^2}$$

8.1.4.9 Differential Theorem of Impulse Train

$$Z\left[kx(k)\right]=-z\frac{dX(z)}{dz} \tag{8.35}$$

Proof: Since $X(z)=\sum_{k=0}^{\infty}x(k)z^{-k}$.

Hence,

$$\frac{dX(z)}{dz}=\sum_{k=0}^{\infty}(-k)x(k)z^{-k-1}$$

Thus,

$$\sum_{k=0}^{\infty}kx(k)z^{-k}=-z\frac{dX(z)}{dz}$$

Therefore,

$$Z[kx(k)] = -z\frac{dX(z)}{dz}$$

8.1.4.10 Z-Transformation of Difference Equations

1. *Forward difference*:

$$Z[\Delta^m x(k)] = (z-1)^m X(z) - \sum_{j=0}^{m-1}(z-1)^{m-j-1}\Delta^j x(0) \tag{8.36}$$

Proof:

$$\begin{aligned} Z[\Delta x(k)] &= Z[x(k+1) - x(k)] \\ &= zX(z) - zx(0) - X(z) \\ &= (z-1)X(z) - zx(0) \end{aligned}$$

and

$$\begin{aligned} Z[\Delta^2 x(k)] &= Z[\Delta x(k+1) - \Delta x(k)] \\ &= (z-1)^2 X(z) - z(z-1)x(0) - z\Delta x(0) \end{aligned}$$

Obviously, when the time m is recursive, we obtain

$$Z[\Delta^m x(k)] = (z-1)^m X(z) - z\sum_{j=0}^{m-1}(z-1)^{m-j-1}\Delta^j x(0)$$

where $\Delta^0 X(0) = X(0)$. When it is a zero-initial condition, Eq. (8.36) can be simplified as

$$Z[\Delta^m x(k)] = (z-1)^m X(z) \tag{8.37}$$

2. *Backward difference*:

$$Z[\nabla^m x(k)] = \left(\frac{z-1}{z}\right)^m X(z) \tag{8.38}$$

Proof: When $m=1$, we get

$$Z[\nabla x(k)]=Z[x(k)-x(k-1)]=X(z)-z^{-1}X(z)=\frac{z-1}{z}X(z)$$

When $m=2$, we get

$$Z[\nabla^2 x(k)]=Z[\nabla x(k)-\nabla x(k-1)]=\left(\frac{z-1}{z}\right)^2 X(z)$$

When the recursive time m is increased, we get

$$Z[\nabla^m x(k)]=\left(\frac{z-1}{z}\right)^m X(z)$$

8.1.4.11 Z-Transformation for Sum of Impulse Train

Let, the Z-transformation of $x(t)$ is $X(z)$. With zero-initial condition, we obtain

$$Z\left[\sum_{j=0}^{k-1} x(j)\right]=\frac{X(z)}{z-1} \tag{8.39}$$

Proof: Since $x(k)=\sum_{j=0}^{k} x(j)-\sum_{j=0}^{k-1} x(j)=\Delta\sum_{j=0}^{k-1} x(j)$.

According to the Z-transformation of difference equations, we get

$$Z\left[\Delta\sum_{j=0}^{k-1} x(j)\right]=(z-1)Z\left[\sum_{j=0}^{k-1} x(j)\right]-zx(0)$$

Since $x(0)=0$.

Hence,

$$Z\left[\sum_{j=0}^{k-1} x(j)\right]=\frac{Z\left[\Delta\sum_{j=0}^{k-1} x(j)\right]}{z-1}=\frac{Z[x(k)]}{z-1}=\frac{X(z)}{z-1}$$

Example 8.16

Given $Z[x(k)]=\frac{z}{(z-1)^2}$, evaluate the Z-transformation of $Z\left[\sum_{j=0}^{k-1} x(j)\right]$ in zero-initial condition.

Solution

$$Z\left[\sum_{j=0}^{k-1} x(j)\right]=\frac{X(z)}{z-1}=\frac{z}{(z-1)^2}\times\frac{1}{z-1}=\frac{z}{(z-1)^3}$$

8.1.4.12 Convolution Theorem

Let

$$X_1(z)=\sum_{m=0}^{\infty} x_1(m)z^{-m} \text{ and } X_2(z)=\sum_{m=0}^{\infty} x_2(m)z^{-m}$$

Then, we get

$$\begin{aligned} X_1(z)X_2(z) &= Z\left[x_1(m)^* x_2(m)\right] \\ &= Z\left[\sum_{m=0}^{k} x_1(m)x_2(k-m)\right] \\ &= Z\left[\sum_{m=0}^{k} x_1(k-m)x_2(m)\right] \end{aligned} \tag{8.40}$$

Proof:

$$Z\left[\sum_{m=0}^{\infty} x_1(m)x_2(k-m)\right]=\sum_{j=0}^{\infty}\left[\sum_{m=0}^{\infty} x_1(m)x_2(j-m)\right]z^{-j}$$

$$=\sum_{m=0}^{\infty} x_1(m)\sum_{j=0}^{\infty} x_2(j-m)z^{-j}$$

Let $j - m = k$, and then we get

$$Z\left[\sum_{m=0}^{\infty} x_1(m)x_2(j-m)\right]=\sum_{m=0}^{\infty} x_1(m)\sum_{k=-m}^{\infty} x_2(k)z^{-m}z^{-k}$$

$$=\sum_{m=0}^{\infty} x_1(m)z^{-m}\sum_{k=0}^{\infty} x_2(k)z^{-k}$$

$$=X_1(z)X_2(z)$$

Similarly, we get

$$Z\left[\sum_{m=0}^{\infty} X_1(k-m)X_2(m)\right]=X_1(z)X_2(z)$$

8.2 THE INVERSE Z-TRANSFORM

To ensure the time response in a continuous system, we need inverse Laplace transform or inverse Fourier transform. Similarly, to ensure the time response in a distributed control system, we need an inverse Z-transform. It can be written as

$$Z^{-1}\left[X(z)\right]=x^{*}(t) \tag{8.41}$$

The inverse Z-transform cannot give out continuous signals $x(t)$, mainly because $x^{*}(t)$ and $x(t)$ are not directly corresponded. There are some methods to evaluate inverse Z-transform that we can see here.

8.2.1 Formulae of Inverse Z-Transform

The given Z-transformation is $X(z)$. When $t=kT$, we can use Eq. (8.42) to prove the instantaneous function value of $x(kT)$ as

$$x(kT)=\frac{1}{2\pi j}\oint_{s} X(z)z^{k-1}\,dz \tag{8.42}$$

We can use the residue theorem to ensure the integral on the right-hand side, where S is a closed curve, and then the surrounding poles of $X(z)z^{k-1}$ can be written as

$$x(kT)=\frac{1}{2\pi j}\oint_{s} X(z)z^{k-1}\,dz=\sum X(z)z^{k-1} \tag{8.43}$$

Example 8.17

Evaluate the inverse Z-transformation of $X(z)=\dfrac{\left(1-e^{-aT}\right)z}{(z-1)\left(z-e^{-aT}\right)}$ by using the inverse Z-transform theorem.

Solution

Since $X(z)=\dfrac{\left(1-e^{-aT}\right)z}{(z-1)\left(z-e^{-aT}\right)}$

Hence,

$$X(z)z^{k-1}=\frac{\left(1-e^{-aT}\right)z^{k}}{(z-1)\left(z-e^{-aT}\right)}$$

Thus, the two poles are $z_1=1$ and $z_2=e^{-aT}$. According to Eq. (8.43), we get

$$\begin{aligned}x(kT)&=X(z)z^{k-1}\Big|_{z=1}+X(z)z^{k-1}\Big|_{z=e^{-aT}}\\&=\frac{\left(1-e^{-aT}\right)z^{k}}{z-e^{-aT}}\Bigg|_{z=1}+\frac{\left(1-e^{-aT}\right)z^{k}}{z-1}\Bigg|_{z=e^{-aT}}\\&=1-e^{-akT}\end{aligned}$$

Therefore,

$$x^*(t)=\sum_{k=0}^{\infty}x(kT)\delta(t-k)=\sum_{k=0}^{\infty}\left(1-e^{-akT}\right)\delta(t-kT)$$
$$=\left(1-e^{-aT}\right)\delta(t-T)+\left(1-e^{-2aT}\right)\delta(t-2T)+\cdots \tag{8.44}$$

8.2.2 The Inverse Z-Transformation with Power Series

Power series is also called Laurent series or synthetic method or division. From the examples of Z-transform, we obtain that the function of the Z-transformation can be expressed by the ratio of the two polynomials like the Laplace transform. Thus, we rewrite as

$$X(z)=\frac{b_oz^m+b_1z^{m-1}+\cdots+b_m}{a_oz^n+a_1z^{n-1}+\cdots+a_n} \tag{8.45}$$

In the engineering control system, $n\geq m$. Expanding Eq. (8.45) in the Laurent series, we can obtain

$$X(z)=C_0+C_1z^{-1}+C_2z^{-2}+\cdots+C_nz^{-n}=\sum_{n=0}^{\infty}C_nz^{-n} \tag{8.46}$$

According to the definition of Z-transform, the coefficient of z^{-n} is also called the coefficient C_n (where $n=0,\ 1,\ 2,\ 3,\ \cdots$); when $t=nT$, C_n is the instantaneous value $x(nT)$ in sampling. Usually, we only need to calculate the limited number of items in engineering calculation. Hence, according to the definition of the coefficient of the Laurent series, we obtain

$$C_n=\frac{1}{2\pi j}\oint_s\frac{X(z)}{(z-z_o)^{k+1}}dz\left(k=0,\ \pm1,\ \pm2,\ \cdots\right) \tag{8.47}$$

We obtain C_n or use synthetic division to calculate the inverse Z-transformation. Substantially, the coefficient of synthetic division is the coefficient of the Laurent series, which is one of the advantages of Z-transform. The method makes it easy to calculate $x^*(t)$.

Example 8.18

Evaluate the inverse Z-transformation of $X(z)=\dfrac{0.5z}{(z-1)(z-0.5)}$.

Solution

$$X(z)=\frac{0.5z}{(z-1)(z-0.5)}=\frac{0.5z}{z^2-1.5z+0.5}$$

By using synthetic division, we can obtain

$$X(z)=0.5z^{-1}+0.75z^{-2}+0.875z^{-3}+0.9375z^{-4}+\cdots$$

Hence,

$$x^{*}(t)=0.5\delta(t-T)+0.75\delta(t-2T)+0.875\delta(t-3T)+0.9375\delta(t-4T)+\cdots$$

Example 8.19

Given $X(z)=\dfrac{\left(1-e^{-aT}\right)z}{(z-1)\left(z-e^{-aT}\right)}$, evaluate $x^{*}(t)$.

Solution

By using Synthetic division, we can obtain

$$X(z)=\left(1-e^{-aT}\right)z^{-1}+\left(1-e^{-2aT}\right)z^{-2}+\left(1-e^{-3aT}\right)z^{-3}+\cdots$$

$$x^{*}(t)=\left(1-e^{-aT}\right)\delta(t-T)+\left(1-e^{-2aT}\right)\delta(t-2T)$$

$$+\left(1-e^{-3aT}\right)\delta(t-3T)+\cdots$$

$$=\sum_{k=0}^{\infty}\left(1-e^{akT}\right)\delta(t-kT)$$

8.2.3 Partial Fraction Expansion Method

We expand the Z-transform function $X(z)$ into the sum of all its partial fractions; then, from the Z-transform pairs (from Table 8.1), we can find the time function $x(t)$ of each part and transfer it into a sampled signal $x^{*}(t)$. From the Z-transform pairs (from Table 8.1), we know that each part on the numerator of the Z-transformation has a factor of z. Thus, when we want to expand it into partial fraction, we should follow the steps listed below:

1. The $X(z)$ should be divided by z;
2. Expand $\dfrac{X(z)}{z}$ into partial fractions; and
3. Multiply by z of each item of the result to obtain the partial fractions of $X(z)$.

Example 8.20

Evaluate the inverse transformation of $X(z)=\dfrac{0.5z}{(z-1)(z-0.5)}$.

Solution

Expand $\frac{X(z)}{z}$ into partial fractions. Thus, we get

$$\frac{X(z)}{z}=\frac{0.5}{(z-1)(z-0.5)}=\frac{1}{z-1}-\frac{1}{z-0.5}$$

Hence,

$$X(z)=\frac{z}{z-1}-\frac{z}{z-0.5}$$

From Table 8.1, we get

$$x(t)=1(t)-e^{-0.693\frac{t}{T}}$$

Therefore, we obtain

$$x^*(t)=\left[1(t)-e^{-0.693\frac{t}{T}}\right]\delta_T(t)$$

$$=0.5\delta(t-T)+0.75\delta(t-2T)+0.875\delta(t-3T)+0.9375\delta(t-4T)+\cdots$$

Example 8.21

Given $X(z)=\frac{\left(1-e^{-aT}\right)z}{(z-1)\left(z-e^{-aT}\right)}$, evaluate $x^*(t)$.

Solution

Since $\frac{X(z)}{z}=\frac{1-e^{-aT}}{(z-1)\left(z-e^{-aT}\right)}=\frac{1}{z-1}-\frac{1}{z-e^{-aT}}$

Hence,

$$x(t)=1(t)-e^{-at}$$

Therefore,

$$x^*(t)=\sum_{k=0}^{\infty}\left(1-e^{-akT}\right)\delta(t-kT)$$

$$=\left(1-e^{-aT}\right)\delta(t-T)+\left(1-e^{-2aT}\right)\delta(t-2T)+\left(1-e^{-3aT}\right)\delta(t-3T)+\cdots$$

From the above examples, the results are the same though the methods are different. That is, when we choose the methods to evaluate the inverse transformation, we should pay attention to the questions. Generally speaking, the Z-transform pairs is more convenient, and it is not easy to make mistakes.

8.3 THE Z-TRANSFORM TO EVALUATE DIFFERENCE EQUATIONS

To evaluate the difference equations in a control system, we should first transfer it into an algebraic equation with z as an independent variable and $X(z)$ as a dependent variable by using Z-transform. We evaluate $X(z)$ and its inverse Z-transformation, then we can obtain $x(k)$.

Example 8.22

Find the solution of the difference equation as given below:

$$x(k+2)-3x(k+1)+2x(k)=u(k)$$

where

$$x(k)=0(k \quad 0) \; u(k)=\begin{cases} 0 & ;k\neq 0 \\ 1 & ;k=0 \end{cases}$$

Solution

Put $k=-1$ into the difference equation, we can obtain

$$x(1)=0.$$

We evaluate the Z-transformation of the difference equation by considering it in the initial condition, and we can obtain

$$\left(z^2-3z+2\right)X(z)=U(z)$$

Since the Z-transformation of $u(k)$ is

$$U(z)=\sum_{k=0}^{\infty}u(k)z^{-k}=1$$

Hence,

$$X(z)=\frac{1}{z^2-3z+2}=\frac{-1}{z-1}+\frac{1}{z-2}$$

Since

$$Z\left[x(k+1)\right]=zX(z)-zx(0)=zX(z)$$

Thus,

$$x(0)=0$$

Therefore,

$$Z[x(k+1)] = -\frac{z}{z-1} + \frac{z}{z-2}$$

whereas since

$$Z\left[1^k\right] = \frac{z}{z-1}; Z\left[2^k\right] = \frac{z}{z-2}$$

Thus,

$$x(k+1) = -1 + 2^k (k = 0,\ 1,\ 2,\ldots)$$

Hence,

$$x(k) = -1 + 2^{k-1} (k = 1,\ 2,\ 3,\ldots)$$

Example 8.23

Find the solution to the difference equation:

$$x(k+2) + 3x(k+1) + 2x(k) = 0; x(0) = 0, \text{ and } x(1) = 1$$

Solution

We evaluate the Z-transformation of the difference equation by considering it in the initial condition, and we can obtain

$$\left(z^2 + 3z + 2\right)X(z) - z = 0$$

Hence,

$$X(z) = \frac{z}{z^2 + 3z + z} = \frac{z}{z+1} - \frac{z}{z+2}$$

From Table 8.1, we get

$$Z\left[(-1)^k\right] = \frac{z}{z+1}$$

and

$$Z\left[(-2)^k\right] = \frac{z}{z+2}$$

Hence,

$$x(k) = (-1)^k - (-2)^k (k = 0,\ 1,\ 2,\ 3,\ldots)$$

8.4 IMPULSE OR PULSE TRANSFER FUNCTION

8.4.1 Definition and Solution for Pulse Transfer Function

Figure 8.3 shows the transient process in a distributed control system. We usually use the equation in the following to express its difference equation as

$$\begin{aligned}&a_0x_o(k)+a_1x_o(k-1)+\cdots+a_{n-1}x_o(k-n+1)+a_nx_o(k-n)\\&=b_0x_i(k)+b_1x_i(k-1)+\cdots+b_{m-1}x_i(k-m+1)+b_mx_i(k-m)\end{aligned} \tag{8.48}$$

We evaluate the Z-transformation of both sides of Eq. (8.48) and obtain

$$\begin{aligned}&\left(a_0+a_1z^{-1}+a_2z^{-2}+\cdots+a_{n-1}z^{-n+1}+a_nz^{-n}\right)X_o(z)\\&=\left(b_0+b_1z^{-1}+b_2z^{-2}+\cdots+b_{m-1}z^{-m+1}+b_mz^{-m}\right)X_i(z)\end{aligned} \tag{8.49}$$

Thus, the transfer function of distributed control system may be written as

$$G(z)=\frac{X_o(z)}{X_i(z)}=\frac{b_0+b_1z^{-1}+b_2z^{-2}+\cdots+b_{m-1}z^{-m+1}+b_mz^{-m}}{a_0+a_1z^{-1}+a_2z^{-2}+\cdots+a_{n-1}z^{-n+1}+a_nz^{-n}} \tag{8.50}$$

Since $x_i(k)$ and $x_o(k)$ are pulse train, $G(z)$ is called the pulse transfer function. To evaluate the solutions of the above Eq. (8.50) can be expressed as follows:

1. Evaluate the transfer function $G(s)$ of the system.
2. Evaluate the impulse response function.

$$x_o(t),\ \left\{x_o(t)=L^{-1}\left[G(s)\right]\right\}$$

3. Evaluate $G(z)=\sum_{k=0}^{\infty}x_o(kT)z^{-k}$.

After evaluating the system transfer function $G(s)$, we can expand it into the sum of its partial fractions. Then from the Laplace transform pairs and the Z-transform pairs (Table 8.1), we can obtain the impulse transfer function of the system.

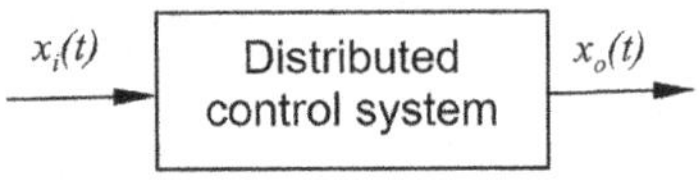

FIGURE 8.3 Transient process in distributed control system.

Example 8.24

Given a system transfer function $G(s)=\dfrac{10}{s(s+10)}$, evaluate its impulse transfer function.

Solution

We expand $G(s)$ into the sum of its partial fractions as

$$G(s)=\frac{1}{s}-\frac{1}{s+10}$$

From Table 8.1, we get

$$G(z)=\frac{z}{z-1}-\frac{z}{z-e^{-10T}}$$

Example 8.25

Evaluate the impulse transfer function of the system as shown in Figure 8.4.

Solution

We expand the transfer function $G(s)$ into the sum of its partial fractions as

$$G(s)=\frac{K}{(s+a)(s+b)}=\frac{K}{b-a}\left(\frac{1}{s+a}-\frac{1}{s+b}\right)$$

From Table 8.1, we get

$$G(z)=\frac{K}{b-a}\left(\frac{z}{z-e^{-aT}}-\frac{z}{z-e^{-bT}}\right)=\frac{K}{b-a}\frac{z\left(e^{-aT}-e^{-bT}\right)}{\left(z-e^{-aT}\right)\left(z-e^{-bT}\right)}$$

8.4.2 The Impulse Transfer Function of Series Connected Open-Loop System

Figure 8.5 shows an open-loop system with ideal switches between two nodes.

The solutions for the system transfer function are as follows:

$$X_{o1}(z)=G_1(z)X_i(z) \tag{8.51}$$

$$X_o(z)=G_2(z)X_{o1}(z) \tag{8.52}$$

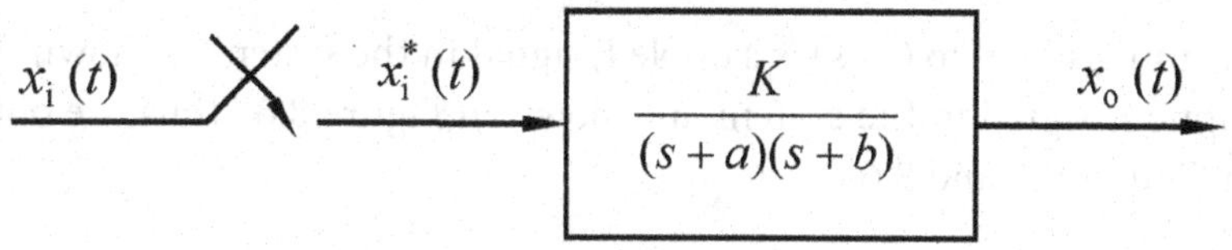

$x_i(t)$ $x_i^*(t)$ $\dfrac{K}{(s+a)(s+b)}$ $x_o(t)$

FIGURE 8.4 A distributed control system.

where $G_1(z)$ and $G_2(z)$ are the impulse transfer function of linear element $G_1(s)$ and $G_2(s)$, respectively. If we substitute Eq. (8.51) into Eq. (8.52), we obtain

$$X_o(z)=G_1(z)G_2(z)X_i(z) \tag{8.53}$$

Thus, the impulse transfer function of the series element is written as

$$G(z)=\frac{X_o(z)}{X_i(z)}=G_1(z)G_2(z) \tag{8.54}$$

From Eq. (8.54), in the open-loop system with two ideal switches between two nodes, the impulse transfer function is equal to the product of the impulse transfer function of the two elements. This conclusion can be extended to the open-loop system with the number of n ideal switches between the number of n nodes. Now, the open-loop impulse transfer function of the whole system is equal to the product of the impulse transfer function of each element, i.e.,

$$G(z)=G_1(z)G_2(z)\ldots G_n(z) \tag{8.55}$$

Figure 8.6 shows an open-loop control system with no ideal switches between two nodes.
Let, $G(s)=G_1(s)G_2(s)$, and then the open-loop impulse transfer function is expressed as

$$G(z)=\frac{X_o(z)}{X_i(z)}=Z\left[G_1(s)G_2(s)\right]=G_1G_2(z) \tag{8.56}$$

where $G_1G_2(z)$ is the Z-transformation of $G_1(s)^* G_2(s)$. In the open-loop system with two linear elements in series and with no ideal switches between two nodes, the impulse transfer function is equal to the Z-transformation of the product of the transfer functions of the two elements. Obviously, when the number of n elements is in series and with no ideal switches between two nodes, the impulse transfer function of the system is equal to the Z-transformation of the product of the transfer functions of the n elements. Thus, we obtain

$$G(z)=G_1G_2\ldots G_n(z) \tag{8.57}$$

Note that the impulse transfer function of the system as shown in Figure 8.5 is different from Figure 8.6. Thus, we can say

$$G_1(z)G_2(z)\neq G_1G_2(z) \tag{8.58}$$

Since the incoming signal to $G_2(s)$ is sampled, signal in the system as shown in Figure 8.5 and its continuous signal in the system are shown in Figure 8.6. Thus, we obtain the following from Figures 8.5 and 8.6

$$G_1(s)=\frac{1}{s},\quad G_2(s)=\frac{10}{s+10}$$

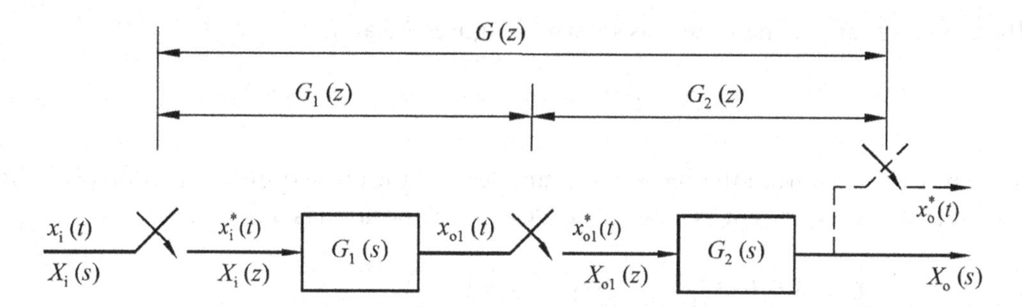

FIGURE 8.5 An open-loop system with ideal switches between two nodes.

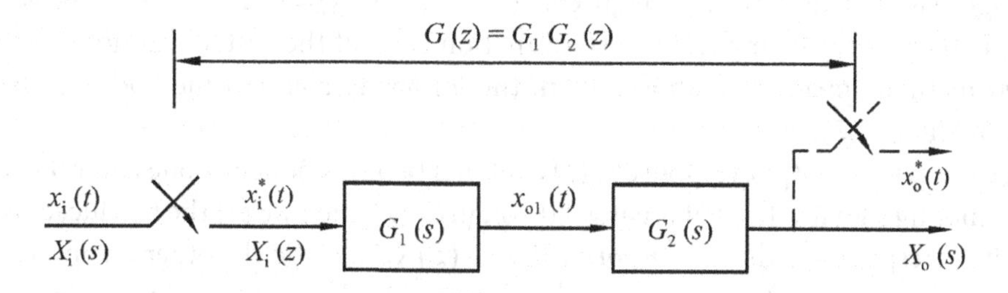

FIGURE 8.6 An open-loop control system with no ideal switches between two nodes.

The impulse transfer function of the open-loop system as shown in Figure 8.5 is written as

$$G(z)=G_1(z)G_2(z)=Z[G_1(s)]\cdot Z[G_2(s)]=\frac{z}{z-1}\cdot\frac{10z}{z-e^{-10T}}$$

And the impulse transfer function of the system as shown in Figure 8.6 is written as

$$G(z)=G_1G_2(z)=Z[G_1(s)G_2(s)]=Z\left(\frac{10}{s(s+10)}\right)=\frac{z\left(1-e^{-10T}\right)}{(z-1)\left(z-e^{-10T}\right)}$$

The impulse transfer functions of the system as shown in Figures 8.5 and 8.6 are expressed as

$$G_1(s)=\frac{1}{s+a},\quad G_2(s)=\frac{1}{s+b}$$

The Z-transform of the system as shown in Figure 8.5 is given by

$$G(z)=G_1(z)G_2(z)=\frac{z}{z-e^{-aT}}\cdot\frac{z}{z-e^{-bT}}$$

The Z-transform of the system as shown in Figure 8.6 is given by

$$G(z)=\frac{1}{b-a}\cdot\frac{z\left(e^{-aT}-e^{-bT}\right)}{\left(z-e^{-aT}\right)\left(z-e^{-bT}\right)}$$

Therefore, we can see the system as shown in Figures 8.5 and 8.6:

$$G_1(z)G_2(z) \neq G_1G_2(z)$$

It can be concluded from the above two examples that the pulse transfer function of all the similar systems is different as long as the ideal switch position is different.

8.5 THE OPEN-LOOP PULSE TRANSFER FUNCTION WITH ZERO-ORDER HOLDER

Suppose the open-loop system as shown in Figure 8.7, in order to analyze simply, the equivalent Figure 8.7a transforms into Figure 8.7b, because $G(s)=1-e^{-Ts}$ is not a rational fraction of s. Therefore, the impulse transfer function $G(z)$ of the system cannot be solved by the two methods mentioned above. Then, the following solution method is adopted, as shown in Figure 8.7.

Since the system output response $x_o^*(t)$ contains two components, one component is the input sampling signal $x_i^*(t)$ that is passed through $G_2(s)$, then we get the produced response $x_{o1}^*(t)$; it corresponds to the *Z-transformation* $G_2(z)X_i(z)$. And the other component $x_i^*(t)$ passes through $e^{-Ts}G_2(s)$, the response $x_{o2}^*(t)$ is produced. Because e^{-Ts} is a delay factor in order to delay one sample period, the response $x_{o2}^*(t)$ is delayed of one sample period than $x_{o1}^*(t)$ according to the real shifting theorem of Z-transformation. The Z-transformation of the response $x_{o2}^*(t)$ is $z^{-1}G_2(z)X_i(z)$. Thus, we get

$$\begin{aligned} X_o(z) &= G_2(z)X_i(z) - z^{-1}G_2(z)X_i(z) \\ &= (1-z^{-1})G_2(z)X_i(z) = \frac{z-1}{z}G_2(z)X_i(z) \end{aligned} \tag{8.59}$$

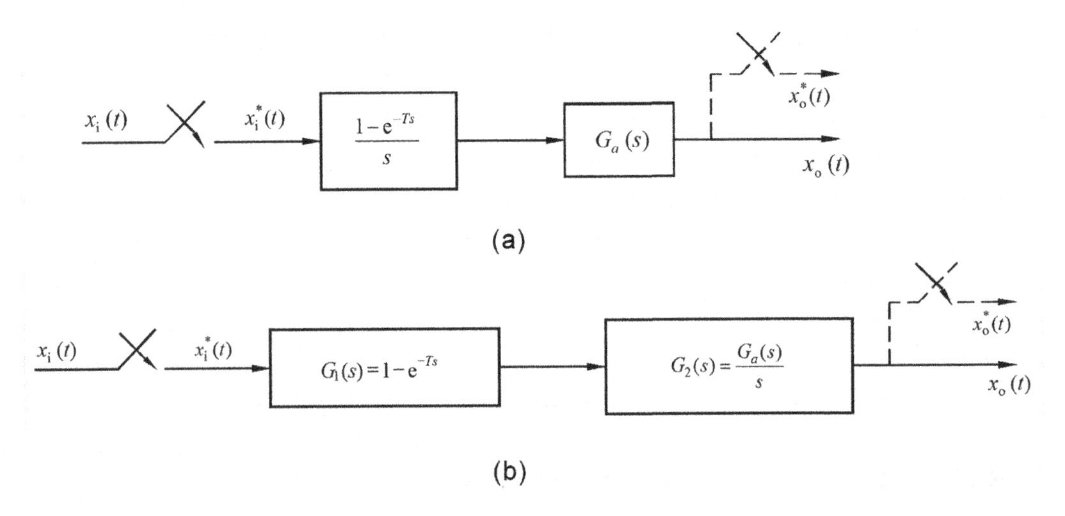

FIGURE 8.7 (a) The open-loop pulse system with zero-order holder and (b) transformed open-loop pulse system with zero-order holder.

Therefore, the pulse transfer function with zero-order holder, we can write as

$$G(z)=\frac{X_o(z)}{X_i(z)}=\frac{z-1}{z}G_2(z)=\frac{z-1}{z}Z\left(\frac{G_a(s)}{s}\right) \tag{8.60}$$

We assume $G_a(s)$ as a rational fraction, and it has no poles, but it includes a pole at $s=0$. Thus, we get

$$G_a(s)=\frac{P(s)}{sQ(s)}$$

At this point, we get

$$\frac{G_a(s)}{s}=\frac{P(s)}{s^2Q(s)}=\frac{c_1}{s}+\frac{c_2}{s^2}+\sum_{k=3}^{n}\frac{c_k}{s-s_k} \tag{8.61}$$

where

$$c_1=\frac{d}{ds}\left(\frac{P(s)}{Q(s)}\right)\Bigg|_{s=0},\quad c_2=\frac{P(0)}{Q(0)},\quad c_k=\frac{P(s_k)}{s_k^2Q'(s_k)}$$

The Z-transformation of Eq. (8.61) is expressed as

$$Z\left(\frac{G_a(s)}{s}\right)=\frac{c_1z}{z-1}+\frac{c_2Tz}{(z-1)^2}+\sum_{k=3}^{n}\frac{c_kz}{z-e^{s_kT}}$$

Thus, the open-loop pulse transfer function can be defined as

$$G(z)=\frac{z-1}{z}\left[\frac{c_1z}{z-1}+\frac{c_2Tz}{(z-1)^2}+\sum_{k=3}^{n}\frac{c_kz}{z-e^{s_kT}}\right]=c_1+\frac{c_2T}{z-1}+\sum_{k=3}^{n}\frac{c_k(z-1)}{z-e^{s_kT}} \tag{8.62}$$

It can be seen as shown in Eq. (8.62) that the transfer function is still a rational fraction of z, and the number of poles and $G(z)$ still is equal to the continuous part of the transfer function $G_a(s)$ on the number of poles and it is one-to-one correspondence. In other words, the number of poles $G(z)$ and their distribution depend on $G_a(s)$, and it has nothing to do with the zero-order holder.

The open-loop transfer function with the zero-order holder can be determined by Eq. (8.62). We can also get output response $x_o(t)$ or $x_o(kT)$ according to $G_a(s)$, and then we get the transfer function of the Z-transform according to the definition of the *Z-transformation as*

$$G(z)=\sum_{k=0}^{\infty}x_o(kT)z^{-k}$$

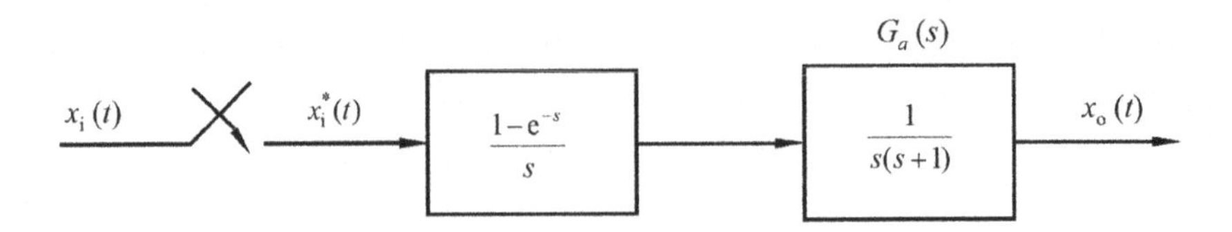

FIGURE 8.8 A discrete control system.

Example 8.26

Determine the pulse transfer function of a discrete control system as shown in Figure 8.8.

Solution

From Figure 8.8, we can get

$$G(s)=\frac{X_o(s)}{X_i^*(s)}=\frac{1-e^{-s}}{s^2(s+1)}=\left(1-e^{-s}\right)\left(\frac{1}{s^2}-\frac{1}{s}+\frac{1}{s+1}\right)$$

Thus,

$$x_o(t)=\left(t-1+e^{-t}\right)-\left(t-1-1+e^{-(t-1)}\right)\cdot 1(t-1)$$

Since $T=1$, $kT=k$; thus, we get

$$x_o(k)=\left(k-1+e^{-k}\right)-\left(k-2+e^{-(k-1)}\right)=e^{-k}+1-e^{-(k-1)}$$

where $k=1, 2, 3, \ldots$, and also $x_o(0)=0$. Therefore, it is possible to have a pulse transfer function of an open-loop system. Thus, we get

$$G(z)=\sum_{k=0}^{\infty}x_o(k)z^{-k}=\sum_{k=1}^{\infty}\left(e^{-k}+1-e^{-(k-1)}\right)z^{-k}$$

$$=\frac{e^{-1}-1}{z-e^{-1}}+\frac{1}{z-1}=\frac{e^{-1}z+1-2e^{-1}}{z^2-\left(1+e^{-1}\right)z+e^{-1}}$$

Example 8.27

An open-loop system shown in Figure 8.7, where $G_a(s)=\frac{10}{s(s+10)}$. Determine the pulse transfer function of the open-loop system.

Solution

Since $G_a(s)=\frac{10}{s(s+10)}$

Thus, we have

$$G_a(s)=\frac{P(s)}{sQ(s)}$$

Hence,

$$P(s)=10,\quad Q(s)=s+10$$

According to Eq. (8.61), we get

$$c_1=\frac{d}{ds}\left(\frac{P(s)}{Q(s)}\right)\Bigg|_{s=0}=\frac{d}{ds}\left(\frac{10}{s+10}\right)\Bigg|_{s=0}=-10\frac{1}{(s+10)^2}\Bigg|_{s=0}=-0.1$$

$$c_2=\frac{P(0)}{Q(0)}=1,\text{and}\, c_3=\frac{P(s_3)}{s_3^2Q'(s_3)}\Bigg|_{s_3=-10}=\frac{10}{s_3^2\times 1}\Bigg|_{s_3=-10}=\frac{10}{(-10)^2}=0.1$$

Substituting c_1, c_2 and c_3 into Eq. (8.62), we get

$$G(z)=-0.1+\frac{T}{z-1}+\frac{0.1(z-1)}{z-e^{-10T}}$$

$$=\frac{\left(T-0.1+0.1e^{-10T}\right)z+\left(0.1-Te^{-10T}-0.1e^{-10T}\right)}{(z-1)\left(z-e^{-10T}\right)}$$

8.6 PULSE TRANSFER FUNCTION OF CLOSED-LOOP SYSTEM

In a closed-loop system, there are different configuration methods due to the different positions of the sampling switch.

1. Determine the transfer function of a closed-loop system as shown in Figure 8.9.

   From the closed-loop system as shown in Figure 8.9, we get

$$e(t)=x_i(t)-b(t) \tag{8.63}$$

   The Z-transformation of Eq. (8.63) can be expressed as

$$E(z)=X_i(z)-B(z) \tag{8.64}$$

   and

$$B(z)=GH(z)\cdot E(z) \tag{8.65}$$

   From Eqs. (8.63) and (8.64), we get

$$GH(z)=Z\left[G(s)H(s)\right] \tag{8.66}$$

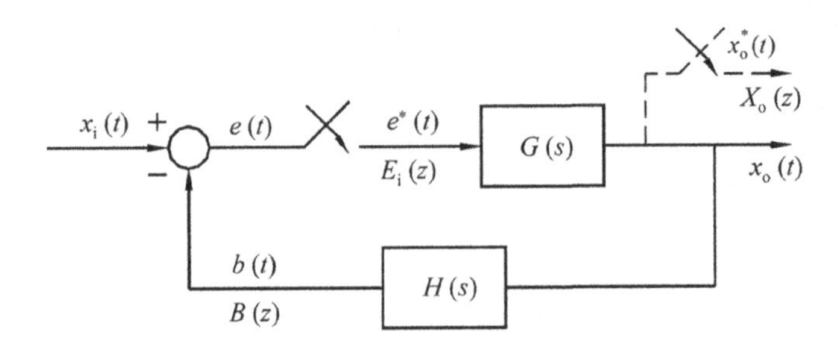

FIGURE 8.9 A closed-loop control system.

Substituting Eq. (8.66) into Eq. (8.65) and then we substitute $B(z)$ into Eq. (8.64). Thus, we get

$$E(z)=X_i(z)-GH(z)E(z) \tag{8.67}$$

The solution of Eq. (8.67) is expressed as

$$E(z)=\frac{X_i(z)}{1+GH(z)}=\Phi_e(z)X_i(z) \tag{8.68}$$

or

$$\Phi_e(z)=\frac{1}{1+GH(z)} \tag{8.69}$$

Eq. (8.69) is called the input error pulse transfer function of a closed-loop sampling system which is referred to as an error pulse transfer function.

Since

$$X_o(z)=G(z)E(z) \tag{8.70}$$

Substituting Eq. (8.68) into Eq. (8.70), we get

$$X_o(z)=G(z)\frac{X_i(z)}{1+GH(z)}=\frac{G(z)}{1+GH(z)}X_i(z) \tag{8.71}$$

The solution of Eq. (8.71) is written as

$$\Phi(z)=\frac{G(z)}{1+GH(z)} \tag{8.72}$$

where $\Phi(z)$ is called the pulse transfer function of the closed-loop system. From Eqs. (8.69) and (8.72), we get the pulse transfer function of the closed-loop sampling system but it is not derived from the Z-transformation of $\Phi(s)$ and $\Phi_e(s)$, i.e.,

$$\Phi(z)\neq Z\left[\Phi(s)\right]$$

$$\Phi_e(z)\neq Z\left[\Phi_e(s)\right]$$

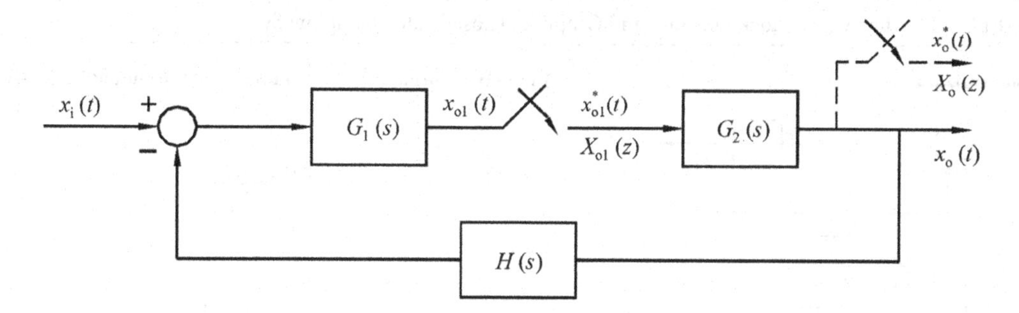

FIGURE 8.10 A closed-loop control system with two series transfer function.

2. Determine the pulse transfer function of the closed-loop system shown in Figure 8.10. From Figure 8.10, we get

$$X_{o1}(z)=X_iG_1(z)-X_{o1}(z)G_2HG_1(z) \tag{8.73}$$

And among them, we get

$$X_iG_1(z)=Z[X_i(s)G_1(s)] \tag{8.74}$$

$$G_2HG_1(z)=Z[G_2(s)H(s)G_1(s)] \tag{8.75}$$

The solution of Eq. (8.73) is written as

$$X_{o1}(z)=\frac{X_iG_1(z)}{1+G_2HG_1(z)} \tag{8.76}$$

Because

$$X_o(z)=X_{o1}(z)G_2(z)=\frac{X_iG_1(z)G_2(z)}{1+G_2HG_1(z)} \tag{8.77}$$

Apparently, we can not get the solution of $\frac{X_o(z)}{X_i(z)}$ from Eq. (8.77). Therefore, we can not determine the pulse transfer function of the closed-loop system as shown in Figure 8.10, but we can find the output response from Eq. (8.77). The output response of the Z-transformation function $X_o(z)$ is shown in Table 8.2 for several common typical sampler various kinds of structural configurations in a closed-loop system. In Table 8.2, G represents $G(s)$, H represents $H(s)$, and so on.

Example 8.28

Determine the unit-step response of the closed-loop system shown in Figure 8.11.

Solution

We get the *Z-transformation* of the output response of the closed-loop system shown in Figure 8.11 based on Table 8.2. Thus,

$$X_o(z)=\frac{G(z)}{1+G(z)}X_i(s)$$

TABLE 8.2 The Typical Block Diagram and Output of Closed-Loop Sampling System

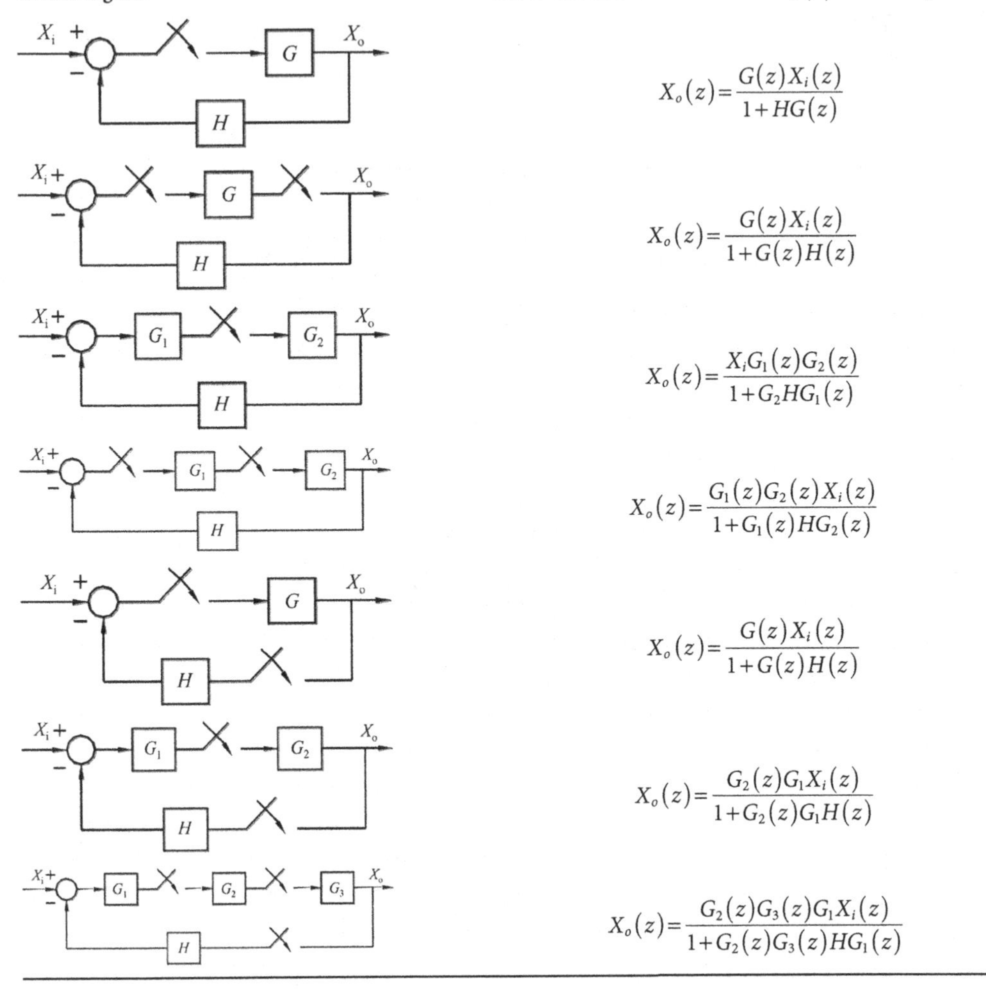

Block Diagram	The Z-Transformation Function $X_o(z)$ of the Output Response
	$X_o(z)=\frac{G(z)X_i(z)}{1+HG(z)}$
	$X_o(z)=\frac{G(z)X_i(z)}{1+G(z)H(z)}$
	$X_o(z)=\frac{X_iG_1(z)G_2(z)}{1+G_2HG_1(z)}$
	$X_o(z)=\frac{G_1(z)G_2(z)X_i(z)}{1+G_1(z)HG_2(z)}$
	$X_o(z)=\frac{G(z)X_i(z)}{1+G(z)H(z)}$
	$X_o(z)=\frac{G_2(z)G_1X_i(z)}{1+G_2(z)G_1H(z)}$
	$X_o(z)=\frac{G_2(z)G_3(z)G_1X_i(z)}{1+G_2(z)G_3(z)HG_1(z)}$

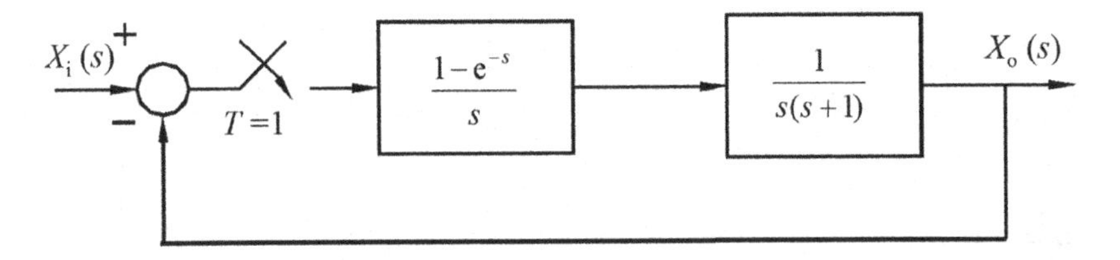

FIGURE 8.11 A closed-loop system with $T = 1$.

Since

$$G(s)=\frac{1-e^{-s}}{s^2(s+1)}$$

According to the results of Figure 8.8, we get

$$G(z)=\frac{e^{-1}z+1-2e^{-1}}{z^2-\left(1+e^{-1}\right)z+e^{-1}}=\frac{0.368z+0.264}{z^2-1.368z+0.368}$$

Thus, we get

$$X_o(z)=\frac{0.368z+0.264}{z^2-z+0.632}X_i(z)$$

Since the input is unit-step response, we get

$$X_i(z)=\frac{z}{z-1}$$

We get the Z-transformation of output as

$$X_o(z)=\frac{(0.368z+0.264)z}{\left(z^2-z+0.362\right)(z-1)}$$

$$=\frac{0.368z^2+0.264z}{z^3-2z^2+1.632z-0.632}$$

$$=0.368z^{-1}+z^{-2}+1.4z^{-3}+1.4z^{-4}+1.147z^{-5}+0.895z^{-6}+0.802z^{-7}+\cdots$$

Therefore, we get

$$x_o(k)=0.368\delta(k-1)+\delta(k-2)+1.4\delta(k-3)+1.4\delta(k-4)$$

$$+1.147\delta(k-5)+0.895\delta(k-6)+0.802\delta(k-7)+\cdots$$

Hence,

$$x_o(0)=0,x_o(1)=0.368,x_o(1)=0.368,x_o(2)=1,x_o(3)=1.4,$$

$$x_o(4)=1.4,x_o(5)=1.147,x_o(6)=0.895,x_o(7)=0.802\text{TM}$$

The operating results are shown in Figure 8.12.

8.7 STABILITY ANALYSIS OF PULSE SYSTEMS

The stability analysis of the system in the S-plane, we should observe the characteristic equation of the transfer function of the closed-loop system in the right half S-plane whether there are poles or not. If there is no pole in the right half S-plane, the system is stable. However, the *Z-transformation* and the *S-transformation* have interrelated; therefore, we should first study the mapping relationship between *Z*-plane and *S*-plane.

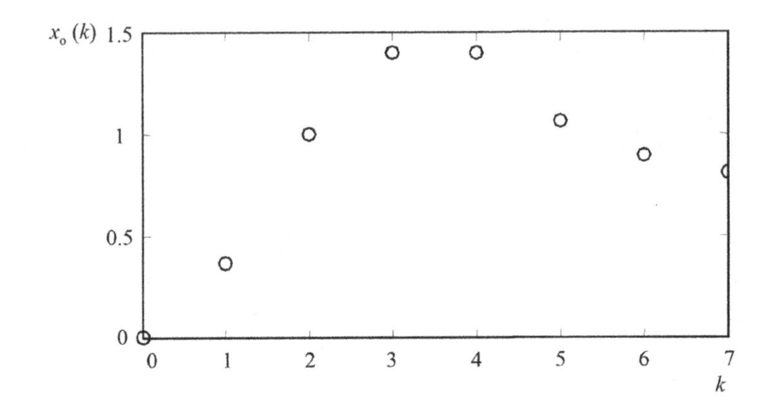

FIGURE 8.12 Discrete points of $x_o(k)$ and k.

According to the definition of the *Z-transformation*, we get

$$z = e^{Ts} \tag{8.78}$$

whereas in the S-plane, we get

$$s = \sigma + j\omega \tag{8.79}$$

Substituting Eq. (8.79) into Eq. (8.78), we get

$$z = e^{(\sigma + j\omega)T} = e^{T\sigma} e^{j\omega T} \tag{8.80}$$

The absolute value of Eq. (8.80) is written as

$$|z| = e^{T\sigma} \tag{8.81}$$

And the angle of Eq. (8.80) can be written as

$$\angle z = \omega T \tag{8.82}$$

According to the Eq. (8.81), when $\sigma < 0$, the correspondence value is in the left half S-plane; thus, the system is stable. Whereas when $|z| < 1$, the correspondence is mapped to the Z-plane, within the corresponding unit circle in the Z-plane, the pulse system is stable. However, when $\sigma > 0$, the correspondence is in the right half S-plane, the system is unstable, and it is mapped to the Z-plane and $|z| > 1$; thus, the correspondence is the unit circle in the Z-plane, and the pulse system is marginally unstable. When $\sigma = 0$, the correspondence is on the imaginary axis of the S-plane, the system is stable, and it is mapped to the Z-plane and $|z| = 1$; thus, the correspondence is the unit circle on the Z-plane, and the pulse system is marginally stable.

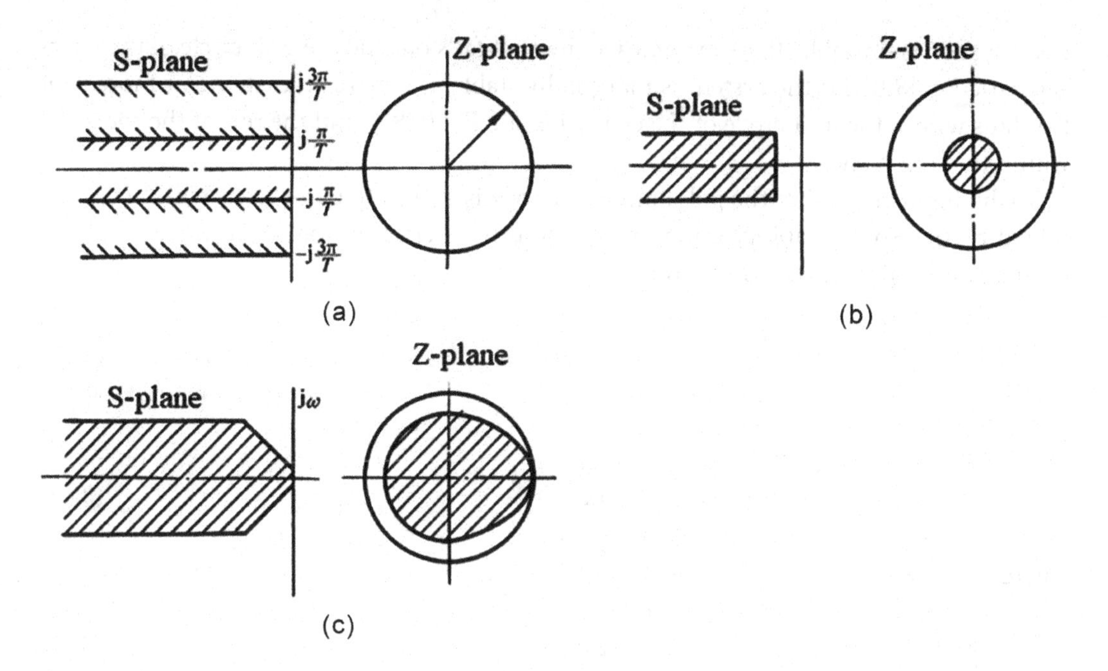

FIGURE 8.13 (a) Band mapping in *S*-plane and zone mapping in *Z*-plane, (b) Band mapping in *S*-plane to the Z-plane, and (c) the area in the *s*-plane is mapped to the Z-plane.

Eq. (8.78) can also be expressed as

$$s = \frac{1}{T}\ln z = \frac{1}{T}\ln\left[e^{T(\sigma+j\omega)}\right] = \frac{1}{T}\left[\ln|z| + j(\omega T + 2k\pi)\right] \tag{8.83}$$

where $k = 0, \pm 1, \pm 2, \ldots$, and in Eq. (8.83), every width of $\frac{2\pi}{T}$ in S-plane is mapped to the unit circle in *Z*-plane, as shown in Figure 8.13. Figure 8.13b and c shows the corresponding region in the *Z*-plane.

8.7.1 Stability Analysis

To discuss the stability analysis of the system in the *Z*-plane, first, we assume the transfer function of the closed-loop pulse as

$$\Phi(z) = \frac{G(z)}{1 + G(z)H(z)} \tag{8.84}$$

To support the system stability, the characteristic equation of the system can be written as

$$1 + G(z)H(z) = 0 \tag{8.85}$$

We can determine the root of the characteristic equation in Eq. (8.85). If the system is stable, all the roots of the characteristic Eq. (8.85) fall within the unit circle. Thus, we say

$$|z_i| < 1 \tag{8.86}$$

If the system is unstable, there are at least one root falls outside the unit circle of the characteristic Eq. (8.85). If the system is marginally stable, there are at least one or more root that falls within the unit circle of the characteristic Eq. (8.85), and the rest of the roots fall within the unit circle.

To distinguish whether the polynomial root of z is within the unit circle which is determined by the Routh stability criterion. The Routh criterion is applied in the *Z*-plane by means of the following w conversion:

$$z = \frac{w+1}{w-1} \tag{8.87}$$

or

$$w = \frac{z+1}{z-1} \tag{8.88}$$

where

$$z = x + jy \tag{8.89}$$

$$w = u + jv \tag{8.90}$$

Eq. (8.88) can be transformed into

$$w = \frac{x + jy + 1}{x + jy - 1} = \frac{\left(x^2 + y^2 - 1\right) - 2jy}{(x-1)^2 + y^2} \tag{8.91}$$

By comparing Eq. (8.90) with (8.91), we get

$$u = \frac{x^2 + y^2 - 1}{(x-1)^2 + y^2} \tag{8.92}$$

and

$$v = \frac{-2y}{(x-1)^2 + y^2} \tag{8.93}$$

Thus,

when $|z| < 1$, we have $x^2 + y^2 < 1$, and then $u < 0$;

when $|z| > 1$, we have $x^2 + y^2 > 1$, and then $u > 0$; and

when $|z| = 1$, we have $x^2 + y^2 = 1$, and then $u = 0$.

For a high-order system, it is difficult to determine u and v. If we can substitute Eq. (8.87) into characteristic Eq. (8.85), then we get the characteristic polynomial of argument w; therefore, we can determine the stability of the system according to the rules of the Routh criterion.

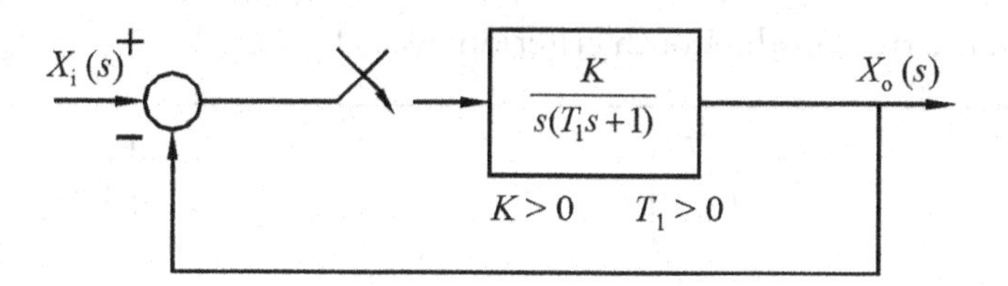

FIGURE 8.14 A unity closed-loop system with $K > 0$ and $T_1 > 0$.

Example 8.29

Determine the stability of the system shown in Figure 8.14.

Solution

Since

$$G(s)=\frac{K}{s(T_1 s+1)};\ H(s)=1$$

Thus, we can say

$$G(z)=\frac{K\left(1-e^{-\frac{T}{T_1}}\right)z}{(z-1)\left(z-e^{-\frac{T}{T_1}}\right)}$$

The characteristic equation of the transfer function can be written

$$1+G(z)=0$$

Hence,

$$z^2+\left[K\left(1-e^{-\frac{T}{T_1}}\right)-\left(1+e^{-\frac{T}{T_1}}\right)\right]z+e^{-\frac{T}{T_1}}=0 \tag{8.94}$$

Applying transformation, we get

$$z=\frac{w+1}{w-1} \tag{8.95}$$

Substituting Eq. (8.95) into Eq. (8.94), we get

$$K\left(1-e^{-\frac{T}{T_1}}\right)w^2+2\left(1-e^{-\frac{T}{T_1}}\right)w+2\left(1+e^{-\frac{T}{T_1}}\right)-K\left(1-e^{-\frac{T}{T_1}}\right)=0$$

According to the rules of the Routh criterion, we get

w^2	$K\left(1-e^{-\frac{T}{T_1}}\right)$	$2\left(1+e^{-\frac{T}{T_1}}\right)-K\left(1-e^{-\frac{T}{T_1}}\right)$
w^1	$2\left(1-e^{-\frac{T}{T_1}}\right)$	0
w^0	$2\left(1+e^{-\frac{T}{T_1}}\right)-K\left(1-e^{-\frac{T}{T_1}}\right)$	

Since

$$\left(1-e^{-\frac{T}{T_1}}\right)>0$$

Hence, it requires if and only if

$$\begin{cases} K>0 \\ 2\left(1+e^{-\frac{T}{T_1}}\right)-K\left(1-e^{-\frac{T}{T_1}}\right)>0 \end{cases}$$

Then, it satisfies the stability criteria, and we get

$$\frac{2\left(1+e^{-\frac{T}{T_1}}\right)}{\left(1-e^{-\frac{T}{T_1}}\right)}>K>0$$

Therefore,

$$0<K<2\coth\left(\frac{T}{2T_1}\right)$$

Example 8.30

Determine the pulse transfer function of the closed-loop system shown in Figure 8.15. When $T = 1s$, determine the discrete value of unit-step response.

Solution

The transfer function of the continuous system is expressed as

$$G(s)=\frac{1-e^{-Ts}}{s^2(s+1)}$$

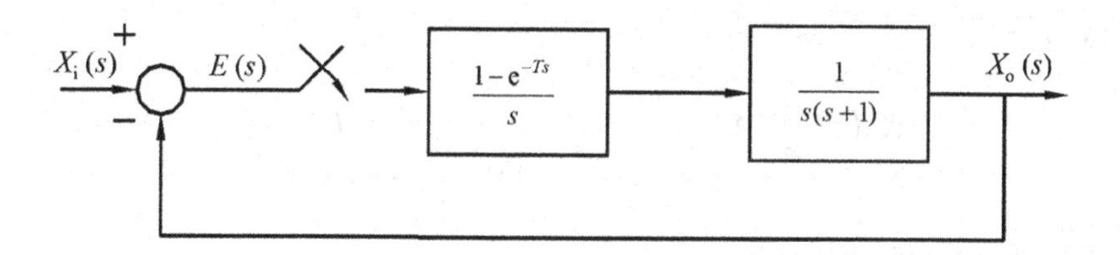

FIGURE 8.15 A discrete control system with unity feedback system.

The transfer function of the open-loop pulse is written as

$$G(z)=\left(1-z^{-1}\right)\left[\frac{Tz}{(z-1)^2}-\frac{\left(1-e^{-T}\right)z}{(z-1)\left(z-e^{-T}\right)}\right]$$

$$=\frac{\left(T+e^{-T}-1\right)z+1-(1+T)e^{-T}}{z^2-\left(1+e^{-T}\right)z+e^{-T}}$$

The pulse transfer function of the closed-loop system is written as

$$\Phi(z)=\frac{G(z)}{1+G(z)}=\frac{\left(T+e^{-T}-1\right)z+1-(1+T)e^{-T}}{z^2-(2-T)z+1-Te^{-T}}$$

The *Z-transformation* of unit-step function is defined as

$$X_i(z)=\frac{z}{z-1}$$

The *Z-transformation* of the unit-step response can be converted as

$$X_o(z)=\Phi(z)X_i(z)=\frac{\left(T+e^{-T}-1\right)z+1-(1+T)e^{-T}}{z^2-(2-T)z+1-Te^{-T}}\cdot\frac{z}{z-1}$$

When $T = 1$ is substituted into the above equation, we get

$$X_o(z)=\frac{\left(1+e^{-1}-1\right)z+1-(1+1)e^{-1}}{z^2-(2-1)z+1-e^{-1}}\cdot\frac{z}{z-1}=\frac{0.368z+0.264}{z^2-z+0.632}\cdot\frac{z}{z-1}$$

$$=\frac{0.368(z+0.717)}{(z-0.5-j0.618)(z-0.5+j0.618)}\cdot\frac{z}{z-1}$$

$$=\frac{0.368(z+0.717)}{\left(z-0.795e^{j0.89}\right)\left(z-0.795e^{-j0.89}\right)}\cdot\frac{z}{z-1}$$

Thus, we get

$$\frac{X_o(z)}{z} = \frac{A}{z-1} + \frac{B_1}{z-0.795e^{j0.89}} + \frac{B_2}{z-0.795e^{-j0.89}}$$

By using the residue theorem, we get

$$\begin{cases} A = 1 \\ B_{1,2} = 0.511e^{\pm j2.93} \end{cases}$$

Therefore,

$$\frac{X_o(z)}{z} = \frac{1}{z-1} + \frac{0.511e^{j2.93}}{z-0.795e^{j0.89}} + \frac{0.511e^{-j2.93}}{z-0.795e^{-j0.89}}$$

and

$$X_o(z) = \frac{z}{z-1} + \frac{0.511e^{j2.93}z}{z-0.795e^{j0.89}} + \frac{0.511e^{-j2.93}z}{z-0.795e^{-j0.89}}$$

The *Z-inverse-transformation* of the above equation can be written as

$$x_o(k) = 1 + 1.022 \times 0.795^k \cos(0.89k + 2.93)$$

where $k = 0, 1, 2, \ldots$. The MATLAB program is performed discrete graphs as shown in Figure 8.16.

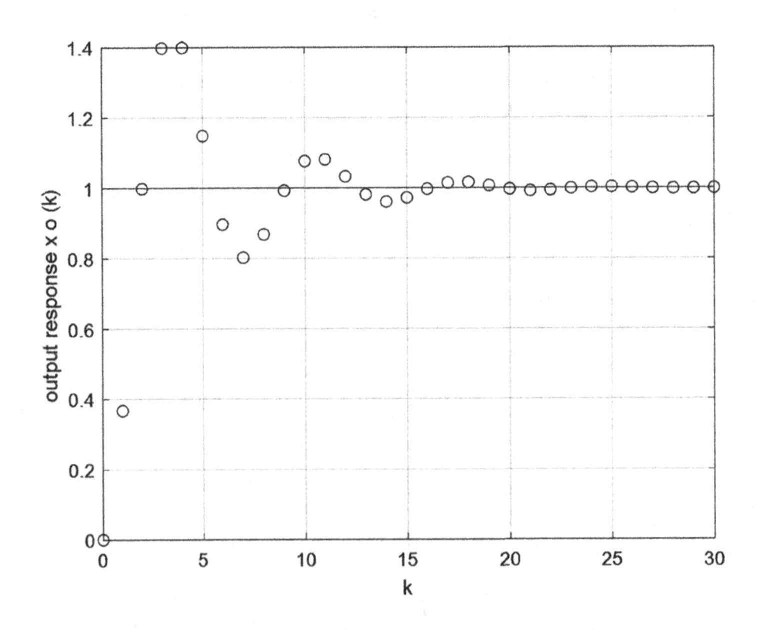

FIGURE 8.16 Discrete responses of step response in Example 8.30.

The MATLAB program:

```
for j = 1: 31
k (j) = j-1;
xo (j) = 1 + 1.022 * 0.795 ^ k (j) * cos (0.89 * k (j) +2.93);
xoo (j) = 1;
end
plot (k, xo, 'o ', k, xoo)
xlabel ('k')
ylabel ('output response x o (k)')
grid
```

8.8 APPLICATION OF MATLAB LANGUAGE IN DISCRETE CONTROL SYSTEM

The application of MATLAB language in the discrete control system is very simple. As long as we learn the application of MATLAB in the continuous system, the parameters of the continuous system change slightly, and we get the application of MATLAB in the discrete control system. The calling format is as follows:

Functions	Meaning
dstep (num, den)	% determine the step response request of system;
dbode (num, den, T)	% determine the Bode diagram of system;
dmargin (num, den, T)	% determine the Bode diagram of system and gain margin and phase margin;
dnyquist (num, den, T)	% determine the Nyquist curve of system;

Example 8.31

Simulation discrete system is shown in Example 8.30 using MATLAB language.

Solution

First, we substitute $T = 1$ into the transfer function of the system, we get

$$\Phi(z)=\frac{G(z)}{1+G(z)}=\frac{\left(T+e^{-T}-1\right)z+1-(1+T)e^{-T}}{z^2-(2-T)z+1-Te^{-T}}$$

Thus, the transfer function of the closed-loop system can be written as

$$\Phi(z)=\frac{ze^{-1}+1-2e^{-1}}{z^2-z+1-e^{-1}}$$

The step response of discrete system can be obtained in accordance with the transfer function. Its MATLAB procedure is as follows:

```
num=[exp(-1), 1-2*exp(-1)]
den=[1-1 1-exp(-1)]
dstep(num, den)
```

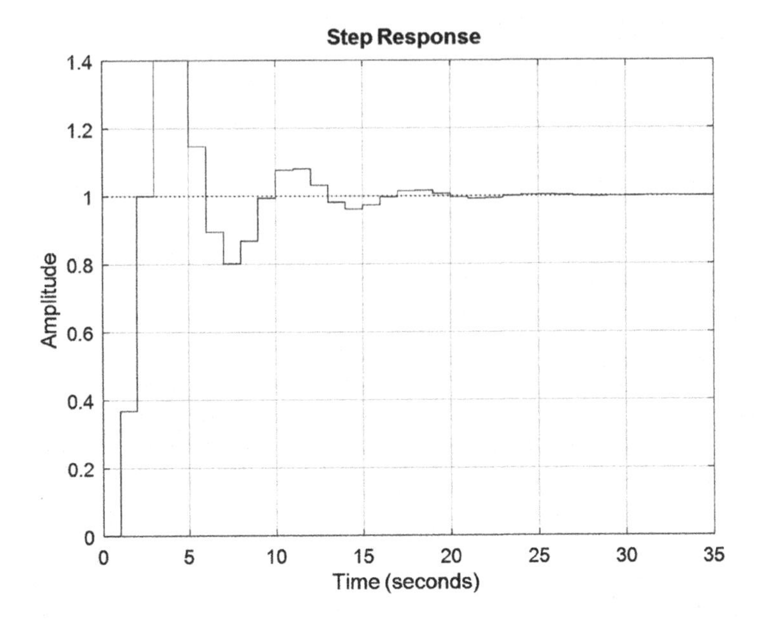

FIGURE 8.17 Discrete coordinates of step response of Example 8.31 using MATLAB functions.

Comparing Figure 8.16 with 8.17, we get that the desired curves are similar in nature and the results are same by programming using MATLAB, and using MATLAB built-in function.

Example 8.32

The known transfer function of the open-loop pulse of a discrete control system is given as

$$G(z)=\frac{0.74z+0.52}{z^2-1.37z\ +0.37}$$

Draw the Bode diagram of the control system using MATLAB language when the sample time $Ts = 1s$.

Solution

The MATLAB program of system is as follows:

```
num=[0.74 0.52]
den=[1-1.37 0.37];
dbode(num, den, 1)
grid
```

The Bode diagram is shown in Figure 8.18.

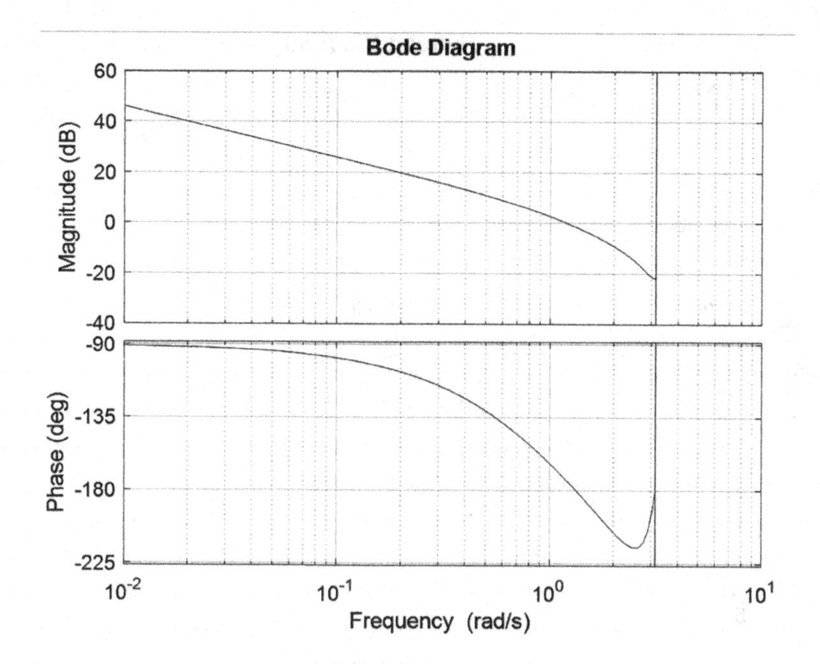

FIGURE 8.18 Bode diagram in Example 8.32.

Example 8.33

The known transfer function of the open-loop pulse of a second-order discrete system is given as

$$G(z)=\frac{0.7z+0.06}{z^2-0.5z+0.43}$$

Determine Nyquist curve of the discrete system when $Ts = 0.1s$.

Solution

```
num=[0.7 0.06];
den=[1-0.5 0.43];
dnyquist(num, den, 1)
grid
```

The Nyquist curve is shown in Figure 8.19.

EXAMPLES

This chapter requires you to understand the basic concepts of discrete control systems, learn mathematical tools of Z-transformation, learn to write a simple transfer function of discrete systems, and determine its stability.

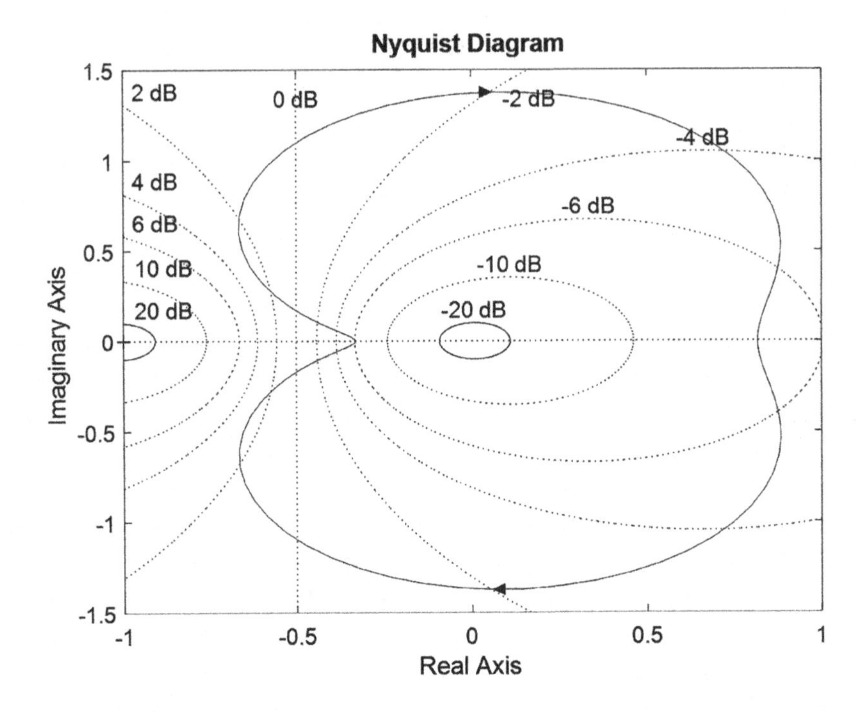

FIGURE 8.19 The Nyquist plot of Example 8.33.

Example 8.34

If the differential equations of system

$$y(k+2)-0.1y(k+1)-0.2y(k)=x(k+1)+x(k)$$

Let input $x(k)=1(k)$, the initial conditions $y(0)=y(1)=0, x(0)=0$. Determine $y(k)$.

Solution

We take the Z-transformation of both sides of the differential equations and get

$$z^2Y(z)-z^2y(0)-zy(1)-0.1zY(z)-0.1zy(0)-0.2Y(z)=zX(z)-zx(0)+X(z)$$

Substituting the initial conditions, we get

$$\left(z^2-0.1z-0.2\right)Y(z)=(z+1)X(z)$$

Thus,

$$Y(z)=\frac{z+1}{z^2-0.1z-0.2}X(z)$$

We know $x(k)=1(k)$, $X(z)=z/(z-1)$. By substituting them into the above equation, we get

$$Y(z)=\frac{(z+1)z}{(z^2-0.1z-0.2)(z-1)}=\frac{(z+1)z}{(z+0.4)(z-0.5)(z-1)}$$

Therefore, we get

$$y(k)=Z^{-1}[Y(z)]=0.476(-0.4)^k-3.333(0.5)^k+2.875\times 1(k)$$

where $K=0, 1, 2,\ldots$

Example 8.35

The block diagram of the sampling system is shown in Figure 8.20, where the sampling period $T = 0.2s$. Determine the stability of the closed-loop system.

Solution

The transfer function of the open-loop continuous system is written as

$$G(s)=\frac{5(1-e^{-Ts})}{s^2(0.1s+1)(0.05s+1)}$$

$$=(1-e^{-Ts})\cdot\frac{1000}{s^2(s+10)(s+20)}$$

$$=(1-e^{-Ts})\left(\frac{5}{s^2}-\frac{0.75}{s}+\frac{1}{s+10}-\frac{0.25}{s+20}\right)$$

From Table 8.1, we get the pulse transfer function of the open-loop system as

$$G(z)=(1-z^{-1})\left[\frac{5Tz}{(z-1)^2}-\frac{0.75z}{z-1}+\frac{z}{z-e^{-10T}}-\frac{0.25z}{z-e^{-20T}}\right]$$

Substituting $T = 0.2s$ into the above equation, we get

$$G(z)=\frac{0.3805z^2+0.449z+0.0198}{z^3-1.1533z^2+0.1558z-0.0025}$$

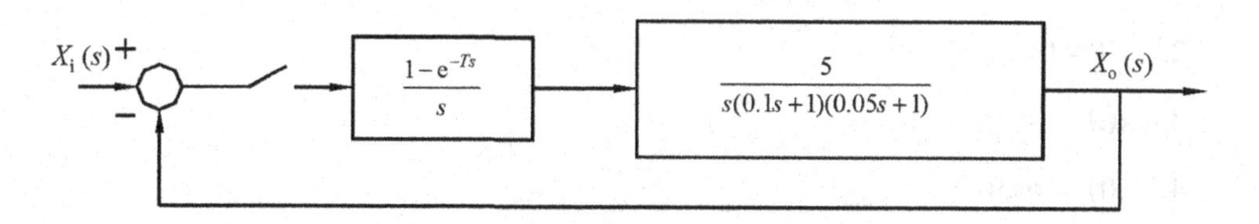

FIGURE 8.20 Block diagram of the sampling system of Example 8.35.

From Table 8.2, we get the pulse transfer function of the closed-loop system as

$$\varphi(z)=\frac{G(z)}{1+G(z)}=\frac{0.3805z^2+0.449z+0.0198}{z^3-0.7728z^2+0.6048z-0.0173}$$

Therefore, the characteristic equation of the closed-loop system is written as

$$z^3-0.7728z^2+0.6048z-0.0173=0$$

Applying w-transformation, we get

$$z=\frac{w+1}{w-1}$$

Substituting it into the characteristic equation of the closed-loop system, we get

$$\left(\frac{1+w}{1-w}\right)^3-0.7728\left(\frac{1+w}{1-w}\right)^2+0.6048\left(\frac{1+w}{1-w}\right)+0.0173=0$$

We multiply by $(1-w)^3$ in both ends of the above equation and get

$$2.3603w^3+3.2199w^2+1.5705w+0.8493=0$$

By using the Routh table in the above equations, we get

w^3	2.3603	1.5705
w^2	3.2199	0.8493
w^1	0.9479	
w^0	0.8493	

Since the numbers in the first column of Routh's table are positive, the closed-loop system is stable.

EXERCISES

1. Determine the *Z-transformation* of the following functions.

   1. $x(t)=1-e^{-at}$
   2. $x(t)=te^{-at}$
   3. $x(t)=\sin 10t$
   4. $x(t)=\cos 3t$
   5. $x(t)=e^{-at}\sin\omega t$

2. Determine the *Z-transformation* of the following Laplace transformation.

1. $X(s)=\dfrac{a}{s(s+a)}$

2. $X(s)=\dfrac{\omega}{s^2-\omega^2}$

3. $X(s)=\dfrac{s+3}{(s+1)(s+2)}$

4. $X(s)=\dfrac{s+1}{s^2}$

5. $X(s)=\dfrac{1}{s(s+1)(s+2)}$

3. Determine the initial value and final value of the following function.

1. $X(z)=\dfrac{z^2}{(z-1)(z-0.5)}$

2. $X(z)=\dfrac{Tz}{(z-1)^2}$

3. $X(z)=\dfrac{2}{1-z^{-1}}$

4. $X(z)=\dfrac{10z^{-1}}{(1-z^{-1})^2}$

5. $X(z)=\dfrac{z^2(z^2+z+1)}{(z^2-0.8z+1)(z^2+z+0.8)}$

4. Determine pulse train $x^*(t)$ corresponding to the following Z-transformations.

1. $X(z)=\dfrac{10z}{(z-1)(z-2)}$

2. $X(z)=\dfrac{z}{(z-1)(z+0.5)^2}$

3. $X(z)=\dfrac{10z(z+1)}{(z-1)(z^2+z+1)}$

4. $X(z)=\dfrac{z}{(z+1)(3z^2+1)}$

5. $X(z)=\dfrac{z^2}{(z-1)0.8(z-0.1)}$

5. Solve the following differential equation and represent it by $x(kT)$.

   1. $x(k+2)+4x(k+1)+3x(k)=2k;\ x(0)=x(1)=0$
   2. $x(k+3)+6x(k+2)+11x(k+1)+6x(k)=0;\ x(0)=x(1)=1, x(2)=0$

6. The known transfer function of a system is given

   1. $G(s)=\dfrac{a\left(1-e^{-Ts}\right)}{s^2(s+a)}$ (T is the sampling period)
   2. $G(s)=\dfrac{10}{s(s+10)}$

   Determine the pulse transfer function of the system.

7. If the differential equations which describe the sampling system is

$$x_o(k+2)-0.7x_o(k+1)-0.1x_o(k)=5x_i(k+1)+x_i(k)$$

   Determine the pulse transfer function.

8. Determine the output *Z-transformation* $X_o(z)$ of the system shown in Figure 8.21.

9. Sampling system block diagram is shown in Figure 8.22. Determine the open-loop pulse transfer function $G(z)$ and closed-loop pulse transfer function $\Phi(s)$.

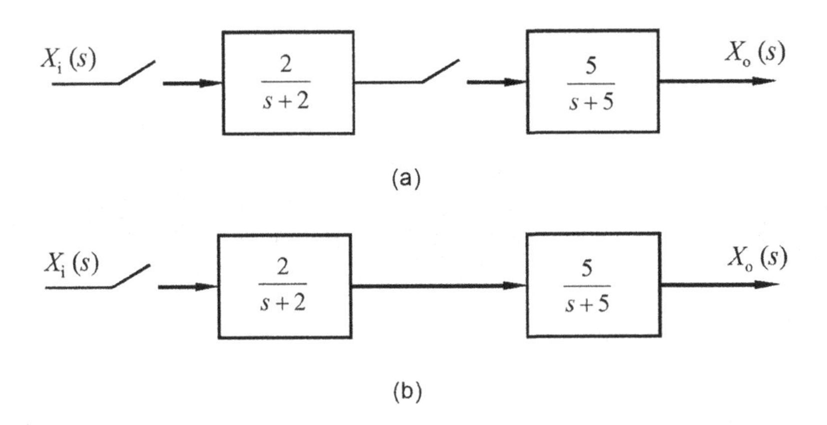

FIGURE 8.21 (a)-(b) Block diagram of open-loop pulse transfer system.

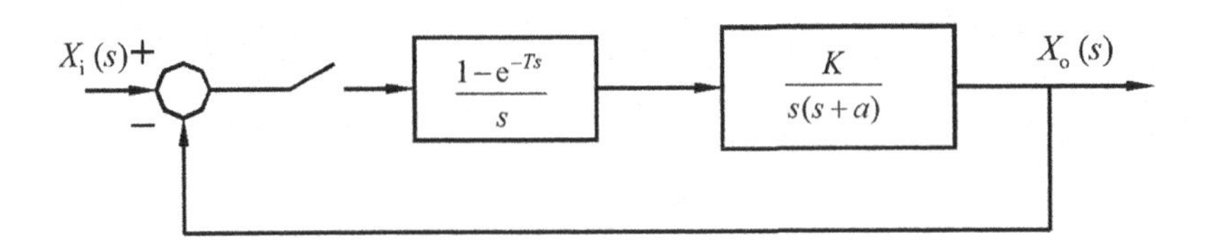

FIGURE 8.22 Block diagram of sampling system.

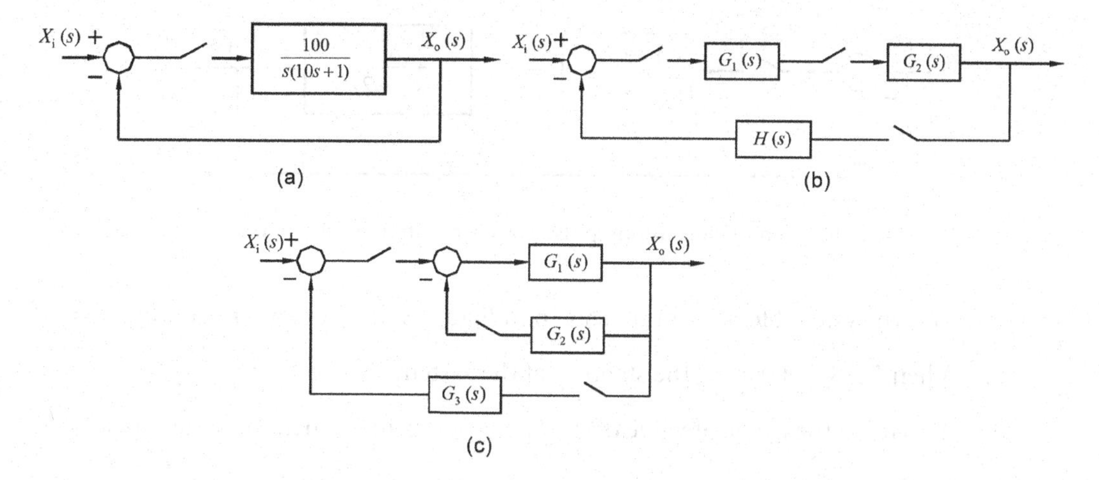

FIGURE 8.23 (a)-(c) Block diagram of closed-loop pulse transfer system.

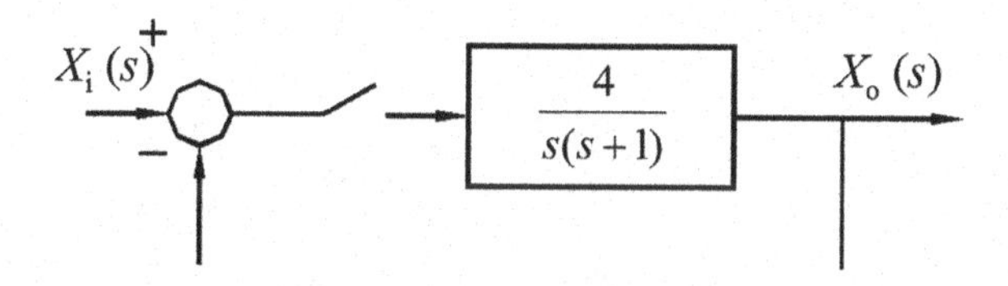

FIGURE 8.24 Block diagram of closed-loop pulse transfer system of Figure 8.23.

10. Determine the closed-loop transfer function of the system shown in Figure 8.23.
11. The known transfer function of the system is given as

$$\frac{X_o(z)}{X_i(z)} = \frac{z(z+1)}{\left(z-\frac{2}{5}\right)\left(z+\frac{1}{2}\right)}$$

Determine the unit impulse response and step response of the system.

12. The known characteristic equations of closed-loop sampling system is the following, determine the stability of the system.

    1. $z^2 - 0.632z + 0.368 = 0$
    2. $(z+1)(z+0.5)(z+2) = 0$
    3. $z^4 + 0.2z^3 + z^2 + 0.36z + 0.8 = 0$

13. Determine the stability of the system shown in Figure 8.24. (T is the sampling *period*, $T = 1$.)

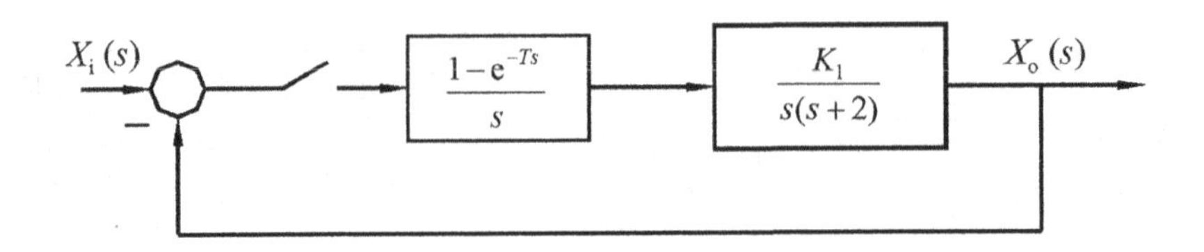

FIGURE 8.25 Block diagram of closed-loop pulse transfer system with $T = 1$s.

14. The known system block diagram shown in Figure 8.25, the sampling period $T = 1s$.
    1. When $K_{1=}8$, determine the stability of the system;
    2. Determine the range of value of the open-loop gain K when the system is stable.

CHAPTER 9

Applications with PID and Motor Control System

A COMPLETE DRIVE SYSTEM IS shown in block diagram form in Figure 9.1. The job of the converter is to draw electrical energy from the main motor at whatever voltage and frequency necessary to achieve the desired mechanical output. Control signals, in the form of low-power analog or digital voltages, tell the converter what it is supposed to be doing, while other low-power feedback signals are used to measure what is actually happening. By comparing the demand and feedback signals and adjusting the output accordingly, the target output is maintained. The simple arrangement shown in Figure 9.1 has only one input representing the desired speed and one feedback signal indicating the actual speed, but most drives will have extra feedback signals as we will see later. Almost all drives employ closed-loop (feedback) control.

A characteristic of power electronic converters which is shared with most electrical systems is that they have very little capacity for storing energy. This means that any sudden change in the power supplied by the converter to the motor must be detected in a sudden increase in the power drawn from the supply. In most cases, this is not a serious problem, but it does have two drawbacks. First, a sudden increase in the current drawn from the supply will cause a momentary drop in the supply voltage, because of the voltage source of the supply impedance. With a single-phase mains supply, for example, there can be no sudden increase in the power supply from the mains at the instant when the main voltage is zero, because instantaneous power is necessarily zero at this point in the cycle because the voltage is itself zero. It would be better if a significant amount of energy could be stored within the converter itself: short-term energy demands could then be met instantly, thereby reducing rapid fluctuations in the power drawn from the mains. But unfortunately, this is just not economic: most converters do have a small store of energy in their smoothing inductors and capacitors.

DOI: 10.1201/9781003293859-9

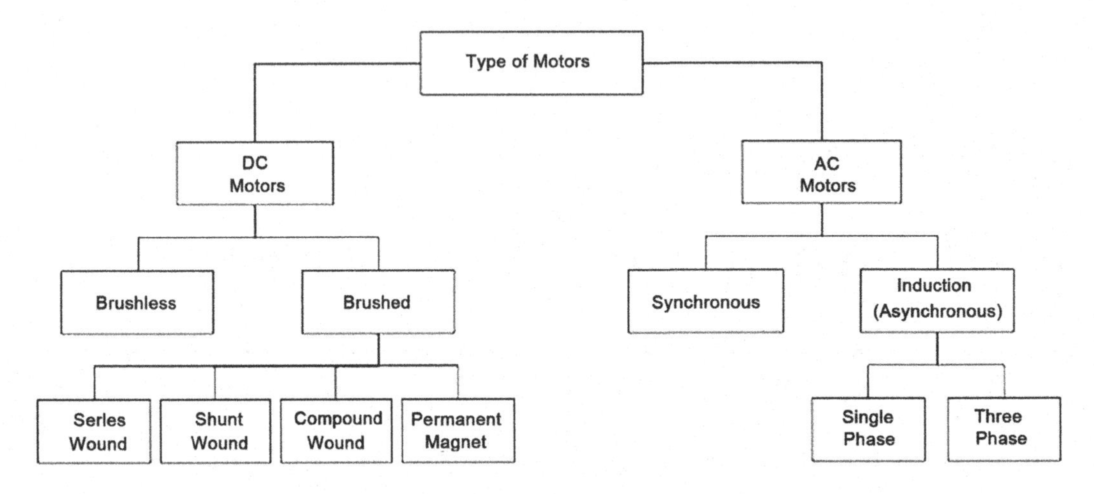

FIGURE 9.1 Types of motors.

9.1 DC AND AC MOTORS

9.1.1 DC Motors

DC motors were the first type of motor widely used and the systems (motors and drive) initial costs tend to be typically less than AC systems for low-power units, but with higher power, the overall maintenance costs increase and would need to be taken into consideration. The DC motors speed can be controlled by varying the supply voltage and are available in a wide range of voltages; however, the most popular type is 12 and 24 V, with some of the advantages being

- Easy installation
- Speed control over a wide range
- Quick starting, stopping, reversing, and acceleration
- High starting torque
- Linear speed-torque curve
- DC motors are widely used and can be used from small tools and appliances to electric vehicles, lifts, and hoists.

1. *DC brushless motors*
   DC brushless motors were first developed to achieve higher performance in a smaller space than DC brushed motors, and they are smaller than comparable AC models, as shown in Figure 9.2. An embedded controller is used to facilitate operation in the absence of a slip ring or commutator.

   Brushless motors alleviate some of the issues associated with the more common brushed motors (short life span for high-use applications) and are mechanically much simpler in design (not having brushes). The motor controller uses Hall effect

FIGURE 9.2 DC brushless motors.

sensors to detect the rotor's position, and using this, the controller can accurately control the motor via current in the rotor coils) to regulate the speed. The advantages of this technology are long life, little maintenance, and high efficiency (85%–90%), whereas the disadvantages are higher initial costs and more complicated controllers. These types of motors are generally used in speed and positional control with applications such as fans, pumps, and compressors, where reliability and ruggedness are required. An example of brushless design is stepper motors, which is primarily used in open-loop position control, with uses from printers through to industrial applications, such as high-speed pick and place equipment.

Brushless motors are also available with a feedback device that allows the control of the speed, torque and position of the motor, and the intelligent electronics control all three. So if more torque is required to accelerate it quicker to a certain speed, then more current is delivered—these are known as brushless servomotors, as shown in Figure 9.3.

Electric motors are now more diverse and adaptable than ever before. When planning a motion control system, the choice of motor is extremely important. The motor must align with the purpose and overall performance goals of the system. Luckily, there is a motor design suitable for any imaginable purpose.

2. *DC brushed motors*

   In a DC brushed motor, brush orientation on the stator determines current flow. In some models, the brush's orientation relative to the rotor bar segments is decisive instead. The commutator is especially important in any DC brushed motor design.

FIGURE 9.3 Brushless servomotors.

These are the more traditional type of motor and are typically used in cost-sensitive applications, where the control system is relatively simple, such as in consumer applications and more basic industrial equipment, these types of motors can be broken down as discussed below.

9.1.1.1 Series Wound

This is where the field winding is connected in series with rotor winding and speed control by varying the supply voltage; however, this type offers poor speed control, and as the torque to the motor increases, the speed falls. Applications include automotive, hoists, lifts, and cranes as it has a high starting torque.

9.1.1.2 Shunt Wound

This type has one voltage supply and the field winding is connected in parallel with the rotor winding and can deliver increased torque, without a reduction in speed by increasing the motor current. It has a medium level of starting torque with constant speed, so it is suitable for applications such as lathes, vacuum cleaners, conveyors, and grinders.

9.1.1.3 Compound Wound

This is a cumulative of Series and Shunt, where the polarity of the shunt winding is such that it adds to the series fields. This type has a high starting torque and runs smoothly if the load varies slightly and is used for driving compressors, variable-head centrifugal pumps, rotary presses, circular saws, shearing machines, elevators, and continuous conveyors.

9.1.1.4 Permanent Magnet

As the name suggests, rather than an electromagnet, a permanent magnet is used in applications with precise control and low torque, such as in robotics and servo systems.

9.1.2 AC Motors

AC motors (shown in Figure 9.4) are highly flexible in many features including speed control (VSD—variable speed drives) and have a much larger installed base compared to DC motors; some of the key advantages are:

- Low power demand on start
- Controlled acceleration
- Adjustable operational speed
- Controlled starting current
- Adjustable torque limit
- Reduced power line disturbances

FIGURE 9.4 AC motors.

The current trend for VSD is to add more features and programmable logic controller (PLC) functionality, which are advantages for the experienced used but require greater technical expertise during maintenance.

9.1.2.1 Types of AC Motors

1. *Synchronous AC motors*
   In this type of motor, the rotation of the rotor is synchronized with the frequency of the supply current and the speed remains constant under varying loads, so is ideal for driving equipment at a constant speed and is used in high precision positioning devices like robots, instrumentation, machines, and process control.

2. *Induction (asynchronous) AC motors*
   This type of motor uses electromagnetic induction from the magnetic field of the stator winding to produce an electric current in the rotor and hence torque. These are the most common type of AC motor and are important in the industry due to their load capacity with single-phase induction motors being used mainly for smaller loads, like in household appliances, whereas three-phase induction motors are used more in industrial applications including like compressors, pumps, conveyor systems, and lifting gear.

3. *AC brushless motors*
   AC brushless motors are some of the most popular motors in motion control. They use induction of a rotating magnetic field, generated in the stator, to turn both the stator and rotor at a synchronous rate. They rely on permanent electromagnets to operate.

9.2 STEPPER AND SERVOMOTORS

Direct drive is a high-efficiency, low-wear technology implementation that replaces conventional servomotors and their accompanying transmissions. In addition to being far easier to maintain over a longer period, these motors accelerate more quickly.

The linear electric motors feature an unrolled stator and motor, producing linear force along the device's length, as shown in Figure 9.5. In contrast to cylindrical models, they have a flat active section featuring two ends. They are typically faster and more accurate than rotatory motors.

9.2.1 Stepper Motors

Stepper motors use an internal rotor, electronically manipulated by external magnets. The rotor can be made with permanent magnets or a soft metal. As windings are energized, the rotor teeth align with the magnetic field. This allows them to move from point to point in fixed increments. Before work begins on any new system, think carefully about the competing properties of the different motors. The selection of the right motor gets any project off to a better start.

Stepper motors offer several advantages over servomotors beyond the larger number of poles and easier drive control. The design of the stepper motor provides a constant holding torque without the need for the motor to be powered. The torque of a stepper motor at low speed is greater than a servomotor of the same size. One of the biggest advantages of stepper motors is their relatively low cost and availability.

9.2.1.1 Stepper Limitations

For all of their advantages, stepper motors have a few limitations which can cause significant implementation and operation issues depending on your application. Stepper motors

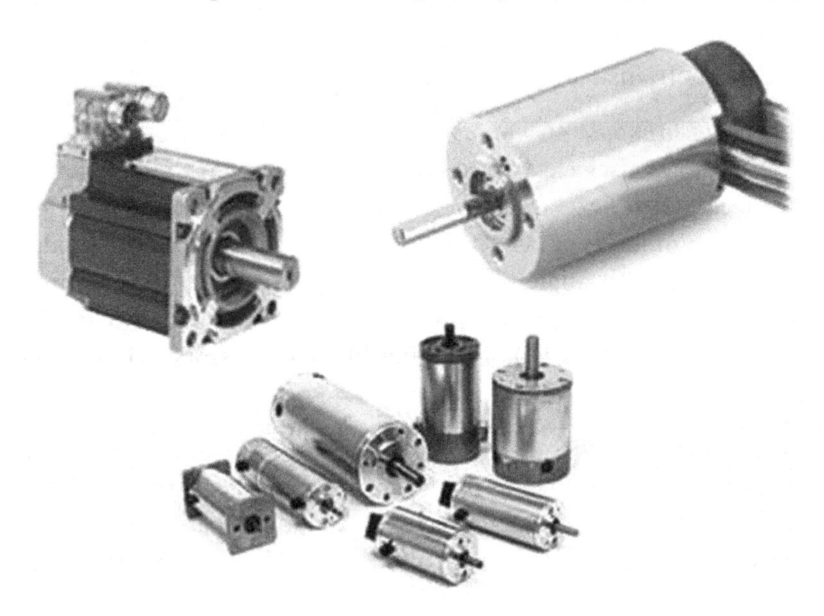

FIGURE 9.5 Linear electric AC motors.

do not have any reserve power. In fact, stepper motors lose a significant amount of their torque as they approach their maximum driver speed. A loss of 80% of the rated torque at 90% of the maximum speed is typical. Stepper motors are also not as good as servomotors at accelerating a load. Attempting to accelerate a load too fast where the stepper cannot generate enough torque to move to the next step before the next drive pulse will result in a skipped step and a loss in position. If positional accuracy is essential, either the load on the motor must never exceed its torque or the stepper must be combined with a position encoder to ensure positional accuracy. Stepper motors also suffer from vibration and resonance problems. At certain speeds, partially depending on the load dynamics, a stepper motor may enter resonance and be unable to drive the load. This results in skipped steps, stalled motors, excessive vibration, and noise.

9.2.2 Servomotors

Servomotors are electrical devices used for converting electrical signals provided as input into a precise angular velocity. It is known as a servo system. Basically, a servo system is the one that generates some form of mechanical variable like velocity, acceleration, or position as its output. A servomotor is designed to generate the mechanical equivalent of the applied electrical signal. It is known to be a low-power motor that finds applications in servomechanisms or position control systems. In servomotor, the applied electric input allows the rotation of the motor at a certain angle. Also, it does so with great precision, which means rotation at a specified angle can be accurately achieved.

A servomotor is any motor coupled with a feedback sensor to facilitate positioning; thus, servomotors are the backbone of robotics. Both rotary and linear actuators are used. Low-cost brushed DC motors are common but are being superseded by brushless AC motors for high-performance applications.

A servomotor is a linear or rotary actuator that provides fast precision position control for closed-loop position control applications. Unlike large industrial motors, a servomotor is not used for continuous energy conversion. Servomotors have a high-speed response due to low inertia and are designed with small diameter and long rotor length. They work on servo mechanism that uses position feedback to control the speed and final position of the motor. Internally, a servomotor combines a motor, feedback circuit, controller, and other electronic circuits.

A servomotor is one of the widely used VSD in industrial production and process automation and building technology worldwide. Although servomotors are not a specific class of motors, they are intended and designed to use in motion control applications that require high accuracy positioning, quick reversing, and exceptional performance. These are widely used in robotics, radar systems, automated manufacturing systems, machine tools, computers, CNC machines, tracking systems, etc.

Servomotor uses an encoder or speed sensor to provide speed feedback and position. This feedback signal is compared with the input command position (desired position of the motor corresponding to a load) and produces the error signal (if there exists a difference between them). The error signal available at the output of the error detector is not enough to drive the motor. So, the error detector followed by a servo amplifier raises the voltage and power level of the error signal and then turns the shaft of the motor to the desired position.

Basically, servomotors are classified into AC and DC servomotors depending upon the nature of supply used for their operation. Brushed permanent magnet DC servomotors are used for simple applications owing to their cost, efficiency, and simplicity. These are best suited for smaller applications. With the advancement of microprocessor and power transistors, AC servomotors are used more often due to their high accuracy control.

A servomotor is a revolving or linear motion actuator that allows the precise control of its linear velocity, acceleration, or angular position. This motor is available in the market with coupled sensor for feedback control. This motor is controlled with a sophisticated controller or a special kind of module is designed for controlling this motion. Servomotor is not a special class of motor, but actually, the servo word is used for those motors that are used for control loop system. Servomotor is usually interfaced with microcontroller for control applications. A simple servomotor with a feedback sensor is shown in Figure 9.6.

There are some special types of applications of electrical motor where rotation of the motor is required for just a certain angle and not continuously for a long period of time. For these applications, some special types of motor are required with some special arrangement that makes the motor rotate at a certain angle for a given electrical input (signal). For this purpose, servomotor comes into the picture. This is normally a simple DC motor that is controlled for specific angular rotation with the help of additional servomechanism (a typical closed-loop feedback control system). Nowadays, servo system has huge industrial applications. Servomotor applications are also commonly seen in remote-controlled toy cars for controlling the direction of motion and are also very commonly used as the motor that moves the tray of a CD or DVD player. Besides these, there are other hundreds of servomotor applications we see in our daily life. The main reason behind using a servo is that it provides angular precision, i.e., it will only rotate as much we want and then stop and wait for the next signal to take further action. This is unlike a normal electrical motor that starts rotating as and when power is applied to it and the rotation continues until we switch

FIGURE 9.6 Servomotors.

off the power. We cannot control the rotational progress of the electrical motor, but we can only control the speed of rotation and can turn it ON and OFF.

Servomotor is a special type of motor that is automatically operated up to a certain limit for a given command with the help of error-sensing feedback to correct the performance. This is nothing but a simple electrical motor, controlled with the help of servomechanism. If the motor as a controlled device, associated with servomechanism as a DC motor, then it is commonly known as DC servomotor. If the controlled motor is operated by AC, it is called AC servomotor.

For applications in which high speed and high torque are needed, servomotors shine. Stepper motors peak around speeds of 2,000 rotation per minute (rpm), while servomotors are available many times faster. Servomotors also maintain their torque rating at high speed, up to 90% of the rated torque is available from a servo at high speed. Servomotors are also more efficient than stepper motors with efficiencies between 80% and 90%. A servomotor can supply roughly twice its rated torque for short periods, providing a well of capacity to draw from when needed. In addition, servomotors are quiet, available in AC and DC drive, and do not vibrate or suffer from resonance issues.

9.2.2.1 Servo Limitations

Servomotors are capable of delivering more power than stepper motors but do require much more complex drive circuitry and positional feedback for accurate positioning. Servomotors are also much more expensive than stepper motors and are often harder to find. Servomotors often require gearboxes, especially for lower speed operation. The requirement for a gearbox and position encoder makes servomotor designs more mechanically complex and increases the maintenance requirements for the system. To top it all off, servomotors are more expensive than stepper motors before adding on the cost of a position encoder.

9.2.3 Servo Motor Mechanism

Servomechanism is the term associated with positioning systems. It is a mechanism that is only applied to the systems where the feedback and error correction signals control the mechanical position or its derivatives like velocity and acceleration. Servomotor uses servomechanism to convert the applied electrical energy into precisely controlled motion by making use of negative feedback. The servo systems are used in position control systems. Thus, they are referred to as the automatic control system that uses error signals for their operation. The amplified form of these signals is used to drive the motors. In the case of servomotors, position feedback is used to control its motion and final position. It is known to be a simple electric motor that is controlled by servomechanism.

The servomotor is basically composed of three basic parts:

1. Controlling device
2. Output sensor
3. Feedback system

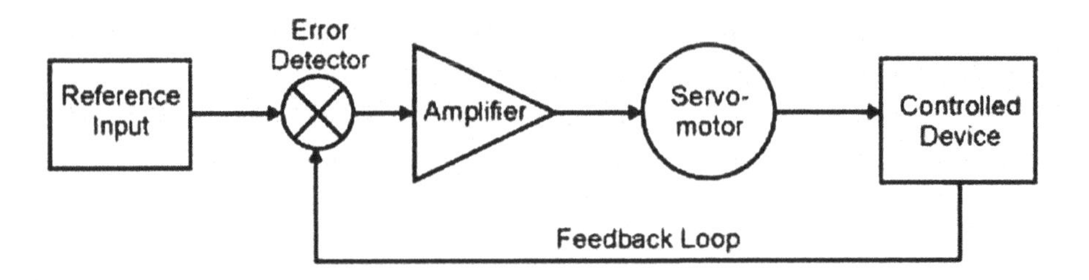

FIGURE 9.7 Servomotor mechanism.

Servomotor mechanism is an automatic closed-loop control system. For this system, a controller is required, which is composed of a comparator and a feedback path, whereas instead of controlling a device by applying the variable input signal, the device is controlled by a feedback signal generated by comparing the output and reference input signals. The controller has two inputs and one output. In the controller, the comparator compares the output signal with the desired reference signal for producing the input for servomotor. The output signal is sensed through the sensor. The input signal for the motor is called the feedback signal, as shown in Figure 9.7. The motor works on the basis of this feedback signal. The comparator signal is basically a logic signal for motor. If the logical difference is high, then the motor would be on for a desired time, and if the logical difference is low, then the motor would be off for a desired time. Actually, the controller the in comparator would be decided the whether the motor would be on or off. A good controller is mandatory for the proper function of the servomotor. Hence, the primary task of a servomechanism is to maintain the output of a system at the desired value in the presence of disturbances.

9.2.3.1 Working of Servomotors

A servomotor has four major components on which its working is dependent. These are as follows:

1. Motor
2. Gear system
3. Position sensor
4. Control circuit

These four units combined comprise the servomotor. We have already discussed that it uses a feedback control system that utilizes the error signal to correct the position of the motor. More specifically, by using a feedback control system, the shaft precisely achieves a position at a particular angle. It operates in a way that initially the motor is excited by externally supplying power to it. It is to be noted here that the speed of the motor depends on the applied voltage to it. This means that to control the motor speed, the voltage must be adjusted accordingly. In servomotor, the potentiometer is generally used as the position sensing and controlling device. Basically, according to the supplied input voltage, the device rotates and attains a specific position.

Further, the position of the shaft is analyzed by the position control device. The feedback potentiometer after analyzing the position of the shaft converts it into a specific voltage level. Later on, a comparison between the actually achieved voltage level and the desired voltage level is done. The difference generated from the two different voltage levels is the error signal that is amplified and is provided to the control unit. The control unit then generates a signal that implies the amplified error signal that controls the voltage provided to the motor. The voltage level of the error signal provided to the motor now changes its position. In this way, the desired position of the motor is achieved with high precision using a feedback potentiometer.

9.2.4 Construction of Servo Motor

The servomotor is a DC motor that consists of the following main parts:

1. Stator winding
2. Rotor winding
3. Bearing
4. Shaft
5. Encoder

The servomotor consists of two winding stators and rotor winding. The stator winding is wound on the stationary part of the motor, and this winding is also called field winding of the motor, this winding could the permanent magnets. The rotor winding is wound on the rotating part of the motor and this winding is also called the armature winding of the motor. The motor consists of two bearings on the front and backside for the free movement of the shaft. Shaft is basically the iron rod on which the armature winding is coupled. The encoder has the approximate sensor for telling the rotational speed and revolution per minute of the motor. The construction of the servomotor is shown in Figure 9.8.

9.2.5 Working Principle of Servo Motor

As we know that a servomotor is basically a DC motor (in some special cases it is AC motor) along with some other special-purpose components that make a DC motor a servo, as shown in Figure 9.9. In a servo unit, you will find a small DC motor, a potentiometer, gear arrangement, and an intelligent circuitry. The intelligent circuitry along with the potentiometer makes the servo rotate according to our wishes. As we know, a small DC motor will rotate at high speed, but the torque generated by its rotation will not be enough to move even a light load. The gear mechanism will take the high input speed of the motor (fast), and at the output, we will get an output speed that is slower than the original input speed but more practical and widely applicable.

Say at the initial position of the servomotor shaft, the position of the potentiometer knob is such that there is no electrical signal generated at the output port of the potentiometer. This output port of the potentiometer is connected with one of the input terminals

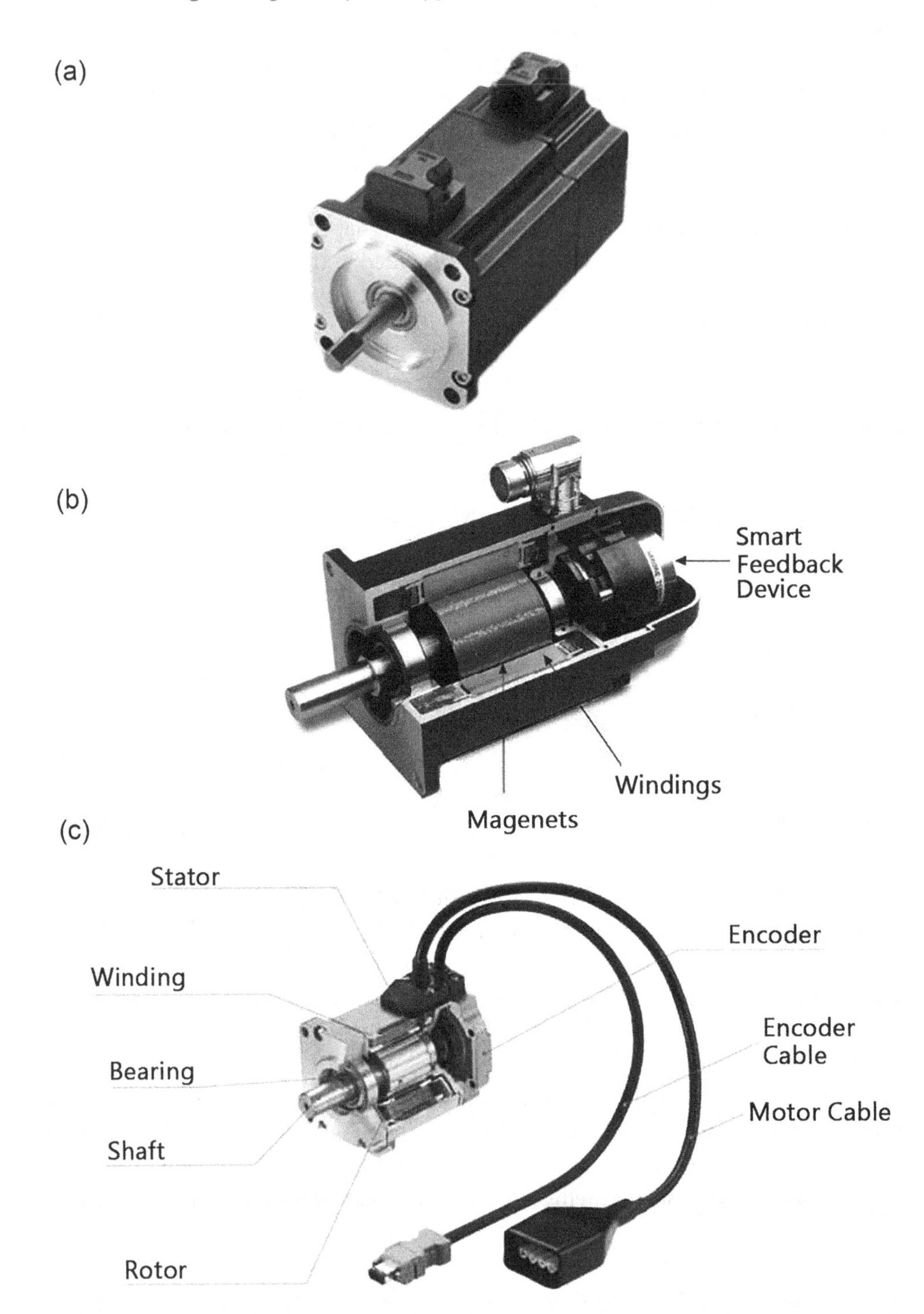

FIGURE 9.8 Structure of the servomotor system.

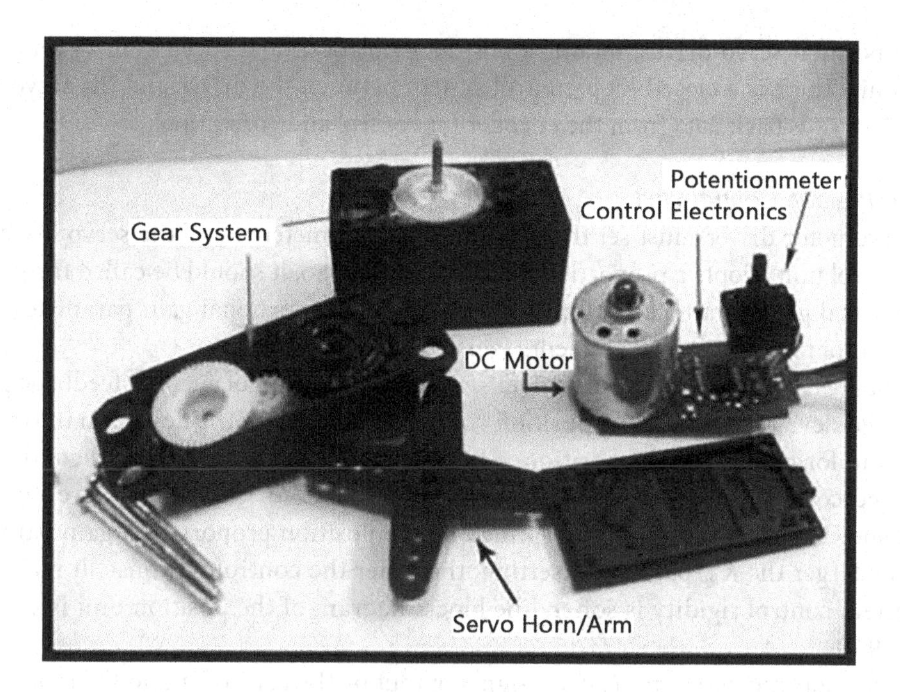

FIGURE 9.9 The gear system inside the servomechanism.

of the error detector amplifier. Now an electrical signal is given to another input terminal of the error detector amplifier. Now the difference between these two signals, one comes from the potentiometer and another from an external source, will be amplified in the error detector amplifier and feeds the DC motor. This amplified error signal acts as the input power of the DC motor, and the motor starts rotating in the desired direction. As the motor shaft progresses, the potentiometer knob also rotates as it is coupled with the motor shaft with help of gear arrangement. As the position of the potentiometer knob changes, there will be an electrical signal produced at the potentiometer port. As the angular position of the potentiometer knob progresses, the output or feedback signal increases. After desired angular position of the motor shaft, the potentiometer knob reaches such a position that the electrical signal generated in the potentiometer becomes the same as of external electrical signal given to the amplifier. At this condition, there will be no output signal from the amplifier to the motor input as there is no difference between the external applied signal and the signal generated at the potentiometer. As the input signal to the motor is nil at that position, the motor stops rotating. This is how a simple conceptual servomotor works.

The working purpose of the servo drive is mainly to work according to the instructions (*P*, *V*, *T*) sent by the servo controller. Synchronous motors are not completely synchronized with the rotating magnetic field. The drive must be corrected to make the motor work stably without losing any step. Therefore, the main task of the servo drive is to drive the motor to follow the control instructions correctly.

The following uses the general servomotor system *P* command (position servo) as an example to illustrate the basic principle of the drive and various necessary adjustment parameters.

The position servo drive contains a position control unit, a speed control unit, and a drive unit. There is a closed-loop control system between the driver and the servomotor. The drive sends back data from the encoder for control and correction.

9.2.5.1 Position Control Unit

The servomotor driver must set the position gain parameter (K_{PP}). The servo drive position control unit adopts a proportional control system. So, it should be called the position proportional gain parameter. Adjusting the position proportional gain parameter is also called servomotor rigidity adjustment.

Compare the number of command pulses with the number of encoder feedback pulses. It is called deviation count. The position control unit converts the deviation into a speed command for correcting the position. After being processed by the speed control unit, the speed command is sent to the drive unit for the motor drive. Therefore, the value of the speed command can be determined by the position proportional gain parameter K_{PP}. The larger the K_{PP} parameter setting, the faster the control response. It means that the system control rigidity is softer. The block diagram of the position unit is shown in Figure 9.10.

The relationship between the drive signal output by the controller and the input is only a proportional gain constant relationship. Therefore, it is called a proportional controller. The input signal is the deviation between the controller command pulse and the feedback pulse generated by the servomotor encoder. The deviation is multiplied by the proportional gain constant P and then sent to the next control unit for processing.

The deviation counter is different from the general function subtractor. It is the action of the number of pulses at the two input terminals offsetting each other, although the offset must be zero in the end. However, there is a delay time difference between the sending of the command pulse and the feedback. This is one of the causes of the deviation. Another part of the deviation is caused by external forces. When the motor stops, the displacement is formed due to the load change, which causes the deviation. It also reflects the necessity of correcting the output. The proportional controller needs to control the amplitude of the correction output.

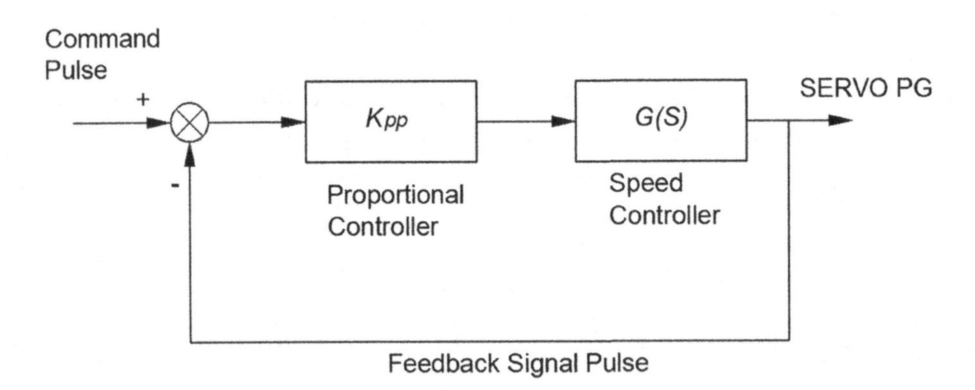

FIGURE 9.10 The block diagram of the position unit.

9.2.5.1.1 The Influence of Position Proportional Gain Parameter K_{PP} Use the instantaneous response of the step input to illustrate the effect of K_{PP}. The unit step response of the proportional controller is shown in Figure 9.11. The distance from the highest point of the curve to the set point is called the maximum overshoot. The time when the output value is 0.1~0.9 is called rise time. The time when the output value changes from 0 to 0.95~1.05 is called stabilization time.

When the K_{PP} value increases, the servomotor has a better response to the position control, generally called rigidity. But it is also prone to vibration and noise. It easily leads to instability.

After the K_{PP} value is adjusted, the effect will be reflected when the servomotor is positioned and stopped. Observed from the step response graph, when the K_{PP} value increases, the rise time decreases. The set point can be reached quickly. The maximum overshoot will increase accordingly. Therefore, the following factors must be considered:

1. Whether the motion mechanism can accept a large overshoot.
2. A shorter rise time does not mean that the stabilization time can be shortened.
3. When the K_{PP} value decreases, the rise time increases. It takes a long time to reach the set point. The maximum overshoot is reduced. But it does not mean that the system stabilization time will be extended.

For the above situation, adjust the K_{PP} value. The shortest stabilization time obtained is the best value. Appropriate instruments are required to measure system stability time. When there is no suitable instrument or tool for assistance, it can only be adjusted manually

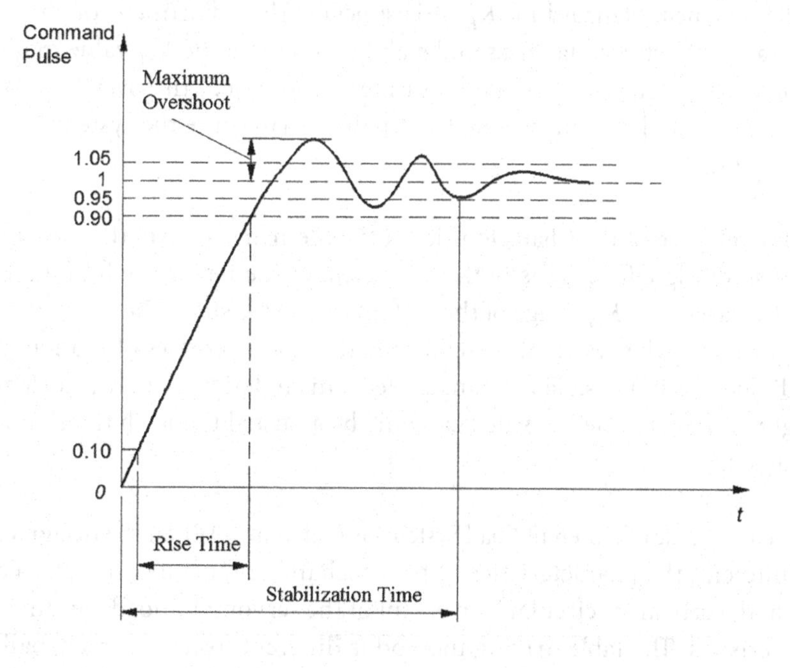

FIGURE 9.11 The unit step response of the proportional controller.

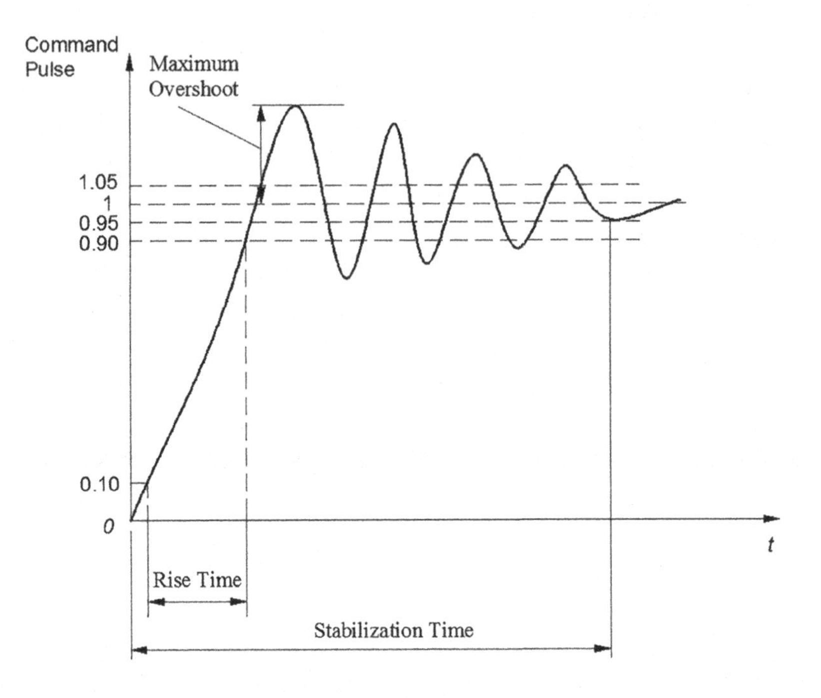

FIGURE 9.12 The unit step response of with larger K_{PP} value.

and judge whether the K_{PP} value is applicable. The unit step response with larger K_{PP} value is shown in Figure 9.12. The unit step response with too large K_{PP} value is shown in Figure 9.13. The unit step response with too small K_{PP} value is shown in Figure 9.14.

9.2.5.1.2 The Judgment Standard for K_{PP} Adjustment The adjustment of the K_{PP} value is actually a trade-off between fastness and stability. Increase the K_{PP} value to quickly reach the control target. It means that the rise time is shortened, the overshoot is increased, and the system instability is increased. Eventually, it will cause the system to oscillate and become unusable.

9.2.5.1.3 The Influence of the Characteristics of the Mechanism Another important factor affecting positioning efficiency is braking efficiency. The better the braking efficiency of the motor, the larger the K_{PP} range of the system in a stable state. Therefore, it can run with a larger K_{PP}, making the system more efficient. The characteristics of motion mechanisms are often difficult or impossible to change. Mechanism design usually prioritizes function and strength. Therefore, the K_{PP} value can only be adjusted to match the characteristics of the mechanism.

9.2.5.1.4 The K_{PP} Value of Different Load Systems Is Different When the design of the mechanism is different, the characteristics of the mechanism must be different. For example, the horizontal, vertical, or circular movement of the servomotor load has different operating characteristics. The table driving method is different from gear rack, ball screw, and belt drives.

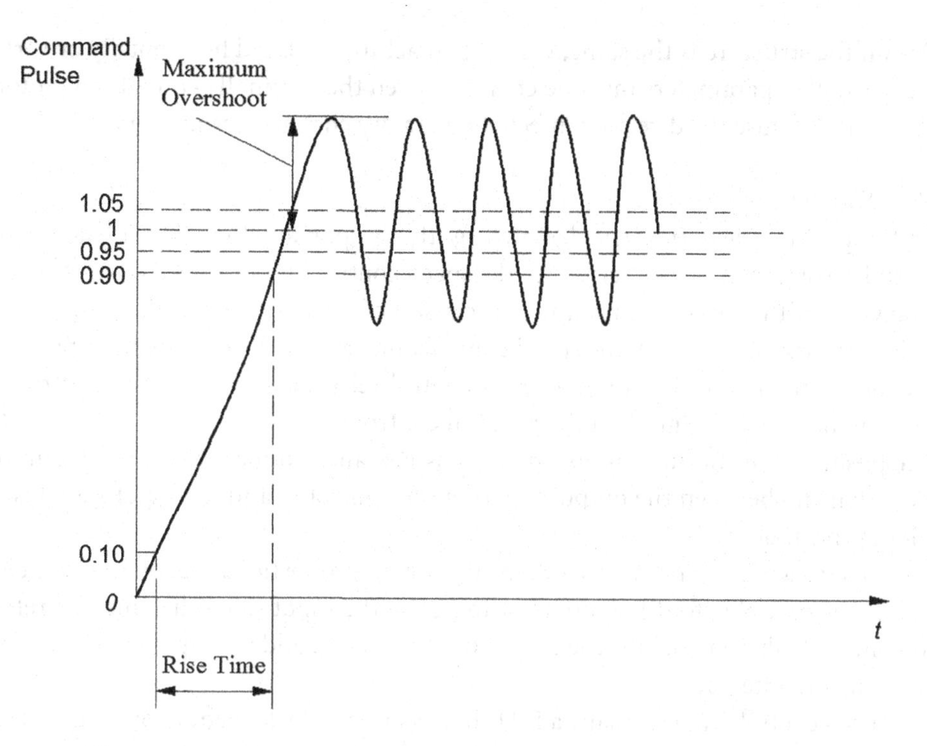

FIGURE 9.13 The unit step response with too large K_{PP} value.

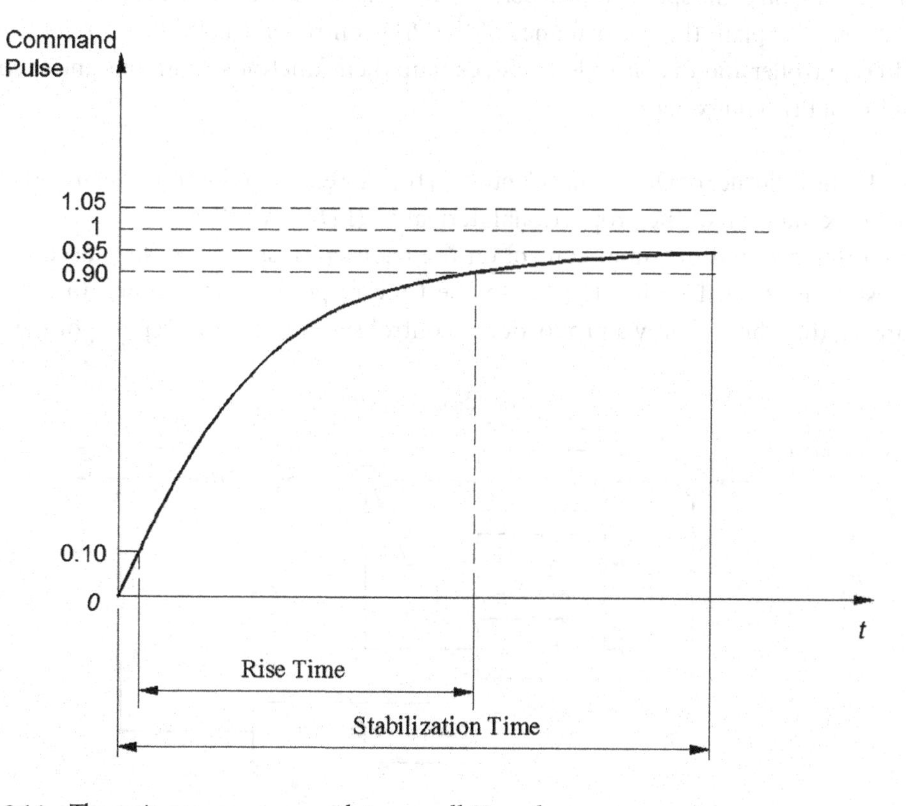

FIGURE 9.14 The unit step response with too small K_{PP} value.

Even if the structure is the same, when the machine is placed horizontally or vertically, the servo system parameters must be changed. Even the originally configured motor cannot be used. Because the direction affected by gravity changes, results vary.

9.2.5.2 Speed Control Unit

After the position deviation is calculated by the proportional controller, the correction amplitude is obtained. Then it is sent to the speed control unit for speed control. The speed setting value of the speed control unit is the result of the operation of the position control unit. For control engineering, the speed control unit can also be regarded as a part of position control. The two units are in a series control relationship. The speed control unit is more complicated in design than the position control unit.

The position control unit discussed above is just an amplifier with a fixed gain of K_{PP}. The relationship between the output signal of the controller and the input signal is a proportional constant.

From a mathematical point of view, in addition to proportional operations, the controller should also be able to differentiate or integrate the input signal in time. Therefore, an ideal controller should contain the following equipment: adder, amplifier, attenuator, differentiator, and integrator.

The speed control unit is actually a PID controller. The block diagram of the speed control unit PID controller is shown in Figure 9.15 Adjusting the K_{VP}, K_{VI}, and K_{VT} values of the PID controller can make the speed control performance of the servo system meet the requirements.

In order to explain the performance of the PID controller, the PID controller is divided into PD controller and PI controller. First, explain their functions separately and then merge them into a PID controller.

9.2.5.2.1 The Influence of Differential Control on the Closed-Loop Control System Figure 9.16 is the block diagram of the proportional derivative (PD) controller.

The influence of differential control on the transient response of the closed-loop control system is studied with reference to the time response curve shown in Figure 9.17. Assuming that there is only a proportional control system, its unit step response is shown

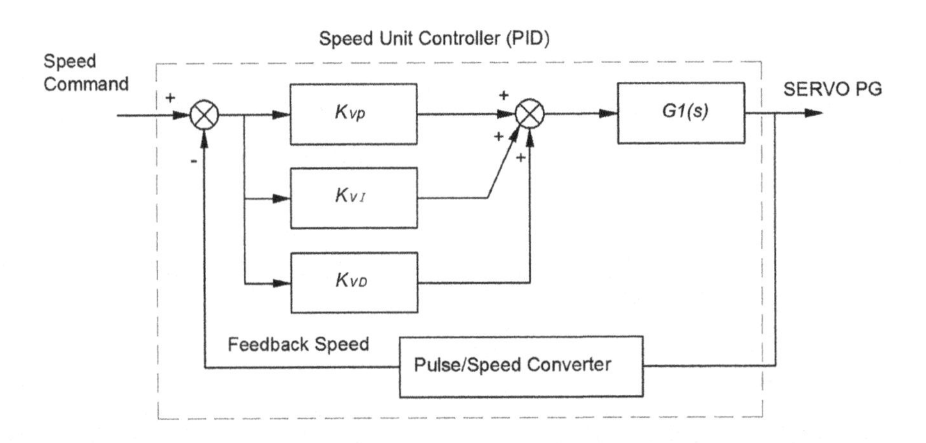

FIGURE 9.15 Block diagram of speed control unit PID controller.

in Figure 9.17a. The corresponding error signal $e(t)$ and time derivative $de(t)/dt$ are shown in Figure 9.17b and c, respectively.

Note: The response peak overshoot in Figure 9.17a is quite high, and there are some oscillations. For many practical control purposes, this is not desirable. This situation is the same as the control method of the position control unit as discussed above. But the speed

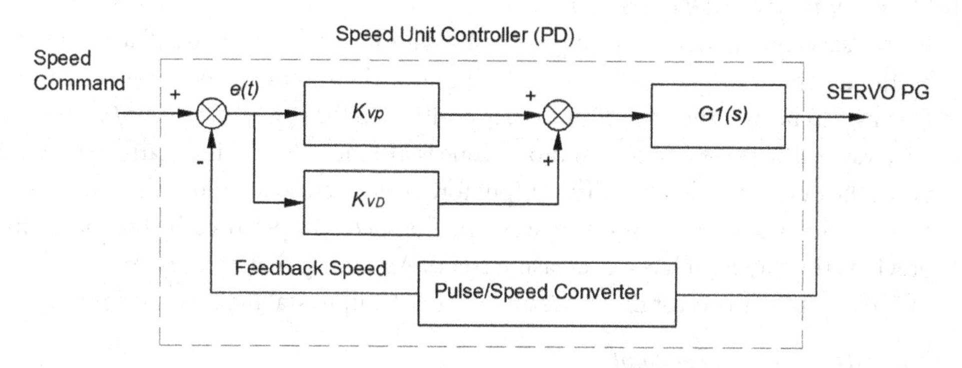

FIGURE 9.16 Block diagram of the proportional derivative (PD) controller.

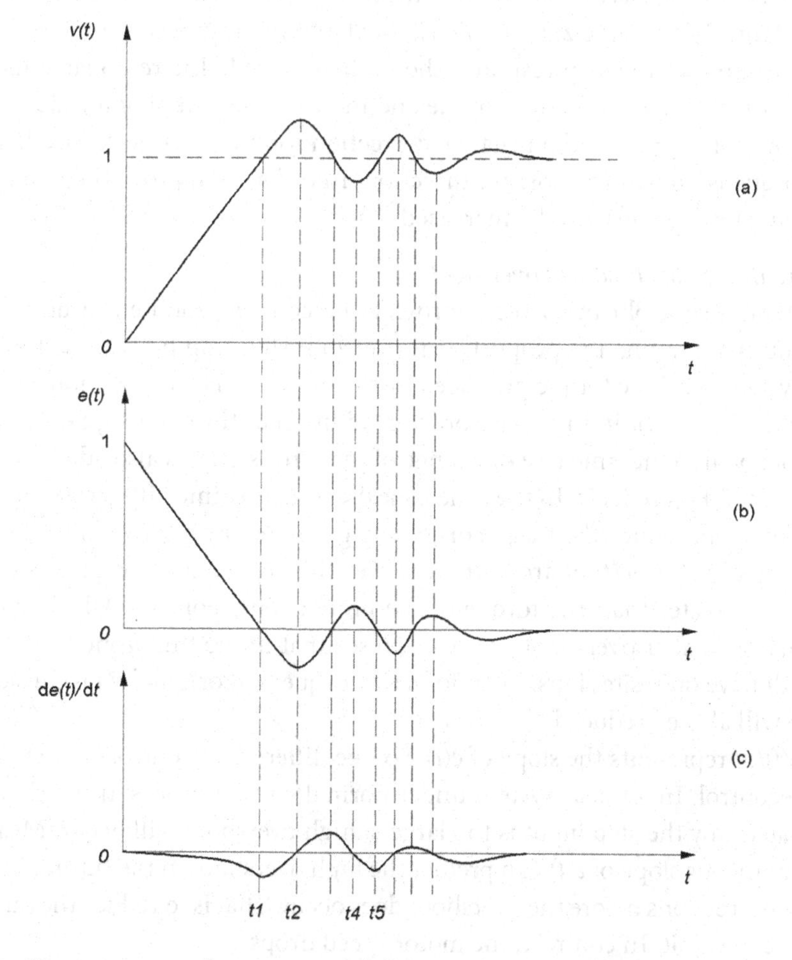

FIGURE 9.17 Time response of proportional differentiation controller (PD).

control unit cannot only use the proportional control system because the physical quantities of its input and output are not the same, and it must be corrected so that the output is the same as the input. Generally, the speed control unit of a servomotor is at least a proportional integral (PI) controller or a PI derivative (PID) controller.

1. *The generation of high overshoot*
   For the servomotor system, the high overshoot and the subsequent oscillation are caused by the excessive torque and lack of damping of the motor in the time zone $0<t<t_1$. During this period, the error signal $e(t)$ is positive. In the time zone $t_1<t<t_3$, $e(t)$ is a negative value. The corresponding motor torque is also negative. This negative torque slows down the output acceleration. The output torque is reversed and down. In the time zone $t_3<t<t_5$, the motor torque is positive again. The speed drop caused by the torque in the previous time interval has a decreasing trend. Assuming that the system is stable, after each oscillation, the error decreases and the final output stabilizes at the desired value.

2. *Methods to reduce overshoot*
   Considering the above description, it can be considered that the factors leading to high overshoot are as follows: In the time zone $0<t<t_1$, the positive correction torque is too large. In the time zone $t_1<t<t_2$, the damping torque is insufficient. Therefore, the overshoot of the step response should be reduced. The reasonable method is to reduce the positive correction torque and increase the damping torque. Similarly, in the time zone $t_2<t<t_4$, the negative correction torque must be reduced. The damping torque is now in the positive direction. In order to improve the down jump, the damping torque must also be increased.

3. *Differential control reduces overshoot*
   Differential control can correctly provide the compensation effect described in the previous paragraph. The proportional control system can be modified to a PD control system. And the torque produced by the motor can be proportional to the signal $K_{VP}\,e(t)+K_{VD}\,de(t)/dt$. In other words, in addition to the error signal, another signal proportional to the time rate of change of the error signal is also added to the motor, as shown in Figure 9.17c. In the time zone $0<t<t_1$, the time differential of $e(t)$ is negative. This will reduce the torque originally generated by $e(t)$ alone. In the time zone $t_1<t<t_2$, $e(t)$ and $de(t)/dt$ are both negative. This means that the generated damping torque is greater than the torque of the proportional control. All these effects will produce a smaller overshoot. It is easy to see that in the time zone $t_2<t<t_3$, $e(t)$ and $de(t)/dt$ have opposite signs. Therefore, the torque that originally produced the down jump will also be reduced.

   $de(t)/dt$ represents the slope of $e(t)$. So, the differential control is essentially a kind of precontrol. In a linear system under normal circumstances, if the slope of $e(t)$ or $v(t)$ caused by the step input is too large, a high overshoot will occur. Measuring the instantaneous slope of $e(t)$ can predict the high overshoot in the future. Make appropriate corrections before the overshoot does occur. That is to reduce the output torque of the drive unit. In contrast, the motor speed drops.

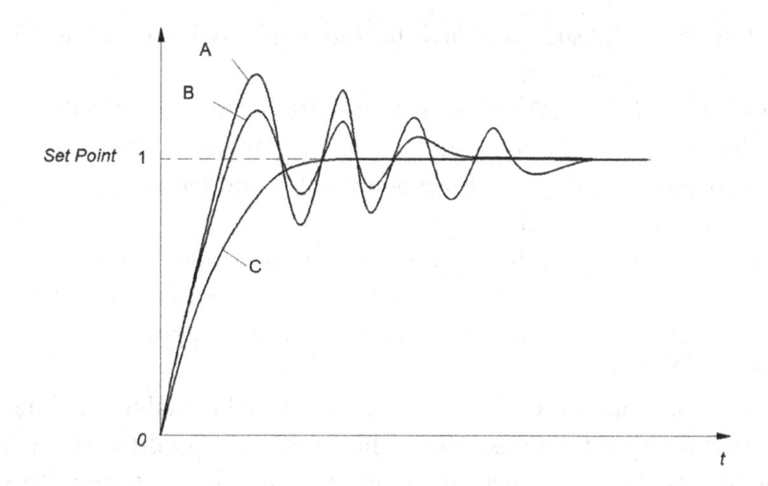

FIGURE 9.18 Different effects of parameters K_{VP} and K_{VD} in the PD controller.

9.2.5.2.2 Conclusion of Proportional Derivative Control The following assumes different values of A and B for comparison.

From Figure 9.18, it can be observed that when the curves $K_{VP=}100$ in groups *A*, *B*, and *C* are the same, the effects of the proportional control part are all the same. The K_{VP} value of the derivative control part adopts different settings. So different effects are produced.

When $K_{VD}=0$, the derivative control part is equal to invalid. There is a considerable amount of overshoot.

When $K_{VD}=0.5$, the differential control part has an effect. The amount of overshoot has decreased.

When $K_{VD}=0.8$, the effect of the differential control part is more obvious. There is almost no overshoot.

Adding the proportional controller to the differential control obviously suppresses the overshoot. It reduces the amount of overshoot. This reduces the number of system oscillations. In this way, the controller is closer to the ideal situation.

Like the K_{PP} value, if the set K_{VP} value exceeds the critical value, the system will become an unstable oscillation state. The rotation speed of the servomotor will be fast and slow, and it will not be able to rotate smoothly.

If the K_{VD} value is set too large, the curve shown in Figure 9.18 will be close to the level. This means that because the correction amount is too large, although the rotation speed of the servomotor rotates smoothly, it affects the efficiency of the system.

Curve A: $K_{VP}= 100$, $K_{VD}=0$

Curve B: $K_{VP}= 100$, $K_{VD}=0.5$

Curve C: $K_{VP}= 100$, $K_{VD}=0.8$

9.2.5.2.3 The Influence of Integral Control on the Closed-Loop Control System Figure 9.19 is a block diagram of a PI controller.

The main purpose of integral control is to modify the amount of output value so that the output value is equal to the command value. The purpose of the speed control unit is to control the rotation speed of the servomotor stably and equal to the speed command according to the speed command.

If the speed control unit is only a proportional (P) or PD controller, assuming that the controller can increase the speed to the command speed, then the command speed = feedback speed, then $e(t) = 0$. In contrast, at this time, the speed output = 0, and the speed curve is shown in Figure 9.20.

However, in actual conditions, the above curve cannot be established. The biggest contradiction is that the speed decreases from the command speed to zero with a negative infinite slope. That is, the infinite negative acceleration decelerates to zero. This violates the laws of physics. Therefore, the curve must be corrected, as shown in Figure 9.21.

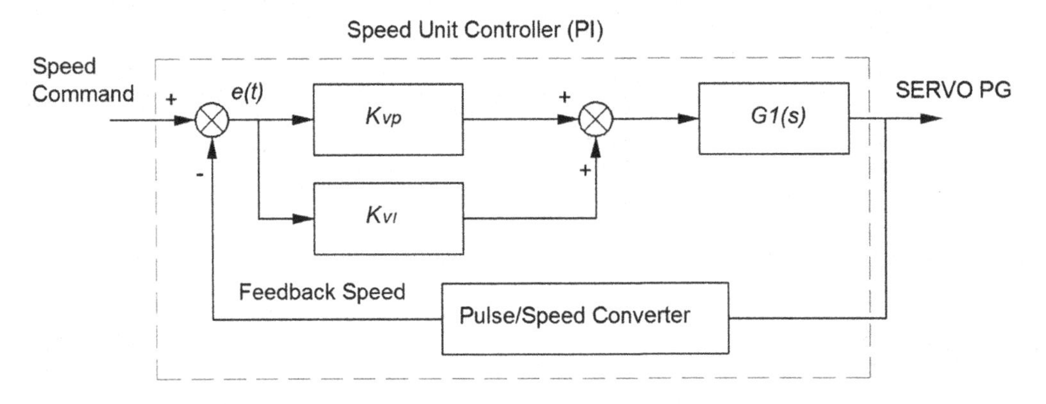

FIGURE 9.19 Block diagram of a proportional integral (PI) controller.

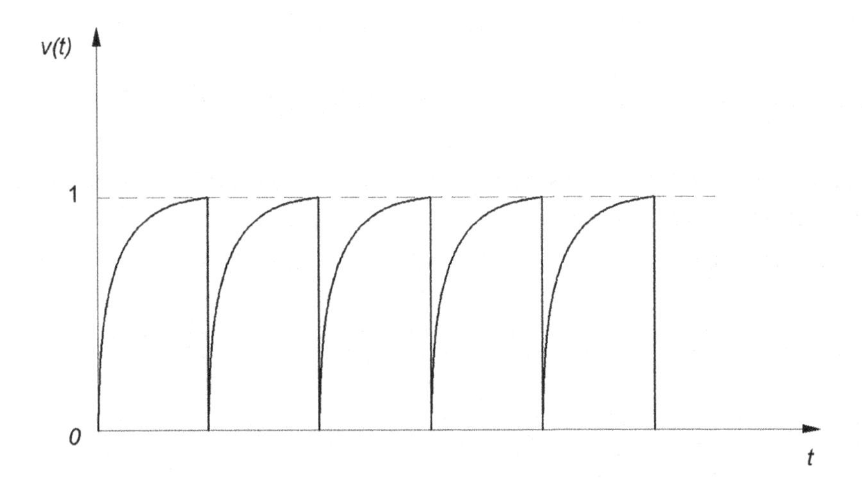

FIGURE 9.20 The impossible curve.

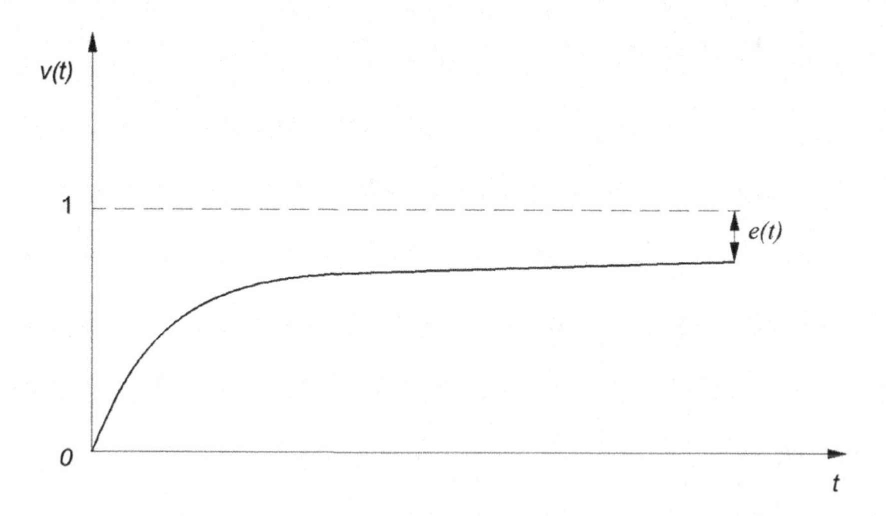

FIGURE 9.21 The speed unit step response without integral controller (I).

9.2.5.2.4 Function of Proportional Integral Controller The function of the PI controller is shown in Figure 9.22.

Figure 9.22c is the time integral curve of the speed error curve shown in Figure 9.22b, that is, the curve for speed error correction. To adjust the K_{VI} parameter value is to adjust the magnitude of the speed correction amount. When the correction range is too large, instability will also occur.

9.2.5.2.5 Proportional-Integral-Derivative Control Combine the integral correction curve and the PD control curve into a PID control curve, as shown in Figure 9.23. Although the output speed changes due to changes in load characteristics, it can also be corrected to be equal to the command speed. It will not be the same as the P or PD controller, resulting in the contradiction that the output value is equal to zero when the error value is zero, and the situation that the command speed cannot be reached occurs. The block diagram of the speed control unit PID controller is shown in Figure 9.24.

For a complete PID speed control unit, different parameter settings will produce different unit response speed control curves, as shown in Figure 9.25. Group C curve is closest to the ideal control curve. It has almost zero overshoot and stable rotation speed. If K_{VI} is adjusted too large, it will cause unstable oscillation. If the K_{VI} value of the group A curve continues to increase, oscillation will also occur. The purpose of adjusting these drive parameters is to hope that the rotation speed can respond to the command speed as quickly as the C group curve, reduce the overshoot, shorten the settling time, and achieve the performance of the fast response of the servomotor.

Curve A: K_{VP}= 100, K_{VD}=0, K_{VI}=0

Curve B: $K_{VP=}$100, K_{VD}=0.5, K_{VI}=1

Curve C: $K_{VP=}$100, K_{VD}=0.8, K_{VI}=0.2

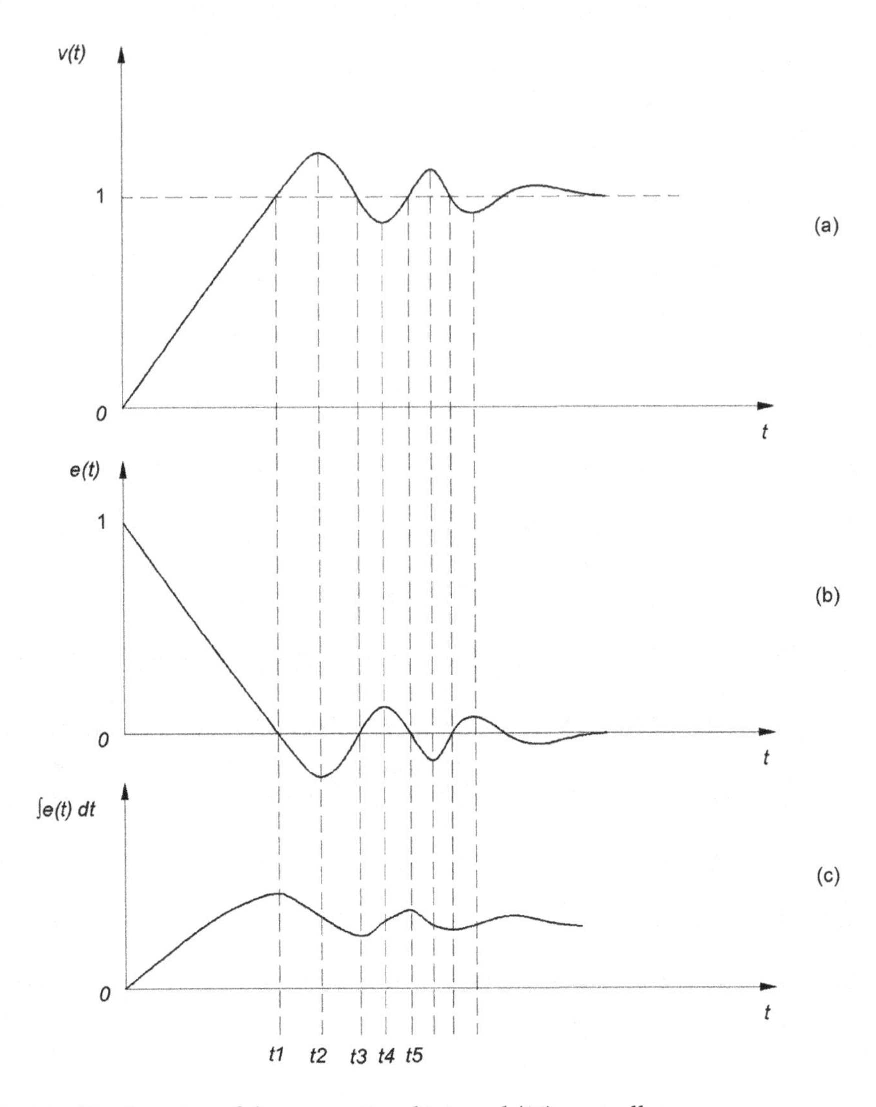

FIGURE 9.22 The function of the proportional integral (PI) controller.

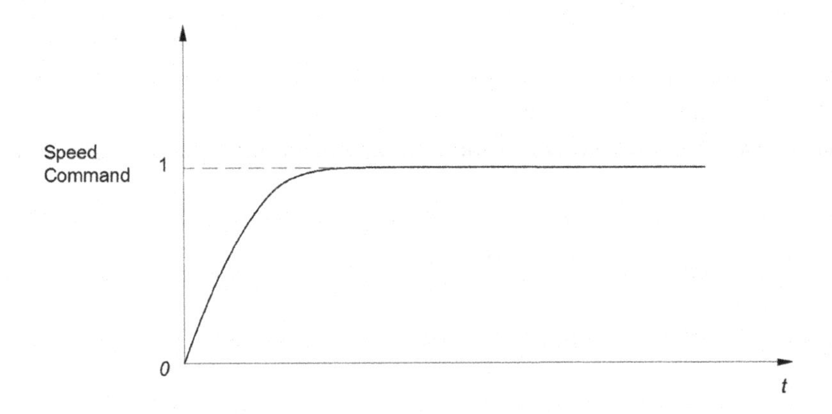

FIGURE 9.23 ID control curve of speed.

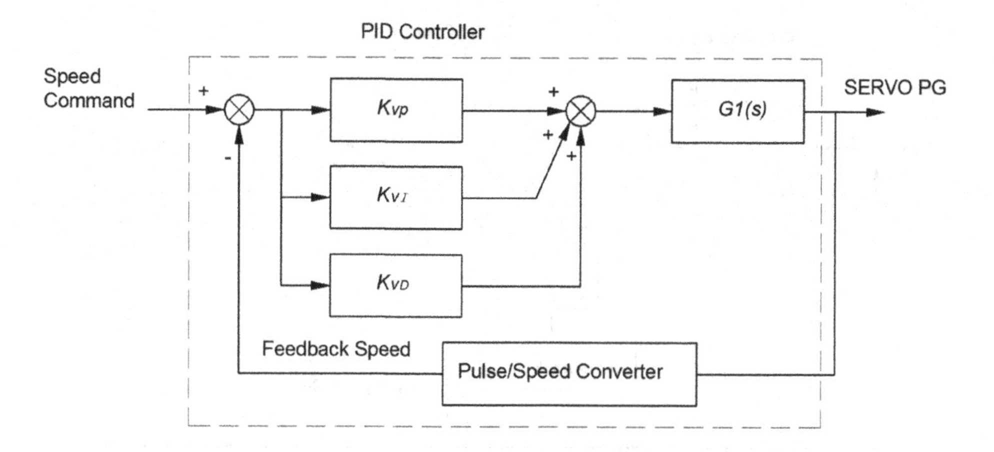

FIGURE 9.24 Block diagram of the speed control unit PID controller.

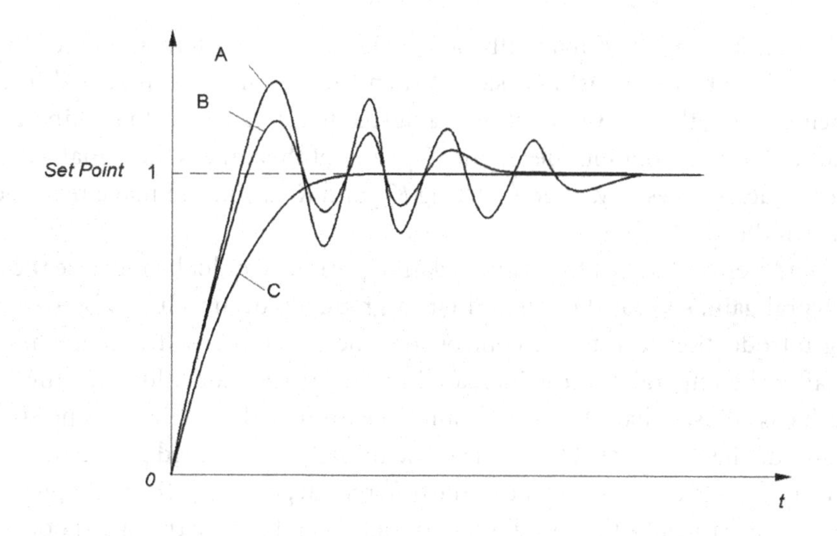

FIGURE 9.25 Different effects of parameters K_{VP}, K_{VD} and K_{VI} in the PID controller.

9.2.5.3 Complete Servo Motor Driver

The complete servomotor driver should include a position control unit, a speed control unit, and a drive unit, as shown in Figure 9.26. The P command (position servo) driver is composed of these units. The user must really understand, in order to adjust and use the servomotor correctly.

9.2.5.4 Adjustment of Drive Gain Parameters

9.2.5.4.1 Manually Adjust the Gain Parameters The first step is to adjust the speed proportional gain K_{VP} value. After the servo system is installed, the parameters must be adjusted to make the system rotate stably. First, adjust the speed proportional gain K_{VP} value. Before adjustment, the integral gain K_{VI} and the derivative gain K_{VD} must be adjusted to zero. Then gradually increase the K_{VP} value. At the same time, observe whether the servomotor

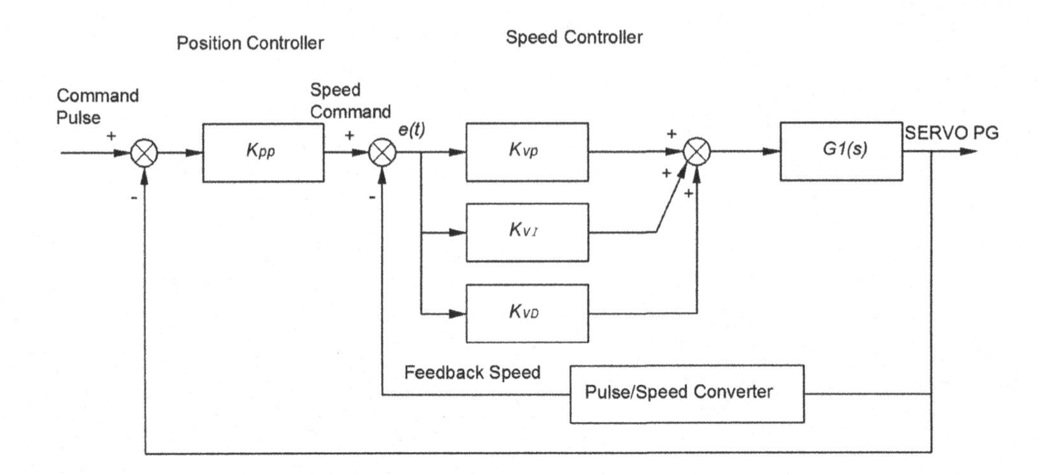

FIGURE 9.26 Block diagram of the P command driver.

oscillates when it stops. And manually adjust the K_{VP} parameters. Observe whether the rotation speed is obviously fast and slow. When the K_{VP} value is increased to cause the above phenomenon, the K_{VP} value must be adjusted to a small value to eliminate the oscillation and stabilize the rotation speed. The K_{VP} value at this time is the initially determined parameter value. If necessary, after adjusting K_{VI} and K_{VD}, you can make repeated corrections to reach the ideal value.

The second step is to adjust the integral gain K_{VI} value. Gradually increase the K_{VI} value of the integral gain, so that the integral effect gradually occurs. It can be seen from the foregoing introduction to integral control that the K_{VP} value will oscillate and become unstable after the integral effect is increased to the critical value. Just like the K_{VP} value, the K_{VI} value is adjusted back to a small value to eliminate the oscillation and stabilize the rotation speed. The K_{VI} value at this time is the initially determined parameter value.

The third step is to adjust the value of the differential gain K_{VD}. The main purpose of the differential gain is to make the speed rotate smoothly and reduce the overshoot. Therefore, increasing the K_{VD} value gradually can improve the speed stability.

The fourth step is to adjust the K_{PP} value of the position proportional gain. If the K_{PP} value is adjusted to a large extent, excessive motor positioning overshoot will occur when the servomotor is positioned, causing instability. At this time, reduce the K_{PP} value, reduce the overshoot, and avoid the unstable area. But it can't be adjusted too small to reduce the positioning efficiency. Therefore, the adjustment should be carefully coordinated.

9.2.5.4.2 Automatic Adjustment of Gain Parameters Most servomotors provide an automatic gain adjustment function, which can cope with most load conditions. When adjusting the parameters, you can use the automatic parameter adjustment function first, and then manually adjust it if necessary.

In fact, there are options for automatic gain adjustment. Generally, the control response is divided into several levels, such as high response, medium response, and low response. Users can set it according to actual needs.

9.2.5.4.3 The Difference between PI and PID Some simpler servomotor drives provide three parameter settings: P, I, and D. But it is not informed that the P parameter is a parameter used by the position control or the speed control unit. According to the control theory, the integral (I) and derivative (D) control items in this situation cannot be used alone and must be used in conjunction with the proportional (P) control item. At this time, it can be judged that the P parameter is used by the speed control unit. But you should ask the supplier how to deal with the P parameter of position control.

The position control unit of some servomotors provides proportional (P) control, and the speed control unit provides proportional (P) and integral (I) control without derivative (D) control items. This is reasonable. Because according to the previous discussion, proportional (P) and integral (I) control are necessary, while the derivative (D) control term is not. Therefore, there are PI combination controllers and PID combination controllers, but no PD combination controllers.

9.2.5.4.4 The Problem of Gain and Time The various parameters mentioned above—proportion (P), integral (D), and derivative (D)—are all expressed in terms of gain. The larger the value, the greater the effect it causes. This is the method used for convenience of explanation and easy identification. In fact, integral (I) often uses time constant instead of gain constant. The gain constant is inversely proportional to the time constant. To increase the effect of integral control, the time constant must be reduced, and vice versa.

9.2.5.4.5 Proportional Control Mode Some manufacturers' servo drive speed control units only use proportional control and cancel integral control. This control method is generally not used in the speed closed-loop control system but used in the servomotor positioning control system with mechanical lock.

As mentioned earlier, the servomotor is a dynamic operation. In fact, there is no stop state except for the system stop. Even if it does not rotate, it is in the position correction state at any time. Therefore, it has been in a state of correction within the minimum range.

If it is to be locked mechanically, a correction stress must be generated on the mechanism. Switching the drive speed control unit to proportional control mode can suppress the correction torque. Before the servomotor rotates, it must be switched to the normal mode.

9.2.6 Servomotor Control

For understanding servomotor control, let us consider an example of servomotor that we have given a signal to rotate by an angle of 45° and then stop and wait for further instruction. The shaft of the DC motor is coupled with another shaft called the output shaft, with help of gear assembly. This gear assembly is used to step down the high RPM of the motor's shaft to low RPM at output shaft of the servo system.

With the experimental point of view, the servomotor is control with the controller or module here for understating the servomotor control we will explain the control with an example. Consider a motor is rotate with an angle 45° by a given signal then stop for some time and wait for the further instruction. The shaft of this motor is coupled to another shaft with a gear assembly for decreasing the high RPM in to the low RPM at the output side and that shaft is called output shaft shown in Figure 9.27.

The voltage adjusting knob of a potentiometer is so arranged with the output shaft by means of another gear assembly, that during rotation of the shaft, the knob also rotates and creates a varying electrical potential according to the principle of potentiometer. This signal, i.e., electrical potential is increased with angular movement of potentiometer knob along with the system shaft from 0° to 45°. This electrical potential or voltage is taken to the error detector feedback amplifier along with the input reference commends, i.e., input signal voltage.

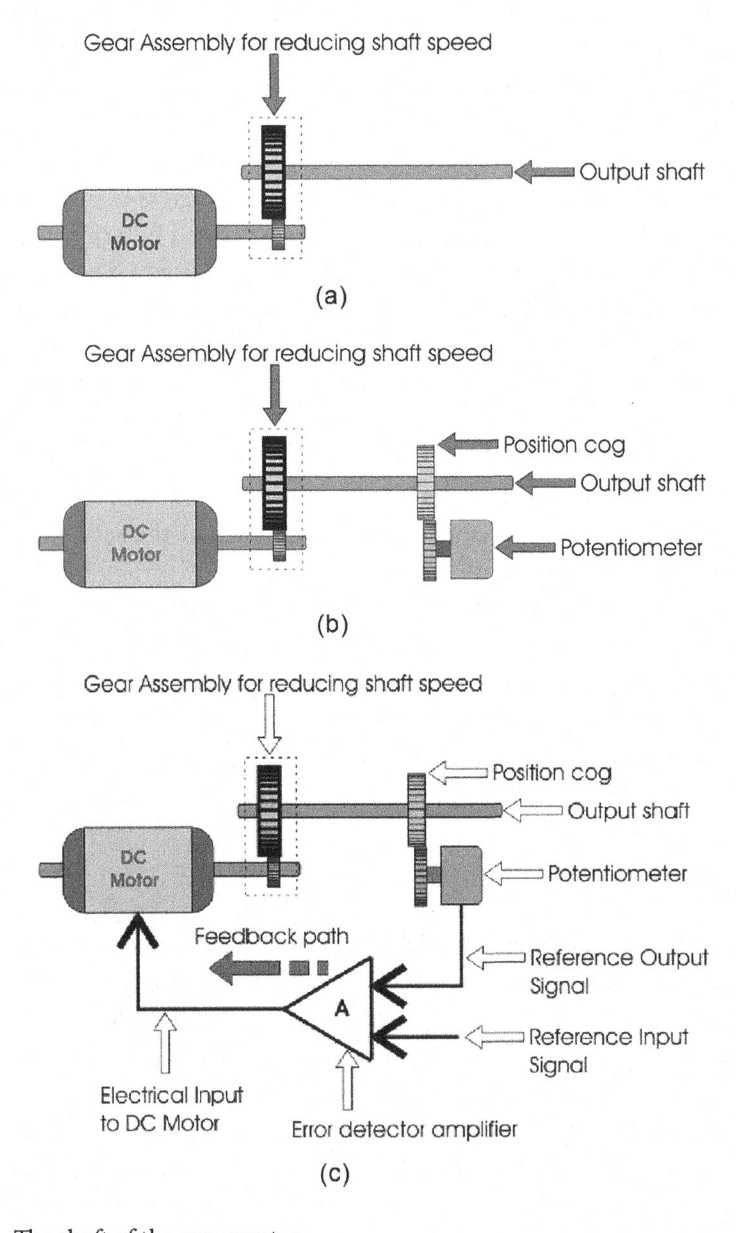

FIGURE 9.27 The shaft of the servomotor.

As the angle of rotation of the shaft increases from 0° to 45°, the voltage from potentiometer increases. At 45° this voltage reaches a value that is equal to the given input command voltage to the system. At this position of the shaft, there is no difference between the signal voltage coming from the potentiometer and reference input voltage (command signal) to the system, the output voltage of the amplifier becomes zero.

As per the picture given above the output electrical voltage signal of the amplifier, acts as input voltage of the DC motor. Hence, the motor will stop rotating after the shaft rotates by 45°. The motor will be at this rest position until another command is given to the system for further movement of the shaft in desired direction. From this example, we can understand the most basic servomotor theory and how servomotor control is achieved.

From this basic working principle of servomotor, it can be concluded. The shaft of the servo is connected to a potentiometer. The circuitry inside the servo, to which the potentiometer is connected, knows the position of the servo. The electric current position will be compared with the desired position continuously with the help of an Error Detection Amplifier. If a mismatch is found, then an error signal is provided at the output of the error amplifier and the shaft will rotate to go to the exact location required. Once the desired location is reached, it stops and waits.

According to Figure 9.28, a potentiometer is adjusted at the voltage adjusting knob and this knob is coupled with another gear assembly at the output shaft. During the rotation of the output shaft, the potentiometer knob is also rotated and the potentiometer creates the electrical potential according to rotating output shaft. This electrical potential is basically the reference output signal. When the shaft rotates 0°–45°, the signal is produced according to this rotation. The reference output signal and reference input signal are given to the error detector amplifier, this error detector amplifier decided when the motor would be on and when would be off. When the error detector amplifier shows the positive difference between the input reference signal and output reference signal, the motor moves in the forward direction to that angle according to the output shaft and when the error detector amplifier shows the negative difference then the motor moves in the reverse direction for that angle. Similarly, when the error detector amplifier shows the zero difference, the servomotor remains stationary for that time until the error detector amplifier becomes positive or negative. These are the simple control logics of servomotor control.

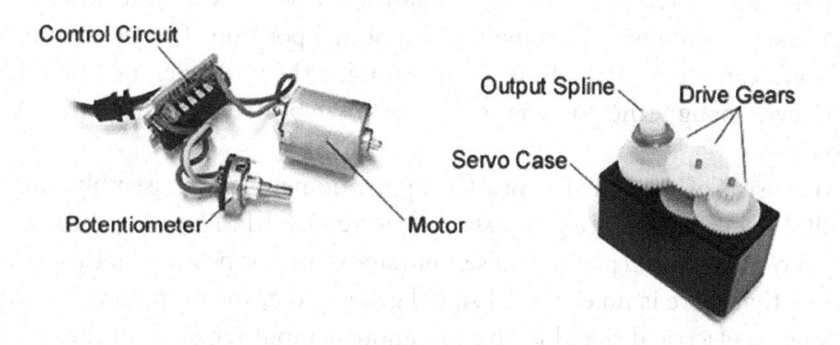

FIGURE 9.28 Potentiometer control circuit with servomotor.

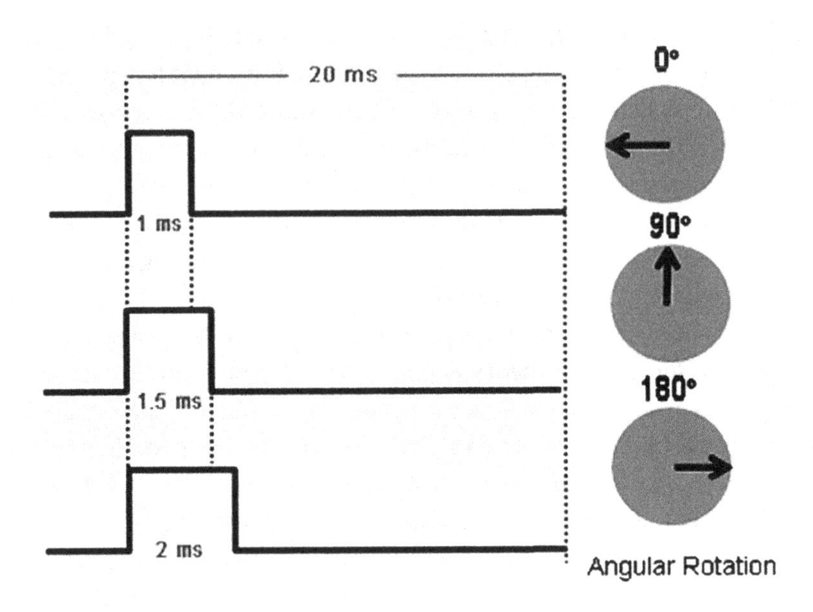

FIGURE 9.29 Pulse width modulation (PWM) for position control.

As shown in Figure 9.29, the servos are controlled by sending an electrical pulse of variable width or pulse width modulation (PWM) through the control wire. There is a minimum pulse, a maximum pulse, and a repetition rate. A servomotor can usually only turn 90° in either direction for a total of 180° movement. The motor's neutral position is defined as the position where the servo has the same amount of potential rotation in both the clockwise or counterclockwise directions. The PWM sent to the motor determines position of the shaft, and based on the duration of the pulse sent via the control wire; the rotor will turn to the desired position. The servomotor expects to see a pulse every 20 ms and the length of the pulse will determine how far the motor turns. For example, a 1.5 ms pulse will make the motor turn to the 90° position. Shorter than 1.5 ms moves it in the counterclockwise direction toward the 0° position, and any longer than 1.5 ms will turn the servo in a clockwise direction toward the 180° position.

When these servos are commanded to move, they will move to the position and hold that position. If an external force pushes against the servo while the servo is holding a position, the servo will resist from moving out of that position. The maximum amount of force the servo can exert is called the torque rating of the servo. Servos will not hold their position forever though; the position pulse must be repeated to instruct the servo to stay in position.

A servo consists of a motor (DC or AC), a potentiometer, gear assembly, and a controlling circuit. First of all, we use gear assembly to reduce RPM and increase the torque of the motor. Say at the initial position of servomotor shaft, the position of the potentiometer knob is such that there is no electrical signal generated at the output port of the potentiometer. Now an electrical signal is given to another input terminal of the error detector amplifier. Now the difference between these two signals, one comes from potentiometer and another comes from another source, will be processed in a feedback mechanism and output will be provided in terms of error signal. This error signal acts as the input for the motor and the motor starts rotating. Now motor shaft is connected with potentiometer and as motor rotates so the potentiometer and it will generate a signal. So as the potentiometer's angular position changes, its output feedback signal changes. After some time, the position of the potentiometer reaches a position where the output of the potentiometer is the same as the external signal provided. At this condition, there will be no output signal from the amplifier to the motor input as there is no difference between external applied signal and the signal generated at the potentiometer, and in this situation, the motor stops rotating.

9.2.6.1 Controlling Servomotor

All motors have three wires coming out of them, out of which two will be used for supply (positive and negative) and one will be used for the signal that is to be sent from the MCU.

Servomotor is controlled by PWM, which is provided by the control wires. There is a minimum pulse, a maximum pulse, and a repetition rate. Servomotor can turn 90° from either direction from its neutral position. The servomotor expects to see a pulse every 20 ms, and the length of the pulse will determine how far the motor turns. For example, a 1.5 ms pulse will make the motor turn to the 90° position, such as if the pulse is shorter than 1.5 ms shaft moves to 0°, and if it is longer than 1.5 ms, then it will turn the servo to 180°.

Servomotor works on PWM (pulse width modulation) principle, which means its angle of rotation is controlled by the duration of applied pulse to its Control PIN. Basically, the servomotor is made up of DC motor that is controlled by a variable resistor (potentiometer) and some gears. High-speed force of DC motor is converted into torque by gears. We know that WORK = FORCE × DISTANCE, in DC motor force is less and distance (speed) is high and in servo, force is high and distance is less. Potentiometer is connected to the output shaft of the servo, to calculate the angle and stop the DC motor at the required angle.

Servomotor can be rotated from 0° to 180°, but it can go up to 210°, depending on the manufacturing. This degree of rotation can be controlled by applying the electrical pulse of proper width, to its Control PIN. Servo checks the pulse every 20 ms. Pulse of 1 ms width can rotate servo to 0°, 1.5 ms can rotate to 90° (neutral position), and 2 ms pulse can rotate it to 180°.

All servomotors work directly with your +5 V supply rails, but we have to be careful about the amount of current the motor would consume. If you are planning to use more than two servomotors, a proper servo shield should be designed.

9.2.7 Continuous Rotation Servomotors

Continuous rotation **servomotors** are actually a modified version of what the servos are actually meant to do, that is, control the shaft position, as seen in Figure 9.30. The 360° rotation servos are actually made by changing certain mechanical connections inside the servo. However, certain manufacturer like parallax sells these servos as well. With the continuous rotation servo, you can only control the direction and speed of the servo, but not the position.

FIGURE 9.30 Continuous rotation servomotors.

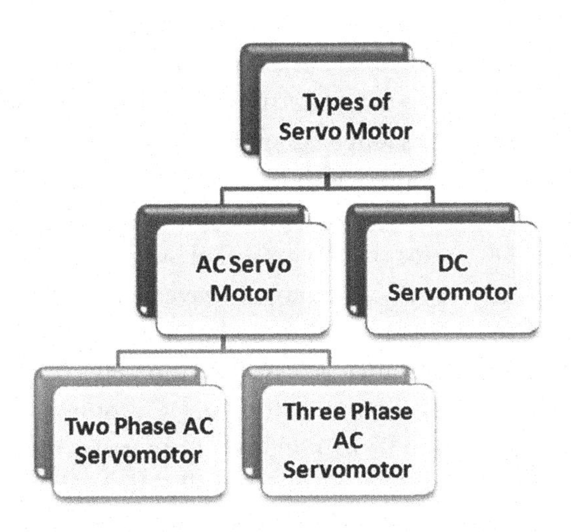

FIGURE 9.31 Types of servomotor.

9.2.7.1 Different Types of Servomotor and Its Applications

The servomotor is a commonly used motor for high technology devices in various industries like automation. This motor is a self-controlled electrical device that switches part of a machine with high productivity and great accuracy. The o/p shaft of this motor can be stimulated to a specific angle. These motors are mainly used in different applications like home electronics, cars, toys, and airplanes.

9.2.8 Types of servomotor

These motors are classified into different types based on their application like Brushless DC, AC, continuous rotation, linear and positional rotation, etc. as shown in Figure 9.31. Typical servomotors contain three wires, such as power control and ground. The outline and dimension of these motors depend on their applications. The most common type of

FIGURE 9.32 DC servomotor.

this motor is an RC servomotor used in interest applications like robotics due to their ease, reliability, and affordability of control by microprocessors.

9.2.8.1 DC servomotor

Generally, this motor has a separate DC source in the winding and the armature winding field, as shown in Figure 9.32. The control can be archived either by controlling the armature current or field current. Field control comprises some benefits over armature control. Similarly, armature control comprises some benefits over field control. Based on the uses, the control should be functional to the DC servomotor. This motor offers very precise and also quickly react to start or stop commands due to the low armature inductive reactance. These motors are used in related equipment's and computerized mathematically controlled machines.

9.2.8.2 DC servomotor Principle and Its Applications

A DC motor along with servomechanism (closed-loop control system) acts as a servomotor that is basically used as a mechanical transducer in the automation industry. Based on its accurate closed-loop control, it has versatile applications used in many industries. The **DC servomotor** is a motor that is used in servo systems. A servo system is a closed-loop system where the feedback signal (position, velocity, acceleration, etc.) drives the motor. This signal acts as an error and based on controller, accurate position or velocity is achieved. The motors are coupled to an output shaft (load) through a gear train for power matching. Servomotor acts as a mechanical transducer as they convert an electrical signal to an angular velocity or position.

9.2.8.3 DC servomotor Working Principle

It is a normal DC motor with its field winding separately excited. Based on the nature of excitation, it can be further classified as field-controlled and armature-controlled servomotors. It is connected to a mechanical shaft of the motor. It could be any industrial load or a simple fan load. It acts as a mechanical transducer to convert the output of the motor in the form of position, acceleration or velocity based on application. It senses the position of

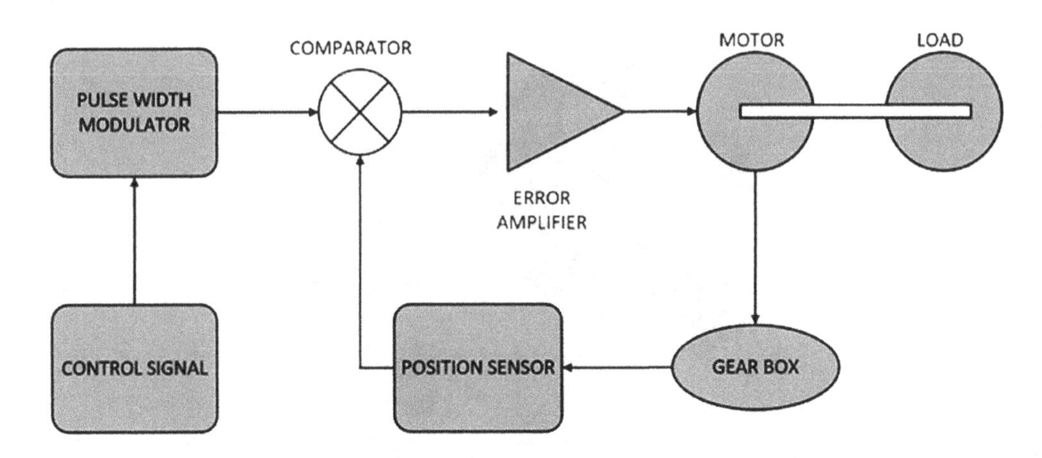

FIGURE 9.33 Block-diagram-of-servo-motor.

the rotor and feeds it to the comparator. Commonly used sensors are hall effect sensors. It compares the output of a position sensor and a reference point to generate the error signal and feeds to the amplifier. If the motor runs with exact control, then no error is zero. The gearbox, position sensor, and comparator make the system closed loop. It amplifies the error from the comparator, for feeding into the motor. It acts like a proportional controller where the gain is amplified for zero steady-state error.

9.2.8.4 Control Signal and Pulse Width Modulator (PWM)

Based on the feedback signal, controlled gives the input to pulse width modulator, which modulates the input of motor (voltage or field excitation) for exact control or zero steady-state error. The pulse width modulator further uses a reference waveform and comparator to generate pulses.

By making the system closed-loop exact position, acceleration or velocity is obtained. The name servomotor implies a controlled motor that gives the desired output because of the effect of feedback and controller. The error signal is amplified and used to drive the servomotor, as shown in Figure 9.33. Based on the nature of producing control signal and pulse width modulator, servomotors can have advanced controlled techniques using digital signal processors or Field Programmable Gate Arrays (FPGA) chips.

A DC reference voltage is set to the value corresponding to the desired output. This voltage can be applied by using another potentiometer, control pulse width to voltage converter, or through timers depending on the control circuitry. The dial on the potentiometer produces a corresponding voltage that is then applied as one of the inputs to error amplifier. In some circuits, a control pulse is used to produce DC reference voltage corresponding to desired position or speed of the motor and it is applied to a pulse width to voltage converter. In this converter, the capacitor starts charging at a constant rate when the pulse is high. Then the charge on the capacitor is fed to the buffer amplifier when the pulse is low and this charge is further applied to the error amplifier. So, the length of the pulse decides the voltage applied at the error amplifier as a desired voltage to produce the desired speed or position. In digital control, microprocessor or microcontroller are used for generating the PWM pluses in terms of duty cycles to produce more accurate control signals.

9.2.8.5 Characteristics of DC Servomotor

The characteristic or in other words the requirements of a servomotor are as follows:

- The inertia of the servomotor should be less for precision and accuracy.
- It should have a fast response which can be obtained by keeping high torque to weight ratio.
- The torque-speed characteristics should be linear.
- Four quadrant operation is desirable by using further converters.
- Stable operation.
- Robust nature.

A DC servomotor consists of a small DC motor, feedback potentiometer, gearbox, motor drive electronic circuit, and electronic feedback control loop, as shown in Figure 9.34. It is

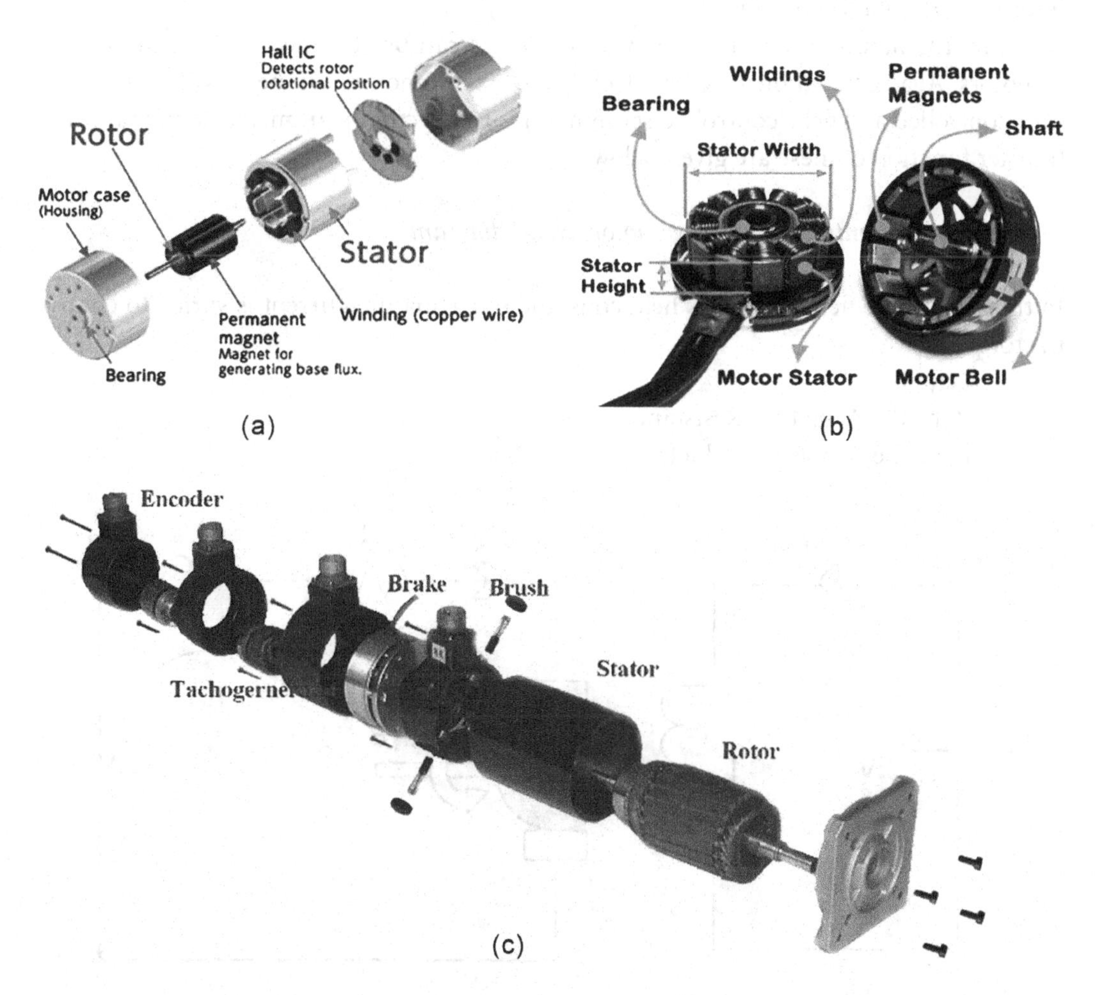

FIGURE 9.34 Block-diagram-of-servo-motor.

more or less similar to the normal DC motor. The stator of the motor consists of a cylindrical frame and the magnet is attached to the inside of the frame.

The rotor consists of brush and shaft. A commutator and a rotor metal supporting frame are attached to the outside of the shaft and the armature winding is coiled in the rotor metal supporting frame. A brush is built with an armature coil that supplies the current to the commutator. At the back of the shaft, a detector is built into the rotor in order to detect the rotation speed. With this construction, it is simple to design a controller using simple circuitry because the torque is proportional to the amount of current flow through the armature. And also, the instantaneous polarity of the control voltage decides the direction of torque developed by the motor. Types of DC servomotors include series motors, shunt control motor, split series motor, and permanent magnet shunt motor.

A DC servomotor is an assembly of four major components, namely a DC motor, a position sensing device, a gear assembly, and a control circuit. The Figure 9.35 shows the parts that consisting in RC servomotors in which small DC motor is employed for driving the loads at precise speed and position.

9.2.8.6 *Types of DC servomotor*

Based on the nature of electric supply servomotors can be classified as AC and DC servomotors. Again, based on speed control, the DC servomotors can be classified as armature-controlled and field-controlled servomotors. The circuit diagram, block diagram, and transfer function of these are given below:

1. *Armature-controlled DC servomotor circuit diagram*

In this motor, the field current is held constant and armature current is varied to control the torque.

Let,
R_a be the Armature Resistance
L_a be the Armature Inductance

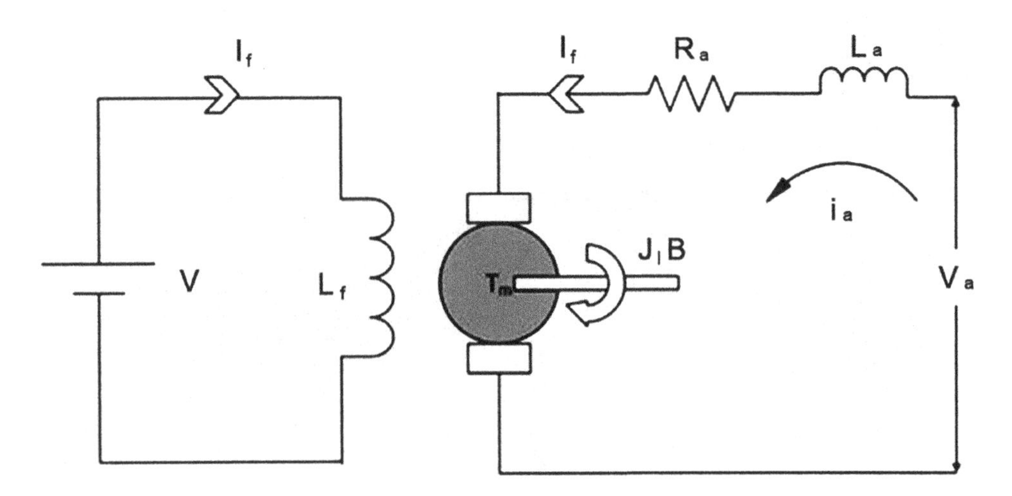

FIGURE 9.35 Circuit-diagram-of-armature-controlled-dc-servo-motor.

I_a be the Armature Current
V_a be the Armature Voltage
W_m be the Angular Velocity
E_b be the Back EMF
J be the Moment of Inertia
I_f be the Field Current
L_f be the Field Inductance
T_m be the Torque Produced
Eb be the Back EMF produced by the motor
Applying KCL in the armature, we get

$$V_a = i_a R_a + L_a * (di_a / dt) + e_b$$

The load torque equation is given by

$$J * \left(d^2\theta / dt^2\right) + B * (d\theta / dt) = T_m = K_1 i_a$$

Substituting i_a in the second equation and converting into the Laplace domain, the transfer function of the DC servomotor can be obtained as below.

9.2.8.7 Transfer Function and Block Diagram

As shown in Figure 9.36, the block diagram of the armature-controlled DC servomotor is shown below.

The transfer function of armature-controlled DC servomotor is shown below.

$$\theta(s) / V_a(s) = \left(K_1 / \left(Js^2 + Bs\right) * (L_a s + R_a)\right) / 1 + (K_1 K_b K_s) / \left(Js^2 + Bs\right) * (L_a s + R_a)$$

2. *Field-controlled DC servomotor circuit diagram*
   In this method of speed control, a variable input voltage is applied to the field winding of DC motor, while keeping the armature current constant, as shown in Figure 9.37.
   Let R_f be the field resistance, L_f be the field inductance, I_f be the field current, V_f be the variable field voltage, θ be the angular displacement of the motor shaft, T_m be

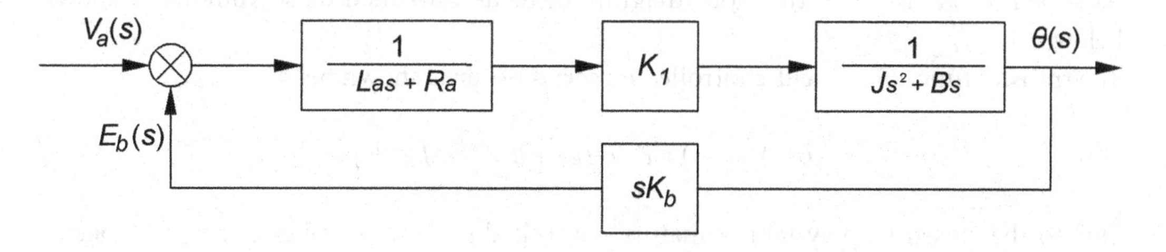

FIGURE 9.36 Block-diagram-of-armature-controlled-DC-servomotor.

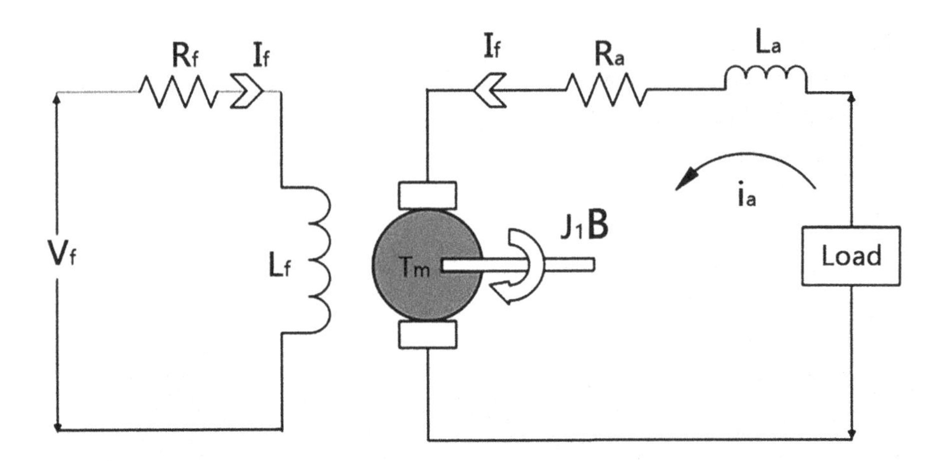

FIGURE 9.37 Circuit-diagram-of-field-controlled-DC-servo-motor.

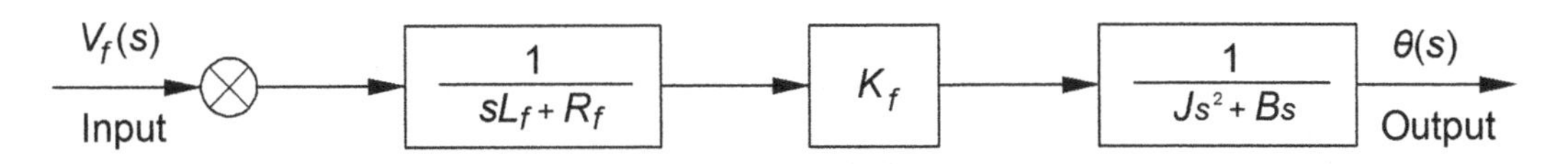

FIGURE 9.38 Block-diagram-of-field-controlled-DC-servo-motor.

the torque developed by the motor, B be the coefficient of viscous friction, J be the moment of inertia.

The field equation is given by

$$V_f = i_f R_f + L_f * di/dt$$

Torque equation is given by

$$J * d^2\theta/dt^2 + B * d\theta/dt = T_m$$

Converting into Laplace domain, and substituting for field current, we get the transfer function as

9.2.8.8 Transfer Function and Block Diagram

As shown in Figure 9.38, the block diagram of field-**controlled** dc servomotor is shown below.

The transfer function of field-controlled dc **servomotor** is shown below:

$$\theta(s)/V_f(s) = K_f/\left(sL_f + R_f\right)*\left(s^2 J + Bs\right)$$

Due to the closed-loop system, armature controlled system gives better performance as compared to field controlled which is the open-loop system. Also, the speed of response is slow in the field control system. In the case of armature controlled, the inductance of the armature is negligible, which is not the case for field control. Infield control, improved damping is not possible, which can be achieved in armature control.

9.2.8.9 Difference between servomotor and DC Motor

The following points may be regarded as the difference between the servomotor and DC motor:

9.2.8.10 DC Servomotor Applications

Because of its precise control and accuracy, servomotor has numerous applications, a few of them have been enlisted below:

- Automation industry
- Robotic industry
- Aviation
- Manufacturing industry
- Pharmacy
- Food services
- Toys and radio-controlled cars.

9.2.8.11 DC servomotor Advantages and Disadvantages

The advantages of DC servomotor are

- Precise control and accuracy
- Stable operation
- Fast response
- Lightweight and portable
- Four quadrant operation possible.

TABLE 9.1 Difference between Servomotor and DC Motor

Servomotor	DC Motor
Servomotor works on the principle of the closed-loop control system, in which DC motor is also one of the components.	A DC motor works on the principle of Lorentz's force law, i.e., when a current-carrying conductor is placed in a magnetic field, it experiences a force.
Due to a closed-loop control system, servomotor would have higher precision and accuracy.	This is not the case with the DC motor.
The inertia of servomotor is less.	The inertia of dc motor is less.
A servomotor is more suited for automation and robotics industry.	The DC motors have versatile applications.
Servomotor needs maintenance.	A DC motor is rugged in nature.
Servomotors are more costly compared to DC motors.	DC motors are not expensive compared to servomotors.

FIGURE 9.39 AC servomotor.

The disadvantages of DC servomotor are as follows:

- Due to the complex circuit, reliability is less,
- Due to closed-loop components i.e. amplifier, gearbox, etc. it is costly.

AC Servomotor

A type of servomotor that uses AC electrical input in order to produce mechanical output in the form of precise angular velocity is known as **AC servomotor**, as shown in Figure 9.39. AC servomotors are basically two-phase induction motors with certain exceptions in designing features.

The output power achieved from ac servomotor ranges between some watt to a few hundred watts, while the operating frequency range is between **50 and 400 Hz**. It provides closed-loop control to the feedback system as here the use of a type of encoder provides feedback regarding speed and position.

This motor includes encoder which is used with controllers for giving closed-loop control and also feedback. This motor can be employed to high accuracy and also controlled exactly as required for the applications. Often these motors have advanced designs of easiness or better bearings and some simple designs also use higher voltages in order to achieve greater torque. AC motor applications, mainly involve in robotics, automation, CNC machinery, etc.

Construction of AC Servomotor

We have already said in the beginning that an AC servomotor is regarded as a two-phase induction motor. However, ac servomotors have some special design features which are not present in normal induction motor, thus it is said that two somewhat differs in construction.

It is mainly composed of two major units, stator and rotor.

- *Stator*: First have a look at the Figure 9.40 shown below, representing stator of ac servomotor:

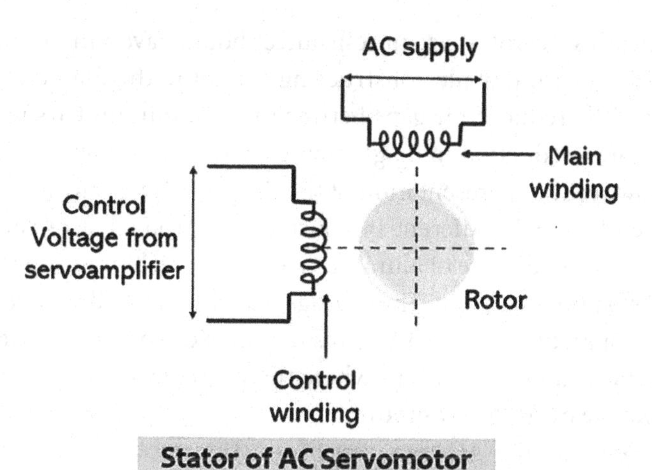

FIGURE 9.40 Stator of AC servomotor.

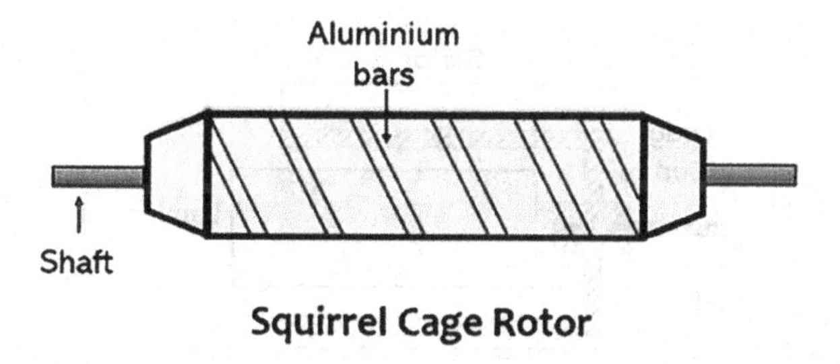

FIGURE 9.41 Rotor of AC servomotor.

The stator of ac servomotor consists of two separate windings uniformly distributed and separated at 90°, in space. Out of the two windings, one is referred as main or fixed winding while the other one is called control winding.

A constant ac signal as input is provided to the main winding of the stator. However, as the name suggests, the control winding is provided with the variable control voltage. This variable control voltage is obtained from the servo amplifier.

It is to be noted here that to have a rotating magnetic field, the voltage applied to the control winding must be 90° out of phase w.r.t the input ac voltage.

- *Rotor*: The rotor is generally of two types; one is squirrel cage type while the other is drag cup type.

  As shown in Figure 9.41, the squirrel cage type of rotor is shown below:

  In this type of rotor, the length is large while the diameter is small and is constructed with aluminum conductors thus weighs less. It is to be noted here that the torque-speed characteristics of a normal induction motor have both positive as well as negative slope regions that represent unstable and stable regions, respectively. However, ac servomotors are designed to possess high stability; thus, its torque-slip characteristics must not have a positive slip region. Along with this the torque developed in the motor must reduce in a linear manner with speed.

To achieve this the rotor circuit resistance should have a high value, with low inertia. Due to this reason, while constructing the rotor, the diameter to length ratio is kept smaller. The reduced air gaps between the aluminum bars in the squirrel cage motor facilitate a reduction in magnetizing current, as shown in Figure 9.42.

Let us now see the representation of the drag cup type rotor:

This type of rotor is different in construction from that of squirrel cage one. It consists of a laminated core of aluminum around which drag cup is present with certain air gaps on both the side. These drag cups are attached with a driving shaft that facilitates its operation. The two air gaps in both sides of the core lead to reducing the inertia thus is used in applications where there is a low power requirement.

Working Principle of AC Servomotor

Figure 9.43 represents the AC two-phase induction motor that uses the principle of servomechanism.

Initially, a constant ac voltage is provided at the main winding of the stator of the ac servomotor. The other stator terminal of the servomotor is connected to the control

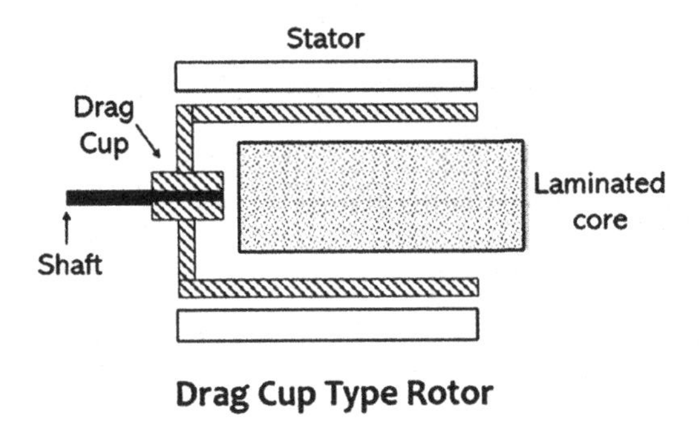

FIGURE 9.42 Drug cup type rotor of AC servomotor.

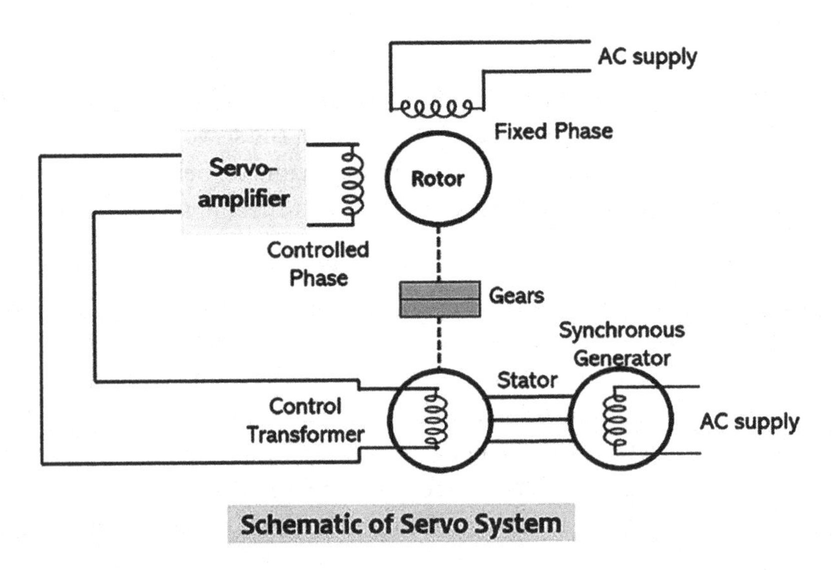

FIGURE 9.43 Schematic diagram of AC servomotor.

transformer through the control winding. Due to the provided reference voltage, the shaft of the synchro generator rotates with a particular speed and attains a certain angular position. Also, the shaft of the control transformer has a certain specific angular position which is compared with the angular position of the shaft of the synchro generator.

Further, the comparison of two angular positions provides the error signal. More specifically, the voltage levels of the corresponding shaft positions are compared which generates the error signal. This error signal corresponds to the voltage level present at the control transformer. This signal is then provided to the servo amplifier which generates variable control voltage. With this applied voltage, the rotor again attains a specific speed and starts rotation and sustains until the value of the error signal reaches 0, thereby attaining the desired position of the motor in the AC servomotors.

The schematic diagram of servo system for AC two-phase induction motor is shown in the figure below. In this, the reference input at which the motor shaft has to maintain at a certain position is given to the rotor of synchro generator as mechanical input theta. This rotor is connected to the electrical input at rated voltage at a fixed frequency.

The three stator terminals of a synchro generator are connected correspondingly to the terminals of control transformer. The angular position of the two-phase motor is transmitted to the rotor of control transformer through gear train arrangement and it represents the control condition alpha. Initially, there exist a difference between the synchro generator shaft position and control transformer shaft position. This error is reflected as the voltage across the control transformer. This error voltage is applied to the servo amplifier and then to the control phase of the motor. With the control voltage, the rotor of the motor rotates in required direction till the error becomes zero. This is how the desired shaft position is ensured in AC servomotors. Alternatively, modern AC servo drives are embedded controllers like PLCs, microprocessors and microcontrollers to achieve variable frequency and variable voltage in order to drive the motor. Mostly, pulse width modulation and Proportional-Integral-Derivative (PID) techniques are used to control the desired frequency and voltage. The block diagram of AC servomotor system using programmable logic controllers, position and servo controllers is given below.

Torque-Speed Characteristics

Figure 9.44 represents the torque-speed characteristics of the two-phase induction motor.

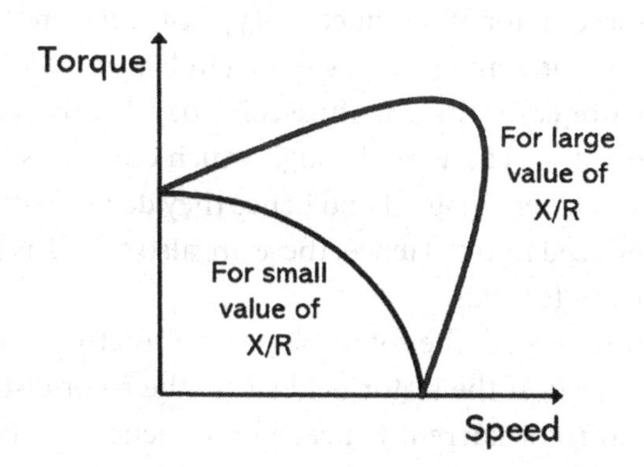

FIGURE 9.44 Torque-speed characteristics of AC servomotor.

We have already discussed that the motor must be designed in a way to provide linear torque-speed characteristics, in which the torque changes in a linear manner with the speed. However, as we have seen in the above figure that the torque-speed characteristics here are not actually linear.

This is so because it depends on the ratio of reactance to resistance. The low value of the ratio of reactance to resistance implies that motor possesses high resistance and low reactance; in such case, the characteristics is more linear that high value of the ratio for reactance to resistance.

Features

- These are low weight devices.
- It offers reliability as well as stability in operation.
- There is not much noise generated at the time of operation.
- It offers almost linear torque-speed characteristics.
- As brushes and slip rings are not present here thus it reduces maintenance cost.

Applications of AC Servomotors

Due to the various advantages offered by the AC servomotors, these majorly find applications in the instruments that operate on servomechanism, in position controlling devices, and computers. Moreover, these also find applications in tracking systems, machine tools, and robotics machinery.

Synchronous-Type AC servomotor

AC servomotors are basically two-phase squirrel cage induction motors and are used for low-power applications. Nowadays, three-phase squirrel cage induction motors have been modified such that they can be used in high-power servo systems. The main difference between a standard split-phase induction motor and AC motor is that the squirrel cage rotor of a servomotor has made with thinner conducting bars, so that the motor resistance is higher.

Based on the construction there are two distinct types of AC servomotors, they are synchronous-type AC servomotor and induction-type AC servomotor. **Synchronous-type AC servomotor** consist of stator and rotor, as shown in Figure 9.45. The stator consists of a cylindrical frame and stator core. The armature coil wound around the stator core and the coil end is connected to with a lead wire through which current is provided to the motor. The rotor consists of a permanent magnet and hence they do not rely on AC induction-type rotor that has current induced into it. Hence, these are also called as brushless servomotors because of structural characteristics.

When the stator field is excited, the rotor follows the rotating magnetic field of the stator at the synchronous speed. If the stator field stops, the rotor also stops. With this permanent magnet rotor, no rotor current is needed and hence less heat is produced. Also, these motors have high efficiency due to the absence of rotor current. In order to know

FIGURE 9.45 Synchronous-type of AC servomotor.

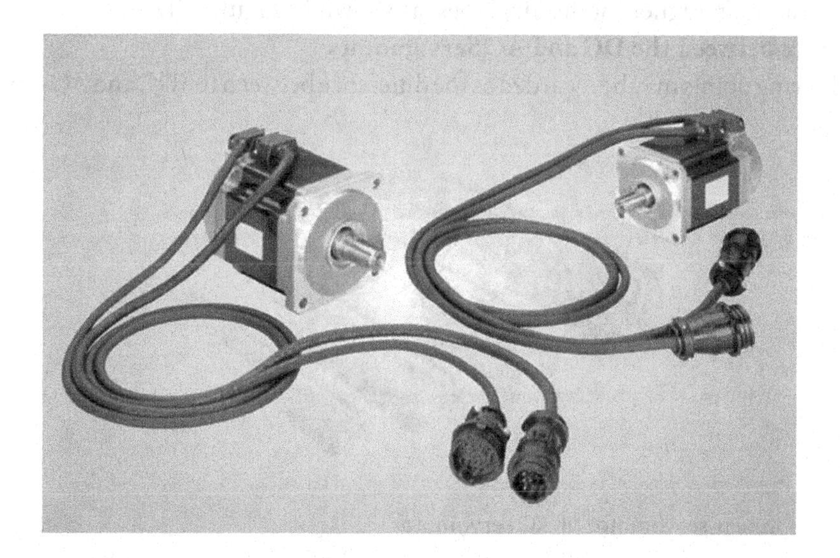

FIGURE 9.46 Induction-type of AC servomotor.

the position of rotor with respect to stator, an encoder is placed on the rotor and it acts as feedback to the motor controller.

Induction-Type AC Servomotor

The induction-type AC servomotor structure is identical with that of the general motor. In this motor, stator consists of stator core, armature winding and lead wire, while rotor consists of shaft and the rotor core that built with a conductor as similar to squirrel cage rotor, as shown in Figure 9.46. The working principle of this servomotor is similar to the normal induction motor. Again, the controller must know the exact position of the rotor using encoder for precise speed and position control.

Positional Rotation Servomotor

This is a most common type of motor and the o/p of the shaft rotates at about 180°. It comprises physical stops situated in the gear device to stop revolving outside limits to protect the rotation sensor. These common servomotors include in radio-controlled cars, radio-controlled water, toys, aircraft, robots, and many other applications.

Continuous Rotation Servomotor

This motor is quite correlated to the common positional rotation servomotor, but it can move in any direction indefinitely. As per the speed and direction of rotation, the control signal rather than set the static position of the servo is assumed. The range of potential commands causes the motor to rotate clockwise or anticlockwise as chosen, at varying speed, depending on the command signal. This kind of motor is used in a radar dish, and it can be used as a drive motor on a mobile robot.

Linear Servomotor

This motor is also similar to the positional rotation servomotor, which is discussed above, but with an extra gear to alter the o/p from circular to back-and-forth. These motors are not simple to find, but sometimes you can find them at hobby stores where they are used as actuators in higher model airplanes, as shown in Figure 9.47.

Difference between the DC and AC Servomotors

The following points may be regarded as the difference between the DC and AC servomotors.

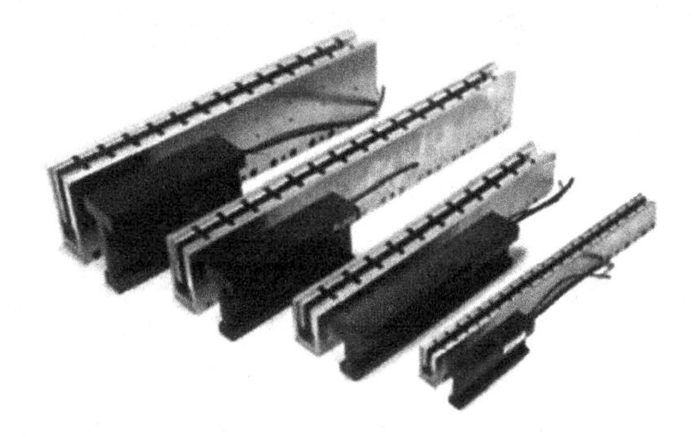

FIGURE 9.47 Linear servomotor of AC servomotor.

TABLE 9.2 Difference between the DC and AC Servomotors

DC Servomotor	AC Servomotor
It delivers high power output.	Delivers low output of about 0.5–100 W.
It has more stability problems.	It has fewer stable problems.
It requires frequent maintenance due to the presence of a commutator.	It requires less maintenance due to the absence of a commutator.
It provides high efficiency.	The efficiency of AC servomotor is less and is about 5%–20%.
The life of DC servomotor depends on the life of brush life.	The life of AC servomotor depends on bearing life.
It includes a permanent magnet in its construction.	The synchronous-type AC servomotor uses permanent magnet while induction type doesn't require it.
These motors are used for high-power applications.	These motors are used for low power.

9.2.9 Servo Motor Interfacing with Microcontroller

The feedback signal corresponding to the present position of the load is obtained by using a position sensor. This sensor is normally a potentiometer that produces the voltage corresponding to the absolute angle of the motor shaft through the gear mechanism. Then the feedback voltage value is applied at the input of the error amplifier (comparator).

The error amplifier is a negative feedback amplifier and it reduces the difference between its inputs. It compares the voltage related to the current position of the motor (obtained by potentiometer) with desired voltage related to the desired position of the motor (obtained by pulse width to voltage converter) and produces the error of either a positive or negative voltage. This error voltage is applied to the armature of the motor. If the error is more, more output is applied to the motor armature. As long as an error exists, the amplifier amplifies the error voltage and correspondingly powers the armature. The motor rotates till the error becomes zero. If the error is negative, the armature voltage reverses and hence the armature rotates in the opposite direction.

Servomotor is one of the most special motors that can be used for precise angular movement. The main advantage of this motor is that if we are controlling this motor with microcontroller then there is no need for any feedback controller or mechanism. This motor is mostly used in industrial and commercial applications. Its working principle and operation are very simple. It consists of three wires; two wires are used to power the motor and the third is used for the control signal, as shown in Figure 9.48.

According to Figure 9.48, one wire is connected to the VCC of microcontroller, second wire is connected to the ground, and the third wire is connected to any output port of microcontroller. PWM (pulse width modulation) technique is used of controlling the servomotor control signal and its angular position. A total of 11.0592 MHz crystal oscillator is used for clock pulse and 22 PF capacitor is used for the stable operation of crystal oscillator.

This tutorial is about dc motor speed control with a pic microcontroller using PWM method. There are many applications of DC motors, where we need a variable speed of DC motor. For example, DC motors have an application in electric cars, trucks, and aircraft. These are three examples where we need a variable speed of DC motor. Although nowadays AC motors have more usage in the market for variable speed application over DC motors

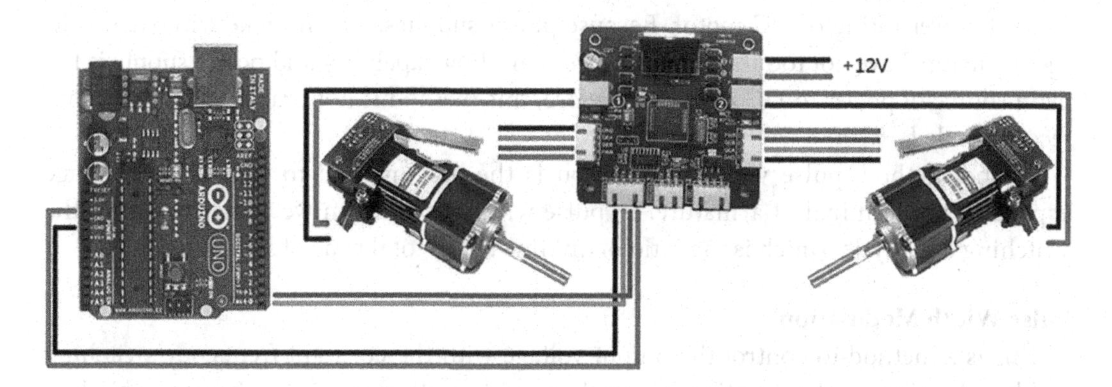

FIGURE 9.48 Block-diagram-of-servo-motor.

due to enhancement in power electronics and solid-state electronics. But DC motors are still used today in many applications for variable speed.

According to armature and field current circuit configuration, there are five types of DC motor:

- DC motor with separate excitation
- DC motors with shunt
- Series DC motors
- Compound DC motors
- DC motors with excitation from permanent magnet

The speed of separately excited DC motor and DC shunt motor can be made variable by changing terminal voltage and by changing the armature resistance. The speed of DC series motor and permanent magnet DC motor can only be changed by changing the armature terminal voltage. Compound DC motor has characteristics of both DC series motor and DC shunt motor.

So, there is one thing common in all types of DC motors. Speed control by changing armature terminal voltage is the same in all types of DC motors. Now to make this project you have to learn a method through which we can apply variable DC voltage to the armature of any DC motor so that we can control the speed of DC motor according to our requirement.

Methods to Generate Variable Voltage

- Variable power supply
- Pulse width modulation technique

In variable power supply, there are some limitations but still speed of very small power DC motors can be controlled by making a variable power supply with the help of voltage regulators, which are used to generate variable voltage within a specified limit provided by the manufacturer. There are many voltage regulators in the market. But there will be an issue of power rating of DC motor. Because power supplies, which make with such voltage regulators have not too much high-power handling capability and power supply is the main source of power to DC motor. Therefore, this method is not practically feasible and recommended.

Second method, pulse width modulation is the best method to control DC voltage applied to the terminal of armature. In pulse width modulation we actually control the switching duty cycle, which is the ratio of on time to the total time of switching.

Pulse Width Modulation

This is a method to control the output voltage with the constant frequency switching and by adjusting on the duration of switching and in other words by changing the duty cycle of switching.

i) Constant switching time period = on time + off time

ii) Duty Cycle = on time / on time + off time %

Duty cycle cannot be greater than 1 or 100%. Because on time will always be less than total timer period of switching frequency. The relationship of input-output voltage and duty cycle is

$$\text{Output Voltage} = \text{Duty Cycle} * \text{Input Voltage}$$

Hence output voltage and duty cycle are directly related to each other but their output also depends on the switching frequency of the switch.

In Figure 9.49, there are four PWM having different duty cycle. So, we need to change the duty cycle according to the requirement of output voltage. But output voltage can never be greater than the input voltage. But now the question which into mind. How to generate PWM and how to use PWM for variable output voltage generation. Check this introduction to pulse width modulation tutorial for further details.

PWM Generation Methods

There are two ways to generate pulse width modulation with a variable duty cycle:

1. Using analog electronics (operational amplifier, comparator, and sawtooth wave, etc.)
2. Using digital electronics (microcontrollers and dedicated PWM controller's ICs)

I will not discuss the first method in this article. But in the future, I will write a complete post on pulse width modulators using analog electronics. But those who don't know about programming and microcontrollers can use the analog electronics method for this purpose.

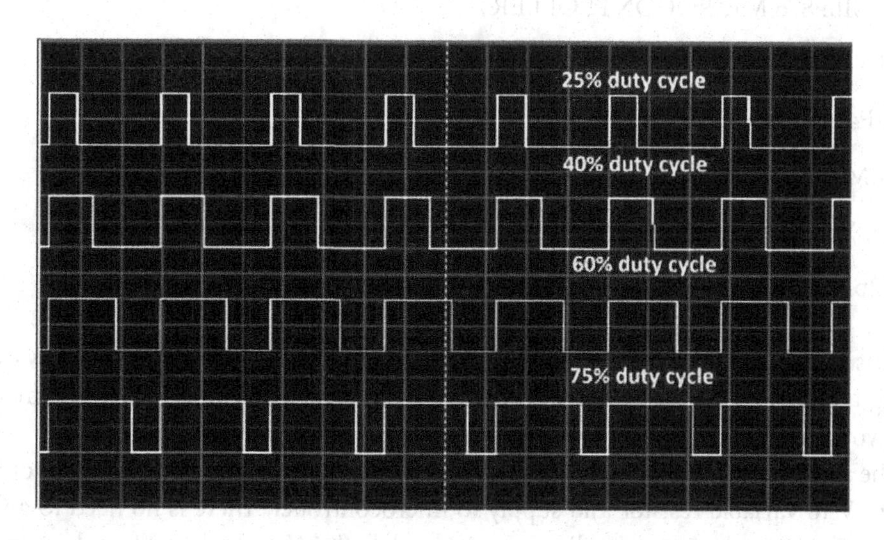

FIGURE 9.49 Different Pulse width modulation having different duty cycles.

9.2.9.1 Circuit Diagram Speed Control of DC Motor

To control DC motor speed with PIC microcontroller, you should know how to generate PWM using PIC microcontroller and how to use the analog channel of PIC microcontroller to read analog voltage. If you don't know about these topics, kindly visit the following link to get an understanding of PIC microcontrollers and PWM programming.

- PWM using pic microcontroller
  I have already mentioned there are two methods to make a pulse width modulator. These are analog electronics and digital electronics method using microcontrollers. There are dedicated PWM controller ICs that are also available in the market like SG3525, UC3842, SG2525, and many others. These PWM controller ICs also use the analog electronics method. Because their internal circuitry consists of operational amplifiers, resistors, and capacitors. I will write also write an article on PWM controllers for those who hate programming and don't know about the use of microcontrollers. But there are a few issues with these ready-to-use PWM controllers ICs. The most important issues are

  1. You will require many extra components to make them ready to use.
  2. Exact frequency is not achievable. 1%–2% error.
  3. There is also a frequency limitation which in some cases generates audible in motors.

  But there are no such issues with pulse width modulators of digital electronics type.

Components List

1. 7805 voltage regulators
2. Resistors
3. PIC16F876 MICROCONTROLLER
4. IRL1004 FET
5. UF4007 DIODE
6. 10MHZ CRYSTAL
7. 1K Variable resistor
8. 22pF capacitors

I have used PIC16F876 microcontroller to generate PWM and to change the duty cycle by reading the analog value of voltage across a variable resistor. You can use any microcontroller you want. But logic will remain the same, as shown in Figure 9.50.

In the above circuit diagram, 12 V is used as a power source. 7805 voltage regulator is used to give 5 V to variable resistor and supply to microcontroller. There is no need to give supply 5 V in Proteus to microcontroller, because by default 5 V and ground have been supplied to microcontroller in Proteus. A variable resistor is used to change the duty cycle of PWM. By moving the knob of the variable resistor, you can change the duty cycle which is directly related to speed of DC motor, as shown in Figure 9.51. So, you can adjust the speed of DC motor by adjusting the variable resistor knob according to your speed requirement.

The PWM frequency is about 8 kHz which will not make any audible noise. You can adjust it according to your motor specifications. IRL1004 is used as a switch. IRL1004 FET requires only 5 V logic level driver to operate FET and a negligible amount of current. Therefore, there is no need of any driver to drive IRL1004 FET. It can handle more than one-ampere current very easily as it is mentioned in its datasheet. Diode 1N4007 is used as

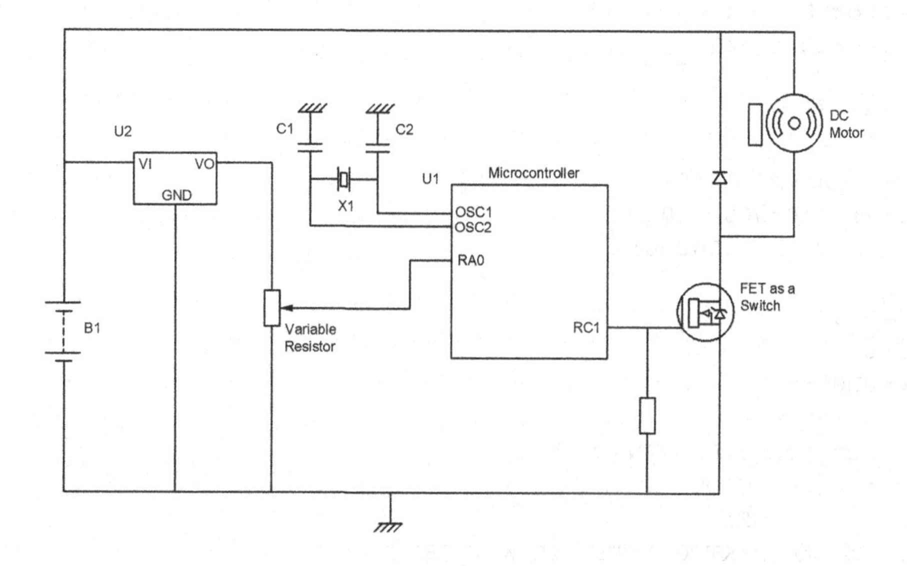

FIGURE 9.50 Circuit diagram of speed control of DC motor.

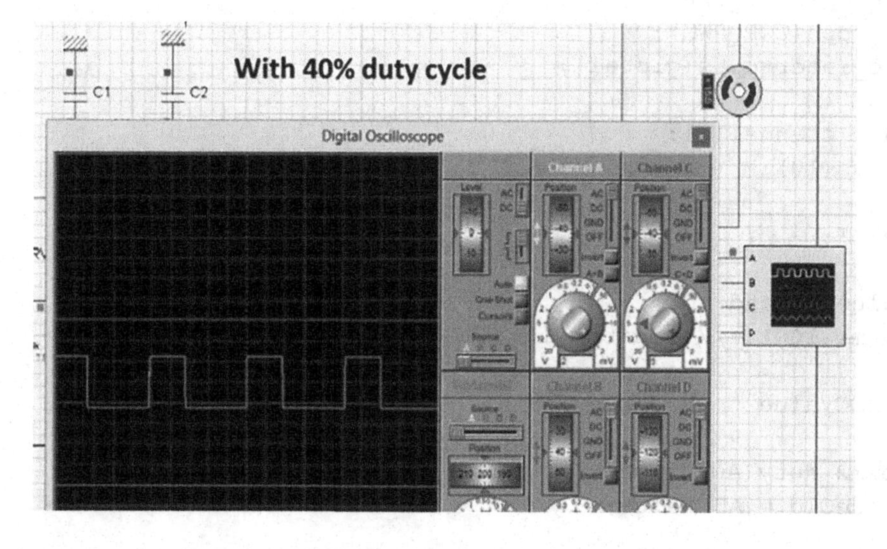

FIGURE 9.51 Output of speed control of DC motor.

a freewheeling fast recovery diode. Because the motor is an inductive load and its back emf may damage the circuit. Freewheel diode will provide a flow path to back emf current and avoid sparking across motor terminals.

Code for Speed Control of DC Motor

```
long ADCValue=0;
long ADCValueOld=1;

#byte portA = 0X05
#byte portB = 0X06
#byte portC = 0X07

void main()
{
set_tris_A(0b00101011);
set_tris_B(0b00000001);
set_tris_C(0b00000000);

portA=0X00;
portB=0X00;
portC=0X00;

setup_adc_ports(ALL_ANALOG);
setup_adc(ADC_CLOCK_INTERNAL);
setup_spi(FALSE);
setup_counters(RTCC_INTERNAL,WDT_288MS);
setup_timer_1(T1_INTERNAL|T1_DIV_BY_1);
setup_timer_2(T2_DIV_BY_1,255,1);
setup_ccp1(CCP_OFF);
setup_ccp2(CCP_PWM);
enable_interrupts(global);

set_adc_channel(0);
delay_us(10);

while(1)
{
ADCValue = Read_ADC();
delay_ms(100); // monitor 10 times a second

if ( ADCValue != ADCValueOld )
{
set_pwm2_duty(ADCValue);
ADCValueOld = ADCValue;
}
}
}
```

9.2.9.2 Line Follower Robot Using Microcontroller

Line follower robot using AVR microcontroller. In this tutorial, you will learn how to design a line follower robot using AVR microcontroller and path sensors. It follows a black line automatically and it also turns its direction according to the black line. Liner follower robot is a very famous project at the university level among electrical, electronics, computer, and software engineering students. There are a lot of competitions held every year in engineering universities related to robotics. But for first-year engineering students, the Liner follower robot is very popular, because it is easy to make and can be easily ready to run within a minimum time. You can also like to check out other robotics projects:

- Obstacle avoidance robot
- Voice-controlled robot
- Metal detector robot
- Wi-Fi–controlled robot using Arduino
- Pick and place robotic arm

Applications of Line Follower Robot

Line follower robot which is usually made at the university level is just to make students familiar with the field of robotics. But actual robots used in fields are much more complex and they can perform very complicated tasks in the industry. It is not possible to make a practical robot at the university level. That's why international robotics club encourages students to make simple robots like Liner follower robot, obstacle avoided robot, metal detection robot to get basic understanding of practical robots. In this article, I have presented you with an idea of a Liner follower robot. How to make line follower robot using microcontroller.

Circuit Working of Line Follower Robot

The block diagram of the line follower using AVR microcontroller is shown in Figure 9.52. As I have clearly mentioned, in a block diagram, two motors are used. One motor is attached to the left type of robot and another motor is attached to the right type of robot. Front-wheel of line follower robot is freely moving wheel. Motor driver IC is used to rotate both motors either clockwise or anticlockwise direction according to the turning direction of the line follower robot. L298 motor driver IC is used as a motor driver IC. Microcontroller

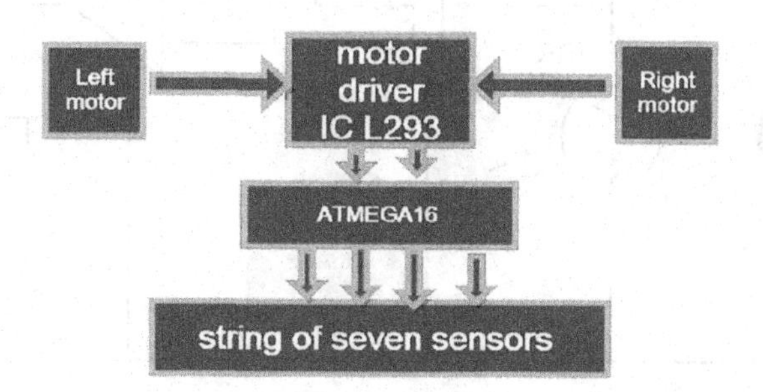

FIGURE 9.52 Block diagram of line follower robot using microcontroller.

Atmega16 is used to give control signals to motor driver IC according to sensors output. A light-dependent resistor in combination with a light-emitting diode is used to sense the black line. In case the robot goes off from the black line, the respective sensor operates and microcontrollers sense its value and take control actions by turning motors either clockwise or anticlockwise. A block diagram of the line follower robot is shown below:

Sensors Circuit Diagram for Line Follower Robot

We have already mentioned sensors are used to sense black line. We have used seven sensors in this project. But you can also use 3, 5 sensors according to the width of the black line. The light-dependent resistor in combination with the light-emitting diode is used to sense whether either line follower robot is on the black line or off the line. Circuit diagram of a single sensor is shown in Figure 9.53, but you should connect all seven sensors on single strip with a distance of 2–3 cm between each set LDR and LED to get better results of line follower robot.

Circuit Diagram of Line Follower Robot Using AVR Microcontroller

The circuit diagram of the line follower robot is shown below. Diodes are used with motors and motor driver IC to avoid sparking due to back emf voltage. Because DC motor is inductive in nature. For simulation purposes, dip switches are used instead of the actual sensors strip, as shown in Figure 9.54. LCD is used to display the direction of both motors. But it is optional to use. It has no link with the performance of the line follower robot.

Coding for Line Follower Robot

The code given below is written in AVR studio compiler and an internal oscillator is used for clock frequency.

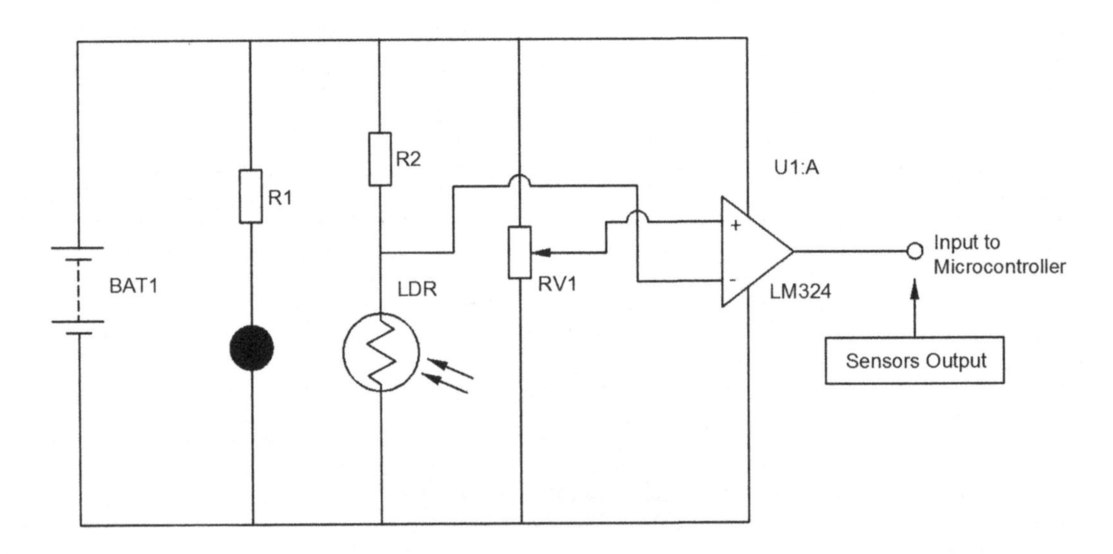

FIGURE 9.53 Sensor for line follower robot.

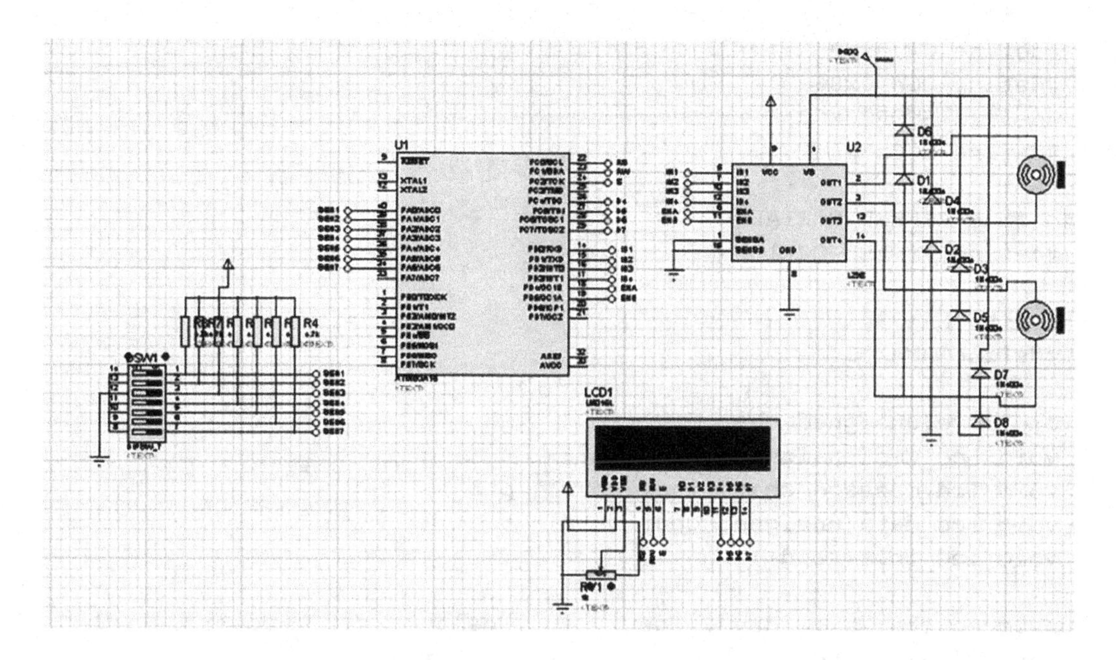

FIGURE 9.54 Circuit diagram of line follower robot using pic microcontroller.

```
// *******************************************************
 // Project: LINE FOLLOWER ROBOT
 // Author: Bilal Malik      "microcontrollerslabhub@gmail.com"
 // Module description:
 // *******************************************************
 #include <avr\io.h> // Most basic include files
 #include <avr\interrupt.h> // Add the necessary ones
 #include <util\Delay.h> // here
 #include <stdio.h>

#define RS 0
 #define RW 1
 #define EN 2
 #define LCD_PORT PORTC
 #define LCD_DIR DDRC

#define CLEAR_LCD 0x01
 #define LCD_HOME 0x02
 #define LCD_LINE_1 0x80
 #define LCD_LINE_2 0xC0
 #define LCD_R_SHIFT 0x1C
 #define LCD_L_SHIFT 0x18
 #define FWD 0x0A
 #define REV 0x05
 #define R 0x02
 #define L 0x08
```

```
#define CW 0x09
#define CCW 0x06
#define STOP 0x00
#define B 0xFF

#define RSPEED OCR1AL
#define LSPEED OCR1BL

#define MAX 3
#define HMAX 1

void move(unsigned char dir);
void Lcd_Cmd(unsigned char cmd);
void lcd_message(char line, char * ptr );
void lcd_data(unsigned char data);
void lcd_init(void);

unsigned char i , rdev , ldev , ip , delay , dir , power , dirl ,
history[MAX] , hcount = 0 , rotpow , val = 0;

unsigned char rep=0, prev=0;

void disp()
{
char mess[] = "LINE FOLLOWING ROBOT CREATED BY";
char mess2[] = "BILAL MALIK";
lcd_message(LCD_LINE_1, mess );

_delay_ms(1000);

for(int i=0;i<20;i++)
{
Lcd_Cmd(LCD_R_SHIFT);
_delay_ms(100);
}
Lcd_Cmd(LCD_HOME);
lcd_message(LCD_LINE_2, mess2);
_delay_ms(1000);

}

int main(void)
{

lcd_init();
disp();
```

```
PORTA = 0x00;
 DDRA = 0x00;

PORTB = 0xFF;
 DDRB = 0xFF;

PORTC=0x00;
 DDRC=0xFF;

PORTD=0xff;
 DDRD=0xff;

TCCR0=0x00;
 TCNT0=0x00;
 OCR0=0x00;

TCCR1A=0xA1;
 TCCR1B=0x0A;
 TCNT1H=0x00;
 TCNT1L=0x00;

OCR1AH=0xFF;
 OCR1AL=0xFF;
 OCR1BH=0xFF;
 OCR1BL=0xFF;

ACSR=0x80;// adc disable
 move(STOP);
 _delay_ms(1000);
 while(1)
 {

if(PINA!=0)
 {

ldev=0;
 rdev=0;
 val = PINA;

if((val & 0x08)==8)
 {
 rdev=1;
 ldev=1;
 }
 if((val & 0x04)==4)
 rdev=2;
```

```
if((val & 0x10)==0x10)
 ldev=2;

if((val & 0x02)==2)
 rdev=3;

if((val & 0x20)==0x20)
 ldev=3;

if((val & 0x01)==1)
 rdev=4;

if((val & 0x40)==0x40)
 ldev=4;
 if(rdev>ldev)
 move(L);
 else if(rdev<ldev)
 move(R);
 else if(rdev==ldev)
 move(FWD);
 }
 else
 {
 for(i=0,dirl=0;i
 {
 if(history[i]==L)
 {
 dirl++;
 }
 }
 if(dirl>HMAX)
 {
 move(CCW);//,0,rotpow);

_delay_ms(1000); // TESTING & DEBUG
 }
 else
 {
 move(CW);//,0,rotpow);

_delay_ms(1000); // TESTING & DEBUG
 }

_delay_ms(1000); // TESTING & DEBUG

}

};
```

```
 }
 void move(unsigned char dir)//,unsigned char delay,unsigned char
power)
 {
 PORTD=dir;

char ch[10];
 itoa(dir,ch,10);
 Lcd_Cmd(CLEAR_LCD);

if(dir == STOP)
 {
 lcd_message(LCD_LINE_1, "STOP");
 }
 else if(dir == FWD)
 {
 lcd_message(LCD_LINE_1, "FORWARD");
 }
 else if(dir == REV)
 {
 lcd_message(LCD_LINE_1, "REVERSE");
 }
 else if(dir == R)
 {
 lcd_message(LCD_LINE_1, "RIGHT");
 }
 else if(dir == L)
 {
 lcd_message(LCD_LINE_1, "LEFT");
 }
 else if(dir == CW)
 {
 lcd_message(LCD_LINE_1, "CW");
 }
 else if(dir == CCW)
 {
 lcd_message(LCD_LINE_1, "CCW");
 }

lcd_message(LCD_LINE_2, ch );
 if(dir==L||dir==R)
 {
 hcount=(hcount+1)%MAX;
 history[hcount]=dir;
 }
 LSPEED=50;
 RSPEED=50;
 _delay_ms(1000);
```

```
}

void Lcd_Cmd(unsigned char cmd)
 {

LCD_PORT &= ~((1<<rs)|(1<<rw)); rs="" low="" for="" command,=""
rw="" <span="" class="hiddenSpellError" pre="for "
data-mce-bogus="1">write
 LCD_PORT |= (1<<EN); // Set enable
 LCD_PORT &= 0x0F;
 LCD_PORT |= (cmd & 0xF0); // Send upper nibble
 LCD_PORT &= ~(1<<EN); // Enable low
 _delay_ms(1);
 LCD_PORT |= (1<<EN); // Set enable
 LCD_PORT &= 0x0F;
 LCD_PORT |= (cmd << 4); // Send lower nibble
 LCD_PORT &= ~(1<<EN); // Enable low
 _delay_ms(1);
 }

void lcd_data(unsigned char data)
 {

LCD_PORT |= (1<<RS); //RS high for data
 LCD_PORT &= ~(1<<rw); rw="" low="" for="" <span=""
class="hiddenSpellError" pre="for " data-mce-bogus="1">write
 LCD_PORT |= (1<<EN); // Set enable
 LCD_PORT &= 0x0F;
 LCD_PORT |= (data & 0xF0); // Send upper nibble
 LCD_PORT &= ~(1<<EN); // Enable low
 _delay_ms(1);
 LCD_PORT |= (1<<EN); // Set enable
 LCD_PORT &= 0x0F;
 LCD_PORT |= (data << 4); // Send lower nibble
 LCD_PORT &= ~(1<<EN); // Enable low
 _delay_ms(1);
 }

void lcd_init(void)
 {
 LCD_DIR = 0xFF;
 _delay_ms(15);
 Lcd_Cmd(0x03);
 _delay_ms(5);
 Lcd_Cmd(0x03);
 _delay_ms(1);
```

```
 Lcd_Cmd(0x03);
 _delay_ms(1);
 Lcd_Cmd(0x02);
 _delay_ms(1); //as said in the book
 Lcd_Cmd(0x28); //Set interface length
 _delay_ms(1);
 Lcd_Cmd(0x08); // Turn off display
 _delay_ms(1);
 Lcd_Cmd(0x01); //Clear display
 _delay_ms(1);
 Lcd_Cmd(0x06); //Set cursor move direction
 _delay_ms(1);
 Lcd_Cmd(0x0F); //Enable display cursor
 _delay_ms(1);
 }
 void lcd_message(char line, char * ptr ) // 1st argument is the
string pointer the other is line No.
 {

//Lcd_Cmd(CLEAR_LCD);
 Lcd_Cmd(line); //move cursor to line

while(ptr != '\0')
 lcd_data((ptr++));
}
```

9.2.9.3 Applications of Servomotor

The servomotor is a small and effective motor and it can be used in some serious applications like precise position control. The controlling of this motor can be done with a PWM (pulse width modulator) signal. The applications of these motors mainly include industrial robotics, computers, toys, CD/DVD players, etc., as shown in Figure 9.55. These motors are widely used in some applications where a specific task is to be done regularly in an exact manner.

This motor is used to activate movements in robotics for giving the arm to its exact angle. This motor is used to start, move and also stop the conveyor belts carrying the product along with many stages. For example, bottling, packaging, and product labeling,

This motor can be used in the camera to set a lens of the camera to enhance the focus of images. This motor is built into the camera to correct the lens of the camera to improve out-of-focus images. This motor can be used to control the robotic vehicle by controlling robot wheels, speed, generating plenty of torque to move and also start and stop the robot.

It can be used in solar tracking system to set the angle of the solar panel so that panel stays to face the sun. This motor is used in milling machines for metal cutting and forming to provide specific motion. This is used in textile industries to control and knit machines like spinning and weaving. This motor is used in automatic door opening and closing in public places like hospitals, theaters, and supermarkets.

FIGURE 9.55 Applications of servomotor.

1. It is used in the robotic industry for position control.
2. It is used in robotic arms.
3. It is used in press and cutting industry for the cutting and pressing the piece precisely.
4. It is used in conveyor belts to start and stop the conveyor belt at every position.
5. It is used in digital cameras for autofocusing.
6. It is used in a solar tracking system for tracking the sun at every precise moment.
7. It is used in the labeling and packing industry for labeling the monogram and packing the things.

Common Industrial Applications for Servomotors

Servomotors are small and efficient but critical for use in applications requiring precise position control. The servomotor is controlled by a signal (data) better known as a pulse width modulator (PWM). Here are several of the more common servomotor applications in use today.

Robotics: A servomotor at every "joint" of a robot is used to actuate movements, giving the robot arm its precise angle.

Conveyor belts: Servomotors move, stop, and start conveyor belts carrying products along to various stages, for example, in product packaging/bottling, and labeling.

Camera autofocus: A highly precise servomotor built into the camera corrects a camera's lens to sharpen out-of-focus images.

Robotic vehicle: Commonly used in military applications and bomb detonation, servomotors control the wheels of the robotic vehicle, generating enough torque to move, stop, and start the vehicle smoothly as well as control its speed.

Solar tracking system: Servomotors adjust the angle of solar panels throughout the day so that each panel continues to face the sun, harnessing maximum energy from sunup to sundown.

Metal cutting and metal forming machines: Servomotors provide precise motion control for milling machines, lathes, grinding, centering, punching, pressing, and bending in metal fabrication for such items as jar lids to automotive wheels.

Antenna positioning: Servomotors are used on both the azimuth and elevation drive axis of antennas and telescopes such as those used by the National Radio Astronomy Observatory (NRAO).

Woodworking/CNC: Servomotors control woodturning mechanisms (lathes) that shape table legs and stair spindles, for example, as well as targeting and drilling the holes necessary for assembling those products later in the process.

Textiles: Servomotors control industrial spinning and weaving machines, looms, and knitting machines that produce textiles such as carpeting and fabrics as well as wearable items such as socks, caps, gloves, and mittens.

Printing presses/printers: Servomotors stop and start the print heads precisely on the page as well as move paper along to print multiple rows of text or graphics in exact lines, whether it's a newspaper, a magazine, or an annual report.

Automatic door openers: Supermarkets and hospital entrances are prime examples of automated door openers controlled by servomotors, whether the signal to open is via a push plate beside the door for handicapped access or by a radio transmitter positioned overhead.

The world would be a much different place without servomotors. Whether they're used in industrial manufacturing or commercial applications, they make our lives better, easier, and in many cases provide us with more affordable products.

Engineering Application of Control Theory

In order to design a control system for engineering applications, we need to use knowledge of control theory. Engineering application is a relatively complex process. In order to design an ideal control system, we need to consider many factors, such as the reliability of the system, the replaceability of equipment components, economic costs, and the difficulty of engineering realization. Reliability is an important consideration in engineering applications. In some industrial fields, such as nuclear power plants, thermal power plants, chemical plants, and aerospace industries, the reliability of their control systems is the most important. When designing the control system, it is necessary to select a suitable design scheme according to the reliability requirements and cost budget.

To design a complete control system, in addition to learning the content in this book, we must also learn the following knowledge, but not limited to the following knowledge.

1. Knowledge of electronic circuits.
2. Knowledge of programming languages, such as Matlab, C, VB, LAD, etc.
3. Knowledge of industrial communication networks, such as Modbus, RS-232, PROFIBUS, etc.

The above learning suggestions are mainly aimed at the application of control systems in the process industry. Nuclear power, thermal power, petrochemical, and food production belong to process industries.

This chapter takes the process control system as a case to introduce the application of control theory in practical engineering. The process control system usually includes the control of multiple parameters such as liquid level, temperature, and flow. It involves system models such as pneumatic, electric, hydraulic and thermal. For the study of control theory, the process control system is a good research object. This chapter takes the control system of the sterilization device as the research object.

9.3 PROCESS FLOW OF STERILIZATION DEVICE

Sterilization devices are commonly used in food, chemical, and pharmaceutical process industries. The sterilization device is a typical process control equipment. The heat exchanger is the core component of the sterilizer. The device uses steam as the heat source. The process fluid is sterilized at a high temperature. The main function of the sterilizer control system realizes the control of sterilization temperature. During the production process, the inlet pressure and temperature of the steam and the temperature of the process fluid will change. The control system dynamically adjusts parameters such as valve opening and pumps speed to control the sterilization temperature. The process flow of sterilization device is shown in Figure 9.56.

The functions of each part of the sterilizer are as follows:

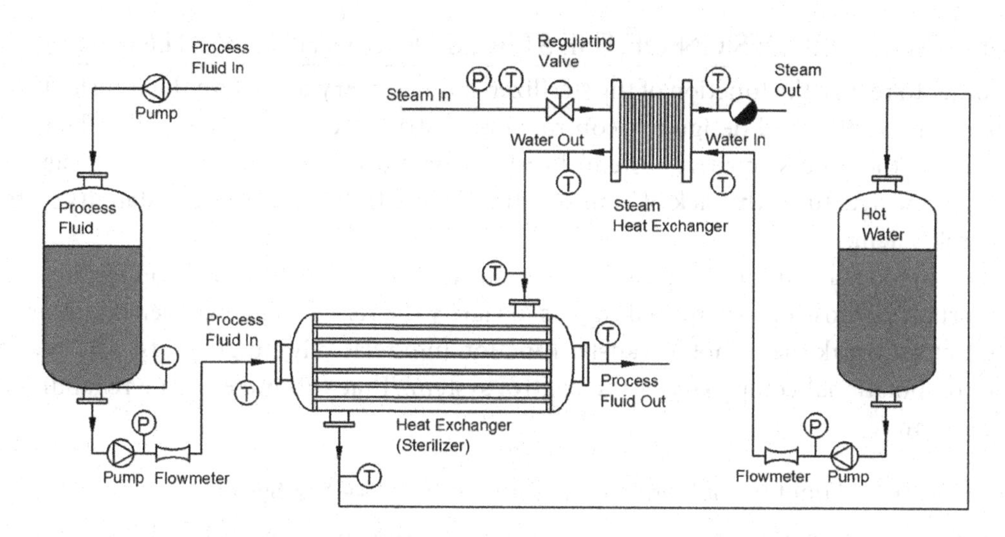

FIGURE 9.56 Process flow of sterilization device.

1. *Main heat exchanger (Sterilizer)*
   The sterilization heat exchanger is the core component of the sterilizer. Its main function is to sterilize the process fluid with hot water. The process fluid flows through the tube side of the heat exchanger. And the hot water flows through the shell side of the heat exchanger.

2. *Steam heat exchanger*
   The heat of the steam is transferred to the hot water through the steam heat exchanger. And the hot water transfers heat to the process fluid through the sterilizer.

3. *Steam regulating valve*
   The steam regulating valve is a linear actuator. The opening of the valve is proportional to the received signal. The system controls the flow of steam by adjusting the opening of the steam regulating valve. And then realize the control of the sterilization temperature of the sterilizer.

4. *Hot water tank*
   The system uses a hot water tank as a buffer for hot water.

5. *Process fluid tank*
   The system uses a process fluid tank as a buffer for the process fluid.

6. *Pumps*
   Centrifugal pumps are used to transport process fluid and hot water. The process fluid pump and the hot water pump are driven by frequency converters.

7. *Sensors*
   The system collects production process signals such as temperature, pressure, flow, speed, and liquid level through sensors.

9.4 STRUCTURE DESIGN OF THE STERILIZER CONTROL SYSTEM

In order to realize the function of the sterilizer, it is necessary to analyze the transfer function of the system and design the control system structure. There are many influencing factors in the process control system. Every factor must be considered when designing the system structure. The lack of consideration of any factor may cause instability in the control system.

Due to too many influencing factors, the transfer function will be very complicated. An important purpose of structural design is to clarify the relationship between all influencing factors. Break the complex transfer function into relatively simple parts. And reduce the computational complexity of the control system. That will increase the reliability of the system.

9.4.1 Transfer Function of Sterilization Temperature Control System

The block diagram of the sterilizer control system is shown in Figure 9.57. Where $R_1(s)\ D_a$ is the temperature input, $H_1(s)\ D_a$ is the transfer function of temperature transducer, $C_1(s)\ D_a$ is the temperature of process fluid out, $R_2(s)\ D_a$ is the process fluid flow input, $H_2(s)\ D_a$ is the transfer function of process fluid flow transducer, $C_2(s)\ D_a$ is the process fluid flow, $R_3(s)\ D_a$ is the hot water flow input, $H_3(s)\ D_a$ is the transfer function of hot water flow transducer, $C_3(s)\ D_a$ is the hot water flow, $G^{**}(s)\ D_a$ is the transfer functions of control system elements, and $D^{**}(s)\ D_a$ is the disturbances.

The target parameter of the control system is the sterilization temperature. The direct effects that affect the sterilization temperature include the following:

1. *Steam side*
   It includes the flow rate and the inlet temperature of the steam.

2. *Hot waterside*
   It includes the flow rate and the inlet temperature of the hot water.

3. *Process fluid side*
   It includes the flow rate and the inlet temperature of the process fluid.

4. *Heat exchanger*
   It includes heat exchange efficiency of the steam heat exchanger and the sterilizer.

The indirect effects that affect the sterilization temperature include the following:

1. *Steam side*
   It includes the steam regulating valve opening and the resistance of the steam pipeline. It will affect the flow of steam.

2. *Hot waterside*
   It includes the speed of the hot water pump and the resistance of the hot water pipe. It will affect the flow of hot water.

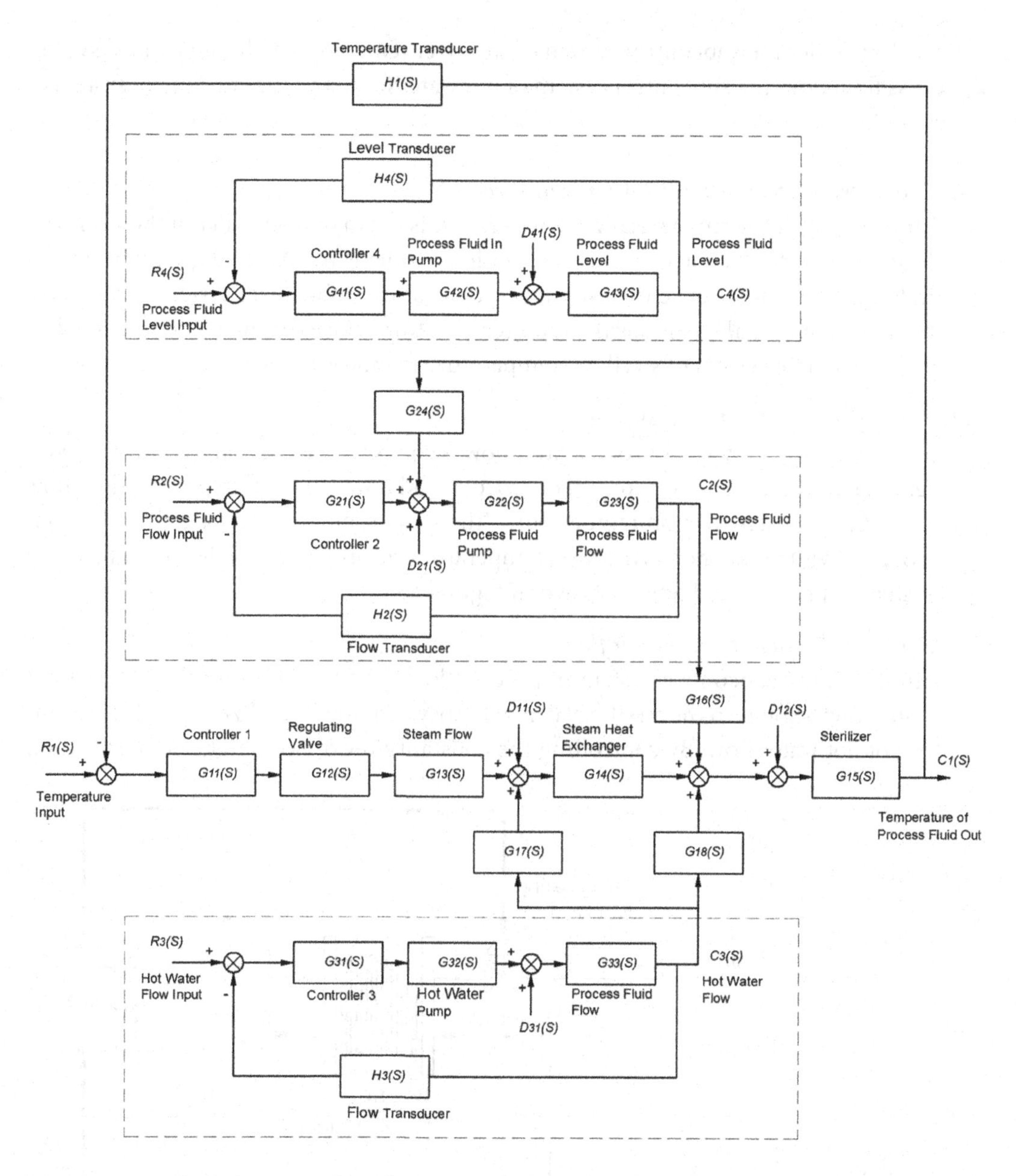

FIGURE 9.57 Block diagram of sterilizer control system.

3. *Product side*

   It includes the speed of the process fluid pump, the resistance of the process fluid pipeline, and the liquid level of the process fluid storage tank. It will affect the flow of the process fluid.

9.4.2 Subcontrol Systems of Sterilizer

In order to reduce the computational complexity of the control system and improve its stability of the system, the control system is divided into four independent subcontrol systems. It includes the subcontrol system of sterilizer temperature, the subcontrol system of process fluid flow, the subcontrol system of hot water flow, and the subcontrol system of process fluid level.

1. *Subcontrol system of sterilizer temperature*
   The subcontrol system of sterilizer temperature is the main controller of the sterilizer. It realizes the final control of the sterilization temperature. All changes in direct and indirect effects will cause the output response of the subcontrol system of sterilizer temperature. It makes the sterilization temperature consistent with the preset value. The subcontrol system of sterilizer temperature is shown in Figure 9.58.
2. *Subcontrol system of process fluid flow*
   The subcontrol system of process fluid flow belongs to the auxiliary control subsystem. It realizes closed-loop control of process fluid flow. Changes in process fluid flow will affect the sterilization temperature. The flow of process fluid needs to be controlled. It will make the sterilization temperature control more stable. The subcontrol system of process fluid flow is shown in Figure 9.59.
3. *Subcontrol system of hot water flow*
   Similar to the subcontrol system of process fluid flow, the subcontrol system of hot water flow belongs to the auxiliary control subsystem too. It realizes closed-loop control of hot water flow. By controlling the constant flow of hot water, the sterilization

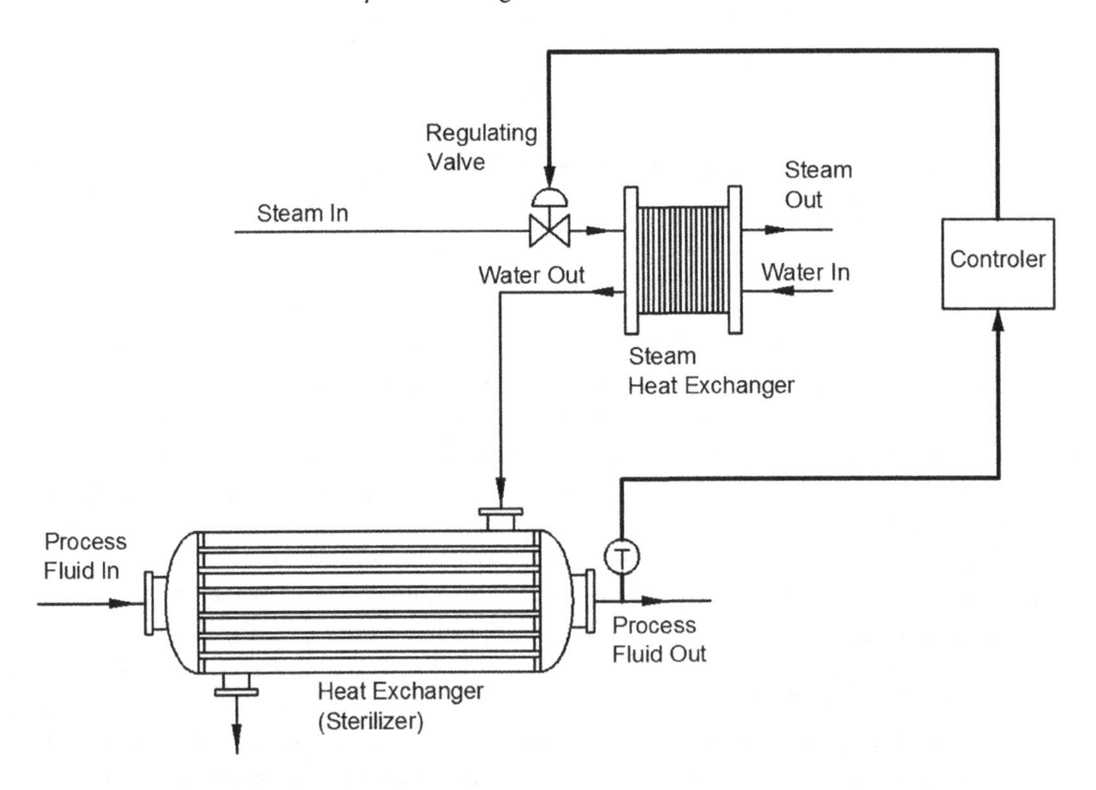

FIGURE 9.58 Subcontrol system of sterilizer temperature.

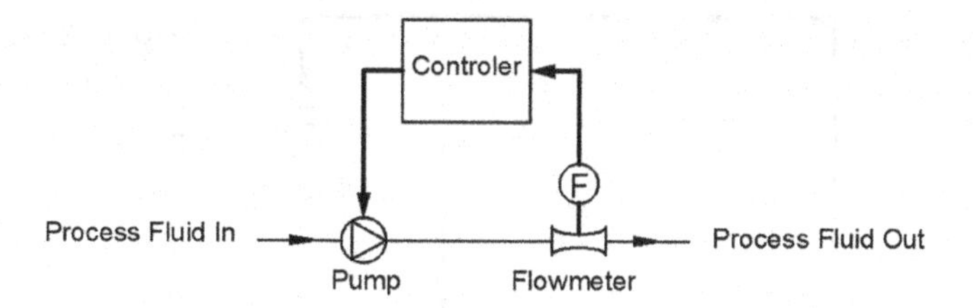

FIGURE 9.59 Subcontrol system of process fluid flow.

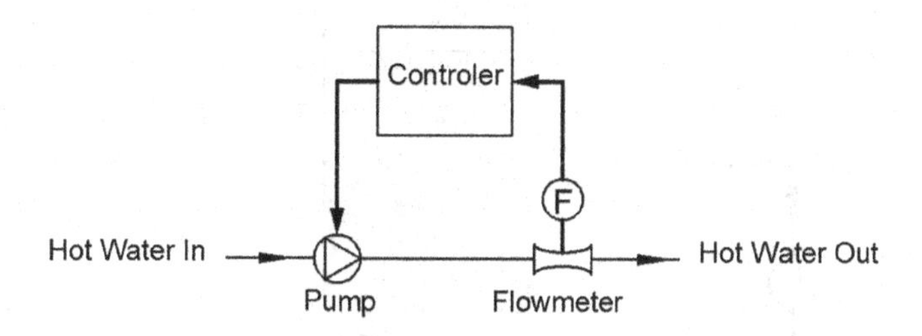

FIGURE 9.60 Subcontrol system of hot water flow.

temperature control tends to be more stable. The subcontrol system of hot water flow is shown in Figure 9.60.

4. *Subcontrol system of process fluid level*
   The subcontrol system of process fluid level can realize two functions. The first function is to ensure continuous production. The second is to achieve closed-loop control of the liquid level of process fluid tank. The liquid level affects the inlet pressure of the process fluid pump. Then it will affect the flow of the process fluid. Through liquid level control, the process fluid level tends to be stable. That will make the process of fluid flow control and sterilization temperature control more stable. The subcontrol system of process fluid level is shown in Figure 9.61.

9.4.3 Hardware Configuration of Sterilizer Control System

Distributed Computer System (DCS) is used as the control system of the sterilizer. DCS is very commonly used in the field of process control. The control system is composed of PC, PLC, HMI, I/O modules, sensors, and actuators. The control system of the sterilizer is shown in Figure 9.62. In industrial production, usually, one PC controls multiple PLCs. Each PLC controls a functionally independent device. This chapter does not give a detailed introduction to the DCS system. Readers interested in DCS can further study in other ways.

The functions of each component of the control system are as follows:

1. *PC*
   PC is used for functions such as data storage, data query, complex calculation, production operation, and parameter setting. The sterilizer is a subsystem of the production line. Through communication with PC, all equipment on the production line can work together.

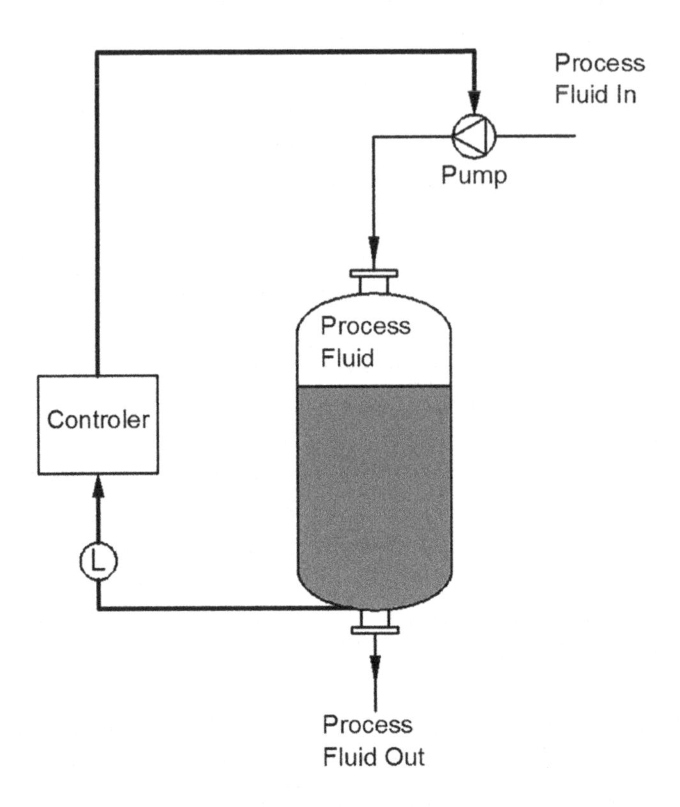

FIGURE 9.61 Subcontrol system of process fluid level.

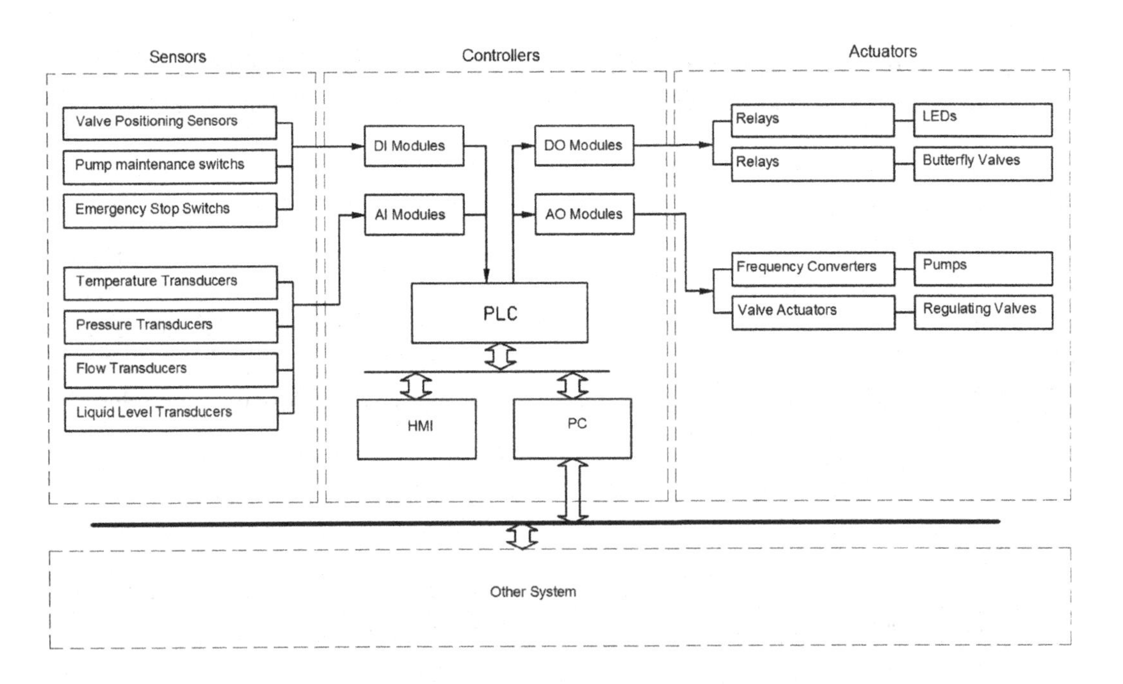

FIGURE 9.62 Configuration of sterilizer control system.

2. *PLC*
   PLC (programmable logic controller) is the main controller of sterilizer. PLC is used to realize sterilization process control. PLC will collect signals such as temperature, pressure, flow, speed, and liquid level in the production process. It is also responsible for controlling the sterilization temperature, the pump speed, and the valve opening.

3. *HMI*
   HMI (human–machine interface) is a user interface that connects a person to the sterilizer device. HMI is placed near the sterilizer. Operators can operate the sterilizer locally.

4. *I/O module*
   I/O modules include AI, AO, DI, and DO modules.
   AI (Analog Input) modules are used to monitor the temperature, pressure, flow, speed, and liquid level. AO (Analog Output) modules are used to set the pump speed and valve opening. DI (Digital input) modules are used to check the status of any devices whether it is ON or OFF. DO (Digital Output) modules are used to operate solenoid valves, relays, indicating lamps or as a command to any other devices.

5. *Sensor*
   Sensors include digital sensors and analog sensors. Valve position sensor and switch are digital sensors. Analog sensors include temperature, pressure, flow, speed, and liquid level sensors, which convert physical quantities into electrical signals. Analog signals can be formatted as 4–20 mA, 0–20 mA, 1–5 VDC, 0–5 VDC, -10-10 VDC.

6. *Actuator*
   Actuators include valves and pumps to regulate the flow of steam, water, and process fluid.

9.5 TRANSFER FUNCTIONS OF THE SUBCONTROL SYSTEMS

1. *PID subcontrol system of sterilizer temperature*
   Sterilization temperature is the target parameter of the control system. A PID controller is used to control the sterilization temperature. The block diagram of sterilizer temperature PID subcontrol system is shown in Figure 9.63.
   where
   $R_1(s)$ is the setting temperature value.
   $C_1(s)$ is the measured temperature value of the outgoing process fluid.
   $E_1(s)$ is the difference value between $R_1(s)$ and $C_1(s)$.
   $U_1(s)$ is the output value of the temperature PID controller.
   $H_1(s)$ is the transfer function of the temperature transducer.
   $G_{11}(s)$ is the transfer function of the temperature PID controller.
   $G_{12}(s)$ is the transfer function of the regulating valve driver.
   $G_{13}(s)$ is the transfer function of the relationship between valve opening and flow.
   $G_{14}(s)$ is the transfer function of the steam heat exchanger.

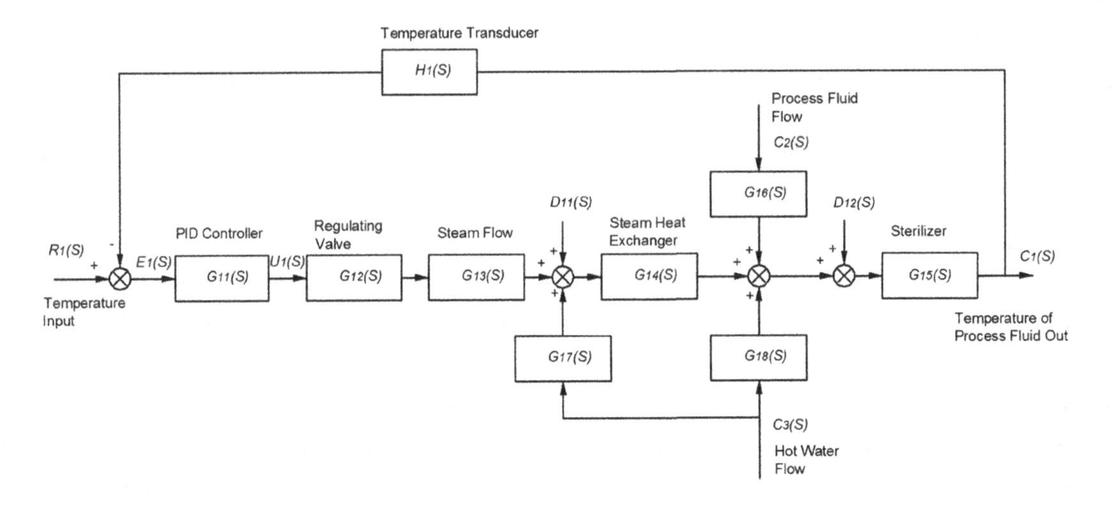

FIGURE 9.63 Block diagram of sterilizer temperature PID subcontrol system.

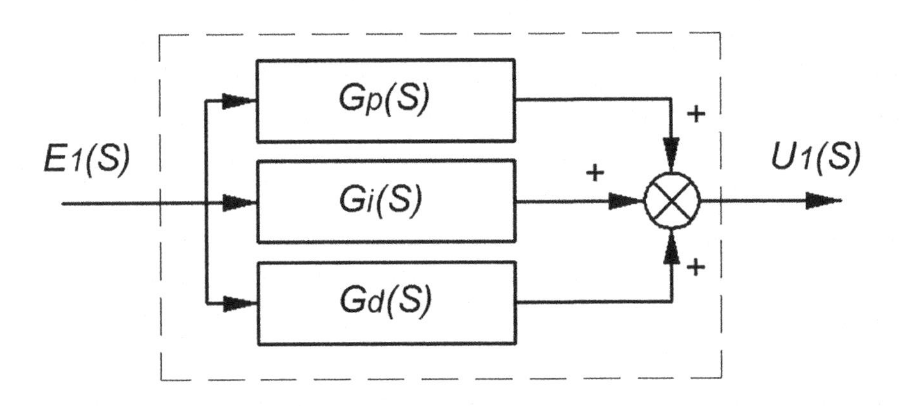

FIGURE 9.64 Block diagram of PID controller.

$G_{15}(s)$ is the transfer function of the sterilizer.
$G_{16}(s)$ is the transfer function of the process fluid flow.
$G_{17}(s)$ is the transfer function of the hot water flow.
$G_{18}(s)$ is the transfer function of the hot water flow.

In this block diagram, classical PID controller is used as the temperature controller. The block diagram of PID controller is shown in Figure 9.64. Many industrial controllers employ a PID regulator arrangement that can be tailored to optimize a particular control system. PID controller is the most commonly used algorithm for controller design and it is the most widely used controller in the industry. PID controller has all the necessary dynamics: fast reaction on change of the controller input (D mode), increase in the control signal to lead error toward zero (I mode), and suitable action inside the control error area to eliminate oscillations (P mode).

$G_p(s)$ is the transfer function of proportional. $G_i(s)$ is the transfer function of integral, and $G_d(s)$ is the transfer function of differential. The PID controller can be implemented by analog circuitry or by microprocessor technology. Practically, most modern PID controllers are designed based on microprocessor technology. The PID equation is

$$u(t)=k_P\left(e(t)+\frac{1}{T_I}\int e(t)dt+T_D\dot{e}(t)\right) \tag{9.1}$$

where

$u(t)$ is the output value of the temperature PID controller.

$e(t)$ is the difference value between the setting temperature and the measured temperature.

K_p is the Proportional gain.

T_i is the Integral time.

T_d is the Differential time.

K_p is the proportional gain. The output varies based on how far you are from your target. T_i is the integral time. The output varies based on how long it's taking you to get to your target. T_d is the differential time. The output varies based on the change in the error.

The following transfer function of PID controller is obtained by the Laplace transformation.

$$G_{11}(s)=\frac{U_{11}(s)}{E_{11}(s)}=K_p\left(1+\frac{1}{T_i s}+T_d s\right) \tag{9.2}$$

2. *PI subcontrol system of process flow*

According to the principle of the heat exchanger, the change of process flow will affect the sterilization temperature of the sterilization heat exchanger. Flow control is a fast response system. The PI controller is used to control the process flow. The constant process flow will improve the stability of sterilization temperature regulation. The block diagram of the process flow PI controller is shown in Figures 9.65 and 9.66.

where

$R_2(s)$ is the setting flow value of the process fluid.

$C_2(s)$ is the measured flow value of the process fluid.

$C_4(s)$ is the measured level value of the process fluid.

$C_2(s)$ is the transfer function of the flow transducer.

$C_{21}(s)$ is the transfer function of the flow PI controller.

$C_{22}(s)$ is the transfer function of the process fluid pump.

$C_{23}(s)$ is the transfer function of the process fluid flow.

$C_{24}(s)$ is the transfer function of the process fluid level.

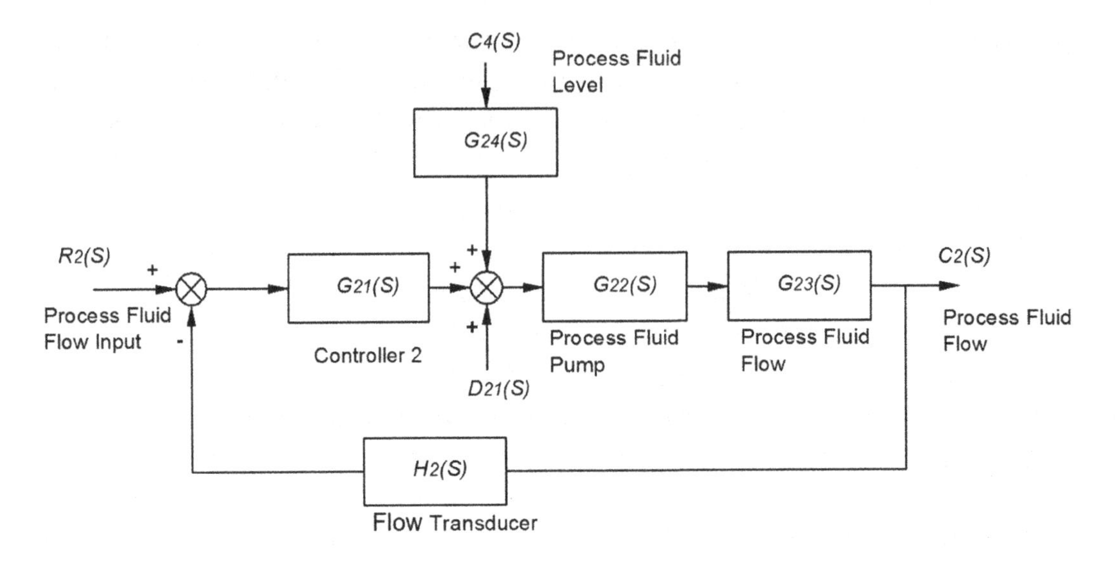

FIGURE 9.65 Block diagram of process flow subcontrol system.

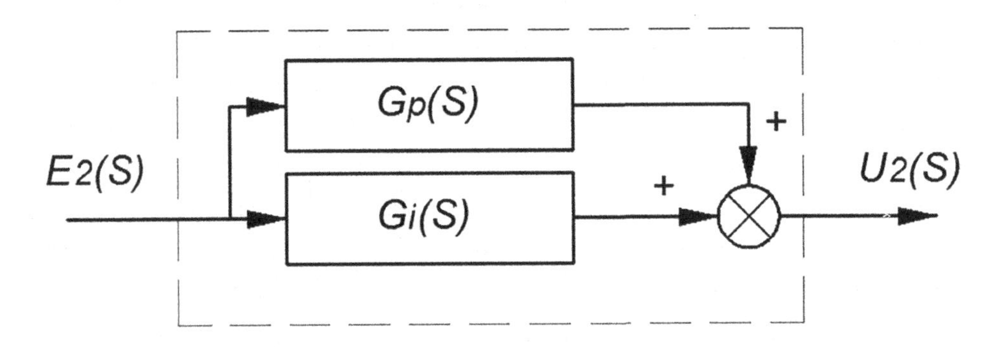

FIGURE 9.66 Block diagram of PI controller.

$G_p(s)$ is the transfer function of proportional. $G_i(s)$ is the transfer function of integral. The PI equations is

$$u(t) = k_P\left(e(t) + \frac{1}{T_I}\int e(t)dt\right) \tag{9.3}$$

where

$u(t)$ is the output value of the flow PI controller.

$e(t)$ is the difference value between the setting flow and the measured flow.

K_p is the Proportional gain.

T_i is the Integral time.

T_d is the Differential time.

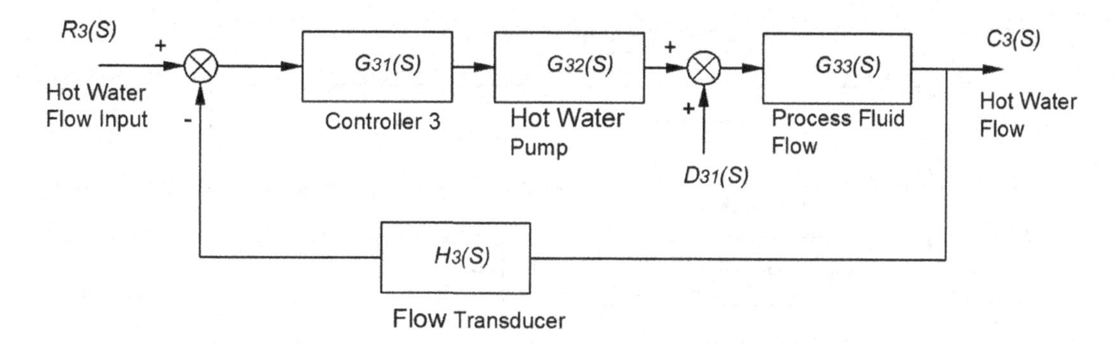

FIGURE 9.67 Block diagram of hot water flow subcontrol system.

The following transfer function of PI controller is obtained by the Laplace transformation.

$$G_{21}(s)=\frac{U_{21}(s)}{E_{21}(s)}=K_p\left(1+\frac{1}{T_i s}\right) \tag{9.4}$$

3. *PI subcontrol system of hot water flow*

The change of hot water flow will affect the hot water out temperature of the sterilization heat exchanger. The PI controller is used to control the hot water flow. The constant hot water flow will improve the stability of sterilization temperature regulation. The block diagram of the hot water flow PI controller is shown in Figure 9.67.

where

$R_3(s)$ is the setting flow value of the hot water.
$C_3(s)$ is the measured flow value of the hot water.
$H_3(s)$ is the transfer function of the hot water flow transducer.
$G_{31}(s)$ is the transfer function of the hot water flow PI controller.
$G_{32}(s)$ is the transfer function of the hot water pump.
$G_{33}(s)$ is the transfer function of the hot water flow.
The PI subcontrol system of hot water flow is written as,

$$u(t)=k_P\left(e(t)+\frac{1}{T_I}\int e(t)dt\right) \tag{9.5}$$

The following transfer function of PI controller is obtained by the Laplace transformation.

$$G_{31}(s)=\frac{U_{31}(s)}{E_{31}(s)}=K_p\left(1+\frac{1}{T_i s}\right) \tag{9.6}$$

4. *PI subcontrol system of process fluid tank liquid level*

The level of the process fluid storage tank will affect the inlet pressure of the process fluid pump. And the inlet pressure affects the process of fluid flow. Although, the PI controller

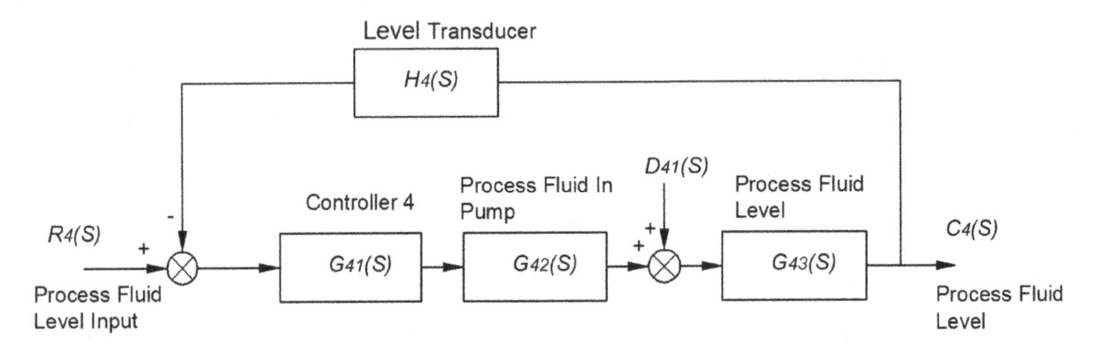

FIGURE 9.68 Block diagram of process fluid tank liquid level PI controller.

has been used to control process fluid flow. A PI controller of the process fluid storage tank liquid level is used to further improve the accuracy of flow control. The block diagram of the process fluid tank liquid level PI controller is shown in Figure 9.68.

where

$R_4(s)$ is the setting liquid level value of the process fluid tank.
$C_4(s)$ is the measured liquid level value of the process fluid tank.
$H_4(s)$ is the transfer function of the liquid level transducer.
$G_{41}(s)$ is the transfer function of the process fluid tank liquid level PI controller.
$G_{42}(s)$ is the transfer function of the pump of incoming process fluid.
$G_{43}(s)$ is the transfer function of the process fluid tank liquid level.

The PI subcontrol system of hot water flow is

$$u(t)=k_P\left(e(t)+\frac{1}{T_I}\int e(t)dt\right) \tag{9.7}$$

The following transfer function of PI controller is obtained by the Laplace transformation.

$$G_{41}(s)=\frac{U_{41}(s)}{E_{41}(s)}=K_p\left(1+\frac{1}{T_i s}\right) \tag{9.8}$$

9.6 REALIZATION OF CONTROL ALGORITHM IN COMPUTER

In practice, controllers are nowadays almost exclusively implemented digitally. This means that the controller operates in discrete time, although the controlled systems usually operate in continuous time. Therefore, the digital controller has to be connected to the system by interfaces that transform the continuous-time system output $u(t)$ to a discrete sequence $u(t_n)$ which the digital controller can process.

1. *Discretization of PID controller*

   In a computer control system, transforming Eq. (9.2) to the time-discrete domain results in Eq. (9.9).

$$\Delta u(t_n)=k_P\left(e(t_n)-e(t_{n-1})\right)+k_I e(t_n)+k_D\left(e(t_n)-2e(t_{n-1})+e(t_{n-2})\right) \tag{9.9}$$

where

$\Delta u(t_n)$ D_a is the increment of the nth output of the PID controller to the previous output, $e(t_n)$ is the deviation signal of the nth sample, k_P is the proportional gain,

k_I is the integral coefficient, $k_I = k_P \frac{T_S}{T_I}$. T_S is the is the sampling period.

k_D is the differential coefficient. $k_D = k_P \frac{T_D}{T_S}$.

2. *Discretization of PI controller*

Transforming Eqs. (9.4), (9.6) and (9.8) to the time-discrete domain results in Eq. (9.10).

$$\Delta u(t_n)=k_P\left(e(t_n)-e(t_{n-1})\right)+k_I e(t_n) \tag{9.10}$$

Here,

$\Delta u(t_n)$ D_a D_a is the increment of the nth output of the PID controller to the previous output.

$e(t_n)$ is the deviation signal of the nth sample.

k_P is the proportional gain.

k_I is the integral coefficient $k_I = k_P \frac{T_S}{T_I}$. T_S is the sampling period.

9.7 TUNING FOR CONTROLLERS

In order to realize PID and PI algorithms in the computer, the parameters K_p, K_I and K_D must be known. K_p, K_I, and K_D are determined by various components such as sensors, drivers, controllers, and fluid delivery pipelines. To accurately calculate all the coefficients, the transfer functions of all components must be obtained first. But in practical applications, it is difficult to obtain the transfer functions of all components. Therefore, it is difficult to obtain the accurate transfer function of the control equation only by calculation.

The Ziegler–Nichols design methods are the most popular methods used in process control to determine the parameters of a PID or PI controllers. The unit step response method is based on the open-loop step response of the system. The unit step response of the plant is shown in Figure 9.69.

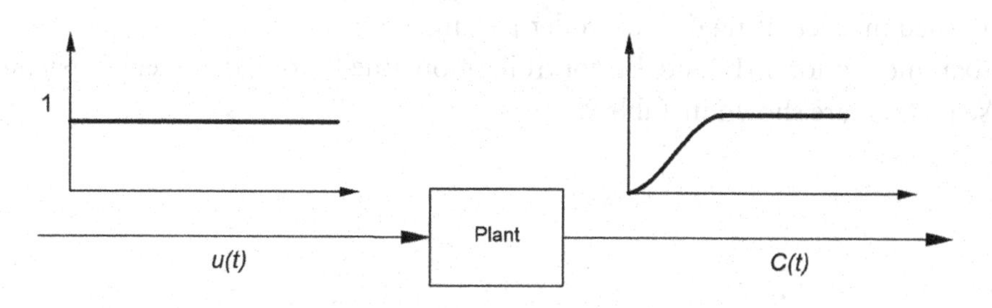

FIGURE 9.69 Unit step response of plant.

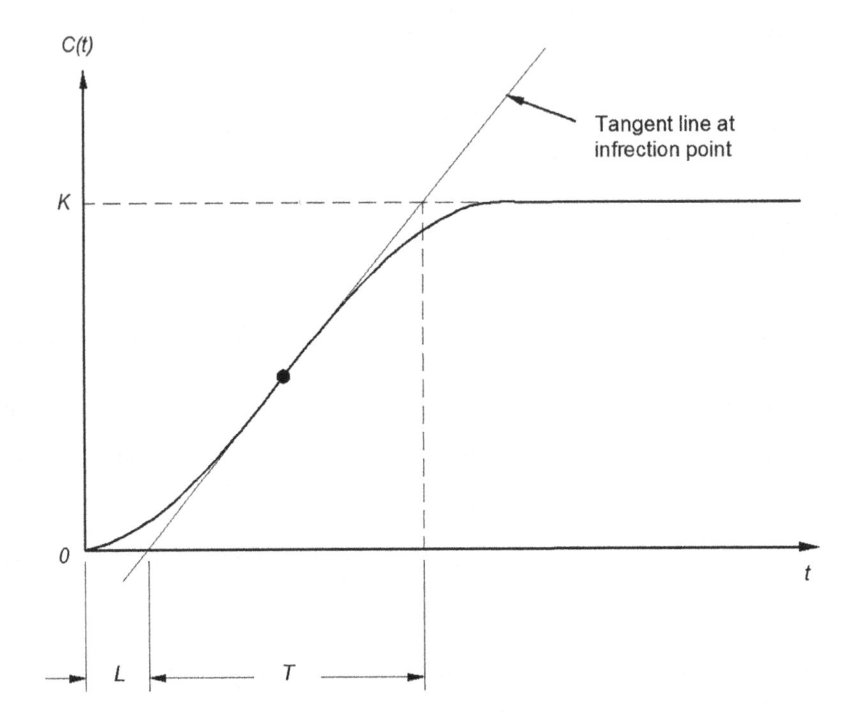

FIGURE 9.70 Response curve of Ziegler–Nichols tuning.

TABLE 9.3 Controller's Parameters for the Ziegler–Nichols Method

Controller	K_p	T_I	T_D
P	$\frac{T}{L}$	∞	0
PI	$0.9\frac{T}{L}$	$\frac{L}{0.3}$	0
PID	$1.2\frac{T}{L}$	$2L$	$0.5L$

The unit step response of the process is characterized by two parameters, delay L and time constant T. These are determined by drawing a tangent line at the inflection point, where the slope of the step response has its maximum value. The intersections of the tangent and the coordinate axes give the process parameters as shown in Figure 9.70, and these are used in calculating the controller parameters.

The parameters for PID and PI controllers obtained from the Ziegler–Nichols step response method are shown in Table 9.3.

Bibliography

1. G. Hu, *Engineering Control Theory*. Beijing: China Machine Press, 1997.
2. W. Shi, *Modern Electromechanical Control Engineering*. Beijing: China Machine Press, 1998.
3. G. Hu, *Fundamentals of Electromechanical Control Engineering Theory and Application*. Beijing: China Machine Press, 1997.
4. S. Yue, T. Weilin, *Mechanical Engineering Control Basic Learning Guidance and Detailed Exercises*. Beijing: China Machine Press, 1999.
5. H. Shousong, *Automatic Control Principles Problem Set*. Beijing: National Defense Industry Press, 1997.
6. Z.K. Shi, L. Chao, *Automatic Control Principle FAQ Analysis and Simulation Questions*. Xi'an: Northwestern Polytechnical University Press, 1999.
7. H. Zhonglin, *Control System MATLAB Calculation and Simulation*. Beijing: National Defense Industry Press, 2001.
8. S. Xiaosheng, *Master MATLAB6.0 and Its Engineering Application*. Beijing: Science Press, 2002.
9. Q. Xuesen, S. Jian, *Engineering Cybernetics*. Beijing: Science Press, 1980.
10. Z. Qi, S. Lin, et al., *Electro Hydraulic Control Engineering Practice*. Xi'an: Northwestern Polytechnical University Press, 1988.
11. Z. Yinglin, *Stability Analysis of Control System*. Lanzhou: Lanzhou University Press, 1987.
12. Z. Lifeng, et al., *Servo Mechanical Structure: Volume 1, Basic Principles of Servo System*. Beijing: National Defense Industry Press, 1980.
13. D. Zhong, *Theoretical Basis of Automatic Control*. Beijing: Tsinghua University Press, 1991.
14. S. Zhisun, *Hydraulic Control System*. Beijing: National Defense Industry Press, 1985.
15. T. Nie, et al., *Numerical Calculation Method*. Xi'an: Northwestern Polytechnical University Press, 1987.
16. W. Lin, *Principle of Automatic Control*. Beijing: Tsinghua University Press, 1990.
17. X. Deling, *Automatic Control Theory Experiments and Problem Sets*. Beijing: China Machine Press, 1994.
18. H. Luo, et al., *Electromechanical Control*. Hangzhou: Zhejiang University Press, 1994.
19. J.E. Johnson, P. Zhu, et al., *Electro-hydraulic Servo System*. Beijing: National Defense Industry Press, 1981.
20. G. Zhu, *Automatic Control Systems*. 4th edition, Beijing: Beijing Science and Technology Press, 1987.
21. W. Chunxing, *Hydraulic Servo Control System*. Beijing: China Machine Press, 1984.
22. B. Zhang, *Control Engineering Foundation*. Beijing: China Machine Press, 1982.
23. N. Wiener, *Cybernetics*. Beijing: Science Press, 1963.
24. S.M. Hinnels, *Modern Control System Theory and Application*. Beijing: China Machine Press, 1980.
25. G. Ruilong, *Control Theory and Electro-hydraulic Control System*. Beijing: China Machine Press, 1984.
26. K. Ogata, *Modern Control Engineering*. 5th Edition, Beijing: Science Press, 1976–2010.

27. S. Hu, *Principle of Automatic Control*. Beijing: National Defense Industry Press, 1984.
28. J. Dong, et al., *Control Engineering Foundation*. Beijing: Tsinghua University Press, 1992.
29. S. Yang, et al., *Fundamentals of Mechanical Engineering Control*. Wuhan: Huazhong Institute of Technology Press, 1984.
30. G.l. Xiong, *Digital Simulation of Control System*. Beijing: Tsinghua University Press, 1982.
31. Y. Yang, *Mechanical Control Engineering*. Beijing: China Machine Press, 1986.
32. C. Taorui, *Control System Analysis Program Design*. Xi'an: Northwestern Polytechnical University Press, 1985.
33. H.Z. Deng, *Computer Control System Analysis and Design*. Beijing: Tsinghua University Press, 1988.
34. D. Lu, et al., *Principle and Design of Automatic Control*. Shanghai: Shanghai Science and Technology Press, 1978.
35. J. Bao, *Theoretical Basis of Automatic Control*. Shanghai: Shanghai Science and Technology Press, 1987.
36. Y.Z. Deng, *Automatic Control System*. Shenyang: Liaoning People's Publishing House, 1982.
37. S.Z. Xiaodeng, *Automatic Control Theory*. Beijing: National Defense Industry Press, 1979.
38. Y. Zihou, *Principle of Automatic Control*. Beijing: Metallurgical Industry Press, 1980.
39. H. Li, *Hydraulic Control System*. Beijing: National Defense Industry Press, 1981.
40. W. Chunxing, *Hydraulic Servo Control System*. Beijing: China Machine Press, 1981.
41. Y. Lu, et al., *Electro-hydraulic Proportional Control Technology*. Beijing: China Machine Press, 1988.
42. M. Te, *Hydraulic Control System*. Beijing: Science Press, 1976.
43. L. Youshan, *Principle of Automatic Control*. Beijing: National Defense Industry Press, 1980.
44. W. jiaqi, *Principle and System of Automatic Control*. Beijing: National Defense Industry Press, 1979.
45. F. Weigetezhuo, *Feedback Control System Details*, Translated by W. Changhui, Beijing: Xiaoyuan Press, 1993.
46. X. Ma, et al., *MATLAB Language*. Beijing: Chinese Academy of Sciences Hope Computer Technology Company, 1980.
47. S.D. Fukefu, *Principles of Self-Regulation*. Volume 1, Translated by W. Zhongtuo, Beijing: Water Resources and Electric Power Press, 1957.
48. X. Chen, *Automatic Principle Problem Set*. Beijing: National Defense Industry Press, 1982.
49. W. Ke, *Fundamentals of Automatic Control*, Translated by A. Zhao. Shanghai: Shanghai Science and Technology Press, 1983.
50. Y. Chu, *Fundamentals of Automatic Control*. Beijing: National Defense Industry Press, 1985.
51. SMC, Electric - gas technology tutorial.
52. Y. Zihou, *Principle of Automatic Control*. Beijing: Metallurgical Industry Press, 1980.
53. W. Erman, *Root Locus Method in Automatic Control Theory*. Shanghai: Shanghai Science and Technology Press, 1966.
54. W. Nuofu, *Disturbance Regulation*. Shanghai: Shanghai Science and Technology Press, 1963.
55. By C. Kai-chung, *Analysis and Synthesis of Linear Control Systems*. Beijing: National Defense Industry Press, 1982.
56. Y.-g. Xi, *Predictive Control*. Beijing: National Defense Industry Press, 1993.
57. H. Wenjiao, *Feedback Control Theory*. Beijing: Guangming Daily Press, 1986.
58. W. Tezhao, *Feedback Control System Details*. Beijing: Xiaoyuan Press, 1993.
59. L. Yao, et al., *Automatic Control System Analysis and Synthesis*. Harbin: Harbin Institute of Ship Engineering Press, 1990.
60. Z. Deng, *Automatic Control System Experimental Technology*. Beijing: China Machine Press, 1986.
61. X. Zhou, et al., *Introduction to Control Engineering*. Xi'an: Northwestern Polytechnical University Press, 1988.

62. S. Hu, et al., *Automatic Control Principle Problem Set.* Beijing: National Defense Industry Press, 1990.
63. G. Hu, *China Fluid Transmission and Control and Mechatronic-Hydraulic Integration Technology Latest Achievements.* Xiamen: Xiamen University Press, 1997.
64. D. Xue, *Computer Aided Design of Control System.* Beijing: Tsinghua University Press 1996.
65. D.K. Anand, *Introduction to Control System.* 2nd Edition, New York: Pergamon Press Ltd., 1984.
66. J.A. Aseltine, *Transform Methods in Linear Analysis.* New York: McGraw-Hill Book Company Inc., 1958.
67. M. Athanassiades, et al., Time optimal control for plants with numerator dynamics. *IRE Trans. Auto. Cont.*, 39, 46–47, 1962.
68. M. Athans, et al., *Optimal Control: An Introduction to the Theory and Its Applications.* New York: McGraw-Hill Book Company Inc., 1965.
69. D.P. Atnerton, *Nonlinear Control Engineering.* London: Van Rein-hold Company Limited, 1975.
70. R. Bateson, *Introduction to Control System Technology.* Columbus: Charles, E. Merrill Publishing, 1980.
71. N.H. Beachley, *Introduction to Dynamic System Analysis.* New York: Harper and Row Publishing, 1978.
72. C.K. Benjamin, *Automation Control Systems.* 4th Edition, Cambridge: Prentice Hall Inc., 1982.
73. B. Friedland, *Control System Design: An Introduction to State-Space Methods.* New York: McGraw-Hill Book Company, 1986.
74. B.S. Blanchard, *System Engineering and Analysis.* Englewood Cliffs, NJ: Prentice Hall Inc., 1981.
75. D. Burghes, *Introduction to Control Including Optimal Control Theory.* New York: EllisHorwood Limited, 1980.
76. J.J. D'Azzo, *Linear Control System Analysis and Design: Conventional and Modern.* New York: Mcgraw Hill Book Co., 1981.
77. E.O. Doeblin, *Control System Principles and Design.* Englewood Cliffs, NJ: Prentice Hall Inc., 1985.
78. J. Gertler, L. Keviczky, *A Bridge Between Control Science and Technology.* Volume 3, Number 3, Oxford: IFAC Proceedings Series, Robert Maxwell Publisher at Pergamon Press, 1985.
79. J.D. Glomb, On the approximation of roots of the nth order phlynomials. *Trans. IRE.* 5, 1–4, 1960.
80. R.G. Jacquot, *Modern Digital Control System.* New York: Marcel Dekker Inc., 1981.
81. M.S. Jasmhidi, Z.M. Malek, *Linear Control Systems: Computer Adided Approach.* Oxford: Oxford Pergamon Press, 1985.
82. J. Van De Vetge, *Feedback Control System.* Englewood Cliffs, NJ: Prentice Hall Inc, 1986.
83. R.E. Kalman, Control system analysis and design via the second method of lyapunov and continuous time system, *ASME J. Basic Eng.*, 82(2), 371–393, 1960.
84. R.E. Kalman, *Automatic and Remote Control.* London: Butterwoths & Company, Ltd., 481–492, 1961.
85. R.E. Kalman, Canonical structure of linear dynamical system. *Proceed. Nat. Sci. Acad.*, 48(4), 96–600, 1962.
86. R.E. Kalman, *Controllability of Linear Dynamic Systems in Contributions to Differential Equation.* Vol. 1. New York: Interscience Publishers Inc., 1962.
87. R.E. Kalman, When is a linear control system optimal? *ASME J. Basic Eng.*, 51–60, 86, 1964.
88. B.C. Kuo, *Analysis and Synthesis of Sample Data Control System.* Englewood Cliffs, NJ: Prentice-Hall, Inc., 1963.
89. G. Lago, *Control System Theory.* New York: Ronald Press Company, 1962.
90. N.E. Leonard, W.S. Levine, *Using MATLAB to Analyse and Design Control Systems.* Amsterdam: Benjamin/Cummings, 1993.

91. S.J. Lewis, The use of feedback to improve the transient response of a servomechanism. *AIEE Trans. Part II*, 71, 449–453, 1952.
92. S.J. Mason, Feedback theory and some properties of signal flow graphs. *Cambrid: Proced. IRE*, 41, 1144–1156, 1953.
93. MathWorks, *MATLAB1*. New York: Springer-Verlag, 1998.
94. MathWorks, *MATLAB2*. New York: Springer-Verlag, 1998.
95. MathWorks, *MATLAB3*. New York: Springer-Verlag, 1999.
96. Mitchel and Gauthier Associate, *Advanced Continuous Simulation Language (ACSL)-User's Manual*. Mitchell & Gauthier Assoc, New York, Mitchell and Gauthier, Assoc., Inc., 1987.
97. R.V. Monopoli, *Controller Design for Non-linear and Time Varying Plants*. Washington, DC: NASA, 1965.
98. I.J. Nagrath, *Control Systems Engineering*. New York: Wiley, 1982.
99. Numerical Algorithm Group, *How to Use the NAG Library and its Documentation*, Oxford: NAG FORTRAN Liberary Manual, 1982.
100. H. Nyquist, *Regeneration Theory*. New York, *Bell System Technical Journal*, 1932.
101. K. Ogata, *Solving Control Engineering Problems with MATLAB*. Englewood Cliffs, NJ: Prentice Hall Inc., 1994.
102. B. Shahian, M. Hassul, *Computer Adided Control System Design Using MATLAB*. Englewood Cliffs, NJ: Prentice-Hall, 1993.
103. W.J. Palm, *Control System Engineering*. Englewood Cliffs, NJ: Prentice Hall Inc., 1986.
104. K. Sattelmeyer, *Mathematica Products Catalog*. Illinois: Wolfram Research Inc., 1995.
105. S. Wolfram, *Mathematica: A System for Doing Mathematics by Computer*. 2nd Edition, New York: Addision-Wesley, 1991.
106. N.S. Nise, *Control System Engineering*. 4th Edition, Charleston, SC: John Wiley & Sons, 2004.

Index

Note: Bold page numbers refer to tables; *italic* page numbers refer to figures.

Printed in the United States
by Baker & Taylor Publisher Services